Advanced Mobile Robotics

Advanced Mobile Robotics

Volume 2

Special Issue Editor

DaeEun Kim

MDPI • Basel • Beijing • Wuhan • Barcelona • Belgrade

Special Issue Editor
DaeEun Kim
Yonsei University
Korea

Editorial Office
MDPI
St. Alban-Anlage 66
4052 Basel, Switzerland

This is a reprint of articles from the Special Issue published online in the open access journal *Applied Sciences* (ISSN 2076-3417) from 2018 to 2019 (available at: https://www.mdpi.com/journal/applsci/special_issues/Advanced_Mobile_Robotics).

For citation purposes, cite each article independently as indicated on the article page online and as indicated below:

LastName, A.A.; LastName, B.B.; LastName, C.C. Article Title. *Journal Name* **Year**, *Article Number*, Page Range.

Volume 2
ISBN 978-3-03921-944-5 (Pbk)
ISBN 978-3-03921-945-2 (PDF)

Volume 1-3
ISBN 978-3-03921-942-1 (Pbk)
ISBN 978-3-03921-943-8 (PDF)

Cover image illustrated by Danho Kim.

Contents

About the Special Issue Editor

DaeEun Kim received his BE and MS from the Department of Computer Science and Engineering of Seoul National University, South Korea, and the University of Michigan at Ann Arbor, USA, respectively. He received his Ph.D. from the University of Edinburgh, UK, in 2002. From 2002 to 2006 he was a research scientist at the Max Planck Institute for Human Cognitive and Brain Sciences in Munich, Germany. Currently he is a professor in Yonsei University in Seoul, Korea. His research interests are in the areas of biorobotics, autonomous robots, artificial intelligence, artificial life, neural networks and neuroethology.

applied sciences

MDPI

Article

Disturbance-Rejection Control for the Hover and Transition Modes of a Negative-Buoyancy Quad Tilt-Rotor Autonomous Underwater Vehicle

Tao Wang, Jianqin Wang, Chao Wu *, Min Zhao and Tong Ge

State Key Laboratory of Ocean Engineering, Collaborative Innovation Center for Advanced Ship and Deep-Sea Exploration, Shanghai Jiao Tong University, Shanghai 200240, China; wangtaonice@sjtu.edu.cn (T.W.); wjq-394805990@163.com (J.W.); min.zhao@sjtu.edu.cn (M.Z.); tongge@sjtu.edu.cn (T.G.)

* Correspondence: wuchaorr@sjtu.edu.cn; Tel.: +86-135-6438-4939

Received: 30 October 2018; Accepted: 24 November 2018; Published: 2 December 2018

Abstract: This paper proposes a Negative-buoyancy Quad Tilt-rotor Autonomous Underwater Vehicle (NQTAUV), for which an attitude-tracking controller is designed for the hover and transition modes based on a disturbance-rejection control scheme. First, the structure of NQTAUV is illustrated, a mathematical model based on the Rodrigues parameters is proposed, and the attitude-tracking error model is derived. To simplify the disturbance-observer design procedure, a disturbance observer with a single parameter was designed to estimate the disturbance torque acting on the vehicle. The controller was designed to track the target attitude, and the stability of the whole system is analyzed. Finally, the performance of the proposed method was validated by three experiments. The primary benefit of the proposed method is the simplicity of its tuning and implementation.

Keywords: disturbance-rejection control; extended state observer (ESO); hover mode; transition mode; negative buoyancy; quad-tilt rotor; autonomous underwater vehicle (AUV); Rodrigues parameters

1. Introduction

In recent decades, as the demand for ocean exploration has increased, an increasing number of Autonomous Underwater Vehicles (AUVs) have been developed to expand the boundary of human knowledge [1,2]. Research on controlling these AUVs has received considerable attention from the robotics and marine research communities [3–5]. Based on the density of a given vehicle, AUVs can be classified as one of three types: positive buoyancy, neutral buoyancy, and negative buoyancy [6,7]. Compared with positive and neutral AUVs, negative buoyancy AUVs have the advantages of high energy efficiency, high speed, and a long cruising range. For the development of a previous version of the Negative-buoyancy Tri-tilt-rotor AUV (NTAUV), the attitude-stabilization control system was designed based on Immersion and Invariance (I and I) methodology [8,9]. However, when the NTAUV changes its mode from hover to transition, its asymmetric arrangement of rotors caused disturbance torque to impact the attitude control. To overcome this disadvantage, a negative-buoyancy quad-tilt-rotor AUV (NQTAUV) is proposed in this paper with the design of a disturbance-rejection attitude-tracking system.

The NQTAUV has many advantages. The symmetric arrangement of rotors makes attitude stabilization more convenient, and this structure provides the NQTAUV with the capability of rotor-failure recovery. In the case of one or two of its rotors failing, it retains the ability to hover using the remaining rotors [10,11]. The other advantage of the NQTAUV is that it maintains a level body attitude while moving forward, producing minimum extra hydrodynamic torque and reducing drag force. Quad-tilt rotors provide AUV pitch control within $0 \sim 10^\circ$ without horizontal motion [12].

Attitude-tracking performance under disturbance is essential for rigid-body vehicles [13]. When the NQTAUV is deployed in water, its attitude control is affected by ocean currents, waves [14–16], and the motion of its rotors, and they produce disturbance torques that affect the attitude. To reduce the influence of this disturbance, a number of methodologies have been studied. Pan designed a robust depth controller for AUVs in the presence of hydrodynamic-parameter uncertainties and disturbances [17]. Healey designed a sliding-mode controller for combined steering, diving, and speed-control functions based on a six-degree-of-freedom model for the maneuvering of an underwater vehicle [18]. Cui designed an adaptive neural-network controller for the problem of trajectory tracking under external disturbances, control input nonlinearities, and model uncertainties [19]. Harun designed a PID back-stepping controller with the help of a particle-swarming optimization approach in optimizing performance [20]. Shen designed a Lyapunov-based model predictive controller for AUVs to utilize computational resources (online optimization) to improve trajectory-tracking performance [21].

Disturbance rejection has attracted wide attention in recent years, for which disturbance-observer design is a widely studied core concept. A disturbance observer is an artificial system that can estimate the complex disturbances acting on a nominal system, and its controller design procedure can be enumerated in three steps: (1) Designing a controller for the nominal system to achieve stability and other performance improvements under the assumption that the disturbance is measurable; (2) designing a nonlinear disturbance observer to estimate the disturbance; (3) integrating the disturbance observer with the controller by replacing the disturbance in the control law with its estimation yielded by the disturbance observer [22,23]. Based on disturbance-observer methodology, Liu designed a guidance controller for a small unmanned aerial vehicle (UAV) to achieve a path following the presence of wind disturbances [24]. Chen derived a nonlinear disturbance observer to compensate for the friction of two-link robotic manipulators [22]. In addition to nonlinear disturbance observers, a Linear Extended State Observer (LESO) is a form of linear disturbance observer that has the advantage of tuning [25–28]. The feedforward control technique also plays an important role in control engineering as a useful and classical disturbance-compensation method, as it can compensate for measurable disturbance as part of control engineering. Bao designed an adaptive feedforward compensation controller to improve disturbance-rejection and tracking performance for jumping disturbances and large noises [29]. Kempf designed and implemented a discrete time-tracking controller for a precision positioning table actuated by direct-drive motors for which the existing disturbance-observer design techniques were extended to account for time delay in an industrial plant, and it performs well in practical scenarios [30].

In this paper, we propose a novel type of NQTAUV by simplifying the LESO design to one parameter and integrating a control scheme that compensates for unmeasurable underwater disturbances. We propose a nonlinear controller and an LESO with a single tuning parameter for attitude tracking the hover and transition modes, and these lead to simplicity in controller design, tuning, and implementation. The NQTAUV encounters various disturbances. During transition mode, the rotors tilt from 0° to 60°, and the tilt process results in disturbance torque. As speed increases, hydrodynamic force and torque play an important role as disturbance to the body of the vehicle [31]. The proposed LESO estimates the disturbance, changing from the nonlinear system to the nominal system without disturbance, and the nonlinear controller was designed to stabilize the nominal system.

This article is structured as follows. Section 2 introduces the attitude system model of the NQTAUV, where the attitude is presented based on the Rodrigues parameters, and the error model of the attitude-tracking system is then established. In Section 3, a disturbance rejection scheme is proposed that contains an LESO and a controller, which was designed to track the target attitude under disturbances, and the stability of the entire system is analyzed. In Section 4, three typical experiments validate the performance of the proposed scheme by comparing the attitude stabilization of the hover mode, the attitude tracking of the hover mode, and the attitude stabilization of the transition mode, both with the LESO on and with the LESO off.

2. Preliminaries

2.1. Mechanical Structure and Three Working Modes

The mechanical configuration of the NQTAUV is illustrated in Figure 1, where the rotors are mounted on the tip of the hydrofoil. The servos tilt the rotors, separately generating vector thrust. Each rotor generates force $f_i, i = 1,2,3,4$, and the servo tilts the rotor to angle $\alpha_i, i = 1,2,3,4$.

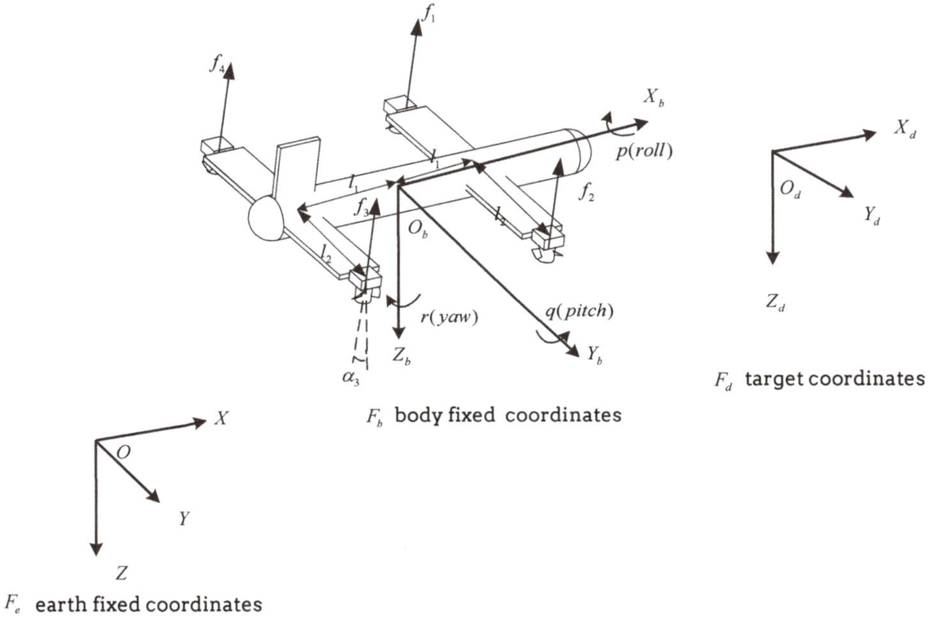

Figure 1. Mechanical configuration and co-ordinates of the negative-buoyancy quad-tilt-rotor autonomous underwater vehicle (NQTAUV).

The NQTAUV has three working modes: hover mode, transition mode, and level cruise mode. These are illustrated in Figure 2. The NQTAUV takes off, operating first in hover mode. In this mode, it can move slowly in a horizontal direction. The rotors then tilt to a certain angle, which is usually 60°, to generate forward thrust to accelerate until it reaches a speed at which the lift force generated by the hydrofoil can balance the weight of the vehicle in the water, which describes the transition mode. After that, the rotors tilt to 90°, and the NQTAUV operates in level-cruise mode. Unlike neutral-buoyancy vehicles, the NQTAUV must overcome less drag force [9], resulting in high efficiency and speed.

2.2. System Model

Three co-ordinates are used for attitude-tracking control: Earth-fixed co-ordinates $\mathcal{F}_e$, body-fixed co-ordinates $\mathcal{F}_b$, and target co-ordinates $\mathcal{F}_d$. These are illustrated in Figure 1.

There are a few methods to describe the orientation of a rigid body, and Euler angles have been widely used to express the attitude, which is convenient and intuitive. However, Euler angles have a singularity angle at $\pm\pi/2$. Rodrigues-parameter representation is a three-dimensional alternative form of Euler representation. Compared with Euler angles, Rodrigues parameters have double the range of angles without singularity, as it has a singularity at $\pm\pi$. In this paper, we chose the Rodrigues parameters to express the attitude.

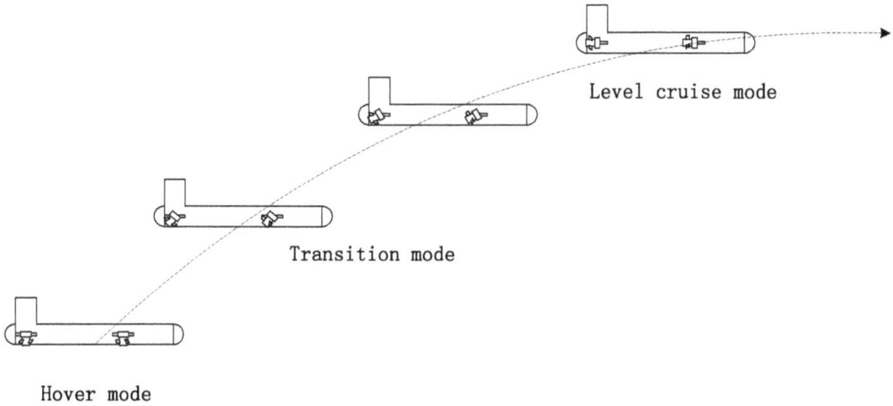

Figure 2. Three working modes of NQTAUV.

The NQTAUV shares the same kinetics equations of the previous version of NTAUV. The kinematics and kinetics equations of NQTAUV are in References [1,9,32]

$$\dot{\xi} = H(\xi)\omega \tag{1}$$

$$J\dot{\omega} = -S(\omega)J\omega - I_A\dot{\omega} - C(\omega)\omega - D(\omega) + \tau \tag{2}$$

where $\xi = \bar{\mathbf{e}}tan(\Phi/2), \bar{\mathbf{e}} \in \mathbb{R}^3, ||\bar{\mathbf{e}}|| = 1$ are the Rodrigues parameters, ω is the body-angle velocity vector, J is the rigid-body inertia tensor with respect to the origin of body-fixed co-ordinates, I_A is the added inertia matrix, $C(\omega) = S(I_A\omega)$ is the hydrodynamic Coriolis and centripetal matrix, $D(\omega)$ is the hydrodynamic damping matrix, and τ is the control torque. $S(\lambda)$ is defined as

$$S(\lambda) = \begin{bmatrix} 0 & -\lambda_3 & \lambda_2 \\ \lambda_3 & 0 & -\lambda_1 \\ -\lambda_2 & \lambda_1 & 0 \end{bmatrix} \tag{3}$$

in which $\lambda = [\lambda_1, \lambda_2, \lambda_3]^T$, which satisfies $\lambda^T S(\lambda) \equiv 0$. $H(\xi)$ is defined as

$$H(\xi) = \frac{1}{2}[I_3 + S(\xi) + \xi\xi^T] \tag{4}$$

which satisfies

$$\begin{aligned} \xi^T H(\xi) &= \frac{1}{2}\xi^T[I_3 + S(\xi) + \xi\rho^T] \\ &= \frac{1}{2}(\xi^T + \xi^T\xi\xi^T) \\ &= \frac{1}{2}(1 + \xi^T\xi)\xi^T \end{aligned} \tag{5}$$

2.3. Problem Formulation

In the attitude-tracking problem, we define the target attitude as $[\xi_d, \omega_d, \dot{\omega}_d]$, so the relative attitude is

$$\tilde{\xi} = \xi \otimes \xi_d^{-1} = \frac{\xi - \xi_d + S(\xi)\xi_d}{1 + \xi^T\xi_d} \tag{6}$$

$$\tilde{\omega} = \omega - \tilde{R}\omega_d \tag{7}$$

where $\tilde{R}$ is the relative attitude matrix, and the detailed derivation of $\tilde{R}$ can be found in Reference [32] as

$$\tilde{R} = \frac{(1 - \tilde{\xi}^T\tilde{\xi})I + 2\tilde{\xi}\tilde{\xi}^T - 2S(\tilde{\xi})}{1 + \tilde{\xi}^T\tilde{\xi}} \tag{8}$$

where the derivative of $\tilde{R}$ is

$$\dot{\tilde{R}} = -S(\tilde{\omega})\tilde{R} \tag{9}$$

and the derivative of $\tilde{\omega}$ is

$$\dot{\tilde{\omega}} = \dot{\omega} - (\dot{\tilde{R}}\omega_d + \tilde{R}\dot{\omega}_d) = \dot{\omega} - [-S(\tilde{\omega})\tilde{R}\omega_d + \tilde{R}\dot{\omega}_d] \tag{10}$$

We then obtain the attitude-tracking error system model

$$\dot{\tilde{\xi}} = H(\tilde{\xi})\tilde{\omega} \tag{11}$$

$$\dot{\tilde{\omega}} = J^{-1}[-S(\omega)J\omega - I_A\dot{\omega} - C(\omega)\omega - D(\omega) + \tau] - (-S(\tilde{\omega})\tilde{R}\omega_d + \tilde{R}\dot{\omega}_d) \tag{12}$$

where $\omega = \tilde{\omega} + \tilde{R}\omega_d$.

We use feedback linearization to cancel the nonlinearities of the system dynamics

$$\tau = v + [S(\omega)J\omega + I_A\dot{\omega} + C(\omega)\omega + D(\omega)] + J(-S(\tilde{\omega})\tilde{R}\omega_d + \tilde{R}\dot{\omega}_d) \tag{13}$$

and System (12) then becomes

$$\dot{\tilde{\omega}} = J^{-1}(v + d) \tag{14}$$

where d is the disturbance, and d comes from various sources. First, the tilting servo generates a disturbance torque to the pitch axis. Second, when the NQTAUV flies in the water, the hydrofoil generates extra force and torque for the vehicle. Without a linear velocity sensor, both of the above torques cannot be calculated, we treat those and other outside disturbances as one combined disturbance. We then estimated combined disturbance d. Noting that $d' = J^{-1}d$, System (14) becomes

$$\dot{\tilde{\omega}} = J^{-1}v + d' \tag{15}$$

3. Disturbance-Rejection Controller Design Based On Linear Extended State Observer (LESO)

The whole control scheme is illustrated in Figure 3. The dynamic model is Equation (2), $H(\xi)$ is Equation (1), the designed LESO is described in Section 3.1, and the designed controller is described in Section 3.2.

3.1. LESO Design

The disturbance acting on the system cannot be measured directly, so the extended states of the system are introduced to estimate the disturbance, and we assume that the disturbance is a constant or is a slowly changing variable generated by a linear system, so we design a corresponding linear extended state system to estimate the disturbance. This system is the LESO.

Assume the change rate of disturbance d' is $\dot{d}' = h(t)$, where $h(t)$ is the first-order derivative of d', so the system can be written as:

$$\dot{\tilde{\omega}} = J^{-1}v + d' \tag{16}$$

$$\dot{d}' = h(t) \tag{17}$$

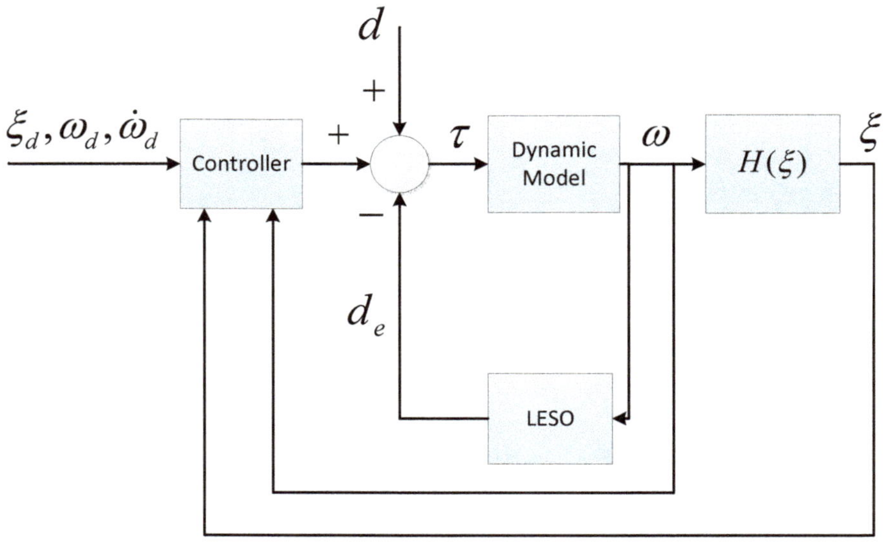

Figure 3. Control scheme. LESO: Linear Extended State Observer.

Based on system models (16) and (17), we designed a second-order disturbance observer. The second-order LESO system is

$$\dot{z}_1 = J^{-1}v + z_2 + l_1(\tilde{\omega} - z_1) \tag{18}$$

$$\dot{z}_2 = l_2(\tilde{\omega} - z_1) \tag{19}$$

where z_1, z_2 are the corresponding states of $\tilde{\omega}, d'$, and l_1, l_2 are positive constants. Insert Equation (14) into (18), and the Laplace transform of LESO is

$$sz_1 = (z_2 - d') + s\tilde{\omega} + l_1(\tilde{\omega}z_1) \tag{20}$$

$$sz_2 = l_2(\tilde{\omega} - z_1) \tag{21}$$

as a characteristic polynomial. We can get the transform function of d' and z_2 as

$$z_2 = \frac{l_2}{s^2 + l_1 s + l_2} d' \tag{22}$$

We needed to design l_1, l_2 to ensure z_2 tends to d'. According to the Routh–Hurwitz stability criterion, $s^2 + l_1 s + l_2$ should be Hurwitz. However, there are countless solutions in the parameter space of l_1, l_2. Considering response time and overshoot, we considered the critical damping of the second-order system, which had damping coefficient 1, and we selected $l_1 = 2\omega_o, l_2 = \omega_o^2, \omega_o > 0$, which balanced the response time and overshoot, and the characteristic polynomial of Equation (22) is Hurwitz. There is only one parameter ω_o for tuning, which simplifies the design, tuning, and implementation.

The estimate error is

$$\begin{aligned} \tilde{d}' &= d' - z_2 \\ &= d' - \frac{l_2}{s^2 + l_1 s + l_2} d' \\ &= \frac{s^2 + l_1 s}{s^2 + l_1 s + l_2} d' \end{aligned} \tag{23}$$

3.2. Controller Design

We used a backstepping method to design the nonlinear attitude-tracking controller. Consider the following Lyapunov candidate function:

$$V_1 = k_1 \ln(1 + \tilde{\xi}^T \tilde{\xi}) + \frac{1}{2} \tilde{\omega}^T J \tilde{\omega} \tag{24}$$

for which the derivative of V_1 is

$$\begin{aligned} \dot{V}_1 &= \frac{2k_1 \tilde{\xi}^T}{1 + \tilde{\xi}^T \tilde{\xi}} H(\tilde{\xi}) \tilde{\omega} + \tilde{\omega} J \dot{\tilde{\omega}} \\ &= k_1 \tilde{\xi}^T \tilde{\omega} + \tilde{\omega}^T (v + d) \\ &= \tilde{\omega}^T (k_1 \tilde{\xi} + v + d) \end{aligned} \tag{25}$$

We can select control input v as

$$v = -k_1 \tilde{\xi} - k_2 \tilde{\omega} - d_e \tag{26}$$

where $d_e = J z_2$ is the estimation of disturbance d. We then get the final control input

$$\tau = -k_1 \tilde{\xi} - k_2 \tilde{\xi} - d_e + [S(\omega) J \omega + I_A \dot{\omega} + C(\omega)\omega + D(\omega)] + J(-S(\tilde{\omega}) \tilde{R} \omega_d + \tilde{R} \dot{\omega}_d) \tag{27}$$

Noting $\tilde{d} = d - d_e$, we then get

$$\dot{V}_1 \leq ||\tilde{\omega}|| ||\tilde{d}|| - k_2 ||\tilde{\omega}||^2 \tag{28}$$

3.3. Stability Analysis

We used Lyapunov stability theory to analyze the stability of the system with disturbance d and control law (27).

Theorem 1. *Given Systems (1) and (2) under disturbance d, for target attitude $[\xi_d, \omega_d, \dot{\omega}_d]$, with feedback control law (27), attitude error $\tilde{\xi}, \tilde{\omega}$ and disturbance estimation error $\tilde{d}$ are globally asymptotically stable.*

Proof. We rewrite System (23) to the state space form

$$\begin{aligned} \dot{x} &= Ax + Bd' \\ \tilde{d}' &= Cx + d' \end{aligned} \tag{29}$$

where $x = \begin{bmatrix} x_1 \\ x_2 \end{bmatrix}, A = \begin{bmatrix} 0 & -\omega_o^2 \\ 1 & -2\omega_o \end{bmatrix}, B = \begin{bmatrix} -\omega_o^2 \\ 0 \end{bmatrix}, C = \begin{bmatrix} 0 & -1 \end{bmatrix}$, and we can see $z_2 = -x_2 = Cx$.

Assume disturbance $d' = J^{-1}d$ is constant. From Equation (22), we know the direct current gain from d' to the estimation of the disturbance z_2 is 1, so we get $-CA^{-1}B = 1$.

We then get

$$\begin{aligned} z_2 &= Cx \\ &= C(e^{At}x(0) + \int_0^t e^{A(t-\tau)} B\, d'\, d\tau) \\ &= C(e^{At}x(0) + e^{At}A^{-1}Bd' - A^{-1}Bd') \\ &= Ce^{At}(x(0) + A^{-1}Bd') - CA^{-1}Bd' \\ &= Ce^{At}(x(0) + A^{-1}Bd') + d' \end{aligned} \tag{30}$$

From Systems (23) and (30), we get

$$\tilde{d}' = d' - z_2 = -Ce^{At}(x(0) + A^{-1}Bd') \tag{31}$$

Assume the upper bound of d' is $||d'|| \leq \bar{d}$, and choose

$$\begin{aligned} c &= a_0 + \frac{a_0 \bar{d}'}{||x(0)||} \\ -\omega_o &< \lambda < 0 \end{aligned} \tag{32}$$

where $a_0 > 0$, which satisfies $||\tilde{d}'|| \leq c||x(0)||e^{\lambda t}$, so $\tilde{d}'$ can exponentially converge to 0. Based on the Lyapunov converse theorem [33], there exists a Lyapunov function V_2, which satisfies

$$\begin{aligned} a_1||\tilde{d}||^2 \leq V_2 \leq a_2||\tilde{d}||^2 \\ \dot{V}_2 \leq -a_3||\tilde{d}||^2 \end{aligned} \tag{33}$$

where $a_1, a_2, a_3 > 0$.

Define a new Lyapunov function $V_3 = V_1 + V_2$, and then we get

$$\dot{V}_3 \leq ||\tilde{\omega}||||\tilde{d}|| - k_2||\tilde{\omega}||^2 - a_3||\tilde{d}||^2 \tag{34}$$

Then, select $k_2 \geq \frac{1}{4a_3}$, and $\dot{V}_3 \leq 0$. □

4. Experiment Results and Discussion

We carried out three experiments to validate the proposed control scheme based on disturbance rejection with the help of a testbed.

4.1. Testbed

The three-degree-of-freedom testbed of the NQTAUV is illustrated in Figure 4, and it is similar to the one in Reference [9]. This testbed was designed to verify the performance of the presented controller. The body of the vehicle was mounted on a ball joint, which allowed the vehicle ±30° of roll and pitch, and free rotation of yaw. A desktop PC was used as the host computer, sending mode-switch commands and receiving data to and from the NQTAUV using serial communication. The controller law was implemented on a Nucleo STM32 F401RE board with 512 KB memory and 84 MHz CPU frequency. The control frequency was 100 Hz. In each control cycle, the control board sends data to the host, including the attitude, control torque, and disturbance observer output $\hat{d}$. An attitude sensor module reads three-axis-accelerometer, gyroscope, and magnetometer data from an MPU-9250 motion-tracking device, and then runs a Kalman filter algorithm to obtain the fusion attitude. The thruster is a brushless DC motor with a mounted propeller, and provides a maximum thrust of 15 N. The thruster curve was obtained by experiment. The thruster was mounted on the servo, which rotated at maximum of 6.9 rad/s.

The mechanical parameters of the NQTAUV are listed in Table 1.

The control law ran at 100 Hz on an STM32 board. The parameter of the LESO was selected as $w_o = 1.5$ rad/s, and the parameters of the controller were selected as $k_1 = 3.2, k_2 = 0.6$. To verify the disturbance-estimation effect, roughly 0.08 N·m torque was applied to both the x and y axes as the disturbance. The disturbance was loaded by hanging a weight on the axes, which resulted in the applied torque. The advantage of this method is its simplicity and accuracy.

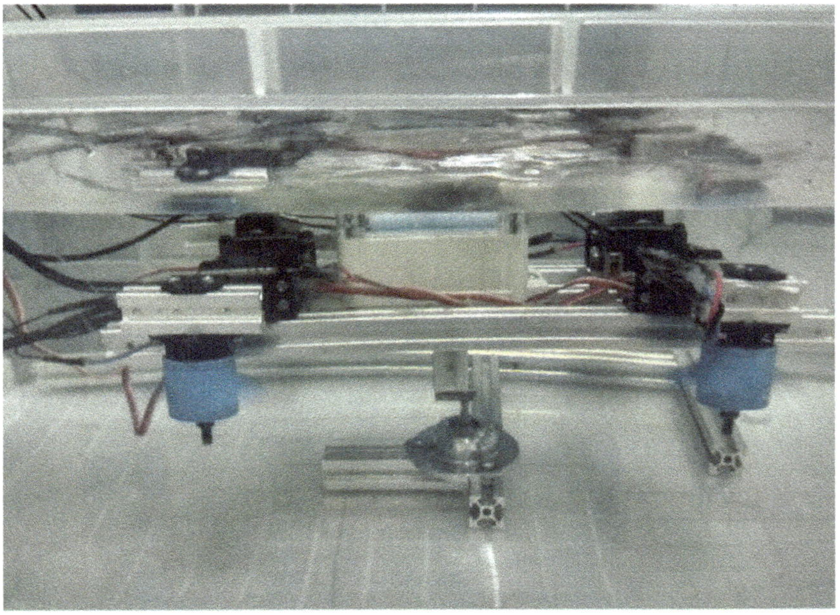

Figure 4. Testbed in the water tank.

Table 1. Mechanical parameters of the NQTAUV.

Parameters	Value	Unit (SI)
l_1	0.11	m
l_2	0.09	m
m	1.5	kg
B	3.15	N
I_{xx}	0.006	kg/m^2
I_{yy}	0.006	kg/m^2
I_{zz}	0.012	kg/m^2

4.2. Attitude Stabilization of Hover Mode

Attitude stabilization of the hover mode is essential for the NQTAUV, and there is a special case when the target attitude is $\xi_d = [0, 0, 0]^T$. To verify the disturbance-rejection effect of the hover mode, we toggled the LESO off and on with the same controller. The LESO being off means the observer states are reset to 0. The LESO takes 0∼30 s to turn off, and 30∼70 s to turn on. Figure 5 shows attitude ξ of the vehicle, and Figure 6 shows control input τ. Figure 7 shows the estimation of disturbance d_e.

From Figure 5, we can see that the disturbances lead to an attitude-stabilization error of 0.02 of the x and y axes. When the LESO is on, the disturbance is estimated within 3 s, and with this compensation, the attitude-stabilization error is eliminated. From Figure 7, we can see the estimation of the disturbance is stable and near the true value of the added torque. However, from Figure 6, we can see that control inputs become more violent when the LESO is on.

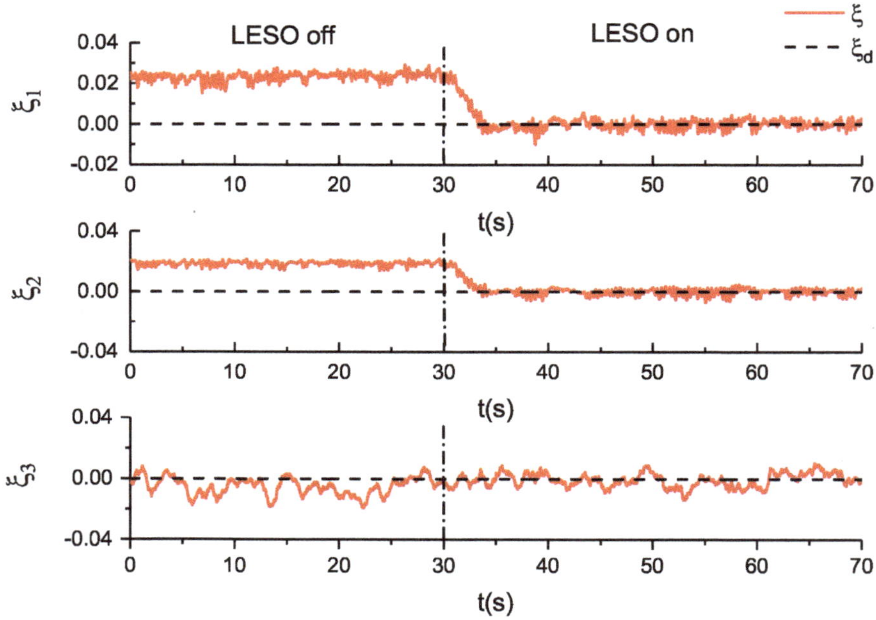

Figure 5. Attitude-stabilization effect with the LESO off and on.

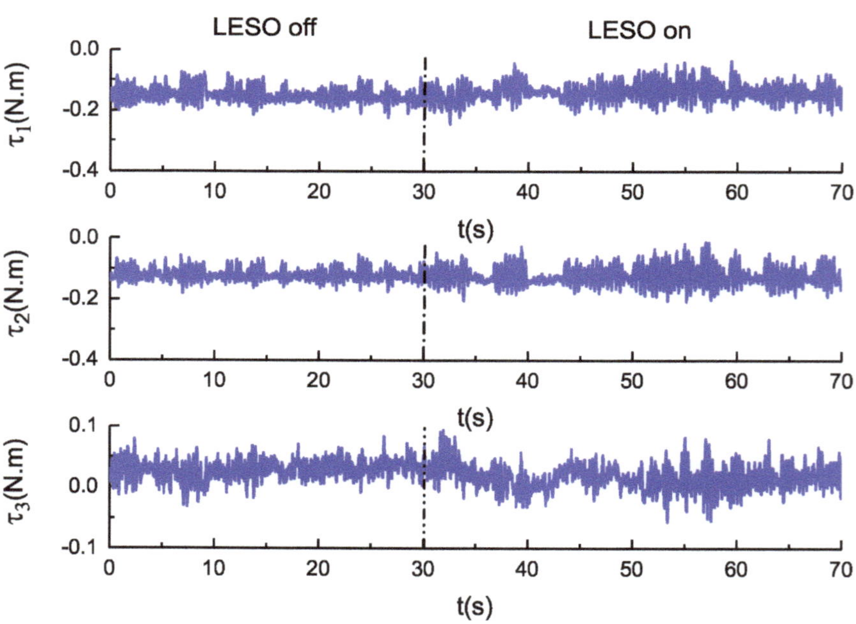

Figure 6. Control torque with the LESO off and on using attitude stabilization.

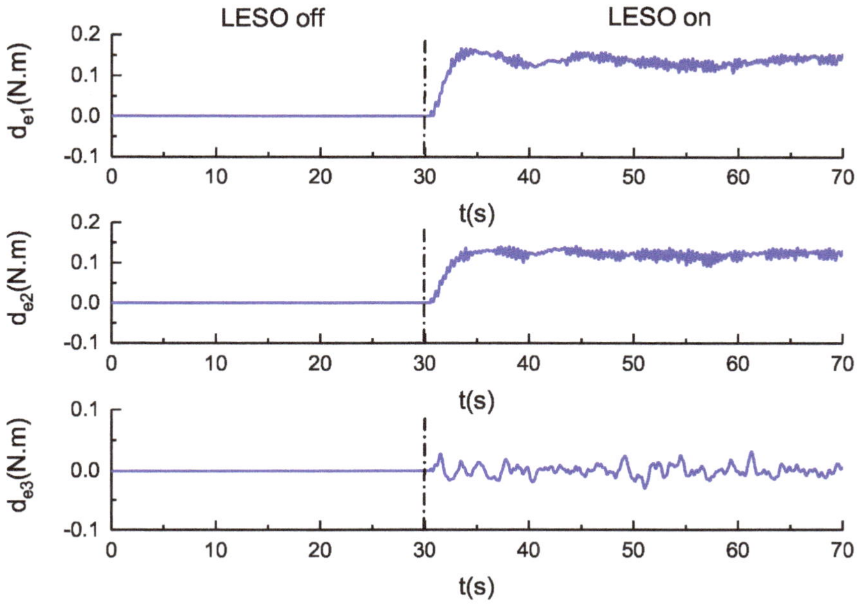

Figure 7. Disturbance estimation with the LESO on using attitude stabilization.

4.3. Attitude Tracking of Hover Mode

Attitude tracking is a fundamental function of horizontal movement. The tracking performance of pitch angle is typical. To verify the effect of attitude tracking, the target attitude was set as

$$\xi_d = \begin{bmatrix} 0 \\ tan(\frac{\pi}{36})sin(\frac{\pi}{10}t) \\ 0 \end{bmatrix} \tag{35}$$

To verify the effect of the proposed disturbance-rejection control effect, we turned the LESO off over 0∼55 s, and turned the LESO on over 55∼117 s. Figure 8 shows the attitude-tracking effect, Figure 9 shows the control torque, and Figure 10 shows the estimation of disturbance.

From Figure 8, we can see that, when the LESO is off, the disturbance resulted in an attitude-tracking error. When the LESO is on, the attitude tracks the desired ξ_d precisely in the amplitude. We can also see the attitude tracking had a small phase delay, and when the LESO was on, the delay was significantly reduced. This delay iswascaused by the design of the disturbance observer, as it is based on the assumption that the disturbance is a constant or slowly changing variable. However, when the attitude is tracking a harmonic signal, the water periodically affects the NQTAUV. This is a source of harmonic disturbance, and the proposed method cannot estimate the frequency of the harmonic disturbance in phase very well. We can see the harmonic disturbance indirectly from Figure 10. Future work will focus on improving the design of the harmonic disturbance observer.

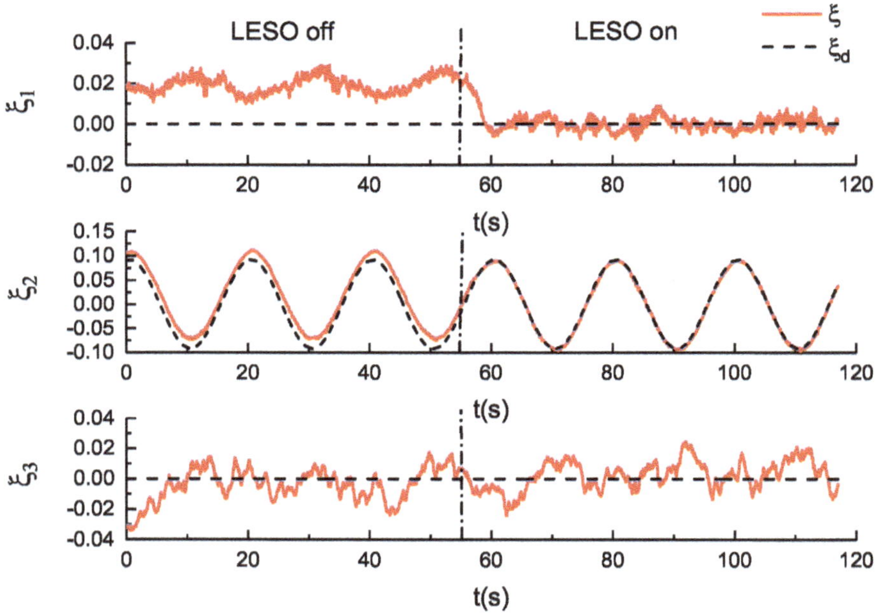

Figure 8. Attitude-tracking effect with the LESO off and on.

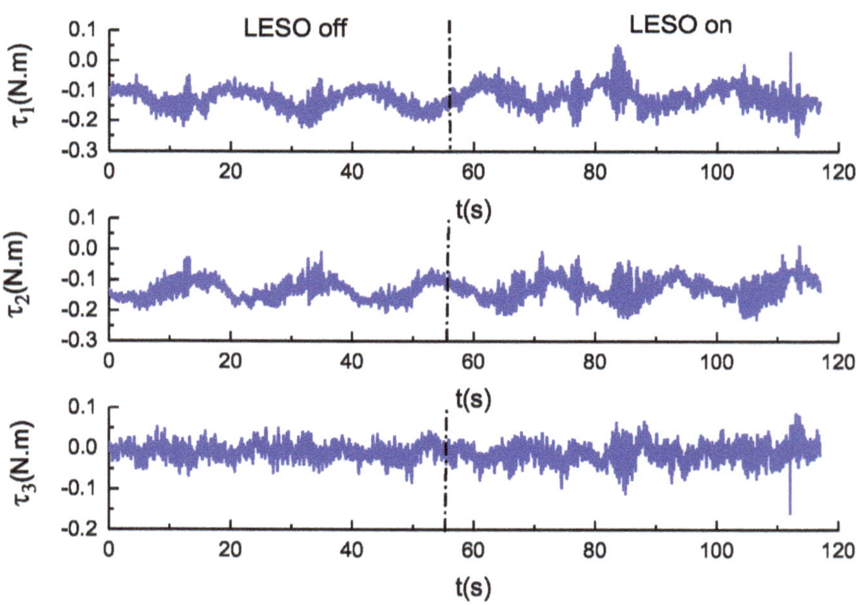

Figure 9. Control torque with the LESO off and on using attitude tracking.

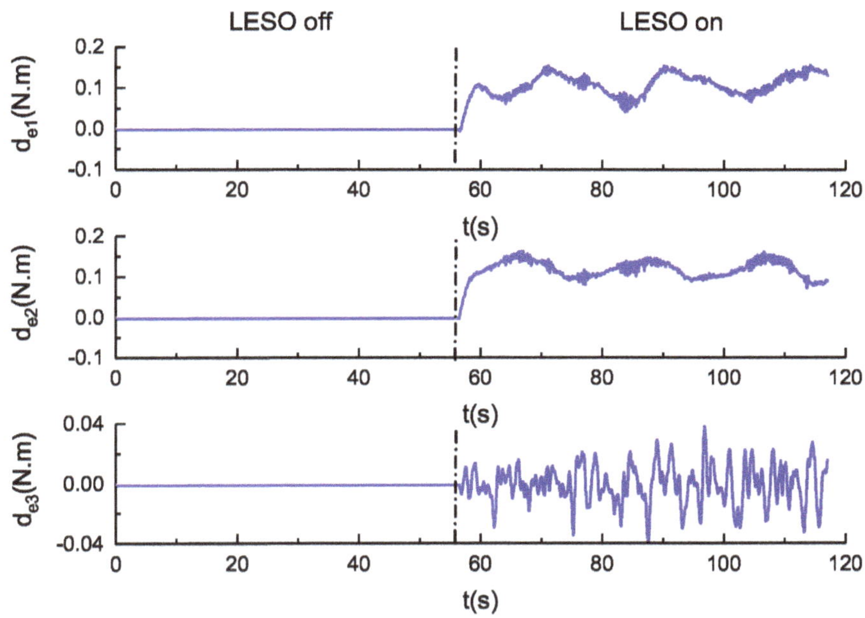

Figure 10. Disturbance estimation with the LESO on using attitude tracking.

4.4. Attitude Stabilization of Transition Mode

Attitude stabilization is a key function during the transition mode. The transition mode requires the attitude to be stabilized to $\xi_d = [0,0,0]^T$, which reduces the hydrodynamic torque as well as disturbance to the depth subsystem, making depth control more stable.

During the transition mode, the rotors tilt to produce a forward force, and as the speed of the vehicle increases, the hydrofoil produces lift force to balance the weight of the vehicle in the water. The hydrofoil also produces torque, which is treated as a disturbance to the attitude stabilization.

To verify the effect of the proposed disturbance-rejection control in transition mode, two experiments were carried out for the rotor tilt with the the LESO on and with the LESO off. The tilt ran for 10 s, tilting from 0° to 60°, performed over the 10∼20 s segment of the experiment. Figure 11 shows the effect of attitude stabilization on transition mode. Figure 12 shows control torque τ. Figure 13 shows estimation of disturbance d_e.

From Figure 11, we can see that the attitude-stabilization error with the LESO on is ±0.005, which is much smaller than with the LESO off, which is ±0.02. With the help of the LESO, the disturbance torque is compensated, and the attitude is stabilized to $[0,0,0]^T$ during transition mode. From Figure 12, we can also see that the input torque vibrated much more violently than when LESO is off, which costs more energy. Figure 13 shows the estimated disturbance torques during transition mode; this is obviously different from the attitude-stabilization and attitude-tracking modes. It is not a constant or harmonic disturbance, and cannot be measured. This experiment shows that the proposed control scheme can compensate for this kind of unmeasured disturbance well.

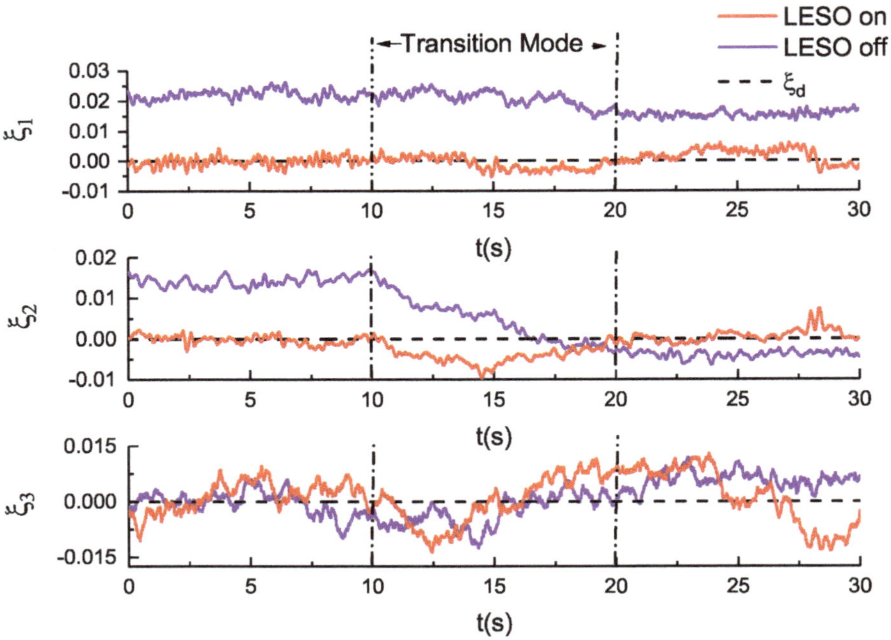

Figure 11. Attitude-stabilization effect with the LESO on and off in transition mode.

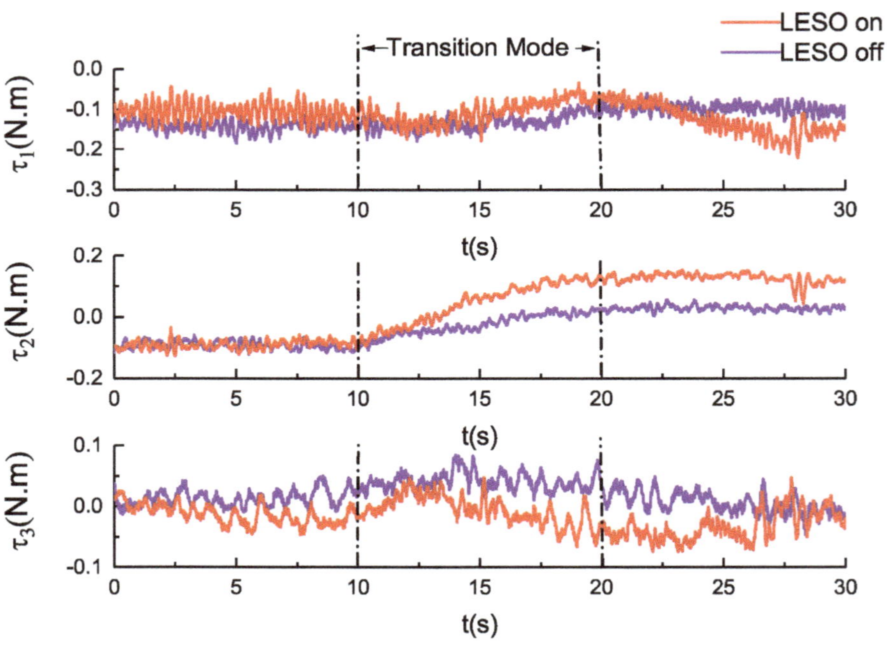

Figure 12. Control torque with the LESO on and off in transition mode.

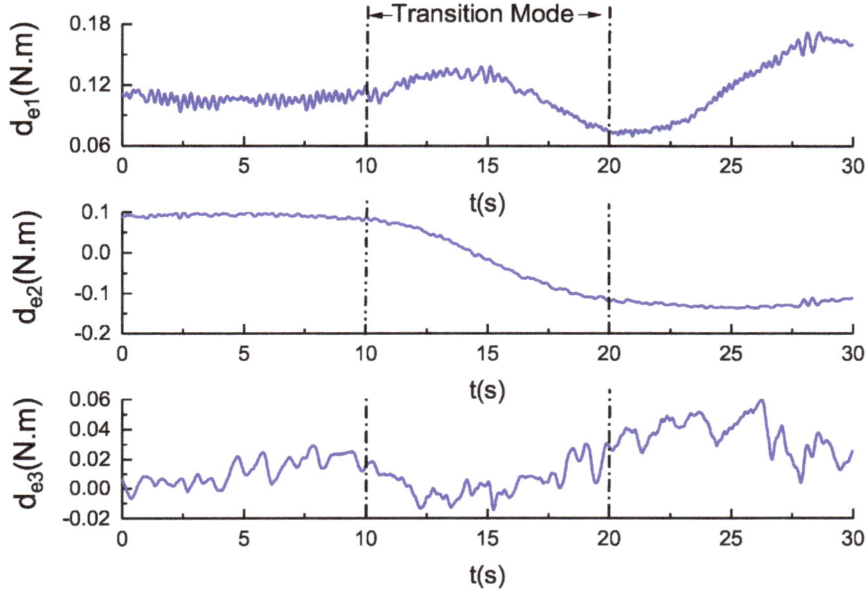

Figure 13. Disturbance estimation with the LESO on in transition mode.

Overall, for most situations, the proposed controller and disturbance observer performs well in attitude-tracking tasks. However, harmonic disturbance cannot be estimated without phase delay, even if it is insignificant. Moreover, input torque vibration costs significant amounts of energy.

5. Conclusions

In this paper, a negative-buoyancy quad-tilt-rotor autonomous underwater vehicle was presented, and a disturbance-rejection control scheme is proposed for attitude tracking of the hover and transition modes. The mathematical model of the NQTAUV was established with attitude using the Rodrigues parameters. An LESO was designed to estimate the disturbance torque. We simplified the LESO to one turning parameter, which is more convenient for practical use. A controller based on the disturbance observer was designed to perform attitude tracking, and the stability of the proposed control scheme was analyzed. The advantage of the proposed method is the simplicity of tuning the LESO with only one parameter, and the proposed controller only has two parameters for tuning. These make the proposed scheme simple to use in practice.

Finally, the performance of the control scheme was validated by three real-time experiments: the attitude stabilization of the hover mode, the attitude tracking of the hover mode, and the attitude stabilization of the transition mode. The results indicated satisfactory performance by comparing the effect of the controller with the LESO on and off. However, the proposed linear disturbance observer cannot estimate the time variance or harmonic disturbance without phase delay, and the input torque seriously vibrates and costs much energy. This harmonic disturbance is usually caused by hydrodynamic sources when the vehicle performs harmonic movements, so the frequency is known, but the amplitude is unknown. Further work will include the design of an exogenous disturbance observer to estimate harmonic disturbances. Additionally, the frequency response of the observer can be optimized. The parameters of the observer should be optimized to avoid the resonant frequency of the mechanical structure.

Author Contributions: T.W. put forward the original concept, proposed the control approach, and wrote the article. C.W., M.Z., J.W., and T.G. gave their valuable suggestions on research design. T.W. and J.W. analyzed and discussed the experimental results.

Funding: This work is supported by the National Natural Science Foundation of the National Research and Development of major scientific instruments (41427806), the National Natural Science Foundation of China (Grant No. 51779139), the National Key Research and Development Program of China (Grant No. 2016YFC0300700), the ROV System For Exploration and Sampling (DY125-21-Js-06) funded by the State Oceanic Administration People's Republic of China and National High Technology Research, and the Development Program of China (863 Program, Grant No. 2012AA092103).

Acknowledgments: The authors would like to express their appreciation to Chunwen Zhang for his considerable help on the experiment.

Conflicts of Interest: The authors declare no conflict of interest.

References

1. Fossen, T. *Guidance and Control of Ocean Vehicles*; John Willey & Sons: New York, NY, USA, 1994.
2. Brown, C.L. Deep sea mining and robotics: Assessing legal, societal and ethical implications. In Proceedings of the 2017 IEEE Workshop on Advanced Robotics and its Social Impacts (ARSO), Austin, TX, USA, 8–10 March 2017; pp. 1–2.
3. Hussain, N.A.A.; Arshad, M.R.; Mohd-Mokhtar, R. Underwater glider modelling and analysis for net buoyancy, depth and pitch angle control. *Ocean Eng.* **2011**, *38*, 1782–1791. [CrossRef]
4. Paull, L.; Saeedi, S.; Seto, M.; Li, H. AUV Navigation and Localization: A Review. *IEEE J. Ocean. Eng.* **2014**, *39*, 131–149. [CrossRef]
5. Palomeras, N.; Vallicrosa, G.; Mallios, A.; Bosch, J.; Vidal, E.; Hurtos, N.; Carreras, M.; Ridao, P. AUV homing and docking for remote operations. *Ocean Eng.* **2018**, *154*, 106–120. [CrossRef]
6. Hui, Y.; Tong, G.; Jia-wang, L.; Qiang, W. Prediction of mode and static stability of negative buoyancy vehicle. In Proceedings of the 2011 Chinese Control and Decision Conference (CCDC), Mianyang, China, 23–25 May 2011; pp. 1903–1909.
7. Wynn, R.B.; Huvenne, V.A.I.; Le Bas, T.P.; Murton, B.J.; Connelly, D.P.; Bett, B.J.; Ruhl, H.A.; Morris, K.J.; Peakall, J.; Parsons, D.R.; et al. Autonomous Underwater Vehicles (AUVs): Their past, present and future contributions to the advancement of marine geoscience. *Mar. Geol.* **2014**, *352*, 451–468. [CrossRef]
8. Astolfi, A.; Karagiannis, D.; Ortega, R. *Nonlinear and Adaptive Control with Applications*; Springer: London, UK, 2008.
9. Wang, T.; Wu, C.; Wang, J.; Ge, T. Modeling and Control of Negative-Buoyancy Tri-Tilt-Rotor Autonomous Underwater Vehicles Based on Immersion and Invariance Methodology. *Appl. Sci.* **2018**, *8*, 1150. [CrossRef]
10. Sarkar, N.; Podder, T.K.; Antonelli, G. Fault-accommodating thruster force allocation of an AUV considering thruster redundancy and saturation. *IEEE Trans. Robot. Autom.* **2002**, *18*, 223–233. [CrossRef]
11. Ji, S.W.; Bui, V.P.; Balachandran, B.; Kim, Y.B. Robust control allocation design for marine vessel. *Ocean Eng.* **2013**, *63*, 105–111. [CrossRef]
12. Benkhoud, K.; Bouallègue, S. Model Predictive Control design for a convertible Quad Tilt-Wing UAV. In Proceedings of the 2016 4th International Conference on Control Engineering & Information Technology (CEIT), Hammamet, Tunisia, 16–18 December 2016; pp. 1–6.
13. Li, X.; Zhao, M.; Ge, T. A Nonlinear Observer for Remotely Operated Vehicles with Cable Effect in Ocean Currents. *Appl. Sci.* **2018**, *8*, 867. [CrossRef]
14. Guo-Qing, X.; Ying, Y.; Wei-Guang, Z. Path-Following in 3D for Underactuated AUV in the Presence of Ocean Current. In Proceedings of the 2013 Fifth International Conference on Measuring Technology and Mechatronics Automation, Hong Kong, China, 16–17 January 2013; pp. 788–791.
15. Hosseini, M.; Seyedtabaii, S. Robust ROV path following considering disturbance and measurement error using data fusion. *Appl. Ocean Res.* **2016**, *54*, 67–72. [CrossRef]
16. Xiang, X.; Yu, C.; Zhang, Q. Robust fuzzy 3D path following for autonomous underwater vehicle subject to uncertainties. *Comput. Oper. Res.* **2017**, *84*, 165–177. [CrossRef]
17. Pan, H.; Xin, M. Depth control of autonomous underwater vehicles using indirect robust control method. In Proceedings of the 2012 American Control Conference (ACC), Montreal, QC, Canada, 27–29 June 2012; pp. 6216–6221.

18. Healey, A.J.; Lienard, D. Multivariable sliding mode control for autonomous diving and steering of unmanned underwater vehicles. *IEEE J. Ocean. Eng.* **1993**, *18*, 327–339. [CrossRef]
19. Cui, R.; Yang, C.; Li, Y.; Sharma, S. Adaptive Neural Network Control of AUVs with Control Input Nonlinearities Using Reinforcement Learning. *IEEE Trans. Syst. Man Cybern. Syst.* **2017**, *47*, 1019–1029. [CrossRef]
20. Harun, N.; Zain, Z.M.; Noh, M.M. PSO approach for a PID back-stepping control method in stabilizing an underactuated X4-AUV. In Proceedings of the 2017 IEEE 7th International Conference on Underwater System Technology: Theory and Applications (USYS), Kuala Lumpur, Malaysia, 18–20 December 2017; pp. 1–6.
21. Shen, C.; Shi, Y.; Buckham, B. Trajectory Tracking Control of an Autonomous Underwater Vehicle Using Lyapunov-Based Model Predictive Control. *IEEE Trans. Ind. Electron.* **2018**, *65*, 5796–5805. [CrossRef]
22. Wen-Hua, C.; Ballance, D.J.; Gawthrop, P.J.; Reilly, J.O. A nonlinear disturbance observer for robotic manipulators. *IEEE Trans. Ind. Electron.* **2000**, *47*, 932–938. [CrossRef]
23. Huang, Y.; Xue, W.; Zhiqiang, G.; Sira-Ramirez, H.; Wu, D.; Sun, M. Active disturbance rejection control: Methodology, practice and analysis. In Proceedings of the 33rd Chinese Control Conference, Nanjing, China, 28–30 July 2014; pp. 1–5.
24. Liu, C.; McAree, O.; Chen, W.H. Path following for small UAVs in the presence of wind disturbance. In Proceedings of the 2012 UKACC International Conference on Control (CONTROL), Cardiff, UK, 3–5 September 2012; pp. 613–618.
25. Qin, W.; Liu, Z.; Chen, Z. Formation control for nonlinear multi-agent systems with linear extended state observer. *IEEE/CAA J. Autom. Sin.* **2014**, *1*, 171–179. [CrossRef]
26. Tatsumi, J.; Gao, Z. On the enhanced ADRC design with a low observer bandwidth. In Proceedings of the 32nd Chinese Control Conference, Xi'an, China, 26–28 July 2013; pp. 297–302.
27. Liu, P.; Sun, Z.; Qiao, Y.; Sun, W. Attitude control and moment management of space station based on LESO. In Proceedings of the 2015 IEEE International Conference on Information and Automation, Lijiang, China, 8–10 August 2015; pp. 918–922.
28. Zhou, Y.; Li, R.; Zhao, D.; Wu, Q. Ship heading control using LESO with wave disturbance. In Proceedings of the 2016 IEEE International Conference on Robotics and Biomimetics (ROBIO), Qingdao, China, 3–7 December 2016; pp. 2081–2086.
29. Bao, Y.; Wang, L.Y.; Wang, C.; Jiang, J.; Jiang, C.; Duan, C. Adaptive Feedforward Compensation for Voltage Source Disturbance Rejection in DC–DC Converters. *IEEE Trans. Control Syst. Technol.* **2018**, *26*, 344–351. [CrossRef]
30. Kempf, C.J.; Kobayashi, S. Disturbance observer and feedforward design for a high-speed direct-drive positioning table. *IEEE Trans. Control Syst. Technol.* **1999**, *7*, 513–526. [CrossRef]
31. Xiang, X.; Lapierre, L.; Jouvencel, B. Smooth transition of AUV motion control: From fully-actuated to under-actuated configuration. *Robot. Autom. Syst.* **2015**, *67*, 14–22. [CrossRef]
32. Xing, G.; Shabbir, A. Alternate forms of relative attitude kinematics and dynamics equations. In *2001 Flight Mechanics Symposium*; Lynch, J.P., Ed.; NASA Goddard Space Flight Center: Greenbelt, MD, USA, 2001; pp. 83–97.
33. Khalil, H.K. *Nonlinear Systems*, 3rd ed.; Prentice-Hall: Upper Saddle River, NJ, USA, 2002.

applied sciences

MDPI

Article

Modeling and Control of Negative-Buoyancy Tri-Tilt-Rotor Autonomous Underwater Vehicles Based on Immersion and Invariance Methodology

Tao Wang, Chao Wu *, Jianqin Wang and Tong Ge

State Key Laboratory of Ocean Engineering, Collaborative Innovation Center for Advanced Ship and Deep-Sea Exploration, Shanghai Jiao Tong University, Shanghai 200240, China; wangtaonice@sjtu.edu.cn (T.W.); wjq-394805990@163.com (J.W.); tongge@sjtu.edu.cn (T.G.)

* Correspondence: wuchaorr@sjtu.edu.cn; Tel.:+86-021-3420-8063

Received: 16 June 2018; Accepted: 13 July 2018; Published: 15 July 2018

Abstract: Spot hover and high speed capabilities of underwater vehicles are essential for ocean exploring, however, few vehicles have these two features. Moreover, the motion of underwater vehicles is prone to be affected by the unknown hydrodynamics. This paper presents a novel negative-buoyancy autonomous underwater vehicle equipped with tri-tilt-rotor to obtain these two features. A detailed mathematical model is derived, which is then decoupled to altitude and attitude subsystems. For controlling the underwater vehicle, an attitude error model is designed for the attitude subsystem, and an adaptive nonlinear controller is proposed for the attitude error model based on immersion and invariance methodology. To demonstrate the effectiveness of the proposed controller, a three degrees of freedom (DOF) testbed is developed, and the performance of the controller is validated through a real-time experiment.

Keywords: negative-buoyancy; tri-tilt-rotor; autonomous underwater vehicle (AUV); immersion and invariance

---

1. Introduction

With the increasing demand for marine equipment over the past decades, different types of marine vehicles have been developed to expand human exploration capabilities [1] from the surface to the bottom of the ocean. Underwater vehicles perform various missions, such as collecting samples [2], acquiring data [3], and repairing marine structures [4].

In accordance with the level of autonomy for self-driving vehicles, underwater vehicles can be classified into remotely operated underwater vehicles (ROV) [5], autonomous underwater vehicles (AUV) [6,7], and human occupied vehicles (HOV) [8]. ROV consist of a cable that is used for power supply and as a communication line, which is used by the operator to remotely control the ROV [2,9]; an AUV is autonomous, and it performs motion control and mission planning [10–12]; an HOV has a life support system, and a pilot inside the vehicle controls the movement of the HOV for precise movement. ROV and AUV are unmanned underwater vehicles, and therefore, they have the advantages of no risk to life and long operation time. However, the densities of these vehicles are similar to that of the water owing to the buoyancy of the material used [13,14]. This increases the size and drag force, which considerably slows down the speed of the vehicle [15]. In some scenarios, a high-speed underwater vehicle is required to perform time-sensitive missions. Moreover, with the development of deep sea mining [16], a spot hover is a necessary capability for missions such as recharging and payload transition.

Traditional underwater vehicles have a cylindrical shape or an open frame. Cylindrical-shaped underwater vehicles have the advantage of a low drag force. They are propelled using fixed thrusters, and some of these vehicles are equipped with vertical and horizon thrusters to provide extra control forces. An ocean glider is an autonomous underwater vehicle used for ocean science [15,17]; it uses a small change of buoyancy in order to ascend and descend. A fixed wing converts the vertical motion to horizontal motion [18], thus acting like a saw tooth pattern [19]. The energy effect is so high that the glider can continually glide over hundreds of kilometers for months. Open frame vehicles such as an ROV can operate at one spot with the help of multiple rotors [20]; the AUV cannot achieve this [21]. However, the cable connecting the ROV to the mother ship limits the work range of the ROV [2]. A kind of negative buoyancy vehicle is designed to achieve high speed and long cruise range, it is more efficient than traditional AUV at high speed. However, it has to fly in the water to maintain depth, besides, spot hover capability is not achieved [22].

The design and control of an underwater vehicle involves many problems such as the nonlinearity of the model [13,21,23], underactuation [24], and the influence of the ocean current, waves, and turbulence [25].

In this paper, we present a negative-buoyancy tri-tilt-rotor autonomous underwater vehicle (NTAUV) to achieve the capability of spot hover and high-speed motion. The NTAUV is illustrated in Figure 1. The NTAUV is heavier than water, has negative buoyancy, and it balances the weight by buoyancy and lift force generated by the fixed wing or thrusters. Further, it operates under three modes: hover, horizontal motion, and transition between them. The NTAUV can hover or slowly cruise by controlling the rotor speed and the tail rotor angle in the hover mode. The hover motion control of NTAUV uses a hierarchical control scheme. The outer layer is position control, and the inner layer is attitude control. Attitude control is the fundamental function of the underactuated system. This paper focuses on hover mode modeling and control, especially attitude control. We design an adaptive nonlinear attitude controller using the immersion and invariance (I&I) methodology.

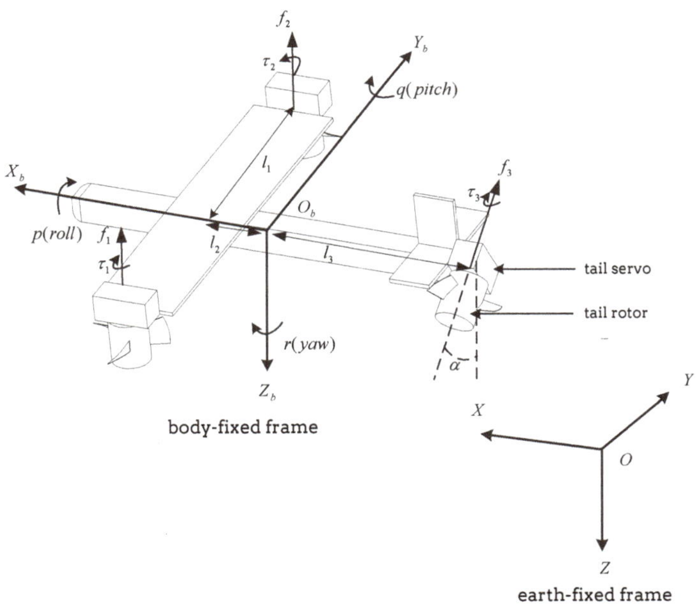

Figure 1. Configuration of NTAUV with Earth-fixed and body-fixed frame.

The article is structured as follows. Section 2 introduces some preliminaries, including the kinematic equations, mechanical structure, mathematical model, and subsystems of altitude and attitude. Section 3.1 presents the attitude error model. In Section 3.2, an adaptive nonlinear I&I

controller is designed for the attitude subsystem. The stability analysis is presented in Section 3.3. A three degree of freedom testbed is designed and the experiment results are shown in Section 4.

2. Preliminaries

2.1. Kinematics and Kinetics

Modeling a marine vehicle involves the study of statics and dynamics. The 6 DOF motion of a marine vehicle is analyzed by defining two coordinate frames, as illustrated in Figure 1. $O_bX_bY_bZ_b$ is fixed to the vehicle and is called the body-fixed frame. $OXYZ$ is fixed to the earth and is called the earth-fixed frame. The origin of the body-fixed frame is the center of gravity (CG). The center of buoyancy locates at CG.

The notations of the frames used in this paper are [1]

$$\eta = [\eta_1^\top, \eta_2^\top]^\top; \eta_1 = [x, y, z]^\top; \eta_2 = [\phi, \theta, \psi]^\top \tag{1}$$

$$\nu = [\nu_1^\top, \nu_2^\top]^\top; \nu_1 = [u, v, w]^\top; \nu_2 = [p, q, r]^\top \tag{2}$$

here η denotes the position and orientation of the vehicle and ν denotes the linear and angular velocity of the vehicle.

The rigid body kinematics of the vehicle are given by [1]

$$\dot{\eta}_1 = J_1(\eta_2)\nu_1 \tag{3}$$

$$\dot{\eta}_2 = J_2(\eta_2)\nu_2 \tag{4}$$

in which

$$J_1(\eta_2) = \begin{bmatrix} c\psi\theta & -s\psi c\phi + c\psi s\theta s\phi & s\psi s\phi + c\psi c\phi s\theta \\ s\psi c\theta & c\psi c\phi + s\phi s\theta s\psi & -c\psi s\phi + s\theta s\psi c\phi \\ -s\theta & c\theta s\phi & c\theta c\phi \end{bmatrix} \tag{5}$$

$$J_2(\eta_2) = \begin{bmatrix} 1 & s\phi t\theta & c\phi t\theta \\ 0 & c\phi & -s\phi \\ 0 & s\phi / c\theta & c\phi / c\theta \end{bmatrix} \tag{6}$$

where $s\cdot = sin(\cdot), c\cdot = cos(\cdot), t\cdot = tan(\cdot)$.

The mathematical model of the 6 DOF rigid body dynamics is

$$M_{RB}\dot{\nu} + C_{RB}(\nu)\nu = \tau_{RB} \tag{7}$$

where

$$\tau_{RB} = \tau_H + \tau_P \tag{8}$$

τ_H denotes the hydrodynamic forces and moments, and τ_P denotes the propulsion forces and moments. τ_H can be calculated as

$$\tau_H = -M_A\dot{\nu} - C_A(\nu)\nu - D(\nu)\nu \tag{9}$$

For underwater vehicles, if the movement is at low speed, it can be assumed that the vehicle performs a non-coupled motion. For simplicity, M_A and $D(\nu)$ have a diagonal structure with only linear damping terms on the diagonal

$$M_A = -diag\{X_{\dot{u}}, Y_{\dot{v}}, Z_{\dot{w}}, K_{\dot{p}}, M_{\dot{q}}, N_{\dot{r}}\} \tag{10}$$

The Coriolis terms of added mass are

$$C_A(\nu) = \begin{bmatrix} 0 & 0 & 0 & 0 & -Z_{\dot{w}}w & Y_{\dot{v}}v \\ 0 & 0 & 0 & Z_{\dot{w}}w & 0 & -X_{\dot{u}}u \\ 0 & 0 & 0 & Y_{\dot{v}}v & X_{\dot{u}}u & 0 \\ 0 & -Z_{\dot{w}}w & Y_{\dot{v}}v & 0 & -N_{\dot{r}}r & M_{\dot{q}}q \\ Z_{\dot{w}}w & 0 & -X_{\dot{u}}u & N_{\dot{r}}r & 0 & -K_{\dot{p}}p \\ Y_{\dot{v}}v & X_{\dot{u}}u & 0 & -M_{\dot{q}}q & K_{\dot{p}}p & 0 \end{bmatrix} \tag{11}$$

The damping terms are

$$D(\nu) = -diag\{X_u, Y_v, Z_w, K_p, M_q, N_r\} \tag{12}$$

For the rigid body, the inertia matrix M_{RB} is

$$M_{RB} = \begin{bmatrix} mI_{3\times 3} & 0 \\ 0 & I_0 \end{bmatrix} \tag{13}$$

where

$$I_0 = \begin{bmatrix} I_x & -I_{xy} & -I_{xz} \\ -I_{yx} & I_y & -I_{yz} \\ -I_{zx} & -I_{zy} & I_z \end{bmatrix} \tag{14}$$

The Coriolis and centripetal terms are

$$C_{RB}(\nu) = \begin{bmatrix} 0_{3\times 3} & -mS(\nu_1) \\ -mS(\nu_1) & -S(I_0\nu_2) \end{bmatrix} \tag{15}$$

Remark 1. *The skew-symmetric matrix $S(\lambda)$ is defined as*

$$S(\lambda) = \begin{bmatrix} 0 & -\lambda_3 & \lambda_2 \\ \lambda_3 & 0 & -\lambda_1 \\ -\lambda_2 & \lambda_1 & 0 \end{bmatrix} \tag{16}$$

where $\lambda = [\lambda_1, \lambda_2, \lambda_3]^\top$. S satisfies $x^\top S(\nu_2)x \equiv 0, x, \nu_2 \in \mathbb{R}^3$.

Based on the mechanical structure of the vehicle, the propulsion forces f and moments τ acting on the vehicle are [26–28]

$$f = \begin{bmatrix} 0 \\ f_3 sin\alpha \\ -f_1 - f_2 - f_3 cos\alpha \end{bmatrix} \tag{17}$$

$$\tau = \begin{bmatrix} l_1(f_1 - f_2) \\ l_2(f_1 + f2) - l_3 f_3 cos\alpha - \tau_3 sin\alpha \\ -l_3 f_3 sin\alpha + \tau_1 - \tau_2 + \tau_3 cos\alpha \end{bmatrix} \tag{18}$$

where the force and torque generated by each rotor is

$$\tau_i = k_\tau \omega^2 \tag{19}$$

$$f_i = k_f \omega^2 \tag{20}$$

The total moments (18) can be considered as the sum of the following two terms

$$\tau_m = \begin{bmatrix} l_1(f_1 - f_2) \\ l_2(f_1 + f_2) - l_3 f_3 cos\alpha \\ -l_3 f_3 sin\alpha \end{bmatrix} \tag{21}$$

$$\tau_{tail} = \begin{bmatrix} 0 \\ -\tau_3 sin\alpha \\ \tau_1 - \tau_2 + \tau_3 cos\alpha \end{bmatrix} \tag{22}$$

Rotor 1 and rotor 2 rotate in the opposite directions; therefore, the moments of the rotors are counteracted. Moreover, $k_\tau \ll k_f$, and τ_{tail} is much smaller compared to the control force, and it can be neglected [26]. Thus, the control force and moment are

$$\tau_P = \begin{bmatrix} f \\ \tau_m \end{bmatrix} \tag{23}$$

Note $M = M_{RB} + M_A, C(\nu) = C_{RB}(\nu) + C_A(\nu)$. Therefore, the whole system can be written as

$$M\dot{\nu} + C(\nu)\nu + D(\nu) = \tau_P \tag{24}$$

The system (24) has four inputs and six outputs; therefore, it is an underactuated system. We can see that the challenge is the lateral force $f_3 sin\alpha$ of f in (17). Consider f as

$$f = [0, 0, f_z]^\top + \Phi\tau_m \tag{25}$$

where

$$\Phi = \begin{bmatrix} 0 & 0 & 0 \\ 0 & 0 & -1/l_3 \\ 0 & 0 & 0 \end{bmatrix} \tag{26}$$

and

$$f_z = -f_1 - f_2 - f_3 cos\alpha \tag{27}$$

Note that (21) and (27) define a diffeomorphism, and therefore, the control inputs $[f_1, f_2, f_3, \alpha]^\top$ can be recovered from $[f_z, \tau_m^\top]^\top$ [27].

2.2. Altitude and Attitude Subsystems

The lateral component $f_3 sin\alpha$ of (17) is the main contributor of the yaw control input; therefore, the 6 DOF system can be decoupled to an altitude subsystem and an attitude subsystem. f_z is the altitude control input and τ_m is the attitude control input.

The altitude equation is

$$(m + Z_{\dot{w}})\ddot{z} - Z_w\dot{z} - (W - B) = f_z cos\phi cos\theta \tag{28}$$

where $W = mg$ is the rigid body weight in the air, $B = \rho g \nabla$ is the buoyancy in the water, g is the gravity constant, and ∇ is the displacement.

The attitude equation is

$$(I_0 + I_A)\dot{\nu_2} + (-S((I_0 + I_A)\nu_2))\nu_2 + D\nu_2 = \tau_m \tag{29}$$

where

$$I_A = -diag\{K_{\dot{p}}, M_{\dot{q}}, N_{\dot{r}}\} \tag{30}$$

$$D = -diag\{K_p, M_q, N_r\} \tag{31}$$

As $S((I_0 + I_A)\nu_2)\nu_2 = S(\nu_2)(I_0 + I_A)\nu_2$, we rewrite the attitude equation

$$(I_0 + I_A)\dot{\nu_2} - S(\nu_2)(I_0 + I_A)\nu_2 + D\nu_2 = \tau_m \tag{32}$$

The whole underactuated system is decoupled into two fully actuated subsystems: altitude subsystem and attitude subsystem. We could design the controller separately to control the whole system. The attitude control is the core function of NTAUV; we design the attitude controller in Section 3.

3. Attitude Controller Design

For the NTAUV, which is heavier than water, the attitude control is the essential function for maneuvering control. In general, an attitude controller is designed as the inner loop of a higher level controller, such as path following or trajectory tracking.

In this section, an adaptive I&I attitude controller is designed. I&I is a nonlinear controller design method, and it yields a stabilizing scheme that counters the effect of the uncertain parameters adopting a robustness perspective [29–31]. The stability of the controller is proved in Section 3.3.

3.1. Attitude Error Model

Without loss of generality, the reference attitude is $\eta_{2d} = \eta_{2d}(t)$, therefore, the tracking error is

$$\tilde{\eta}_2 = \eta_2 - \eta_{2d}, \tilde{\nu}_2 = \nu_2 - J_2^{-1}(\eta_2)\dot{\eta}_{2d} \tag{33}$$

We define the energy function, which is the sum of the potential and kinetic energy, as

$$H(\tilde{\eta}_2, \tilde{\nu}_2) = \frac{1}{2}\tilde{\eta}_2^\top K_1 \tilde{\eta}_2 + \frac{1}{2}\tilde{\nu}_2^\top I^2 \tilde{\nu}_2 \tag{34}$$

where $K_1 = K_1^\top > 0$ (positive definite matrix), $I = I_0 + I_A = [I_1^\top, I_2^\top, I_3^\top]^\top$.

The partial derivative of H are

$$\frac{\partial H^\top}{\partial \tilde{\eta}_2} = K_1 \tilde{\eta}_2 \tag{35}$$

$$\frac{\partial H^\top}{\partial (I\tilde{\nu}_2)} = I\tilde{\nu_2} \tag{36}$$

Then, the derivatives of the reference attitude (33) are

$$\begin{aligned} \dot{\tilde{\eta}}_2 &= \dot{\eta}_2 - \dot{\eta}_{2d} \\ &= J_2(\eta_2)\nu_2 - J_2(\eta_2)(\nu_2 - \tilde{\nu_2}) \\ &= J_2(\eta_2)\tilde{\nu_2} \\ &= J_2(\eta_2)I^{-1}I\tilde{\nu_2} \\ &= J_2(\eta_2)I^{-1}\frac{\partial H^\top}{\partial \tilde{\nu}_2} \end{aligned} \tag{37}$$

$$\dot{\tilde{\nu}}_2 = \dot{\nu}_2 - \dot{J}_2(\eta_2)\dot{\eta}_{2d} - J_2(\eta_2)\ddot{\eta}_{2d} \tag{38}$$

The product of I and $\dot{\tilde{\nu}}_2$ is

$$\begin{aligned}
I\dot{\tilde{\nu}}_2 =& I\dot{\nu}_2 - I\dot{J}_2(\eta_2)\dot{\eta}_{2d} - IJ_2(\eta_2)\ddot{\eta}_{2d} \\
=& S(\nu_2)I\nu_2 - D\nu_2 + \tau_m - I\dot{J}_2(\eta_2)\dot{\eta}_{2d} - IJ_2(\eta_2)\ddot{\eta}_{2d} \\
=& S(\nu_2)I(\tilde{\nu}_2 + J_2^{-1}(\eta_2)\dot{\eta}_{2d}) - D\nu_2 + \tau_m - I\dot{J}_2(\eta_2)\dot{\eta}_{2d} - IJ_2(\eta_2)\ddot{\eta}_{2d} \\
=& S(\nu_2)I\tilde{\nu}_2 + S(\nu_2)IJ^{-1}(\eta_2)\dot{\eta}_{2d} - D\nu_2 + \tau_m - I\dot{J}_2(\eta_2)\dot{\eta}_{2d} - IJ_2(\eta_2)\ddot{\eta}_{2d} \\
& + I^{-1}J_2^\top(\eta_2)K_1\tilde{\eta}_2 - I^{-1}J_2^\top(\eta_2)K_1\tilde{\eta}_2 \\
=& S(\nu_2)\frac{\partial H^\top}{\partial \tilde{\nu}_2} - I^{-1}J_2^\top(\eta_2)\frac{\partial H^\top}{\partial \tilde{\eta}_2} + \tau_m - \mathcal{M} + \kappa
\end{aligned} \tag{39}$$

where

$$\mathcal{M} = D\nu_2 \tag{40}$$

$$\kappa = S(\nu_2)IJ_2^{-1}(\eta_2)\dot{\eta}_{2d} - I\dot{J}_2(\eta_2)\dot{\eta}_{2d} - IJ_2(\eta_2)\ddot{\eta}_{2d} + I^{-1}J_2^\top(\eta_2)K_1\tilde{\eta}_2 \tag{41}$$

in which $\dot{J}_2(\eta_2)$ can be obtained as follows

$$\dot{J}_2(\eta_2) = -S(\nu_2)J_2(\eta_2) \tag{42}$$

Then, the attitude error system is

$$\begin{bmatrix} \dot{\tilde{\eta}}_2 \\ I\dot{\tilde{\nu}}_2 \end{bmatrix} = \begin{bmatrix} 0_{3\times3} & J_2(\eta_2)I^{-1} \\ -I^{-1}J_2^\top \eta_2 & S(\nu_2) \end{bmatrix} \begin{bmatrix} \frac{\partial H^\top}{\partial \tilde{\eta}_2} \\ \frac{\partial H^\top}{\partial (I\tilde{\nu}_2)} \end{bmatrix} + \begin{bmatrix} 0 \\ \tau_m - \mathcal{M} + \kappa \end{bmatrix} \tag{43}$$

The system states are $\dot{\tilde{\eta}}_2$ and $I\dot{\tilde{\nu}}_2$.

The η_2 state will follow η_{2d} if the system (43) converges to the zeros.

3.2. Controller and Estimator Design

Define unknown parameters $\vartheta = -[K_p, M_q, N_r]^\top$, which are the diagonal of D. The estimator error is

$$z_i = \hat{\vartheta}_i - \vartheta_i + \beta_i(\nu_{2i}) \tag{44}$$

where β_i is a continuous function.

For the convenience of controller design, we rewrite $\mathcal{M}$ as

$$\mathcal{M} = D\nu_2 = \begin{bmatrix} D_{11}\nu_{21} \\ D_{22}\nu_{22} \\ D_{33}\nu_{23} \end{bmatrix} = \begin{bmatrix} \vartheta_1\rho_1(\nu_{21}) \\ \vartheta_2\rho_2(\nu_{22}) \\ \vartheta_3\rho_3(\nu_{23}) \end{bmatrix} \tag{45}$$

where ρ_i is a continuous function.

The controller can be constructed as

$$\tau_m = \begin{bmatrix} (\hat{\vartheta}_1 + \beta_1(\nu_{21}))\rho_1(\nu_{21}) \\ (\hat{\vartheta}_2 + \beta_2(\nu_{22}))\rho_2(\nu_{22}) \\ (\hat{\vartheta}_3 + \beta_3(\nu_{23}))\rho_3(\nu_{23}) \end{bmatrix} - \kappa - K_2 I\tilde{\nu}_2 \tag{46}$$

where K_2 is a positive-defined matrix valued function.

With the control input (46), the closed-loop system (43) can be written as

$$\begin{bmatrix} \dot{\tilde{\eta}}_2 \\ I\dot{\tilde{\nu}}_2 \end{bmatrix} = \begin{bmatrix} 0_{3\times3} & J_2(\eta_2)I^{-1} \\ -I^{-1}J_2^\top(\eta_2) & S(\nu_2) - K_2 \end{bmatrix} \begin{bmatrix} \frac{\partial H^\top}{\partial \tilde{\eta}_2} \\ \frac{\partial H^\top}{\partial I\tilde{\nu}_2} \end{bmatrix} - \begin{bmatrix} 0 \\ \Delta \end{bmatrix} \tag{47}$$

where

$$\Delta_i = z_i \rho_i(\nu_{2i}) \tag{48}$$

The estimator can be designed as

$$\dot{\hat{\vartheta}}_i = -\frac{\partial \beta_i}{\partial(\nu_{2i})}[\dot{\nu}_{2i} + \Delta_i] \tag{49}$$

We can see that Δ_i can be obtained from (47), which is

$$\begin{aligned} I\dot{\nu}_2 + \Delta &= I\dot{\nu}_2 + \begin{bmatrix} -I^{-1}J_2^\top \eta_2 & S(\nu_2) - K_2 \end{bmatrix} \begin{bmatrix} \frac{\partial H^\top}{\partial \tilde{\eta}_2} \\ \frac{\partial H^\top}{\partial I\tilde{\nu}_2} \end{bmatrix} - I\dot{\tilde{\nu}}_2 \\ &= \begin{bmatrix} -I^{-1}J_2^\top \eta_2 & S(\nu_2) - K_2 \end{bmatrix} \begin{bmatrix} K_1\tilde{\eta}_2 \\ I\tilde{\nu}_2 \end{bmatrix} + \dot{J}_2(\eta_2)\dot{\eta}_{2d} + J_2(\eta_2)\ddot{\eta}_{2d} \end{aligned} \tag{50}$$

As a result, $[I_i\dot{\nu}_2 + \Delta_i]$ in (49) is a function of $\nu_2, \eta_2, \dot{\eta}_{2d}, \ddot{\eta}_{2d}$, and they are measurable or can be calculated from the given reference signal.

The continuous function β_i can be selected as

$$\beta_i(I_i\nu_2) = \gamma_i \int_0^{\nu_{2i}} \rho_i(\zeta)d\zeta = \frac{1}{2}\gamma_i\nu_{2i}^2 \tag{51}$$

where $\gamma_i > 0$, which implies that

$$\frac{\partial \beta_i}{\partial \nu_{2i}} = \gamma_i \rho_i(\nu_{2i}) \tag{52}$$

The control scheme is illustrated in Figure 2.

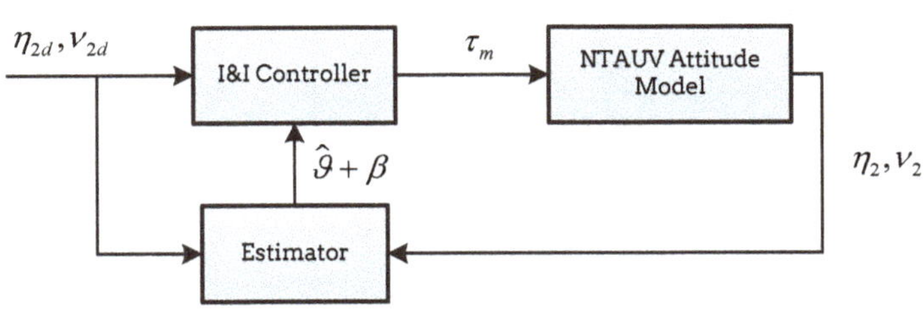

Figure 2. I&I control scheme.

3.3. Stability Analysis

The derivative of (44) is

$$\begin{aligned} \dot{z}_i &= \dot{\hat{\vartheta}}_i - \dot{\vartheta}_i + \dot{\beta}_i(\nu_{2i}) \\ &= -\frac{\partial \beta_i}{\partial(\nu_{2i})}[\dot{\nu}_{2i} + z_i\rho_i(\nu_{2i})] + \frac{\partial \beta_i}{\partial(\nu_{2i})}(\dot{\nu}_{2i}) \\ &= -\gamma_i\rho_i(\nu_{2i})\rho_i(\nu_{2i})z_i \end{aligned} \tag{53}$$

where ϑ_i is assumed to be a constant.

The Lyapunov function is $W(z) = \Sigma_{i=1}^{3} \frac{1}{\gamma_i} |z_i|^2$, then

$$\begin{aligned} \dot{W}(z) &= \Sigma_{i=1}^{3} \frac{2}{\gamma_i} z_i \dot{z}_i \\ &= \Sigma_{i=1}^{3} \frac{2}{\gamma_i} z_i (-\gamma_i \rho_i(\nu_{2i}) \rho_i(\nu_{2i}) z_i) \\ &= -2\Sigma_{i=1}^{3} z_i \rho_i(\nu_{2i}) \rho_i(\nu_{2i}) z_i \\ &= -2\Delta^\top \Delta \\ &= -2|\Delta|^2 \leq 0 \end{aligned} \tag{54}$$

which means that $\hat{\vartheta} \to \vartheta$ as $t \to \infty$.

We want the whole system to be stable, which means $(\tilde{\eta}_2, \tilde{\nu}_2, z)^\top$ has the equilibrium point $(0,0,0)^\top$ that is stable.

Consider the Lyapunov function $V = H(\tilde{\eta}_2, \tilde{\nu}_2) + W(z)$. The derivative of V is

$$\begin{aligned} \dot{V} &= \frac{\partial H}{\partial \tilde{\eta}_2} \dot{\tilde{\eta}}_2 + \frac{\partial H}{\partial (I\tilde{\nu}_2)} (I\dot{\tilde{\nu}}_2) + \dot{W}(z) \\ &= \frac{\partial H}{\partial \tilde{\eta}_2} J_2(\eta_2) I^{-1} \frac{\partial H^\top}{\partial \tilde{\nu}_2} + \frac{\partial H}{\partial (I\tilde{\nu}_2)} (-I^{-1} J_2^\top(\eta_2) \frac{\partial H^\top}{\partial \tilde{\eta}_2} + (S(\nu_2) - K_2) \frac{\partial H^\top}{\partial \tilde{\nu}_2} - \Delta) - 2\Delta^\top \Delta \\ &= \frac{\partial H}{\partial (I\tilde{\nu}_2)} (-K_2 \frac{\partial H^\top}{\partial \tilde{\nu}_2} - \Delta) - 2\Delta^\top \Delta \\ &= -\frac{\partial H}{\partial (I\tilde{\nu}_2)} K_2 \frac{\partial H^\top}{\partial \tilde{\nu}_2} - \frac{\partial H}{\partial (I\tilde{\nu}_2)} \Delta - 2\Delta^\top \Delta \\ &= -\frac{\partial H}{\partial (I\tilde{\nu}_2)} (K_2 - \frac{1}{2\sqrt{2}} I_{3\times3}) \frac{\partial H^\top}{\partial \tilde{\nu}_2} - \frac{\partial H}{\partial (I\tilde{\nu}_2)} (\frac{1}{2\sqrt{2}} I_{3\times3}) \frac{\partial H^\top}{\partial \tilde{\nu}_2} - \frac{\partial H}{\partial (I\tilde{\nu}_2)} \Delta - 2\Delta^\top \Delta \end{aligned}$$

select K_2 such that $K_2 - \frac{1}{2\sqrt{2}} I_{3\times3} > 0$; then, $\dot{V} \leq 0$.

4. Experiment Results

4.1. Testbed

A 3 DOF testbed is designed for verifying the performance of the presented controller. The testbed includes three parts: the unmovable base, 3-DOF ball joint, and NTAUV. The ball joint enables a maximum $\pm 40^\circ$ roll and pitch angle and 360° yaw angle. The 3 DOF ball joint of the testbed is illustrated in Figure 3.

The mechanical parameters of the NTAUV are listed in Table 1.

We designed our own control system. A desktop PC running ground station software was used as the host computer. This computer sends commands and receives attitude data via a serial port. An STM32 Nucleo F401RE board, with 512 KB memory and 84 MHz CPU frequency is used as the controller board. The computation power guarantees the capability to apply an advanced control algorithm, dealing with complex matrix calculation running at 100 Hz.

The attitude sensor module reads raw three-axis accelerometers, gyroscopes, and magnetometers data from MPU9250, runs a Kalman filter algorithm, and sends the attitude and angular speed to the controller board. The rotor is a brushless DC motor; a propeller is mounted on the top of the motor, and it can provide maximum thrust of 15 N. The servo provides a maximum torque of 0.15 N· m, and the maximum speed of rotation is 6.9 rad/s .

Table 1. Mechanical parameters of NTAUV.

Parameters	Value	Unit (SI)
l_1	0.13	m
l_2	0.075	m
l_3	0.15	m
m	1	kg
B	3	N
I_{xx}	0.0061	kg/m^2
I_{yy}	0.006	kg/m^2
I_{zz}	0.0118	kg/m^2

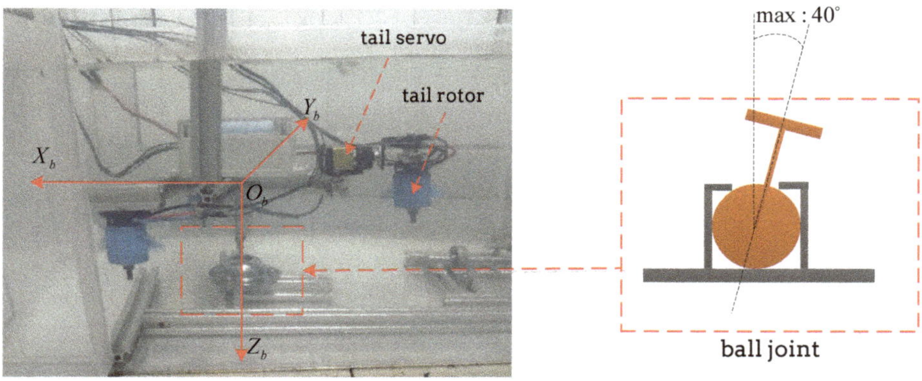

Figure 3. Testbed.

4.2. Experiment Results and Discussion

Attitude control is essential for under actuated rigid body vehicles, such as airplanes, surface vessels, helicopters, and multi-rotor aerial vehicles. Attitude control is often designed as the inner loop of path or trajectory control, named as hierarchical control. η_{2d} is the output of the higher layer controller. As a fundamental function of attitude control, $\eta_{2d} = [0,0,0]^\top$ is an essential state that needs to be stabilized under the influence of disturbance, such as hydrodynamic moments generated by constant fluent or turbulence, and collision.

4.2.1. I&I Control Experiment

The parameters of the I&I controller are selected as follows: $K_1 = diag\{0.02, 0.1, 0.03\}$, $K_2 = diag\{5,5,5\}$, $\gamma = [0.4, 0.4, 0.4]^\top$.

To validate the anti-disturbance performance of I&I controller, disturbances are applied to each axis. Each axis is disturbed by a collision. The I&I controller regulates the state to $\eta_2 = [0,0,0]^\top$. The roll, pitch, and yaw control results of I&I control are shown in Figure 4.

The experiment results show that: (1) the attitude error generated by the collision is near 10°; (2) the attitude converges to η_{2d} in less than 0.8 s; (3) the roll and pith control have shorter adjust time than the yaw control, which is because the yaw axis has a higher moment of inertia and a lower control moment, resulting from the mechanical design of the rotor arrangement; and (4) when no disturbance is applied, the control accuracy of roll and pitch is better than yaw; this is mainly for the unmodeled dynamics of the tail servo.

The control torque is shown in Figure 5. The roll and pitch control toques are generated by the change in the speed of the rotors, which is quick, whereas the yaw control torque is generated by the servo sway, which is slow and has a slight chattering.

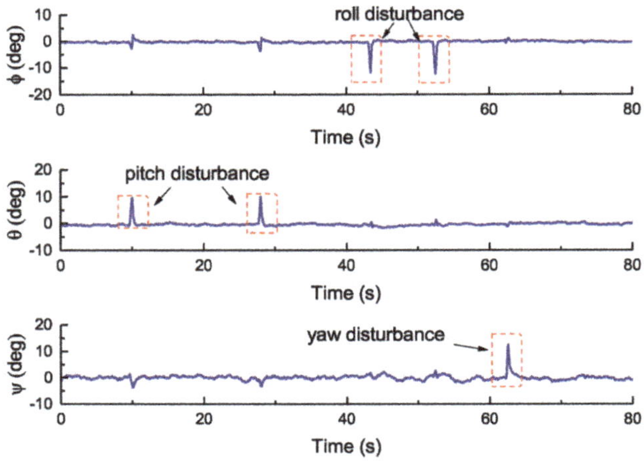

Figure 4. Attitude of NTAUV under I&I control with disturbance on each axis.

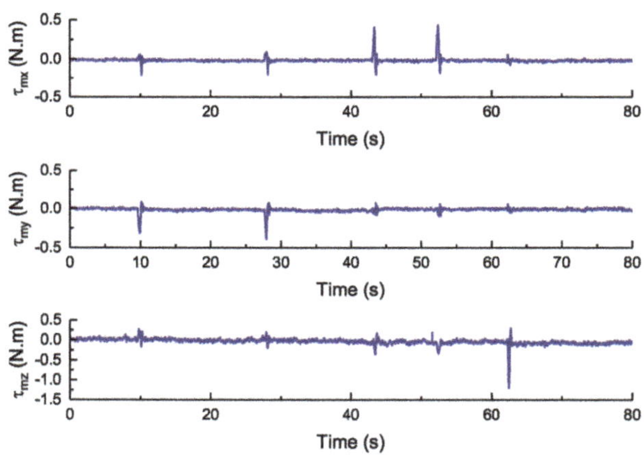

Figure 5. I&I control torque.

4.2.2. PI Control Experiment

We compare experiment results generated by the I&I controller with the ones generated by the typical cascaded proportional integral (PI) controller. The parameters of PI controller are selected carefully to get good performance. The parameters of PI approach are as follows: angular velocity loop controller $k_p^v = 0.2, k_i^v = 1$, angular position loop controller $k_p^p = 0.08, k_i^p = 0.001$. The experiment results of PI control are shown in Figures 6 and 7. It shows that it takes more than 1.3 s to converge to desired attitude, and the yaw control takes even more than 3 s.

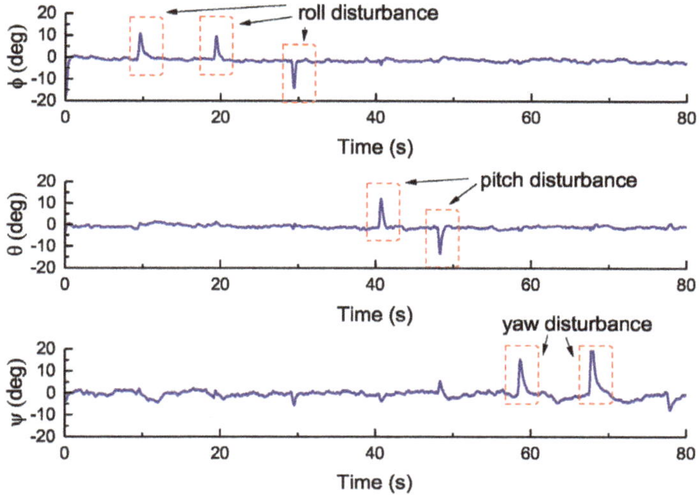

Figure 6. Attitude of NTAUV under PI control with disturbance on each axis.

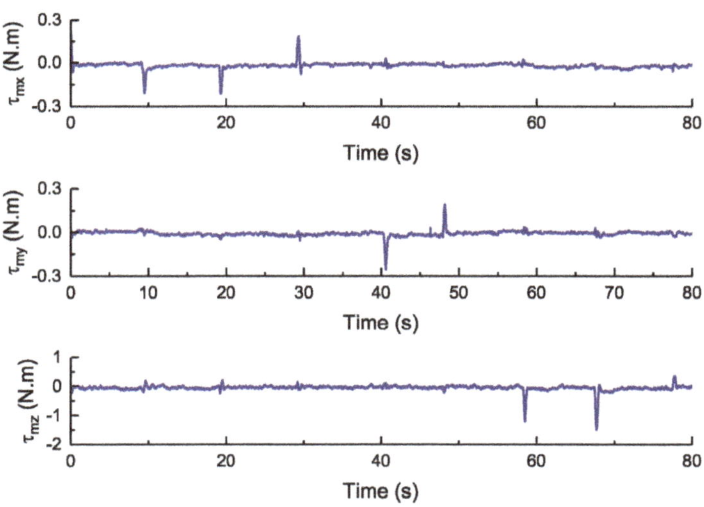

Figure 7. PI control torque.

4.2.3. Comparison Between I&I and PI Control

The comparisons of the roll, pitch and yaw control performance between I&I and PI control are shown separately in Figures 8–10. The I&I control performs higher accuracy than PI control under disturbance.

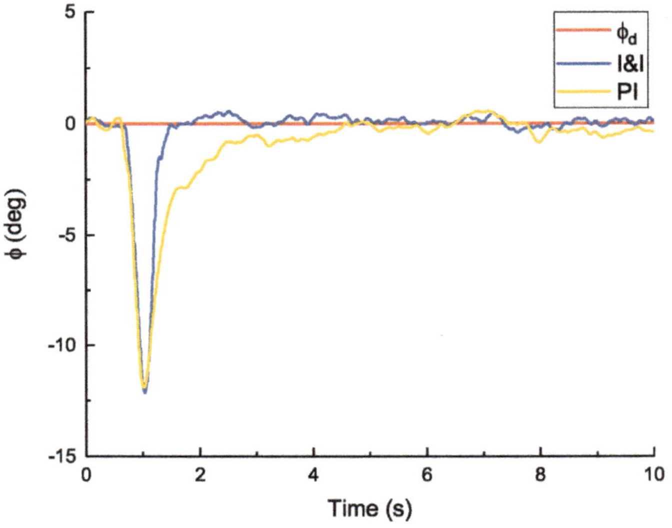

Figure 8. Roll Control Comparison between I&I and PI Control.

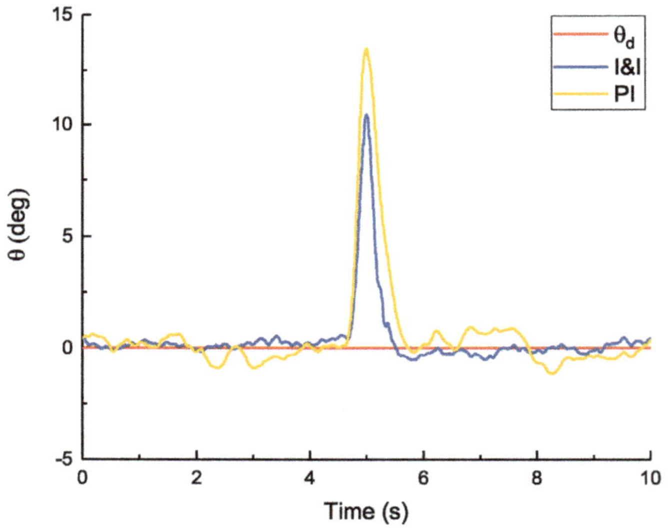

Figure 9. Pitch Control Comparison between I&I and PI Control.

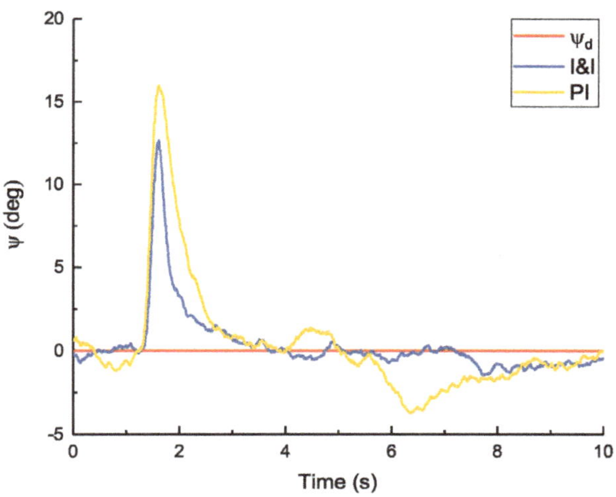

Figure 10. Yaw Control Comparison between I&I and PI Control.

The hydrodynamic force and moment are complex as the underwater vehicle works at a low Reynolds number (Re) condition, especially when hovering and for low-speed horizontal moving. The actual hydrodynamic force is very complex, because there is no stable flow field around. The disturbance rejection performance of I&I control is validated.

5. Conclusions

In this paper, a negative-buoyancy tri-tilt-rotor autonomous underwater vehicle was presented and an attitude controller was designed for attitude stabilization. The full mathematical model of the NTAUV was established, and it was decoupled to attitude and altitude subsystems. Then, the attitude subsystem was investigated, and an adaptive attitude controller was designed based on the I&I theory. A parameter estimator was applied to estimate the unknown parameters. The global stability of the controller was proved. Finally, the performance of the proposed controller was validated through a real-time attitude stabilization experiment. The experimental results indicated a satisfactory performance compared with a well designed PID controller.

Author Contributions: T.W. put forward the original concept, proposed the control approach, and wrote the article. C.W., T.G., J.W. and T.G. gave their valuable suggestions on research design. Further, T.W., C.W. and J.W., analyzed and discussed the experimental results.

Funding: This work is supported by the National Natural Science Foundation of the National Research and Development of major scientific instruments (41427806), the National Natural Science Foundation of China (Grant No. 51779139), the National Key Research and Development Program of China (Grant No. 2016YFC0300700), ROV System For Exploration and Sampling (DY125-21-Js-06) funded by the State Oceanic Administration People's Republic of China and National High Technology Research and Development Program of China (863 Program, Grant No. 2012AA092103).

Acknowledgments: The authors would also like to express their appreciation to Chunwen Zhang and Han Liu for their considerable help on the experiment.

Conflicts of Interest: The authors declare no conflict of interest.

References

1. Fossen, T. *Guidance and Control of Ocean Vehicles*; John Willey & Sons: Hoboken, NJ, USA, 1994.

2. Li, X.; Zhao, M.; Ge, T. A Nonlinear Observer for Remotely Operated Vehicles with Cable Effect in Ocean Currents. *Appl. Sci.* **2018**, *8*, 867. [CrossRef]
3. Wu, N.; Wu, C.; Ge, T.; Yang, D.; Yang, R. Pitch Channel Control of a REMUS AUV with Input Saturation and Coupling Disturbances. *Appl. Sci.* **2018**, *8*, 253. [CrossRef]
4. Salgado-Jimenez, T.; Gonzalez-Lopez, J.L.; Pedraza-Ortega, J.C.; García-Valdovinos, L.G.; Martínez-Soto, L.F.; Resendiz-Gonzalez, P.A. Design of ROVs for the Mexican power and oil industries. In Proceedings of the 2010 1st International Conference on Applied Robotics for the Power Industry, Montreal, QC, Canada, 5–7 October 2010; pp. 1–8.
5. Soylu, S.; Proctor, A.A.; Podhorodeski, R.P.; Bradley, C.; Buckham, B.J. Precise trajectory control for an inspection class ROV. *Ocean Eng.* **2016**, *111*, 508–523. [CrossRef]
6. Wynn, R.B.; Huvenne, V.A.I.; Le Bas, T.P.; Murton, B.J.; Connelly, D.P.; Bett, B.J.; Ruhl, H.A.; Morris, K.J.; Peakall, J.; Parsons, D.R.; et al. Autonomous Underwater Vehicles (AUVs): Their past, present and future contributions to the advancement of marine geoscience. *Mar. Geol.* **2014**, *352*, 451–468. [CrossRef]
7. Palomeras, N.; Vallicrosa, G.; Mallios, A.; Bosch, J.; Vidal, E.; Hurtos, N.; Carreras, M.; Ridao, P. AUV homing and docking for remote operations. *Ocean Eng.* **2018**, *154*, 106–120. [CrossRef]
8. Liu, K.; Huang, Y.; Zhao, Y.; Cui, S.; Wang, X.; Wang, G. Research on error correction methods for the integrated navigation system of deep-sea human occupied vehicles. In Proceedings of the OCEANS 2015—MTS/IEEE Washington, Washington, DC, USA, 19–22 October 2015; pp. 1–5.
9. Hosseini, M.; Seyedtabaii, S. Robust ROV path following considering disturbance and measurement error using data fusion. *Appl. Ocean Res.* **2016**, *54*, 67–72. [CrossRef]
10. Cui, R.; Zhang, X.; Cui, D. Adaptive sliding-mode attitude control for autonomous underwater vehicles with input nonlinearities. *Ocean Eng.* **2016**, *123*, 45–54. [CrossRef]
11. Cui, R.; Li, Y.; Yan, W. Mutual Information-Based Multi-AUV Path Planning for Scalar Field Sampling Using Multidimensional RRT. *IEEE Trans. Syst. Man Cybern. Syst.* **2016**, *46*, 993–1004. [CrossRef]
12. Cui, R.; Yang, C.; Li, Y.; Sharma, S. Adaptive Neural Network Control of AUVs With Control Input Nonlinearities Using Reinforcement Learning. *IEEE Trans. Syst. Man Cybern. Syst.* **2017**, *47*, 1019–1029. [CrossRef]
13. Nouri, N.M.; Valadi, M. Robust input design for nonlinear dynamic modeling of AUV. *ISA Trans.* **2017**, *70*, 288–297. [CrossRef] [PubMed]
14. Paull, L.; Saeedi, S.; Seto, M.; Li, H. AUV Navigation and Localization: A Review. *IEEE J. Ocean. Eng.* **2014**, *39*, 131–149. [CrossRef]
15. Hussain, N.A.A.; Arshad, M.R.; Mohd-Mokhtar, R. Underwater glider modelling and analysis for net buoyancy, depth and pitch angle control. *Ocean Eng.* **2011**, *38*, 1782–1791. [CrossRef]
16. Brown, C.L. Deep sea mining and robotics: Assessing legal, societal and ethical implications. In Proceedings of the 2017 IEEE Workshop on Advanced Robotics and its Social Impacts (ARSO), Austin, TX, USA, 8–10 March 2017; pp. 1–2.
17. Javaid, M.Y.; Ovinis, M.; Hashim, F.B.M.; Maimun, A.; Ahmed, Y.M.; Ullah, B. Effect of wing form on the hydrodynamic characteristics and dynamic stability of an underwater glider. *Int. J. Naval Archit. Ocean Eng.* **2017**, *9*, 382–389. [CrossRef]
18. Li, Y.; Pan, D.; Zhao, Q.; Ma, Z.; Wang, X. Hydrodynamic performance of an autonomous underwater glider with a pair of bioinspired hydro wings—A numerical investigation. *Ocean Eng.* **2018**, *163*, 51–57. [CrossRef]
19. Tchilian, R.D.S.; Rafikova, E.; Gafurov, S.A.; Rafikov, M. Optimal Control of an Underwater Glider Vehicle. *Procedia Eng.* **2017**, *176*, 732–740. [CrossRef]
20. Bessa, W.M.; Dutra, M.S.; Kreuzer, E. An adaptive fuzzy sliding mode controller for remotely operated underwater vehicles. *Robot. Auton. Syst.* **2010**, *58*, 16–26. [CrossRef]
21. Campos, E.; Monroy, J.; Abundis, H.; Chemori, A.; Creuze, V.; Torres, J. A nonlinear controller based on saturation functions with variable parameters to stabilize an AUV. *Int. J. Naval Archit. Ocean Eng.* **2018**. [CrossRef]
22. Yan, H.; Ge, T.; Li, J.W.; Wang, Q. Prediction of mode and static stability of negative buoyancy vehicle. In Proceedings of the 2011 Chinese Control and Decision Conference (CCDC), Mianyang, China, 23–25 May 2011; pp. 1903–1909.
23. Ji, S.W.; Bui, V.P.; Balachandran, B.; Kim, Y.B. Robust control allocation design for marine vessel. *Ocean Eng.* **2013**, *63*, 105–111. [CrossRef]

24. Xiang, X.; Lapierre, L.; Jouvencel, B. Smooth transition of AUV motion control: From fully-actuated to under-actuated configuration. *Robot. Auton. Syst.* **2015**, *67*, 14–22. [CrossRef]
25. Xiang, X.; Yu, C.; Zhang, Q. Robust fuzzy 3D path following for autonomous underwater vehicle subject to uncertainties. *Comput. Oper. Res.* **2017**, *84*, 165–177. [CrossRef]
26. Salazar-Cruz, S.; Kendoul, F.; Lozano, R.; Fantoni, I. Real-time stabilization of a small three-rotor aircraft. *IEEE Trans. Aerosp. Electron. Syst.* **2008**, *44*, 783–794. [CrossRef]
27. Papachristos, C.; Tzes, A. Modeling and control simulation of an unmanned Tilt Tri-Rotor Aerial vehicle. In Proceedings of the 2012 IEEE International Conference on Industrial Technology, Athens, Greece, 19–21 March 2012; pp. 840–845.
28. Ta, D.A.; Fantoni, I.; Lozano, R. Modeling and control of a tilt tri-rotor airplane. In Proceedings of the 2012 American Control Conference (ACC), Montreal, QC, Canada, 27–29 June 2012; pp. 131–136.
29. Astolfi, A.; Ortega, R. Immersion and invariance (I2): A new tool in nonlinear control design. In Proceedings of the 2001 European Control Conference (ECC), Porto, Portugal, 4–7 September 2001; pp. 2854–2859.
30. Astolfi, A.; Ortega, R. Immersion and invariance: A new tool for stabilization and adaptive control of nonlinear systems. *IEEE Trans. Autom. Control* **2003**, *48*, 590–606. [CrossRef]
31. Astolfi, A.; Karagiannis, D.; Ortega, R. *Nonlinear and Adaptive Control with Applications*; Springer: Berlin/Heidelberg, Germany, 2008.

applied sciences

MDPI

Article

A Nonlinear Observer for Remotely Operated Vehicles with Cable Effect in Ocean Currents

Xiang Li, Min Zhao * and Tong Ge

State Key Laboratory of Ocean Engineering, Collaborative Innovation Center for Advanced Ship and Deep-Sea Exploration, Shanghai Jiao Tong University, Shanghai 200240, China; lixiang.sjtu@hotmail.com (X.L.); tongge@sjtu.edu.cn (T.G.)

* Correspondence: min.zhao@sjtu.edu.cn; Tel.: +86-021-3420-8063

Received: 28 April 2018; Accepted: 23 May 2018; Published: 25 May 2018

Featured Application: Applying on Observer and Controller for Remotely Operated Vehicles.

Abstract: A nonlinear observer for a remotely operated vehicle (ROV) is investigated, and a four-degree-of-freedom nonlinear sliding state observer is designed in this study. An ocean current model and a simplified umbilical cable disturbing force model of ROVs were set up; the simplified cable force model characterized the cable disturbing force. The velocity information and the cable force were observed and estimated both online and in real time. We proved that the observation error was uniformly ultimately bounded. The modeling of the disturbing force and the compensation for the observer was an effective method to improve the observation precision and to reduce the chattering of the observer outputs.

Keywords: remotely operated vehicle; ocean current; cable disturbance modeling; lumped parameter method; sliding mode observer

1. Introduction

The improvement of marine resources exploitation technology has become important for the development of the marine industry. The design and development of new types of marine engineering equipment to enhance the technological level of the ocean engineering industry is of considerable significance. Because of the complexity of the marine environment, underwater vehicles have become an important and valuable tool for the exploration and development of marine resources. Unmanned underwater vehicles, including remotely operated vehicles (ROVs), autonomous underwater vehicles (AUVs), and underwater gliders, have been extensively and increasingly used in marine environments for exploration, inspection, and engineering operations [1]. They can provide safe and effective access to deep sea and hadal environments without physically entering them.

With advantages like functional diversity, strong operational ability, and a high safety factor, ROVs have been widely used and have become important technical equipment in areas such as ocean resource development, exploration, marine scientific research, and underwater engineering [2].

To guarantee the operation quality, high precision, and safety of ROVs, a high-performance controller for the trajectory tracking or station-keeping of ROVs is required. In practice, there are a number of technical challenges in the control of an underwater vehicle, such as model uncertainties and unknown external disturbances [3–7]. The model uncertainties of ROVs are usually caused by inaccurate hydrodynamic coefficients, which are obtained using towing tank experiments or computational fluid dynamics methods. The unknown disturbances in practical oceanic environments often include currents, waves, and tides. For the control design of ROVs, the external disturbing force caused by the cable that connects with the support ship should also be considered. The currents and the umbilical cable simultaneously affect the ROV motions as the umbilical cable is connected to the ROV.

In many works on the control of underwater dynamic systems, the state vector is assumed to be measured. The quality of feedback signals from the ROV sensor system plays an important role in the control performance of the vehicle as the signals affected by noises can debase the control quality and even lead to system instability [8]. However, in practice, it is difficult or impossible to measure all of the state variables of an ROV system with sensors for technical or economic reasons. For example, cable disturbing force cannot be easily obtained with normal sensors, and there is not enough money to buy high-precision but expensive sensors. Hence, effective state observers for ROVs need to be developed to provide accurate and robust signals. Moreover, it is necessary and is of considerable significance to design an observer to fully know the state variables of the system. The state observers for vehicles have been studied by a number of researchers in the past [9–12]. The use of a complementary observer is one of the most popular techniques for sensor combination. The algorithm allows for straightforward implementation without requiring high computational resources and is suitable for small and low-cost autonomous vehicles with limited onboard power. The linear observer has been solved by Kalman and Luenberger, but the nonlinear case is still an active domain of research. Gauthier et al. developed the high-gain observer approach, which is closely related to the triangular structure and is derived from the uniform observability of nonlinear systems [13]. Fridman et al. proposed a higher-order sliding-mode observer to estimate precisely the observable states and asymptotically the unobservable ones in a multi-input-multi-output nonlinear system with unknown inputs and stable internal dynamics [14]. Rezazadegan and Chatraei derived an adaptive control law for a six-degree-of-freedom (6-DOF) model for the trajectory tracking problem of an underactuated underwater vehicle in the presence of a parametric uncertainty [15]. Khadhraoui et al. proposed to control ROVs for exploration in sub-sea historical sites and designed a nonlinear observer to estimate the linear and angular velocity of an ROV [16]. Chu et al. developed a new adaptive neural network control approach for a class of ROVs and introduced an adaptive observer for the velocity state and angular velocity state estimation with a local recurrent neural network [17].

In complex missions, an ROV often requires a high degree of autonomy, precision, and maneuverability. Moreover, a single sensor is not sufficient to obtain the position and velocity information of the ROV because each type of sensor has its own disadvantages. Thus, ROVs are usually equipped with different types of sensors, such as the inertial measurement unit (IMU) and the velocity log [18]. The IMU can measure the linear accelerations and the rotational velocity of the vehicle. Further, the velocity log, usually a Doppler velocity log (DVL), can measure the linear velocities of the vehicle. The IMU and the DVL measurements are usually combined to provide stable linear velocity, angular velocity, and orientation signals. Depending on the quality of the onboard IMU, the position of the vehicle can be computed from the linear acceleration and velocity measurements. However, because of the inevitable drift in the IMU and DVL, the position estimate will not be stable over time. When the ROV swims through the water, ocean currents always have an influence on the motion of the ROV. To increase the performance, accuracy, and the situation awareness of the ROV, the estimation of the ocean current velocity is often desired. However, the IMU can only measure the total acceleration experienced by the vehicle and cannot distinguish between the motion induced by the ocean current and that induced by the onboard propulsion system. Moreover, the DVL cannot measure the ocean current velocity in its usual configuration. As the hydrodynamic forces acting on the ROV depend on the velocity of the vehicle with respect to the water, but not the total velocity of the vehicle with respect to the fixed coordinate frame, obtaining the ocean current velocity information can considerably improve the performance of the vehicle control system.

For ocean currents, researchers have developed a variety of observers to estimate the current velocity. Refsnes et al. designed an exponentially stable current observer for a reduced-order dynamic vehicle model, and used the estimated current velocity to compute the hydrodynamic forces and moments more accurately for inclusion in a feedback control strategy. Further, the observer only required position measurements [19]. Aguiar and Pascoal proposed an estimation scheme and designed a kinematic observer for ocean current estimation. The proposed observer provides exponential convergence to the true ocean current velocity, but requires the knowledge of both the position and the relative velocity of the vehicle [20]. Børhaug et al. took a model-based approach and designed a six-degree-of-freedom (6-DOF) observer for the ocean current velocity on the basis of the dynamic model and the measurement of the linear velocity of the vehicle [21].

The deep sea ROV systems typically consist of a large support vessel, a winch, umbilical cable, and ROV. Thus far, most of the research has focused on the numerical simulation and prediction of the cable configuration and the ROV motion with the cable effect [22–24]. However, few studies have dealt with motion control models for ROVs considering the cable disturbing force and the current effects because of the corresponding complexity and difficulty.

As the ocean current velocity and the cable disturbing force cannot be measured by the standard onboard sensors, the development of an estimation scheme is required. In this paper, a nonlinear observer for the state estimation of a tethered ROV system in slowly varying ocean currents is proposed. The observer will estimate the velocity state, the external current velocity, and the cable disturbance.

The rest of this paper is organized as follows. In Section 2, the kinematic and dynamic models of ROVs are introduced and formulated. In Section 3, numerical methods for discrete model solutions are given. In Section 4, a nonlinear observer is proposed for estimating velocities, unknown ocean currents, and the cable disturbing force, and the stability properties of the observer are analyzed. Then, in Section 5, the performance of the proposed observer is illustrated through case studies. Finally, the conclusions are presented in Section 6.

2. ROV Description

The deep sea ROV "Sea Dragon" is a large-scale working class ROV developed and built by the Underwater Engineering Research Institute of Shanghai Jiao Tong University (Figure 1). The results presented in this paper are based on the Sea Dragon ROV. The main physical data of the ROV are reported in Table 1. The ROV is equipped with seven thrusters (three vertical and four horizontal). The position of the thrusters on the vehicle is shown in Figure 2.

Figure 1. The Sea Dragon remotely operated vehicle (ROV) at sea.

Table 1. Main parameters of the ROV.

Parameter	Value
Mass in the air (kg)	3450
Mass in the water (kg)	0
Working depth (m)	3500
Length × width × height (m)	3.17 × 1.81 × 1.76
Coordinate of buoyancy center (m)	0, 0, −0.4
Coordinate of gravity center (m)	0, 0, 0
Moment of inertia I_{xx}, I_{yy}, I_{zz} (kgm^2)	2200, 710, 652

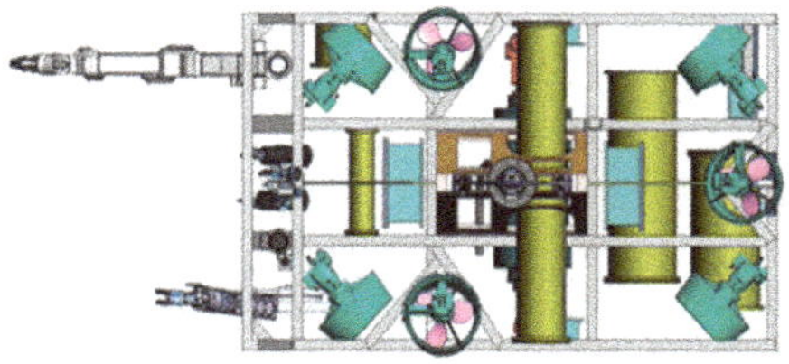

Figure 2. Thruster configuration of ROVs.

The sensor set available for the working class ROV includes:

- The global positioning system (GPS);
- Ultra-short baseline (USBL);
- Altimeter: 200 kHz transmit frequency standard, depth rating 6000 m;
- Depth sensor: 6000 m rating, 0.01% accuracy;
- The inertial measurement unit (IMU);
- Doppler Velocity Log (DVL): working frequency 600 kHz, 0.3% full scale accuracy.

An Ordinary Compact PCI industrial control computer is adopted as the upper computer and PC104 is used for the lower computer.

3. Dynamic Model of ROV

3.1. Coordinate Systems

The dynamic model of an underwater vehicle is established and analyzed in two orthogonal coordinate systems, as shown in Figure 3, namely the earth-fixed frame O-XYZ and the body-fixed frame o-xyz.

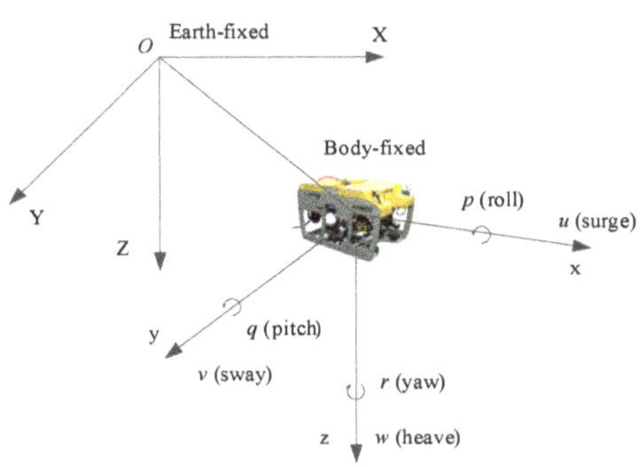

Figure 3. Earth-fixed and body-fixed reference frames.

The earth-fixed frame O-XYZ is a global inertial coordinate system fixed at the ocean surface ship with the origin at O. The OZ axis points vertically down to the water, and the OX and OY axes are in two mutually perpendicular horizontal directions.

The body-fixed frame o-xyz is a local coordinate system fixed on the vehicle with the origin at o. Here, the ox axis points to the front of the vehicle, the oz axis points downward, and the oy axis completes the right-hand system with the other two axes.

Underwater remotely operated vehicles have 6-DOF motions, namely, the surge, sway, heave, roll, pitch, and yaw. Three translation displacements, namely X (surge), Y (sway), and Z (heave), and three Euler angles, namely φ (roll), θ (pitch), and ψ (yaw), represent the position and the attitude of the vehicle with respect to the inertial frame, respectively. The instantaneous velocity and the angular velocity with respect to body-fixed frame are represented by (u, v, w) and (p, q, r), respectively.

The transformation of the forces and motions from the global to the local coordinate system can be fulfilled by using the transformation matrix through the Euler angles $\phi\ \theta\ \psi$. The orientation of the vehicle in the global coordinates can be specified by the vector r_o from O to o.

The position of the vehicle is denoted as $\begin{bmatrix} X & Y & Z \end{bmatrix}^T$ in the inertial coordinates (earth-fixed). Linear velocities $\begin{bmatrix} u & v & w \end{bmatrix}^T$ and angular velocities $\begin{bmatrix} p & q & r \end{bmatrix}^T$ are expressed in the body-fixed coordinates. There exists

$$\begin{bmatrix} \dot{X} & \dot{Y} & \dot{Z} \end{bmatrix}^T = J_1 \begin{bmatrix} u & v & w \end{bmatrix}^T \tag{1}$$

where

$$J_1 = \begin{bmatrix} \cos\theta\cos\psi & \sin\varphi\sin\theta\cos\psi - \cos\varphi\sin\psi & \cos\varphi\sin\theta\cos\psi + \sin\varphi\sin\psi \\ \cos\theta\sin\psi & \sin\varphi\sin\theta\sin\psi + \cos\varphi\cos\psi & \cos\varphi\sin\theta\sin\psi - \sin\varphi\cos\psi \\ -\sin\theta & \sin\varphi\cos\theta & \cos\varphi\cos\theta \end{bmatrix} \tag{2}$$

J_1 is the coordinate transformation matrix from the body-fixed to the earth-fixed frame, ${J_1}^{-1} = {J_1}^T$, and J_1 and ${J_1}^{-1}$ are both units of an orthogonal array.

Similarly, the relationship between the angular velocities and the attitude angles can be obtained as follows:

$$\begin{bmatrix} \dot{\varphi} & \dot{\theta} & \dot{\psi} \end{bmatrix}^T = J_2 \begin{bmatrix} p & q & r \end{bmatrix}^T \tag{3}$$

where

$$J_2 = \begin{bmatrix} 1 & \sin\varphi\tan\theta & \cos\varphi\tan\theta \\ 0 & \cos\varphi & -\sin\varphi \\ 0 & \sin\varphi\sec\theta & \cos\varphi\sec\theta \end{bmatrix} \tag{4}$$

3.2. ROV Dynamic Model

In the 6-DOF motion equations of the ROV, $v_1 = \begin{bmatrix} u & v & w \end{bmatrix}^T$ denotes the ROV's velocity in the body-fixed frame, $v_2 = \begin{bmatrix} p & q & r \end{bmatrix}^T$ represents the angular velocity, and $r_g = \begin{bmatrix} x_g & y_g & z_g \end{bmatrix}^T$ is the coordinate of the center of gravity. The momentum equation and the moment of momentum equation of the ROV are expressed as follows [25]:

$$m\left[\dot{v}_1 + v_2 \times v_1 + \dot{v}_2 \times r_G + v_2 \times (v_2 \times r_G)\right] = F \tag{5}$$

$$I_0\dot{v}_2 + v_2 \times (I_0 v_2) + m r_G \times (\dot{v}_1 + v_2 \times v_1) = M \tag{6}$$

where m denotes the mass and I_0 represents the inertial mass moment matrix of the ROV in the body-fixed frame. F and M denote the forces and the moment acting on the center of the ROV, respectively. The two abovementioned equations can be combined as follows:

$$M_{RB}\dot{x} + C_{RB}(v)x = \tau \tag{7}$$

where $x = \begin{bmatrix} v_1 & v_2 \end{bmatrix}^T$, M_{RB} denotes the general mass matrix, and C_{RB} represents the Coriolis matrix. If $r_g = \begin{bmatrix} 0 & 0 & 0 \end{bmatrix}^T$, then

$$M_{RB} = \begin{bmatrix} m & 0 & 0 & 0 & 0 & 0 \\ 0 & m & 0 & 0 & 0 & 0 \\ 0 & 0 & m & 0 & 0 & 0 \\ 0 & 0 & 0 & I_x & -I_{xy} & -I_{xz} \\ 0 & 0 & 0 & -I_{xy} & I_y & -I_{yz} \\ 0 & 0 & 0 & -I_{xz} & -I_{yz} & I_z \end{bmatrix} \tag{8}$$

$$C_{RB}(x) = \begin{bmatrix} 0 & 0 & 0 & 0 & mw & -mv \\ 0 & 0 & 0 & -mw & 0 & mu \\ 0 & 0 & 0 & mv & -mu & 0 \\ 0 & mw & -mv & 0 & -I_{yz}q - I_{xz}p + I_z r & I_{yz}r + I_{xy}p - I_y q \\ -mw & 0 & mu & I_{yz}q + I_{xz}p - I_z r & 0 & -I_{xz}r - I_{xy}q + I_x p \\ mv & -mu & 0 & -I_{yz}r - I_{xy}p + I_y q & I_{xz}r + I_{xy}q - I_x p & 0 \end{bmatrix} \tag{9}$$

$\tau = \begin{bmatrix} F & M \end{bmatrix}^T$, τ includes all of the external forces, such as gravity, buoyancy, inertia, and viscous forces due to the fluid, propulsion, and the cable tension.

3.3. Added Mass Force on ROV

The added mass force acting on the ROV is defined as follows:

$$F_A = -M_A\dot{x} - C_A(x)x$$

where M_A denotes the general mass matrix and C_A represents the Coriolis matrix corresponding to the added mass of the ROV. M_A and C_A can be further expressed as follows:

$$M_A = -\begin{bmatrix} X_{\dot{u}} & X_{\dot{v}} & X_{\dot{w}} & X_{\dot{p}} & X_{\dot{q}} & X_{\dot{r}} \\ Y_{\dot{u}} & Y_{\dot{v}} & Y_{\dot{w}} & Y_{\dot{p}} & Y_{\dot{q}} & Y_{\dot{r}} \\ Z_{\dot{u}} & Z_{\dot{v}} & Z_{\dot{w}} & Z_{\dot{p}} & Z_{\dot{q}} & Z_{\dot{r}} \\ K_{\dot{u}} & K_{\dot{v}} & K_{\dot{w}} & K_{\dot{p}} & K_{\dot{q}} & K_{\dot{r}} \\ M_{\dot{u}} & M_{\dot{v}} & M_{\dot{w}} & M_{\dot{p}} & M_{\dot{q}} & M_{\dot{r}} \\ N_{\dot{u}} & N_{\dot{v}} & N_{\dot{w}} & N_{\dot{p}} & N_{\dot{q}} & N_{\dot{r}} \end{bmatrix} \tag{10}$$

$$C_A(x) = \begin{bmatrix} 0 & 0 & 0 & 0 & -a_3 & a_2 \\ 0 & 0 & 0 & a_3 & 0 & -a_1 \\ 0 & 0 & 0 & -a_2 & a_1 & 0 \\ 0 & -a_3 & a_2 & 0 & -b_3 & b_2 \\ a_3 & 0 & -a_1 & b_3 & 0 & -b_1 \\ -a_2 & a_1 & 0 & -b_2 & b_1 & 0 \end{bmatrix} \tag{11}$$

where

$$\begin{cases} a_1 = X_{\dot{u}}u + X_{\dot{v}}v + X_{\dot{w}}w + X_{\dot{p}}p + X_{\dot{q}}q + X_{\dot{r}}r \\ a_2 = X_{\dot{v}}u + Y_{\dot{v}}v + Y_{\dot{w}}w + Y_{\dot{p}}p + Y_{\dot{q}}q + Y_{\dot{r}}r \\ a_3 = X_{\dot{w}}u + Y_{\dot{w}}v + Z_{\dot{w}}w + Z_{\dot{p}}p + Z_{\dot{q}}q + Z_{\dot{r}}r \\ b_1 = X_{\dot{p}}u + Y_{\dot{p}}v + Z_{\dot{p}}w + K_{\dot{p}}p + K_{\dot{p}}q + K_{\dot{r}}r \\ b_2 = X_{\dot{q}}u + Y_{\dot{q}}v + Z_{\dot{q}}w + K_{\dot{q}}p + M_{\dot{q}}q + M_{\dot{r}}r \\ b_3 = X_{\dot{r}}u + Y_{\dot{r}}v + Z_{\dot{r}}w + K_{\dot{r}}p + M_{\dot{r}}q + N_{\dot{r}}r \end{cases} \tag{12}$$

3.4. Viscous Hydrodynamic Damping Forces

Viscous resistance is a function of the velocity with respect to the current velocity. The relative velocity state vector of the ROV gives $x_r = [u_r, v_r, w_r, p, q, r]^T$, $x_r = x - x_c$, and $x_c = [u_c, v_c, w_c, 0, 0, 0]^T$, where x_c is the ocean current velocity. Moreover, the viscous resistance F_V can be written in a simplified form as follows:

$$F_V = -Dx_r \tag{13}$$

where $D = D_L + D_Q$, $D_L x_r$ denotes the linear damping term, and $D_Q x_r$ represents the quadratic damping term. D_Q is very complex, and only its diagonal elements are kept as the influence of the coupling term is small. This is applicable to the modeling and simulation of a slow underwater vehicle. The hydrodynamic coefficients owing to the accelerations and angular accelerations of the vehicle can be obtained by using the planar motion mechanism model tests. D can be expressed as follows:

$$D = -\begin{bmatrix} X_u + X_{u|u|}|u_r| & X_v & X_w & X_p & X_q & X_r \\ Y_u & Y_v + Y_{v|v|}|v_r| & Y_w & Y_p & Y_q & Y_r \\ Z_u & Z_v & Z_w + Z_{w|w|}|w_r| & Z_p & Z_q & Z_r \\ K_u & K_v & K_w & K_p + K_{p|p|}|p| & K_q & K_r \\ M_u & M_v & M_w & M_p & M_q + M_{q|q|}|q| & M_r \\ N_u & N_v & N_w & N_p & N_q & N_r + N_{r|r|}|r| \end{bmatrix} \tag{14}$$

3.5. Weight and Buoyancy

The gravity and the buoyancy of the ROV in the earth-fixed coordinate system give W and B, respectively. The gravity and the buoyancy in the body-fixed coordinates W_b, B_b can be obtained as follows:

$$W_b = J_1^{-1}\begin{bmatrix} 0 \\ 0 \\ W \end{bmatrix}, B_b = -J_1^{-1}\begin{bmatrix} 0 \\ 0 \\ B \end{bmatrix} \tag{15}$$

Thus, the restoring force F_R has the following form:

$$F_R = \begin{bmatrix} W_b + B_b \\ r_G \times W_b + r_B \times B_b \end{bmatrix} \tag{16}$$

The gravitational force and the buoyant force are defined in the global coordinate system, and therefore, they should be transformed to the local coordinate system as follows:

$$F_R = -g = -\begin{bmatrix} (W-B)\sin\theta \\ -(W-B)\cos\theta\sin\varphi \\ -(W-B)\cos\theta\cos\varphi \\ -(y_G W - y_B B)\cos\theta\cos\varphi + (z_G W - z_B B)\cos\theta\sin\varphi \\ (z_G W - z_B B)\sin\theta + (x_G W - x_B B)\cos\theta\cos\varphi \\ -(x_G W - x_B B)\cos\theta\sin\varphi - (y_G W - y_B B)\sin\theta \end{bmatrix} \tag{17}$$

where $r_G = [x_G, y_G, z_G]^T$ and $r_B = [x_B, y_B, z_B]^T$ denote the locations of the gravitational and the buoyancy center of the ROV in the body-fixed frame, respectively.

3.6. Thruster Force and Moment

The ROV model considered in this study is equipped with seven thrusters. That is, T_1, T_2, T_3, and T_4 are installed horizontally and are responsible for the forward and sideward motions, while T_5, T_6, and T_7 are installed vertically at an inclined angle to induce the ascending and descending motions. The resultant forces and moments induced by all of the thrusters acting on the centroid in the body-fixed coordinate frame can be expressed as follows:

$$F_T = \begin{bmatrix} F_{Tx} & F_{Ty} & F_{Tz} & M_{Tx} & M_{Ty} & M_{Tz} \end{bmatrix}^T \tag{18}$$

where F_{Tx}, F_{Ty}, F_{Tz} are the three axial components of the resultant thrust force in the body-fixed coordinate system, and M_{Tx}, M_{Ty}, M_{Tz} are the three axial components of the resultant thrust moment.

3.7. Umbilical Cable Force

The umbilical cable plays an important role in facilitating the power supply and the communication function between the ROV and the support vessel. However, the attachment of the cable and the drag relative to the current places some restrictions on the maneuverability of the ROV. Therefore, the estimation of the corresponding effect caused by the umbilical cable and the current will be helpful for the controller design of the ROV. However, thus far, most of the researchers have neglected the effect of the umbilical cable because of the complexity involved, particularly in including the current effect.

There are two types of modeling techniques available to predict the response of tethered systems: continuous analytical methods and discrete numerical models [26]. Discrete numerical models are valid for some nonlinear properties such as the quadratic drag and the spatially varying properties of cables. The nonlinear coupling motion principle between the tether and the vehicle can be included in these models. The most prevalent numerical approaches used nowadays for determining the hydrodynamic performance of an underwater tethered system are the lumped mass method [27], the finite difference method [28,29], and the finite element method [30,31]. The lumped mass method is adopted in this study for cable simulation because of its simplicity and effectiveness [26].

In the present study, for simplifying the problem, the following assumptions are made to solve the configuration and tension of the umbilical cable attached to the ROV:

1. The umbilical cable is incompressible.
2. The cable surface is relatively smooth, ignoring the attachments on the cable.
3. The bending stiffness of the cable is ignored. The umbilical cable can only resist the tension force, but not the bending moment and the compression force.
4. The torsional rigidity and the quality of cable point rotation effect, which do not consider the torque, are ignored.

The umbilical cable force is one of the most important nonlinear disturbance forces on the ROV when the current is strong and the cable length is sufficiently long. The movement of a flexible cable moving in a fluid can be described by the following lumped parameter equations.

The cable model is described using the lumped parameters in the earth-fixed frame of O-XYZ. The coordinate system for analyzing the umbilical cable is shown in Figure 4. The origin O coincides with the end point of the umbilical cable. As shown in the figure, the cable is divided into n segments by $(n + 1)$ nodes. The node number is sorted from the bottom to the top. The first node is on the ROV, and the last node is on the tether management system (TMS) or the surface ship. The jth segment refers to the segment between the jth node and the $(j + 1)$th node. Moreover, the segment is considered a linear elastic unit with the same tension directing along the tangent of the segment as that shown in Figure 4.

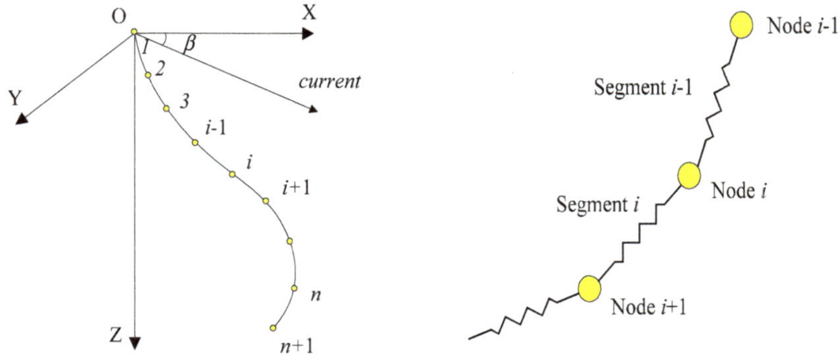

Figure 4. Discrete cable obtained using the lumped mass method.

According to the lumped mass model, the mass of each segment is distributed on the two boundary nodes equally and the nodes are connected with a light spring. Therefore, the first node bears half of the mass of the first segment as well as the last node in particular.

For the nodes except the first and the last, the kinetics equations of each node can be obtained by using Newton's laws of motion as follows:

$$m_i \frac{du_i}{dt} = W_i - B_i + F_{Ai} + D_i + T_i - T_{i-1} \quad (19)$$

$$\frac{dr_i}{dt} = u_i \quad (20)$$

where m_i is the ith lumped mass of the cable, u_i and r_i are the velocity and position vectors of the ith lumped mass with respect to the fixed inertial reference, and W_i and B_i are the gravity and buoyancy of the ith lumped mass. D_i represents the drag force, and T_i represents the tension force of the ith segment unit. F_{Ai} represents the additional mass force of the lumped mass with the following form:

$$F_{Ai} = m_a \left(\frac{du_c}{dt} - \frac{du_i}{dt} \right) \quad (21)$$

where m_a is the additional mass, $m_a = k_a \rho A l_0$, ρ is the sea water density, k_a is the additional mass coefficient, A is the cross-sectional area, l_0 is the initial segment length of the cable, and u_c is the current velocity.

As the cable shape is cylindrical, the drag force can be obtained according to the resistance formula of cylindrical objects in the current, and the resistance of each segment unit is distributed to the nodes at both the ends as follows:

$$D = -\frac{1}{2} \rho C S |v'| v' \quad (22)$$

where ρ is the sea water density; C is the cable resistance coefficient, which is often determined by tests; S is the incident flow area of the cylinder; and v' is the relative velocity to the flow. The drag force can be decomposed into the tangential and the normal components. The drag force lumped onto the ith node can be expressed as follows, and the hydrodynamic force on the umbilical cable can be resolved into the tangential component and the normal component.

$$D_{ti} = -\rho d_l C_t (|l_{i-1}| + |l_i|) |u_{ri} \tau_i| [(u_{ri} \tau_i) \tau_i] / 4 \quad (23)$$

$$D_{ni} = -\rho d_l C_n(|l_{i-1}| + |l_i|)|u_{ri} - (u_{ri}\tau_i)\tau_i|[u_{ri} - (u_{ri}\tau_i)\tau_i]/4 \tag{24}$$

where C_t, C_n are the tangential and the normal drag coefficients of the cable in the ith link, respectively, and l_i, u_{ri} can be calculated as follows:

$$l_i = r_{i+1} - r_i,\ (i = 1 \cdots n) \tag{25}$$

$$u_{ri} = u_i - u_c,\ (i = 2 \cdots n) \tag{26}$$

Thus, the lumped drag force on each node can be expressed as follows:

$$D_i = D_{ti} + D_{ni} \tag{27}$$

In fact, the cable is curved in the sea, and resistances at different points usually have different directions. To reduce the error due to the resistance direction, a tangential vector of average resistance is introduced. The unit tangential vector at the ith node is defined as follows:

$$\tau_i = (r_{i+1} - r_{i-1})/|r_{i+1} - r_{i-1}| \tag{28}$$

As the cable is equivalent to a discrete segmented spring model, the tension can be calculated according to Hooke's law when the tensile deformation of the spring unit occurs, as follows:

$$T = EA\varepsilon \tag{29}$$

where E denotes the elastic modulus of the piecewise cable unit, A is the cross-sectional area of the cable, and ε is the longitudinal strain. For an actual cable that can only be stretched, but not be compressed, the strain ε can only be positive. That is, there exists tension inside the cable when the segment unit is stretched more than the initial segmented length; otherwise, the tension is zero.

The tension force exerted in the ith link can be calculated by using the position of the nodes as follows:

$$T_i = \begin{cases} \pi {d_i}^2 E_i(|l_i| - l_0)l_i/(4l_0|l_i|), & |l_i| \geq l_0 \\ 0, & |l_i| < l_0 \end{cases} \tag{30}$$

3.8. Interaction Between the ROV and Umbilical Cable

When the ROV moves, it is pulled by the cable with tensile force. In turn, the movement of the ROV changes the position and the velocity of the end of the cable, thus changing the shape and the internal tension of the cable. The tension variation also affects the movement of the ROV and is a cyclic process. The interaction between the ROV and the umbilical cable can be described as follows:

$$u_{n+1} = J_1(v_1 + v_2 \times r_c) \tag{31}$$

$$r_{n+1} = r_{rov} + J_1 r_c \tag{32}$$

where J_1 is the transformation matrix and r_c is the position vector of the cable's tying point on the ROV in the body coordinate system.

For the ith node, u_i and r_i represent the displacement and the velocity, respectively. For the last node on the TMS or surface vessel without motion,

$$u_1 = u_s(t) \tag{33}$$

$$r_1 = r_s(t) \tag{34}$$

Because the cable is built on a fixed coordinate system, the tension is relative to the fixed system. The conversion matrix transformation is still required to calculate the ROV cable force in the O-xyz coordinate system as follows:

$$F_{bcable} = -J_1^{-1} T_{n+1} \tag{35}$$

$$M_{bcable} = r_c \times \left(-J_1^{-1} T_{n+1}\right) \tag{36}$$

Considering all of the abovementioned forces acting on the ROV, we can express the vector form of the dynamic model of the ROV in still water as follows:

$$M\dot{x} + C(x)x + D(x)x + g = F_T + F_{bcable} \tag{37}$$

where

$$\begin{cases} M = M_{RB} + M_A \\ C(x) = C_{RB}(x) + C_A(x) \end{cases} \tag{38}$$

$C(x)$ is the Coriolis and centripetal force matrix, $D(x)$ is the damping term, g is gravity and buoyancy, F_T is the control thrust, and F_{bcable} is the cable disturbing force in the body-fixed frame. When the existing currents, $\dot{x}_r = \dot{x} - \dot{x}_c$, $\dot{x}_c = 0$, are calculated using the relative velocity vector x_r to replace the ROV velocity x in the equations of motions, the general vector expressions for the dynamics equation of the ROV in the current can be expressed as follows:

$$M\dot{x}_r + C(x_r)x_r + D(x_r)x_r + g = F_T + F_{bcable} \tag{39}$$

where

$$\begin{cases} M = M_{RB} + M_A \\ C(x_r) = C_{RB}(x_r) + C_A(x_r) \end{cases} \tag{40}$$

4. Numerical Methods for the Solution

4.1. Numerical Integration Method

A set of coupled differential equations of the multi-body system can be solved simultaneously with dynamic equilibrium at each time step. The static equilibrium position of the system can be used to provide the initial condition. All of the nonlinearities (material, geometric, explicit loads, and hydrodynamic loads) are treated in a consistent manner. These governing equations are integrated by using the Runge–Kutta method.

4.2. Parameters and Calculation Steps

Because Equations (11) to (16) are a set of first-order ordinary differential equations with two-point boundary values, the Runge–Kutta method is applied to solve these equations. On the basis of the dynamic model of the ROV and the umbilical cable, the programming flow chart of cable simulation with the lumped parameter method is as follows (Figure 5):

For solving the nonlinear equations, an initial value must be given. In the previous studies, most researchers considered the straight line running steadily to be the initial configuration, and the vertical line of the cable in the still water was assumed as its initial state in this study. Further, the program allows running steadily in advance as the initial configuration.

An ROV with the principal dimensions of 3.17 m (length) × 1.81 m (beam) × 1.76 m (depth) is adopted as the numerical model for the calculations. The ROV is neutrally buoyant in the water. All the corresponding hydrodynamic coefficients of the maneuvering characteristics of the ROV can be obtained from previous work [32]. The connected point at the free surface near the supported vessel is assumed to be fixed at (0 m, 0 m, 0 m), and the other point of the cable connected to the ROV is set at the gravitational center of the ROV. Parameters of the lumped parameter method for ROV simulation are shown in Table 2.

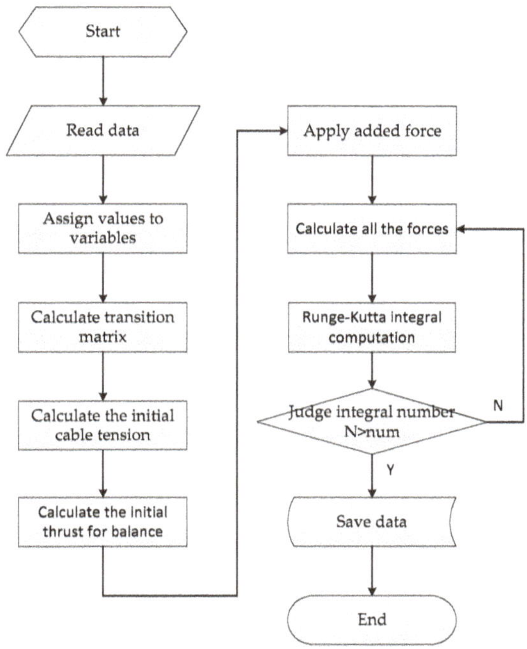

Figure 5. Flow chart of tethered ROV motion simulation.

Table 2. Parameters of the lumped parameter method (LPM).

Parameters	Values	Parameters	Values
Seawater density (kg/m^3)	1025	Cable length (m)	150
Drag coefficient in transverse direction C_n	1.2	Cross-sectional area (cm^2)	12.56
Drag coefficient in longitudinal direction C_t	0.024	Segment length (m)	15
Relative working depth (m)	100	Number of links	10
Diameter (m)	0.04	Segment mass (kg)	18.3
Young's modulus (N/m^2)	1.0×10^9	Mass in the water (kg/m)	0
Mass in the air (kg/m)	1.22	Minimum fracture strain (kN)	90

4.3. Comparison with Experiment

A small tank experiment is taken for comparison. The cable is fully immersed and allowed to reach the initial static configuration at the beginning. The experimental cable is initially suspended vertically and statically in the tank. The flow velocity is 1.543 m/s (3 kn), pulling a ball weighing 8.9 N in the water. The cable is 3.66 m long, and its diameter is 3.05 mm. The simulation time is 12 s.

The comparisons indicate that the simulated results are in fairly good agreement with the experiment (Figure 6). The slight discrepancy may be attributed to the inaccurate drag coefficients or certain non-modeled effects such as the bending stiffness of the cable.

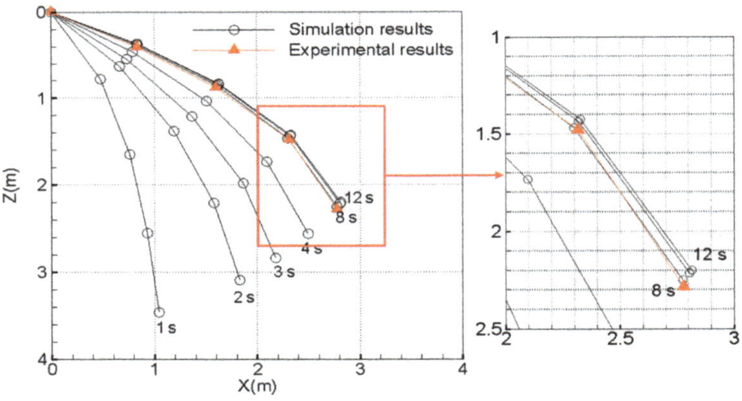

Figure 6. Comparison of the simulation results with the experimental results.

5. Nonlinear Observer

5.1. Nonlinear Observer Design

The ROV is a highly nonlinear and time-varying system with coupling between the degrees of freedom. For cases of weak maneuvering, the ROV's motion can be divided into horizontal and vertical motion, which can obtain satisfactory results. The control system of the ROV is often the integration of the multiple function modules, including the observer, controller units, and dead reckoning units.

Because the number of sensors that the ROV carries onboard is limited, only some of the state information can be measured. The nonlinear observer uses the measurable variables to estimate the velocities and the cable disturbing force information that cannot be measured online and in real-time. The structure diagram of the control system is shown below (Figure 7).

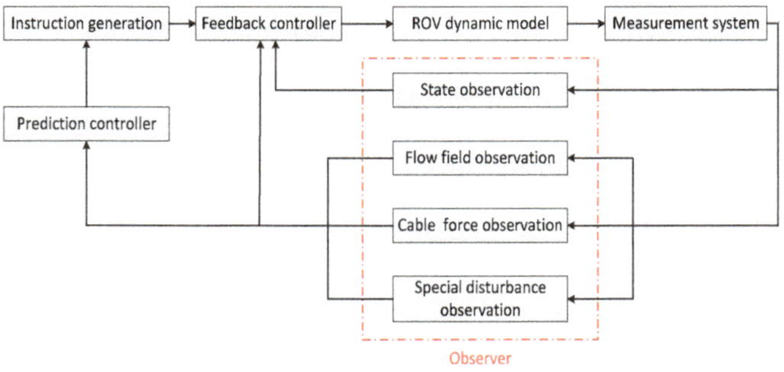

Figure 7. Control system structure of the ROV.

The observer structure (Figure 8) includes two parts: (1) The ROV dynamic model represents the dynamic and kinematic characteristics of the ROV in the water, including the ideal ROV model and the external disturbing cable force model, which is generated by using the lumped parameter method; (2) The observer model includes the ideal ROV model and a simplified model of the cable disturbing force.

The observer inputs are the thruster outputs $F_{Tx}, F_{Ty}, F_{Tz}, M_{Tz}$ in the body-fixed coordinates. The observer also needs measurements $u_{em}, v_{em}, w_{em}, r_{em}$ provided by the dead reckoning units. $\hat{u}, \hat{v}, \hat{w}, \hat{r}$ represent the observed values of u_e, v_e, w_e, r_e. e_u, e_v, e_w, e_r represent the observation error. The observer outputs are the velocity estimation $\hat{u}_c, \hat{v}_c, \hat{w}_c$ and the cable disturbing force $F_{cablex}, F_{cabley}, F_{cablez}$.

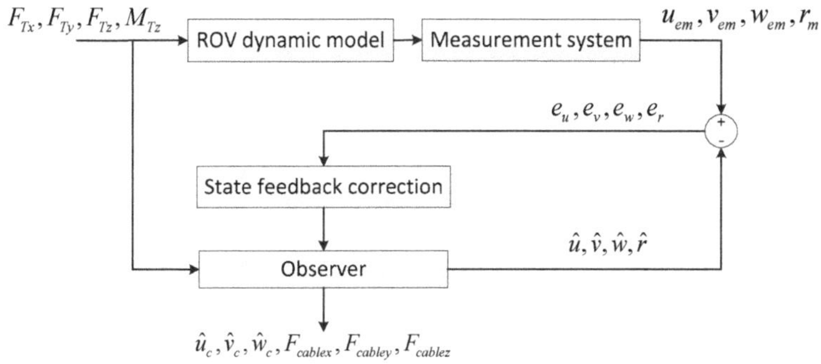

Figure 8. Observer structure.

Different types of sensors are installed on the ROV, such as the gyroscope, three-axis accelerometer, compass, Doppler velocity log, and the baseline system. Considering the equipment cost and the installation space, the number of sensors installed on ROV is limited. Some information cannot be obtained with sensors, such as the flow rate and the cable disturbing force. The nonlinear observer uses the measurable information to estimate the unknown state variables without any increase in the peripheral sensor equipment, and improves the overall performance of the system effectively.

An ROV always has a bilateral symmetrical structure, and its rolling and pitching motion velocities are small. Therefore, the influence of the rolling and pitching motion terms is often ignored. For the convenience of discussion, the ROV model can be simplified into a 4-DOF expression as follows:

$$\begin{cases} \dot{u}_r = \frac{1}{m-X_{\dot{u}}}\left[(m-Y_{\dot{v}})v_r r + X_u u_r + X_{u|u|}u_r|u_r| + (F_{Tx}+F_{bcablex})\right] \\ \dot{v}_r = \frac{1}{m-Y_{\dot{v}}}\left[-(m-X_{\dot{u}})u_r r + Y_v v_r + Y_{v|v|}v_r|v_r| + \left(F_{Ty}+F_{bcabley}\right)\right] \\ \dot{w}_r = \frac{1}{m-Z_{\dot{w}}}\left[Z_w w_r + Z_{w|w|}|w_r|w_r + (W-B+F_{Tz}+F_{bcablez})\right] \\ \dot{r} = \frac{1}{I_{zz}-N_{\dot{r}}}\left[(Y_{\dot{v}}-X_{\dot{u}})u_r v_r + N_r r + N_{r|r|}r|r| + (M_{Tz}+M_{bcablez})\right] \end{cases} \tag{41}$$

The motion function can be written as follows:

$$\begin{cases} \dot{X} = u_r\cos\psi - v_r\sin\psi + u_c \\ \dot{Y} = u_r\sin\psi + v_r\cos\psi + v_c \\ \dot{Z} = w_r + w_c \\ \dot{\psi} = r \end{cases} \tag{42}$$

The current model can be expressed as follows:

$$\begin{cases} \dot{u}_c = -\mu_{cx}u_c + \omega_{cx} \\ \dot{v}_c = -\mu_{cx}v_c + \omega_{cy} \\ \dot{w}_c = -\mu_{cx}w_c + \omega_{cz} \end{cases} \tag{43}$$

where $\mu_{cx}, \mu_{cy}, \mu_{cz} > 0$, $\omega_{cx}, \omega_{cy}, \omega_{cz}$ are the white noise with a mean value of zero. (41)–(43) express the 4-DOF model. Next, a reasonable state observer is designed to observe the unmeasurable parameters with sensors such as u_c, v_c, w_c.

For linear systems, the observer theory is almost complete and mature. Research on the nonlinear system state observer began in the 1970s, and various types of nonlinear observers have been developed and applied to different fields since then. In 1974, Utkin proposed a sliding mode observer using a switch function, which is often used in sliding mode control [33]. Its advantage is that when the nonlinear term and the disturbance of the system are bounded, it exhibits good robustness. This observer has invariance for the disturbance and uncertain factors when the matching conditions are satisfied. In view of the above advantages, sliding mode observers have been widely used and studied.

As the exact mathematical model of an ROV is difficult to build and will be affected by the external disturbance, the sliding mode observer is considerably suitable for such a system. Therefore, (41)–(43) can be rewritten as follows:

$$\begin{cases} \dot{\chi} = f(\chi) + BF_T + B\sigma \\ \eta = J\chi \end{cases} \tag{44}$$

The observation function is as follows:

$$y = \eta - \omega \tag{45}$$

where ω represents the measurement noise. $\chi = [u_r, v_r, w_r, r, u_c, v_c, w_c]^T$, $\eta = [\dot{X}, \dot{Y}, \dot{Z}, \dot{\psi}]^T$, F_T represents the control input, and σ represents the sum of the model error and the external disturbance. Further, σ is assumed to be bounded, that is, $\|\sigma\| \leq \beta_1$.

A nonlinear observer is designed as follows:

$$\begin{cases} \dot{\hat{\chi}} = f(\hat{\chi}) + BF_T + L\tilde{y} + \Gamma \\ \hat{\eta} = J\hat{\chi} \\ \hat{y} = \hat{\eta} \end{cases} \tag{46}$$

where $\tilde{y} = y - \hat{y}$, $\Gamma = \rho \mathrm{sgn}(\tilde{y})$ is the switching compensation term, $\rho \geq \beta_1$. The switching compensation term is used to represent the effects of the system modeling error and the external disturbance to enhance the robustness of the observer.

Taking (44) and (45) minus (46), and $e = \chi - \hat{\chi}$, $\tilde{\eta} = \eta - \hat{\eta}$, we obtain the following error functions:

$$\begin{cases} \dot{e} = f(\chi) - f(\hat{\chi}) + B\sigma - L\tilde{y} - \Gamma \\ \tilde{\eta} = Je \\ \tilde{y} = \tilde{\eta} - \omega = Je - \omega \end{cases} \tag{47}$$

Then:

$$\dot{e} = f(\chi) - f(\hat{\chi}) + B\sigma - LJe + L\omega - \Gamma \tag{48}$$

Let $\|L\omega\| \leq \beta_2$, as $e \approx 0$; then, $f(\chi) - f(\hat{\chi})$ can be expressed as follows:

$$f(\chi) - f(\hat{\chi}) = Ae + \Delta(e) \tag{49}$$

where Ae represents the linear part and $\Delta(e)$ represents the nonlinear part. Assuming $\|\Delta(e)\| \leq \beta_3$ and substituting (49) into (48) yields the following equation:

$$\begin{aligned} \dot{e} &= Ae + \Delta(e) + B\sigma - LJe - \Gamma + L\omega \\ &= (A - LJ)e - \Gamma + \Delta(e) + B\sigma + L\omega \\ &= A_0 e - \Gamma + \Delta(e) + B\sigma + \omega_1 \end{aligned} \tag{50}$$

where $\omega_1 = L\omega$, $A_0 = A - LJ$, and (50) is the differential equation of the observed error.

The following assumptions are put forward:

(1) If (A, J) is observable, then there exists L making A_0 stable.

(2) There exists a positive definite symmetric matrix P and a positive definite matrix Q, satisfying Lyapunov's equation as follows:

$$A_0^T P + PA_0 = -Q \tag{51}$$

The stability of the nonlinear observer as shown in (49) is proven as follows:
Let Lyapunov's function $V(e) = e^T Pe$; then,

$$\begin{aligned} \dot{V} &= \dot{e}^T Pe + e^T P\dot{e} \\ &= (A_0 e - \Gamma + \Delta(e) + B\sigma + \omega_1)^T Pe + e^T P(A_0 e - \Gamma + \Delta(e) + B\sigma + \omega_1) \\ &= e^T (A_0^T P + PA_0)e + 2e^T P(\Delta(e) + B\sigma + \omega_1 - \Gamma) \\ &= -e^T Qe + 2e^T P(\Delta(e) + B\sigma + \omega_1 - \rho \mathrm{sgn}(\widetilde{y})) \end{aligned} \tag{52}$$

Let $\lambda_{\min}(Q)$ and $\lambda_{\max}(P)$ be the minimum characteristic root of Q and the maximum characteristic root of P, respectively; then,

$$\dot{V} \leq -\lambda_{\min}(Q)\|e\|^2 + 2\|e\|\lambda_{\max}(P)(\beta_1 + \beta_2 + \beta_3 + \rho) \tag{53}$$

As $\dot{V} \leq 0$ and $\|e\| \geq 0$,

$$\|e\| \geq R_e \tag{54}$$

where $R_e = 2\frac{\lambda_{\max}(P)}{\lambda_{\min}(Q)}(\beta_1 + \beta_2 + \beta_3 + \rho)$. Therefore, when (54) exists, the observer is stable and the observation error decreases gradually until it enters the spherical domain with a radius of R_e centered on the origin in an n-dimensional space. After entering the spherical domain, the state estimation error reduces to a certain degree and oscillates, meeting the following condition:

$$\|e\| \leq R_e \tag{55}$$

From (55), we infer that the observer error is uniformly bounded. To reduce the observer error, when given P, Q, β_2, β_3, an effective method is to model the disturbing force and the disturbance compensation to reduce β_1. Even though the disturbance model cannot fully represent the actual disturbance, it can effectively reduce the upper bound of the observation error and considerably weaken the chatter of the observer.

5.2. Cable Disturbance Force Simplified Model

To improve the precision of the observer, there are usually two methods. One is to improve the structure of the observer and design an optimization algorithm and an adaptive method to improve the robustness of the observer. The advantage of this method is that the algorithm robustness for the parameter perturbation can be proven in theory for a particular system. However, its limitation is that it is difficult to find a universal observer satisfying the requirements of different systems.

The other method is to model the unknown disturbing force. Modeling the disturbing force can reflect the main factors of the disturbing force and make the feedforward compensation for the observer, thus reducing the upper bound of the unknown disturbance force and improving the precision of the observation. The advantage of this approach is that it is applicable to any observer for performance improvement.

The observer state variables are $\chi = [u_r, v_r, w_r, r, u_c, v_c]^T$, and the observer formula can be expressed as follows:

$$\begin{cases} \dot{\hat{\chi}} = f(\hat{\chi}) + B(F_T + J^{-1}F_{cable}(\hat{\chi})) + L\widetilde{y} + \Gamma \\ \hat{\eta} = C\hat{\chi} \\ \hat{y} = \hat{\eta} \end{cases} \tag{56}$$

6. Results and Discussion

6.1. Simulation of Cable Disturbing Force

The spatial changing rule of the cable disturbing force is analyzed and discussed, and a simplified model of the cable disturbing force is attempted for use in the design of the observer to reduce the influence of the cable disturbing force on the observer.

In the numerical calculation model, the cable disturbing force is calculated using the lumped mass method iteratively and is regarded as the true value of the cable force in a real circumstance. A simplified model of the cable disturbing force also needs to be built for the observer as the disturbing compensation. The cable disturbing force is generated by the relative motion of the cable in the water and is influenced by various parameters, such as the current velocity, displacement and velocity boundary conditions of the cable ends, cable length, and the cable material properties.

The simulation results are given below for different operating conditions, and the simplified model structure of the cable disturbing force is discussed.

Case one: The top end and the bottom end of cable are static at the initial time; the top end coordinates of the cable are (0 m, 0 m, 0 m), and the bottom end coordinates of the cable are (0 m, 0 m, 100 m). Further, the current velocity is (u_c, 0 m/s, 0 m/s). Tension results under different conditions of cable length and working depth are shown in Table 3.

Table 3. Tension on the ROV at the end of the cable under different conditions.

Cable Length (m)	Working Depth (m)	u_c (m/s)	F_{cablex} (N)	F_{cabley} (N)	F_{cablez} (N)
120	100	0	0	0	0
		−0.25	−55.45	0	−45.58
		−0.3	−79.44	0	−68.01
		−0.5	−220.21	0	−190.68
		−0.75	−495.29	0	−428.54
		−1	−880.45	0	−761.07
		−1.25	−1375.20	0	−1186.70
		−1.5	−1979.45	0	−1704.94
150	100	0	0	0	0
		−0.25	−44.96	0	−19.66
		−0.5	−179.66	0	−82.12
		−0.75	−404.16	0	−184.76
		−1	−719.00	0	−329.74
		−1.25	−1123.27	0	−514.93
		−1.5	−1617.16	0	−740.76
150	125	0	0	0	0
		−0.25	−68.47	0	−58.53
		−0.5	−275.12	0	−237.54
		−0.75	−619.05	0	−535.34
		−1	−1100.30	0	−950.30
		−1.25	−1718.30	0	−1480.80
		−1.5	−2472.94	0	−2125.86

Case two: The top end of the cable is static with the coordinates (0 m, 0 m, 0 m), and the bottom end of the cable moves with the absolute velocity (u, 0 m/s, 0 m/s); the relative current velocity is (u_r, 0 m/s, 0 m/s), and the current velocity is (u_c, 0 m/s, 0 m/s). Tension results under different relative velocity conditions are shown in Table 4.

Table 4. Tension on the ROV under different relative velocity conditions.

Cable Length (m)	Working Depth (m)	u_c (m/s)	u_r (m/s)	u (m/s)	F_{cablex} (N)	F_{cablez} (N)
150	100	−1.5	1.5	0	−1617	−741
		−1.5	1.6	0.1	−1731	−790
		−1.5	1.8	0.3	−1970	−895
		−1.5	2	0.5	−2230	−1000
		−1.5	2.2	0.7	−2500	−1125
		−1.5	2.4	0.9	−2800	−1250

It can be seen that when the top end of the cable is static and the bottom end moves, the tension components F_{cablex} and F_{cablez} at the end of the cable on the ROV have a linear relationship with the relative velocity with respect to the current, and have the opposite direction of the relative velocity.

To conclude, the tension at the end of the cable on the ROV is a complex function of the relative velocity to the current, displacement, and velocity boundary conditions of the cable ends, the cable length, and so on. When the cable length is 150 m and the top end of the cable is static, the expressions of the cable terminal tension versus the relative velocity and the spatial position under the typical velocity conditions can be expressed as follows.

When the current velocity is (−1.5 m/s, 0 m/s, 0 m/s), the top end of the cable is static with the coordinates (0 m, 0 m, 0 m) and the bottom end of the cable or the ROV moves. The ROV often swims against the current in the water, and the lateral and vertical velocities are relative small. Therefore, the effect of the longitudinal velocity on the cable tension is mainly considered here. The tension components at the bottom end of the cable to the ROV can be expressed as follows:

$$\begin{aligned} F_{cablex} &= 366 - 1308u_{re} + \left[-0.3381X - 0.0164Y^2 - 0.025(Z-100)^2 - 3.1634(Z-100)\right] \times 9 \\ F_{cabley} &= -0.9126Y \times 9 \\ F_{cablez} &= 114 - 564u_{re} + \left[-0.0076X^2 + 0.5277X - 0.0125Y^2 - 0.077(Z-100)^2 - 3.6955(Z-100)\right] \times 9 \end{aligned} \tag{57}$$

where u_{re} is the relative current velocity component along the x direction in earth coordinates, $u_{re} = u_e - u_c$. Further, u_e is the absolute velocity in earth coordinates and u_c is the current velocity. Equation (57) is mainly used for estimating the disturbance compensation in the observer.

The tension components at the end of the cable F_{cablex}, F_{cablex}, F_{cablez} are obtained in the earth coordinate system. The cable tension in the body-fixed coordinate system of the ROV can be obtained with a coordinate transformation.

6.2. Observer Simulation under Ideal Condition without Cable Disturbing Force

The state variables adopt $q = [u_r, v_r, w_r, r, u_c, v_c]^T$, the observer variables adopt $\eta = [\dot{X}, \dot{Y}, \dot{Z}, \dot{\psi}]^T$, and the observer parameters adopt $L = diag\{5,5,5,5,5\}$, $\rho = diag\{0.5, 0.5, 0.5, 5, 5\}$. When there is no cable disturbing force, σ is zero in (44). The parameters of the slowly varying current flow field model are $\mu_{cx} = 1$, $\mu_{cy} = 1$. ω_{cx} and ω_{cy} are composed of the Gaussian white noise signal and the step signal. The amplitudes of the step signals are −0.4 and −0.3, respectively, and the power spectrum density of the white noise is 10^{-5}. The current velocity generated by (43) is regarded as the true value of the current velocity, and the current velocity averages (−0.4 m/s, −0.3 m/s, 0 m/s). The effects of the measurement noise are ignored, and the propeller thrust acting on the ROV is $F_T = (2000N, 500N, 0N)$. The simulation time is 100 s, and the simulation step size is 0.001 s.

From the above figures, we can infer that if the effect of the sensor noise is not considered, when the system model does not have the current velocity and the cable disturbing force, the observer can rapidly track the observation state variables, including the ROV's position, attitude angle, velocity, and current velocity (Figures 9–11). The tracking error is small, as the velocity observation error is less than 0.02 m/s, and the displacement observation error is less than 0.03 m (Figure 12). This shows that the designed state observer has good observation performance without divergence in the ideal circumstance.

6.3. Observer Simulation under Conditions of Cable Disturbing Force without Compensation

To observe the effects of the cable disturbing force on the performance of the observer, the cable force is considered in the ideal ROV model. The cable disturbing force mentioned above is generated by using the lumped parameter method, regarded as the true value in practical environments. The parameters adopted by the lumped parameter method are presented in Table 2. The observer simulation is conducted under the conditions of a cable disturbing force without the disturbance compensation and depends only on the observer's own robustness. The parameters of the observer remain unchanged. The effects of the measurement noise are ignored. The current velocity averages (−0.5 m/s, 0 m/s, 0 m/s), and the propeller thrust acting on the ROV is $F_T = (2000N, 0N, 0N)$. The simulation time is 50 s, and the simulation step size is 0.001 s. The simulation results are presented below.

As can be seen from the observation results, the observation errors of the current velocity and the relative velocity of the ROV increase obviously (Figure 13). Although the observer has certain robustness, the added cable disturbing force (Figure 14) affects the accuracy of the observation (Figures 15 and 16). Therefore, when there exist external disturbances in the model or the actual system, the external disturbance has a considerable influence on the observation accuracy, and part of the observation information may possibly exhibit a divergence trend.

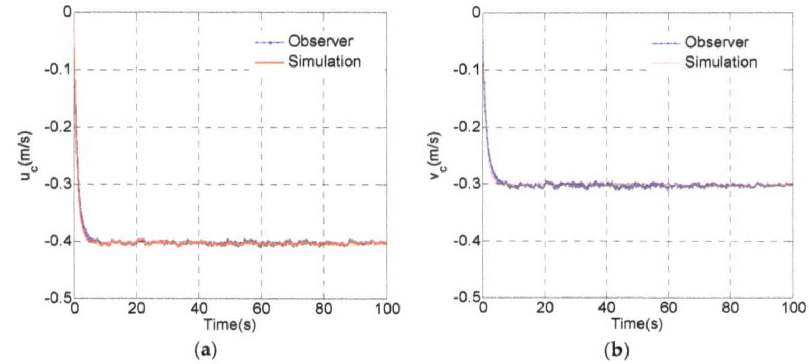

Figure 9. Observed results of current velocity: (**a**) current velocity in the X direction of earth-fixed frame; (**b**) current velocity in the Y direction of earth-fixed frame.

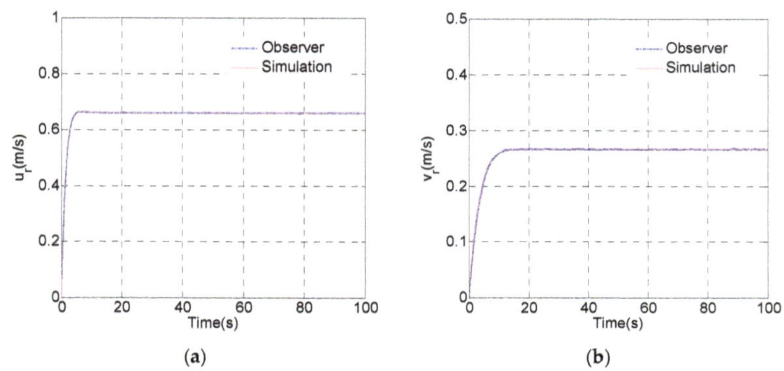

Figure 10. Observed results of relative velocity: (**a**) relative velocity of ROV in the x direction of body-fixed frame; (**b**) relative velocity of the ROV in the y direction of body-fixed frame.

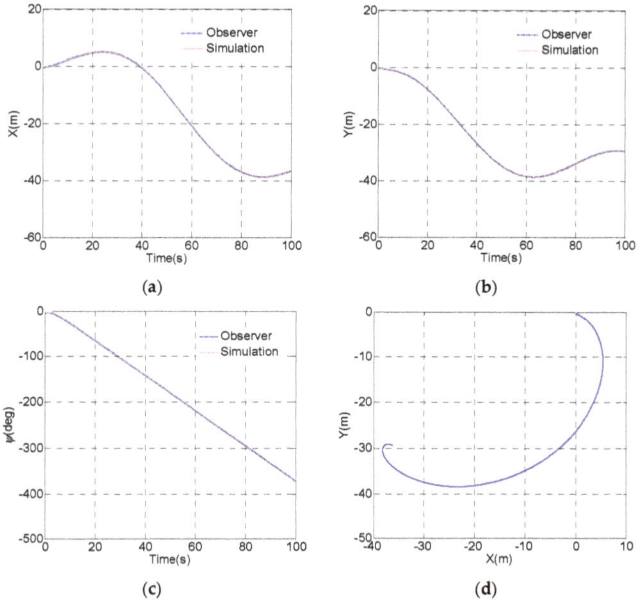

Figure 11. Observed results of position and heading of ROV: (**a**) X; (**b**) Y; (**c**) heading; (**d**) trajectory of ROV in the XY plane.

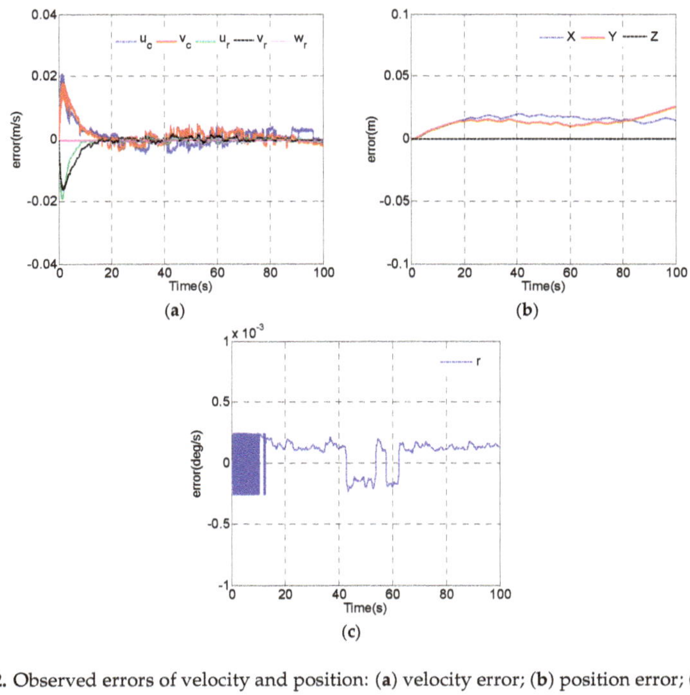

Figure 12. Observed errors of velocity and position: (**a**) velocity error; (**b**) position error; (**c**) angular velocity error.

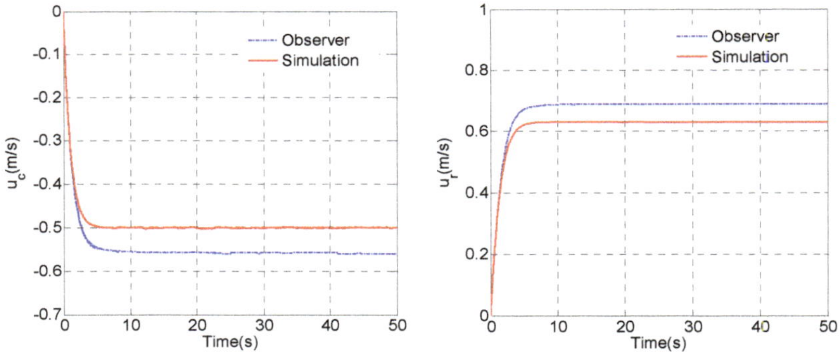

(**a**) (**b**)

Figure 13. Observed result of velocity: (**a**) current velocity in the X direction; (**b**) relative velocity of ROV in the x direction.

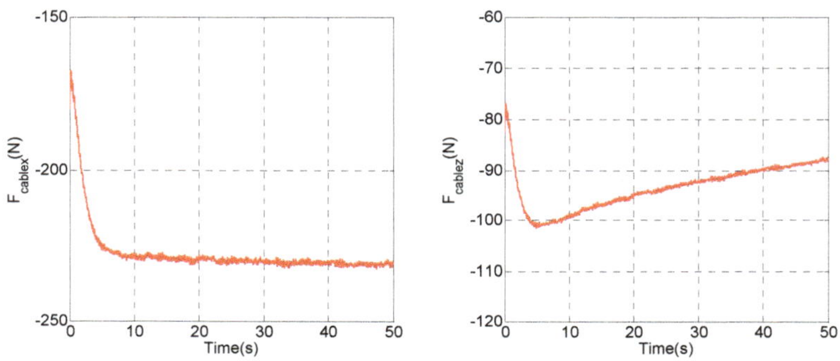

(**a**) (**b**)

Figure 14. Cable disturbing force (LPM): (**a**) cable force in the X direction; (**b**) cable force in the Z direction.

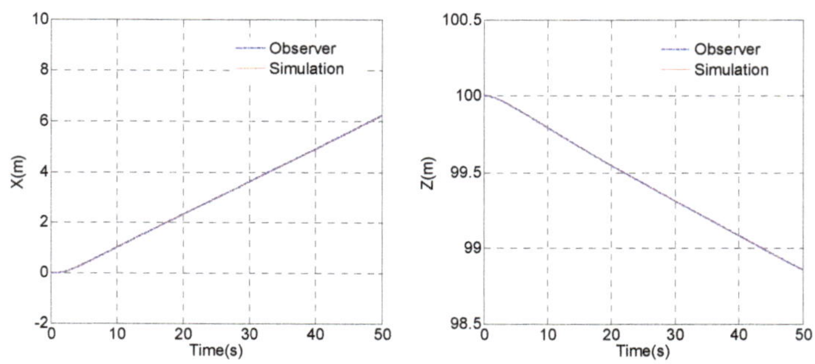

(**a**) (**b**)

Figure 15. Observed result of ROV position: (**a**) X; (**b**) Z.

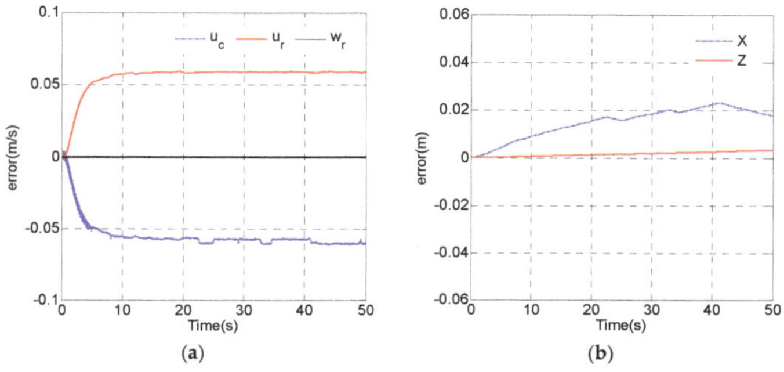

Figure 16. Observed error of velocity and position: (a) velocity error; (b) position error.

6.4. Observer Simulation under Conditions of Cable Disturbing Force with Compensation

To observe the effects of the cable disturbing force compensation on the performance of the observer, the cable force is considered in the ideal ROV model. The cable disturbing force mentioned above is generated by using the lumped parameter method, regarded as the true value in practical environments. The initial cable shape in the vertical plane is shown in Figure 17. The observer simulation is conducted under the conditions of a cable disturbing force with disturbance compensation. The parameters of the observer remain unchanged, $L = diag\{5, 5, 5, 5, 5\}$, $\rho = diag\{0.5, 0.5, 0.5, 5, 5\}$. The effects of the measurement noise are ignored. The current velocity averages (-0.5 m/s, 0 m/s, 0 m/s). The simulation time is 100 s, and the simulation step size is 0.001 s.

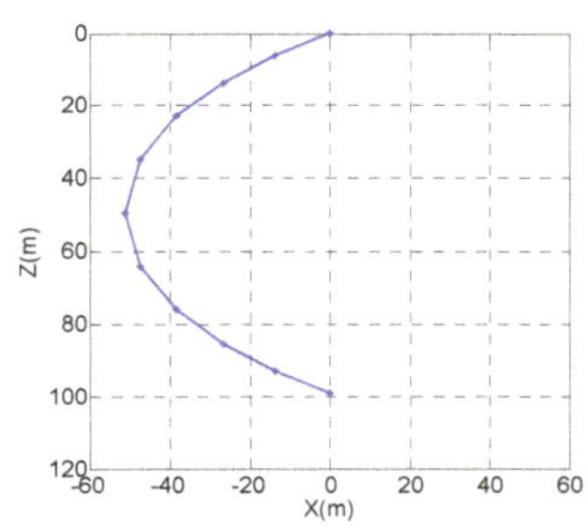

Figure 17. Initial cable shape in the vertical plane.

When the propeller thrust acting on the ROV is $F_T = (2000N, 0N, 0N)$, the following simulation results are obtained.

The simulation results show that when the longitudinal thrust acts on the ROV system, the observation accuracy of state variables u_r, w_r, u_c is improved considerably by the addition of the cable disturbing force compensation to the observer model compared with the observations without compensation (Figures 18 and 19). The absolute value of the velocity observation error decreases from 0.05 m/s to 0.01 m/s, and the vertical displacement observation error becomes 0.12 m after 100 s (Figure 20), which is acceptable. The cable disturbing force error between the observed value and the true value is small, and the observed result can accurately reflect the variation in the disturbance force (Figure 21).

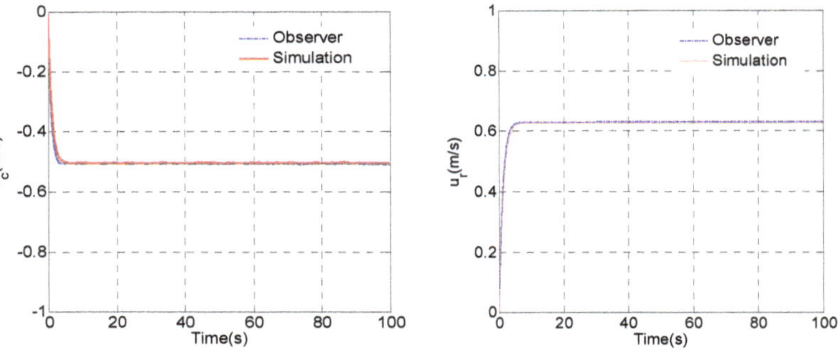

(**a**) (**b**)

Figure 18. Observed result of velocity: (**a**) current velocity in the X direction; (**b**) relative velocity of ROV in the x direction.

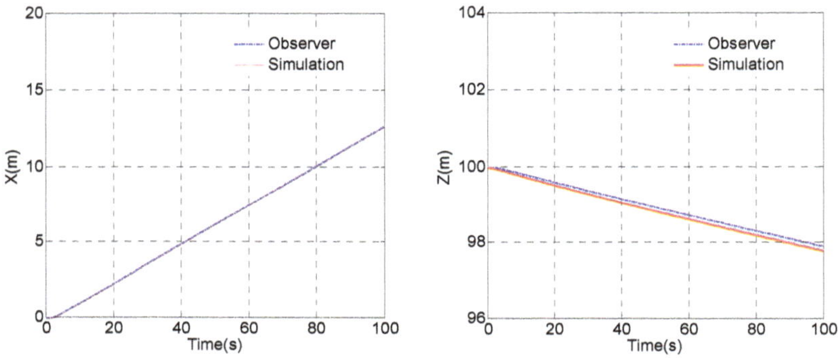

(**a**) (**b**)

Figure 19. Observed result of ROV position: (**a**) position in the X direction; (**b**) position in the Z direction.

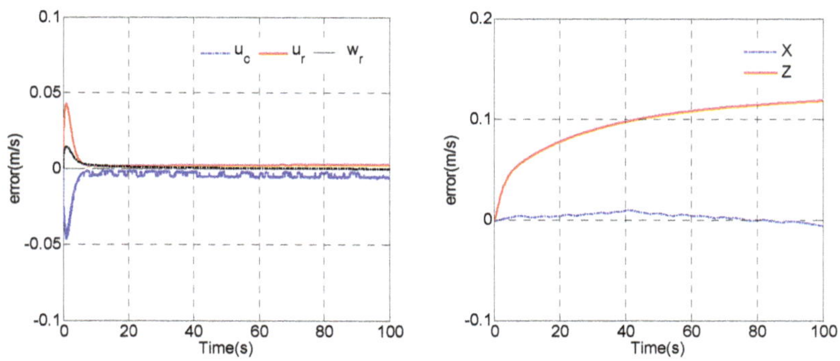

(**a**) (**b**)

Figure 20. Observed error of velocity and position: (**a**) velocity error; (**b**) position error.

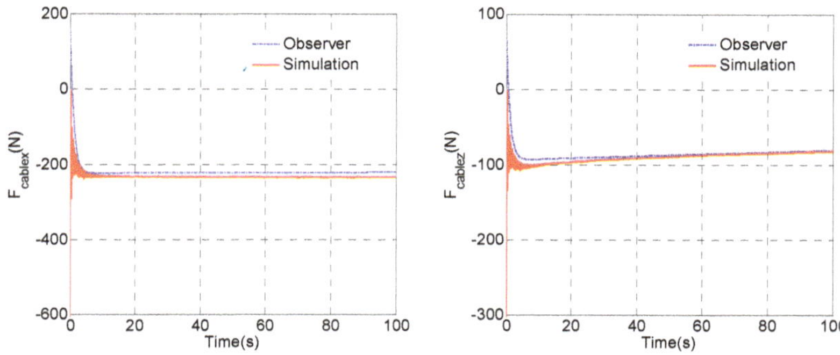

(a) (b)

Figure 21. Cable disturbing force (LPM): (**a**) cable force in the X direction; (**b**) cable force in the Z direction.

It is concluded that as the existence of the disturbing force deteriorates the observer performance, modeling the disturbance force and compensating for the observer is an effective method to increase the accuracy of the observer without modifying its parameters, although disturbing force modeling cannot fully characterize the cable disturbing force.

When the ROV's resultant thrust is $F_T = (3000N, 0N, 0N)$, the simulation results are as follows.

When the ROV's vertical thrust increases from 2000 N to 3000 N, the observation results of the state variables u_r, w_r, u_c are still effective and have a high observation precision (Figures 22 and 23). The absolute value of the velocity observation error converges within 0.01 m/s (Figure 24). The vertical displacement observation error is relatively large, mainly because the vertical component error between the observed value and the true value of the cable disturbing force is relatively large, of which the relative error is around 10% (Figure 25). It is illustrated that the established simplified model of the cable disturbing force can accurately reflect the change in the disturbing force at different velocities, which is effective and universal.

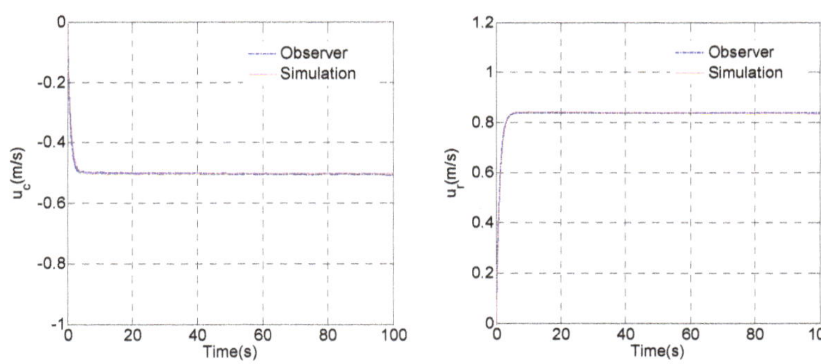

(a) (b)

Figure 22. Observed result of velocity: (**a**) current velocity in the X direction; (**b**) relative velocity of ROV in the x direction.

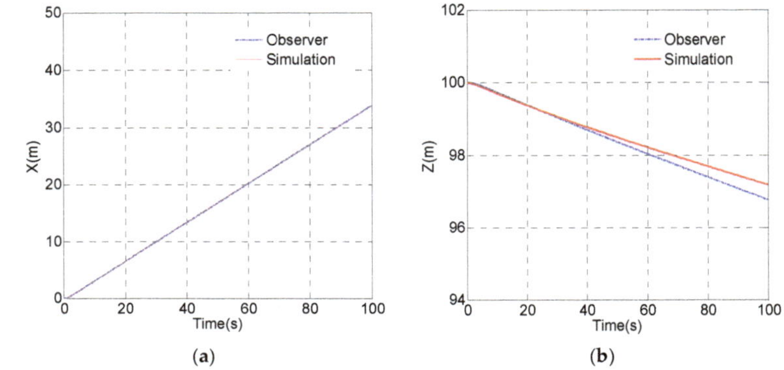

Figure 23. Observed result of ROV position: (**a**) position in the X direction; (**b**) position in the Z direction.

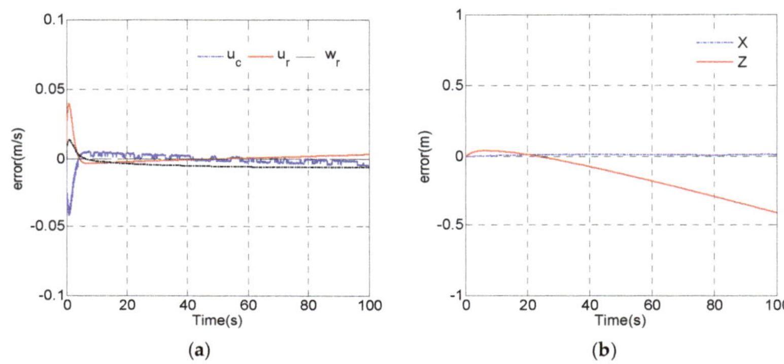

Figure 24. Observed error of velocity and position: (**a**) velocity error; (**b**) position error.

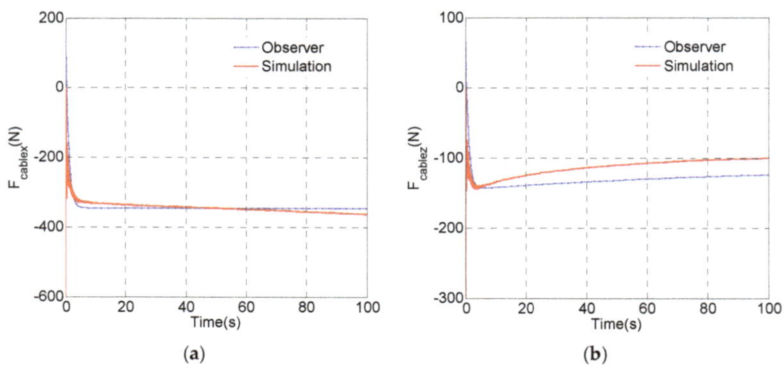

Figure 25. Cable disturbing force: (**a**) cable force in the X direction; (**b**) cable force in the Z direction.

7. Conclusions

A nonlinear observer for the ROV was investigated, and a 4-DOF nonlinear sliding state observer was designed in this study. A 6-DOF dynamic model was set up under the condition of ocean current.

The lumped mass method was adopted for the dynamic simulation of the umbilical cable and verified by a comparison with the experimental data. A coupling dynamic model of the ROV and the umbilical cable was established. Further, a 4-DOF nonlinear sliding mode observer for the ROV system was set up, and the ocean current model and the simplified flexible cable disturbance force model were established. The convergence and the stability of the observer were proven, and the applicability and the performance of the observer were verified by simulation under different working conditions. Unmeasured states such as the velocity state, current velocity, and cable disturbance were observed with the designed observers. We concluded that establishing a simplified disturbance model that could effectively estimate the actual disturbing force, and adopting the disturbing force compensation method effectively improved the observation precision and reduced the chattering of the observer outputs; doing so can also provide a basis and a reference for the design and application of other tethered ROV systems.

Author Contributions: X.L. designed the disturbance estimator and the integration with the main observer, and wrote the manuscript. M.Z. conceived and designed the nonlinear observer. T.G. analyzed the results and wrote the manuscript.

Acknowledgments: This work was supported by the Natural Science Foundation of China (Grant No. 51779139), the National Key Research and Development Program of China (Grant No. 2016YFC0300700), the 12th Five-Year Plan Project of International Sea Area Resources Survey and Development of China Ocean Mineral Resources R & D Association of the People's Republic of China (DY125-21-Js-06).

Conflicts of Interest: The authors declare that there is no conflict of interest.

References

1. Do, K.D.; Pan, J. *Control of Ships and Underwater Vehicles: Design for Underactuated and Nonlinear Marine Systems*; Springer: Berlin, Germany, 2009.
2. Cohan, S. Trends in ROV development. *Mar. Technol. Soc. J.* **2008**, *42*, 38–43. [CrossRef]
3. Yoerger, D.; Slotine, J. Robust trajectory control of underwater vehicles. *IEEE J. Ocean Eng.* **1985**, *10*, 462–470. [CrossRef]
4. Bessa, W.M.; Dutra, M.S.; Kreuzer, E. Depth control of remotely operated underwater vehicles using an adaptive fuzzy sliding mode controller. *Robot. Auton. Syst.* **2008**, *56*, 670–677. [CrossRef]
5. Bagheri, A.; Moghaddam, J.J. Simulation and tracking control based on neural-network strategy and sliding-mode control for underwater remotely operated vehicle. *Neurocomputing* **2009**, *72*, 1934–1950. [CrossRef]
6. Cui, R.; Zhang, X.; Cui, D. Adaptive sliding-mode attitude control for autonomous underwater vehicles with input nonlinearities. *Ocean Eng.* **2016**, *123*, 45–54. [CrossRef]
7. Soylu, S.; Proctor, A.A.; Podhorodeski, R.P.; Bradley, C.; Buckham, B.J. Precise trajectory control for an inspection class ROV. *Ocean Eng.* **2016**, *111*, 508–523. [CrossRef]
8. Cui, R.; Chen, L.; Yang, C.; Chen, M. Extended state observer-based integral sliding mode control for an underwater robot with unknown disturbances and uncertain nonlinearities. *IEEE Trans. Ind. Electron.* **2017**, *64*, 6785–6795. [CrossRef]
9. Wang, Y.; Gu, L.; Gao, M.; Zhu, K. Multivariable output feedback adaptive terminal sliding mode control for underwater vehicles. *Asian J. Control* **2016**, *18*, 247–265. [CrossRef]
10. Kinsey, J.C.; Whitcomb, L.L. Model-based nonlinear observers for underwater vehicle navigation: Theory and preliminary experiments. In Proceedings of the IEEE International Conference on Robotics and Automation, Roma, Italy, 10–14 April 2007; pp. 4251–4256.
11. Song, B.K.; An, J.H.; Choi, S.B. A New Fuzzy Sliding Mode Controller with a Disturbance Estimator for Robust Vibration Control of a Semi-Active Vehicle Suspension System. *Appl. Sci.* **2017**, *7*, 1053. [CrossRef]
12. Fernandes, D.D.A.; Sørensen, A.J.; Pettersen, K.Y.; Donha, D.C. Output feedback motion control system for observation class ROVs based on a high-gain state observer: Theoretical and experimental results. *Control Eng. Pract.* **2015**, *39*, 90–102. [CrossRef]
13. Gauthier, J.P.; Kupka, I.A.K. Observability and observers for nonlinear systems. *SIAM J. Control Optim.* **1994**, *32*, 975–994. [CrossRef]

14. Fridman, L.; Shtessel, Y.; Edwards, C.; Yan, X.G. Higher-order sliding-mode observer for state estimation and input reconstruction in nonlinear systems. *Int. J. Robust Nonlin. Control* **2008**, *18*, 399–412. [CrossRef]
15. Rezazadegan, F.; Shojaee, K.; Chatraei, A. Design of an adaptive nonlinear controller for an autonomous underwater vehicle. *Int. J. Adv. Electr. Electron. Eng.* **2013**, *2*, 1–8.
16. Khadhraoui, A.; Beji, L.; Otmane, S.; Abichou, A. Observer-based controller design for remotely operated vehicle ROV. In Proceedings of the IEEE International Conference on Informatics in Control, Automation and Robotics, Vienna, Austria, 1–3 September 2014; pp. 200–207.
17. Chu, Z.; Zhu, D.; Jan, G.E. Observer-based adaptive neural network control for a class of remotely operated vehicles. *Ocean Eng.* **2016**, *127*, 82–89. [CrossRef]
18. Kinsey, J.C.; Eustice, R.M.; Whitcomb, L.L. A survey of underwater vehicle navigation: Recent advances and new challenges. In Proceedings of the IFAC Conference of Manoeuvering and Control of Marine Craft, Lisbon, Portugal, 20–22 September 2006.
19. Refsnes, J.E.; Pettersen, K.Y.; Sorensen, A.J. Control of slender body underactuated AUVs with current estimation. In Proceedings of the 45th IEEE Conference on Decision and Control, San Diego, CA, USA, 13–15 December 2006; pp. 43–50.
20. Aguiar, A.P.; Pascoal, A.M. Dynamic positioning and way-point tracking of underactuated AUVs in the presence of ocean currents. *Int. J. Control* **2007**, *80*, 1092–1108. [CrossRef]
21. Børhaug, E.; Pivano, L.; Pettersen, K.Y.; Johansen, T.A. A model-based ocean current observer for 6DOF underwater vehicles. *IFAC Proc.* **2007**, *40*, 169–174. [CrossRef]
22. Lueck, R.G.; Driscoll, F.R.; Nahon, M. A wavelet for predicting the time-domain response of vertically tethered systems. *Ocean Eng.* **2000**, *27*, 1441–1453. [CrossRef]
23. Fang, M.C.; Chen, J.H.; Luo, J.H.; Hou, C.S. On the behavior of an underwater remotely operated vehicle in a uniform current. *Mar. Technol.* **2008**, *45*, 241–249.
24. Zhu, K.Q.; Zheng, D.C.; Cai, Y.; Yu, C.L.; Wang, R.; Liu, Y.L.; Zhang, F. Nonlinear hydrodynamic response of marine cable-body system under random dynamic excitations. *J. Hydrodyn. Ser. B* **2009**, *21*, 851–855. [CrossRef]
25. Fossen, T.I. *Guidance and Control of Ocean. Vehicles*; John Wiley & Sons: Chichester, UK, 1994.
26. Driscoll, F.R.; Lueck, R.G.; Nahon, M. Development and validation of a lumped-mass dynamics model of a deep-sea ROV system. *Appl. Ocean Res.* **2000**, *22*, 169–182. [CrossRef]
27. Walton, T.S.; Polachek, H. Calculation of transient motion of submerged cables. *Math. Comput.* **1960**, *14*, 27–46. [CrossRef]
28. Sanders, J.V. A three dimensional dynamic analysis of a towed system. *Ocean Eng.* **1982**, *9*, 483–499. [CrossRef]
29. Ablow, C.M.; Schechter, S. Numerical simulation of undersea cable dynamics. *Ocean Eng.* **1983**, *10*, 443–457. [CrossRef]
30. Leonard, J.W. Nonlinear dynamics of cables with low initial tension. *J. Eng. Mech.* **1972**, *98*, 293–309.
31. Ma, D.C.C.; Chu, K.H.; Leonard, J.W. Slack-elasto-plastic dynamics of cable system. *J. Eng. Mech. Div.* **1979**, *105*, 207–222.
32. Xu, G.; Ge, T.; Zhu, J.M.; Lian, L. Motion modelling and simulation of varying length cable for "Sea Dragon 3500" ROV. *Mar. Technol.* **2005**, *5*, 22–26. (In Chinese)
33. Utkin, V.I. Sliding mode and their application in discontinuous systems. *Automat. Rem. Control* **1974**, *1*, 1898–1907.

applied sciences

MDPI

Article

The Effects on Muscle Activity and Discomfort of Varying Load Carriage With and Without an Augmentation Exoskeleton

Huaixian Li [1,*], Wenming Cheng [1,*], Fang Liu [1], Mingkui Zhang [1] and Kun Wang [2]

[1] School of Mechanical Engineering, Southewest Jiaotong University, Chengdu 610031, Sichuan Province, China; liufang@swjtu.edu.cn (F.L.); sunson1688@163.com (M.Z.)
[2] School of Sports Medicine and Health, Chengdu Sport University, Chengdu 610041, Sichuan Province, China; Michellebaobei@163.com
* Correspondence: ynlihuaixian@163.com (H.L.); wmcheng@home.swjtu.edu.cn (W.C.); Tel.: +86-833-519-8015 (H.L.)

Received: 27 October 2018; Accepted: 11 December 2018; Published: 16 December 2018

Abstract: Load carriage is a key risk factor for Muscular Skeletal Disorders (MSDs). As one way to decrease such injuries, some exoskeletons have been developed for regular load carriage. We examined the ergonomic potential of an augmentation exoskeleton. Nine subjects completed eight trials of carrying tasks, using four loading levels (0, 15, 30, and 45 kg) and two carrying conditions (with and without the exoskeleton). Electromyography (EMG) and the extended NASA-TLX rating scales were investigated and analyzed by linear mixed modeling and two-way ANOVA methods. Noraxon MR3.8, SPSS19.0, and MATLAB R2014b software were adapted. The results show that most of the muscle mean activities increased significantly ($p < 0.05$) with exoskeleton assistance. However, the interactive effects illustrate a decreasing trend with increase of load level. The mean discomfort rating scale values were generally higher, but subjects generally preferred using the exoskeleton in heavier loading tasks. The exoskeleton can effectively augment the performance of humans in heavy load carriage. The main reasons for higher muscle activity are from inflexible structures and inharmonious human–robot interactions. In order to decrease the MSD risks and increase comfort, optimal human–robot control strategies and adaptable kinematic design should be improved.

Keywords: exoskeleton; load carriage; muscle activities; human–robot interaction; discomfort

1. Introduction

Regular load carriages are criteria risk factors for Muscular Skeletal Disorders (MSDs) in hikers, backpackers, and soldiers [1,2]. The prevalence of MSDs is significantly related to the weight and mode of carrying a backpack among the load carrying population [3]. For military populations, high injury rates and high stress fracture rates have been associated with load carriage [1]. Knapik reported that while carrying heavy loads, 79 out of 335 infantry soldiers suffered from marching-related injuries in one road march [4].

Various preventive measures have been proposed for decreasing load carriage-related MSDs, such as mechanical aids like cranes [5]. However, with the development of new technologies, some potentially preventive strategies have emerged. One of these could be the use of wearable exoskeletons [6–8]. A wearable exoskeleton is defined as an active mechanical device, which is "worn" by an operator and fits closely to their body and works in concert with the operator's movements [9]. These newly developed exoskeletons proved to some extent helpful in decreasing the feeling of burden of the operators. Some positive study results reported the benefits of specific exoskeletons in reducing the internal muscle forces in corresponding body regions [8] and reducing high physical demands in standing status [10]. These exoskeletons have been designed to make load carriage easier by providing a parallel load path to the ground. These research results represent remarkable achievements in the field of load carrying magnitudes. However, the mean metabolic consumption and muscle force of human beings significantly increased despite the assistance of an exoskeleton [11]. This situation might be caused by kinematic misalignments between the complex human body structure and the simplified exoskeleton, since the augmentation devices were designed to adapt to the movement of human beings and to be "worn" by an operator. The design features determined that the augmentation devices could not self-stablilize and would start work in concert with the operator when they form a closed system [12]. Therefore, load carriage processes with an augmentation exoskeleton resulted in more load on human beings because of the adaptability to the movement of the human beings. The incompatibilities might actually increase the risks of muscular skeletal injuries, disorders, and discomforts as well. These potential problems are enormous challenges because they are influenced by some serious factors. For example, there are compromises to mobility, agility, and cognitive and physical performance in the human–exoskeleton interaction system with exoskeleton assistance, which may arise from aligning the exoskeleton to biological joints of human beings, the inertia compensation of the exoskeleton system [13], and adaptable control strategies of human–robot interactions. Therefore, in human–exoskeleton interaction processes, human beings have to spend energy and power to start, overcome hindrances, and stabilize the whole system. This might cause added load and fatigue.

Recently, a new series of studies reported a kind of elastic multi-joint soft exosuit which can help to decrease 7.3 ± 5.0% of the metabolic power during carriage of a load equivalent to 30% human body mass. However, the lower limbs' electromyography (EMG) activity reduction in the experiment was not significantly different ($p < 0.05$). The small benefits have no clear effect on stress fractures and risk of sustaining injuries from heavy load carriage [8,14]. The soft exosuit needs no aligning to the human joints and overcomes most defects of the rigid exoskeletons. However, the device only augments the power of the operators' muscles and less is considered about the skeleton of the human beings and the injuries and MSDs which may happen in conditions of heavier loads and longer time of carriage.

Both of the above mentioned augmentation exoskeletons may form new kinds of risks of increasing Musculoskeletal Disorders and discomfort. To the authors' knowledge, however, there is still a lack of a readily available exoskeleton device for heavy load carriage to reduce muscular discomfort and the risk of MSDs. Most experiment studies were focused on oxygen consumption and muscle activity, with no further consideration to the discomfort, muscle activity patterns, and risk factors of MSDs [11].

Actual and virtual experiments in assessing the performance of the exoskeletons have already been undertaken [15,16]. In order to investigate the ergonomic performance of the augmentation exoskeleton developed by Southwest Jiaotong University, the effects on subjective discomfort and objective leg muscle activity are studied by a series of actual load carrying tasks. We investigated the subjective performance based on an extended questionnaire for discomfort and all six indices of the NASA TLX rating scale. We measured the surface electromyography signal of the knee extensor and ankle planter flexor to study the muscle activity patterns in the with-EXO condition. The results should display the ergonomic potentials of the augmentation exoskeleton. At the same time, the results might inspire kinematic design choices and optimal human–robot control strategies for exoskeletons in improving the usability and ergonomic comfort and decreasing the MSDs risk factors from the exoskeleton itself.

2. Materials and Methods

2.1. Subjects and Prototypes

In this study, nine healthy male participants (years: 25 ± 8; body mass: 71 ± 12.4 kg; height: 176 ± 10 cm) volunteered to take part in the study. None of the participants reported muscular-skeletal disorders in the previous three months. All subjects gave their informed consent for inclusion before they participated in the study. The study was conducted in accordance with the Declaration of Helsinki, and the protocol was reviewed and approved by the Ethics Committee of the Southwest Jiaotong University (approval number 2017_001).

A powered augmentation exoskeleton prototype, which was developed by the Southwest Jiaotong University, was used (as shown in Figure 1). The exoskeleton was described in detail by F.Liu and M.K.Zhang et al. [17,18]. The experimental prototype is the third generation prototype. Its mechanical structures consist of a backpack and two powered legs which power the knee joints via two hydraulic actuating cylinders. Using the servo valve to realize the servo-control for displacement of the hydraulic cylinder, the feedback Proportion-Integral-Derivative (PID) control law of the hydraulic cylinder displacement was designed to carry out force-assisted demands in a portable system under changing conditions of different weights and different poses. Each link could be adjusted to suit for the length of the human extremities. Additionally, the whole exoskeleton could be adjusted to suit human dimensions ranging from 165–185 cm, covering 95% of the anthropometry population. Each leg has 6 degrees of freedom (DOFs): three hip joint DOFs, one knee joint DOF, and two ankle joint DOFs. The backpack is connected to the human trunk by shoulder girdles and a flexible waist bandage. The lower exoskeleton links are connected to the thighs by flexible bandages. Additionally, the operator's feet are put into the exoskeleton shoes.

Figure 1. The argumentation exoskeleton prototype for load carriage experiments.

The exoskeleton is composed of rigid parts, two hydraulic cylinders, and some damping components, which are implanted in the exoskeleton to reduce vibration impacts. The Rigid–Flexible Coupling System has intrinsic compliance due to damping parts which allow for, and passively counteract, deviations from a target position in a safe way. Subjects were strapped to the exoskeleton by flexible, adaptable bandages, allowing free movement at the human joints [17].

The whole system control strategy was a kind of combination of position control and ZMP (zero moment point) control. The control scheme was shown in Figure 2. There are five different control strategies corresponding to five walking modes: level walking, level running, climbing slope, ladder walking, and standing. In order to study the basic muscle activities and subjective discomfort, we selected the level walking control mode to mimic free treadmill walking while carrying a load.

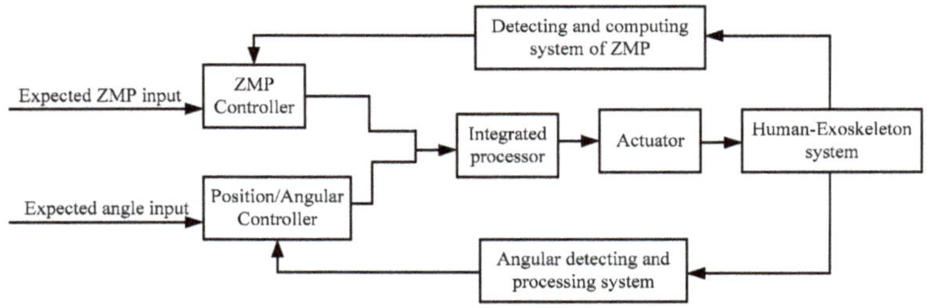

Figure 2. The overall control scheme.

2.2. Testing Procedures

Subjects completed two groups of contrast tasks, carrying 0 kg, 15 kg, 30 kg, and 45 kg weight in each group of trials, with and without the exoskeleton prototypes. All subjects had recently completed about 30 min basic training on the treadmill, walking with the exoskeleton and without loads in order to familiarize them with the device. Each participant performed 3 trials of 8 min walking in succession with 2 min rest between the trials. The training section had been carried out a week before the formal experiment. All volunteers were healthy and without muscular-skeletal injuries or disorders. We started with the group of trials without the exoskeleton, followed by the group of tasks with exoskeleton; the order of the two groups (with and without exoskeleton) within the tasks was assigned and based on a Latin square approach. The order of presentation of the load conditions was randomly assigned and based on a Latin square approach, as shown in Figure 3b. In order to keep the consistency of the trials, the subjects completed all of the trials with the help of the treadmill handlebars in case of falling down when carrying heavier loads. All sessions were performed in a laboratory at a constant ambient temperature of 22 °C.

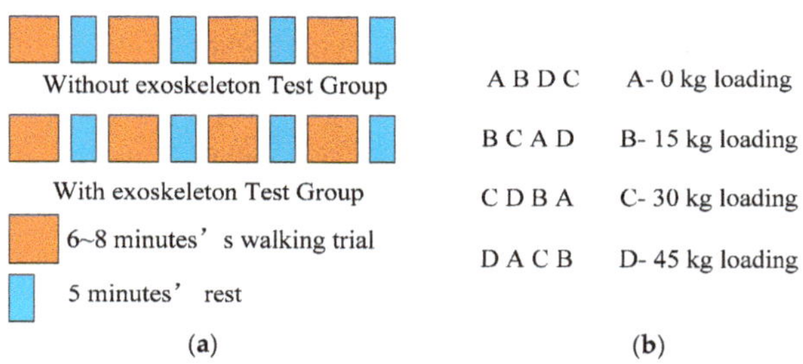

Figure 3. Process of the experiments. (**a**) Testing procedure: the orange square block means 6–8 min walking trial and the blue rectangle means 5 min rest; (**b**) the Latin square arrangement: the capital letters A, B, C, and D are used to represent the load level 0 kg, 15 kg, 30 kg, and 45 kg separately.

Prior to testing, the exoskeleton was adjusted for each participant, ensuring that an acceptable fit was achieved, as determined independently by an exoskeleton engineer. The criteria was to keep the exoskeleton links fitting the length of the shank and thigh of each subject, as measured by a ruler and the engineer's experience. Each volunteer then completed a total of 6–8 min warm-up with the exoskeleton. According to the completed training session, the volunteers become familiar with carrying each load while wearing the exoskeleton whilst walking on the treadmill at 1 m/s (3.6 km/h). Participants walked on the treadmill for 6 min in each trial at a speed of 3.6 km/h. At the beginning of the fourth minute of trial, 30 s data of human kinematics and muscle activity was collected; the signals used for controlling the exoskeleton were collected as well. It concerns about 12 walking cycles. After each trial was completed, all load components were removed and the volunteer was allowed approximately 5 min to rest between trials.

After each test section, the participants were asked to rate a subjective questionnaire on a visual rating scale form. They marked their rating on seven linear rating scales that ranged from 0 (low) to 100 (high) points, as shown in Figure 4.

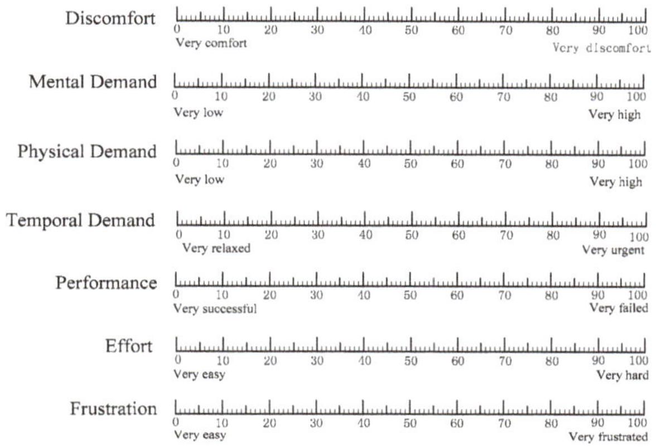

Figure 4. The extended NASA-TLX rating scales [19].

2.3. Measurements

The Noraxon Desktop DTS 8-channel wireless electromyography (EMG) system (Noraxon Ltd., Scottsdale, AZ, USA) was used in combination with bi-polar Al/AgCl surface electrodes. EMG data were recorded through analysis software MR 3.8 (Noraxon Ltd., USA) at a sample rate of 2000 Hz. EMG data of seven muscles on the right leg was recorded. The seven muscles are rectus femoris (RF), vastus medialis (VM), vastus lateralis (VL), semitendinosus (ST), biceps femoris (BF), gastrocnemius lateralis (GL), and peroneus longus (PL). Maximal voluntary contraction values (MVC) were also recorded. The MVC was realized by Isometric contraction. The participants sat in a self-contained chair designed by the Isomed2000 testing system. Knee joint angles of 105° and 150° were selected. Subjects performed a 6 s maximal voluntary contraction. The bi-polar surface electrodes were located in the positions shown in Figure 5, which complied with SENIAM recommendations (http://www.seniam.org). Raw EMG signals were first filtered using a notch filter (50 Hz) to reduce effects of alternating current. Then a band-pass (10–500 Hz) was used to reduce potential effects of transitional adjustments. The integration was expressed as a percentage of the peak voltage of the MVCs. In each of these trials, EMG root mean square (RMS) values were obtained. Similar processing was done for raw EMG values obtained during reference contractions. Mean RMS values obtained during the task were then normalized (nRMS) to the corresponding reference RMS values.

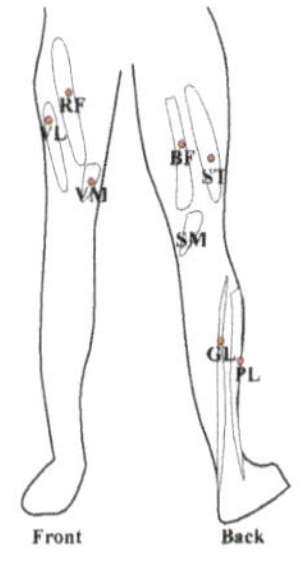

Figure 5. The body parts and points where the bi-polar surface electrodes were located.

An extended NASA TLX rating scale is used to record the discomfort of the size indices [19]. After each subject complete the eight trials, the TLX rating scales were acquired.

2.4. Subjective Performance and Discomfort

The rating scale is an extended NASA TLX questionnaire with an extra scale of discomfort, as introduced by Hart and Stavenland. The discomfort index and the six standard indices were defined as the following [17]:

(1) Discomfort (C): How comfortable was the appointed loading during walking?
(2) Mental Demand (MD): How mentally demanding was the task?
(3) Physical Demand (PD): How physically demanding was the task?
(4) Temporal Demand (TD): How hurried was the pace of the task?
(5) Performance (OP): How successful were you in accomplishing the task?
(6) Effort (EF): How hard did you have to work to accomplish the task?
(7) Frustration (FR): How insecure, discouraged, irritated, stressed, and annoyed were you?

After all eight trials, the weighting factors for the standard rating scales were acquired for each subject by pair-wise comparison and computed into an overall set of group weights. At the same time, the subjects would be asked by the experimenter which body part experienced the most discomfort in the trials.

2.5. Data Analysis

Linear mixed modeling analyses were applied to explain differences in mean EMG activity over time by the factored type of walking ("with- or without-EXO") and load carriage (four-level factor "loading"). Differences in average EMG activation, subjective performance between the two walking types, and four loading tasks were therefore analyzed using a two-way ANOVA for repeated measures with independent factors for the exoskeleton (with or without) and loading (four-level factor "loading"). To evaluate the effects of the exoskeleton in the EMG gait patterns, unilateral gait pattern differences in two types of walking, four-level loading, and two-level factor intervals (stance and swing) were analyzed. A paired sample *t*-test was carried out for the loading tasks.

After carefully inspecting data, the Kolmogorov-Smirnov and Shapiro-Wilk test were performed. The NASA-TXL rating scores were not normally distributed. Differences between the two walking conditions (with or without exoskeleton) during the loading tasks were analyzed using Wilcoxon signed rank tests (i.e., using participants as their own controls). Significance was accepted at $p < 0.05$ and all statistical analyses were performed using SPSS (IBM SPSS Statistics 19.0).

3. Results

The experimental results should be presented and briefly discussed in the following order: (1) Overall Stance and Swing durations, with- or without-Exo load carriage. (2) Overall gait pattern and characteristics, with- or without-EXO load carriage through a case study. (3) For the two independent variables, load and with- or without-Exo, the mean and maximum muscle activities from EMG data will be presented for the knee extensors (RF, VM, VL), knee flexors (ST, BF), and ankle planter flexors (PL, GL). (4) The two-way ANOVA interaction effects of load $\times$ with- or without-EXO are presented. (5) The NASA-TXL discomfort and standard subjective performance.

3.1. The Overall Stance and Swing Duration for with- or without-Exo Carriage Condition

As shown in Figure 6, the investigation of the stance and swing duration shows that the stance duration was prolonged with increased loading in the without-Exo condition. And in the with-Exo condition, the stance and swing durations were almost the same. However, the statistic results illustrated that there were no significant difference ($p > 0.05$) between with-Exo and without-Exo condition. The reason for the increase of the stance phase with the EXO might be the trigger time of the EXO since the exoskeleton must be started by human beings.

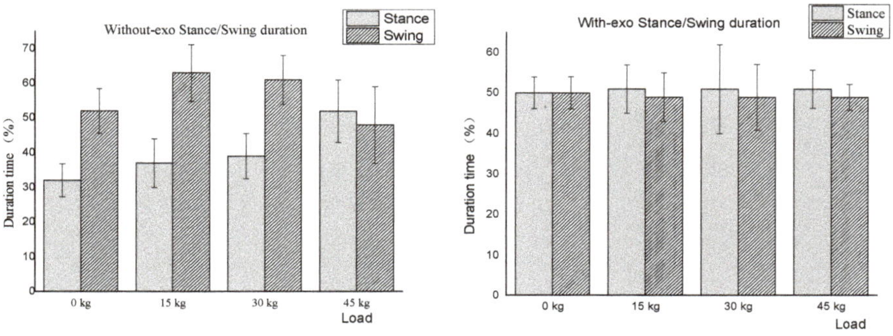

(a) (b)

Figure 6. Comparison of gait phase stance and swing duration time. (**a**) without-Exo; (**b**) with-Exo.

3.2. Unilateral Gait Report of with-Exo and without-Exo in a 45 kg Load Carriage Trial

The unilateral gait report is a kind of time normalized gait cycle EMG pattern which can show the typical muscle activity characteristics and coordination of muscle groups during a walking trial. With appropriate test standardization, the averaged EMG pattern of gait cycles in a trial is highly reproducible. The averaged gait cycle data could be used to analyze the overall patterns and characteristics of the muscle activity within one gait cycle. It could be also used to compare the muscle activations under loaded (stance) and unloaded (swing) gait phases between with-Exo and without-Exo trials in the same load carriage condition.

The exoskeleton prototype was designed to augment human beings' performance of heavy load carriage. Therefore, a comprehensive case study was carried out on the heaviest load carriage condition, which was 45 kg load level, in order to access the loading performance of the exoskeleton prototype. The comparison was carried out between the conventional 45 kg carriage walking trials and exoskeleton-assisted 45 kg carriage walking trials.

The results are shown in Figure 7. Under the 45 kg carriage, the overall patterns and characteristics of the muscle activities were almost the same for the both with- or without-Exo load carriage conditions. In the loaded (stance) stage, the averaged activity of exoskeleton augmenting conditions were a little bit lower than the ones for conventional load walking conditions, except for the planter flexor PL. However, in the unloaded (swing) conditions, the curves' trends were almost contrary to the loaded (stance) phase. This situation complied with the poor adaptability and impedance effect from inertia. However, the load carriage performance was good. Considering the standard deviations, the conventional load carriage gait graphs are smoother than the assistive ones, as shown in Figure 7b,c.

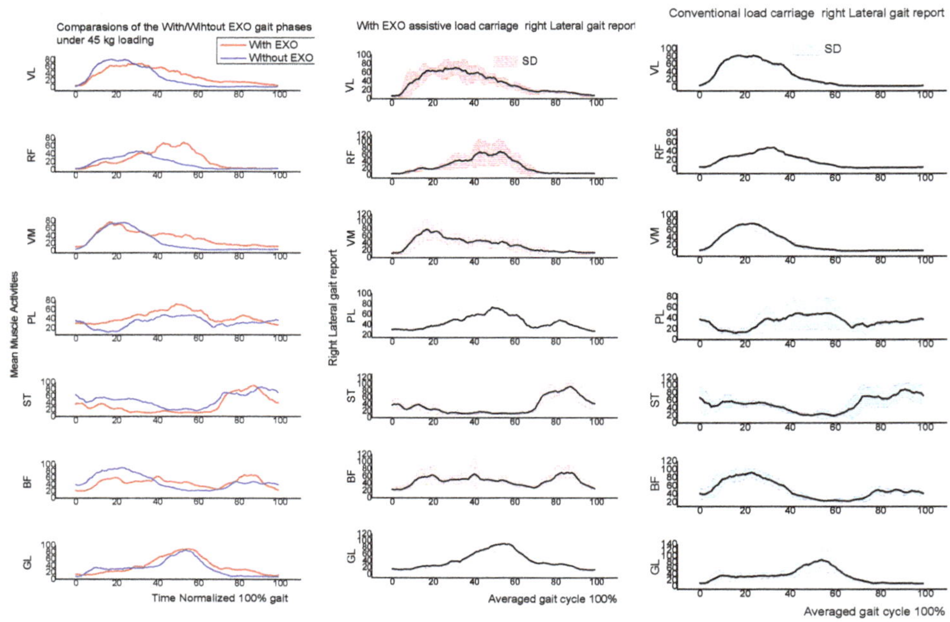

(**a**) (**b**) (**c**)

Figure 7. Time normalized EMG gait patterns for the measured right leg muscle groups during with-Exo and without-Exo walking in the case of 45 kg load carriage.

3.3. Mean Muscle Activity

As shown in Table 1, in the conventional carriage, the mean muscle activity of knee extensors RF, VM, and VL increased almost linearly with increasing of loads. The R^2 were 0.917, 0.992, and 0.991, respectively. However, there was no obvious change for knee flexor BF and ST. Additionally, the R^2 were 0.829 and 0.562, respectively. Relating to the load level, the changes of the planter flexor PL is linear.

Table 1. Mean Muscle Activities During 0 kg, 15 kg, 30 kg, and 45 kg Load Levels Without the Exoskeleton. (**Note:** The determination coefficient R^2 and the statistics F of Fischer used to confirm the linearity in whole cases are shown. The bold and italic style indicates linearity.)

Muscles	Load (Mean Activities)				F	R^2	*p*
	0 kg	15 kg	30 kg	45 kg			
RF	7.315	8.265	11.16	16	***22.15***	***0.917***	***0.042***
VM	10.02	13.95	18.5	23.9	***393.00***	***0.992***	***0.003***
VL	18.65	22.25	28.15	33.3	***233.17***	***0.991***	***0.004***
BF	4.45	4.35	4.94	5.215	9.71	0.829	0.09
ST	6.61	4.755	5.41	4.645	2.56	0.562	0.25
PL	7.675	8.935	10.28	14.4	***19.24***	***0.906***	***0.048***
GL	5.045	4.155	5.535	7.49	3.51	0.637	0.20

However, in Table 2, the with-Exo conditions, RF, VM, and VL activities are not increasing with load level, as shown in Table 2. Some muscle's mean activities even decrease with heavier load level, such as the knee extensor muscles RF and VM. The PL muscle activities are far higher than the other muscles.

Table 2. Mean Muscle Activities During 0 kg, 15 kg, 30 kg, and 45 kg Load Levels With the Exoskeleton. (Note: The determination coefficient R^2 and the statistics F of Fischer used to confirm the linearity in whole cases are shown. The bold and italic style indicates linearity.).

Muscles	Load (Mean Activities)				F	R^2	p
	EXO + 0	EXO + 15	EXO + 30	EXO + 45			
RF	16.35	23.15	25.8	24.35	4.17	0.68	0.17
VM	19.2	18.15	21.45	26.2	6.68	0.77	0.12
VL	26.1	20.45	22.4	22.2	0.78	0.28	0.47
BF	4.81	6.28	6.57	8.285	***32.40***	***0.94***	***0.03***
ST	4.39	4.94	7.96	8.81	***24.12***	***0.92***	***0.04***
PL	22.8	31.5	53.75	65.5	***72.34***	***0.97***	***0.014***
GL	10.75	11.915	10.81	12.76	1.54	0.43	0.34

Most of the muscles' mean activities increased when using the assistive exoskeletons, especially in the planter flexor PL ($p < 0.001$), as shown in Figure 8. As far as knee flexors BF and ST were concerned, although the mean activities are higher than the conventional condition, there were no obvious change to the muscle activities with increasing loads. According to the experiment, it could be concluded that the flex process was seriously affected by the exoskeleton. For the plantar flexor muscles PL and GL, much more activity was needed when using the EXO, especially for PL. It may be caused by the exoskeleton self-weight and adaptability.

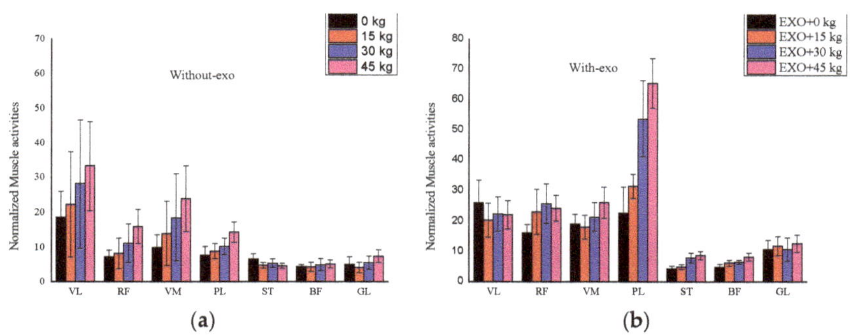

Figure 8. Normalized averaged muscle activities of with-Exo and without-Exo load carriage walking.

3.4. Main and Two-Way ANOVA Interaction Effects of with or without * Load Level of EMG Values

A summary of ANOVA for the effects of load (L) with- or without-Exo (W) for nRMS values of nEMG data are given in Table 3. The two-way ANOVA method was used in the calculations. Significant changes of the load main effects in nRMS are found for almost all the measured muscles ($p < 0.05$), except for the muscle BF ($p > 0.05$). nRMS values greatly increased over load levels in the knee extensors RF, VM, and VL ($p < 0.001$). The main effects of with- or without-EXO are significantly different for almost all the muscles ($p < 0.05$), except for the muscle VM ($p > 0.05$). Load * with or without interaction effects were significant in some muscles such as RF, PL, and GL ($p < 0.05$).

The main and interactive effects could also be seen from plots of Figure 9. Although the overall characteristic is that the mean nRMS from the exoskeleton assisted load walking was higher than those from the conventional load walking, almost all the rates of nRMS that increased during the exoskeleton assisted load walking slowed down with the increasing of load level. Even in some muscles the rates are decreasing with the increase of the load level, such as muscles BF, ST, and GL. Especially for the loaded muscle VM, when the load level is over 30 kg, the exoskeleton augmentation potential appears.

Table 3. Summary of Two-Way ANOVA Results for the Main and Interaction Effects of With or Without and Load Final EMG Values. W × L stands for with or without and load interaction. F means F test value. Lower-case *p* means statistical significance and the significant differences were bolded.

Muscles	With- or Without-EXO (Main Effects) F (*p*)	Load Level (Main Effects) F (*p*)	W × L (Interaction Effects) F (*p*)
RF	**6.418 (p = 0.017)**	**26.124 (p = 0.000)**	**3.356 (p = 0.04)**
VM	0.058 (p = 0.810)	**16.320 (p = 0.000)**	0.734 (p = 0.427)
VL	**6.373 (p = 0.017)**	**19.836 (p = 0.000)**	1.185 (p = 0.312)
BF	**22.653 (p = 0.000)**	1.407 (p = 0.250)	2.472 (p = 0.078)
ST	**12.637 (p = 0.001)**	**5.240 (p = 0.005)**	2.767 (p = 0.062)
PL	**15.496 (p = 0.000)**	**6.728 (p = 0.005)**	**4.158 (p = 0.031)**
GL	**21.391 (p = 0.000)**	**9.287 (p = 0.000)**	**6.729 (p = 0.003)**

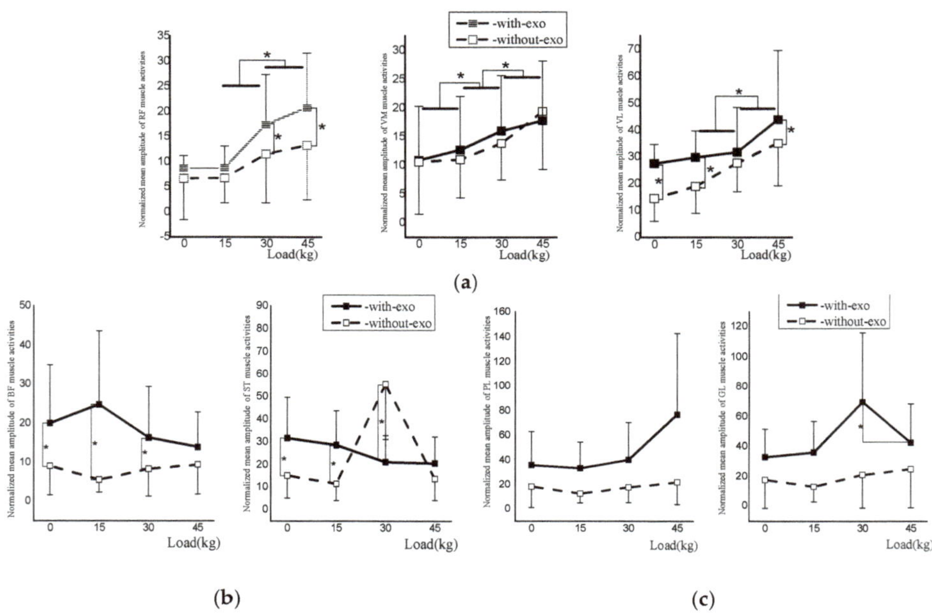

Figure 9. With-Exo or without-Exo and four loading levels effects on initial values of normalized RMS (nRMS) for (**a**) the right knee extensors (RF, VM, VL); (**b**) for the right knee flexors (BF, ST); (**c**) the right plantar flexors (PL, GL). The star (*) indicates the significant difference ($p < 0.001$) between two related factors.

3.5. The Discomfort Index and Six Indices of NASA-TXL Rating Scales

For the subjective performance, the two primary independent factors were still: (1) without-EXO and with-EXO carrying condition and (2) the four loading levels. The dependent experiment variables were: (1) the discomfort rating C and (2) the group weighted ratings of the NASA TLX scales (MD, PD, TD, OP, EF, FR). Because there were just nine investigated samples, the data abnormality was tested by boxplot. The Shapiro-Wilk method was used to test the data normality. A Levene variance homogeneity test was used. The results indicated that there were no abnormal data. The residual errors were normal ($p > 0.05$).

According to the literature [19], this experiment is a kind of Continuous SINGLE-axis MANUAL task. Considering the fact that the TLX index was developed for significantly more complex tasks, the combined numeric value of the TLX is too coarse to be sensitive to the relatively small variations in the task elements. Therefore, the effect of "with or without" variation on the individual mean discomfort rating scales is shown in Figure 10a. Additionally, the individual mean MD, PD, TD, OP, EF, and FR rating scales are shown in Figure 10b. It is noticed that in the conventional load carriage, the workload demands for all effort required and the extended discomfort C for 0 kg level were defined as "1". With increasing load level, the subjective discomfort increased linearly and significantly in the conventional carriage, while in the exoskeleton assisted carriage condition, the discomfort increased much slower. Similar trends for the standard six rating scales could also be observed in Figure 10b.

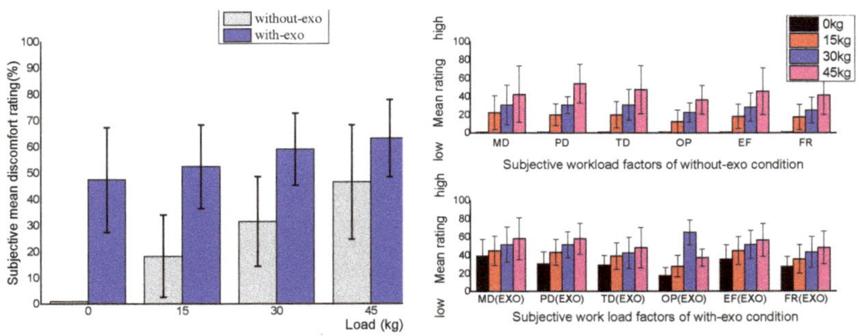

(a) (b)

Figure 10. (**a**) The subjective discomfort performance. (**b**) The comparisons of mean rating value plots of the six workload factors.

3.6. The Weighting Factor and the Results

The mean group weighting factors of the two carriage conditions were determined for the NASA TLX rating scales after the experiment, as shown in Figure 11. The results of the paired t-test illustrate that the MD and PD weights are significantly different in the two carriage conditions. The MD weight in the with-EXO condition is significantly higher ($p = 0.037$) than the one which is in the conventional carriage. It means the assistive exoskeleton increases the mental demand of the subjects. According to the experiment process, the reason may be the subjects have to stabilize the human–robot system not only by providing force and torque, but by providing mental control strategies. As far as PD was concerned, the PD weight in the with-EXO condition is significantly lower ($p = 0.005$) than the one in the conventional carriage. It means the assistive exoskeleton successfully shared the loadings in the experiment process. There were no significant differences for the weighting factors TD, OP, and EF. For the weighting factor FR, it indicated that subjects were more insecure, discouraged, irritated, stressed, and annoyed with exoskeleton assisted loading walking. This indicates a poor ergonomic user interface for the exoskeleton.

According to the comparison of weight factors, the discomfort C, mental demand MD, physical demand PD, and frustration FR were selected into the two-way ANOVA analysis. The results are shown in Table 4. Although there is no interaction effect in the statistical results of Discomfort, Mental Demand, and Physical Demand with carrying condition and load level, the ordinal interactions existed in the interaction plot, as shown in Figure 12. In reality, soldiers often carry loads over 55 kg. If the load level was increased, a disordinal interaction may have appeared. The with- or without-EXO main effects on the three workload factors were significantly different ($p < 0.0001$). The load level main effects on the Physical Demand were significantly different ($p < 0.0001$), and also on Discomfort and Mental Demand. The "with or without" main effects were pair-wise compared. The unweight marginal mean values were 31.208 ± 5.502, $p < 0.0001$. The with-Exo discomfort level was higher 21.438 (95%CI is from 20.890–42.328) than the without-Exo.

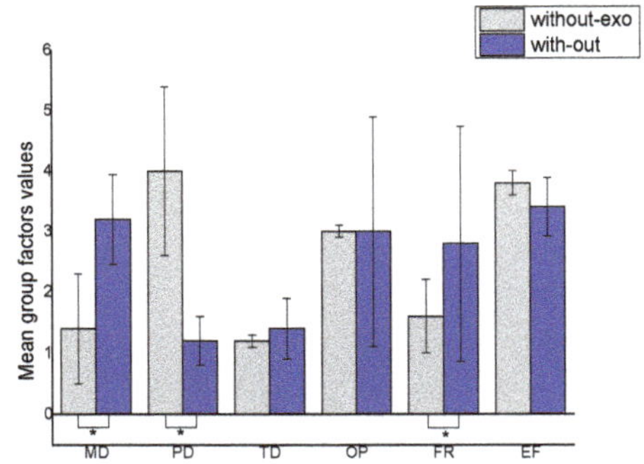

Figure 11. The comparison of weight factors between without-exo and with-exo walking conditions.

Table 4. Summary of Two-Way ANOVA Results for the Main and Interaction Effects on the Discomfort, Mental Demand and Physical Demand. (**Note:** The significant differences were indicated as bold and italic style. F means F test value. Lower-case *p* indicates statistical significance.)

NASA-TLX Rating Scale Values	With- or Without-EXO (Main Effects) F (*p*)	Load Levels (L) (Main Effects) F (*p*)	W × L (Interaction Effects) F (*p*)
Discomfort	***32.177 (<0.0001)***	***5.764 (<0.05)***	1.288 (=0.292)
Mental Demand	***14.513 (<0.0001)***	***4.030 (<0.05)***	0.570 (=0.638)
Physical Demand	***20.006 (<0.0001)***	***15.008(<0.0001)***	1.564 (=0.213)

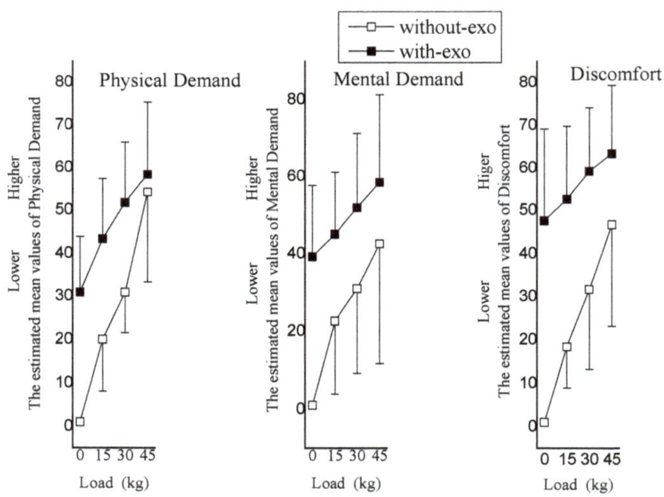

Figure 12. The interaction trend plot of with-Exo or without-Exo × load level.

4. Discussion

Regular load carriage often happens in specific situations, such as for soldiers, school children, hikers, and assembly line workers. It was one of the most frequent risk factors for MSDs which have been confirmed by EU-OSHA. Although a few exoskeletons have been developed to decrease the loading effects on the human body, there is still lacking evidence that these assistive devices allow human beings to move or walk freely or make carriage-loaded MSDs decrease. In this study, one exoskeleton prototype was examined as a potential ergonomic device for application in regular loading tasks. Some advices should be provided to improve the ergonomic performance and human–robot interaction performance so as to decrease carriage-loaded MSDs.

A few assistive exoskeletons for the lower extremities are already commercially available. Some have been introduced into rehabilitation centers, industry assembly lines, and military training [20–22]. Some measures and analyses on the metabolic effects, muscle activities, and human gait patterns have been carried out on the related exoskeletons [11,23,24]. This study remains one of the first to quantify muscle activities for which the exoskeletons are not specifically designed to decrease the metabolic cost. In particular, this investigation aims to quantify muscle amplitude and activities that result from performing four levels of load carriage walking on a treadmill, with and without an exoskeleton. At the same time, this study examined the subjective performance of the exoskeleton using the extended NASA-TLX questionnaire.

Most research has examined muscle activities of the lower limb, comparing the mean and peak amplitude from free treadmill walking to exoskeleton treadmill walking, and have reported a significant increase in muscle activity with the exoskeleton [24]. Few studies have reported decreased muscle activity when using an exoskeleton [25]. It is important to note that the existing augmenting exoskeletons have reported no decrease in muscle activity and oxygen consumption [11,20]. However, this is not to say the assistive exoskeletons pose no use to human beings. It is reported that any real metabolic advantage imparted by the EXO may be dependent on the magnitude of the load carried, such that the metabolic energy cost per unit of mass becomes lower than non-EXO assisted load carriage only when heavy loads (greater than 50% of body mass) are carried by the device [11]. The results of this study are consistence with the former mentioned results. As shown in Figure 9, the interaction of the Load level × with or without-EXO should happen with increasing of load level.

When comparing the load level main effects on muscle activity, it is illustrated that the structure of the effects are different for the with- and without-Exo walking conditions. By looking further into the research results, we can see in the loaded (stance) gait stage, muscle activities decrease with same load level for the with-Exo walking condition. However, in the unloaded (swing) gait stage, the muscle activities increase. This result suggests that the assistive exoskeleton could effectively share the load with human beings. However, exoskeleton weight and adaptability are not yet good enough. The structure design and control strategy should be reconsidered from an ergonomic point of view to improve the agility, ease of push-off, and adaptability of human–robot systems, or the assistive devices could form a new risk factor for MSDs from loaded carriage.

Loading of the lower limb is increased when carrying load. Therefore, a longer ground contact time would increase the amount of time the limb is subjected to increased loading. This may result in increased loading of lower limb structures, such as increased metatarsal compression [24]. Alternatively, a longer ground contact time may be a protective mechanism to lower the rate of loading on the internal structures in order to reduce the potential negative effects of increased load magnitude. It is therefore unclear whether a longer ground contact time is associated with the development of injury, or if it is a mechanism to minimize the risk of injury. Therefore, interventions for increasing muscular strength and endurance of the plantar flexor and knee extensor muscles may be beneficial [2].

EMG data indicated increased ankle plantar, flexor, and knee extensor muscle activity whilst carrying load, in support of the hypotheses. The use of the exoskeleton most significantly increased max and mean muscle activities in the ankle plantar flexor PL in loaded carriage. A significant increase in muscle activation also happened for the knee extensor and hip flexor RF, and the knee flexor and hip extensor ST. However, these significant differences could not be found for the other four pair-wise comparisons (VL, VM, BF, and GL).

The very high rates of PL (peroneus longus) muscle activity which were observed may have been the result of a greater contribution of the plantar flexor muscles while knee extensor moments were reduced. Plantar flexor muscle activation has previously been associated with both tibia and metatarsal stress fractures [26]. The suggested interventions to reduce fatigability of the plantar flexor muscles during load carriage activity should be considered, since there is no activation in the exoskeleton ankle joint. Furthermore, the observed significant increases in PL muscle activity which appeared in the exoskeleton assisted load carriage is more likely a result of the loose fastening, rigidity, and inertia of the exoskeleton shoes. Therefore, the structure of exoskeleton shoes and the human–robot shoe interface should be redesigned.

In the early stance, the quadriceps muscles act to control knee flexion, while the plantar flexors act at the end of stance to contribute to push-off [27]. In theory, carrying heavier loads is more demanding on the knee joint musculature than that of the ankle. Therefore, the exoskeleton prototype is designed to be activated only in the knee joints. However, the increased GL and PL muscle activity suggests a greater plantar flexor force production is required for push-off when carrying a load of 45 kg, as shown in Figure 4. The activation of the ankle joints is significant.

It is clear from the results that load level has a strong influence on the subject's discomfort. The exoskeleton itself also seriously affects the subject's discomfort. However, during the trials with assistance of the exoskeleton, the subject's discomfort increased more slowly with increasing load level. When asked directly, the participants preferred to use the exoskeleton at the heavier load levels of 30 kg and 45 kg. Although no more than a 45 kg load level was tested, ordinal interactions existed in the interaction plot, as shown in Figure 10. That means if the load level was added, the disordinal interaction may appear.

Considering the six standard NASA-TLX indices, the load level also has stronger influence on the subject's performance metrics than the with- or without-Exo has. However, during the with-EXO load carriage, the MD weight is higher and PD weight is lower than during the without-EXO load carriage. This means the exoskeleton increased the mental demand and decreased the physical demand. This situation is consistent with the experiment experience. The subjects reported they had to pay attention to the push-off of the swing for the exoskeleton and the stabilization of the overall system. Furthermore, there existed an interaction force about the lower limb which the subjects had to counteract. This again verified that the adaptability and human–robot interaction performance have to be improved, or the extra risk factors for MSDs may arise for the exoskeleton wearers.

In order to actually help the human being in heavy load carriage and decrease muscle activities, a better ergonomic mechanical structure should be proposed for the tested prototype. Especially for the ankle or foot parts, power should be supplied and the human–exoskeleton connections should be reconsidered. At that point, human beings should be included in the control scheme of the human–exoskeleton system.

5. Conclusions

Although the activities of the knee extensors VM and VL decreased with the help of the powered exoskeleton under the heaviest carriage level, this study indicated much stronger muscular activities of most tested muscles when carrying loads using the assistive exoskeleton, especially the planter flexor PL. These increased muscular responses may result in new muscular skeletal injuries. The subjective comfort performance was lower than that in the conventional carriage condition. The findings show that the argumentation exoskeleton can help reduce the burden of the load, while stabilization and recruitment are severely mentally and physically demanding. This is consistent with the investigation results of more mental demands and lower physical demands in the NASA-TLX factors. Such investigations may have important implications related to the construction design and human–robot control strategies of an augmentation exoskeleton. Some of the results in this article may possibly be used as empirical data for sensorimotor interactions when walking with different types of exoskeletons.

Author Contributions: H.L. and W.C. conceived and designed the experiments. H.L. performed the experiments, analyzed the data, and wrote the paper. F.L. provided suggestions for the experiments. M.Z. contributed the analysis tools. K.W. provided management of the experiments.

Funding: This work was supported by Chinese National Science Foundation Grant Number 51675450. We are especially grateful to the participant for the volunteers.

Conflicts of Interest: The authors declare no conflict of interest.

References

1. Hamzat, T.K.; Abdulkareem, T.A.; Akinyinka, O.O.; Fatoye, F.A. Backpack-related musculoskeletal symptoms among nigerian secondary school students. *Rheumatol. Int.* **2014**, *34*, 1267–1273. [CrossRef] [PubMed]
2. Rice, H.; Fallowfield, J.; Allsopp, A.; Dixon, S. Influence of a 12.8-km military load carriage activity on lower limb gait mechanics and muscle activity. *Ergonomics* **2017**, *60*, 649–656. [CrossRef] [PubMed]
3. Anderson, A.M.; Meador, K.A.; Mcclure, L.R.; Makrozahopoulos, D.; Brooks, D.J.; Mirka, G.A. A biomechanical analysis of anterior load carriage. *Ergonomics* **2007**, *50*, 2104–2117. [CrossRef] [PubMed]
4. Knapik, J.; Reynolds, K.; Staab, J.; Vogel, J.A.; Jones, B. Injuries associated with strenuous road marching. *Mil. Med.* **1992**, *157*, 64–67. [CrossRef] [PubMed]
5. Lavender, S.A.; Ko, P.L.; Sommerich, C.M. Biomechanical evaluation of the eco-pick lift assist: A device designed to facilitate product selection tasks in distribution centers. *Appl. Ergon.* **2013**, *44*, 230–236. [CrossRef]
6. Kazerooni, H. The berkeley lower extremity exoskeleton. *J. Dyn. Syst. Meas. Control* **2006**, *128*, 9–15. [CrossRef]
7. Kawamoto, H.; Taal, S.; Niniss, H.; Hayashi, T.; Kamibayashi, K.; Eguchi, K. Voluntary motion support control of robot suit HAL triggered by bioelectrical signal for hemiplegia. In Proceedings of the International Conference of the IEEE Engineering in Medicine and Biology, Buenos Aires, Argentina, 31 August–4 September 2010; pp. 462–466.
8. Panizzolo, F.A.; Ignacio, G.; Asbeck, A.T.; Christopher, S.; Kai, S.; Holt, K.G. A biologically-inspired multi-joint soft exosuit that can reduce the energy cost of loaded walking. *J. Neuroeng. Rehabil.* **2016**, *13*, 43. [CrossRef]
9. Dollar, A.M.; Herr, H. Lower extremity exoskeletons and active orthoses: Challenges and state-of-the-art. *IEEE Trans. Robot.* **2008**, *24*, 144–158. [CrossRef]
10. Schiffman, J.M.; Gregorczyk, K.N.; Bensel, C.K.; Hasselquist, L.; Obusek, J.P. The effects of a lower body exoskeleton load carriage assistive device on limits of stability and postural sway. *Ergonomics* **2008**, *51*, 1515–1529. [CrossRef]
11. Gregorczyk, K.N.; Obusek, J.P.; Hasselquist, L.; Bensel, J.S.M.; Carolyn, K.; Gutekunst, D. *The Effects of a Lower Body Exoskeleton Load Carriage Assistive Device on Oxygen Consumption and Kinematics during Walking with Loads*; U.S. Army Natick Soldier Center: Natick, MA, USA, 2006.
12. Zanotto, D.; Akiyama, Y.; Stegall, P.; Agrawal, S.K. Knee joint misalignment in exoskeletons for the lower extremities: Effects on user's gait. *IEEE Trans. Robot.* **2015**, *31*, 978–987. [CrossRef]
13. Aguirre-Ollinger, G.; Colgate, J.E.; Peshkin, M.A.; Goswami, A. Inertia compensation control of a one-degree-of-freedom exoskeleton for lower-limb assistance: Initial experiments. *IEEE Trans. Neural Syst. Rehabil. Eng.* **2012**, *20*, 68–77. [CrossRef]
14. Ye, D.; Galiana, I.; Asbeck, A.; Rossi, S.D.; Bae, J.; Santos, T. Biomechanical and physiological evaluation of multi-joint assistance with soft exosuits. *IEEE Trans. Neural Syst. Rehabil. Eng.* **2017**, *25*, 119–130.
15. Hansen, C.; Gosselin, F.; Ben, K.M.; Devos, P.; Marin, F. Design-validation of a hand exoskeleton using musculoskeletal modeling. *Appl. Ergon.* **2018**, *68*, 283–288. [CrossRef] [PubMed]
16. Michalos, G.; Karvouniari, A.; Dimitropoulos, N.; Togias, T.; Makris, S. Workplace analysis and design using virtual reality techniques. *Cirp Ann.-Manuf. Technol.* **2018**, *67*, 141–144. [CrossRef]
17. Liu, F.; Cheng, W.; Wu, Q. Research on human exoskeleton based on rigid-flexible coupling system. *Mech. Sci. Technol. Aerosp. Eng.* **2013**, *32*, 15.
18. Zhang, M.K.; Cheng, W.; Li, H.X. Drive of the powered exoskeleton and driving compensation with the human-machine coupling interaction while squatting. *Robot* **2017**, *39*, 514–522.
19. Hart, S.G.; Staveland, L.E. Development of nasa-tlx (task load index): Results of empirical and theoretical research. *Adv. Psychol.* **1988**, *52*, 139–183.
20. Hidler, J.W.; Wisman, N. Kinematic trajectories while walking within the Lokomat robotic gait-orthosis. *Clin. Biomech.* **2008**, *23*, 1251–1259. [CrossRef]
21. Stegall, P.; Winfree, K.; Zanotto, D.; Agrawal, S.K. Rehabilitation exoskeleton design: Exploring the effect of the anterior lunge degree of freedom. *IEEE Trans. Robot.* **2013**, *29*, 838–846. [CrossRef]
22. Suzuki, K.; Mito, G.; Kawamoto, H.; Hasegawa, Y.; Sankai, Y. Intention-based walking support for paraplegia patients with robot suit hal. *Adv. Robot.* **2007**, *21*, 1441–1469.

23. Knaepen, K.; Beyl, P.; Duerinck, S.; Hagman, F.; Lefeber, D.; Meeusen, R. Human–robot interaction: Kinematics and muscle activity inside a powered compliant knee exoskeleton. *IEEE Trans. Neural Syst. Rehabil. Eng.* **2014**, *22*, 1128–1137. [CrossRef] [PubMed]
24. Del-Ama, A.J.; Asín-Prieto, G.; Piñuela-Martín, E.; Pérez-Nombela, S.; Lozano-Berrio, V.; Serrano-Muñóz, D. Muscle Activity and Coordination During Robot-Assisted Walking with H2 Exoskeleton. In *Converging Clinical and Engineering Research on Neurorehabilitation II*, 1st ed.; Jaime, I., José, G.V., José, M.A., Metin, A., José, L.P., Eds.; Springer International Publishing: Segovia, Spain, 2017; Volume 15, pp. 349–353, ISBN 978-3-319-46668-2.
25. Collins, S.H.; Wiggin, M.B.; Sawicki, G.S. Reducing the energy cost of human walking using an unpowered exoskeleton. *Nature* **2015**, *522*, 212–215. [CrossRef] [PubMed]
26. Arndt, A.; Ekenman, I.; Westblad, P.; Lundberg, A. Effects of fatigue and load variation on metatarsal deformation measured in vivo during barefoot walking. *J. Biomech.* **2002**, *35*, 621–628. [CrossRef]
27. Nindl, B.C.; Harman, E.A.; Marx, J.O.; Gotshalk, L.A.; Frykman, P.N.; Lammi, E. Regional body composition changes in women after 6 months of periodized physical training. *J. Appl. Physiol.* **2000**, *88*, 2251–2259. [CrossRef] [PubMed]

applied sciences

MDPI

Article

Combining Subgoal Graphs with Reinforcement Learning to Build a Rational Pathfinder

Junjie Zeng, Long Qin *, Yue Hu, Cong Hu and Quanjun Yin

College of System Engineering, National University of Defense Technology, Changsha 410073, Hunan, China; zengjunjie13@nudt.edu.cn (J.Z.); huyue.nudt@gmail.com (Y.H.); hccz95@163.com (C.H.); yin_quanjun@163.com (Q.Y.)

* Correspondence: qldbx2007@sina.com; Tel.: +86-136-4744-5281

Received: 9 August 2018; Accepted: 3 September 2018; Published: 17 January 2019

Abstract: In this paper, we present a hierarchical path planning framework called SG–RL (subgoal graphs–reinforcement learning), to plan rational paths for agents maneuvering in continuous and uncertain environments. By "rational", we mean (1) efficient path planning to eliminate first-move lags; (2) collision-free and smooth for agents with kinematic constraints satisfied. SG–RL works in a two-level manner. At the first level, SG–RL uses a geometric path-planning method, i.e., Simple Subgoal Graphs (SSG), to efficiently find optimal abstract paths, also called subgoal sequences. At the second level, SG–RL uses an RL method, i.e., Least-Squares Policy Iteration (LSPI), to learn near-optimal motion-planning policies which can generate kinematically feasible and collision-free trajectories between adjacent subgoals. The first advantage of the proposed method is that SSG can solve the limitations of sparse reward and local minima trap for RL agents; thus, LSPI can be used to generate paths in complex environments. The second advantage is that, when the environment changes slightly (i.e., unexpected obstacles appearing), SG–RL does not need to reconstruct subgoal graphs and replan subgoal sequences using SSGs, since LSPI can deal with uncertainties by exploiting its generalization ability to handle changes in environments. Simulation experiments in representative scenarios demonstrate that, compared with existing methods, SG–RL can work well on large-scale maps with relatively low action-switching frequencies and shorter path lengths, and SG–RL can deal with small changes in environments. We further demonstrate that the design of reward functions and the types of training environments are important factors for learning feasible policies.

Keywords: subgoal graphs; reinforcement learning; hierarchical path planning; uncertain environments; mobile robots

---

1. Introduction

In this paper, we focus on the problem of planning rational paths in continuous and uncertain environments. By "rational", we mean that, firstly, computational costs, mainly the time consumed, brought by the planning algorithm must be low enough, so requirements such as avoiding first-move lags and providing human users with an excellent experience can be met; secondly, resultant paths must be collision-free and smooth, and thus, feasible to follow given kinematic constraints. One typical application of this problem lies with agents who maneuver in time-sensitive scenarios (e.g., service robots working in the real world and Non-Player Characters (NPCs) engaging movement tasks in video games). Briefly, application fields such as robotics and video games strongly require time-saving path-planning methods which can quickly generate realistic paths.

The problem of path planning is extensively researched in the literature [1–6]. A* on grid maps [6] is regarded as a fundamental and standard path-planning algorithm based on search. It is complete and optimal, which means A* can find an optimal path if there is at least one from the start point to the goal point. There are at least two shortcomings of A* on grid maps: (1) path symmetry [3], i.e., the existence of symmetrical paths forces A* to generate same states multiple times along different paths, which consumes more time; (2) cell-wise expansion, i.e., since successors of a cell which are expanded are just its adjacent cells, this short-ranged expansion brings about a huge open list and a large number of operations. These two problems cause many unnecessary cell generations and expansions, and, as a result, A* is so slow that it cannot meet the first requirement for rational paths.

To overcome the above disadvantages, methods for map abstraction attract many researchers' attention. Subgoal graphs (SG) [5] is an archetypical map-abstraction-based algorithm, whose preprocessing stage is responsible for building subgoal graphs. Similar to visibility graphs, its simple version—SSG—places subgoals at the corners of obstacles. SSG links cells, between which a reachability relationship called direct-h-reachability is satisfied, and all symmetrical paths are traversable. Then, over subgoal graphs, A* is used to obtain the abstract path. Finally, depth-first search is safely utilized to refine it into a ground-level path. As a result, A* on subgoal graphs works far better than that on grids and satisfies the first requirement mentioned above. As for the second requirement, however, the optimal ground-level path is constrained to grid edges (i.e., the heading changes are artificially constrained to specific angles) [7], which causes sharp turns in the generated paths, and cannot meet kinematic constrains. Moreover, it is time-consuming for SG to deal with changing environments by reconstructing subgoal graphs.

To plan kinematically feasible paths, interpolating curve planners [8] construct and insert a new set of data within the range of a previously known set. For example, in Reference [9], hypocycloidal curves were used to smooth sharp turns, which can maintain a safe clearance in relation to the obstacles. However, when unexpected obstacles often appear in the detection range of robots, these kinds of planners need to recalculate a new path frequently, causing more time consumption. It is also difficult for sampling-based planners, such as Probabilistic Roadmaps (PRM) [10] and rapidly exploring Random Trees (RRT) [11], to deal with unexpected obstacles. Furthermore, the generated paths are random and suboptimal.

To deal with uncertain environments and improve the path quality, Reinforcement Learning (RL) can be used to learn near-optimal policies to find paths in the real world under certain constrains. RL is learning what to do—how to map situations to actions—to maximize a numerical reward signal [12]. Specifically, RL is an agent that interacts with the environment, and learns an optimal policy by trial and error [13]. RL emerged as a practical method for robot control [14] and was successfully applied to solve complex robot manipulation problems [15] and learn complex skills like playing Go [16] and Atari games [17]. There are two categories of RL (i.e., model-based RL and model-free RL). Since we cannot provide complete models of uncertain environments for RL, model-free RL is more suitable for our scenarios. However, RL cannot be used directly to plan paths in complex environments due to two limitations: (1) sparse reward, i.e., reward is sparsely distributed in large state spaces, causing difficulties for learning feasible policies; (2) local minima trap, i.e., RL agent may be trapped in local minima, like goals on the other side of box canyons.

In this paper, to overcome the abovementioned limitations of SG and RL methods, we present a hierarchical path planning framework, called SG–RL, to integrate the geometric path-planning method (i.e., SG) and the ground-level motion planning method (i.e., RL), which can plan rational paths in uncertain and continuous environments. SG–RL solves this path-planning problem in a two-phase manner. The first phase is a global abstract path-planning phase that focuses on the first requirement for rational paths. The second phase is a feasible trajectory-planning phase that focuses on the second requirement. In the first phase, SG–RL uses SSG to find global abstract paths, also called subgoal sequences, from start points to goal points with high computational efficiency. In the second phase, considering kinematic constraints, SG–RL uses an RL method, i.e., Least-Squares Policy Iteration (LSPI) [18], to learn near-optimal motion-planning policies which can generate kinematically feasible and collision-free trajectories between adjacent subgoals.

Compared with RL methods, SG–RL can use SSG to plan paths more efficiently so as to eliminate first-move lags, and owing to the direct-h-reachability of adjacent subgoals, SG–RL can support RL agents to realize long-range navigation on complex maps by overcoming sparse reward and local minima trap. Then, compared with SG methods, SG–RL can use LSPI to generate smooth paths that follow kinematic equations and preserve G1 continuity [9], and due to the generalizability of LSPI, SG–RL can deal with small changes (i.e., unexpected obstacles) over maps.

The rest of this paper is organized as follows: Section 2 discusses some related work on path-planning methods based on A*, RL, and hierarchical path planning. Section 3 presents the hierarchical path-planning approach based on SG and RL. In Section 4, the performance of the proposed approach is evaluated by simulation experiments. Finally, this paper is concluded in Section 5.

2. Related Work

In this section, studies of three kinds of methods (i.e., path-planning methods based on A*, RL, and hierarchical path planning) in the path-planning field are introduced.

2.1. Path-Planning Methods Based on A*

Hart, Nilsson, and Raphael jointly proposed the A* algorithm, which is a widely used best-first search algorithm and can search an optimal path over grid maps [6]. A* plays an important role in areas that do not require real-time response. However, with many real-time applications getting prevalent in fields such as real-time strategy games and robotics, A* faces severe challenges (e.g., searching in large scale and dynamic environments). A* tends to scale poorly as it must compute complete and optimal paths for agents before agents can move, which brings the first-move lag problem. In brief, A* on grid maps has two mentioned weaknesses (i.e., path symmetry and cell-wise expansion). These two weaknesses cause a huge search space and unnecessary cell generations, which renders A* very slow.

To solve these problems, some recent studies focusing on map representations were proposed. Such kinds of studies mainly construct particular map representations via abstracting topological structures and key information of original maps to reduce the search space. Jump point search (JPS) [2] proposed by Harabor and Grastien uses the canonical ordering method to solve the path symmetry problem and identifies jump points over grid maps to reduce the search space. Then, the optimal canonical path is searched among jump points. However, in JPS, jump points do not really "jump" to one another; instead, they just "roll" between jump points, namely to check cells one by one online. To increase the search speed between jump points, JPS+ [3] builds jump-point maps via preprocessing. Since the distance information regarding adjacent jump points exists in the jump-point map, the path between jump points can be directly found. There is another kind of method named Contraction Hierarchy (CH) [4] that can search over grid maps faster than JPS+ [19], due to its distinctive preprocessing. The preprocessing of CH involves the contraction of one node at a time out of the graph and adds shortcut edges to the remaining graph. However, CH costs more preprocessing time and memory space than JPS+. SSG [5], proposed by Tansel Uras and Sven Koenig, constructs subgoal graphs by identifying subgoals and checking direct-h-reachability in the preprocessing stage. Compared with the original map, the subgoal graph abstracts the information of key points, so as to reduce the search space. The direct-h-reachability ensures that any shortest paths between adjacent subgoals are not blocked by obstacles, allowing the path symmetry problem can be solved. To reduce the size of subgoal graphs, the two-level subgoal graphs method [5] relaxes connection conditions between subgoals and adds a subgoal-pruning strategy. To further increase the search speed, the n-level subgoal graphs approach [20] was proposed. These above path-planning methods based on A* can meet the first requirement of rational paths. However, these methods cannot solve the ground-level motion-planning problem under certain constraints and cannot deal with changes of environments.

2.2. Reinforcement Learning

In the field of ground-level motion planning, RL recently gained prevalence for systems with unknown dynamics [14]. Q-learning and Sarsa [12] can be useful for dealing with discrete state spaces. For example, Mihai Duguleana et al. [21] combined Q-learning with the artificial neural network for solving the problem of autonomous movement of robots in environments that contain both static and dynamic obstacles. However, in large or continuous state spaces, the abovementioned tabular RL methods are inefficient or impractical for applications. Function approximation methods that estimate value function can be used to tackle this problem [22]. There are many function approximators, such as artificial neural networks, decision trees, linear basis functions, and so on. In this paper, we focus on linear-basis-function approximators. The least-squares temporal-difference learning (LSTD) algorithm [23] proposed by Bradtke and Barto, an ideal method for prediction problems, uses linear basis functions to approximate the value function of a fixed policy. Though LSTD has many good properties like data efficiency and fast convergence, it cannot be straightforwardly used to learn good control policies. To solve this problem, the LSPI [18] algorithm proposed by Lagoudakis and Ronald introduces LSTDQ, an algorithm similar to LSTD that learns approximate state-action value functions of fixed policies. LSPI adds LSTDQ into the approximate policy-iteration structure, allowing it to get an optimal policy quickly by combining the policy-search efficiency of policy iteration with the data efficiency of LSTDQ. Due to its generalizability, LSPI can handle changes in environments. In this paper, we choose LSPI as the RL method. However, RL cannot be used directly in path planning in complex environments due to two limitations: (1) sparse reward; and (2) local minima trap.

2.3. Hierarchical Path Planning

To overcome limitations of path-planning methods based on A* and ground-level motion-planning methods, hierarchical path-planning approaches are widely researched in the literature [24–28]. In Reference [24], a two-level-based, goal-driven architecture is used to solve mobile robot navigation in real life with vision systems and infrared sensors. In Reference [25], a two-stage path-planning algorithm uses a variant of A* search to obtain a kinematically feasible trajectory in the first stage and improves the quality of the solution via numeric non-linear optimization in the second stage, for an autonomous vehicle operating in an unknown semi-structured (or unstructured) environment where obstacles are detected online by a robot's sensors. In Reference [26], to guarantee the consistency between high and low levels of planning, Cowlagi and Tsiotras proposed a hierarchical motion-planning framework with a novel mode of interaction between the geometric path planner and the vehicle trajectory planner. In Reference [27], a novel hierarchical global path-planning approach for mobile robots in a cluttered environment with multi-objectives was proposed. This approach has a three-level structure which combines triangular decomposition, Dijkstra's algorithm, and a proposed particle swarm optimization called constrained multi-objective particle swarm optimization. In Reference [28], to find an optimal maneuver that moves a car-like vehicle between two configurations in minimum time, a two-phase algorithm firstly solves a geometric optimization problem and then finds the optimal maneuver with system dynamics and its constraints satisfied.

In this paper, to overcome the abovementioned disadvantages of SSG and LSPI, we designed a suitable hierarchical framework to combine a global abstract path planner based on SSG with a feasible trajectory planner based on LSPI, which can improve the computational efficiency of hierarchical path-planning methods by optimizing the relationship between high and low levels of planning.

3. Materials and Methods

In this section, the problem description and the tracked robot used in the simulation experiments are briefly introduced in the first subsection. Next, the architecture of the proposed hierarchical path-planning method—SG–RL—is presented. Then, the SG-based global abstract path planner and the LSPI-based feasible trajectory planner are described in detail.

3.1. Problem Description

The path-planning problem investigated in this paper can be stated as follows: in continuous and uncertain environments, the problem is to plan rational paths for tracked robots from a start position to a goal position.

In this paper, we used the Cartesian coordinate system. The schematic diagram of the tracked robot is shown in Figure 1. The position state of the tracked robot in this paper is defined as $s = (x_r, y_r, \theta)$ by the global coordinate related to the map, where x_r and y_r are the tracked robot's abscissa and vertical coordinates, respectively, and θ is the angle between the forward orientation of the tracked robot and abscissa axis. We equipped six ultrasonic sensors at the front of the tracked robot to perceive the external environment. The detection angle of each ultrasonic sensor was 30 degrees. The max detection distance of each sensor was D_{max}. The tracked robot was driven by the left and right tracks. The kinematic equations of the tracked robot were

$$\begin{pmatrix} \dot{x} \\ \dot{y} \\ \dot{\theta} \end{pmatrix} = R \begin{bmatrix} \frac{cos\theta}{2} & \frac{cos\theta}{2} \\ \frac{sin\theta}{2} & \frac{sin\theta}{2} \\ \frac{1}{l_B} & \frac{-1}{l_B} \end{bmatrix} \begin{pmatrix} \omega_l \\ \omega_r \end{pmatrix}, \tag{1}$$

where R is the radius of the driving wheels in the tracks, ω_l and ω_r are the angular velocity of left tracks and right tracks (rad$\cdot s^{-1}$), respectively, and l_B is the distance between left tracks and right tracks.

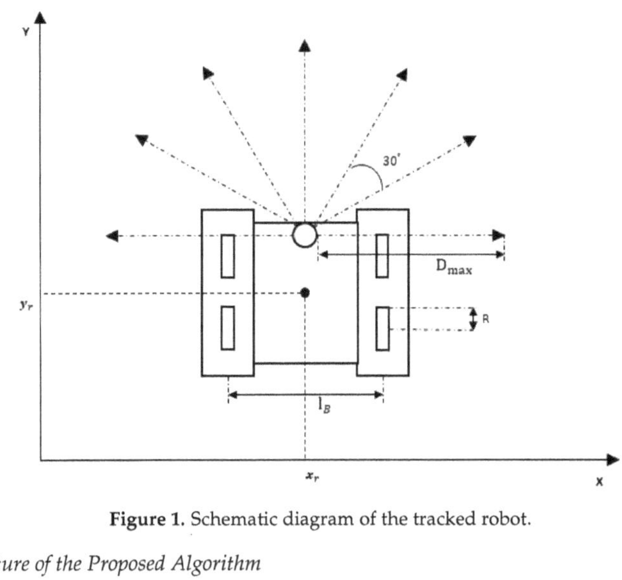

Figure 1. Schematic diagram of the tracked robot.

3.2. The Architecture of the Proposed Algorithm

As mentioned above, we decomposed the complex path-planning problems into two phases: the first for the global abstract path planning and the second for the feasible trajectory planning. In the first phase, we used SSG to find global abstract paths with low computation costs. In the second phase, we chose LSPI as the RL method. Then, a feasible trajectory planner based on LSPI took subgoals, which are cells of abstract paths, as input, and planned collision-free trajectories between subgoals with kinematic constraints satisfied. Moreover, there were online stages and offline stages in both phases. Figure 2 shows an overview of this approach.

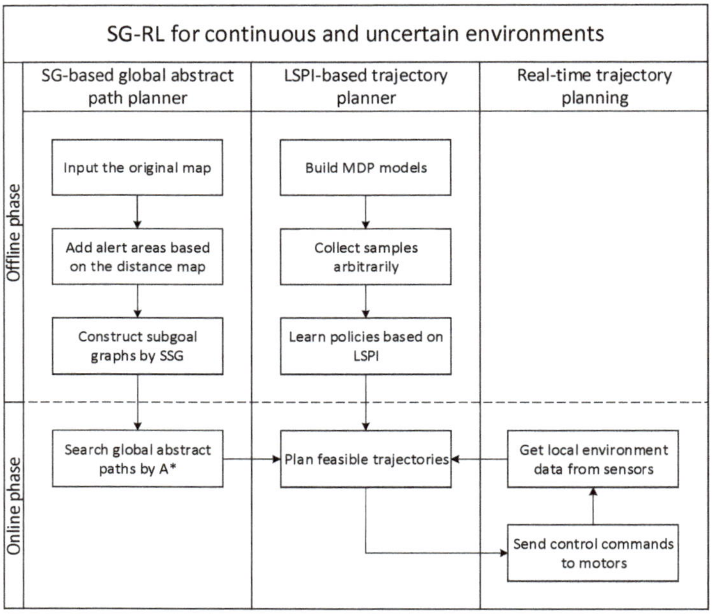

Figure 2. Subgoal graphs–reinforcement learning (SG–RL) flowchart.

The main task of the first phase was providing subgoal sequences which began with the start point and ended with the goal point. During the offline stage, firstly, considering sizes of robots and user-defined safety distance, we added alert areas into original maps based on the distance map to plan safe and feasible paths. Then, we used SSG to construct subgoal graphs from modified maps by placing subgoals at the corners of obstacles and adding edges between related subgoals. During the online stage, we could use A* to search global abstract paths over the subgoal graph. It was not necessary to obtain the optimal ground-level paths between subgoals and assemble them into a complete one. Since basic search methods can hardly generate smooth and feasible solutions, SG–RL passes the task to the RL phase.

The main task of the second phase was planning feasible trajectories between subgoals. During the offline stage, firstly, this task can be divided into two sub-tasks which involve approaching subgoals with kinematic constraints and avoiding initially known or unknown obstacles. By analyzing the key information abstracted from sub-tasks, two Markov Decision Processes (MDPs), which were built separately, formally describe an environment for the RL method, LSPI. Next, according to MDPs, we collected samples arbitrarily from training environments which were smaller than test environments. Then, by carefully tuning parameters, LSPI could train feasible trajectory planners for robots. During the online stage, taking subgoal sequences and local sensor data as input, trajectory planners could choose the best action to generate collision-free and kinematically feasible paths. Note that LSPI uses the original map as the working environment, since alert areas were built by us, and the robot cannot recognize alert areas from local sensor data.

The first strength of SG–RL is that the excellent performance of SSG can be fully utilized, and some limitations of applying LSPI to plan paths in complex environments can be solved by SSG. (1) SSG uses A* to efficiently plan subgoal sequences over subgoal graphs constructed based on maps with alert areas, which can eliminate first-move lags. (2) SG–RL does not select subgoals from the optimal geometrical path according to certain standards [24]; however, it utilizes the intermediate, the abstract path regarded as subgoal sequences, which greatly increases the speed of providing subgoal sequences.

Subgoals are placed at the exit of the local minima area, and the adjacent subgoals are direct-h-reachable, which ensures that all symmetrical paths between them are traversable, so as to eliminate the local minima trap for LSPI. (3) SG–RL uses SSG to generate subgoal sequences that are the input of LSPI. Since the distance between adjacent subgoals is shorter than the distance between the start and goal point, the reward distribution becomes relatively dense, which solves the limitation of sparse reward for LSPI. In conclusion, SSG used in SG–RL not only meets the first requirement of rational paths, but also overcomes the two abovementioned limitations of LSPI.

The second strength is that, based on optimal abstract paths over discretized grid maps, LSPI can train near-optimal feasible trajectory-planning policies which can plan collision-free and smooth continuous paths, following kinematic equations. Owing to linear-basis-function approximators, policies can take continuous state spaces as input to deal with continuous environments. Consequently, SG–RL satisfies the second demand for rational paths.

The third strength is that the SG–RL method can quickly adapt to uncertain environments via the generalizability of LSPI. When the map changes slightly, SG needs to reconstruct all subgoal graphs and replan paths. The constructed subgoal graphs cannot be directly applied to changing environments. Until now, there is no research regarding locally reconstructing subgoal graphs. Therefore, SG–RL uses the generalizability of LSPI to deal with small changes in environments based on original subgoal sequences without reconstructing subgoal graphs.

In conclusion, SG–RL not only makes good use of outstanding advantages of SG and RL, but also overcomes each limitation, allowing it to plan rational paths in continuous and uncertain environments.

3.3. The SG-Based Global Abstract Path Planner

Figure 3 shows the offline work concerning building alert areas and constructing subgoals. Since paths found by SSG over the original map are close to obstacles, they are not safe for robots to follow. To tackle this problem, we added alert areas into original-map-based distance maps.

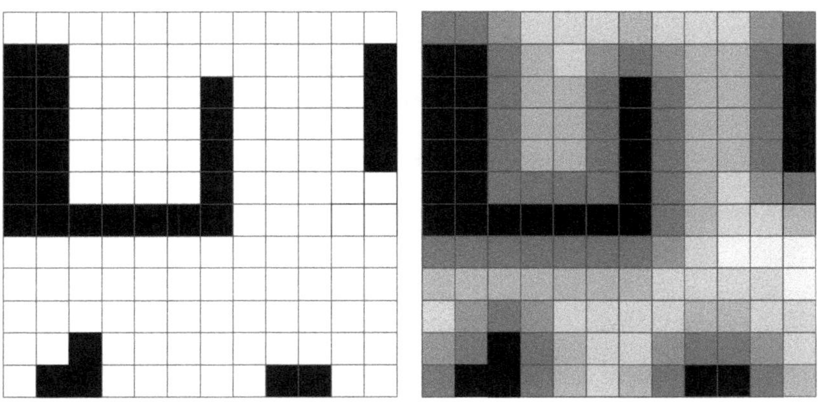

(a) (b)

Figure 3. *Cont.*

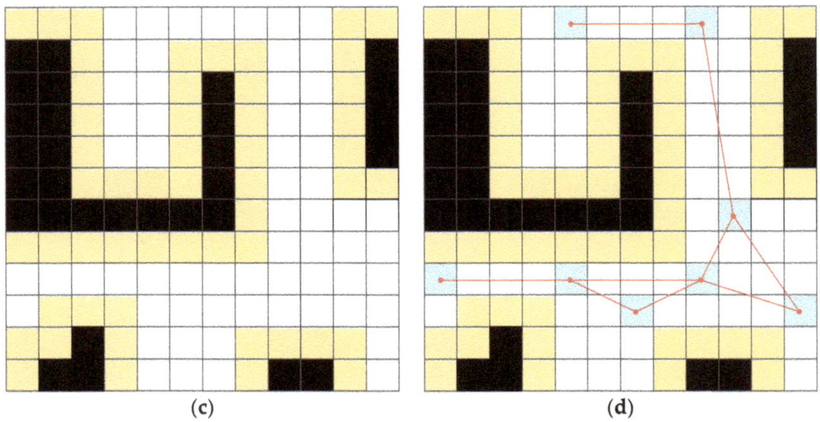

Figure 3. (a)This is the original map. (b) The Euclidean distance map. Black represents occupied cells. The brightness of cells increases with distance. (c) The map with alert areas. The light-yellow cells denote the alert areas. (d) The subgoal graph. The light-blue cells represent subgoals, and the red lines signify the direct-h-reachability of adjacent subgoals.

During the phase of building alert areas, the dynamic brushfire algorithm [29] was used to build the Euclidean distance map (Figure 3b) based on the original map (Figure 3a). Then, the map with alert areas was constructed based on the Euclidean distance map. The size of alert areas was determined by the sizes of the robot and user-defined safety distance. One distinctive advantage of using a distance map to build alert areas is that one distance map can be used to construct various sizes of alert areas.

Then, we used the SG algorithm to find abstract paths on maps with alert areas and provide subgoal sequences for the feasible trajectory planner. There are three kinds of SG algorithm, namely simple, two-level, and n-level subgoal graphs [5,20]. We chose SSG as the method at this phase because, in SSG, all symmetrical paths between adjacent subgoals are not blocked by obstacles, which can provide better subgoal sequences for the trajectory planner.

First of all, in Reference [5], there are some key definitions.

Heuristic $h(s, s')$ is the octile distance between cells s and s'.

Definition 1. *An unblocked cellsis a subgoal iff there are two perpendicular cardinal directions c_1 and c_2 such that $s + c_1 + c_2$ is blocked and $s + c_1$ and $s + c_2$ are not blocked.*

Definition 2. *Two cellssand s' are h-reachable iff there is a path of length $h(s, s')$ between them. Two h-reachable cells are safe-h-reachable iff all shortest paths between them are not blocked by obstacles. Two h-reachable cells s and s' are direct-h-reachable iff none of the shortest paths between them contains a subgoal $s'' \notin \{s, s'\}$.*

Definition 3. *A subgoal graph$G_S = (V_S, E_S)$ is an undirected graph where V_S is the set of subgoals and E_S is the set of edges connecting direct-h-reachable subgoals. The lengths of edges are the octile distances between subgoals they connect.*

In SSG, there are two steps to find the shortest paths from a start cell to a goal cell. The preprocessing stage abstracts subgoal graphs from the underlying grid maps with alert areas, by identifying subgoals and checking direct-h-reachability. This sparse space representation can eliminate the path symmetry with the strict reachability check and generate subgoals which are not close to each other, as successors. As for the online phase, given the start and goal point, it applies a connect–search–refine approach to return ground-level paths. For our scenarios, the refinement was replaced by LSPI to obtain a smooth and collision-free path. Thus, a connect–search method remained, as shown in Algorithm 1.

Algorithm 1 shows how to find the shortest paths over subgoal graphs. Firstly, the start cell and the goal cell are regarded as subgoals and are connected to subgoal graphs. Then, A* is used to search global abstract paths over modified subgoal graphs. As proven in Reference [5], SSG is a complete and optimal global search algorithm.

In Figure 3d, the grid map with alert areas is 12 × 12. However, the size of V_S of the subgoal graph is 7, and the size of E_S is 7. Obviously, the search space of the subgoal graph is smaller than the grid map's. Moreover, subgoals are placed at the exit of the box canyon on the left side of Figure 3d, and adjacent subgoals are direct-h-reachable, which brings greater security for the trajectory-planning phase.

Algorithm 1 Searching subgoal graphs Pseudo-code of simple subgoal graph (SSG).

```
function ConnectToGraph(cell s):
if s ∉ V_S then
  V_S = V_S ∪ {s};
  S ← GetDirectHReachable(s);
  for all s' ∈ S do
      E_S = E_S ∪ {(s, s')};
function FindAbstractPath(cells s, s'):
ConnectToGraph(s)
ConnectToGraph(s')
Π ← find a shortest path from s to s' over the modified subgoal graph
return Π;
function FindPath(cells s, s')
Π ← TryDirectPath(s, s');
if Π ≠ nopath then
     return Π;
Π ← FindAbstractPath(s, s');
return Π;
```

3.4. The LSPI-Based Feasible Trajectory Planner

3.4.1. LSPI Method

LSPI can deal with continuous state spaces and be more data-efficient and time-efficient than other conventional temporal-different RL methods. Figure 4 shows the overview of the entire LSPI framework [18]. It is a completely off-policy and offline algorithm, and can use sample sets collected arbitrarily from the simulation environment or the real world. At the policy evaluation step, it uses the linear structure to obtain the approximate state-action value function, which is easy to implement and use. At the policy improvement stage, it obtains greedy policies by maximizing state-action values, based on approximate value functions. Furthermore, the transparent inner mechanism is beneficial for users to see how it works and why failures occur.

Theoretically, samples from any policy can be collected arbitrarily. Even during policy iteration, samples from the current policy can also be collected and added to train policies. LSPI provides great flexibility for collecting samples, which brings much convenience in practical uses. However, similar to using samples to approximate linear or nonlinear functions, the sample distribution in the state space affects the speed and result of approximating. Therefore, when sample sets are of poor quality, LSPI causes a worse effect via fitting biased distribution based on the data-efficient feature. To solve these problems, in this paper, we not only used random policies to collect samples, but we also tried using random environments to collect samples. Details on how samples are collected are given in Section 4.1.

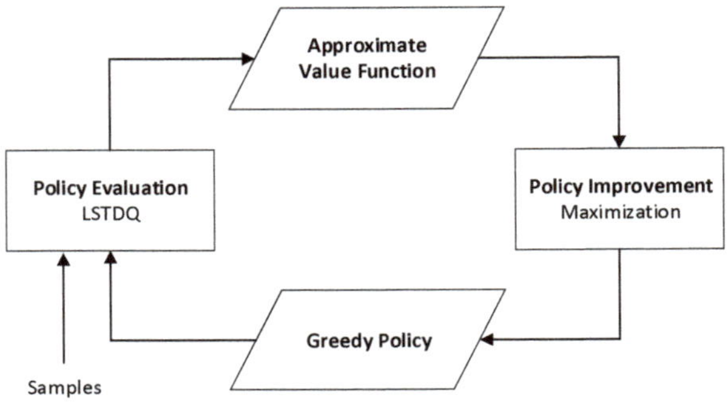

Figure 4. The flowchart of least-squares policy iteration. LSTDQ: least-squares temporal-difference learning for the state-action value function.

The prominent advantage of LSPI is its ability to deal with continuous or large state spaces, while tabular RL methods (e.g., Q-learning and Sarsa) cannot solve them, since it is impractical and computationally expensive to use the table structure to store state-action values, (i.e., Computer Go: 10^{170} states and Helicopter Control: continuous state spaces). The key of LSPI is that, during the policy evaluation stage, it uses linear architectures to approximate value functions (i.e., uses $\hat{Q}^{\pi}(s,a;\omega)$ to approximate $Q^{\pi}(s,a)$ via linear basis functions with free parameters ω). $Q^{\pi}(s,a)$ is the state-action function under policy π. The design of basis functions is fundamental work in LSPI. It is popular to use radial basis functions or polynomial basis functions, and we chose polynomial basis functions, since, compared with radial basis functions, polynomial basis functions have clear structures for designers. The definition of $\hat{Q}^{\pi}(s,a;\omega)$ is

$$\hat{Q}^{\pi}(s,a;\omega) = \sum_{j=1}^{k} \varphi_j(s,a)\omega_j, \tag{2}$$

where ω_j is the weight parameter, $\varphi_j(s,a)$ is the fixed basis function that is linearly independent, s is the state, a is the action, and k is the number of basis functions.

Two kinds of value-function approximation projections (i.e., Bellman residual minimizing approximation and least-squares fixed-point approximation) are used to approximate value functions. Because learning the Bellman approximation requires "doubled" samples [18] which are impossibly collected from the model-free system, we chose the second approximation that is more practical for the learning task and evaluates the state-action value more accurately.

In the policy-evaluation step, LSTDQ was used to find the solution of least-squares fixed-point approximation which is equivalent to learning ω. Since the model was unknown, samples were used to learn ω in an incremental way [18].

$$A\omega = b, \tag{3}$$

where matrix A and vector b are

$$A^{t+1} \approx \widetilde{A}^{t+1} = \widetilde{A}^{t} + \varphi(s_t,a_t)\big(\varphi(s_t,a_t) - \gamma * \varphi(s_t', \pi(s_t'))\big)^{T}, \tag{4}$$

$$b^{t+1} \approx \widetilde{b}^{t+1} = \widetilde{b}^{t} + \varphi(s_t,a_t)r_t, \tag{5}$$

where γ is the discounted factor, π is the policy function, and r_t is the immediate reward.

As seen in the pseudo-code of LSPI [18] in Algorithm 2, during the policy iteration, the same sample sets are used in every iteration, and the iteration ends until ω is stable. Since sample sets are important for LSPI, the collected sample sets should cover the entire state-action space as uniformly as possible to improve the training effect.

Algorithm 2 Pseudo-code of Least-Squares Policy Iteration (LSPI).

```
function LSPI(D, k, φ, γ, ϵ, π0)
//  D:    set of samples (s, a, r, s')
//  k:    The number of basis functions
//  φ:    Basis functions
//  γ:    Discount factor
//  ϵ:    Stopping criterion
//  ω0:   initial weight parameters
Ã ← 0
b̃ ← 0
ω' ← ω0
Repeat
  ω ← ω'
  for each (s, a, r, s') ∈ D
      Ã ← Ã + φ(s, a)(φ(s, a) − γ ∗ φ(s', π(s')))^T
      b̃ ← b̃ + φ(s, a)r
  end
  ω' ← Ã^{-1} b̃
until (|ω − ω'| < ϵ)
return ω
```

3.4.2. Definitions of Subgoal-Approaching (SA) and Obstacle-Avoiding (OA) MDPs

In the feasible-trajectory-planning phase, to increase training efficiency, we decomposed the problem of finding feasible trajectories between subgoals into two sub-problems—approaching subgoals and avoiding obstacles. Since LSPI is used to find a solution to discrete-time MDPs with continuous state spaces, given as a tuple of states, action, rewards, and next states, (s, a, r, s'), we separately built two MDPs via abstracting key information from the two sub-problems, to formally describe environments for LSPI. One MDP, named Subgoal-Approaching (SA), describes the process of approaching subgoals with kinematic constraints satisfied. The other MDP, named Obstacle-Avoiding (OA), describes the process of preventing collisions with obstacles. Definitions of SA and OA are described in Table 1. In these two MDPs, we used kinematic equations to update trajectories of the robot, so that the trajectories could preserve G1 continuity. Figure 5 shows the process of planning feasible trajectories. At first, we check the sensor data, and then decide which MDP to execute.

Table 1. Definitions of Subgoal-Approaching (SA) and Obstacle-Avoiding (OA) Markov Decision Processes (MDPs).

Type of MDP	SA MDP		OA MDP
States	(d_g, a_g)		$(S_1, S_2, S_3, S_4, S_5, S_6)$
Actions	move forward: (0.5, 0.5) turn left: (0.5, 0) turn right: (0, 0.5)		move forward: (0.5, 0.5) turn left: (0.5, 0) turn right: (0, 0.5)
Reward	+10	if $d_g < d_n$	Comparative value reward in Table 4, where P = 0.9
	$-\tilde{a}_g$	if $\tilde{a}_g > 0$	
	$1 - \tilde{d}_g - \tilde{a}_g$	else	

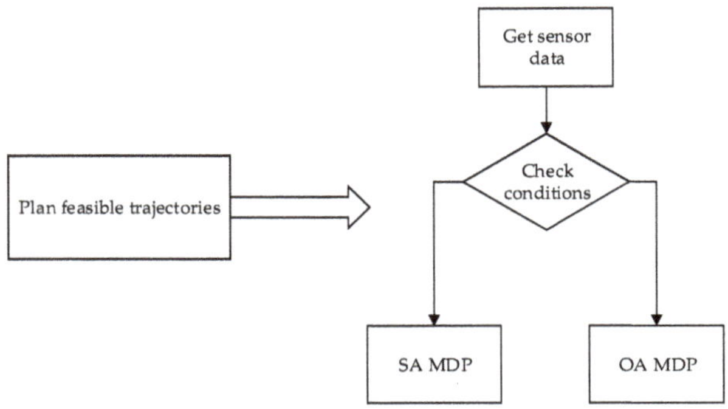

Figure 5. The flowchart of the trajectory planner.

In the SA MDP, state d_g denotes the distance between the robot and the goal, state a_g denotes the angle between the robot's orientation and the direction from the robot heading to the goal, and the range of a_g is $[-\pi, \pi]$. When the goal is on the left side of the robot's orientation, the value of a_g is positive. Otherwise, the value is negative. There are three actions (i.e., going forward, turning left, and turning right) that the robot can execute. Each action corresponds to values of ω_l and ω_r. For example, the action of going forward is equal to setting ω_l and ω_r to 0.5. And the action's execution is through the use of kinematic equations to calculate the next positions of the robot via inputting ω_l and ω_r (i.e., the angular velocity of left and right tracks). The definition of reward plays an important role in MDP, which is directly related to the training speed and the validity of the resulting policy. When arriving into the predefined acceptable area of subgoals, the robot gets a +10 scalar reward; d_n is the goal tolerance. If $\widetilde{a}_g$ is more than 0, the reward function is $-\widetilde{a}_g$, where $\widetilde{a}_g$ is the absolute value of normalization of a_g. Otherwise, the reward function is $1 - \widetilde{d}_g - \widetilde{a}_g$, where $\widetilde{d}_g$ is the normalization of d_g.

In the OA MDP, state S_i $(i \in (1,2,3\ldots,6))$ denotes the reading of sensor i. The range of S_i is $[0, 5]$. The first three sensors belong to left-side sensors and the others belong to right-side sensors. The setting of action is the same as the SA MDP's. Because of complexities of the obstacle-avoiding process, in general, it is tempting to add prior experience into reward functions, but it may not improve the training effect, which is discussed in Section 3.4.3. To explore the effects of adding prior experience into reward functions, we designed two kinds of reward functions—concise reward and comparative value reward. The first is a simple reward function (i.e., colliding with obstacles means receiving −4, otherwise 0). The comparative value reward adds prior experience about the risk of the current state based on the concise reward, which is shown in Table 2. $S_{min} = min(S_1, S_2, \ldots, S_6)$. S_{li} $(i \in (1,2,3))$ denotes the i-th minimum reading of left-side sensors. S_{ri} $(i \in (1,2,3))$ denotes the i-th minimum reading of right-side sensors. P is a constant value to tune the reward function. D_{ur} denotes the brake distance for tracked robots. D_{sa} denotes the ideal safe distance for tracked robots keeping away from obstacles. These conditions are checked in order, as shown in Table 2. To increase the performance of learned policies, an action-changing penalty (i.e., when the current action is different from last one, the agent obtains −0.2) can be combined with abovementioned reward functions, which is verified in Section 4.2.1.

Table 2. Comparative value reward in obstacle-avoiding MDP.

Order	Conditions	Annotation	Reward
1	$S_{min} < D_{ur}$	The minimum reading of all sensors is less than the brake distance.	-4
2	$S_{min} > D_{sa}$	The minimum reading of all sensors is greater than the safe distance.	+0
3	$S_{min} < D_{sa}$ $S_{l1} \neq S_{r1}$	The minimum reading of all sensors is less than the safe distance, and the minimum reading of left-side sensors is not equal to the right-side sensor's minimum reading.	$-P*$ $(D_{sa} - \min(S_{l1}, S_{r1}))$
4	$S_{min} < D_{sa}$ $S_{l2} \neq S_{r2}$	The minimum reading of all sensors is less than the safe distance, and the second minimum reading of left-side sensors is not equal to the right-side sensor's second minimum reading.	$-P*$ $(D_{sa} - \min(S_{l2}, S_{r2}))$
5	$S_{min} < D_{sa}$	The minimum reading of all sensors is less than the safe distance.	$-P*$ $(D_{sa} - \min(S_{l3}, S_{r3}))$

3.4.3. Improvements of Training with LSPI

To make the proposed feasible trajectory planner more efficient, we focused on the factors described below, including training environments and reward functions.

Firstly, from the view of input of LSPI, one significant element of successful training is whether sample sets can cover the entire state-action space. It affects the accuracy of the transition probability function, $P : S \times A \times S \to \mathbb{R}$. P denotes given current action a, the probability of transferring current state s to next state s'.

Sample sets depend on two key factors, which are ways of collecting samples, and training environments. For better sampling, we use not only random policies to ensure that every action is selected with the same probability, but also random positions to make all kinds of different situations happen. When starting the new sampling epoch, the goal position and the robot position are set randomly. We designed two kinds of training environments including simplified office maps and maps with random obstacles. In maps with random obstacles, various collected sample sets contribute to learning the transition probability function, which is proven in Section 4.2.2.

Secondly, scalar reward is important for RL methods. In reinforcement learning, there is a reward hypothesis: realizing goals means the maximization of the expected value of the cumulative sum of a received scalar signal (called reward). The hypothesis is described as follows: [12]

$$G_t = R_{t+1} + \gamma R_{t+2} + \gamma^2 R_{t+3} + \cdots = \sum_{k=0}^{\infty} \gamma^k R_{t+k+1}. \quad (6)$$

It is critical that rewards truly indicate what we want agents to achieve, and generating a reward signal does not depend on knowledge of how the agent chose correct actions. Therefore, the reward signal is not the place to impart to the agent prior knowledge about how to achieve the goal state. When we design the reward function, we should model this function from the brain of intelligent agents, which can realize what designers desire. If environments are well defined, designing the reward function is simple, such as winning +1 and losing 0 in chess. However, if environments are unknown and tasks are complex, it is tempting to add prior experience into reward functions and give the agent a supplemental reward for completing sub-tasks. The reward signal with well-intentioned supplemental rewards may lead agents to behave very differently from what is intended, and agents may end up not achieving the overall goal at all [12]. For example, in OA MDP, the comparative value reward is a reward with well-intentioned supplemental rewards, which is compared with the concise reward in Section 4.2.1, to show the effects of prior experience in reward functions.

4. Results

In this section, firstly, we evaluate the performance of SG–RL on static and dynamic maps. Static maps are constructed by laser sensor data from robots that are deployed with the Robot Operating System (ROS). Dynamic maps mean that some uncertainties such as unexpected obstacles appear along the abstract path. Note that unexpected obstacles mean static obstacles which are not known initially because of errors from sensors or environment models. Secondly, we evaluate the effect of reward function and different types of training maps in the obstacle-avoiding MDP. This paper implements the proposed algorithm based on the source code from *http://www2.cs.duke.edu/research/AI/LSPI* and *http://movingai.com/GPPC*. We run experiments on a 3.4-GHz Intel Core i7-6700 (Intel, Santa Clara, CA, USA) central processing unit (CPU) with 16 GB of random-access memory (RAM).

In the simulation experiments, we exploited numerical methods (i.e., solved equations in small steps) to calculate trajectories via kinematic equations. If the step approximates zero, the computed trajectory (a series of discrete trajectory points) will be equivalent to a continuous trajectory computed by analytical methods. However, small steps require more computational resources and also slow down the computation speed. Considering our experiment platform, we chose the step of simulation time as 0.1 s, which is relatively large in order to achieve higher speed and less memory consumption.

4.1. Performance of the SG–RL Method

SG–RL is a method used to find rational paths from a start position to a goal position, which decomposes complex path-planning problems into two phases: the first for the global abstract path planning and the second for the feasible trajectory planning.

In the first phase, we used the SSG method to generate subgoal sequences. The time to find subgoal sequences, called H-time, is a key assessment indicator, which can evaluate the ability to eliminate first-move lags. SSG has constrained heading changes, which cannot generate smooth paths for agents moving in continuous environments. By contrast, the proposed hierarchical method, SG–RL, has free heading changes. We evaluated its ability to deal with continuous environments by comparing its path lengths of it with those of SSG.

In the second phase, we used a feasible trajectory planner based on LSPI to find safe paths between subgoals, and all trajectory segments between subgoals could be linked to obtain complete feasible trajectories from the start position to the goal position. In the LSPI method, some parameter values came from Reference [18] and others came from trials. The discounted factor γ was 0.9. Basis function φ was polynomial. In OA MDP, the order of basis function φ was 3, and in SA MDP, the order was 4. The subgoal tolerance was 1.5 and the final goal tolerance was 0.5. The subgoal tolerance was bigger since subgoals only play auxiliary roles in the trajectory planner. The time interval of action choice was 0.5 s. The way of collecting samples was to choose actions randomly per second. If robots collided with obstacles or boundaries, this sampling epoch ended and robots started from random positions.

Action-switching frequency is a significant factor which can evaluate the performance of planned trajectories. Its definition is the ratio of the sum of the current action that is different from the last action to all actions. For robots, it cannot be high because of limitations of hardware systems and network latency. Realizing fast and frequent responses requires more precise chassis gears and more energy. Meanwhile, in areas with signal interference, network latency is relatively high.

4.1.1. Performance on Static Maps

We evaluated the SG–RL method with four maps described in Figure 6. We trained the robot to learn to avoid obstacles with LSPI in Figure 6a. The three maps for navigation tasks (Figure 6b–d) were between 250 and 521 times larger than the training map. Since the resolution of these three original maps (i.e., 2.5 cm per pixel) was too high for doing experiments on them, a down-sampling method was used to compress them. These original maps were composed of grayscale information, and we transformed them into grid maps (if there existed an obstacle in a cell, then the value of this cell was set to 1, otherwise 0).

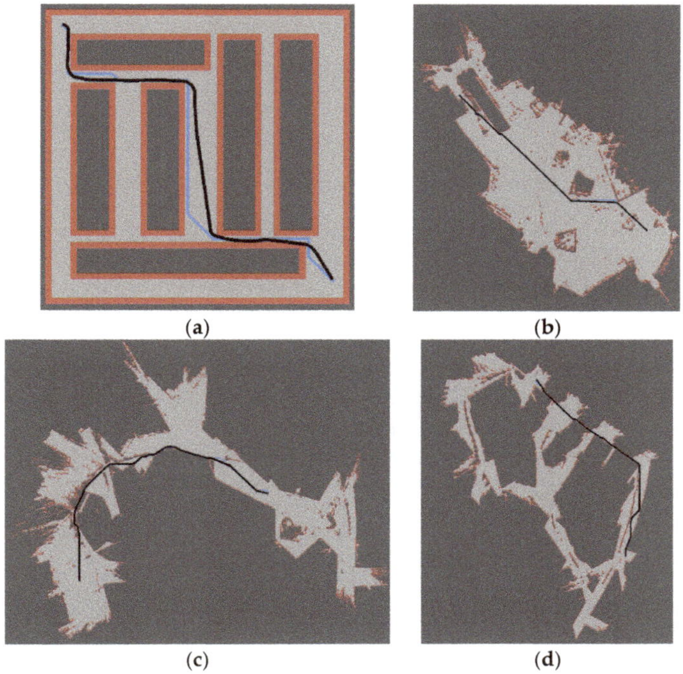

Figure 6. (**a**) Training environment—50 by 50. The training environment was a simplified office. (**b–d**) Given the start position and the goal position, the blue line is the path planned by the simple subgoal graph (SSG), and the black line is the trajectory planned by SG–RL. The dark-gray region is unpassable while the light-gray region is traversable for robots. Red regions deemed close to obstacles denote alert areas. (**b**) Building 1—820 by 820, (**c**) Building 2—1142 by 1142, (**d**) Building 3—792 by 792.

Table 3 shows the action-switching frequency of the robot completing the whole path on different maps. Since the learned policy was deterministic, this indicator was also fixed given the same start–goal location pair. All these values were below 10%, which indicates that the robot maneuvered by SG–RL can work well in large and realistic environments with low action-switching frequencies. In Table 4 the path length of SG–RL in Building 1 was almost equal to that in Building 3; however, in Table 3, the action-switching frequency of Building 3 was larger than that of Building 1, since there were more narrow passages in Building 3, where the robot needed to quickly change its actions for adapting to the complex environment (i.e., the robot changed its action to traverse many left and right turns). As a consequence, as the complexity of test maps increased, the action changes that were essential for the environment accordingly increased in number, which finally, increased the action-switching frequency. Action-switching frequency is also related to the performance of learned policy. If the performance of learned policies improves, the evaluation factor will decrease. Therefore, in Section 4.2, we present two ways to improve the performance of learned policies (i.e., by modifying reward functions and by choosing suitable training maps).

Table 3. Action-switching frequency of SG-RL.

Method	Building 1	Building 2	Building 3
SG-RL	0.0492	0.0849	0.0827

Table 4. Path length of SSG and SG–RL.

Method	Building 1	Building 2	Building 3
SSG	563.23	958.72	587.58
SG–RL	558.18	931.38	574.68

As shown in Table 4, although SSG could obtain an optimal path over the grid map which was the discretization of continuous environments, SG–RL could plan even shorter paths, showing that SG–RL can deal with continuous environments and plan trajectories following kinematical equations. Briefly, SG–RL without constrained heading changes can work well in continuous environments, which is suitable for robots navigating in the real world.

Table 5 shows, given the start position and the goal, the average value of H-time of A* and SG–RL on different maps. The H-time of A* is 195 times that of SG–RL on the Building 2 map and 107 times that of SG–RL on the Building 3 map. The H-time of SG–RL was below 1 millisecond on large-scale maps, which can meet the practical requirements and eliminate first-move lag. SG–RL used SSG with great success to obtain subgoal sequences, which benefitted from reducing the search space by abstracting topological structures of maps.

Table 5. H-time of A* and SG–RL.

Method	Building 1 (ms)	Building 2 (ms)	Building 3 (ms)
SG–RL	0.117	0.541	0.371
A*	20.828	105.582	39.860

4.1.2. Performance on Dynamic Maps

To evaluate SG–RL's ability to deal with uncertainties in environments, we added some random shapes of obstacles along the subgoal sequence. Certainly, by reconstructing the subgoal graphs and replanning abstract paths, SG–RL could still cope with the new environment. However, the preprocessing stage plus the path replanning needs some time, which is not suitable for use online. To tackle the problem elegantly, SG–RL continues using the initial subgoal sequence, but deals locally with unexpected situations. Owing to the capacity of generalizability, SG–RL behaves well enough in uncertain environments, as demonstrated in Figure 7. Even though there were some obstacles blocked in the path between subgoals, the robot controlled by SG–RL could achieve non-collision moves.

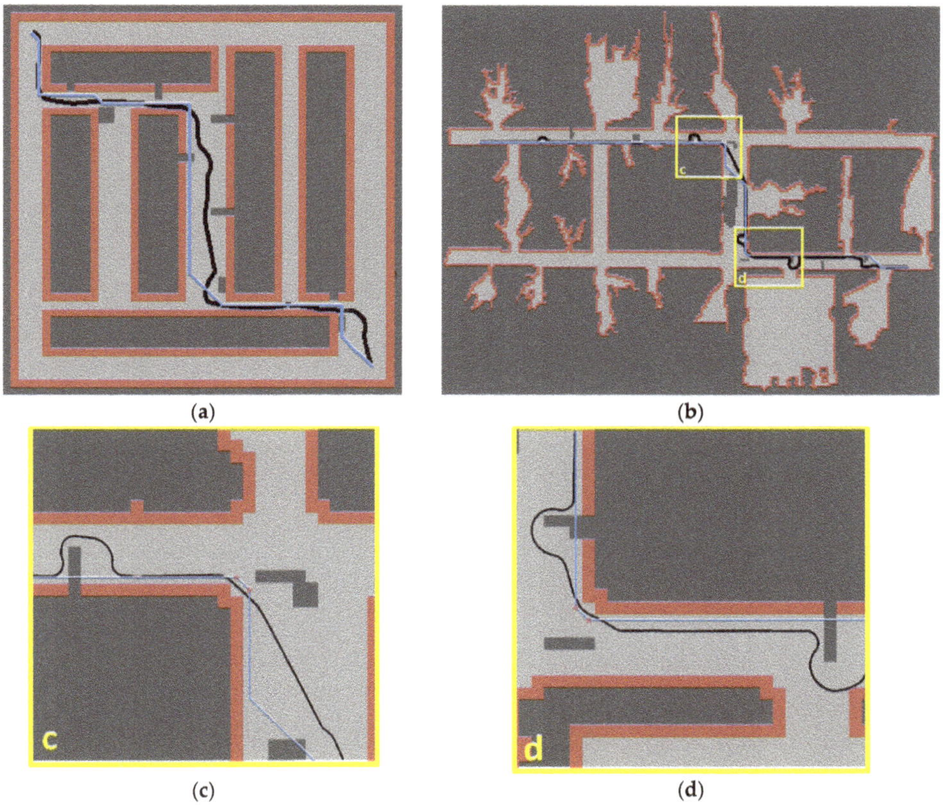

Figure 7. (**a**,**b**) The blue line is the path planned by SSG based on the map without unexpected obstacles. The black line is the trajectory planned by SG–RL without replanning the subgoal sequence. The dark-gray region is unpassable, while the light-gray region is traversable for robots. Red regions deemed close to obstacles denote alert areas. (**a**) Simplified office map with unexpected obstacles, (**b**) real office map with unexpected obstacles; (**c**,**d**) Two magnified details. (**c**) Magnified detail 1, (**d**) Magnified detail 2.

4.2. Influence Factors of the Obstacle-Avoiding MDP

During the second phase, SG–RL formalizes the problem of finding smooth and collision-free trajectories into two MDPs. OA MDP is an important part of the feasible trajectory planner, and it is difficult to train this MDP since it needs to deal with complex states. Therefore, it is essential to research its influence factors like reward functions and training environments.

4.2.1. Reward Functions

Firstly, to explore the effects of prior experience in reward functions, we compared the concise reward with the comparative value reward. The comparative value reward consists of the concise reward and some supplemental rewards concerning the risk of the current state. Secondly, to reduce action-switching frequencies, we added action-changing punishment into the original reward functions.

To reduce the randomness of training, we trained 100 times and randomly collected 60,000 samples each time. To evaluate the learned obstacle-avoiding policy, we let the robot move from the start position to the finish line on the test map (Figure 8). If the robot collided with obstacles, the policy was considered as a failure, otherwise it was considered a success.

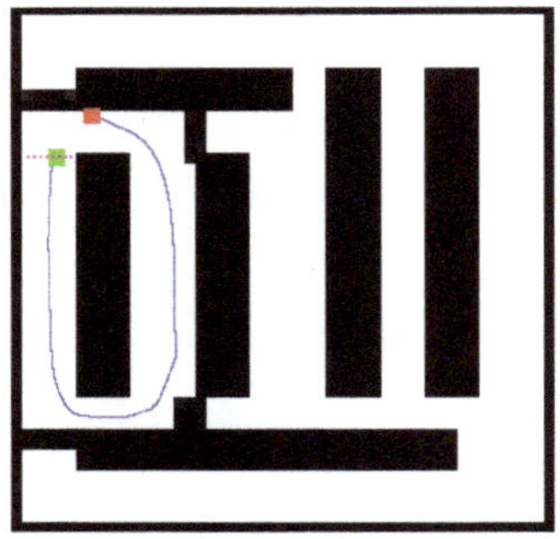

Figure 8. This is a successful sample of one robot with learned obstacle-avoiding policy completing the test. The red square is the start position, and the pink dotted line is the finish line. The green square is the position that the robot arrived at the finish line.

As seen in Table 6, the ratio of successful OA policies with the concise reward is five times higher than that with the comparative value reward. The other evaluation indicator of concise reward is over five times higher than that of the comparative value reward. Obviously, training with the concise reward can more easily get an obstacle avoidance policy than training with the comparative value reward. The comparative value reward including prior experience makes rewards dense; however, meanwhile, it brings difficulties for training. Therefore, training with the concise reward can perform better, and adding prior experience into reward functions may bring some negative effects on training. The better place to impart to the agent prior experience may be the initialization.

Table 6. Comparison of concise reward and comparative value reward.

Reward Function	Ratio of Successful OA Policies	Ratio of Policies with Action-Switching Frequency Below 30%
concise reward	45%	38%
comparative value reward	9%	7%

As seen in Table 6, the learned policies did not perform well in the aspect of action-switching frequencies; thus, we brought the punishment concerning action changes into reward functions to reduce action-switching frequencies while avoiding obstacles.

The punishment concerning action changes suggests that when current action is different from last one, the agent obtains a value of −0.2. Then, the entire combined reward has "strong" requirements (i.e., avoid obstacles) and "weak" requirements (i.e., reduce action changes) which means that, after "strong" requirements are satisfied, "weak" requirements can be satisfied as much as possible.

Shown in Table 7, the evaluation factor is the ratio of policies with action-switching frequency below 30%. Obviously, original reward functions with action-changing punishment can increase this evaluation factor. Therefore, action-changing punishment does a great favor to both the concise reward and the comparative value reward, reducing the action-switching frequency by increasing the performance of learned policies, thereby also demonstrating that the combined rewards work better than separate ones.

Table 7. Effect of adding action-changing punishment.

Reward Function	Without Action-Changing Punishment	With Action-Changing Punishment
concise reward	38%	49%
comparative value reward	7%	14%

4.2.2. Different Types of Training Maps

In this section, there are two types of training maps. The first type is the simplified office map (Figure 7a). The second type is the map with random obstacles, whose size is the same as the simplified office map. To build the second-type map, we firstly set 0.05 as the ratio of obstacles in the whole map. We randomly set cells as obstacles one by one under the same probability until the ratio of obstacles reached 0.05. Random obstacles were uniformly distributed in the space represented in the map. The test map was the same as the first type. The reward function we used was a concise reward with action-changing punishment.

As shown in Table 8, using the second type of map can more easily obtain obstacle-avoiding policies than using the first type. On the map with random obstacles, the robot chose action randomly, which realized not only randomness in action selection, but also in training environments. It can be seen that sample sets collected from the second-type map could cover more state-action spaces than sample sets from the first type, which could help LSPI to converge useful policies. Applying the second-type map could decrease the action-switching frequency as a whole. Using the second type meant fewer iteration numbers, which could speed up the training process. The experimental result shows that using the second type of environment could not only collect more different sample sets, but could also increase the accuracy of the learned transition probability function, so as to increase the performance of learned policies and decrease the action-switching frequency.

Table 8. Comparison of different types of training maps.

Reward Function	Ratio of Successful OA Policies	Ratio of Policies with Action-Switching Frequency Below 30%	Average Iteration Numbers of LSPI
simplified office map	53%	48%	7.94
map with random obstacles	89%	57%	6.01

5. Conclusions

In this paper, we proposed a hierarchical path-planning framework called SG–RL in continuous and uncertain environments. By "rational", we mean (1) efficient path planning to eliminate first-move lag; (2) collision-free and smooth for agents with kinematic constraints satisfied. SG–RL plans rational paths in a two-phase manner (i.e., the first phase for global abstract path planning and the second for feasible trajectory planning). In the first phase, SG–RL uses SSG to generate subgoal sequences efficiently for eliminating first-move lag, and overcome sparse reward and local minima trap for LSPI. In the second phase, following kinematical equations, SG–RL uses LSPI to plan feasible trajectories for agents between adjacent subgoals, and handles unexpected obstacles without replanning subgoal sequences.

In simulation experiments, firstly, SG–RL was evaluated on realistic and large-scale maps constructed by laser sensor data. The action-switching frequency was below 10% and the time to obtain subgoal sequences was between 195 times and 107 times faster than A* on grid maps. The path length of SG–RL was shorter than that of SSG. Secondly, evaluated on dynamic maps, SG–RL can deal with uncertainties based on original subgoal sequences without reconstructing subgoal graphs from the changed map.

We demonstrated that the design of reward functions and the type of training environment are important factors for learning feasible policies. A concise reward with action-changing punishment achieved the best results in the obstacle-avoiding MDP, which means that reward functions with prior experience may bring difficulties for agents to learn effective policies. The map with random obstacles increased the accuracy of the learned transition probability function, whose ratio of getting an effective obstacle avoidance policy was 89%. SG–RL, as a hierarchical path-planning method, combines the strengths of SSG and LSPI to overcome their limitations, greatly improving the efficiency of finding rational paths.

In future work, continuous action spaces can be taken into consideration. Specifically, the range of angular velocity can be $[-0.5, 0.5]$, which means that robots can do more free and complex actions, such as going backward. In the second phase of SG–RL, if the number of state-action pairs is not huge, LSPI can keep the relative size of Q-value for every state-action pair to ensure the right action choice. However, since adding continuous action space will greatly increase the state-action space, it is difficult for linear basis functions to evaluate the Q-value precisely. Inspired by the idea of asynchronous advantage Actor-Critic [30], we can use an artificial neural network to approximate state-action value functions, and use another network to approximate policy functions, which can increase the accuracy of evaluating the Q-value.

Author Contributions: J.Z. and L.Q. conceived and designed the paper structure and the experiments; J.Z. performed the experiments; Y.H., Q.Y., and C.H. contributed with materials and analysis tools.

Funding: This work was sponsored by the National Science Foundation (NSF) project 61473300, China.

Acknowledgments: The work described in this paper was sponsored by the National Natural Science Foundation of China under Grant No. 61473300. We appreciate the fruitful discussion with the Sim812 group: Qi Zhang and Yabing Zha.

Conflicts of Interest: The authors declare no conflicts of interest. The founding sponsors had no role in the design of the study; in the collection, analysis, or interpretation of data; in the writing of the manuscript, and in the decision to publish the results.

References

1. Korf, R.E. Real-Time Heuristic Search. *Artif. Instell.* **1990**, *42*, 189–211. [CrossRef]
2. Harabor, D.D.; Grastien, A. Online Graph Pruning for Pathfinding On Grid Maps. In Proceedings of the 25th AAAI Conference on Artificial Intelligence, AAAI 2011, San Francisco, CA, USA, 7–11 August 2011.
3. Rabin, S.; Silva, F. JPS+ An Extreme A* Speed Optimization for Static Uniform Cost Grids. In *Game AI Pro*, 2nd ed.; Rabin, S., Ed.; CRC Press: Boca Raton, FL, USA, 2015; Volume 3, pp. 131–143. ISBN 9781482254792.
4. Geisberger, R.; Sanders, P.; Schultes, D.; Vetter, C. Exact Routing in Large Road Networks Using Contraction Hierarchies. *Transport. Sci.* **2012**, *46*, 388–404. [CrossRef]
5. Uras, T.; Koenig, S.; Hernandez, C. Subgoal graphs in for optimal pathfinding in eight-neighbour grids. In Proceedings of the 23rd International Conference on Automated Planning and Scheduling (ICAPS '13), Rome, Italy, 10–14 June 2013.
6. Hart, P.E.; Nilsson, N.J.; Raphael, B. A Formal Basis for the Heuristic Determination of Minimum Cost Paths. *IEEE Trans. Syst. Sci. Cybern.* **1972**, *4*, 28–29. [CrossRef]
7. Nash, A.; Koenig, S. Any-Angle Path Planning. *AI Mag.* **2013**, *34*, 85–107. [CrossRef]
8. González, D.; Pérez, J.; Milanés, V.; Nashashibi, F. A Review of Motion Planning Techniques for Automated Vehicles. *IEEE Trans. Intell. Transp. Syst.* **2016**, *17*, 1135–1145.
9. Ravankar, A.; Ravankar, A.A.; Kobayashi, Y.; Takanori, E. SHP: Smooth Hypocycloidal Paths with Collision-Free and Decoupled Multi-Robot Path Planning. *Int. J. Adv. Robot. Syst.* **2016**, *13*, 1. [CrossRef]
10. Kavraki, L.E.; Švestka, P.; Latombe, J.C.; Overmars, M.H. Probabilistic roadmaps for path planning in high-dimensional configuration spaces. *IEEE Trans. Robot. Autom.* **1994**, *12*, 566–580. [CrossRef]
11. Lavalle, S.M. Rapidly-exploring random trees: Progress and prospects. In Proceedings of the 4th International Workshop on Algorithmic Foundations of Robotics, Hanover, NH, Germany, 16–18 March 2000.
12. Sutton, R.S.; Barto, A.G. *Reinforcement Learning: An Introduction*, 2nd ed.; MIT Press: Boston, MA, USA, 2017. (In Preparation)
13. Li, Y. Deep Reinforcement Learning: An Overview. *arXiv* **2017**; arXiv:1701.07274.

14. Kober, J.; Bagnell, J.A.; Peters, J. Reinforcement Learning in Robotics: A Survey. *Int. J. Robot. Res.* **2013**, *32*, 1238–1274. [CrossRef]
15. Yahya, A.; Li, A.; Kalakrishnan, M.; Chebotar, Y.; Levine, S. Collective robot reinforcement learning with distributed asynchronous guided policy search. In Proceedings of the International Conference on Intelligent Robots and Systems, Vancouver, BC, Canada, 24–28 September 2017; pp. 79–86.
16. Silver, D.; Schrittwieser, J.; Simonyan, K.; Antonoglou, I.; Huang, A.; Guez, A.; Hubert, T.; Baker, L.; Lai, M.; Bolton, A.; et al. Mastering the game of Go without human knowledge. *Nature* **2017**, *550*, 354–359. [CrossRef] [PubMed]
17. Mnih, V.; Kavukcuoglu, K.; Silver, D.; Rusu, A.; Veness, J.; Bellemare, M.G.; Graves, A.; Riedmiller, M.; Fidjeland, A.K.; Ostrovski, G.; et al. Human-level control through deep reinforcement learning. *Nature* **2015**, *518*, 529. [CrossRef] [PubMed]
18. Lagoudakis, M.G.; Parr, R. Least-Squares Policy Iteration. *J. Mach. Learn. Res.* **2003**, *4*, 1107–1149.
19. Sturtevant, N.R.; Traish, J.M.; Tulip, J.R.; Uras, T.; Koenig, S.; Strasser, B.; Botea, A.; Harabor, D.; Rabin, S. The Grid-Based Path Planning Competition: 2014 Entries and Results. In Proceedings of the Annual symposium on combinatorial search, Ein Gedi, Israel, 11–13 June 2015.
20. Uras, T.; Koenig, S. Identifying hierarchies for fast optimal search. In Proceedings of the Twenty-Eighth AAAI Conference on Artificial Intelligence, Quebec City, QC, Canada, 27–31 July 2014.
21. Duguleana, M.; Mogan, G. Neural networks based reinforcement learning for mobile robots obstacle avoidance. *Expert Sys. Appl.* **2016**, *62*, 104–115. [CrossRef]
22. Blekas, K.; Vlachos, K. RL-based path planning for an over-actuated floating vehicle under disturbances. *Robot. Auton. Syst.* **2018**, *101*, 93–102. [CrossRef]
23. Bradtke, S.J.; Barto, A.G. Linear Least-Squares algorithms for temporal difference learning. *Mach. Learn.* **1996**, *22*, 33–57. [CrossRef]
24. Singh, N.N.; Chatterjee, A.; Chatterjee, A.; Rakshit, A. A two-layered subgoal based mobile robot navigation algorithm with vision system and IR sensors. *Measurement* **2011**, *44*, 620–641. [CrossRef]
25. Dolgov, D.; Thrun, S.; Montemerlo, M.; Diebel, J. Path Planning for Autonomous Vehicles in Unknown Semi-structured Environments. *Int. J. Robot. Res.* **2010**, *29*, 485–501. [CrossRef]
26. Cowlagi, R.V.; Tsiotras, P. Hierarchical Motion Planning With Dynamical Feasibility Guarantees for Mobile Robotic Vehicles. *IEEE Trans. Robot.* **2012**, *28*, 379–395. [CrossRef]
27. Mac, T.T.; Copot, C.; Tran, D.T.; Keyser, R.D. A Hierarchical Global Path Planning Approach for Mobile Robots based on Multi-Objective Particle Swarm Optimization. *Appl. Soft Comput.* **2017**. [CrossRef]
28. Frego, M.; Bertolazzi, E.; Biral, F.; Fontanelli, D.; Palopoli, L. Semi-analytical minimum time solutions with velocity constraints for trajectory following of vehicles. *Automatica* **2017**, *86*, 18–28. [CrossRef]
29. Lau, B.; Sprunk, C.; Burgard, W. Efficient grid-based spatial representations for robot navigation in dynamic environments. *Robot. Auto. Sys.* **2013**, *61*, 1116–1130. [CrossRef]
30. Mnih, V.; Badia, A.P.; Mirza, M.; Graves, A.; Lillicrap, T.P.; Harley, T.; Silver, D.; Kavukcuoglu, K. Asynchronous methods for deep reinforcement learning. In Proceedings of the International Conference on Machine Learning, New York NY, USA, 19–24 June 2016.

applied sciences

MDPI

Article

Multi-Criteria Decision Making for Efficient Tiling Path Planning in a Tetris-Inspired Self-Reconfigurable Cleaning Robot

Maryam Kouzehgar [1,*], Mohan Rajesh Elara [1], Mahima Ann Philip [1], Manimuthu Arunmozhi [1,2] and Veerajagadheswar Prabakaran [1]

1 Engineering Production and Development Pillar, Singapore University of Technology and Design, Singapore 487372, Singapore; rajeshelara@sutd.edu.sg (M.R.E.); philpsmahima@gmail.com (M.A.P.); maniasaldhinesh@gmail.com (M.A.); prabakaran@sutd.edu.sg (V.P.)

2 College of Engineering Guindy, Anna University, Chennai 600025, India

* Correspondence: maryam_kouzehgar@sutd.edu.sg; Tel.: +65-8776-9026

Received: 26 September 2018; Accepted: 21 December 2018; Published: 25 December 2018

Abstract: In this study, we aim to optimize and improve the efficiency of a Tetris-inspired reconfigurable cleaning robot. Multi-criteria decision making (MCDM) is utilized as a powerful tool to target this aim by introducing the best solution among others in terms of lower energy consumption and greater area coverage. Regarding the Tetris-inspired structure, polyomino tiling theory is utilized to generate tiling path-planning maps which are evaluated via MCDM to seek a solution that can deliver the best balance between the two mentioned key issues; energy and area coverage. In order to obtain a tiling area that better meets the requirements of polyomino tiling theorems, first, the whole area is decomposed into five smaller sub-areas based on furniture layout. Afterward, four tetromino tiling theorems are applied to each sub-area to give the tiling sets that govern the robot navigation strategy in terms of shape-shifting tiles. Then, the area coverage and energy consumption are calculated and eventually, these key values are considered as the decision criteria in a MCDM process to select the best tiling set in each sub-area, and following the aggregation of best tiling path-plannings, the robot navigation is oriented towards efficiency and improved optimality. Also, for each sub-area, a preference order for the tiling sets is put forward. Based on simulation results, the tiling theorem that can best serve all sub-areas turns out to be the same. Moreover, a comparison between a fixed-morphology mechanism with the current approach further advocates the proposed technique.

Keywords: self-reconfigurable robot; cleaning robot; Tetris-inspired; polyomino tiling theory; coverage path planning; area decomposition; multi-criteria decision making

---

1. Introduction

Robots are deployed in a wide range of applications and are, therefore, rapidly becoming an integral component of our everyday life. In particular, with several million units sold worldwide, robots developed and implemented for floor cleaning tasks will become an elemental part of the household in the near future. Such a massive market in robotic floor cleaning products involves dominant market players such as IRobot, Dyson, Samsung, Xiaomi, and Neato. These robots can autonomously navigate through a given area using onboard sensors and are typically characterized by circular, square, and D-shaped morphologies. The suitability of current robots to assist in daily domestic tasks is investigated in [1] from the human user viewpoint. Additionally, the study presented in [2], analyses and models floor-cleaning coverage performances of some commercial domestic mobile robots. Moreover, not only in the commercial market but also in the academic literature, there exist numerous cleaning robot studies within a wide spectrum from mechanical design to autonomy. A Swedish wheeled mechanism for a floor cleaning robot was presented in [3] in order to attain efficient locomotion in a populous area. In terms of robot autonomy, an innovative neural network approach is presented in [4] which is utilized for path planning and obstacle avoidance for a cleaning robot in a random environment. On the other hand, since numerous robotic floor cleaning products are available in the marketplace, it is critical to have benchmark schemes to validate the robot's cleaning performance and coverage efficacy [5]. In addition, multi-robot schemes have been proved to be useful in cleaning applications [6]. Although there are numerous studies that highlight the benefits of cleaning robots, the efficiency of traditional cleaning robots is still affected by factors such as fixed morphology. Since such a rigid structure highly limits the accessible spaces like room corners and narrow corridors, one viable approach to overcome this restriction in a traditional cleaning system is to develop next-generation robots with shape-shifting abilities to maximize the area coverage performance.

So far, in the field of reconfigurable robotics, three different architectures are introduced, namely, intra-reconfiguration, inter-reconfiguration, and nested reconfiguration. Intra-reconfiguration deals with a single robotic system that could change its morphology on its own without any external supports [7]. Robots proposed under the inter-reconfigurable principle are basically modular robots that can possess different morphologies by undergoing assembling and disassembling processes [8]. Modular robots can be investigated at higher levels to benefit from swarm intelligence. In this regard, modular swarm robots are utilized to gain advantages on self-reconfigurability, self-replication, and self-assembly [9]. On the other hand, nested reconfiguration can perform both inter and intra reconfigurations. Hinged Tetro presented in [10], is an example of such nested reconfigurable robots. Although numerous studies in the literature address reconfigurable robotics, they are primarily limited to mechanism design and are hardly implemented to an area coverage task like floor cleaning. In order to bridge this gap, in our previous work, we presented a novel reconfigurable floor cleaning robot called hTetro [11,12]. This Tetris-inspired structure can change its morphology to any of the one-sided Tetris pieces and as a result, could achieve superior coverage performance through its shapeshifting ability. In addition, in a couple of our previous works, we utilized polyomino tiling theory to introduce a coverage path planning technique for Tetris-inspired polyomino robots [13,14].

Polyomino tiling theory has been extensively studied in combinatorial mathematics as a distinct branch of mathematics with Golomb as the pioneering inventor in this field. The applications of this theory span a wide variety of fields including medicine [15], graphics, and gaming [16]. However, polyomino tiling theory has rarely been applied to the field of robotics and, therefore, presents a scope for tremendous research. As briefly mentioned above, in [13,14], a novel approach is introduced to harness the polyomino tiling theory as a way of suggesting navigation strategies for reconfigurable cleaning robots. These tiling theorems equip the reconfigurable robots with superlative navigation abilities and make corners and narrow spaces more accessible. In pursuance of [14], this paper utilizes the polyomino tiling theory to propose tiling sets as navigation strategies of the Tetris-inspired cleaning robot in a static environment but goes a step further to definitively select the optimal tiling set for each area under cleaning. This will suggest the navigation strategy that improves the overall efficiency from the viewpoint of area and energy. This way, we will be entering the borders of optimal navigation for a Tetris-inspired reconfigurable cleaning robot.

In our previous experiments on shape-shifting cleaning robots, we only considered area coverage as a single parameter for assessing the robot's performance, which is actually insufficient when the robot is functioning in real-time scenarios [13,14]. Since the amount of energy consumption is also a significant issue in any electromechanical system, the current paper aims at balancing a trade-off between energy consumption and area coverage. Thus, in this study, we consider these two key factors as decision criteria in a Multi-Criteria Decision Making (MCDM) process in order to choose the best navigation strategy (defined by the tiling set) for achieving higher efficiency in terms of less energy with superior area coverage.

MCDM is a general term referring to all methods that help to make decisions based on preferences where there is more than one conflicting criterion [17]. The reason why we apply MCDM is that generally, real applications require the consideration of several possibly conflicting objectives to be fulfilled, thus a multi-objective problem is required to be solved. On the other hand, usually a trade-off between the objectives exists and we rarely have an alternative that can best satisfy all the objectives simultaneously. Accordingly, the intelligent agent must efficiently balance the facing trade-off.

Despite the fact that MCDM methods may be widely diverse, many of them share certain aspects in common [18], such as the concepts of alternatives, criteria, and the weights of importance.

- *Alternatives* represent the different choices available to the decision maker. Usually, the set of alternatives is assumed to be finite.
- *Criteria* represent the different factors on the basis of which the alternatives can be investigated.
- *Importance factors* (weights of importance) are considered as a measure of the significance of each criterion in the decision-making process. Usually, these weights are normalized to add up to one. It is also assumed that the decision maker has determined the weights of the decision criteria based on relative significance.

If the number of alternatives is infinite, then we will face a multi-objective optimization (MOO). Here, since we have a finite set of tiling sets proposed based on the tiling theorems, we are supposed to solve a MCDM regarding the area coverage and energy as the decision-making criteria. Moreover, there are many ways to classify MCDM methods:

- Classification based on the data type: deterministic, stochastic, or fuzzy MCDM methods [19]. However, some situations require combinations of all the above [20].
- Classification based on the number of decision makers: Single vs. group MCDM [21].
- Classification based on the approach applied on the data: The WSM (Weighted Sum Model) or SAW (Simple Additive Weighting), WPM (Weighted Product Model), AHP (Analytic Hierarchy Process), revised AHP, ANP (Analytic Network Process), TOPSIS (Technique for Order Preference by Similarity to Ideal Solution), ELECTRE (Elimination and Choice Translating Reality), PROMETHEE (Preference Ranking Organization Method for Enrichment Evaluations); VIKOR (VlseKriterijumska Optimizacija I Kompromisno Resenje), and LINMAP (Linear Programming for Multidimensional Analysis of Preference) are the methods which are of more use in practice [17,18,22].

In addition, MCDM plays a significant role in many intelligent systems and is widely utilized in many fields such as operation research and soft computing [21,23]. Moreover, regarding the links between MCDM and robotics, it is mostly utilized for the allocation of resources and task scheduling in robotic swarms [24], robot selection in automated industry operations [25], and defining exploration strategies for search and rescue robots [26]. Similarly, in this paper, MCDM will be applied to choose between navigation strategies defined by given tiling sets. Here, we will be focused on single decision-making deterministic MCDM applying concepts of the SAW (Simple Additive Weighting) method [22].

In summary, all aspects of this research are quite novel. In better words, all aspects of it have recently emerged: The mechanical design of the Tetris-inspired cleaning robot is proposed in 2017 [11,12] and the path-planning idea is raised in 2018 in terms of a tiling theory approach [13,14]. However, the previously raised tiling-path planning approach only considers area coverage as a single performance metric with absolutely no consideration of energy consumption issues. The main contributions of this study in comparison to our previous works are as follows:

- In this study, energy consumption is considered as a second significant performance metric in tiling path planning.
- This paper proposes an energy estimation scheme for the novel self-reconfigurable robotic platform.

- The current work suggests the optimal path-planning approach by means of MCDM, i.e., creating an optimal balance between maximum area coverage and minimum energy. In other words, however the tiling-theoretic path planning approach is introduced in [13,14], this paper goes beyond previous works to some degree by taking into account the energy factor and aims at adding some optimality utilizing MCDM.
- In addition, in this study, a more recent tiling theorem has been utilized that is not investigated in previous studies [13,14] and surprisingly, based on simulation results, this specific theorem (here illustrated as Theorem 1 [27]) turns out to be the most-promising tiling theorem that can best serve all the areas targeted for optimal cleaning.

On the whole, *tiling-path-planning* is a novel concept only related to the novel hTetro, and it is our belief that expanding a novel idea (by considering more parameters) and adding optimality to a novel concept is invaluable.

The rest of the paper is organized as follows. Section 2 introduces the experimental environment and explains the utilized Tetris-inspired robot structure. Section 3 casts light on the concepts of Polyomino Tiling Theory applied to our robotic platform and elaborates the four utilized tiling theorems. Moreover, this section is also concerned with the reason (and also technique) for the area decomposition process and finally presents five distinguished tiling sets under each theorem applied on each sub-area. Section 4 is concerned with explaining the method of extracting the data for energy and area coverage as the two decision criteria. Section 5 clarifies the general concepts of a MCDM method called Simple Additive Weighing (SAW) and then applies concepts of this method on the tiling-based path planning for the robot by means of defining a special fitness function. Finally, Section 6 is dedicated to a thorough analysis of the results in terms of fitness function values. In addition, a preference order for different theorems applied to different sub-areas is presented along with the best possible combination of tiling sets for the whole area. Section 7 presents an evaluation of the robot performance with and without the tiling. Consequently, Section 8 concludes the results and suggests possible works in the future.

2. An Overview of the Experimental Environment

An image of the experimental environment can be seen in Figure 1 which includes the hTetro robot and five obstacles. The outline of hTetro robot is shown in Figure 2, where all the four square-blocks are linked together with hinges. These blocks are driven by actuating the hinges with the smart servo motors mounted on it. The dimension of each hTetro block has a length, width, and height of 140 mm, 140 mm, and 55 mm, respectively. Owing to the smaller and compact design structure of hTetro, each box is assigned to perform a specific function: Block 1 manages the locomotion of the robot, while control and power modules are located on block 2. Meanwhile, all the blocks are equipped with cleaning functions, but while the robot is performing reconfigurations, the cleaning function is turned off in order to devote the actuation energy fully to the servo motors controlling hinges. The aggregate weight of hTetro, including all its peripheral devices is approximately 3 kg. For the mobility and transformation, hTetro is equipped with a set of six geared DC motors and three actuator servo motors, respectively. The three actuated servo motors, placed at hinges, are controlled to switch the robot between the seven possible configurations illustrated in Figure 3. For the actuation of hinge points during reconfiguration, Herkulex DRS-0101-7 V smart servos are utilized with a maximum rotation angle of 270°. The microcontroller used in hTetro is an Arduino Atmega 2560 16-Bit which performs three major functions:

- Receiving the command from the user by means of a smartphone which is connected wirelessly through a Bluetooth serial communication protocol. Arduino uses UART (Universal Asynchronous Receiver-Transmitter) for this communication.

- Generating control signals to the motor driver which controls multiple DC motors; in addition, using PWM (Pulse Width Modulation), it controls the speed of the motor by providing the required current for the motor to run.
- Establishing full duplex serial communication with the servomotors where Arduino sends the control signals to the servo motors and simultaneously receives feedback from them.

For a thorough explanation of the mechanical design of hTetro and a detailed report on the content and functionality of each block, the reader is referred to [11,12]. The detailed system architecture of electrical conventions is given in Figure 4.

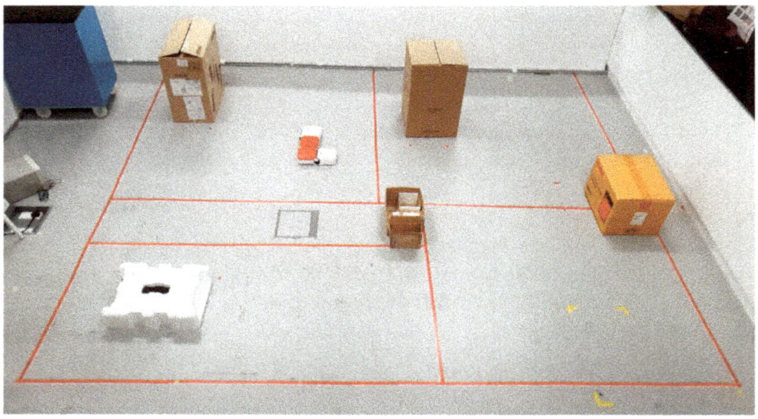

Figure 1. The experimental environment.

Figure 2. The hTetro architecture.

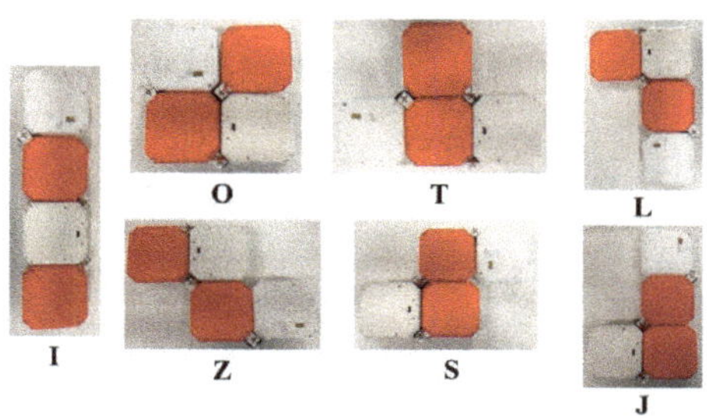

Figure 3. Seven one-sided tetrominoes.

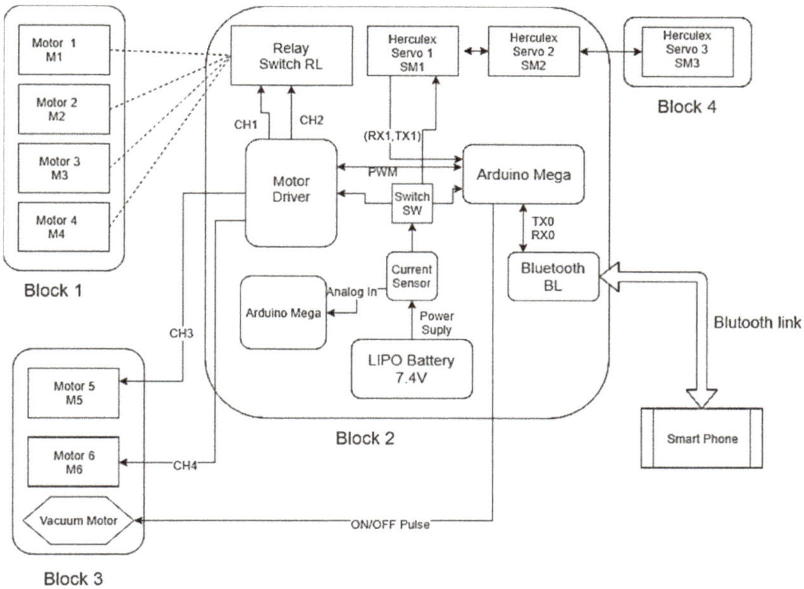

Figure 4. Architecture of electrical conventions in hTetro.

3. Polyomino Tiling Theory Applied to Our Robotic Platform

Polyominoes are two-dimensional geometric figures that have equal sized constituent squares joined to each other by the edges [28]. They are also defined as n-ominoes, depending upon the number of squares that form the polyomino. The robotic platform used in this paper is hTetro that resembles a tetromino, i.e., a polyomino with four squares. As mentioned earlier, each of these hTetro square blocks moves about a hinge connection making it capable of reconfiguring into all seven one-sided shapes of a tetromino, as illustrated in Figure 3.

Polyomino tiling theory recommends mathematical theorems that allow for complete tiling with polyomino tiles without any overlap, via translations, rotations, and reflections, of tiles in the regions that fall within the constraints of the theorem. If these regions strictly adhere to the conditions of the theorems, tiling sets would not produce any gaps and should result in complete (exact) area coverage [28]. As with [13,14], in this paper, polyomino tiling theory is utilized to define the navigation strategies for reconfigurable floor-cleaning robots. To this end, initially, the entire space under cleaning is visualized as a large grid divided into congruent squares with dimensions that match the dimensions of robot blocks. These theorems can provide multiple tiling sets depending upon the dimensions of the region, obstacles contained, and the shapes permitted to be utilized in the theorem. Eventually, by reconfiguring from one shape to another, hTetro follows these tiling sets as a map for its navigation.

3.1. Elaborating on the Applied Tiling Theorems

The theorems enlisted below are applied to the experimental environment regarding the physical characteristics of hTetro cleaning robot. With the aim of applying the results to the experimental environment, only the concepts of the theorems are illustrated in this work. For further illustration on the proofs, the reader is referred to the references mentioned alongside each theorem statement.

In addition, before delving into theorem concepts, it should be mentioned that each theorem employs specific shapes of tetrominoes. Moreover, it should be clarified that, if the space is regarded as a grid divided into congruent square cells:

- A *modified* rectangle is simply an a × b rectangle with both the upper-left and lower-right corner cells removed [29].
- A *deficient* rectangle means a rectangle in which some of its square cells are ignored or occupied by obstacles [30].

Theorem 1. *A deficient square of area $3^N \times 3^N$, with one square removed, can be tiled by set A [27] where Set A includes tiles O, T, and J, as illustrated in Figure 3.*

Theorem 2. *A conventional rectangle a × b can be tiled by the set B if, and only if, either of the following is true [29] where Set B includes tiles Z, S, and T, as illustrated in Figure 3.*

1. *One side is divisible by 4,*
2. *$a, b \equiv 2 \ (mod\ 4)$ and $a + b > 16$.*

Theorem 3. *A modified rectangle a × b can be tiled by the set C, if and only if, either of the following is true [29] where Set C includes tiles Z, S, and T, as illustrated in Figure 3.*

1. *$a \equiv 2 \ (mod\ 4)$ and b is odd,*
2. *$b = 2$ and $a \equiv 1 \ (mod\ 4)$*

Theorem 4. *A deficient rectangle a × b can be tiled by set D, if and only if [30]:*

1. *a and b are odd and the rectangle is sized $(2n + 1) \times (2n + 1)$ when $n \geq 1$, and,*
2. *The missing cell in the deficient rectangle will have*
    - *odd coordinates if the rectangle is sized $(4m + 1) \times (4m + 1)$ when $m \geq 1$, or*
    - *even coordinates if the rectangle is sized $(4m + 3) \times (4m + 3)$ with $m \geq 1$.*

 where Set D includes O, L, and J tiles, as illustrated in Figure 3.

Furthermore, it should be mentioned that in order to reach a complete and exact tiling, the dimensions of the entire area must exactly follow the statement of the theorem and if they do not, a smaller rectangle inside the entire area could be considered that better fits the statement of the theorem. Also, in order to follow the requirements of the theorem, the entire area can be split into several smaller rectangles in a process called area decomposition.

3.2. Area Decomposition Based on Furniture Layout

As mentioned before, reconfigurability allows the robot to navigate around and under obstacles with different shapes. This results in accessing narrow spaces and inaccessible corners. However, this is possible while adhering to a tiling map as the navigation strategy; therefore, we can utilize the polyomino tiling theory. In order to apply these theorems perfectly for full area coverage, any space must satisfy the conditions of the theorem accurately, which is an unrealistic claim. That is why in this study, the entire area is manually decomposed into five sub-areas such that each sub-area will better satisfy the conditions of a theorem.

This decomposition depends on the position of each obstacle (furniture layout). In ideal circumstances, the robot must autonomously detect the presence, shape, and accessibility of the obstacles and tiny or unusual corners that require more attention. This experimental aspect is currently under investigation applying SLAM (Synchronous Localization And Mapping) and sensor fusion techniques. Consequently, depending upon the theorems in the robot's database, it will accordingly decompose the entire floor space into sub-areas for optimal area coverage. However, such decompositions do not always guarantee a perfect coverage of the sub-areas, resulting in untiled zones. That is why, optimizing the area decomposition based on furniture layout would serve as a future scope for this paper. Furthermore, the area decomposition will help to reduce the on-board computational cost working in each individual area with a restricted number of cells to be covered, which is the second reason supporting an area decomposition approach.

In this paper, the positions of randomly-placed obstacles are considered to manually divide the entire space into five sub-areas, as depicted in Figure 5. Notice that, in Figure 5, the sub-areas are numbered, and the yellow sections represent the obstacles. The floor space chosen for this paper is an area of 11.54 m^2, such that the entire area can be covered exactly by 598 square cells. In correlation with the numbering on Figure 5, the respective areas of these distinct sub-areas are 2.058 m^2, 0.7056 m^2, 3.3124 m^2, 3.3124 m^2, and 2.156 m^2 without subtracting the area occupied by obstacles. After subtracting the obstacle-occupied area, the entire area to be cleaned is 10.5448 m^2 which is the sum of 1.8228 m^2, 0.7056 m^2, 3.0184 m^2, 2.9988 m^2, and 1.9992 m^2 respectively for each numbered sub-area.

The general approach is to apply all four previously introduced theorems reasonably in each sub-area and propose consequent tiling sets that cover each sub-area. Finally, the resulting tiling sets will be analyzed in a Multi-Criteria Decision Making (MCDM) framework to test the effectiveness of these tiling theorems. Based on the MCDM results, the most efficient theorem for each sub-area is introduced that can result in a better balance between energy and area coverage.

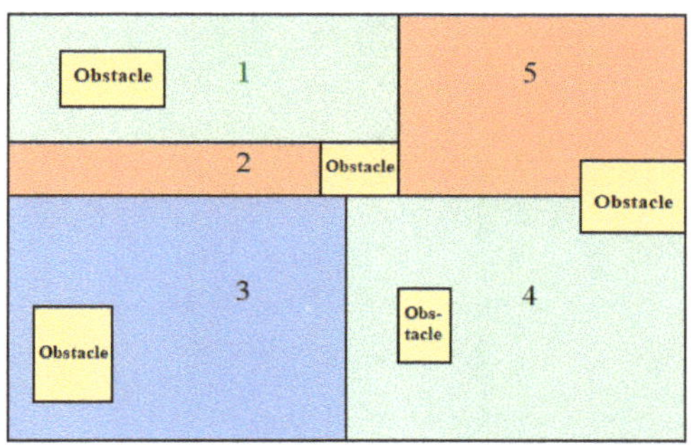

Figure 5. Area Decomposition based on furniture layout.

3.3. *The Resulting Tiling Sets (Navigation Strategies)*

Finally, all the aforementioned theorems are applied in each sub-area to generate five tiling sets which are indeed the robot's navigation strategies.

The proposed tiling sets are summarized in Figure 6, in which yellow rectangles represent obstacles and pink areas represent untiled zones. Each column of Figure 6 contains the results of applying a specific theorem on the whole area, and each row (under each column) represents a sample tiling solution (TS), applying the theorem mentioned in the header. Having four theorems with five tiling samples for each, we will have 20 tiling solutions for the whole area (100 tiling solutions to analyze for all sub-areas). In order to make it highly descriptive, the coloring of each sub-area in Figure 6 goes with Figure 5.

Due to the conditions imposed by each theorem, if the area in question does not perfectly match the conditions, the tiling set does not tile the entire area, resulting in distinct untiled zones within each sub-area. Realistically, the area covered by the tiled zones (tiled area) as shown in Figure 6 differs from the actual area spanned by the robot even when it is strictly following the tiling set for navigation. This is because, every time the robot reconfigures from one shape to another, some area may be traversed that is not tiled and does not exist in the tiling set (fake coverage), or some area will be traversed more than once during the reconfiguration process (re-coverage). However, as mentioned previously, the cleaning function of the robot is switched off while performing reconfigurations in order to devote the actuation energy fully to the servo motors controlling the hinges. This way, any unnecessary area that is not traversed for cleaning is disregarded.

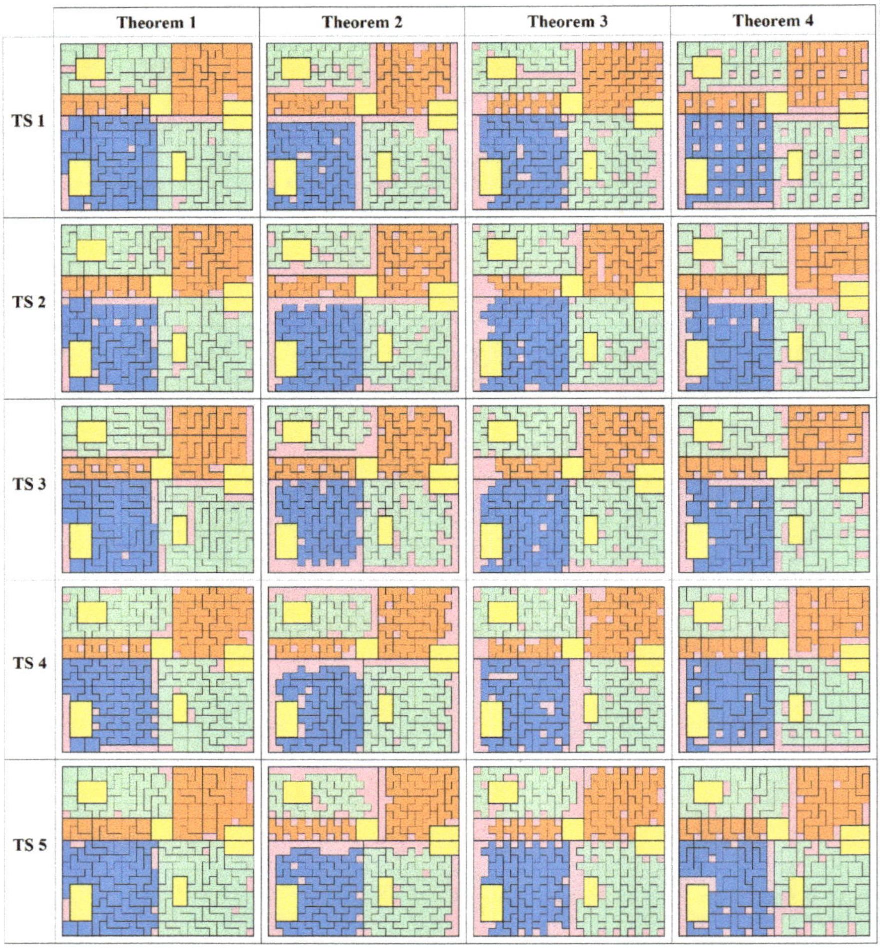

Figure 6. Five sets of tiling sets for all four theorems.

4. Extracting the Data for Area Coverage and Energy

The aim is to use the amounts of area coverage and energy consumption as decision criteria in a multi-criteria decision-making (MCDM) process.

On one hand, since all the blocks of hTetro are equipped with cleaning functions, tiling sets, as depicted in Figure 6, will represent area coverage values.

On the other hand, since the effect of time is not considered as a decision-criteria in this study, the consumed energy can be reported in terms of the consumed current [31,32]. Moreover, the consumed current by the robot can be estimated using a current estimation algorithm developed based on Newton–Raphson's algorithm [33,34] which is illustrated by the pseudo code in Figure 7.

In each sub-area, the starting point of the reconfigurable robot is set as a fixed position for all the tiling patterns to avoid the difference in the starting voltage and current. When the robot begins to navigate, the energy values can be obtained with uniformity by fixing the common starting point and utilizing the same starting current for all the individual tiling sets [33].

```
begin{recording}
caption {Current Estimation for Individual tiling pattern}
Input: Sensor current Value
start= motor: record (rotation: torque: speed ;)
                    %identifying the Locomotion type%
setup_DC Motors status ON;
                              %fixing the test bench%
Use case:: PWM::H-bridge::(Axial rotation);
                              %fixing voltage and speed %
          Clock (timer ON)
Control mechanism initialization;
                    setup:: no load~voltage ();
                              set~duty~cycle ();
                                        load inertia and frequency values
Request to send next step input;
                    {Start~route ();
Control to receive and store data for next gait pattern;
                              if {data delivery status is yes;
                    Check_signal (strength<passband and frequency);
                              Check_battery(threshold level)
else
          halt motor off ();
                    stop operation.
Analyse local workspace for current values;
                    Repeat (next tiling combination)
Next step function;
          Check for obstacle and reconfiguration;
Check for untiled area;
If (perimeter is same)
          Repeat (next gait movement)
                    else {route alteration}
          save_time_slot=0;
                    save_locomotion pattern;
                              save_current and area covered data;
Output: Data sheet with current and area covered;
end{recording}
```

Figure 7. Pseudo code for the energy estimation scheme.

Using the pre-defined tiling set, given in Figure 6, the robot's energy consumption is calculated for individual gaits, while the amount of energy consumed by all the motors for a single-step movement containing all types of locomotion, shapeshifting, and angular movements, overall, is called the gait pattern [33]. Supported by the experimental results, and considering only gait patterns as input, the motor torque, no-load speed, current, and the rated DC voltages are obtained by allowing the robot to traverse within the given tiling set. In the experiments, the output voltage amplitude is taken as the same as the offset value of the DC motor and the duty cycle, while applying the PWM technique, is calculated for the full 12 V DC supply. The total current ratio is set as 7.4 A before starting the area coverage.

Generally, based on the Newton–Raphson algorithm, we developed an energy estimation scheme that characterizes the energy cost in terms of the current consumed with respect to specific locomotion gaits across all valid morphologies. This approach was adopted in our work to compute the current consumed by our hTetro robot for covering a given area.

Tables 1–5 summarize the area coverage and energy values (in both percentages and real values) for each sub-area applying different tiling sets given by the introduced theorems. In addition, it should be mentioned that the *obs-free* in the Tables 1–5 stand for the area which is the result of subtracting the area occupied by obstacles from the entire sub-area.

Table 1. Data for area coverage and energy for sub-area 1.

			Area Coverage		Energy	
			Area %	**Real Value (m^2)**	**Energy %**	**Real Value (A)**
Sub-Area 1 **2.058 m^2** 1.8228 m^2 (*obs-free*)	**Theorem 1**	Tiling set 1	90.322	1.6464	16.11	1.192
		Tiling set 2	94.623	1.7248	15.09	1.117
		Tiling set 3	86.021	1.568	14.90	1.103
		Tiling set 4	90.322	1.6464	15.92	1.178
		Tiling set 5	94.623	1.7248	15.01	1.111
	Theorem 2	Tiling set 1	68.817	1.2544	15.60	1.419
		Tiling set 2	68.817	1.2544	16.78	1.423
		Tiling set 3	64.52	1.2548	14.31	1.431
		Tiling set 4	68.817	1.2544	16.30	1.428
		Tiling set 5	60.215	1.0926	15.76	1.428
	Theorem 3	Tiling set 1	77.419	1.4112	16.29	1.442
		Tiling set 2	86.02	1.568	16.01	1.433
		Tiling set 3	86.021	1.568	15.37	1.440
		Tiling set 4	86.021	1.568	14.98	1.466
		Tiling set 5	73.118	1.3328	15.24	1.441
	Theorem 4	Tiling set 1	73.118	1.3328	15.75	1.449
		Tiling set 2	88.172	1.6072	14.23	1.440
		Tiling set 3	86.021	1.568	14.76	1.480
		Tiling set 4	86.021	1.568	15.90	1.466
		Tiling set 5	90.32	1.6464	15.12	1.473

Table 2. Data for area coverage and energy for sub-area 2.

			Area Coverage		Energy	
			Area %	**Real Value (m^2)**	**Energy %**	**Real Value (A)**
Sub-Area 2 **0.7056 m^2** No Obstacle	**Theorem 1**	Tiling set 1	88.888	0.6266	11.98	0.866
		Tiling set 2	88.888	0.6266	12.55	0.929
		Tiling set 3	88.888	0.6266	12.21	0.903
		Tiling set 4	88.888	0.6266	12.02	0.889
		Tiling set 5	100	0.7056	11.45	0.847
	Theorem 2	Tiling set 1	88.888	0.6266	12.91	1.101
		Tiling set 2	77.777	0.5488	12.11	1.126
		Tiling set 3	88.888	0.6266	11.52	0.998
		Tiling set 4	77.777	0.5488	12.10	1.012
		Tiling set 5	66.666	0.4704	11.00	1.110
	Theorem 3	Tiling set 1	55.555	0.3914	11.78	0.989
		Tiling set 2	55.555	0.3914	12.41	0.961
		Tiling set 3	66.666	0.4704	11.56	1.003
		Tiling set 4	66.666	0.4704	12.30	1.120
		Tiling set 5	55.555	0.3914	12.23	0.890
	Theorem 4	Tiling set 1	88.888	0.6266	12.90	0.994
		Tiling set 2	88.888	0.6266	11.89	1.102
		Tiling set 3	88.888	0.6266	12.32	1.119
		Tiling set 4	88.888	0.6266	12.12	1.103
		Tiling set 5	88.888	0.6266	11.95	0.988

Table 3. Data for area coverage and energy for sub-area 3.

			Area Coverage		Energy	
			Area %	Real Value (m^2)	Energy %	Real Value (A)
Sub-Area 3 **3.3124 m^2** 3.0184 m^2 (*obs-free*)	**Theorem 1**	Tiling set 1	82.825	2.8224	19.81	1.466
		Tiling set 2	83.116	2.5088	19.06	1.410
		Tiling set 3	90.909	2.744	19.59	1.449
		Tiling set 4	83.116	2.5088	18.36	1.358
		Tiling set 5	93.506	2.8224	19.48	1.442
	Theorem 2	Tiling set 1	77.92	2.352	18.43	1.710
		Tiling set 2	80.519	2.4304	17.78	1.702
		Tiling set 3	70.12	2.1168	18.01	1.691
		Tiling set 4	72.227	2.1952	17.44	1.699
		Tiling set 5	75.32	2.2736	18.07	1.710
	Theorem 3	Tiling set 1	81.8181	2.4696	19.91	1.701
		Tiling set 2	83.116	2.5088	18.21	1.699
		Tiling set 3	85.714	2.5872	19.35	1.701
		Tiling set 4	83.116	2.5088	19.71	1.717
		Tiling set 5	77.922	2.352	19.22	1.729
	Theorem 4	Tiling set 1	75.32	2.2736	17.41	1.711
		Tiling set 2	80.519	2.4304	19.09	1.737
		Tiling set 3	80.519	2.4304	19.45	1.712
		Tiling set 4	80.519	2.4304	19.32	1.724
		Tiling set 5	80.519	2.4304	19.88	1.718

Table 4. Data for area coverage and energy for sub-area 4.

			Area Coverage		Energy	
			Area %	Real Value (m^2)	Energy %	Real Value (A)
Sub-Area 4 **3.3124 m^2** 2.9988 m^2 (*obs-free*)	**Theorem 1**	Tiling set 1	91.503	2.744	18.23	1.349
		Tiling set 2	94.117	2.8224	16.79	1.242
		Tiling set 3	86.2745	2.5872	17.10	1.265
		Tiling set 4	91.503	2.744	17.39	1.287
		Tiling set 5	96.732	2.9008	16.58	1.227
	Theorem 2	Tiling set 1	77.124	2.3128	16.89	1.779
		Tiling set 2	80.392	2.4108	17.66	1.767
		Tiling set 3	73.202	2.1952	15.16	1.795
		Tiling set 4	84.313	2.5284	16.78	1.801
		Tiling set 5	81.045	2.4304	17.09	1.780
	Theorem 3	Tiling set 1	81.04	2.4304	16.51	1.771
		Tiling set 2	83.66	2.5088	16.66	1.770
		Tiling set 3	86.274	2.5872	17.71	1.799
		Tiling set 4	75.816	2.2736	15.39	1.787
		Tiling set 5	78.431	2.352	17.41	1.770
	Theorem 4	Tiling set 1	73.202	2.1952	15.89	1.799
		Tiling set 2	83.66	2.5088	17.55	1.769
		Tiling set 3	83.66	2.5088	16.70	1.767
		Tiling set 4	81.045	2.4304	17.45	1.775
		Tiling set 5	83.66	2.5088	16.10	1.778

Table 5. Data for area coverage and energy for sub-area 5.

			Area Coverage		Energy	
			Area %	**Real Value (m^2)**	**Energy %**	**Real Value (A)**
Sub-Area 5 2.156 m^2 1.9992 m^2 (*obs-free*)	**Theorem 1**	Tiling set 1	98.039	1.96	18.30	1.354
		Tiling set 2	94.117	1.8816	18.68	1.382
		Tiling set 3	90.196	1.8032	19.46	1.440
		Tiling set 4	94.117	1.8816	20.04	1.483
		Tiling set 5	94.117	1.8816	19.23	1.4237
	Theorem 2	Tiling set 1	81.372	1.6268	19.09	1.489
		Tiling set 2	86.27	1.7248	20.31	1.491
		Tiling set 3	82.352	1.6364	20.80	1.511
		Tiling set 4	82.352	1.6464	18.31	1.534
		Tiling set 5	86.28	1.7248	21.03	1.487
	Theorem 3	Tiling set 1	90.19	1.8032	20.53	1.512
		Tiling set 2	94.117	1.8816	19.41	1.490
		Tiling set 3	90.196	1.8032	19.05	1.506
		Tiling set 4	86.27	1.7248	19.69	1.493
		Tiling set 5	86.28	1.7248	21.10	1.510
	Theorem 4	Tiling set 1	78.43	1.568	17.26	1.497
		Tiling set 2	78.43	1.568	17.87	1.487
		Tiling set 3	86.274	1.7248	20.33	1.503
		Tiling set 4	78.431	1.568	17.65	1.480
		Tiling set 5	82.352	1.6464	19.77	1.501

5. Multi-Criteria Decision Making

5.1. Simple Additive Weighing (SAW) Method—A Very Brief Review

Owing to its simplicity, the SAW method is probably the most popular and most commonly used method for multiple-criteria decision making (MCDM), especially in single-dimensional problems. If there are M alternatives and N criteria then, the best alternative is the one that satisfies Equation (1).

$$A^* = \{A_i | \max_i \sum_{j=1}^{N} w_j r_{ij}\} i = 1, 2, \ldots, M \tag{1}$$

where, N is the number of decision criteria, r_{ij} is the normalized value of the i-th alternative in terms of the j-th criterion, and w_j is the weight of importance of the j-th criterion. The values are normalized in order to make it possible to add the measurements with different units.

5.2. SAW Concepts Applied to Tiling-Based Path Planning

The energy consumption level is a key factor of any electromechanical system, and for cleaning applications, the amount of area coverage signifies another aspect of the performance. Here, we want to apply MCDM regarding these key factors (area and energy) as the decision-making criteria. As mentioned above, an area decomposition is defined based on the obstacle layout and the whole environment is divided into five sub-areas, as shown in Figure 5. Then, the four introduced tiling theorems are applied to each sub-area and five tiling sets are devised for each theorem in each sub-area. Regarding the given tiling sets for each sub-area, the results for the decision criteria are summarized within Tables 1–5. The final aim is to introduce a superior theorem for each sub-area; the theorem which could best tile the area in terms of area coverage and energy consumption. In order to do this, a fitness function for each given tiling set is described, as in Equation (2), in which c_1 is a positive weighting factor and c_2 is a negative weighting factor since we would like to increase the area coverage and decrease the energy utilization and create a balance between the two. For now, we assume that the area coverage and energy are of the same importance for the user and assume the importance weights c_1 and c_2 are equally 0.5. In order to make the concepts of area and energy commensurable and make the addition possible, first we must normalize the values or we can simply use the values in percentage, which is representative of the real value regardless of units.

$$f(s) = c_1 A(s) - c_2 E(s) \tag{2}$$

6. Result Analysis and Discussion

In order to apply the proposed approach, Tables 1–5 are considered as the input data. For each sub-area, we have 20 given tiling sets, i.e., 20 alternatives: Four theorems that each give five tiling sets. In each sub-area, the value of the fitness function for each of the alternatives is calculated through (2), and consequently, making a comparison amongst the fitness values gives the best tiling set for each sub-area. In addition, not only a preference order between the individual tiling sets is suggested, but also a theorem that can better handle the environment is introduced based on sum of rankings for the tiling sets. Tables 6–10 summarize the ranking of the tiling sets based on the fitness function.

Consequently, based on the sum of rankings summarized in Tables 6–10, the preference order of the Theorems for each sub-area is given by the following (the option with the least sum of rankings appears to be the best).

- For sub-area 1: Theorem 1 > Theorem 4 > Theorem 3 > Theorem 2
- For sub-area 2: Theorem 1 > Theorem 4 > Theorem 2 > Theorem 3
- For sub-area 3: Theorem 1 > Theorem 3 > Theorem 4 > Theorem 2
- For sub-area 4: Theorem 1 > Theorem 3 > Theorem 4 > Theorem 2
- For sub-area 5: Theorem 1 > Theorem 3 > Theorem 2 > Theorem 4

Table 6. Ranking of the tiling sets based on the fitness value for sub-area 1.

Sub-Area 1		Rank	Rank Sum
Theorem 1	Tiling set 1	5	20
	Tiling set 2	2	
	Tiling set 3	8	
	Tiling set 4	4	
	Tiling set 5	1	
Theorem 2	Tiling set 1	16	90
	Tiling set 2	18	
	Tiling set 3	19	
	Tiling set 4	17	
	Tiling set 5	20	
Theorem 3	Tiling set 1	13	58
	Tiling set 2	12	
	Tiling set 3	10	
	Tiling set 4	9	
	Tiling set 5	14	
Theorem 4	Tiling set 1	15	42
	Tiling set 2	6	
	Tiling set 3	7	
	Tiling set 4	11	
	Tiling set 5	3	

Table 7. Ranking of the tiling sets based on the fitness value for sub-area 2.

Sub-Area 2		Rank	Rank Sum
Theorem 1	Tiling set 1	5	30
	Tiling set 2	10	
	Tiling set 3	8	
	Tiling set 4	6	
	Tiling set 5	1	
Theorem 2	Tiling set 1	12	56
	Tiling set 2	14	
	Tiling set 3	2	
	Tiling set 4	13	
	Tiling set 5	15	
Theorem 3	Tiling set 1	18	90
	Tiling set 2	20	
	Tiling set 3	16	
	Tiling set 4	17	
	Tiling set 5	19	
Theorem 4	Tiling set 1	11	34
	Tiling set 2	3	
	Tiling set 3	9	
	Tiling set 4	7	
	Tiling set 5	4	

Table 8. Ranking of the tiling sets based on the fitness value for sub-area 3.

Sub-Area 3		Rank	Rank Sum
Theorem 1	Tiling set 1	8	22
	Tiling set 2	6	
	Tiling set 3	2	
	Tiling set 4	5	
	Tiling set 5	1	
Theorem 2	Tiling set 1	15	81
	Tiling set 2	9	
	Tiling set 3	20	
	Tiling set 4	19	
	Tiling set 5	18	
Theorem 3	Tiling set 1	10	40
	Tiling set 2	4	
	Tiling set 3	3	
	Tiling set 4	7	
	Tiling set 5	16	
Theorem 4	Tiling set 1	17	67
	Tiling set 2	11	
	Tiling set 3	13	
	Tiling set 4	12	
	Tiling set 5	14	

Table 9. Ranking of the tiling sets based on the fitness value for sub-area 4.

Sub-Area 4		Rank	Rank Sum
Theorem 1	Tiling set 1	4	15
	Tiling set 2	2	
	Tiling set 3	5	
	Tiling set 4	3	
	Tiling set 5	1	
Theorem 2	Tiling set 1	18	73
	Tiling set 2	15	
	Tiling set 3	19	
	Tiling set 4	8	
	Tiling set 5	13	
Theorem 3	Tiling set 1	12	60
	Tiling set 2	9	
	Tiling set 3	6	
	Tiling set 4	17	
	Tiling set 5	16	
Theorem 4	Tiling set 1	20	62
	Tiling set 2	11	
	Tiling set 3	10	
	Tiling set 4	14	
	Tiling set 5	7	

Table 10. Ranking of the tiling sets based on the fitness value for sub-area 5.

Sub-Area 5		Rank	Rank Sum
Theorem 1	Tiling set 1	1	18
	Tiling set 2	2	
	Tiling set 3	7	
	Tiling set 4	5	
	Tiling set 5	3	
Theorem 2	Tiling set 1	16	69
	Tiling set 2	10	
	Tiling set 3	17	
	Tiling set 4	14	
	Tiling set 5	12	
Theorem 3	Tiling set 1	8	40
	Tiling set 2	4	
	Tiling set 3	6	
	Tiling set 4	9	
	Tiling set 5	13	
Theorem 4	Tiling set 1	18	83
	Tiling set 2	20	
	Tiling set 3	11	
	Tiling set 4	19	
	Tiling set 5	15	

On the other hand, a graphical analysis of the results is illustrated in Figures 8–12 in which the different tiling sets given by different tiling theorems are depicted based on the ranking of their corresponding fitness function. In the given bar graphs, for each sub-area, the results given by Theorem 1, Theorem 2, Theorem 3, and Theorem 4 are shown in red, green, blue and yellow bars, respectively. Meanwhile, regarding the vertical axis, there is a labeling applied to the real fitness values to increase the visibility for comparison: The first ranking tiling set (with highest fitness function) is given the value 20 and the last ranking one (with lowest fitness function) is given the value of 1. Moreover, in the horizontal axis, the expressions Th i-Ts j stands for Theorem i-Tiling set j. It is clear that Figures 8–12 introduce a precise preference order between each given tiling set, regardless of the theorem applied. However, the overall performance of each theorem is visualized through the total area occupied by the corresponding colored bars.

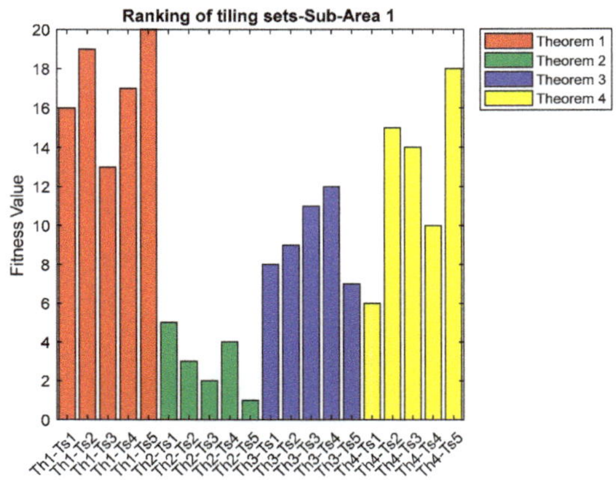

Figure 8. Ranking of fitness values for tiling sets under each theorem-sub-area 1.

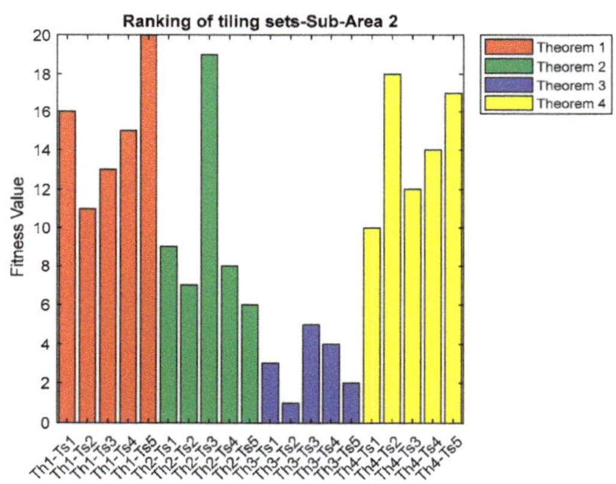

Figure 9. Ranking of fitness values for tiling sets under each theorem-sub-area 2.

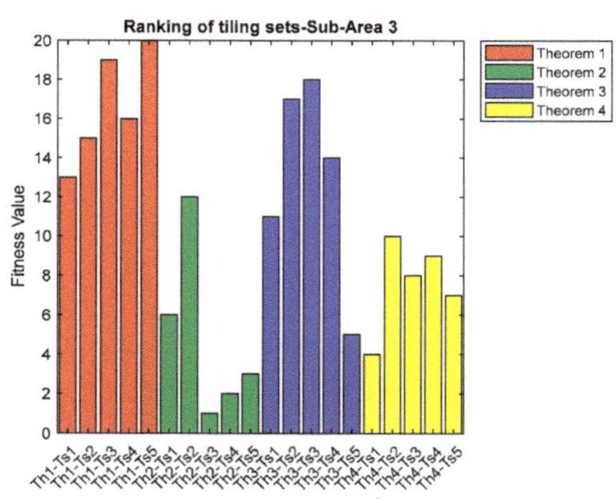

Figure 10. Ranking of fitness values for tiling sets under each theorem-sub-area 3.

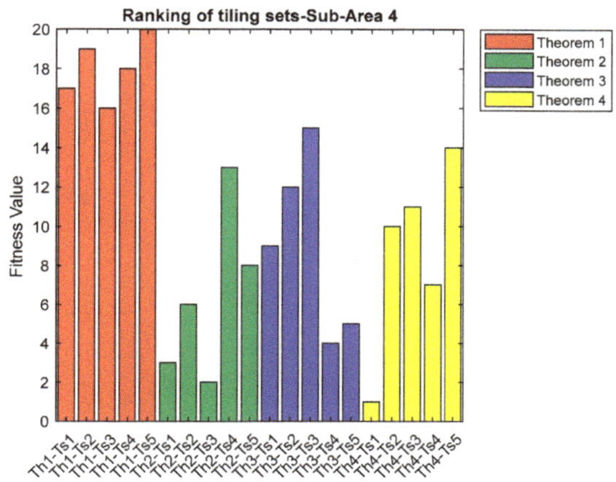

Figure 11. Ranking of fitness values for tiling sets under each theorem-sub-area 4.

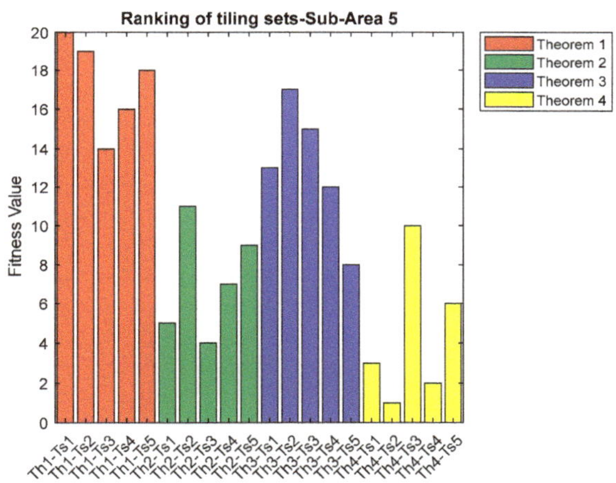

Figure 12. Ranking of fitness values for tiling sets under each theorem-sub-area 5.

It is evident that in all five sub-areas, Theorem 1 has been performing better than other theorems with regard to fulfilling the balance between area coverage and energy consumption. To add a further illustration, a sketch of the best given tiling set is shown in Figure 13, which shows the following tiling sets together:

- Tiling set 5 given by Theorem 1 for sub-area 1,
- Tiling set 5 given by Theorem 1 for sub-area 2,
- Tiling set 5 given by Theorem 1 for sub-area 3,
- Tiling set 5 given by Theorem 1 for sub-area 4,
- Tiling set 1 given by Theorem 1 for sub-area 5,

With reference to Tables 1–5, the whole additive energy value for this composition of the highest-ranking tiling sets is 5.981 A and the whole area covered by this superior tiling set is 10.1136 m^2 (95.9% of the entire area required to be cleaned). As it can be seen, the decision goal is fulfilled to a satisfying extent in terms of the defined criteria since the entire energy can be provided by a one-time battery charge (less than 7.4 A), and the area coverage is the highest available one.

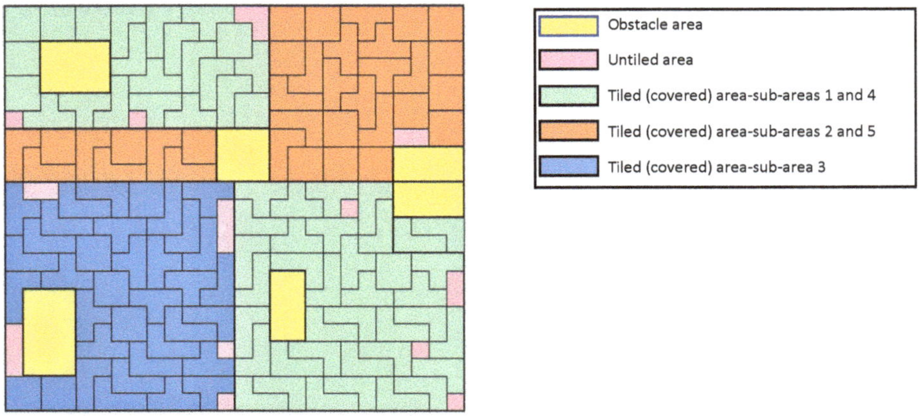

Figure 13. A composition of best tiling sets for each sub-area.

7. Evaluating the Robot Performance with and without the Tiling

Providing access to narrow and difficult spaces, self-reconfiguration is totally reasonable for maximizing the area coverage. However, shape-shifting may sound a waste of energy. In this section, a comparison is illustrated with a fixed-morphology robot with the same size (O reconfiguration in Figure 3) that follows no tiling theory but tries to cover the area like the current commercialized cleaning robots. Figure 14 shows nine such sweeping solutions for the entire experimental environment in which each of the solutions appears with the corresponding numbering and each yellow square is an O morphology, as depicted in Figure 3. In addition, the gray area is the area totally inaccessible and the purple area is the area that is accessible at the expense of re-coverage. The total area coverage percentage and utilized energy amount in each sweeping solution is summarized in Table 11 in which the reported area is the covered area with avoidance of re-coverage, i.e., the whole area subtracted by the gray and purple areas and the area occupied by obstacles.

Figure 14. Sample sweeping solutions with fixed morphology and no tiling theory.

Table 11. Energy and area coverage for fixed-morphologies with no tiling theorem.

Sweeping Solution	Area-Coverage (%)	Energy (A)
1	496/538 = 92.19%	5.23
2	496/538 = 92.19%	5.44
3	496/538 = 92.19%	5.87
4	496/538 = 92.19%	5.91
5	484/538 = 89.96%	6.67
6	484/538 = 89.96%	6.312
7	484/538 = 89.96%	6.18
8	496/538 = 92.19%	6.22
9	484/538 = 89.96%	6.48

Considering Table 11, it is obvious that the fixed-morphology performance was not as energy-efficient as expected and its energy consumption for sweeping the entire space is almost the same as the best alternative selected by MCDM (5.981 A) or even higher (worse) than that in some cases (sweeping solution 5, 6, 7, 8, 9). However, the shape-shifting morphology evidently results in higher area coverage. The results in Table 11 also justify this fact: In a fixed-morphology scheme with no tiling-path-planning, we have, at most, 92% area coverage, but the approach presented in this paper results in almost 96% area coverage, see Figure 13.

To summarize, when a higher area coverage is guaranteed by self-reconfiguration, and the energy consumption in a fixed-morphology scheme is at the level of shape-shifting morphology (slightly better: $5.98 - 5.23 = 0.75$ (A) or even worse: $5.98 - 6.67 = -0.7$ (A)), it sounds reasonable to apply the MCDM-supported tiling path-planning even though it requires a slight compromise in energy consumption.

8. Conclusions

By providing self-reconfiguration abilities for the cleaning robot, it can cope with limitations associated with sweeping narrow corridors and far-to-access corners, making self-reconfigurable robots very interesting for use as cleaners. In this regard, a Tetris-inspired mechanism has already been proposed in our previous studies and, additionally, polyomino tiling theory has been introduced as a means of path planning for such a mechanism. Previous studies in this field, have only considered area coverage as the performance metric for the cleaning robot. Moreover, besides the amount of cleaned area, the amount of battery usage (energy) is a significant issue. In this paper, we applied a MCDM approach aimed at efficient cleaning in terms of energy and area coverage. This study has moved a step forward regarding the quest to find the best alternative among given tiling maps in order to orient the robot towards the optimal navigation.

To this end, in this study, four tiling set theorems are utilized to define the navigation maps for the robot. Before applying the tiling theorems, the entire area is decomposed to five sub-areas based on furniture layout. Then, the tiling theorems are applied to each sub-area separately to generate several distinguished tiling sets. These are the alternatives analyzed from the area and energy viewpoint. To balance the facing trade-off, a MCDM approach is applied by defining a fitness function in terms of the decision criteria and finally, all the alternatives are ranked in terms of the resulting fitness function. Based on the results of this ranking, one of the most recent theorems stood out, showing a more efficient performance throughout all the sub-areas. Furthermore, a comparison between the performance of a fixed-morphology and a shape-shifting mechanism is presented that further suggests the MCDM-supported tiling path planning.

As a scope of future work, the area decomposition method can be improved based on cellular decomposition supported by SLAM. Additionally, here, only the effect of area coverage and energy are investigated as the decision criteria, while time is also an influencing factor in the hectic timetables of people nowadays. Considering the required time as another aspect of the efficiency of a cleaning application could be another aspect of future work. On the other hand, in this study, both the decisive factors in a MCDM process are regarded to be of the same value to the user. Defining the partial importance of the decision criteria based on user preferences is actually not difficult to imagine. This can also open an opportunity for incorporating fuzzy logic.

Author Contributions: Conceptualization, M.K.; Formal analysis, M.K.; Investigation, M.K. and M.A.P.; Methodology, M.K., M.R.E. and M.A.P.; Project administration, M.K. and M.R.E.; Resources, M.R.E.; Software, M.K. and M.A.; Supervision, M.K. and M.R.E.; Writing—original draft, M.K., M.R.E., M.A.P., M.A. and V.P.

Acknowledgments: This research was funded by the National Robotics R & D Program Office, Singapore, under the Grant No. RGAST1702, Singapore University of Technology and Design (SUTD) which are gratefully acknowledged to conduct this research work.

Conflicts of Interest: The authors declare no conflict of interest.

References

1. Vaussard, F.; Fink, J.; Bauwens, V.; Rétornaz, P.; Hamel, D.; Dillenbourg, P.; Mondada, F. Lessons learned from robotic vacuum cleaners entering the home ecosystem. *Robot. Autom. Syst.* **2014**, *62*, 376–391. [CrossRef]
2. Palleja, T.; Tresanchez, M.; Teixido, M.; Palacin, J. Modeling floor-cleaning coverage performances of some domestic mobile robots in a reduced scenario. *Robot. Autom. Syst.* **2010**, *58*, 37–45. [CrossRef]
3. Gao, X.; Li, K.; Wang, Y.; Men, G.; Zhou, D.; Kikuchi, K. A floor cleaning robot using Swedish wheels. In Proceedings of the IEEE International Conference on Robotics and Biomimetics (ROBIO), Sanya, China, 15–18 December 2007.
4. Yang, S.X.; Luo, C. A neural network approach to complete coverage path planning. *IEEE Trans. Syst. Man Cybern. Part B* **2004**, *34*, 718–724. [CrossRef]
5. Wong, S.C.; Middleton, L.; MacDonald, B.A.; Auckland, N. Performance metrics for robot coverage tasks. In Proceedings of the Australasian Conference on Robotics and Automatio, Auckland, New Zealand, 27–29 November 2002.
6. Ahmadi, M.; Stone, P. A multi-robot system for continuous area sweeping tasks. In Proceedings of the IEEE International Conference on Robotics and Automation (ICRA), Orlando, Fl, USA, 15–19 May 2006.
7. Nansai, S.; Rojas, N.; Elara, M.R.; Sosa, R. Exploration of adaptive gait patterns with a reconfigurable linkage mechanism. In Proceedings of the 2013 IEEE/RSJ International Conference on Intelligent Robots and Systems, Tokyo, Japan, 3–7 November 2013.
8. Moubarak, P.; Ben-Tzvi, P. Modular and reconfigurable mobile robotics. *Robot. Autom. Syst.* **2012**, *60*, 1648–1663. [CrossRef]
9. Patil, M.; Abukhalil, T.; Sobh, T. Hardware Architecture Review of Swarm Robotics System: Self-Reconfigurability, Self-Reassembly, and Self-Replication. *ISRN Robot.* **2013**, *2013*. [CrossRef]
10. Kee, V.; Rojas, N.; Elara, M.R.; Sosa, R. Hinged-Tetro: A self-reconfigurable module for nested reconfiguration. In Proceedings of the 2014 IEEE/ASME International Conference on Advanced Intelligent Mechatronics, Besançon, France, 8–11 July 2014.
11. Prabakaran, V.; Elara, M.R.; Pathmakumar, T.; Nansai, S. hTetro: A tetris inspired shape shifting floor cleaning robot. In Proceedings of the 2017 IEEE International Conference on Robotics and Automation (ICRA), Singapore, 29 May–3 June 2017.
12. Prabakaran, V.; Elara, M.R.; Pathmakumar, T.; Nansai, S. Floor cleaning robot with reconfigurable mechanism. *Autom. Constr.* **2018**, *91*, 155–165. [CrossRef]
13. Prabakaran, V.; Mohan, R.E.; Sivanantham, V.; Pathmakumar, T.; Kumar, S.S. Tackling Area Coverage Problems in a Reconfigurable Floor Cleaning Robot Based on Polyomino Tiling Theory. *Appl. Sci.* **2018**, *8*, 342. [CrossRef]
14. Veerajagadheswar, P.; Elara, M.R.; Pathmakumar, T.; Ayyalusami, V. A Tiling-Theoretic Approach to Efficient Area Coverage in a Tetris-Inspired Floor Cleaning Robot. *IEEE Access* **2018**, *6*, 35260–35271. [CrossRef]
15. Twarock, R. The Architecture of Viral Capsids Based on Tiling Theory. *J. Theor. Med.* **2005**, *6*. [CrossRef]
16. Jho, C.W.; Lee, W.H. Video Puzzle Game Application of Polyomino Re-tiling. In Proceedings of the Embedded and Multimedia Computing Technology and Service, Lecture Notes in Electrical Engineering, Dordrecht, The Netherlands, 13 May 2012.

17. Mardani, A.; Jusoh, A.; MD Nor, K.; Khalifah, Z.; Zakwan, N.; Valipour, A. Multiple criteria decision-making techniques and their applications—A review of the literature from 2000 to 2014. *Econ. Res. Ekonomska Istraživanja* **2015**, *28*, 516–571. [CrossRef]
18. Triantaphyllou, E.; Shu, B.; Sanchez, S.N.; Ray, T. *Multi-Criteria Decision Making: An Operations Research Approach. Encyclopedia of Electrical and Electronics Engineering*; Webster, J.G., Ed.; John Wiley & Sons: New York, NY, USA, 1998; pp. 175–186.
19. Mardani, A.; Jusoh, A.; Zavadskas, E.K. Fuzzy multiple criteria decision-making techniques and applications—Two decades review from 1994 to 2014. *Expert Syst. Appl.* **2015**, *42*, 4126–4148. [CrossRef]
20. Zarghami, M.; Szidarovszky, F. Stochastic-fuzzy multi criteria decision making for robust water resources management. *Stoch. Environ. Res. Risk Assess.* **2009**, *23*, 329–339. [CrossRef]
21. Chen, T.-Y. The inclusion-based TOPSIS method with interval-valued intuitionistic fuzzy sets for multiple criteria group decision making. *Appl. Soft Comput.* **2015**, *26*, 57–73. [CrossRef]
22. Tzeng, G.-H.; Huang, J.-J. *Multiple Attribute Decision Making, Methods and Applications*; CRC Press: Boca Raton, FL, USA, 2011.
23. Roszkowska, E.; Wachowicz, T. Application of fuzzy TOPSIS to scoring the negotiation offers in ill-structured negotiation problems. *Eur. J. Oper. Res.* **2015**, *242*, 920–932. [CrossRef]
24. Landa-Torres, I.; Manjarres, D.; Bilbao, S.; Del Ser, J. Underwater Robot Task Planning Using Multi-Objective Meta-Heuristics. *Sensors* **2017**, *17*, 762. [CrossRef] [PubMed]
25. Devi, K. Extension of VIKOR method in intuitionistic fuzzy environment for robot selection. *Expert Syst. Appl.* **2011**, *38*, 14163–14168. [CrossRef]
26. Basilico, N.; Amigoni, F. Exploration strategies based on multi-criteria decision making for searching environments in rescue operations. *Auton. Robots* **2011**, *31*, 401. [CrossRef]
27. Befumo, A.; Lenchner, J. Extensions of Golombs Tromino Theorem. In Proceedings of the Fall Workshop in Computational Geometry, University of Connecticut, Storrs, CT, USA, 31 October–1 November 2014.
28. Golomb, S.W. Checker Boards and Polyominoes. *Am. Math. Mon.* **1954**, *61*, 675–682. [CrossRef]
29. Lester, C. Tilings with T and Skew Tetrominoes. *Quercus Linfield J. Undergrad. Res.* **2012**, *1*.
30. Nitica, V. The tilings of deficient squares by ribbon L-tetrominoes are diagonally cracked. *arXiv* **2017**, arXiv:1701.00419.
31. Plonski, P.A.; Tokekar, P.; Isler, V. Energy-Efficient Path Planning for Solar-Powered Mobile Robots. In *Experimental Robotics*; Springer International Publishing: Heidelberg, Germany, 2013; pp. 717–731.
32. Patil, M.; Abukhalil, T.; Patel, S.; Sobh, T. UB robot swarm—Design, implementation, and power management. In Proceedings of the 12th IEEE International Conference on Control and Automation (ICCA), Kathmandu, Nepal, 1–3 June 2016.
33. Barili, A.; Ceresa, M.; Parisi, C. Energy-saving motion control for an autonomous mobile robot. In Proceedings of the IEEE International Symposium on Industrial Electronics, Dubrovnik, Croatia, 10–14 July 1995.
34. Duleba, I.; Sasiadek, J.Z. Nonholonomic motion planning based on Newton algorithm with energy optimization. *IEEE Trans. Control Syst. Technol.* **2003**, *11*, 355–363. [CrossRef]

applied sciences

MDPI

Article

Mobile Robot Path Planning with a Non-Dominated Sorting Genetic Algorithm

Yang Xue

Department of Mechanics, Tianjin University, Tianjin 300350, China; xueyang216@tju.edu.cn

Received: 18 October 2018; Accepted: 12 November 2018; Published: 15 November 2018

Abstract: In many areas, such as mobile robots, video games and driverless vehicles, path planning has always attracted researchers' attention. In the field of mobile robotics, the path planning problem is to plan one or more viable paths to the target location from the starting position within a given obstacle space. Evolutionary algorithms can effectively solve this problem. The non-dominated sorting genetic algorithm (NSGA-II) is currently recognized as one of the evolutionary algorithms with robust optimization capabilities and has solved various optimization problems. In this paper, NSGA-II is adopted to solve multi-objective path planning problems. Three objectives are introduced. Besides the usual selection, crossover and mutation operators, some practical operators are applied. Moreover, the parameters involved in the algorithm are studied. Additionally, another evolutionary algorithm and quality metrics are employed for examination. Comparison results demonstrate that non-dominated solutions obtained by the algorithm have good characteristics. Subsequently, the path corresponding to the knee point of non-dominated solutions is shown. The path is shorter, safer and smoother. This path can be adopted in the later decision-making process. Finally, the above research shows that the revised algorithm can effectively solve the multi-objective path planning problem in static environments.

Keywords: mobile robot; static environments; path planning; multi-objective optimization; NSGA-II; evolutionary operators

1. Introduction

Path planning (PP) has been fundamental in many areas in recent decades, for instance mobile robots [1–3], unmanned surface vehicles (USVs) [4,5], wireless sensor networks (WSNs) [6–8] and video games [9]. For mobile robots, path planning is devised to find one or more feasible routes from the initial location to the target in a given workspace. The related algorithms in the field of path planning are reviewed in [10,11]. In [12], several path planning and navigation algorithms commonly employed in the domain of unmanned aerial vehicles (UAVs) were studied. The commonly-used methods are the probabilistic roadmap method (PRM) [13], the artificial potential field method (ARF) [14], the rapidly-exploring random tree method (RRT) [15], A* and its variants [16,17], and so on. In [18], the fast marching method (FMM) was applied to the path planning problem. Moreover, the research revealed that the algorithm can successfully attain the collision-free shortest route in many static environments. Most conventional approaches attain available paths and seldom optimize several objectives concurrently.

However, optimization problems in most disciplines should study multiple objectives simultaneously, not just a single one. For example, in the product design process, the cost and quality of the product are often contradictory, that is and decrease in the price of the product and the quality has to decrease, and vice versa. How to decrease the cost of the product and enhance the quality comprise a dual-objective optimization problem. Path planning problems are no exception. In this study, three objectives are presented. Path length is related to the operation time of the mobile robot.

Path safety represents the length from the path to the nearest obstacle. Path smoothness indicates the degree of bending of the path. In general, at least two objectives are conflicting.

At present, based on the complexity of the path planning problem, this issue is identified as NP-hard [19]. Multiple competing goals need to be considered concurrently. Subsequently, common multi-objective weighting methods are employed to solve problems [20,21]. The principle of the weighting method is to set the weighting factors of the objectives and combine the targets into a new one. However, because there is no apparent measurable relationship between these objectives, it is difficult to set their weighting factors. In other words, the transformation process itself is also a multi-objective optimization problem. Therefore, the weighting method is not proper. On the other hand, some researchers extended the deterministic heuristic A* algorithm to multi-objective cases [22–24]. In these algorithms, the heuristic function of any node in the graph search space is a vector.

In later years, evolutionary algorithms (EA) were universally employed for path planning. A practical framework of multi-objective evolutionary algorithm (MOEA) was designed in [25]. Besides, several practical evolution operators were presented. To further advance the computational efficiency, in the initial step, some individuals were generated by the Dijkstra algorithm. In [26], the traditional method was first employed to generate the roadmap in the workspace, and then, the Hopfield neural network was applied to improve path length and safety. In [27], the modified rapidly-exploring random tree was presented to obtain the path. Then, the path length and smoothness were improved based on the neural network curve post-processing strategy. In [28], the Q-learning method was developed to optimize the path. A fast two-stage ant colony optimization algorithm was shown in [29]. This algorithm contained two phases: map preprocessing and ant colony optimization. In the map preprocessing, they calculated the minimum number of steps from all free nodes to the target node. Next, the path length was optimized by the modified ant colony optimization. In [30], the modified tabu search algorithm was designed for the optimal path length in grid environments. In [31], path length and smoothness were enhanced by two multi-objective memetic algorithms. In [32], an intelligent water drop algorithm was proposed to increase the length and safety. An available path was found by the artificial bee colony algorithm in [33], then the length and smoothness of the obtained path were optimized by evolutionary programming. In [34], a model of path length and danger degree was solved by the improved particle swarm algorithm. The degree of risk and path length were minimized by particle swarm optimization (PSO) in unknown circumstances [35]. In [36], the chaotic particle swarm optimization algorithm was used to enhance the control points of the Bezier curve to reach a short and smooth route. The hierarchical global path planning method was introduced in [37]. The method designed a three-stage structure to obtain an optimal route. First, a free geometric configuration space was determined with the triangular decomposition method, and then based on the configuration space, the Dijkstra algorithm was utilized to obtain a viable route. Finally, constrained particle swarm optimization was designed to optimize length and smoothness. Similar hierarchical strategies could be found in [38]. Individuals within the initial population were generated by the surrounding point set algorithm. Then, the length and smoothness were enhanced by the particle swarm optimization algorithm.

In this study, the multi-objective path planning is solved with the improved NSGA-II. The work is shown below:

- The framework of the improved NSGA-II is introduced. Several practical evolutionary operators are presented to enhance the feasibility of the route and optimize three objectives (length, smoothness and safety). They can enhance the local search capabilities of the improved NSGA-II.
- The parameters in the algorithm are systematically studied. The results show that larger population sizes, larger numbers of generations and high operator probabilities are indispensable in complex environments.

- The improved NSGA-II is tested with an existing evolutionary algorithm and different quality metrics. The comparisons show that the non-dominated solutions received via the improved NSGA-II have good characteristics.
- The path corresponding to the knee point of non-dominated solutions is shown. The route is shorter, smoother and safer.

The rest of the work is designed below. The associated work is introduced in Section 2. Section 3 defines environmental modeling, representation of the path and the objectives that need to be optimized. Section 4 details NSGA-II and evolutionary operators. The parametric study is presented in Section 5. Section 6 exhibits the comparison results. Lastly, the conclusions are displayed in Section 7.

2. Related Work

Since the advent of the genetic algorithm, it has been favored by many scholars due to its simple operation process and powerful search ability [39,40]. Genetic algorithms demonstrate robust optimization performance in various areas [41–43]. In path planning, genetic algorithms have also received extensive attention. In [44], the genetic algorithm was designed to shorten the path length in the grid space. In [45], the efficiency of probabilistic roadmap (PRM) and genetic algorithm (GA) for attaining a viable route was studied. NSGA-II was employed to optimize path length and clearance in the grid environments [46]. In [47], multiple techniques of representing a path in the grid environments were presented. Moreover, NSGA-II was used to improve length, smoothness and safety. A new selection operator was designed to avoid falling into a local trap and premature convergence [48]. Then, the adaptive method based on GA was designed to optimize path length. In [49], control points of the Bezier curve were enhanced by NSGA-II to reach the Pareto-optimal solutions. In [50], an efficient initialization technique based on directed acyclic graphs was proposed. This technique can provide multiple feasible minimum paths for the genetic algorithm, which enhances the computational efficiency of the entire genetic algorithm. In [51], a viable route was given with the genetic algorithm and then smoothed by the piecewise cubic Hermite interpolation polynomial. In [52], the environment was converted with a matrix-binary code-based genetic algorithm (MGA). Then, the navigation time and the path length were optimized. Control points of the Bezier curve were optimized with an improved genetic algorithm in [53]. Then, the optimum smooth path could be selected by choosing these control points. In [54], an improved crossover operator was designed in static environments. In [55], NSGA-II was used to improve the clearance and smoothness of the path by optimizing the three parameters obtained by the potential field method. Moreover, the position of the virtual obstacle was redefined for the case where the end point of the path was farther away from the obstacle. This can help the robot safely drive away from obstacles. Finally, a method for identifying obstacles that affect the robot in cluttered environments was presented. As far as we know, genetic algorithms show great vitality in path planning problems. However, due to the existence of obstacles, after using the traditional operators (crossover, mutation), the newly-generated individuals are generally no longer feasible paths. More practical evolutionary operators are needed to further enhance evolution efficiency.

In this study, the path planning problem is resolved with the improved NSGA-II in static environments. Multiple objectives are considered. More practical evolutionary operators are presented. In the remainder of this article, the improved algorithm is proposed in detail.

3. Path Planning Problem

Path planning is to find one or more available paths in the workspace. In this article, statically known environments are considered. That is, all obstacles within the workspace are static, and their location information is entirely known to the robot. The starting location and target of the robot are signified as S and T, respectively. Then, path planning is devised to find one or more paths from S to T that do not collide with obstacles. Next, the environment modeling, path representation and the three optimization objectives are defined. In this work, two-dimensional space is adopted. Any obstacle is supposed to be an arbitrary polygonal shape. Moreover, a polygon robot can be converted into a single point by using the Minkowski sums in the field of computational geometry [56]. A path is signified with $p = [S = p_0, p_1, p_2, ..., p_n, p_{n+1} = T]$. p_i is the i-th rotation points (RPs). p_i and p_{i+1} are connected by straight segments. The representation of a path is shown in Figure 1. The coordinate of p_i is denoted as (X_i, Y_i).

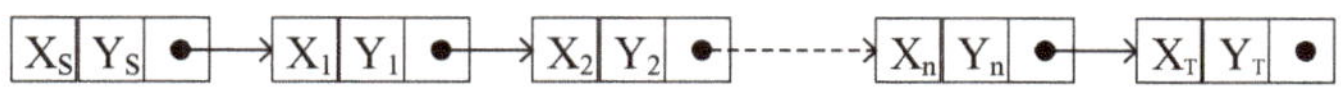

Figure 1. The representation of a path is given.

3.1. Objectives

In this paper, three objectives are introduced: smoothness, safety and length. The energy loss is associated with path smoothness and length. In addition to considering energy loss, the safety of the driving path cannot be ignored. When the distance of the robot from the obstacles during driving is less than the safe range of the sensors, this can be a critical situation, even damaging the robot and the items that are seen as obstacles. Therefore, the farther away from the nearest obstacle, the safer the route. Considering the above factors, the three objectives are designed to enhance the driving route of the robot. This is the multi-objective optimization problem. For convenience, it can be converted into a minimization problem. Next, the mathematical definition of these objectives is presented.

3.1.1. Path Length

The line segment formed by any two points in the space can be calculated based on the Euclidean distance. Thus, for two consecutive rotation points ($p_i = (x_i, y_i)$, $p_{i+1} = (x_{i+1}, y_{i+1})$), the line segment of the points can be computed. Then, the path length can be given by summing all the ordered line segments. The mathematical expression is presented in Equation (1):

$$\begin{cases} dis(p_i, p_{i+1}) = \sqrt{(x_i - x_{i+1})^2 + (y_i - y_{i+1})^2} \\ Length(p) = \sum\limits_{i=0}^{n} dis(p_i, p_{i+1}). \end{cases} \tag{1}$$

3.1.2. Path Smoothness

Path smoothness denotes the degree of sleekness of the path. The smoother the route, the less energy the robot consumes as it travels along the way. In this paper, it is defined by the average turning angle of the path. The corner formed by two consecutive line segments can be determined. Then, the average turning angle of the route can be calculated by Equation (2).

$$\begin{cases} Angle[p_i, p_{i+1}, p_{i+2}] = \pi - \cos^{-1}\left(\frac{(x_{i+1}-x_i)(x_{i+2}-x_{i+1})+(y_{i+1}-y_i)(y_{i+2}-y_{i+1})}{dis(p_i,p_{i+1})\times dis(p_{i+1},p_{i+2})}\right) \\ Smoothness(p) = \frac{1}{n}\sum\limits_{i=0}^{n-1}\{Angle[p_i, p_{i+1}, p_{i+2}]\}. \end{cases} \tag{2}$$

3.1.3. Path Safety

Firstly, the secure interval is defined. It is the minimum distance of the path from the nearest obstacle. The barrier set ($O = O_1, O_2, ..., O_m$) contains all the obstacles in the space; the number is m. $MinDis(\overline{p_i p_{i+1}}, O_j)$ is the minimum length between the line segment ($\overline{p_i p_{i+1}}$) and the obstacle (O_j). Thus, $\underset{0\le i\le n}{\mathrm{Min}}\,\underset{1\le j\le m}{\mathrm{Min}}\,\{MinDis(\overline{p_i p_{i+1}}, O_j)\}$ is the safe interval. For convenience, the maximum safety interval problem is converted to a minimum problem. The general operation is to add a negative sign before the safe distance to convert the function to a negative number. The safety objective of the path can be expressed by Equation (3):

$$Safety(p) = -\underset{0 \leq i \leq n}{\text{Min}} \underset{1 \leq j \leq m}{\text{Min}} \{MinDis(\overline{p_i p_{i+1}}, O_j)\}. \tag{3}$$

At this time, a model of path planning is presented below:

$$\begin{cases} p = [S = p_0, p_1, p_2, \ldots, p_n, p_{n+1} = T] \\ \text{Minimize } f_1(p) = Length(p) \\ \text{Minimize } f_2(p) = Smoothness(p) \\ \text{Minimize } f_3(p) = Safety(p). \end{cases} \tag{4}$$

In the above model, the only constraint is that the path does not collide with the obstacle, but it is allowed to reach the edge of the obstacle. If the path intersects the obstacle, then it is not feasible. The edge of the workspace is also regarded as an obstacle.

In general, at least two goals are contradictory. That is to say, one or more other objectives have to be sacrificed while increasing one objective. This problem is a typical multi-objective optimization problem. There is no longer one best solution, but a set of solutions. Moreover, these solutions are non-dominated. In other words, when one or more objectives of one solution are better than the other, there must be at least one or more poor objectives. In recent years, some surpassing evolutionary algorithms have emerged for solving path planning problems. However, due to the complexity and practicality of this issue, many scholars remain interested.

4. NSGA-II for the Multi-Objective Path Planning Problem

NSGA-II is one of the most popular multi-objective genetic algorithms [40]. It has the advantage of fast running speed and good convergence of solutions. Moreover, it has become the benchmark for evaluating numerous optimization algorithms. NSGA-II is the second generation non-dominated sorting genetic algorithm, and its improvements mainly include three aspects:

- A fast non-dominated sorting algorithm is designed, which stratifies the population according to the non-domination level of the individuals. Individuals within the same layer are non-dominated.
- The crowding distance of individuals in the same layer is calculated. Then, the individuals with higher values are preferentially selected, so that the same layer individuals can be more evenly distributed in the objective space to preserve the population diversity.
- The elite strategy combines the parental population with the offspring population. First, infeasible solutions within the collection are eliminated. Secondly, the remaining individuals are ranked according to the non-dominated sorting algorithm. From the low to high levels, the individuals in each layer are placed sequentially in the new population until the number of individuals exceeds the capacity of the population. Finally, according to the crowding distance metric, the individuals in the specific layer are placed into the new population in descending order until the population is full. The elite strategy preserves all the good individuals in the parental and offspring populations, thereby improving the accuracy and robustness of the optimization results.

Due to the above characteristics, NSGA-II has strong optimization capabilities. Therefore, the path planning is solved with NSGA-II. However, individuals formed by the traditional operators are generally not feasible paths. It is not enough to rely solely on traditional operators. This requires a larger population capacity and more evolutionary algebra. In the work, several practical evolutionary operators are introduced to solve this problem. These evolutionary operators are more purposeful to reduce the length and improve the smoothness and safety. Next, NSGA-II and all the operators are described in detail.

4.1. Flowchart

The process of NSGA-II is as follows:

1 The population (*POP*) is initialized.
2 In the population (*POP*), the objectives of each individual are evaluated.
3 An empty set (*NDPOP*) is generated to save the found feasible non-dominated individuals.
4 While the termination condition has not been reached, DO:

(1) Two empty populations (*NEWPOP*, *POPc*) are generated.
(2) The individuals are chosen from the population (*POP*) by the selection operator and put into the population (*POPc*).
(3) Individuals within the population (*POPc*) are paired. Then, the crossover operator is performed on the two individuals. Next, new individuals are generated and stored in the population (*NEWPOP*).
(4) For any individual in the population (*POPc*), the practical operators are sequentially executed. Then, the generated individuals are put into the population (*NEWPOP*).
(5) Each individual within the population (*NEWPOP*) is evaluated, and the feasible solutions are put into the set (*NDPOP*). Then, the non-dominated individuals of the set (*NDPOP*) are retained.
(6) Parental and children populations (*POP*) and (*NEWPOP*) are merged. Individuals in the combined population are classified according to the individual feasibility: feasible solutions' set and infeasible solutions' set. Then, the non-dominated sorting and crowding distance metric are performed on the two sets respectively. Next, individuals are picked from the two sets, and a new population (*POP*) is formed. The size of the population (*POP*) remains the same.
(7) The loop counter is incremented by one.

In the entire algorithm, a viable individual is a collision-free path. Steps 1–3 are the initialization process. The initial population (*POP*) is generated and evaluated. An empty set (*NDPOP*) is then made to store the found non-dominated solutions. This set is only used for storage and does not participate in the evolution of the population. Step 4 is the core of the algorithm, and the total iterative process is completed in this step until the termination condition is reached.

In each iteration, two empty populations (*POPc* and *NEWPOP*) are created. The population (*POPc*) is used to store the selected individuals of the parental population (*POP*). The population (*NEWPOP*) stores the children individuals. First, the selection operator is applied to choose individuals from the population (*POP*), and the selected individuals are stored in the population (*POPc*). Individuals of the population (*POPc*) are then paired. New individuals are generated by using the crossover operator and then placed in the child population (*NEWPOP*). Next, a series of operators is executed for each within the population (*POPc*). When each operator is applied, a new individual is generated and stored in the child population (*NEWPOP*). After the individuals in the population (*POPc*) are treated with the evolutionary operators, the individuals in the children population (*NEWPOP*) are evaluated, and the feasible individuals are put into the set (*NDPOP*). Then, individuals of the population (*NDPOP*) are updated; in other words, the non-dominated solutions of the population (*NDPOP*) are retained; other individuals are excluded.

Then, the elite strategy is executed. The parental population (*POP*) and the offspring population (*NEWPOP*) are merged, and the individuals in the merged set are classified according to individual feasibility: feasible solutions' set and infeasible solutions' set. Then, the non-dominated sorting and crowding distance metric are performed on the two sets, respectively. Then, individuals are picked from the two sets to form a new population (*POP*). In this step, the infeasible individuals are not entirely discarded, but some better individuals are preserved. Besides, the size of the population (*POP*) is unchanged. Then, the loop counter is incremented by one.

Step 4 is iterated until the end condition is satisfied. At this point, the entire algorithm ends, and the set (*NDPOP*) is output. Furthermore, the operators are introduced. Each operator acts on an individual, and a new individual is generated and then stored in the child population (*NEWPOP*). Besides the traditional operators, these evolutionary operators can improve the length, smoothness and safety of the path more effectively. Next, these operators and the initialization process are presented in detail.

4.2. Selection Operator

The selection operator is to elect individuals from the parental population (*POP*). The constrained tournament selection operator is applied. It selects two solutions at a time, then the tournament between them is run, and the winner participates in the succeeding evolution. For two chosen individuals, if one is not attainable and the other is available, the victor is the viable one. If none are feasible or viable, the non-dominated regulation is applied to pick the victor. If both are non-dominated, the individual with a larger crowding distance is picked as the winner. Finally, the winners are stored in the population (*POPc*).

4.3. Crossover Operator

For two randomly-selected individuals in the population (*POPc*), the crossover operator is used to form new individuals by swapping parts of the individuals. The newly-generated individuals are stored in the child population (*NEWPOP*). Suppose that $p = [S, p_1, p_2, ..., p_{i-1}, p_i, ..., p_j, p_{j+1}, ..., p_n, T]$ and $q = [S, q_1, q_2, ..., q_{k-1}, q_k, ..., q_m, q_{m+1}, ..., q_{n'}, T]$ are two randomly-chosen paths, p_i and p_j are two randomly-selected rotation points (RPs) on path (p). Similarly, q_k and q_m are two randomly-selected rotation points on path (q). The line segments connecting p_i with p_j and the line segments linking q_k with q_m are exchanged. Finally, two new paths ($p' = [S, p_1, p_2, ..., p_{i-1}, q_k, ..., q_m, p_{j+1}, ..., p_n, T]$, $q' = [S, q_1, q_2, ..., q_{k-1}, p_i, ..., p_j, q_{m+1}, ..., q_{n'}, T]$) are generated. Figure 2 shows the result of exchanging line segments with the crossover operator.

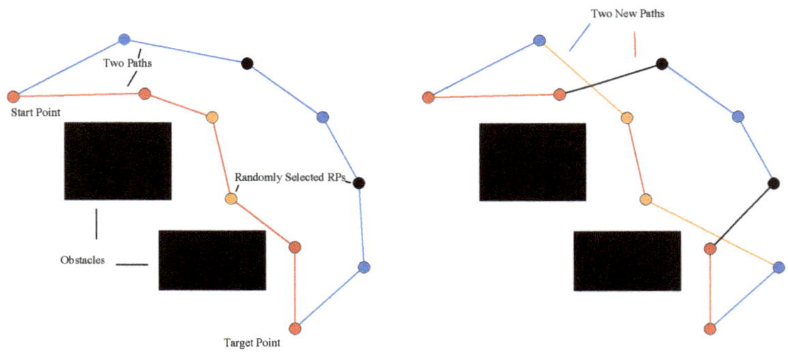

Figure 2. The crossover operator is shown. In the left sub-figure, the yellow circles are the two rotation points randomly selected on the path. Similarly, the black circles are the two rotation points on the other path. The line segments between the yellow circles are exchanged with the line segments between the black circles. Finally, two new paths are generated and displayed in the right sub-figure.

4.4. Invalid Solution Operator

The invalid solution operator is used to convert the path ($p = [S, p_1, p_2, ..., p_i, p_{i+1}, ..., p_n, T]$) that collides with the obstacles ($O = O_1, O_2, ..., O_m$) into a feasible path. The design process of this operator is expressed as follows:

Along the path, it is determined in turn whether the line segment ($\overline{p_i p_{i+1}}$) of the path collides with the obstacles. If so, the intersection points and the obstacles ($O(i)$) that collide with the line segment are found.

Then, the vertices of the obstacles ($O(i)$), the two rotation points (p_i, p_{i+1}) and the intersection points form the vertices of the graph. Any two vertices within the set of vertices are linked to form the set of edges. Meanwhile, the edges that collide with all obstacles (O) are deleted. Then, based on the vertices and edges of the graph, the Dijkstra algorithm can be applied to attain a feasible shortest path ($q = [p_i, q_1, q_2, ..., q_k, p_{i+1}]$) from p_i to p_{i+1}. Next, the line segment ($\overline{p_i p_{i+1}}$) of path (p) can be replaced with the path (q), and the path ($p' = [S, p_1, p_2, ..., p_{i-1}, q, p_{i+1}, ..., p_n, T]$) is generated. Finally, the viable path (p') is made by repeating the above operation from $i = 0$ to n. The invalid solution operator is presented in Figure 3.

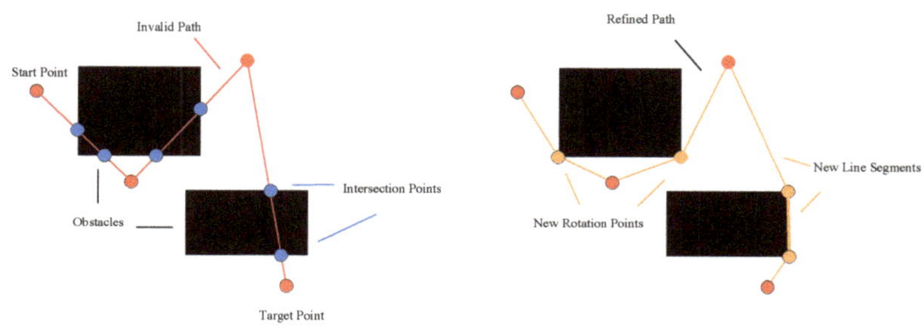

Figure 3. The invalid solution operator is presented. In the sub-figure on the left, the blue circles represent the intersection points of the path and the obstacle. The right sub-figure shows the new path generated by the Dijkstra algorithm.

4.5. Mutation Operator

The single point variation is used in this operator. A rotation point on the path is arbitrarily chosen and displaced by an arbitrary free point. It should be noted that the path after the mutation may be worse or better than before the variation. The mutation operator is displayed in Figure 4.

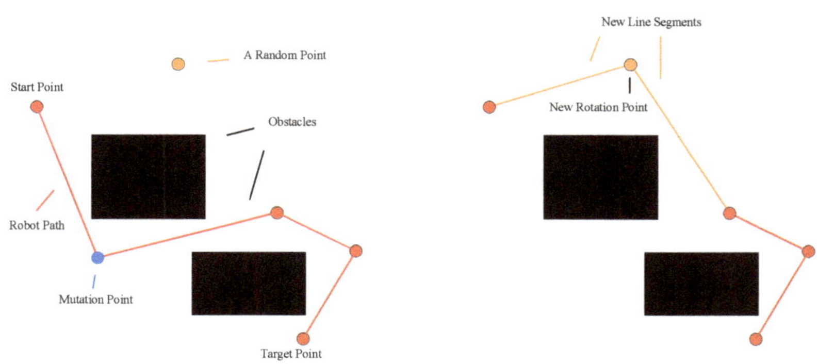

Figure 4. The mutation operator is presented. In the left sub-figure, a randomly selected rotation point (indicated by a blue circle) on the path is replaced by a free point (indicated by a yellow circle) randomly generated in the space. The resulting new path is shown in the right sub-figure.

4.6. Shortness Operator

The path is changed in arbitrary deleting manner in this operator. It randomly removes a rotation point. Figure 5 displays the shortness operator.

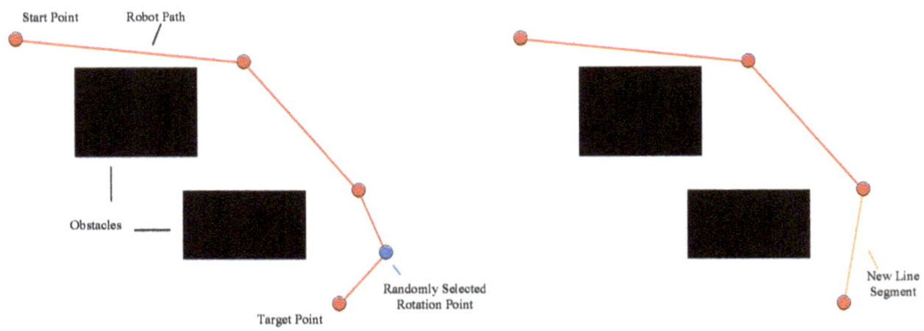

Figure 5. The shortness operator is presented. In the sub-figure on the left, a rotation point (indicated by a blue circle) on the path is randomly selected and removed. Then the new path is shown in the right sub-figure.

4.7. Insertion Operator

A single point insertion is adopted to design the insertion operator. A line segment on the path is randomly selected, and then, a random free point is inserted on the segment. The practice of the insertion operators is similar to the mutation operator. The random points in free space are used in both operators. The difference is that the mutation operator changes a rotation point and two line segments. The insertion operator increases a rotation point and converts one line segment into two new line segments. Figure 6 is an illustration of the insertion operator.

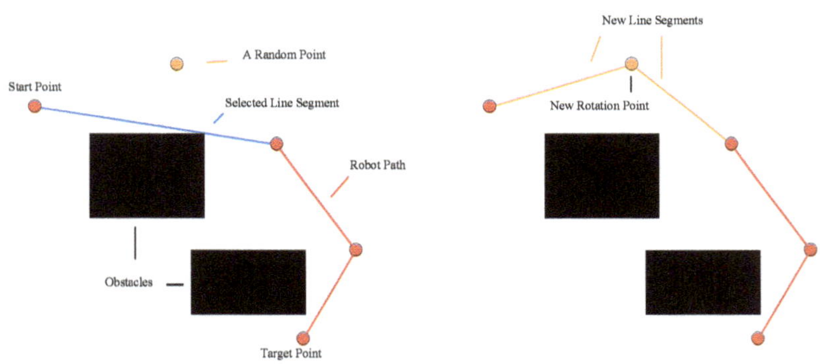

Figure 6. The insertion operator is shown. In the left sub-figure, a line segment on the path is randomly selected and indicated by a blue line. A free point is randomly generated in space and inserted into the blue line segment. The generated path is shown in the right sub-figure.

4.8. Safety Operator

The path safety is improved with the safety operator. On each line segment, the nearest point to the obstacles can be found. The point is named the critical point. Next, the region near the critical point can be meshed. Eight nearby lattices of the critical point are found. By comparing the safety distances of the lattices, the lattice with the largest safety distance is the winner. The center point of the lattice is chosen as a rotation point and then inserted into the line segment. The above method is iterated until all line segments on the route have been considered. At this point, the safety operator is over. Figure 7 shows the safety operator.

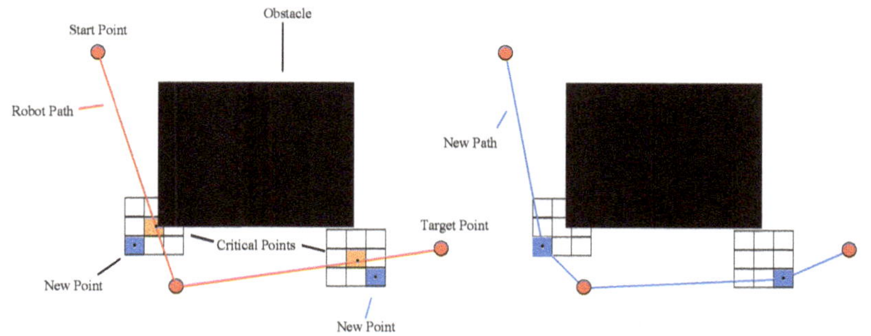

Figure 7. The safety operator is shown. In the left sub-figure, the critical point (indicated by a yellow circle) of each line segment on the path can be replaced by a point with a larger safe distance near the critical point. The sub-figure on the right shows the generated path.

4.9. Smoothness Operator

A smoother route can be found by the smoothness operator. In this operator, it is improved by decreasing the maximum turning angle formed by two consecutive segments on the path. Two free points are randomly generated on the two line segments and replace the intersection of the two line segments. Therefore, the maximum corner of the route is reduced. Figure 8 exhibits the smoothness operator.

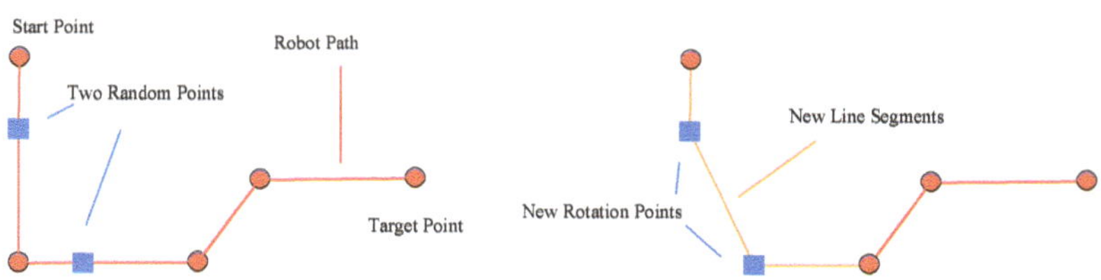

Figure 8. The smoothness operator is presented. In the left sub-figure, two free points are randomly generated on the two line segments where the maximum turning angle is located and are represented by blue squares. Then the intersection of the two segments is replaced with these two points. The right sub-figure shows the generated path.

4.10. Shortest Operator

The length of the path ($p = [S = p_0, p_1, p_2, \ldots, p_i, \ldots, p_n, p_{n+1} = T]$) can be minimized by the shortest operator. For the rotation point (p_i), another rotation point (p_j) is sequentially selected as a candidate from the target point (p_{n+1}) to the rotation point (p_{i+2}). If the new line segment ($\overline{p_i p_j}$) is feasible, other rotation points between the rotation points (p_i, p_j) are deleted. The above operation is repeated from $i = 0$ to $n - 1$. The shortest operator can remove extra rotation points to obtain the shortest path. Figure 9 presents this operator.

4.11. Position Update Operator

The idea of this operator comes from particle swarm optimization (PSO). Based on the positions of three continuing rotation points (p_{i-1}, p_i and p_{i+1}), the position of the middle point (p_i) is updated by Equation (5). This operator is shown in Figure 10.

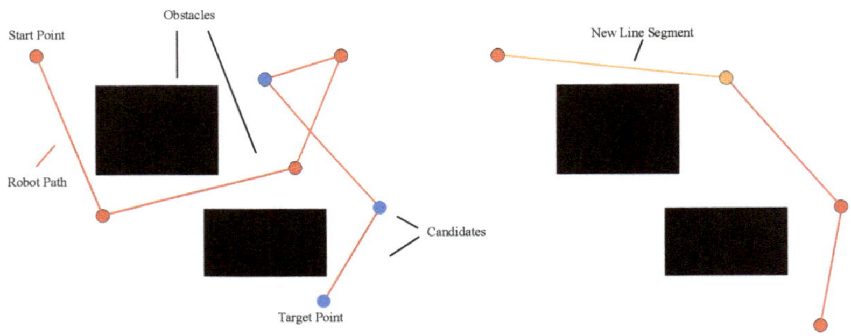

Figure 9. The shortest operator is presented. In the left sub-figure, blue circles indicate the sequentially selected candidate points. The resulting path is shown in the right sub-figure.

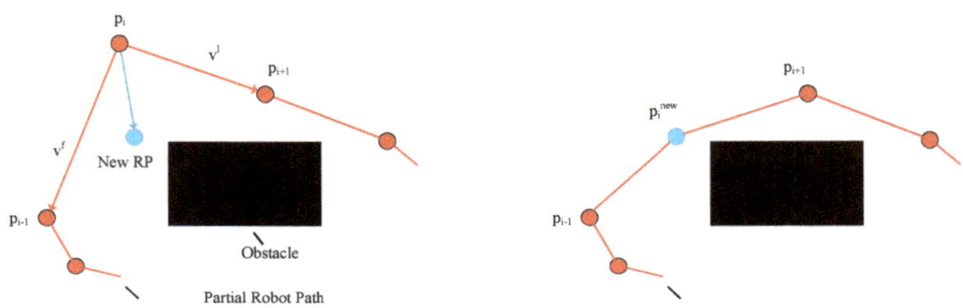

Figure 10. The position update operator is displayed. In the left sub-figure, some line segments of a path are shown. For three consecutive rotation points on the path, the new position of the middle point can be generated using Equation (5) and represented by a blue circle. The right sub-figure presents some line segments of the new path.

$$\begin{aligned} v^f &= p_{i-1} - p_i \\ v^l &= p_{i+1} - p_i \\ v &= r_1 \times v^f + r_2 \times v^l \\ p_i^{new} &= p_i + v. \end{aligned} \tag{5}$$

In Equation (5), the vector (v) should meet a restraint. The restraint is set to be $\pm 1\%$ of the workspace coordinates. In addition, the position (p_i^{new}) and the line segments ($\overline{p_{i-1}^{new} p_i^{new}}$ and $\overline{p_i^{new} p_{i+1}}$) should be feasible. The position (p_i^{new}) is adjusted continually by using Equation (6) until the above conditions are satisfied. Furthermore, r_1, r_2 and r_3 are set to random numbers from zero to one.

$$p_i^{new} = r_3 \times p_i + (1 - r_3) \times p_i^{new}. \tag{6}$$

4.12. Initialization

In general, the initialization process is essential. In the initial phase, the initial population is generated. Each within the population is a randomly generated path. That is, the rotation points on the path are random points in free space. The number of rotation points is set to a random number from one to three. Of course, these initial individuals are generally not feasible in a complicated space. Three objectives are designed to evaluate the path. They are the path length, smoothness and safety, respectively. Additionally, all parameters involved are displayed. The maximum number of generations and population size are 100 and 80 individually. The probabilities of selection, crossover and shortest operators are 1, 0.8 and 0.1. The probabilities of the remaining operators are set to 0.5. Next, through the parameter study, the parameters of the genetic algorithm are reasonably set.

5. Parametric Study

In this section, the parameters studied are population size, the number of generations, crossover rate, mutation rate, insertion rate, invalid solution operation rate, safety operator rate, smoothness rate, position update operator rate, short operator rate and shortest operator rate. The likelihood of optimality ($Lopt$) is used to evaluate the non-dominated solutions obtained by different parameter selections [57]. Based on the same parameter setting, the algorithm is run n times independently. If the optimal solution is found m times, the likelihood of optimality ($Lopt$) of this set of parameters is the estimated probability m/n. This measurement technique was originally used to evaluate the parameter setting of the single-objective optimization algorithm. For multi-objective problems, the optimal solution can be understood as the optimal values of each objective. In the d-dimensional solution space, the $Lopt_k$ of any parameters setting k is the estimated probability $(m_1 + m_2 + ... + m_d)/(d \times n)$. In this work, Let n be 100. Figure 11A is considered. The starting and target points of this map is displayed in Table 1. These maps will be used to test the algorithm in detail in Section 6.

$$Lopt_k = \frac{m_1 + m_2 + ... + m_d}{d \times n}. \tag{7}$$

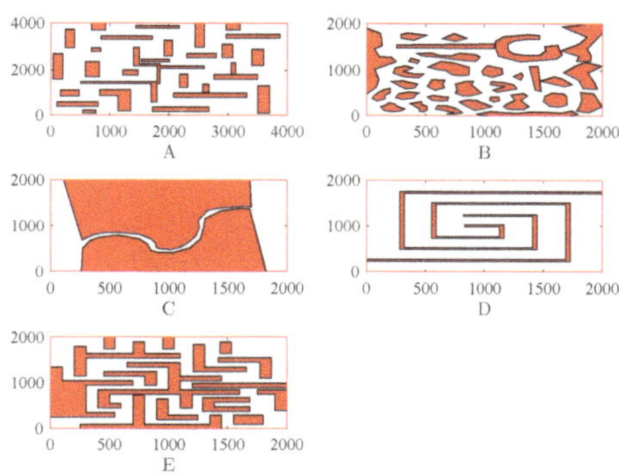

Figure 11. Five maps are used to test the algorithm. These maps have various obstacles.

Table 1. The start and destination of the maps.

Maps	Start	Destination
Figure 11A	(1500, 1200)	(3900, 3900)
Figure 11B	(400, 100)	(1450, 1450)
Figure 11C	(253.6, 403.5)	(1895.3, 1206.9)
Figure 11D	(1007.2, 1400)	(1908.6, 1008.1)
Figure 11E	(120, 120)	(1800, 1800)

Figures 12–22 present the effects of these parameters. Figure 12 presents the effect of population size. Other parameters remain the same as mentioned before. This figure shows that when the population size exceeds 60, the algorithm can run stably. For more safety, the population size is set to 80. The effect of the number of generations is shown in Figure 13. When the number of generations exceeds 80, $Lopt$ is already good. In view of this, the maximum generation of the population is set to 80. The effect of the invalid solution operator rate on $Lopt$ is given in Figure 14. The figure shows that the higher the invalid solution operator rate, the more stable the algorithm. In the next parametric study, the invalid solution operator rate is set to 0.5. Next, Figure 15 presents the effect of the safety rate.

The higher the safety operator rate, the better the *Lopt*. The value is also set to 0.5. Figure 16 shows that an excessively low crossover rate can affect the stability of the algorithm. Therefore, the crossover rate is set to 0.8. Figure 17 implies that when the shortness operator rate changes from 0.4 to one, *Lopt* is basically unchanged. The shortness operator rate is set to 0.5. The effect of smoothness operator rate is shown in Figure 18. We set the smoothness operator rate to 0.5. Figures 19–21 display the effects of the position update operator rate, insert rate and mutation rate, respectively. The effect of the shortest operator rate is presented in Figure 22. When the shortest operator rate changes from 0.1 to one, the algorithm is robust. This is different from other parameters.

The above parameter study also confirms the setting of the parameters in the initialization process. For more complex scenarios, it may be necessary to choose a higher population size, a larger number of generations, a higher invalid solution operator rate, a higher safety operator rate, a higher crossover rate and a higher smoothness operator rate.

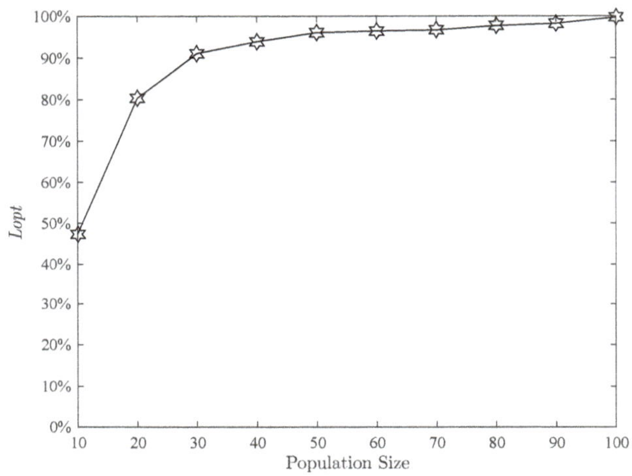

Figure 12. Effect of the population size.

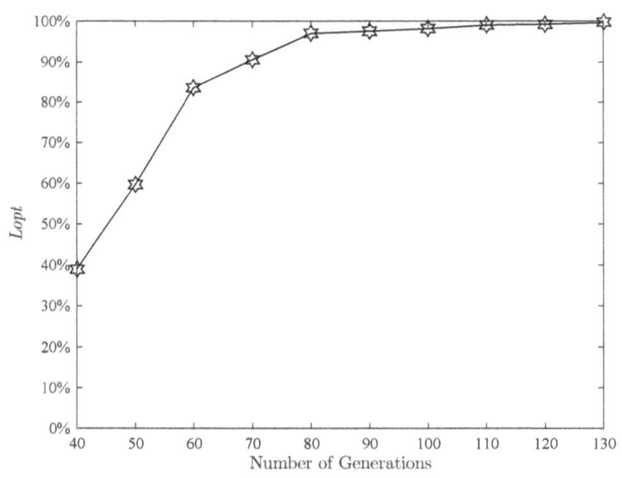

Figure 13. Effect of the number of generations.

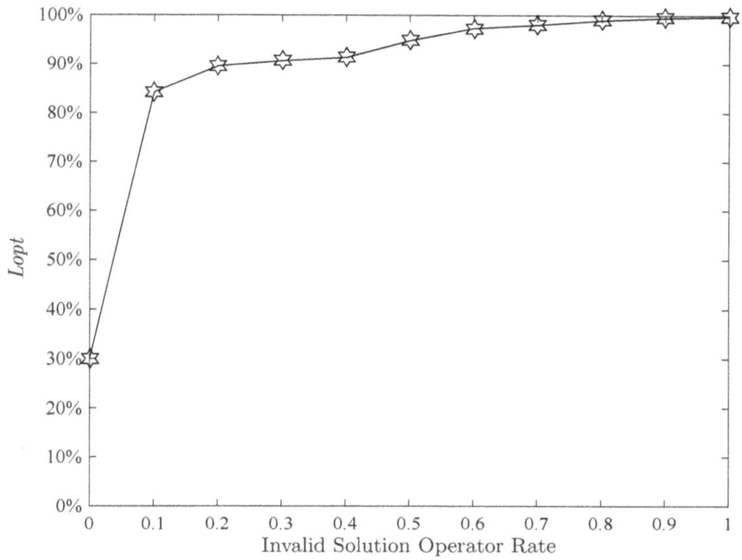

Figure 14. Effect of the invalid solution operator rate.

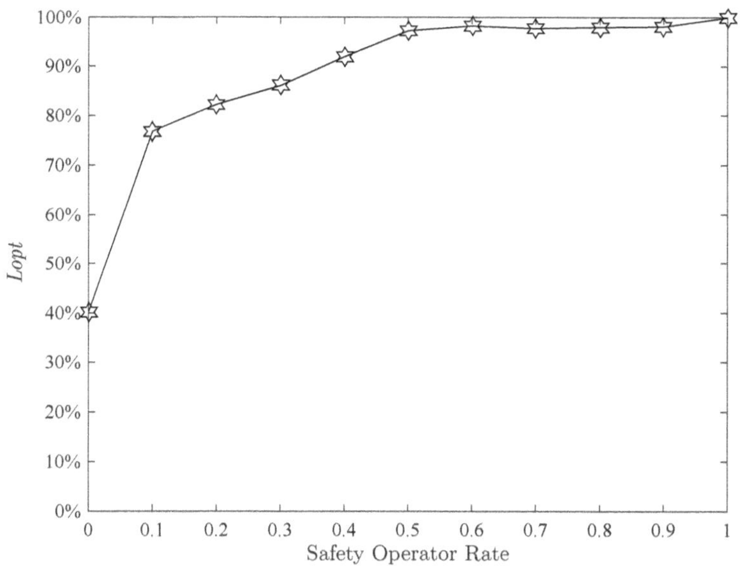

Figure 15. Effect of the safety operator rate.

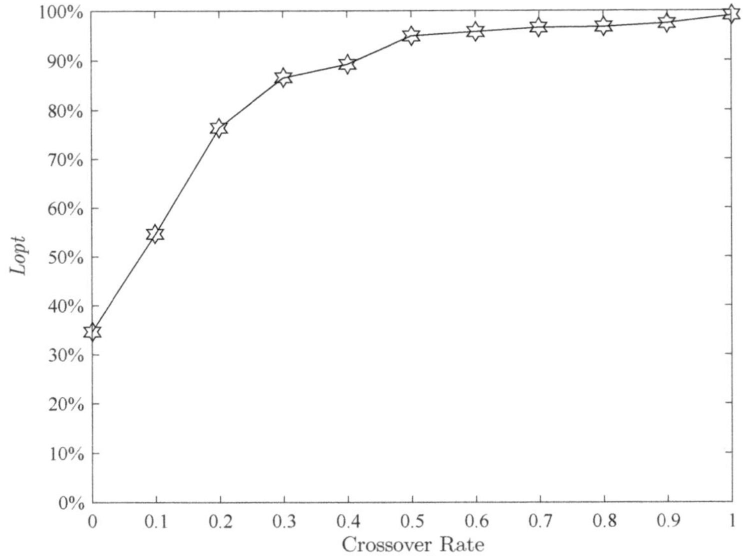

Figure 16. Effect of the crossover rate.

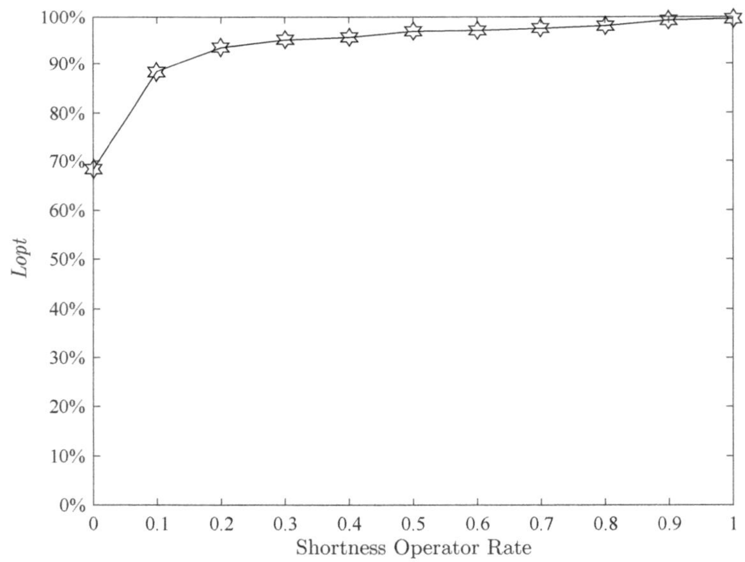

Figure 17. Effect of the shortness operator rate.

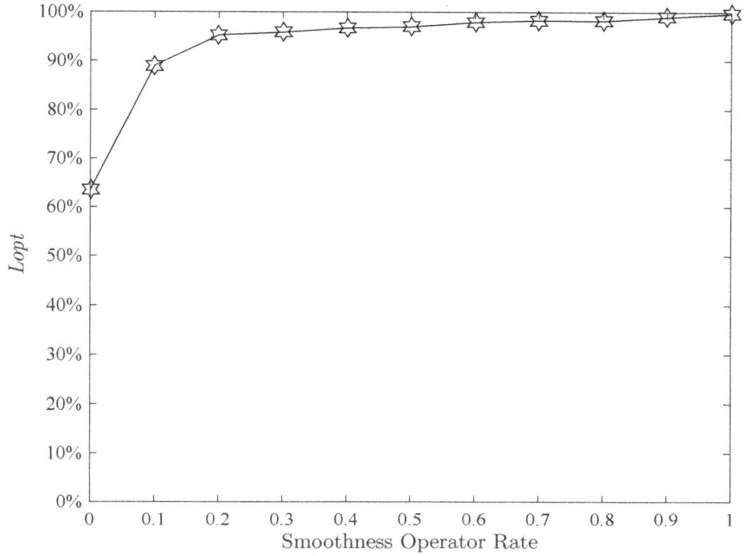

Figure 18. Effect of the smoothness operator rate.

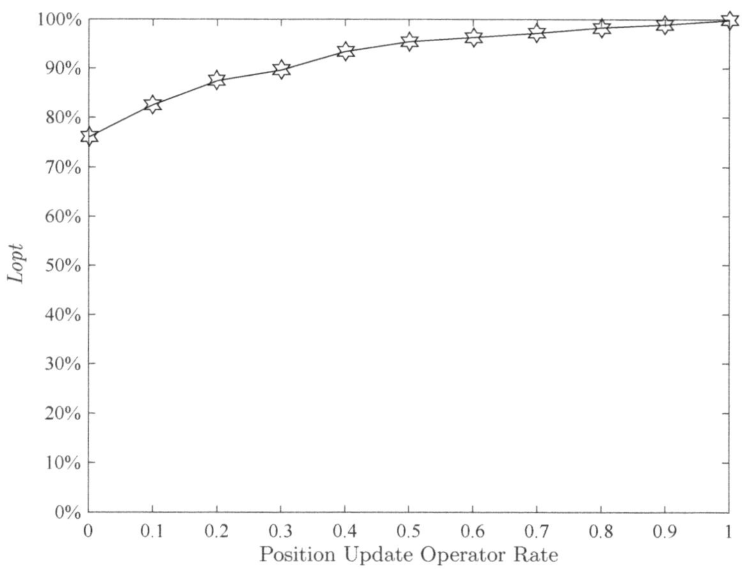

Figure 19. Effect of the position update operator rate.

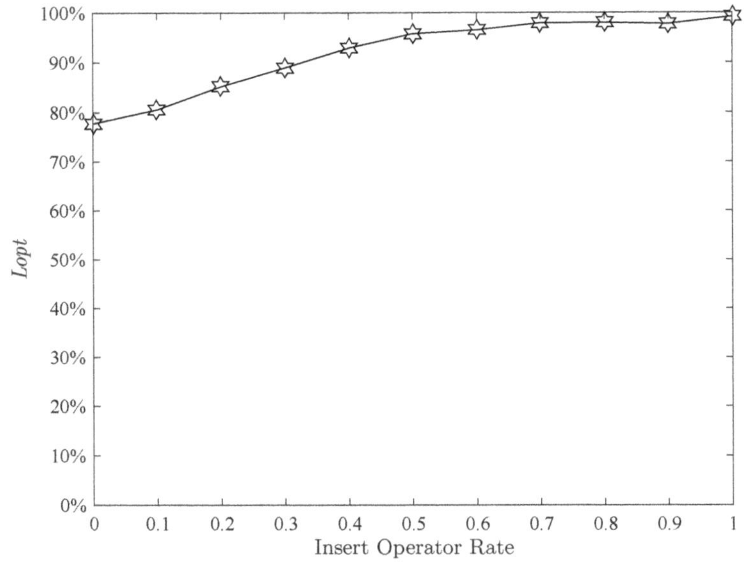

Figure 20. Effect of the insert operator rate.

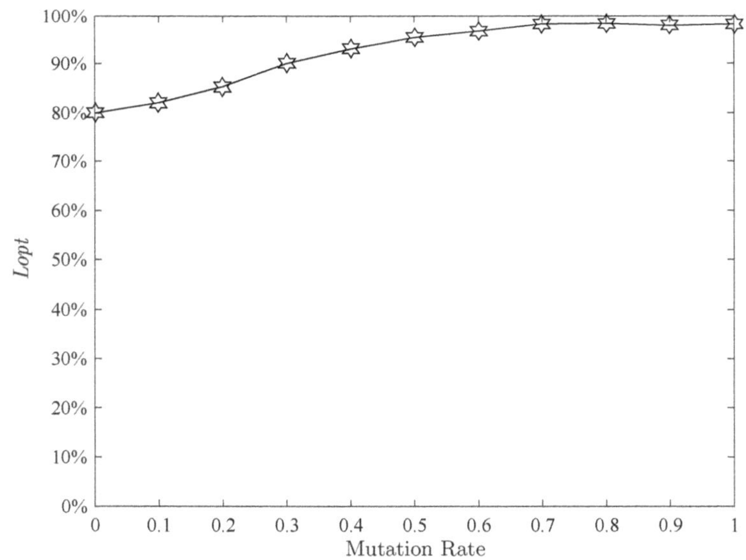

Figure 21. Effect of the mutation rate.

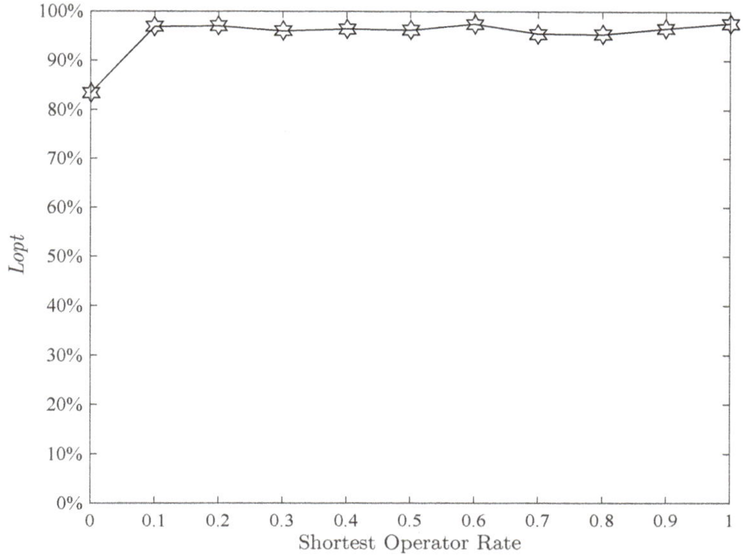

Figure 22. Effect of the shortest operator rate.

6. Results

In this work, the results of the improved non-dominated sorting genetic algorithm (NSGA-II) in various spaces are presented. To better examine the results, different quality metrics [58,59] and the multi-objective evolutionary algorithm (MOEA) algorithm [25] were employed.

The hypervolume indicator (HV) [58] can be used to evaluate the volume of non-dominated solutions in the solution space. In the d-dimensional solution space, the related concepts are defined. $A = \{a_1, a_2, ..., a_n\}$ is the non-dominated solutions, and $r = (r_1, r_2, ..., r_d)$ denotes the reference point, respectively. The Lebesgue measure [60] can be used to compute the hypervolume area of the non-dominated solutions bound by the reference point. The calculation of the HV is shown in Equation (8).

$$\begin{aligned} h(a_i) &= [a_{i_1}, r_1] \times [a_{i_2}, r_2] \times ... \times [a_{i_d}, r_d] \\ HV(A, r) &= L(\bigcup_{i=1}^{|A|} h(a_i) | a_i \in A). \end{aligned} \tag{8}$$

Furthermore, the set coverage metric (SCM) [59] can be applied for computing the dominance ratio of the two sets of non-dominated solutions. $A = \{a_1, a_2, ..., a_n\}$ and $B = \{b_1, b_2, \ldots, b_m\}$ denote the two sets. Consequently, $scm(A, B)$ can be computed with Equation (9). As the relationship ($\preceq$) represents a weak dominance relationship and is asymmetrical, it is also essential to determine $scm(B, A)$.

$$scm(A, B) = \frac{|\{b \in B | \exists a \in A : a \preceq b\}|}{|B|}. \tag{9}$$

In this work, several maps are considered in Figure 11. There are different shapes of obstacles on each map. The starting and target points for these maps are displayed in Table 1. On each map, both algorithms run 30 times. In each run, the two sets of non-dominated solutions obtained by the two algorithms are compared with the quality metrics. The results of the hypervolume are given in Table 2. In contrast, these two sets are dense. Besides, in Table 3, the results of the set coverage metric (SCM) are exhibited. The average of scm (NSGA-II,MOEA) and SCM (MOEA,NSGA-II) is 95.01% and 94.47%, respectively. It is evident that the most solutions within any set are non-dominated relative to another set. This also implies that the solutions obtained by these two algorithms are close to the Pareto front.

Table 2. Hypervolume results of NSGA-II and MOEA.

Maps	NSGA-II(Median$_{\text{Interquartile}}$)	MOEA(Median$_{\text{Interquartile}}$)
Figure 11A	$0.8873_{0.0577}$	$0.8905_{0.0495}$
Figure 11B	$0.8967_{0.0339}$	$0.8840_{0.0391}$
Figure 11C	$0.8975_{0.0837}$	$0.8934_{0.0885}$
Figure 11D	$0.8931_{0.0386}$	$0.9000_{0.0422}$
Figure 11E	$0.8963_{0.0301}$	$0.8950_{0.0322}$

Table 3. The results of the set coverage metric.

Maps	scm(NSGA-II, MOEA)	scm(MOEA, NSGA-II)
Figure 11A	0.9077	0.9032
Figure 11B	0.8974	0.8882
Figure 11C	0.9947	0.9894
Figure 11D	0.9878	0.9680
Figure 11E	0.9630	0.9747
Average	0.9501	0.9447

Additionally, in each map, the reference points need to be calculated. The ideal reference point is optimal in every dimension, and the nadir reference point is worst. Table 4 shows the reference points. An ideal shortest route is the line segment from the start to the destination. The smoothness of the ideal smoothest path should be zero. For the minimum safety of path, this value can be estimated with enhancing the safety of the safest route by 10% in the non-dominated solutions. At this point, the three values of the ideal reference point have been set. Next, the three values of the nadir reference point are set. For the length and smoothness of the nadir reference point, the two values can be calculated by referring to the above means of estimating the minimum safety. The maximum safety of path is zero.

Table 4. Reference points.

Maps	Ideal	Nadir
Figure 11A	$(3612.5, -60.5, 0)$	$(6602.0, 0, 48.5)$
Figure 11B	$(1710.3, -35.4, 0)$	$(8445.0, 0, 59.1)$
Figure 11C	$(1827.7, -15.4, 0)$	$(3545.2, 0, 31.8)$
Figure 11D	$(982.9, -88, 0)$	$(6062.9, 0, 82.5)$
Figure 11E	$(2375.9, -52.3, 0)$	$(4931.6, 0, 63.2)$

Figures 23–27 present the approximate Pareto front of maps. The yellow circle is the ideal reference point. Red triangles and black asterisks are the non-dominated solutions with the improved NSGA-II and MOEA, respectively. In non-dominated solutions, the knee spot is a solution closest to the ideal reference point. Figures 28–32 show the path of the knee point obtained via NSGA-II. Comparison results are analyzed as follows. From the two different quality metrics of hypervolume and set coverage, the improved NSGA-II demonstrates a characteristic that is no worse than MOEA. Moreover, the non-dominated solutions generated via NSGA-II are denser. Besides, the path corresponding to the knee point also takes into account several designed objectives. The path is shorter, smoother and safer.

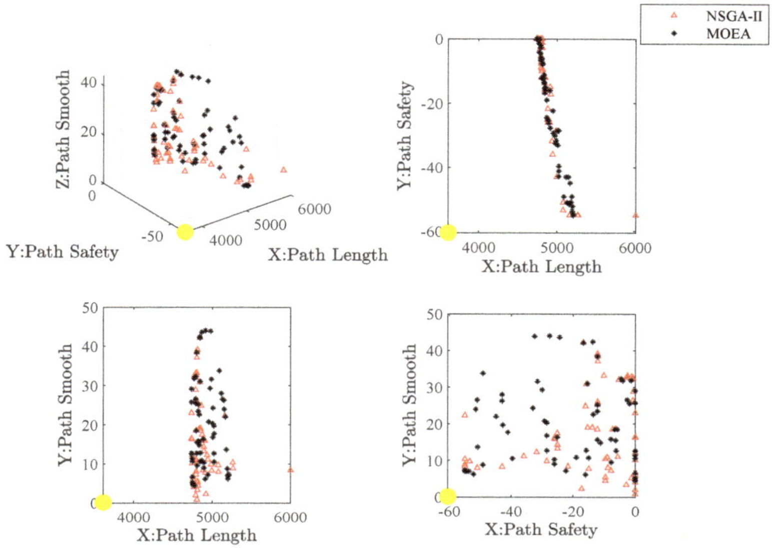

Figure 23. Based on Figure 11A, the approximate Pareto fronts generated by the two algorithms NSGA-II and MOEA. The sub-figure of the upper left corner shows the image of the approximate Pareto front in three-dimensional space. The remaining three sub-figures are the projections of the approximate Pareto front in the two-dimensional space.

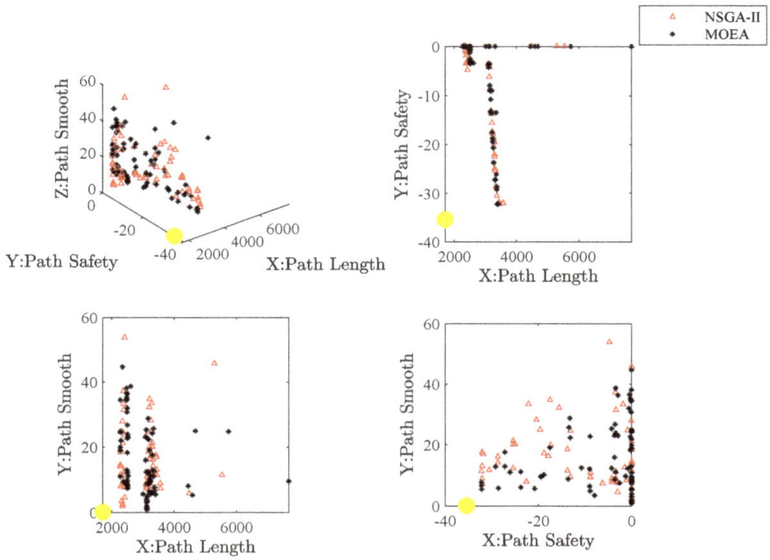

Figure 24. Based on Figure 11B, the approximate Pareto fronts generated by the two algorithms NSGA-II and MOEA. The sub-figure of the upper left corner shows the image of the approximate Pareto front in three-dimensional space. The remaining three sub-figures are the projections of the approximate Pareto front in the two-dimensional space.

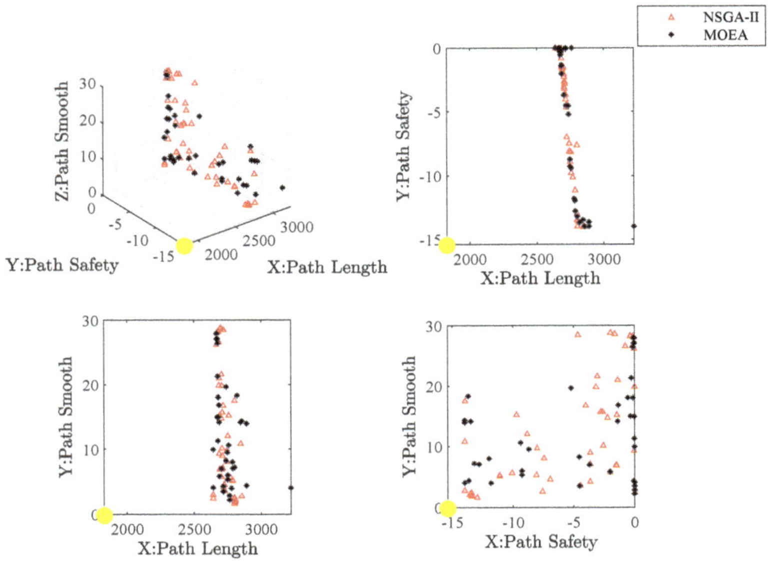

Figure 25. Based on Figure 11C, the approximate Pareto fronts generated by the two algorithms NSGA-II and MOEA. The sub-figure of the upper left corner shows the image of the approximate Pareto front in three-dimensional space. The remaining three sub-figures are the projections of the approximate Pareto front in the two-dimensional space.

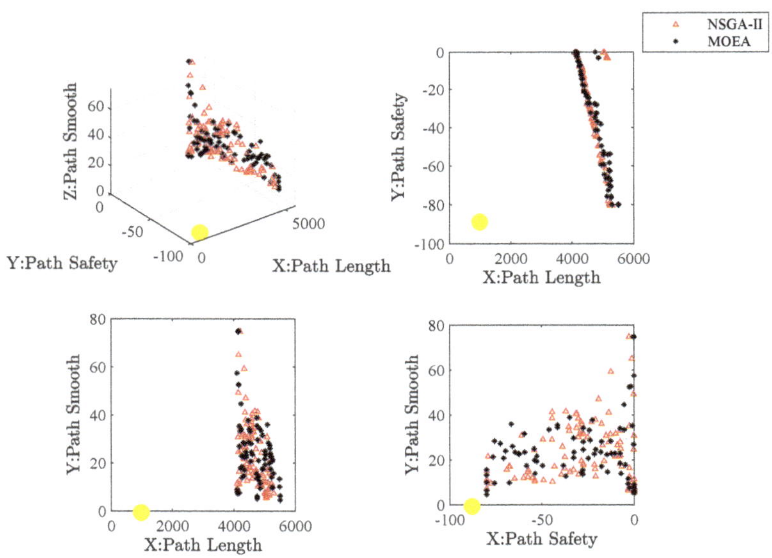

Figure 26. Based on Figure 11D, the approximate Pareto fronts generated by the two algorithms NSGA-II and MOEA. The sub-figure of the upper left corner shows the image of the approximate Pareto front in three-dimensional space. The remaining three sub-figures are the projections of the approximate Pareto front in the two-dimensional space.

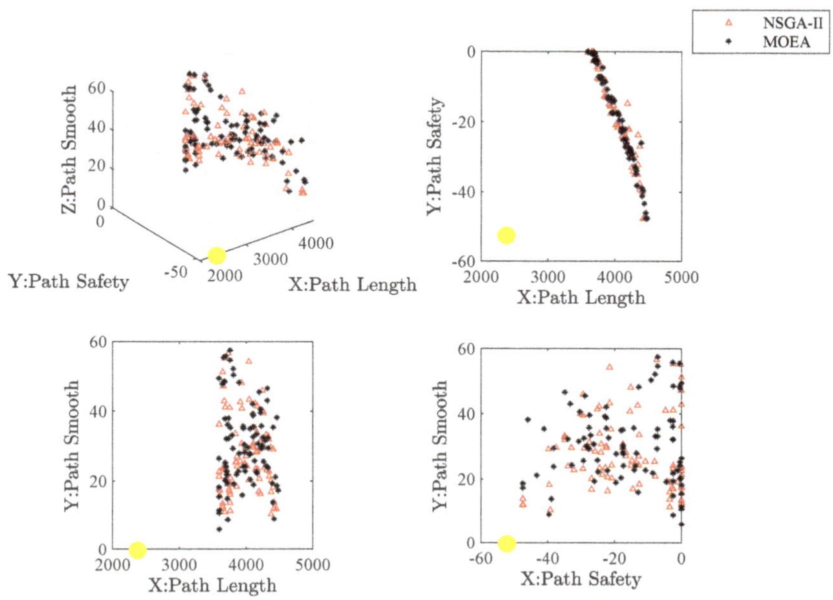

Figure 27. Based on Figure 11E, the approximate Pareto fronts generated by the two algorithms NSGA-II and MOEA. The sub-figure of the upper left corner shows the image of the approximate Pareto front in three-dimensional space. The remaining three sub-figures are the projections of the approximate Pareto front in the two-dimensional space.

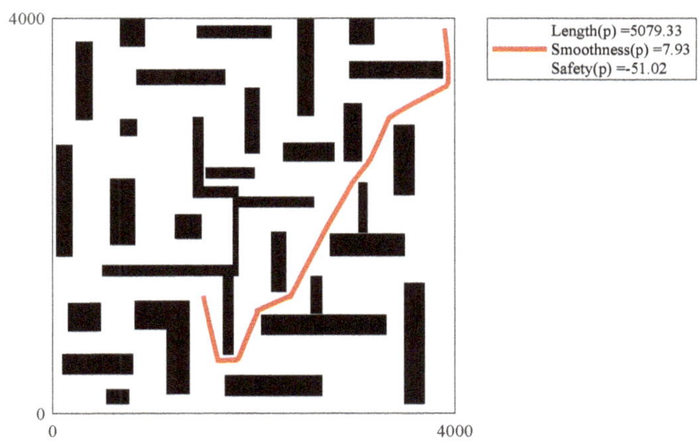

Figure 28. Path of the knee point generated via NSGA-II for Figure 11A.

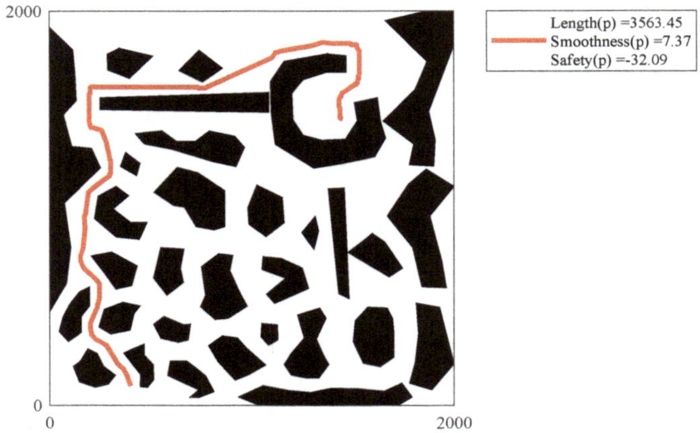

Figure 29. Path of the knee point generated via NSGA-II for Figure 11B.

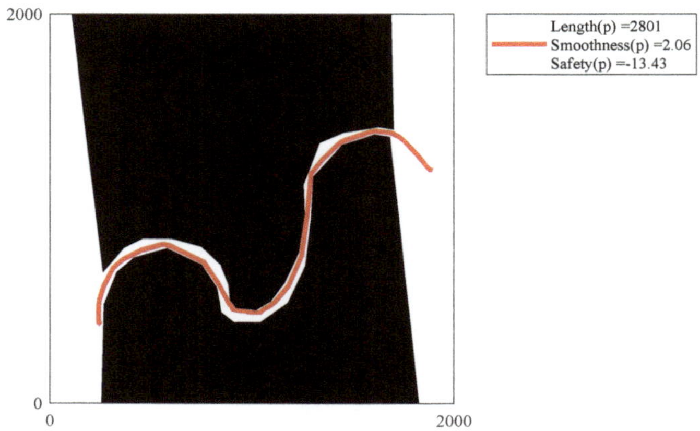

Figure 30. Path of the knee point generated via NSGA-II for Figure 11C.

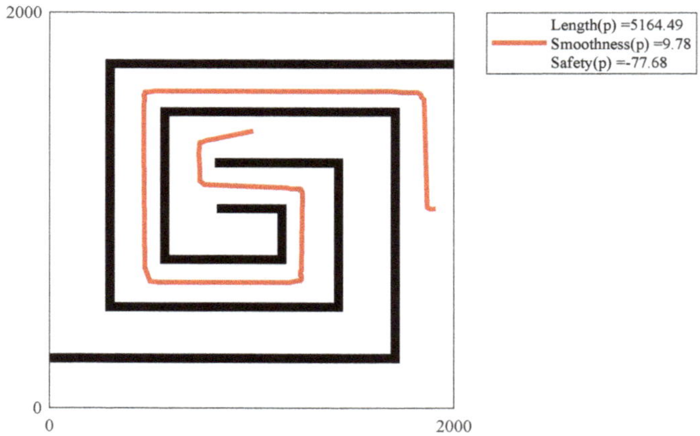

Figure 31. Path of the knee point generated via NSGA-II for Figure 11D.

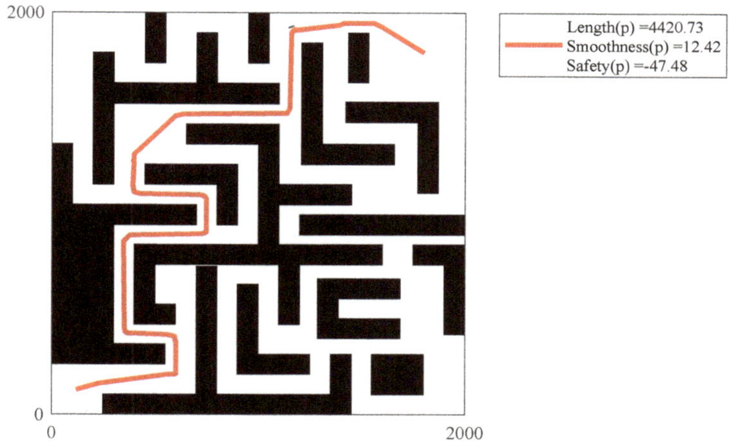

Figure 32. Path of the knee point generated via NSGA-II for Figure 11E.

7. Concluding Remarks

In this work, an improved non-dominated sorting genetic algorithm (NSGA-II) is presented to solve the multi-objective path planning problem in statically known environments. The model and the three objectives to be optimized are introduced. Based on the framework of NSGA-II, more practical operators are proposed to accelerate the evolutionary speed of individuals. These operators help individuals become shorter, safer and smoother. Then, the parameters in the algorithm are systematically studied. To discern the capabilities of the algorithm, an effective multi-objective evolutionary algorithm is employed for comparison. The set coverage metric and hypervolume are adopted. Comparison results demonstrate that the non-dominated solutions generated by the improved NSGA-II have excellent characteristics in the solution space. Finally, the path corresponding to the knee point is displayed. The path is shorter, smoother and safer. It can be adopted as the tracking path of the mobile robot in the later decision.

Due to the speed of the operators in the algorithm, it is currently only applicable to offline path planning. Moreover, the kinematics of the robot is not considered. Besides, the improved algorithm only consider the statically known environments, so the situation with dynamic obstacles in the unknown environments deserves further research.

Author Contributions: Y.X. investigated the relevant literature, proposed and implemented the algorithm, confirmed the feasibility of the algorithm, wrote and revised the paper.

Funding: This research received no external funding.

Conflicts of Interest: The author declares no conflict of interest.

References

1. Choset, H.M.; Hutchinson, S.; Lynch, K.M.; Kantor, G.; Burgard, W.; Kavraki, L.E.; Thrun, S. *Principles of Robot Motion: Theory, Algorithms, and Implementation*; The MIT Press: Cambridge, MA, USA, 2005.
2. LaValle, S.M. *Planning Algorithms*; Cambridge University Press: Cambridge, UK, 2006.
3. Mac, T.T.; Copot, C.; Tran, D.T.; De Keyser, R. Heuristic approaches in robot path planning: A survey. *Robot. Auton. Syst.* **2016**, *86*, 13–28. [CrossRef]
4. Singh, Y.; Sharma, S.; Sutton, R.; Hatton, D.; Khan, A. A constrained A* approach towards optimal path planning for an unmanned surface vehicle in a maritime environment containing dynamic obstacles and ocean currents. *Ocean Eng.* **2018**, *169*, 187–201. [CrossRef]

5. Ma, Y.; Hu, M.; Yan, X. Multi-objective path planning for unmanned surface vehicle with currents effects. *ISA Trans.* **2018**, *75*, 137–156. [CrossRef] [PubMed]
6. Alomari, A.; Phillips, W.; Aslam, N.; Comeau, F. Swarm intelligence optimization techniques for obstacle-avoidance mobility-assisted localization in wireless sensor networks. *IEEE Access* **2018**, *99*, 22368–22385. [CrossRef]
7. Alomari, A.; Phillips, W.; Aslam, N.; Comeau, F. Dynamic fuzzy-logic based path planning for mobility-assisted localization in wireless sensor networks. *Sensors* **2017**, *17*, 1904. [CrossRef] [PubMed]
8. Chen, Y.; Lu, S.; Chen, J.; Ren, T. Node localization algorithm of wireless sensor networks with mobile beacon node. *Peer-to-Peer Netw. Appl.* **2017**, *10*, 795–807. [CrossRef]
9. Algfoor, Z.A.; Sunar, M.S.; Kolivand, H. A comprehensive study on pathfinding techniques for robotics and video games. *Int. J. Comput. Games Technol.* **2015**, *2015*, 7. [CrossRef]
10. Zafar, M.N.; Mohanta, J.C. Methodology for Path Planning and Optimization of Mobile Robots: A Review. *Procedia Comput. Sci.* **2018**, *133*, 141–152. [CrossRef]
11. Zhang, H.Y.; Lin, W.M.; Chen, A.X. Path Planning for the Mobile Robot: A Review. *Symmetry* **2018**, *10*, 450. [CrossRef]
12. Radmanesh, M.; Kumar, M.; Guentert, P.H.; Sarim, M. Overview of Path-Planning and Obstacle Avoidance Algorithms for UAVs: A Comparative Study. *Unmanned Syst.* **2018**, *6*, 95–118. [CrossRef]
13. Wang, Z.; Cai, J. Probabilistic roadmap method for path-planning in radioactive environment of nuclear facilities. *Prog. Nucl. Energy* **2018**, *109*, 113–120. [CrossRef]
14. Sudhakara, P.; Ganapathy, V.; Priyadharshini, B.; Sundaran, K. Obstacle Avoidance and Navigation Planning of a Wheeled Mobile Robot using Amended Artificial Potential Field Method. *Procedia Comput. Sci.* **2018**, *133*, 998–1004. [CrossRef]
15. Wang, W.; Deng, H.; Wu, X. Path planning of loaded pin-jointed bar mechanisms using Rapidly-exploring Random Tree method. *Comput. Struct.* **2018**, *209*, 65–75. [CrossRef]
16. Duchoň, F.; Babinec, A.; Kajan, M.; Beňo, P.; Florek, M.; Fico, T.; Jurišica, L. Path Planning with Modified a Star Algorithm for a Mobile Robot. *Procedia Eng.* **2014**, *96*, 59–69. [CrossRef]
17. Ammar, A.; Bennaceur, H.; Châari, I.; Koubâa, A.; Alajlan, M. Relaxed Dijkstra and A* with linear complexity for robot path planning problems in large-scale grid environments. *Soft Comput.* **2016**, *20*, 4149–4171. [CrossRef]
18. Mrudul, K.; Mandava, R.K.; Vundavilli, P.R. An Efficient Path Planning Algorithm for Biped Robot using Fast Marching Method. *Procedia Comput. Sci.* **2018**, *133*, 116–123. [CrossRef]
19. Canny, J. *The Complexity of Robot Motion Planning*; The MIT Press: Cambridge, MA, USA, 1988.
20. Elshamli, A.; Abdullah, H.A.; Areibi, S. Genetic algorithm for dynamic path planning. In Proceedings of the Electrical and Computer Engineering, Niagara Falls, ON, Canada, 2–5 May 2004; pp. 677–680. [CrossRef]
21. Chen, X.; Li, Y. Smooth Path Planning of a Mobile Robot Using Stochastic Particle Swarm Optimization. In Proceedings of the IEEE International Conference on Mechatronics and Automation, Luoyang, China, 25–28 June 2006; pp. 1722–1727. [CrossRef]
22. Mandow, L.; De La Cruz, J.L.P. Multiobjective A* search with consistent heuristics. *J. ACM* **2010**, *57*, 27. [CrossRef]
23. Lavin, A. A pareto optimal D* search algorithm for multiobjective path planning. *arXiv* **2015**, arXiv:1511.00787.
24. Oral, T.; Polat, F. MOD* Lite: An incremental path planning algorithm taking care of multiple objectives. *IEEE Trans. Cybern.* **2016**, *46*, 245–257. [CrossRef] [PubMed]
25. Xue, Y.; Sun, J.Q. Solving the Path Planning Problem in Mobile Robotics with the Multi-Objective Evolutionary Algorithm. *Appl. Sci.* **2018**, *8*, 1425. [CrossRef]
26. Ghatee, M.; Mohades, A. Motion planning in order to optimize the length and clearance applying a Hopfield neural network. *Expert Syst. Appl.* **2009**, *36*, 4688–4695. [CrossRef]
27. Zhang, P.; Xiong, C.; Li, W.; Du, X.; Zhao, C. Path planning for mobile robot based on modified rapidly exploring random tree method and neural network. *Int. J. Adv. Robot. Syst.* **2018**, *15*. [CrossRef]
28. Konar, A.; Goswami, I.; Singh, S.J.; Jain, L.C.; Nagar, A.K. A Deterministic Improved Q-Learning for Path Planning of a Mobile Robot. *IEEE Trans. Syst. Man Cybern. Syst.* **2013**, *43*, 1141–1153. [CrossRef]
29. Chen, X.; Kong, Y.; Fang, X.; Wu, Q. A fast two-stage ACO algorithm for robotic path planning. *Neural Comput. Appl.* **2013**, *22*, 313–319. [CrossRef]

30. Châari, I.; Koubâa, A.; Bennaceur, H.; Ammar, A.; Trigui, S.; Tounsi, M.; Shakshuki, E.; Youssef, H. On the Adequacy of Tabu Search for Global Robot Path Planning Problem in Grid Environments. *Procedia Comput. Sci.* **2014**, *32*, 604–613. [CrossRef]
31. Zhu, Z.; Xiao, J.; Li, J.Q.; Wang, F.; Zhang, Q. Global path planning of wheeled robots using multi-objective memetic algorithms. *Integr. Comput.-Aided Eng.* **2015**, *22*, 387–404. [CrossRef]
32. Salmanpour, S.; Monfared, H.; Omranpour, H. Solving robot path planning problem by using a new elitist multi-objective IWD algorithm based on coefficient of variation. *Soft Comput.* **2017**, *21*, 3063–3079. [CrossRef]
33. Contreras-Cruz, M.A.; Ayala-Ramirez, V.; Hernandez-Belmonte, U.H. Mobile robot path planning using artificial bee colony and evolutionary programming. *Appl. Soft Comput.* **2015**, *30*, 319–328. [CrossRef]
34. Gong, D.W.; Zhang, J.H.; Zhang, Y. Multi-objective particle swarm optimization for robot path planning in environment with danger sources. *J. Comput.* **2011**, *6*, 1554–1561. [CrossRef]
35. Zhang, Y.; Gong, D.W.; Zhang, J.H. Robot path planning in uncertain environment using multi-objective particle swarm optimization. *Neurocomputing* **2013**, *103*, 172–185. [CrossRef]
36. Tharwat, A.; Elhoseny, M.; Hassanien, A.E.; Gabel, T.; Kumar, A. Intelligent Bézier curve-based path planning model using Chaotic Particle Swarm Optimization algorithm. *Clust. Comput.* **2018**, *2018*, 1–22. [CrossRef]
37. Mac, T.T.; Copot, C.; Tran, D.T.; De Keyser, R. A hierarchical global path planning approach for mobile robots based on multi-objective particle swarm optimization. *Appl. Soft Comput.* **2017**, *59*, 68–76. [CrossRef]
38. Han, J.; Seo, Y. Mobile robot path planning with surrounding point set and path improvement. *Appl. Soft Comput.* **2017**, *57*, 35–47. [CrossRef]
39. Goldberg, D.E. *Genetic Algorithms in Search, Optimization and Machine Learning*; Addison-Wesley: Boston, MA, USA, 1989.
40. Deb, K.; Pratap, A.; Agarwal, S.; Meyarivan, T.A.M.T. A fast and elitist multiobjective genetic algorithm: NSGA-II. *IEEE Trans. Evol. Comput.* **2002**, *6*, 182–197. [CrossRef]
41. Yao, L.; Lim, W.; Tiang, S.; Tan, T.; Wong, C.; Pang, J. Demand Bidding Optimization for an Aggregator with a Genetic Algorithm. *Energies* **2018**, *11*, 2498. [CrossRef]
42. Martínez-Bahena, B.; Cruz-Chávez, M.; Ávila-Melgar, E.; Cruz-Rosales, M.; Rivera-Lopez, R. Using a Genetic Algorithm with a Mathematical Programming Solver to Optimize a Real Water Distribution System. *Water* **2018**, *10*, 1318. [CrossRef]
43. Mahmood, A.; Khan, S.A.; Bahlool, R.A. Correction: Mahmood et al. Hard Real-Time Task Scheduling in Cloud Computing Using an Adaptive Genetic Algorithm. *Computers* **2017**, *6*, 15. *Computers* **2018**, *7*, 35. [CrossRef]
44. Ismail, A.T.; Sheta, A.; Al-Weshah, M. A Mobile Robot Path Planning Using Genetic Algorithm in Static Environment. *J. Comput. Sci.* **2008**, *4*, 341–344. [CrossRef]
45. Santiago, R.M.C.; De Ocampo, A.L.; Ubando, A.T.; Bandala, A.A.; Dadios, E.P. Path planning for mobile robots using genetic algorithm and probabilistic roadmap. In Proceedings of the Humanoid, Nanotechnology, Information Technology, Communication and Control, Environment and Management, Manila, Philippines, 1–3 December 2017; pp. 1–5. [CrossRef]
46. Davoodi, M.; Panahi, F.; Mohades, A.; Hashemi, S.N. Multi-objective path planning in discrete space. *Appl. Soft Comput.* **2013**, *13*, 709–720. [CrossRef]
47. Ahmed, F.; Deb, K. Multi-objective optimal path planning using elitist non-dominated sorting genetic algorithms. *Soft Comput.* **2013**, *17*, 1283–1299. [CrossRef]
48. Karami, A.H.; Hasanzadeh, M. An adaptive genetic algorithm for robot motion planning in 2D complex environments. *Comput. Electr. Eng.* **2015**, *43*, 317–329. [CrossRef]
49. Mittal, S.; Deb, K. Three-dimensional offline path planning for UAVs using multiobjective evolutionary algorithms. In Proceedings of the IEEE Congress on Evolutionary Computation, Singapore, 25–28 September 2007; pp. 3195–3202. [CrossRef]
50. Lee, J.; Kim, D.W. An effective initialization method for genetic algorithm-based robot path planning using a directed acyclic graph. *Inf. Sci.* **2016**, *332*, 1–18. [CrossRef]
51. Bakdi, A.; Hentout, A.; Boutami, H.; Maoudj, A.; Hachour, O.; Bouzouia, B. Optimal path planning and execution for mobile robots using genetic algorithm and adaptive fuzzy-logic control. *Robot. Auton. Syst.* **2017**, *89*, 95–109. [CrossRef]
52. Patle, B.K.; Parhi, D.R.K.; Jagadeesh, A.; Kashyap, S.K. Matrix-Binary Codes based Genetic Algorithm for path planning of mobile robot. *Comput. Electr. Eng.* **2017**, *67*, 708–728. [CrossRef]

53. Elhoseny, M.; Tharwat, A.; Hassanien, A.E. Bezier Curve Based Path Planning in a Dynamic Field using Modified Genetic Algorithm. *J. Comput. Sci.* **2018**, *25*, 339–350. [CrossRef]
54. Lamini, C.; Benhlima, S.; Elbekri, A. Genetic Algorithm Based Approach for Autonomous Mobile Robot Path Planning. *Procedia Comput. Sci.* **2018**, *127*, 180–189. [CrossRef]
55. Shehata, H.H.; Schlattmann, J. Non-dominated sorting genetic algorithm for smooth path planning in unknown environments. In Proceedings of the IEEE International Conference on Autonomous Robot Systems and Competitions, Espinho, Portugal, 14–15 May 2014; pp. 14–21. [CrossRef]
56. De Berg, M.; Van Kreveld, M.; Overmars, M.; Schwarzkopf, O. Computational geometry. In *Computational Geometry*; Springer: Berlin, Germany, 1997; pp. 1–17.
57. Sugihara, K. Measures for performance evaluation of genetic algorithms. In Proceedings of the Joint Conference on Information Sciences, Durham, NC, USA, 1–5 March 1997; pp. 172–175.
58. Bader, J.; Zitzler, E. HypE: An algorithm for fast hypervolume-based many-objective optimization. *Evol. Comput.* **2011**, *19*, 45–76. [CrossRef] [PubMed]
59. Zitzler, E.; Deb, K.; Thiele, L. Comparison of multiobjective evolutionary algorithms: Empirical results. *Evol. Comput.* **2000**, *8*, 173–195. [CrossRef] [PubMed]
60. Bartle, R.G. *The Elements of Integration and Lebesgue Measure*; Wiley Classics Library: New York, NY, USA, 1995.

applied sciences

MDPI

Article

A $\mathbb{G}^3$-Continuous Extend Procedure for Path Planning of Mobile Robots with Limited Motion Curvature and State Constraints

Tomasz Gawron * and Maciej Marcin Michałek

Institute of Automation and Robotics, Poznań University of Technology (PUT), Piotrowo 3A, 60-965 Poznań, Poland; maciej.michalek@put.poznan.pl

* Correspondence: tomasz.gawron@put.poznan.pl

Received: 19 October 2018; Accepted: 30 October 2018; Published: 2 November 2018

Featured Application: The proposed method can be applied to planning of reference paths for autonomous wheeled robots and intelligent vehicles maneuvering in cluttered environments.

Abstract: Provably correct and computationally efficient path planning in the presence of various constraints is essential for autonomous driving and agile maneuvering of mobile robots. In this paper, we consider the planning of $\mathbb{G}^3$-continuous planar paths with continuous and limited curvature in a motion environment that is bounded and contains obstacles modeled by a set of (non-convex) polygons. In practice, the curvature constraints often arise from mechanical limitations for the robot, such as limited steering and articulation angles in wheeled robots, or aerodynamic constraints in unmanned aerial vehicles. To solve the planning problem under those stringent constraints, we improve upon known path primitives, such as Reeds–Shepp (RS) and CC-steer (curvature-continuous) paths. Given the initial and final robot configuration, we developed extend-procedure computing paths that can approximate RS paths with arbitrary precision, but guaranteeing $\mathbb{G}^3$-continuity. We show that satisfaction of all stated path constraints is guaranteed and, contrary to many other methods known from the literature, the method of checking for collisions between the planned path and obstacles is given by a closed-form analytic expression. Furthermore, we demonstrate that our approach is not conservative, i.e., it allows for precise maneuvers in tight environments under the assumption of a rectangular robot footprint. The presented extend procedure can be integrated into various motion-planning algorithms available in the literature. In particular, we utilized the Rapidly exploring Random Trees (RRT*) algorithm in conjunction with our extend procedure to demonstrate its feasibility in motion environments of nontrivial complexity and low computational cost in comparison to a $\mathbb{G}^3$-continuous extend procedure based on η^3-splines.

Keywords: path planning; mobile robots; curvature constraints; state constraints; extend procedure; $\mathbb{G}^3$-continuity; car-like kinematics

1. Introduction

In this paper, we focus on the development of a path primitive and the so-called extend procedure (i.e., a local planning algorithm generating a path connecting two robot configurations), which is crucial for many path-planning algorithms utilized in the navigation of mobile robots. Despite a large body of work concerning path planning for mobile robots and autonomous vehicles (e.g., References [1,2]), this problem remains a challenge, especially in the presence of various constraints arising in practical scenarios. On one hand, mechanical limitations of the robot, such as limited steering and articulation angles in wheeled robots, or aerodynamic constraints in unmanned aerial vehicles, result in path curvature limits. On the other hand, the presence of a bounded-motion environment with obstacles and forbidden areas leads to state constraints imposed on the robot, which reduces the set of feasible paths. It is also important to maintain a high degree of path continuity to achieve smooth control of the robot and increase the comfort of passengers or the safety of the payload. To account for all these constraints, we built upon our approach introduced in Reference [3], where we proposed the extend procedure generating $\mathbb{G}^3$-continuous planar paths (that is, paths with continuous curvature derivative with respect to curve arc length) taking into account a limited curvature of motion in cluttered environments. In contrast to Reference [3], we present the new extend procedure for the carlike kinematics taking into account vehicle-body dimensions in planning collision-free paths, admitting the nonconvex polygonal obstacles present in the operational space. As a consequence, the motion-planning strategy presented in the current paper inherits beneficial properties of the original approach presented in Reference [3], but extends its potential applications to more practical path-planning scenarios.

2. Prerequisites and Problem Statement

While our considerations are quite general, meaning that the planar paths planned with our approach can be applied to various systems and planning tasks, let us consider a rear-driven carlike-vehicle kinematics as an illustratory example used throughout this paper. This is shown in Figure 1. Using results from Reference [4], one can model this system by decomposition into unicycle vehicle-body kinematics $\dot{\bar{\boldsymbol{q}}} = \boldsymbol{G}(\theta)\boldsymbol{v}$ and steering dynamics $\dot{\beta} = u_1$ as follows:

$$\dot{\bar{\boldsymbol{q}}} = \begin{bmatrix} 1 & 0 \\ 0 & \cos\theta \\ 0 & \sin\theta \end{bmatrix} \begin{bmatrix} v_1 \\ v_2 \end{bmatrix} = \boldsymbol{G}(\bar{\boldsymbol{q}})\boldsymbol{v}, \tag{1}$$

$$\dot{\beta} = u_1, \quad \boldsymbol{v} = [v_1\ v_2] \triangleq \left[\tfrac{u_2}{L}\tan\beta \quad u_2\right]^\top \in \mathbb{R}^2, \tag{2}$$
$$\boldsymbol{u} = [u_1\ u_2]^\top \in \mathbb{R}^2,$$

where $\boldsymbol{q} = [\beta\ \theta\ x\ y]^\top = [\beta\ \bar{\boldsymbol{q}}^\top]^\top = [\beta\ \theta\ \tilde{\boldsymbol{q}}^\top]^\top \in \mathcal{Q} = [-\beta_m, \beta_m] \times \mathbb{R} \times \mathcal{P}$ denotes a configuration vector with $\beta_m < \pi/2$ being a steering angle limit, $\mathcal{P} \subseteq \mathbb{R}^2$ denotes a position space, $\boldsymbol{v}$ corresponds to vehicle-body kinematics-control input, with v_1 being vehicle-body angular velocity, and v_2 corresponding to longitudinal velocity of the guidance point $\tilde{\boldsymbol{q}} = [x\ y]^\top$ illustrated in Figure 1, while $\boldsymbol{u}$ is the control input of carlike kinematics comprising steering angle rate u_1 and longitudinal velocity $u_2 \equiv v_2$. As a result of limited steering angle $\beta \in [-\beta_m, \beta_m]$, the following constraint is present:

$$|\kappa| \leq \frac{1}{L}\tan\beta_m =: \kappa_B, \tag{3}$$

where

$$\kappa \triangleq \frac{v_1}{v_2} = \frac{1}{L}\tan\beta \tag{4}$$

is the motion curvature of the robot, whereas L denotes the distance between the rear and the front axle (see Figure 1).

Let us assume that control input $\boldsymbol{u}$ is continuous, which is often desirable in practical applications. According to Equation (2), this implies that $\dot{\beta}$ is continuous. Since $\dot{\beta}$ is related to curvature $\kappa(t)$ by Equation (4), one concludes that admissible trajectories of the carlike kinematics must have curvature $\kappa(t)$ of class C^1 (with a continuous time derivative), which satisfies Equation (3). As a consequence of such requirements, admissible paths for the carlike kinematics must be $\mathbb{G}^3$-continuous, that is, for a path $\tilde{\boldsymbol{q}}_d(s) = [x_d(s)\ y_d(s)]^\top$, its curvature $\kappa_d(s)$ must be at least of class C^1. In this paper we consider three problems of planning such $\mathbb{G}^3$-continuous paths. They shall be solved under the following assumptions:

A1. Planned positional path $\tilde{\boldsymbol{q}}_d(s)$ starts at $s = 0$ and finishes at $s = s_{fin}$, i.e., $s \in [0, s_{fin}]$, $s_{fin} \geq 0$, where s corresponds to the arc-length parameter.
A2. Initial reference steering angle $\beta_d(s = 0)$ and final reference steering angle $\beta_d(s = s_{fin})$ are fixed to 0, i.e., $\beta_d(s = 0) = \beta_d(s = s_{fin}) = 0$.
A3. Initial steering rate $\frac{d\beta_d}{ds}(s = 0)$ and final steering rate $\frac{d\beta_d}{ds}(s_{fin})$ are fixed to 0, i.e., $\frac{d\beta_d}{ds}(0) = \frac{d\beta_d}{ds}(s_{fin}) = 0$.
A4. $\mathcal{P}_f \subseteq \mathcal{P}$ is a free subset of position space bounded by a single (nonconvex) polygon and containing (nonconvex) polygonal obstacles.

Assumption A1 is a consequence of not knowing the resultant path length in advance. Assumptions A2 and A3 have been taken for simplicity of considerations since they help to fix the path structure. Assumption A2 can be lifted by simply changing the initial order of path segments, as explained in the description of the proposed extend procedure. Assumption A4 is used to efficiently check for satisfaction of state constraints by the reference path.

Problem 1. *Feasible path planning in free space. Given a collision-free initial and vehicle-body configurations $\bar{\boldsymbol{q}}_{dinit}$ and $\bar{\boldsymbol{q}}_{dfin}$ plan a $\mathbb{G}^3$-continuous path $\tilde{\boldsymbol{q}}_d(s)$ admissible for the car-like kinematics, such that $\bar{\boldsymbol{q}}_d(s = 0) = \bar{\boldsymbol{q}}_{dinit}$ and $\exists s_{fin}\ \bar{\boldsymbol{q}}(s_{fin}) = \bar{\boldsymbol{q}}_{dfin}$. Curvature $\kappa_d(s)$ of path $\tilde{\boldsymbol{q}}_d(s)$, is limited as follows*

$$\forall s \in [0, s_{fin}]\ |\kappa_d(s)| \leq \kappa_B. \tag{5}$$

Problem 2. *Feasible path planning. Solve Problem 1 by planning a collision-free path, that is, the additional constraint*

$$\forall s \in [0, s_{fin}]\ \tilde{\boldsymbol{q}}_d(s) \in \mathcal{P}_f$$

shall be satisfied.

Problem 3. *Feasible path planning with rectangular footprint. Solve Problem 1 by planning a collision-free path for a robot with rectangular footprint, that is, the reference path shall additionally satisfy*

$$\forall s \in [0, s_{fin}]\ \mathcal{V}(\bar{\boldsymbol{q}}_d(s)) \in \mathcal{P}_f, \tag{6}$$

where $\bar{\boldsymbol{q}}_d(s) = [\theta_d(s)\tilde{\boldsymbol{q}}_d(s)]^\top$ is the reference vehicle-body configuration along the path, and $\theta_d(s)$ is the reference robot orientation tangent to path $\tilde{\boldsymbol{q}}_d(s)$, while $\mathcal{V}(\bar{\boldsymbol{q}}_d(s)) \subsetneq \mathbb{R}^2$ is a position space subset occupied by rectangular footprint of the robot (see Figure 1). Rectangular footprint $\mathcal{V}(\bar{\boldsymbol{q}}_d(s))$ can be expressed in local coordinate frame $\{L\}$ of reference vehicle-body configuration $\bar{\boldsymbol{q}}_d(s)$ as follows:

$$\mathcal{V}^L \triangleq \{x^L, y^L : -c \leq x^L \leq a \wedge \left|y^L\right| \leq b/2\}.$$

Note that even though we only plan a position path $\tilde{\boldsymbol{q}}_d(s)$, orientation component $\theta_d(s)$ is known due to the differential flatness of vehicle-body kinematics. The foundations for solving Problem 1 are given in Sections 4.1–4.3, whereas the final solution is presented in Section 4.4. In Section 4.5, we extend our solution to Problem 1 with collision checking, such that it is capable of solving Problem 2 when coupled with a global motion-planning algorithm (e.g., sampling-based planner). Similarly, in Section 4.6 we extended the collision-checking method to rectangular robots. In Section 4.6 we discuss applications of the proposed extend procedure, and show how it can be applied to solve Problem 3. Before fully explaining our approach, let us briefly survey current path-planning primitives in the next section.

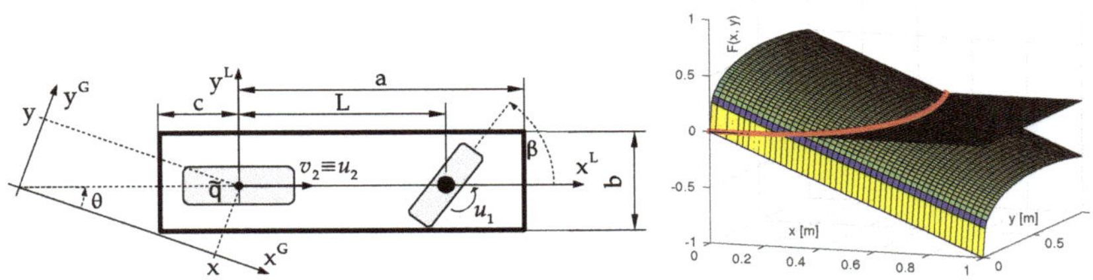

Figure 1. (**a**) A carlike robot with a rectangular footprint. (**b**) A level-curve representation of a transition path segment T, with path shown in red.

3. Related Work

Planning paths of limited curvature has been addressed with various methodologies. The most relevant properties of selected known approaches have been gathered in Table 1. A fundamental result from Reference [5] characterizes the shortest paths of bounded curvature as particular combinations of line segments and circle arcs, i.e., Reeds–Shepp (RS) paths. Note that this result does not consider obstacles, which have been partially taken into account for Dubins paths in Reference [6]. The drive toward paths with continuous curvature and a continuous-curvature arc-length derivative has led to the development of the CC-steer method [7] and its $\mathbb{G}^3$-continuous variant introduced in Reference [8]. Our work can also be viewed as a $\mathbb{G}^3$-continuous extension of the CC-steer method. However, contrary to the approach from Reference [8], our transition segments are not represented by curvature profiles. We propose transition segments with explicit representation of x as a function of y, which can easily be used with known path-following controllers (e.g., Reference [9]). We also devised a closed-form expression utilized to check for collisions between our $\mathbb{G}^3$-continuous path primitive and an obstacle, which leads to efficient and exact collision checking. There has also been some work on improving the computational cost of finding RS-like paths in Reference [10]; however, path-continuity and collision-checking issues for paths of high continuity were not explicitly addressed, to the best of our knowledge.

One can also find various methods of generating paths with a different structure to RS paths. For example, polynomial spline-based path primitives such as η^3-splines [11] and η^4-splines [12] have been developed. Their advantages are generality, conceptual simplicity, and high continuity. However, collision and curvature-constraint checking must usually be done numerically in the case of such primitives, which leads to significant computational cost. Furthermore, the parameters of such primitives can be hard to tune, even though this was partially addressed in Reference [13]. Another group of path primitives and extend procedures relies on B-splines [14–16] providing a limited curvature, curvature continuity (but not $\mathbb{G}^3$-continuity), and parameters that are easy to tune. Collision checking is done numerically in this case as well.

There are also various application-specific methods using the path primitives mentioned above, such as planning algorithms for environments with a roadlike structure similar to References [17,18], a method used to design curvature profiles of paths passing through waypoints [19], a spline-based approach exploiting sum-of-squares optimization to handle exact collision checking, or an elastic-band-like algorithm [20]. Those methods share the various benefits and drawbacks of different path primitives. However, thanks to the proposed path primitive, our approach combines $\mathbb{G}^3$-path continuity with analytically guaranteed curvature limits, as well as fast and exact analytic collision checking. To the best of our knowledge, such a combination of features is not exhibited by most known path primitives, as shown in Table 1.

Table 1. Comparison of the proposed $\mathbb{G}^3$-continuous path primitive with known path primitives.

Path Primitive	Continuity	Bounded κ	Collision Checking	Length Computation
Reeds–Shepp [5]	$\mathbb{G}^1$	yes	analytic	analytic
η^3-splines [11]	$\mathbb{G}^3$	no	numerical	numerical
η^4-splines [12]	$\mathbb{G}^4$	no	numerical	numerical
Clothoid-based	$\mathbb{G}^2$	yes	numerical	analytic
Fermat's spiral [21]	$\mathbb{G}^2$	yes	numerical	numerical
Low-order B-Splines [14,15]	$\mathbb{G}^2$	yes	numerical	numerical
Cubic curvature splines [8]	$\mathbb{G}^3$	yes	numerical	numerical
High-order B-splines	$\mathbb{G}^{3+}$	no	approx. analytic	numerical
HC-Steer [10]	$\mathbb{G}^1$	yes	analytic	numerical
$\mathbb{G}^3$-continuous path primitive	$\mathbb{G}^3$	**yes**	**analytic**	**numerical**

4. The $\mathbb{G}^3$-Continuous Extend Procedure

4.1. Main Concept

We introduce the so called $\mathbb{G}^3$-continuous path primitive, which consists of a transition segment (further explained in Section 4.2), a circle arc, a reversed transition segment, and a line segment. Similarly to the form of path encoding taken from Reference [5], we introduce a language for encoding paths with three words corresponding to path segments:

- $T(\boldsymbol{w}_1, \boldsymbol{w}_2, \mu)$ denotes a transition segment connecting $\boldsymbol{w}_1$ with $\boldsymbol{w}_2$ (defined in Section 4.2),
- $C(\boldsymbol{w}_1, \boldsymbol{w}_2)$ is a circular arc of radius $1/\kappa_c$ connecting $\boldsymbol{w}_1$ with $\boldsymbol{w}_2$,
- $S(\boldsymbol{w}_1, \boldsymbol{w}_2)$ corresponds to a straight line connecting $\boldsymbol{w}_1$ with $\boldsymbol{w}_2$,

where $\boldsymbol{w}_k \triangleq [w_{k\theta}\ w_{kx}\ w_{ky}]^\top = [w_{k\theta}\ \boldsymbol{w}_k^\top]^\top$ for $k = 1,2,3,4,5$ correspond to the reference vehicle-body configurations at endpoints of path segments. Using this encoding, one can describe the proposed $\mathbb{G}^3$-continuous path primitive connecting $\boldsymbol{w}_1$ with $\boldsymbol{w}_5$ as

$$T(\boldsymbol{w}_1, \boldsymbol{w}_2, \mu_1)C(\boldsymbol{w}_2, \boldsymbol{w}_3)T(\boldsymbol{w}_4, \boldsymbol{w}_3, \mu_2)S(\boldsymbol{w}_4, \boldsymbol{w}_5).$$

The geometric interpretation of our $\mathbb{G}^3$-continuous path primitive is shown in Figure 2 (please, note the intentionally reversed order of the arguments in $T(\boldsymbol{w}_4, \boldsymbol{w}_3, \mu_2)$ in the above formula; it simply corresponds to a reversal of the segment's endpoints). Note that, by taking transition segments of zero length, one can obtain an RS path. Depending on a choice of parameters μ_1, μ_2, one can compromise between path length and smoothness resulting from longer transition segments. Since properties of RS paths are well known and problems such as collision checking or curvature limit checking are trivial in their case, we focused on transition segments in the sequel, and show how solutions to Problems 1–3 can be obtained with their help.

The extend procedure utilizing the $\mathbb{G}^3$-continuous path primitive consists of the following main stages:

1. Choose parameters μ_1, μ_2, and curvature $\kappa_c \neq 0$.
2. Find a sequence of $\mathbb{G}^3$-continuous path primitives connecting two prescribed vehicle-body configurations.
3. Check for collisions.
4. Return computed path or a collision signal.

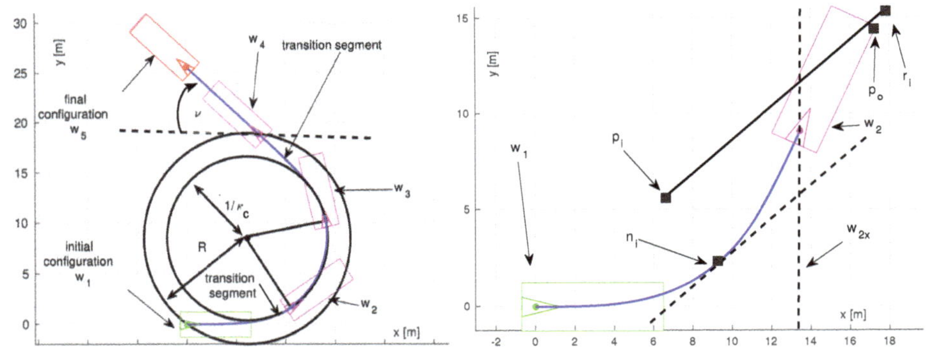

Figure 2. (**a**) Structure and parameters of the proposed $\mathbb{G}^3$-continuous path primitive. (**b**) Collision checking between transition segment and a line segment connecting $\boldsymbol{p}_i$ and $\boldsymbol{r}_i$. Only points $\boldsymbol{p}_i$ and $\boldsymbol{n}_i$ have to be checked for collisions because point $\boldsymbol{r}_i$ lies outside the domain of interest.

4.2. Transition Segments and $\mathbb{G}^3$-Continuous Path Primitives

Since computing the endpoints of a line segment and a circle arc given its radius and length is straightforward, we explicitly only define relations concerning transition segments $T(\boldsymbol{w}_1, \boldsymbol{w}_2, \mu)$. Let us denote variables expressed in a local coordinate frame fixed at point $\boldsymbol{w}_1$ by $(\cdot)^1$, that is, $\boldsymbol{w}_2^1$ corresponds to $\boldsymbol{w}_2$ expressed in coordinates of $\boldsymbol{w}_1$. A transition segment connecting vehicle-body configuration $\boldsymbol{w}_1$ with $\boldsymbol{w}_2$ is defined by the following curve, expressed in the coordinates of endpoint $\boldsymbol{w}_1$:

$$x^1 = f(y^1) = \begin{cases} \frac{-\operatorname{sgn}(w_{2x}^1)|y^1|}{2}\left[\left(\frac{y^1}{p}\right)^{\mu} - \left(\frac{y^1}{p}\right)^{-\mu}\right] & \text{for} \quad y^1 \neq 0, \\ 0 & \text{for} \quad y^1 = 0, \end{cases} \tag{7}$$

$$p \triangleq w_{2y}^1 \exp\left(\frac{\left|\operatorname{arsinh}\left(w_{2x}^1/w_{2y}^1\right)\right|}{\mu}\right), \quad w_{2y}^1 = \frac{K}{\kappa_c}y^*, \quad w_{2x}^1 = \frac{K}{\kappa_c}x^*, \tag{8}$$

with

$$K = \frac{y^*\left(-\mu + \mu^2\frac{x^*}{r}\right)}{-(r)^2\left(1+\mu^2 - 2\mu\frac{x^*}{r}\right)^{3/2}}, \quad r \triangleq \sqrt{(x^*)^2 + (y^*)^2}, \quad x^* = f(y^*),$$

$$y^* = \left(\left(\frac{4}{3}p_5 - p_3\right)^{1/3} - \frac{-6\mu^3 + \mu^2 + 2\mu + 3}{(6\mu+3)(\mu+1)^2} + p_2\right)^{1/2\mu},$$

whereas

$$
\begin{aligned}
p_1 &= -6\mu^4 - 5\mu^3 + 3\mu^2 + 5\mu + 3, \\
p_2 &= \frac{4\mu^2\left(18\mu^4 + 3\mu^3 - 17\mu^2 - 11\mu + 7\right)}{9p_1\left(\mu+1\right)^4\left(2\mu+1\right)^2}, \\
p_3 &= \frac{(p_1)^3}{27(2\mu+1)^3(\mu+1)^9} + \frac{(2\mu-1)(\mu-1)^3}{(4\mu+2)(\mu+1)^3} - p_4, \\
p_4 &= \frac{p_1(-6\mu^4 + 5\mu^3 + 3\mu^2 - 5\mu + 3)}{6(2\mu+1)^2(\mu+1)^6}, \\
p_5 &= \sqrt{\frac{\mu^6(\mu-1)^3(-36\mu^4 + 33\mu^2 - 29)}{(2\mu+1)^4(\mu+1)^9}},
\end{aligned}
$$

where y^* is a rational function of $\mu \in (0.5, 1)$, $\kappa_c \in [-\kappa_B, \kappa_B] \setminus \{0\}$ denotes the curvature of adjacent circle arc C, while $\mu \in (0.5, 1)$ is a design parameter influencing the supremum value of curvature arc-length derivative $|d\kappa_d(s)/ds|$ during the transition segment and its length, as shown in Figures 3 and 4.

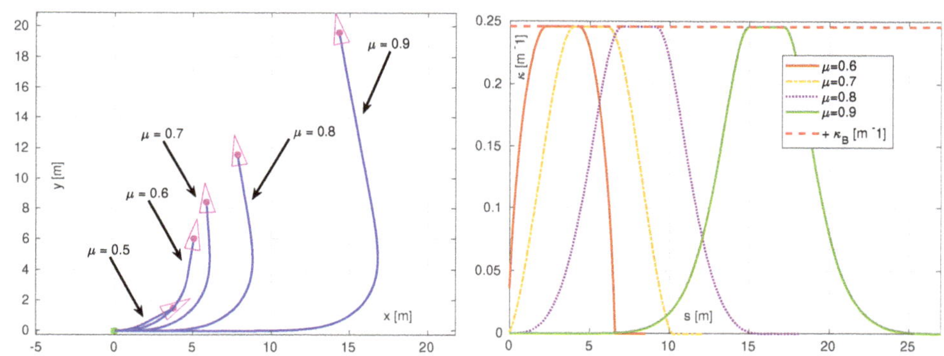

Figure 3. (**a**) $\mathbb{G}^3$-continuous path primitives for selected values of parameter μ. (**b**) Curvature profiles of selected $\mathbb{G}^3$-continuous path primitives.

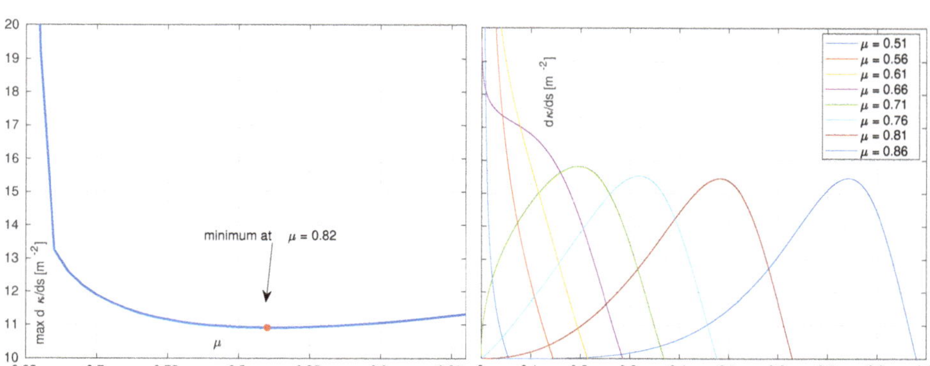

Figure 4. (**a**) Maximal arc-length derivative of transition segment curvature $\kappa_d(s)$ as a function of parameter μ. (**b**) Evolution of transition segment curvature arc-length derivative for selected values of μ.

Given a particular value of μ and transition segment endpoint $\boldsymbol{w}_1$, one can compute coordinates of the other endpoint $\boldsymbol{w}_2$ from Equation (8). The curve representing a transition segment is given by Curve (7) with particular value of parameter p resulting from Equation (8). Curve (7) was derived and analyzed in Reference [22]. It corresponds to an integral curve of the convergence vector field from the Vector Field Orientation (VFO) control law for the waypoint-following task. Its beneficial properties, proved in Reference [22], are instrumental in the construction and analysis of the $\mathbb{G}^3$-continuous path primitive. Given such a $\mathbb{G}^3$-continuous path primitive structure, we analyze its properties in the sequel.

Remark 1. *Transition segments are not explicitly parameterized by arc-length s, but they can immediately be used with path-following controllers utilizing level curves, such as the one from Reference [9].*

4.3. Path Continuity and Curvature Limit Satisfaction Analysis

Let us begin by showing that Curvature Limit (5) is satisfied. Since $\kappa_c \in [-\kappa_B, \kappa_B] \setminus \{0\}$, circle arcs and line segments satisfy the path curvature limit by construction. The curvature limit is also satisfied by both transition segments, since their curvature is limited by κ_c, as proven in Property 2 in Reference [22].

We now turn to path continuity. To ensure $\mathbb{G}^3$-continuity of the proposed primitive, one must guarantee that, for every transition segment T, the following relations hold:

$$\lim_{s\to s_1} \kappa_d(s) = 0 \quad \lim_{s\to s_1} \frac{d\kappa_d}{ds}(s) = 0, \quad \kappa_d(s_2) = \kappa_c, \quad \frac{d\kappa_d}{ds}(s_2) = 0,$$

where s_1 and s_2 are the values of s, such that $\bar{\boldsymbol{q}}_d(s_1) = \boldsymbol{w}_1$ and $\bar{\boldsymbol{q}}_d(s_2) = \boldsymbol{w}_2$. Note that a limit is sufficient to ensure $\mathbb{G}^3$-continuity in the case of point s_1, since at this point only a connection with line segments or boundary conditions of the path can occur, which guarantees continuity on the other side of the transition segment connection.

Condition $\kappa_d(s_2) = \kappa_c$ is immediately satisfied since point w^1_{2y} is defined in such a way, that it corresponds to the point of maximal curvature for the transition segment (assuming $\mu \in (0.5, 1)$, which is satisfied in our case) as shown in the proof of Property 2 in Reference [22]. Since a curvature maximum occurs at s_2, it also implies that $\frac{d\kappa_d}{ds}(s_2) = 0$ holds, because s_2 is a stationary point of $\kappa_d(s)$. In the proof of Property 2 from Reference [22], we have also shown that $\lim_{y^1\to 0} \kappa_f(y^1) = 0$ for $\mu \in (0.5, 1)$, where κ_f denotes curvature of the transition segment as a function of y^1. We conclude that this implies $\lim_{s\to s_1} \kappa(s) = 0$.

Therefore, it remains to show that $\lim_{s\to s_1} \frac{d\kappa_d}{ds}(s) = 0$. Let us recall that, by the definition of curve's curvature, a transition segment curve T has a curvature that can be represented as

$$\kappa_f(y^1) = \frac{\frac{d^2 f^1}{d(y^1)^2}}{\left(1 + \left(\frac{df}{dy^1}\right)^2\right)^{3/2}}. \tag{9}$$

One can also compute the arc-length derivative of curvature $\frac{d\kappa_f}{ds}$ as follows:

$$\frac{d\kappa_f}{ds} = \frac{d\kappa_f(y^1)}{dy^1}\frac{dy^1}{ds} = \frac{d\kappa_f(y^1)}{dy^1} sin\theta_d(s) = \frac{d\kappa_f(y^1)}{dy^1}\frac{m\,\mathrm{sgn}\,(\mu)\,y^1}{\sqrt{f(y^1)^2 + (y^1)^2}}, \tag{10}$$

where $m = \text{const} > 0$, whereas the second equality results from the consideration of Equation (1) expressed in terms of parameter s (i.e., $t = s$), and the last equality results from the definition of $\theta_d(s)$ tangent to the transition segment T, which can be found from the definition of the VFO convergence vector field as shown in, e.g., Reference [22]. Then, differentiation of Equation (9) with respect to s and substitution of the result into the final form of Equation (10) results in a relation, which, after some tedious algebra, can be written as follows

$$\begin{aligned}\frac{d\kappa_f}{ds} &= \frac{8\,\mu\,\mathrm{sgn}\,(w^1_{2x})\,(1-5\mu+(y^1)^{2\mu}\,h_1+(y^1)^{4\mu}\,h2+(y^1)^{6\mu}\,h_3)}{((y^1)^{1-2\mu}((y^1)^{2\mu}+1)^{7/2}(1-\mu^2)^5)},\\ h_1 &= -6\mu^4+6\mu^3+3\mu^2-5\mu+3,\\ h_2 &= -6\mu^4-5\mu^3+3\mu^2+5\mu+3,\\ h_3 &= 2\mu^4+7\mu^3+9\mu^2+5\mu+1.\end{aligned}$$

This clearly implies that, for $\mu > 0.5$, limit $\lim_{y^1\to 0}\frac{d\kappa}{ds}(s)=0$. As a consequence, $\lim_{s\to s_1}\frac{d\kappa}{ds}(s)=0$, which concludes our analysis and proves that the proposed primitive is $\mathbb{G}^3$-continuous.

4.4. Computing Reeds–Shepp-Like Paths Using a $\mathbb{G}^3$-Continuous Path Primitive

In our extend procedure, we assume that parameters μ_1, μ_2, and κ_c are selected or randomly sampled by the global planning algorithm. For simplicity of considerations, we also assume $\mu_1 = \mu_2$, but the proposed procedure can be trivially adapted for the case of $\mu_1 \neq \mu_2$. Finding parameters of two $\mathbb{G}^3$-continuous primitives $T(\boldsymbol{w}_1,\boldsymbol{w}_2,\mu_1)C(\boldsymbol{w}_2,\boldsymbol{w}_3)T(\boldsymbol{w}_4,\boldsymbol{w}_3,\mu_2)S(\boldsymbol{w}_4,\boldsymbol{w}_5)$ and $T(\boldsymbol{w}_5,\boldsymbol{w}_6,\mu_3)C(\boldsymbol{w}_6,\boldsymbol{w}_7)T(\boldsymbol{w}_8,\boldsymbol{w}_7,\mu_4)S(\boldsymbol{w}_8,\boldsymbol{w}_9)$ connecting prescribed vehicle-body configurations $\boldsymbol{w}_1$ and $\boldsymbol{w}_5$ is a nontrivial task. However, one can leverage extensions of the procedure devised by Reeds and Shepp, which were presented in Reference [7] and later Reference [8]. This path-construction procedure requires knowledge of auxiliary circles, on which endpoints $\boldsymbol{w}_4$ and $\boldsymbol{w}_8$ of the transition segments must lie. One must also know the difference between an orientation tangent to this circle and orientation tangent to the transition segment at endpoints $\boldsymbol{w}_4$ and $\boldsymbol{w}_8$. This difference is denoted by ν in Figure 2.

The reference path is computed during an extend procedure as follows:

1. The four possible transition segments T_1, T_2, T_3, T_4 starting at the prescribed initial vehicle body configuration are computed (Curve (7)). They correspond to forward motion with curvature κ_c, forward motion with curvature $-\kappa_c$, backward motion with curvature κ_c, and backward motion with curvature $-\kappa_c$. See Figure 5 for visual interpretation.
2. Step 1 is repeated for the four possible transition segments, T_5, T_6, T_7, T_8, ending at a prescribed final vehicle-body configuration.
3. For every transition segment $T(\boldsymbol{w}_1,\boldsymbol{w}_2,\mu)$ computed up to this step, find center $\tilde{\boldsymbol{q}}_c = [x_c\, y_c]^\top$ of the auxiliary circle, on which the next transition segment must lie according to a simple geometric formula:
$$\tilde{\boldsymbol{q}}_c = [w_{2x}\, w_{2y}]^\top + 1/\kappa_c[-\sin w_{2\theta}\, \cos w_{2\theta}]^\top.$$
4. For every transition segment $T(\boldsymbol{w}_1,\boldsymbol{w}_2,\mu)$ computed up to this step, find auxiliary circle radius R as follows:
$$R = \left\| \tilde{\boldsymbol{q}}_c - [w_{1x}\, w_{1y}]^\top \right\|.$$
5. For every transition segment $T(\boldsymbol{w}_1,\boldsymbol{w}_2,\mu)$ without a fixed value of $\boldsymbol{w}_1$, find ν, which is straightforward given the knowledge of transition segment Curve (7) and the auxiliary circle.
6. For every circle segment $C(\boldsymbol{w}_1,\boldsymbol{w}_2)$, compute its remaining unknown endpoint using the algorithm from Reference [7].
7. If motion cost J is defined, compute the cost for all the paths and choose the optimal path. Otherwise, return a random path, or all found paths (depending on the utilized global planning algorithm).

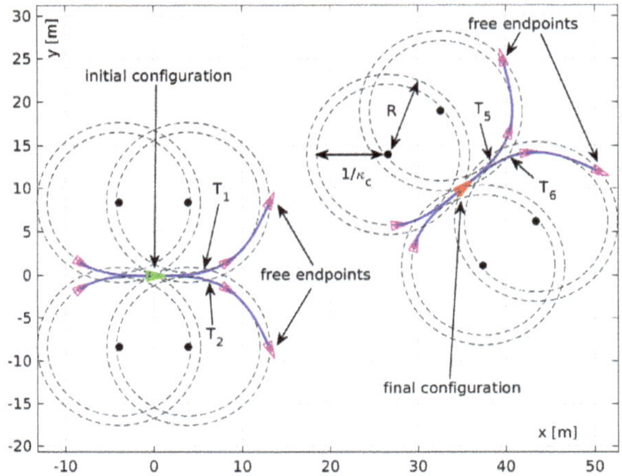

Figure 5. Visualization of the path computation procedure. Free endpoint positions must lie on auxiliary circles of radius R. Such circle arc lengths can be computed that allow for the connection of two free endpoints (i.e., endpoints of paths connected to the initial and final configuration, respectively) by a line segment without discontinuity in the reference orientation.

Note that the choice of μ should depend on the assumed motion costs. For example, if smooth paths are desired, then our computational studies summarized in Figure 4 show that choosing $\mu = 0.82$ leads to the paths with minimal $|d\kappa_d(s)/ds|$. On the other hand, for the shortest paths, choose μ close to 0.5. The proposed extend procedure solves Problem 1; however, to solve the other problems one must account for obstacles. This is solved in the next sections.

4.5. Satisfaction of State Constraints for Point Robots

Thanks to Assumption A4, set $\mathcal{P}_f$ can be described by all the obstacle edges and environment boundary edges (see Figure 2) gathered in the set of N line segments:

$$\mathcal{E} \triangleq \{(\boldsymbol{p}_i, \boldsymbol{r}_i)\}_{i=0}^{N}, \quad \boldsymbol{p}_i = [p_{ix}\, p_{iy}]^\top, \quad \boldsymbol{r}_i = [r_{ix}\, r_{iy}]^\top, \tag{11}$$

where $\boldsymbol{p}_i$ and $\boldsymbol{r}_i$ are endpoints of an i-th line segment. Condition (2) from Problem 2 is satisfied if no line segments from set $\mathcal{E}$ are crossed by the reference path. It is straightforward to analytically check this condition for circle arcs C and line segments S. We now show how to check this condition analytically for a transition segment $T(\boldsymbol{w}_1, \boldsymbol{w}_2, \mu)$ with $w_{2x}^1 > 0$, since the proposed approach is easy to generalize for other cases.

After expressing edge $\boldsymbol{p}_i$ and $\boldsymbol{r}_i$ in the coordinates of a transition segment endpoint $\boldsymbol{w}_1$, one concludes that a transition segment does not collide with line segment $(\boldsymbol{p}_i, \boldsymbol{r}_i)$ if

$$\forall k \in [0,1] \text{ and } d_x^1 \text{ such that } d_x^1 \in [0, w_{2x}^1] \quad \operatorname{sgn}\left(f(d_y^1) - d_x^1\right) = \text{const}, \tag{12}$$

where

$$\boldsymbol{d} = [d_x\, d_y]^\top = k\boldsymbol{p}_i + (1-k)\boldsymbol{r}_i.$$

This condition is illustrated in Figure 2. To explain, we consider curve f as a function and conclude that all points from the i-th line segment must lie either entirely above or below its graph. However, since the transition segment is bounded by its endpoints $\boldsymbol{w}_1$ and $\boldsymbol{w}_2$, we only consider such points from the i-th line segment that lie in subdomain $[0, w_{2x}^1]$ of curve f corresponding to the transition segment.

Condition (12) can be checked for the whole i-th line segment by simply checking it just for the maximal and minimal value of d_x^1 satisfying the left-hand-side conditions from Equation (12) and an additional critical point $\boldsymbol{n}_i \triangleq [n_{ix}\ n_{iy}]^\top$ computed according to Equation (42) from Reference [22]. At critical point $\boldsymbol{n}_i$, the orientation tangent to curve f is also tangential to the i-th line segment; thus, $\boldsymbol{n}_i$ is the point closest to or farthest from a line containing the i-th line segment. Note that, contrary to the approach from Reference [22], we check point $\boldsymbol{n}_i$ only if $\exists k \in [0, 1]$, such that $d_x = n_{ix}$.

4.6. Satisfaction of State Constraints for Robots with Rectangular Footprint

Following Reference [15], we only perform collision checking for the key points of a rectangular robot footprint. Namely, it was proven in Reference [15] that, apart from initial and final configuration, it is sufficient to check for the clearance of 0.5b m (see Figure 1) around the path and check for collisions of point $\boldsymbol{p}_o$ on the robot footprint (see Figure 2). The collision-checking procedure is performed as follows:

1. Using a simple algebra check if all footprint edges in the initial and final vehicle-body configurations are collision free.
2. Inflate the obstacles by 0.5b m. Perform the collision checking using our proposed fast method verifying condition (12).
3. Upon the instantaneous center of rotation compute the orientation θ_o tangent to instantaneous velocity of the point $\boldsymbol{p}_o$ (see Figure 2). Check for collisions of circle arcs and transition segments connected to a vehicle-body configuration $[\theta_o\ \boldsymbol{p}_o^\top]^\top$ with modified curvature $\kappa_a = \kappa_o(\kappa_c)$ instead of κ_c, where
$$\kappa_o(\kappa) = 1/\sqrt{(1/\kappa + b/2)^2 + L^2}, \tag{13}$$
where κ_o is the motion curvature of the point $\boldsymbol{p}_o$ which is furthest from the path, whereas κ is the curvature of robot motion.

As shown in Figure 6, the approach taken in Step 3 is conservative for the transition segments because by taking κ_a for collision checking, one assumes the motion curvature of point $\boldsymbol{p}_O$ to change linearly with respect to the curvature of the robot motion. However, it is also computationally efficient, and, as shown by our computational examples, its conservativeness does not hinder maneuverability of the robot in tight environments due to the ability to plan short transition segments.

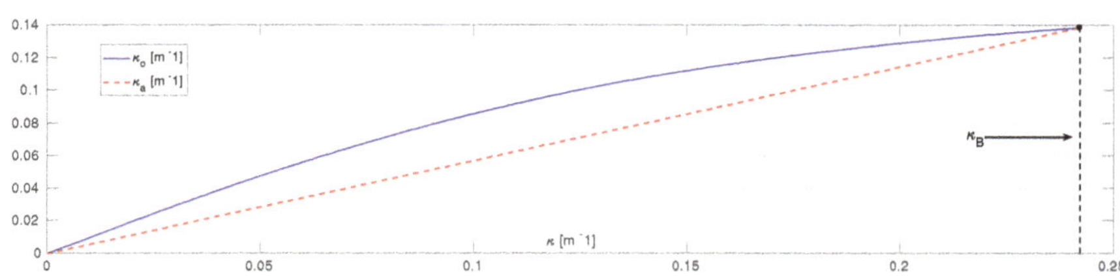

Figure 6. Curvature κ_O of motion for the outer vehicle point $\boldsymbol{p}_O$ (see Equation (13)), which is furthest from the path, expressed as a function of path curvature κ_d for $a = 5.9$ m and $b = 2.5$ m. Dashed line corresponds to a conservative inner approximation of this relation utilized during collision checking.

One can reason about the completeness of the presented extend procedure as follows. The proposed $\mathbb{G}^3$-continuous extend procedure leads to the solution of Problem 3 for its application to every complete global planning algorithm if there exists a collision-free RS path with ϵ-clearance for $\epsilon > 0$ solving this problem without the constraint of $\mathbb{G}^3$-continuity of the reference path. To illustrate why this is the case, let us consider that Equation (13) is not conservative for $\kappa(s) = 0$ and $\kappa(s) = \kappa_B$, that is, $\kappa_o = \kappa_a = 0$ and $\kappa_o = \kappa_a = \kappa_B$, respectively. Furthermore, due to Property 2 from Reference [22] and continuity of y^*, one concludes that $y^* \to 0$ as $\mu \to 0.5$. This means that, as $\mu \to 0.5$, the length of the transition segments tends to 0, and paths obtained from the concatenation of our $\mathbb{G}^3$-continuous path primitives approximate RS paths arbitrarily closely. Therefore, if there exists a feasible RS path that is no closer to obstacles than ϵ, one can always find such μ sufficiently close to 0.5, so that the maximal distance between the RS path and corresponding $\mathbb{G}^3$-continuous path is less than ϵ, meaning that the $\mathbb{G}^3$-continuous path is feasible.

5. Computational Results

To verify feasibility and effectiveness of planning with the proposed $\mathbb{G}^3$-continuous path primitive, it was investigated how it impacts computation times for the most crucial primitive operations in path planning, such as collision checking, checking of curvature-constraint satisfaction (κ checking), and extension procedure computation. Similarly, we tested the computational performance of our approach in path-planning scenarios S1–S3, shown in Figures 7–9. The proposed extend procedure was integrated with the RRT* motion-planning algorithm (see Reference [23] for details). All simulations were performed with the following parameters: $L = 5.7$ m, $b = 2.5$ m, $a = 5.9$ m, $\beta_m = 0.96$ rad. We used motion cost $J \triangleq \bar{l} + \int_0^{s_{fin}} \left(\frac{d\kappa}{ds}(s)\right)^2 ds$, where $\bar{l}$ corresponds to the total path length. Note that the planning procedure was finished when the obtained motion cost was within 5% of the value precomputed over the time of 600 s.

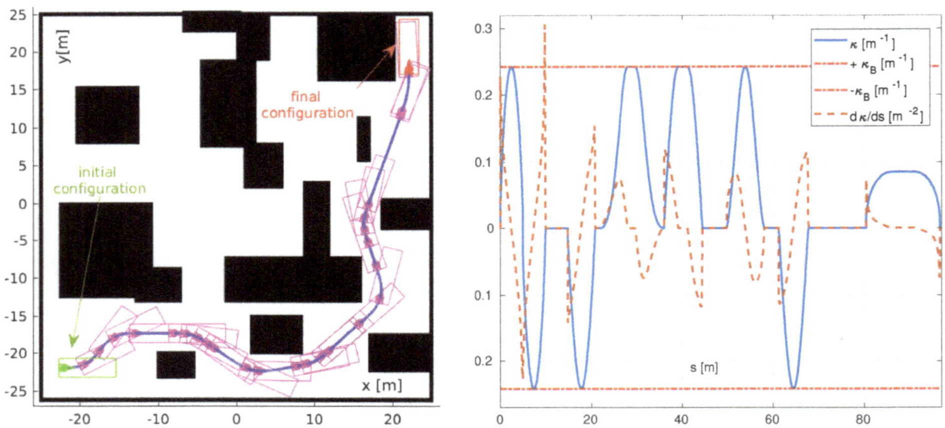

Figure 7. Planned path and curvature profile for forward parking Scenario S1. Only forward robot motion was allowed.

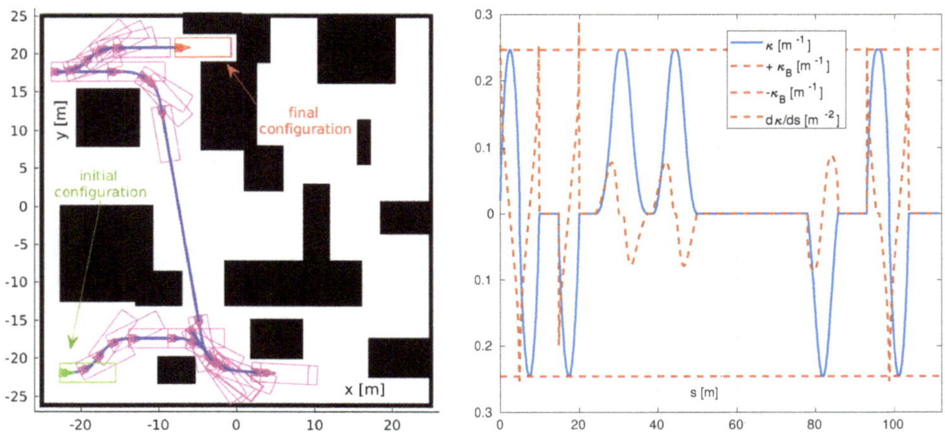

Figure 8. Planned path and curvature profile for parking Scenario S2. Both forward and backward robot motion were allowed.

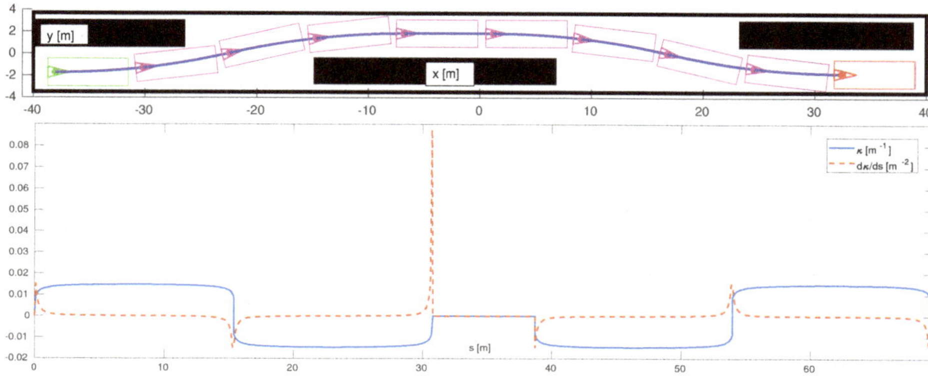

Figure 9. Planned path and curvature profile for Scenario S3 corresponding to a lane-change-like maneuver.

The computation times obtained in MATLAB are presented in Tables 2 and 3. One can observe that the $\mathbb{G}^3$ path continuity of our approach comes with the price of increased computational cost in comparison to the classic RS paths with curvature discontinuities. This was expected, since one has to specifically account for additional transition segments during collision checking, and also check a larger amount of paths due to availability of additional combinations of transition segments and circle arcs. However, we note that the computational cost for the more general η^3-splines, which are also $\mathbb{G}^3$-continuous, is significantly higher than in our approach. This observation holds even if one assumes that numerical checking of curvature constraints and collision checking can be performed in parallel. Unsurprisingly, since collision-checking procedures can account for over 90% of planning time, those computational performance tendencies propagate to computation times of a full path-planning procedure for example Scenario S2. Such results suggest that our approach can represent a viable alternative when the planning of $\mathbb{G}^3$-continuous paths is necessary due to task-specific constraints or the mechanical construction of the robot.

Table 2. Average computation times of primitive operations for 1000 random scenarios.

Path Primitive	Collision Checking (μs)	κ Checking (μs)	Extend Procedure (μs)
Reeds–Shepp	24	0	27
η^3-splines	1397	1264	63
$\mathbb{G}^3$-continuous path primitive	124	0	67

Table 3. Average computation times in MATLAB for 10 planning trials in Scenario S2.

Path Primitive	Planning Time of S2 (S)
Reeds-Shepp	27
η^3-splines	86
$\mathbb{G}^3$-continuous path primitive	39

Figures 7 and 8 illustrate the results of path planning with the proposed $\mathbb{G}^3$-continuous extend procedure for parking Scenarios S1 and S2, whereas in Figure 9 we show a challenging lane-change-like maneuver in Scenario S3. During Scenario S3, only forward robot motion was assumed admissible. Obstacles are shown in black, whereas the planned path is in blue. The magenta rectangles correspond to robot footprints at the endpoints of path segments comprising $\mathbb{G}^3$-continuous path primitives, whereas the green and red rectangles illustrate the initial and final robot configuration, respectively. It can be seen that planning is successfully performed in a severely constrained environment despite the conservative approximation utilized in our collision-checking algorithm. Curvature profiles and curvature arc-length derivative profiles are continuous; however, in some cases, e.g., in Scenario S3, relatively low μ values corresponding to relatively high curvature arc-length derivative values have been planned. This is due to the environment boundaries, which prohibited smoother curvature profiles, since those would lead to collision between the environment boundaries and the robot footprint, specifically point $\boldsymbol{p}_0$ of the footprint. One can also find that, in some cases, paths contain more reversals (changes in motion strategy), because additional space is needed for transition segments, which allow smooth evolution of path curvature. Such a tendency can be eliminated by putting a bigger emphasis on path length in the planning motion cost; however, this inevitably leads to low μ values and less smooth paths.

6. Conclusions

The $\mathbb{G}^3$-continuous path primitive proposed in this paper allows for an extension of the well-known RS paths, and constitutes an easy-to-implement component for various path planners available in the literature. The proposed method guarantees that planned paths are collision-free, satisfy curvature constraints, and preserve continuity of the curvature arc-length derivative. It is worth emphasizing the computational efficiency of the method due to the fact that distance between the robot with a rectangular footprint and obstacles can be effectively checked in a continuous domain using the derived analytical formulas. As opposed to other solutions (e.g., those using clothoid-based approaches), the introduced $\mathbb{G}^3$-continuous path primitives are represented by closed-form expressions, which can be conveniently utilized by path-following feedback controllers. Upon the results included in the paper, one may conclude that the proposed planning strategy provides all the mentioned beneficial properties under a reasonable computational cost when compared to other methods known from the literature.

Author Contributions: Conceptualization, T.G. and M.M.M.; methodology, T.G. and M.M.M.; software, T.G.; validation, T.G. and M.M.M.; writing—original draft preparation, T.G.; writing—review and editing, M.M.M.; visualization, T.G.; supervision, M.M.M.

Funding: This work was financially supported by the National Science Center, Poland via research grant No. 2016/21/B/ST7/02259.

Conflicts of Interest: The authors declare no conflict of interest. The founding sponsors had no role in the design of the study; in the collection, analyses, or interpretation of data; in the writing of the manuscript; or in the decision to publish the results.

References

1. LaValle, S.M. *Planning Algorithms*; Cambridge University Press: Cambridge, UK, 2006.
2. Paden, B.; Čáp, M.; Yong, S.Z.; Yershov, D.; Frazzoli, E. A Survey of Motion Planning and Control Techniques for Self-Driving Urban Vehicles. *IEEE Trans. Intell. Veh.* **2016**, *1*, 33–55. [CrossRef]
3. Gawron, T.; Michałek, M.M. Planning $\mathbb{G}^3$-continuous paths for state-constrained mobile robots with bounded curvature of motion. In *Trends in Advanced Intelligent Control, Optimization and Automation*; Mitkowski, W., Kacprzyk, J., Oprzędkiewicz, K., Skruch, P., Eds.; Springer International Publishing: Cham, Switzerland, 2017; pp. 473–482.
4. Michałek, M.; Kozłowski, K. Feedback control framework for car-like robots using the unicycle controllers. *Robotica* **2012**, *30*, 517–535. [CrossRef]
5. Reeds, J.; Shepp, R. Optimal paths for a car that goes both forward and backwards. *Pac. J. Math.* **1990**, *145*, 367–393. [CrossRef]
6. Agarwal, P.K.; Biedl, T.; Lazard, S.; Robbins, S.; Suri, S.; Whitesides, S. Curvature-Constrained Shortest Paths in a Convex Polygon. *SIAM J. Comput.* **2002**, *31*, 1814–1851. [CrossRef]
7. Fraichard, T.; Scheuer, A. From Reeds and Shepp's to continuous-curvature paths. *IEEE Trans. Robot.* **2004**, *20*, 1025–1035. [CrossRef]
8. Oliveira, R.; Lima1, P.F.; Cirillo, M.; Martensson, J.; Wahlberg, B. Trajectory Generation using Sharpness Continuous Dubins-like Paths with Applications in Control of Heavy-Duty Vehicles. In Proceedings of the European Control Conference (ECC), Limassol, Cyprus, 12–15 June 2018.

9. Michałek, M.M.; Gawron, T. VFO Path following Control with Guarantees of Positionally Constrained Transients for Unicycle-Like Robots with Constrained Control Input. *J. Intell. Robot. Syst.* **2018**, *89*, 191–210. [CrossRef]
10. Banzhaf, H.; Palmieri, L.; Nienhüser, D.; Schamm, T.; Knoop, S.; Zöllner, J.M. Hybrid curvature steer: A novel extend function for sampling-based nonholonomic motion planning in tight environments. In Proceedings of the IEEE 20th International Conference on Intelligent Transportation Systems, Yokohama, Japan, 16–19 October 2017; pp. 1–8.
11. Lini, G.; Piazzi, A.; Consolini, L. Multi-optimization of η^3-splines for autonomous parking. In Proceedings of the Decision and Control and European Control Conference (CDC-ECC), Orlando, FL, USA, 12–15 December 2011; pp. 6367–6372.
12. Ghilardelli, F.; Lini, G.; Piazzi, A. Path Generation Using η^4-Splines for a Truck and Trailer Vehicle. *IEEE Trans. Autom. Sci. Eng.* **2014**, *11*, 187–203. [CrossRef]
13. Bianco, C.G.L.; Gerelli, O. Generation of Paths with Minimum Curvature Derivative with η^3-Splines. *IEEE Trans. Autom. Sci. Eng.* **2010**, *7*, 249–256. [CrossRef]
14. Elbanhawi, M.; Simic, M.; Jazar, R. Continuous Path Smoothing for Car-Like Robots Using B-Spline Curves. *J. Intell. Robot. Syst.* **2015**, *80*, 23–56. [CrossRef]
15. Yoon, S.; Lee, D.; Jung, J.; Shim, D.H. Spline-based RRT* Using Piecewise Continuous Collision-checking Algorithm for Car-like Vehicles. *J. Intell. Robot. Syst.* **2018**, *90*, 537–549. [CrossRef]
16. Yang, K.; Moon, S.; Yoo, S.; Kang, J.; Doh, N.; Kim, H.; Joo, S. Spline-Based RRT Path Planner for Non-Holonomic Robots. *J. Intell. Robot. Syst.* **2014**, *73*, 763–782. [CrossRef]
17. Bertolazzi, E.; Bevilacqua, P.; Biral, F.; Fontanelli, D.; Frego, M.; Palopoli, L. Efficient Re-planning for Robotic Cars. In Proceedings of the European Control Conference (ECC), Limassol, Cyprus, 12–15 June 2018.
18. Gim, S.; Adouane, L.; Lee, S.; Dérutin, J. Clothoids Composition Method for Smooth Path Generation of Car-Like Vehicle Navigation. *J. Intell. Robot. Syst.* **2017**, *88*, 129–146. [CrossRef]
19. Parlangeli, G.; Ostuni, L.; Mancarella, L.; Indiveri, G. A motion planning algorithm for smooth paths of bounded curvature and curvature derivative. In Proceedings of the 17th Mediterranean Conference Control and Automation, Makedonia Palace, Thessaloniki, Greece, 24–26 June 2009; pp. 73–78.
20. Ahmadzadeh, A.; Jadbabaie, A.; Pappas, G.J.; Kumar, V. Elastic multi-particle systems for bounded-curvature path planning. In Proceedings of the 17th American Control Conference, Seattle, WA, USA, 11–13 June 2008; pp. 5035–5040.
21. Candeloro, M.; Lekkas, A.M.; Sørensen, A.J.; Fossen, T.I. Continuous Curvature Path Planning using Voronoi diagrams and Fermat's spirals. *IFAC Proc. Vol.* **2013**, *46*, 132–137. [CrossRef]
22. Gawron, T.; Michałek, M.M. The VFO-Driven Motion Planning and Feedback Control in Polygonal Worlds for a Unicycle with Bounded Curvature of Motion. *J. Intell. Robot. Sys.* **2018**, *89*, 265–297. [CrossRef]
23. Karaman, S.; Frazzoli, E. Sampling-based optimal motion planning for non-holonomic dynamical systems. In Proceedings of the IEEE International Conference Robotics and Automation, Karlsruhe, Germany, 6–10 May 2013; pp. 5041–5047.

applied sciences

MDPI

Article

Bio-Inspired Structure and Behavior of Self-Recovery Quadruped Robot with a Limited Number of Functional Legs

Sarun Chattunyakit [1,*], Yukinori Kobayashi [2], Takanori Emaru [2] and Ankit A. Ravankar [2]

1 Division of Human Mechanical Systems and Design, Graduate School of Engineering, Hokkaido University N13W8, Kita-ku, Sapporo, Hokkaido 060-8628, Japan

2 Division of Human Mechanical Systems and Design, Faculty of Engineering, Hokkaido University N13W8, Kita-ku, Sapporo, Hokkaido 060-8628, Japan; kobay@eng.hokudai.ac.jp (Y.K.); emaru@eng.hokudai.ac.jp (T.E.); ankit@eng.hokudai.ac.jp (A.A.R.)

* Correspondence: mr.sarun.ch@gmail.com

Received: 14 December 2018; Accepted: 20 February 2019; Published: 25 February 2019

Abstract: In this study, the authors focus on the structural design of and recovery methods for a damaged quadruped robot with a limited number of functional legs. Because the pre-designed controller cannot be executed when the robot is damaged, a control strategy to avoid task failures in such a scenario should be developed. Not only the control method but also the shape and structure of the robot itself are significant for the robot to be able to move again after damage. We present a caterpillar-inspired quadruped robot (CIQR) and a self-learning mudskipper inspired crawling (SLMIC) algorithm in this research. The CIQR is realized by imitating the prolegs of caterpillars and by using a numerical optimization technique. A reinforcement learning method called Q-learning is employed to improve the adaptability of locomotion based on the crawling behavior of mudskipper. The results show that the proposed robotic platform and recovery method can improve the moving ability of the damaged quadruped robot with a few active legs in both simulations and experiments. Moreover, we obtained satisfactory results showing that a damaged multi-legged robot with at least one leg could travel properly along the required direction. Furthermore, the presented algorithm can successfully be employed in a damaged quadruped robot with fewer than four legs.

Keywords: fault recovery; reinforcement learning; gait adaptation; legged robot; bio-inspired robot

1. Introduction

In recent years, legged robots have been widely utilized in several applications due to the fact that legged robots are more flexible than wheel-based robots in terms of mobility and energy efficiency [1]. Despite the agility and complex maneuverability of legged robots over wheeled robots, one major drawback is their inability to operate when they are damaged. In general, legged robots can function properly with predesigned controllers. However, there are some failures that occur when some parts of robots are not working, such as encoder drifting, broken legs, and joint failure [2,3]. Prior control strategies cannot be employed efficiently with the transferred models of damaged robots. Researchers have discovered and developed solutions to overcome this problem. The ability of system that can continue functioning in the presence of faults is known as fault tolerance [4]. In fault tolerant system, there are four processes to be considered which are fault detection, fault location, fault containment, and fault recovery. Fault detection is a process to determine that a fault has occurred. Fault location is a process to determine where a fault has occurred. Fault containment is a process to isolate a fault from the system. Fault recovery is a process to make the system recover from a fault. Fault-tolerant gait planning is the recovery method used with multi-legged robots after a failure has occurred and hampers its ability to walk or maintain stability [5]. According to the study in the literature, we can categorize recovery methods for damaged multi-legged robots into four main groups, namely, evolutionary-assist method, gait transition method, task-based method, and learning method [6]. Josh Bongard et al. proposed an algorithm that help damaged robot walk again [7] . In their work, the robot could start the process of identifying its current model and employing an evolutionary algorithm to determine the behavior (self-learning method) that provided the best movement. They showed that a robot with the broken legs can travel forward. However, this method was not successful in all scenarios. Other similar methods have been reported by Qiu G. al. and Liang et al. [8,9]. They employed the evolutionary methods, such as a Genetic Algorithm (GA), to create sequential actions that can guide the robot to move forward. However, these methods are time-consuming tasks. The robots are required to execute actions repeatedly until they receive the maximum outcome. The problems include not just the time spent but also the gap between the results of simulation and real experiments. Typically, the evolutionary methods require a level of numerical processing that is tough enough to perform with the hardware available on the robot body. Therefore, the researchers used a simulation model on a computer to discover the maximum-distance-traveled actions, which is the cause of the aforementioned mismatch between the simulation and the experimental results. Koos et al. presented a method that works in simulation as well as in reality [10]. Using this method, the robot can identify suitable actions in a simulation and then execute such actions on real robots. This method provided very good results in terms of moving ability and time efficiency. However, given that the method required some processing time, the same research group came up with a new idea called "the robot that can adapt like an animal" that was published by Cully et al. [11]. They used the trial-and-error technique combined with a large search space. The robot performed many possible actions in the simulation beforehand (which lasted approximately one week) and these actions were stored in the search space to be picked up when the robot needed to adapt. By using this algorithm, the robot spent less than two minutes for recovery. Even though this method was applied successfully with a hexapod robot and manipulator systems, implementation with a quadruped robot was not reported. In another work, gait transition using a central pattern generator or CPG was presented [12]. Changes in frequency are applied to CPG to help a robot overcome the limitation on movement due to damage to its body. This method uses multiple chaotic CPGs in conjunction with online learning to let a robot execute fault tolerant movement. The result shows that the robot can move along the desired path and reach its destination. However, the method uses multiple infrared and force sensors for feedback sensing. As a result, it can be applied only in specific scenarios, as reported by Ren et al. in [12]. Most existing methods do not consider legged robots with fewer than four legs. Only the algorithms proposed by Qiu et al.and Liang et al. were tested in scenarios in which the robot had only three legs [8,9]. These methods suffer from a number of pitfalls as mentioned earlier. Moreover, not only the recovery algorithm but also the robot hardware can ensure that the robot remains movable even after sustaining damage, as reported by Zhang, in which it was suggested that, with hardware improvement, the robot would become more resilient [13]. Because a self-reconfigurable robot can change its structure when it is damaged, structural design is important from the view point of ensuring that the robot can move even after it is damaged.

In this study, we focus on two key problems associated with self-recovery robots, namely, robot structure (inspired by caterpillar) and recovery method (inspired by mudskipper). We propose a novel structure of quadruped robots legs based on the behavior of a caterpillar. Caterpillar-like robots have been developed because they can perform not only crawling but also climbing. Caterpillars employ multiple prolegs, to perform crawling, as shown in Figure 1. Because a caterpillar can move forward by using its prolegs to propel its body, this concept is applied to a quadruped robot by using only one leg for motion in the case of damage. However, the structure of the leg must be designed considering the fact that the proleg can limit the reachable space of the robot legs when operating in the normal quadruped gait. In this study, a new shape of robotic legs is designed with the inspiration from caterpillar legs. The Particle Swarm Optimization (PSO) algorithm is employed to optimize the design because the optimized parameter is floating numbers. The fitness function of the PSO is set as the distance that the robot can travel with both crawling and trotting gait. The proposed robotic platform is called Caterpillar-inspired Quadruped Robot or CIQR. The process of design and performance testing are conducted in a simulation environment. The parts of the CIQR are printed using a three-dimensional (3D) printer. Afterward, the performance of the proposed quadruped platform is tested in both simulation and real experiment.

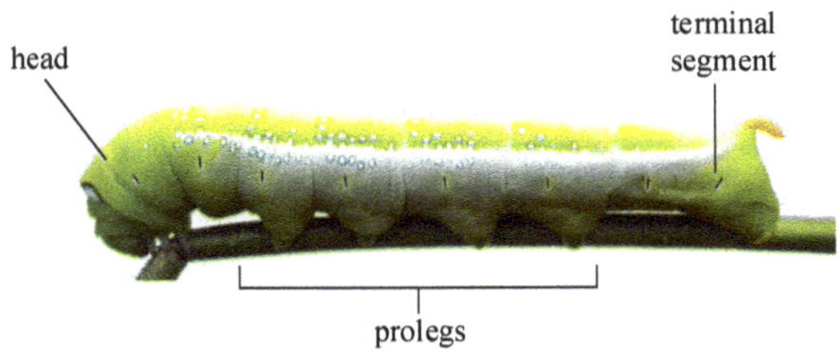

Figure 1. Caterpillar anatomy.

Moreover, a new recovery algorithm is also developed in this study. We attempted to use an evolutionary method to find the best action for the damaged robot. A similar approach published by Qiu et al. [8] was adopted, but, instead of GA, we applied the PSO algorithm. The reason for using PSO algorithm is that, unlike GA, encoding and decoding processes are not necessary for PSO algorithms. Hence, a new control method based on predefined tasks with a learning method is developed herein. We have designed our algorithm based on mudskippers' behavior. Mudskippers are fish that can crawl on mud with only two fins. We define the mudskipper model to have two degrees of freedom (2-DOFs) and create an action loop with sine and cosine functions. Additionally, the Q-learning method is integrated with mudskipper-inspired behavior to enable the robot to move faster in a more efficient manner. Both numerical simulation and practical experiment with a CIQR are conducted to test the performance of proposed recovery algorithm.

The rest of the paper is organized as follows. Section 2 describes the development of a new structural design of a quadruped robot. The system description and robot model, including forward kinematics, inverse kinematics, and robot component, is explained in this section. Afterward, optimization of robot structure inspired by caterpillar is presented. In Section 3, we present the self-recovery methods for a damaged quadruped robot. The conventional self-recovery method is presented along with the proposed method based on mudskipper-inspired behavior. Section 4 presents the results and discussion of both robotic structure design and self-recovery algorithm. The results of the optimization of the robotic structure followed by the simulation and experiments of CIQR are discussed. The evaluation of the proposed method (SLMIC) is presented with simulation and experimental results. Finally, Section 5 provides the conclusion of this study.

2. Development of Quadruped Robots

2.1. System Description and Robot Model

In this study, a new quadruped robot is developed to increase the maneuverability of ordinary-legged robots after sustaining damage. Recently, several self-recovery algorithms have been investigated successfully and implemented in robots with at least six legs. However, in most of the test cases, a maximum of two legs lost was considered, meaning that the robots still had four legs to perform any movement. By contrast, in this study, we focus on robots that have a fewer legs to begin with. Figure 2 shows the quadruped robot model considered in this study. The quadruped robot developed herein has four legs and each leg contains three links, which means that the robot has 12-DOFs.

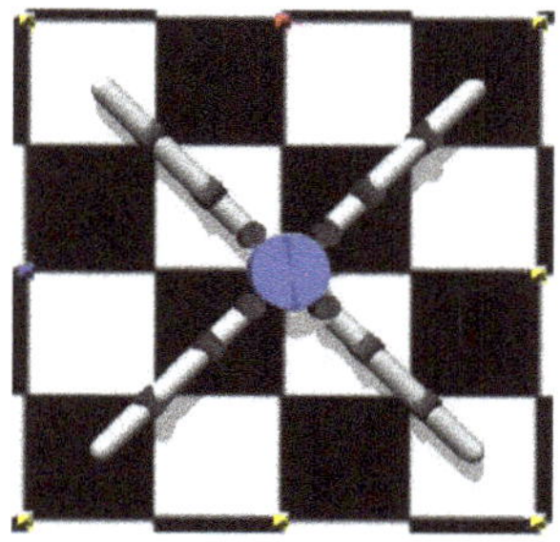

Figure 2. Quadruped robot model used in this study.

2.1.1. Forward Kinematics of Robot Legs

The forward kinematics of robot legs was analyzed using Denavit–Hartenberg parameters (also called DH parameters) [14], which are used widely for describing robotic systems and other mechanical problems [15]. In the set of DH parameters, four parameters are defined as transformation parameters, namely, d_i, θ_i, a_i, and α_i. The notation of each of the parameter can be described as follows [16]:

- d_i (joint displacement):length between the two joints.
- θ_i (joint angle): angle measured between the orthogonal of the common normals. This parameter is variable for revolute joint while the other parameters remain constant.
- a_i (link length): mathematical link length (distance between common normals).
- α_i (link twist): angle measured between the orthogonal of the joint axes. For a prismatic joint, the other parameters are fixed, but this parameter is variable.

The top and side views of the geometrical model of the robot's leg are shown in Figure 3. Moreover, the link coordinate frame is illustrated herein. As aforementioned, each leg comprises three links, namely, link1, link2, and link3. Joint 1 rotates in the horizontal direction, and joints 2 and 3 move the leg upward and downward along the vertical direction. The joint angles θ_1, θ_2, and θ_3 are the angles of link1, link2, and link3, respectively. According to the parameters shown in Figure 3, we can define the DH parameters of the systems as summarized in Table 1.

Table 1. DH parameter for robot legs.

Link	d_i	θ_i	a_i	α_i
1	0	θ_1	a_1	90
2	0	θ_2	a_2	0
3	0	θ_3	a_3	0

Position of the robot leg end-effector can be written as follows:

$$x = a_3 \cdot \cos(\theta_1)\cos(\theta_2+\theta_3) + a_2 \cdot \cos(\theta_1)\cos(\theta_2) + a_1 \cdot \cos(\theta_1), \quad (1)$$

$$y = a_3 \cdot \sin(\theta_1)\cos(\theta_2+\theta_3) + a_2 \cdot \sin(\theta_1)\cos(\theta_2) + a_1 \cdot \sin(\theta_1), \quad (2)$$

$$z = a_3 \cdot \sin(\theta_2+\theta_3) + a_2 \cdot \sin(\theta_3). \quad (3)$$

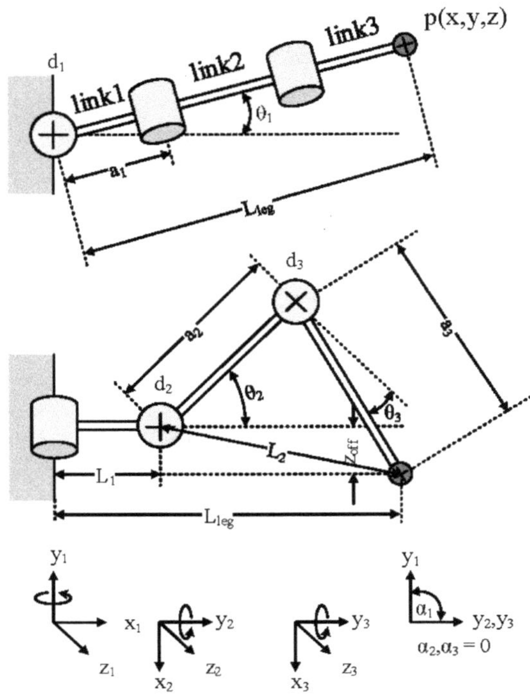

Figure 3. Geometrical model of robot legs.

2.1.2. Inverse Kinematics of Robot Legs

Forward kinematics, which is used to transform joint information into end-effector position, is explained in the previous subsection. Here, we explain how to control the joint angles of a robot leg to achieve the desired goal. In robotics and animation, this method is generally called "inverse kinematic". Several methods to obtain the inverse kinematics of a system are available, such as numerical calculation, Jacobian transpose method, and geometric approach. In this study, the geometric approach is used to solve the inverse kinematic because the robot leg has only 3-DOFs. As shown in Figure 3, we have L_{leg}, L_1, and L_2 as follows:

$$L_{leg} = \sqrt{x^2 + y^2}, \tag{4}$$

$$L_1 = a_1 \cdot \cos(\theta_1), \tag{5}$$

$$L_2 = \sqrt{(L_{leg} - L_1)^2 + Z_{off}^2}, \tag{6}$$

and the angles of the joints can be described as

$$\theta_1 = \tan^{-1}\left(\frac{y}{x}\right), \tag{7}$$

$$\theta_2 = \tan^{-1}\left(\frac{L_{leg} - L_1}{Z_{off}}\right) + \tan^{-1}\left(\frac{\sqrt{1-A^2}}{A}\right) - 90, \tag{8}$$

$$\theta_3 = 90 - \tan^{-1}\left(\frac{\sqrt{1-B^2}}{B}\right), \tag{9}$$

where A and B are $\frac{a_2^2+L_2^2-a_3^2}{2 \cdot a_2 \cdot L_2}$ and $\frac{a_2^2-L_2^2+a_3^2}{2 \cdot a_2 \cdot a_3}$, respectively.

2.1.3. Robot Components

The design of the quadruped robot is geared towards the concept of modularity so that the robot can be assembled easily. The proposed robotic platform comprises of five main components, namely, actuator, controller board, power supply, sensor, and control station. Figure 4 shows the block diagram of the overall system and the details of the connections among the components. Three components are mounted on the robot body, namely, the motors, inertial measurement unit (IMU), and controller board.

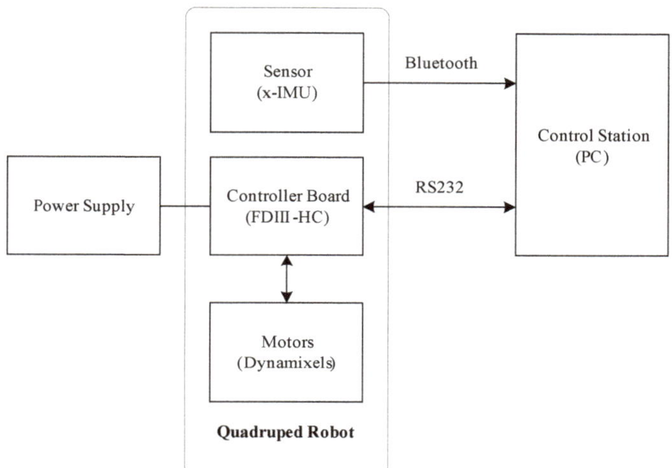

Figure 4. Block diagram of robotic system.

In this study, we used 3D printed parts for fabrication and making the robot allowing for faster prototyping and design changes. The final design was later fabricated using higher grade materials. The ABS filament was employed as the material supply in the 3D printer owning to its favorable characteristics. It is beneficial for the components that will be assembled together. Moreover, the original parts provided by Robotis™ company are integrated on the robot body.

2.2. Caterpillar-Inspired Structure

In nature, an animal can adapt itself extraordinarily for survival. Flexibility is the main key to survival and reproduction. We believe that a bio-inspired robot mechanism would be more robust. According to Trimmer et al. [17], caterpillars are excellent at climbing with their body and prolegs, and they can achieve fault-tolerant maneuverability. As a result, caterpillar-inspired robots have been researched and developed for use in many applications, for instance, wall-climbing [18]. Because the prolegs and body movements can be used to achieve forward propulsion, this concept is applied to help a quadruped robot move by using only one leg in the case of damage. In this study, the design of the CIQR is based mainly on the structure of a caterpillar's prolegs. Each robot limb is designed to have triangular shape to imitate the caterpillar's prolegs. The proleg mounted on the robot leg will increase the number of contact points between the robot and environment. Hence, the movement of robot becomes more flexible such that the robot can use prolegs or end-effector for locomotion. The parameter l is the distance from the beginning of the limb to the foot of the altitude, and h is the altitude (height) of the triangle. The quadruped robot used in this study has 3-DOFs per leg. Additionally, the prolegs are added only on the upper and lower limbs, as shown in Figure 5. The reason for using triangular prolegs is to ensure smooth robot motion.

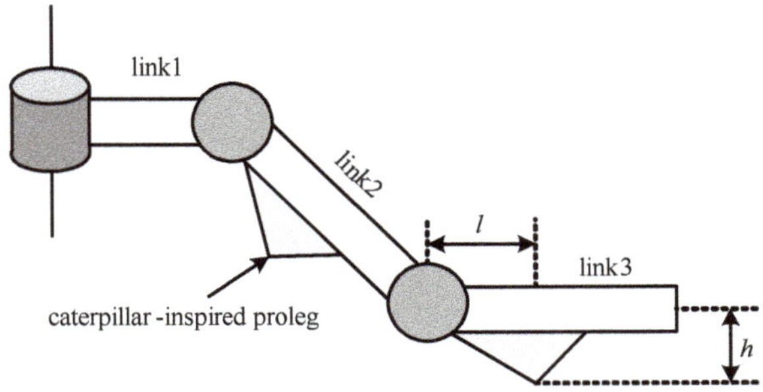

Figure 5. Leg structure of quadruped robot.

2.3. Optimization of Robot Structure

To ensure that the CIQR can perform well with both normal quadruped gait and caterpillar-inspired crawling gait, the robot structure was optimized by operating with two types of locomotion. The first type of locomotion is trotting gait to control the robot because it is the fastest gait as reported by Darici et. al. in [19]. Two pairs of diagonal legs are moved back and forth between two states, as shown in Figure 6. For the second type of locomotion, a sinusoidal generator is used to produce the rhythmic motion of caterpillar-like locomotion (crawling gait). This method can ensure smooth motion and easy control over the motion [20]. The joint rotating angle can be calculated as follows:

$$y_i = Amp_i \sin(\frac{2\pi}{T}t + \phi_i) + O_i, \tag{10}$$

where y_i is the rotation angle of joint i, Amp_i the amplitude, T the control period, t the time, ϕ_i the phase, and O_i the initial offset. Because the limb of robot is designed to be united, two parameters must be optimized, as expressed by Equation (11),

$$P = \{l, h\}, \tag{11}$$

where l and h are the parameters of the prolegs of link2 and link3, as illustrated in Figure 5. The quadruped robot is programmed to execute the crawling and the trotting gaits to ensure that the designed model can move properly when crawling using one leg and when performing the normal quadruped gait. Therefore, the fitness function is set as the traveling distance, as follows:

$$D = \mu_1 \cdot d_c + \mu_2 \cdot d_t, \tag{12}$$

where μ_1 and μ_2 are weight parameters and d_c and d_t are the distance traveled by robot with caterpillar-crawling and trotting gait, respectively.

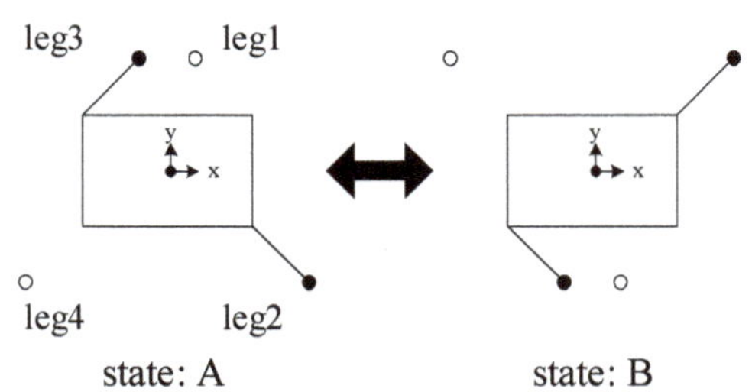

Figure 6. Trotting gait.

3. Self-Recovery Method

3.1. Conventional Self-Recovery Method

We adopted a modified version of the work described in [8,9]. PSO was employed over GA due to its simplistic modeling and optimization convergence. The method can be described into three main parts, namely, movement sequence coding, objective function, and evolutionary process.

3.1.1. Movement Sequence Coding

As aforementioned, we employed smart electric actuators, Dynamixel™(ROBOTIS, Seoul, Korea), to control the joints to the programmed angles. The PSO particles were designed based on the idea of sequence actions. Each motor is required to perform five sequential actions. The robot performs each action for t seconds ($t = 0.2$). Every joint is driven by a self-desired action at the same time. Hence, we can define the actions of each motor as follows:

$$S_i = [Ang_{i1}, Ang_{i2}, \ldots, Ang_{i5}], \tag{13}$$

where S is the sequence of actions, i the number id of joint motor, and Ang the control motor angle. Then, the PSO particles (P) can be described using Equation (14):

$$P = [S_1, S_2, \ldots, S_N], \tag{14}$$

where N is the number of joints of the legged robot. The proposed robot has 12-DOFs so N is equal to 12. The overall parameters of one particle are $5 \times 12 = 60$ values. An example of robot movement is shown in Figure 7. The robot performs five actions iteratively until it completes the desired task.

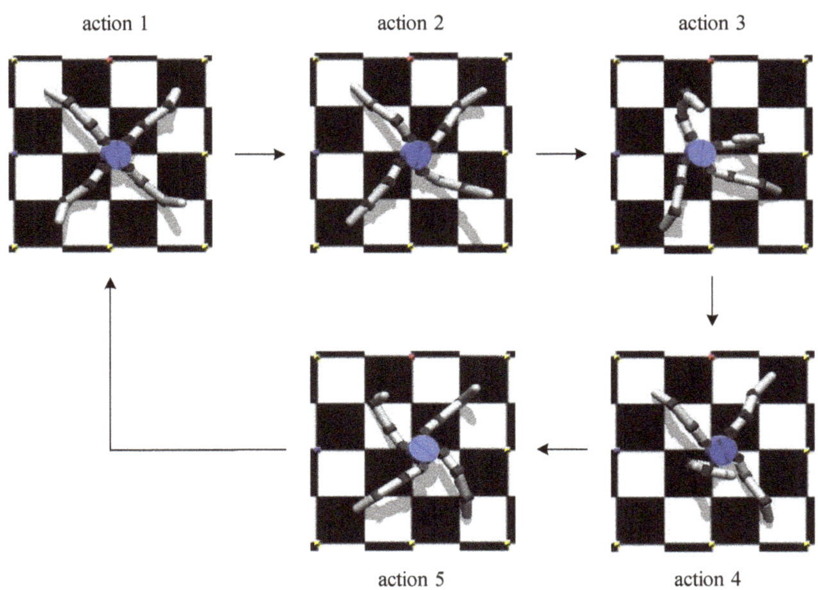

Figure 7. Example of robot movement with sequential actions. Robot will start executing from action 1 to action 5 continuously in round-robin fashion.

3.1.2. Objective Function

The criteria to achieve the recovery movement of a damaged robot can be set in a form of an objective function. This function is used to judge which solutions can provide the best result. To overcome damage, movability is the main objective of the recovery process. If the robot cannot move properly, it will be impossible to fix the robot. By contrast, if the robot can move to the desired position, we can repair the broken parts of the robot and can send it back to complete its mission. Accordingly, we set movability as the travel distance in a given direction. The damaged robot that can move faster to the required position is said to have the maximum of movability. The distance traveled by the robot can be calculated after the robot moves for T seconds. Figure 8 shows the performance of the broken robot traveling from the original position to the final position. The distance traveled D can be determined using a basic geometrical approach as follows:

$$D = \sqrt{(x_f - x_o)^2 + (y_f - y_o)^2}, \quad (15)$$

where x_o and x_f are the position of the robot along the x-axis of the original and the final position, respectively. In addition, y_o and y_f are the robot position along the y-axis of the original and the final positions, respectively. To follow the required direction, the parameter θ_f is put mathematically in the objective function F to decrease its value in case if the robot travels the wrong distance. In addition, we focus on straight line motion in this study. If θ_f becomes 0, it means that the robot has traveled straight and F is maximized:

$$F = e^{-\theta_f^2} \cdot D, \quad (16)$$

where θ_f denotes the different angles between the original and the final positions. Moreover, the calculated value of F is used to process the evolutionary method.

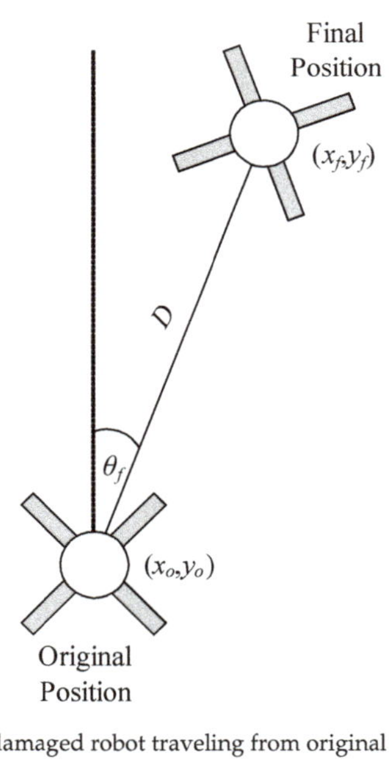

Figure 8. Trajectory of damaged robot traveling from original position to final position.

3.1.3. Evolutionary Process

The evolutionary process is similar to the PSO process, as discussed in the previous sections. A flowchart of the overall process is shown in Figure 9. The process starts with the generation of random populations of n particles (P). Thereafter, all particles execute the actions for time T. Next, the performance of each particle is judged based on the value of the objective function F. The evolutionary method is executed under the following conditions:

- Normal: If a particle is allocated to this state, the action parameter will be updated according to the PSO rule.
- Capsizing: In practice, the robot attitude should be considered because sensors and loads, e.g., camera, are generally integrated on top of the robot. If the robot flips over during recovery, the sensor or loads may break. Therefore, the particles that cause the robot to flip over should reset all of their parameters randomly.
- Moving Backward: If the particles cause the robot to move backwards, their parameters should be set as random values, likewise.
- Mutant: Given the probability $prob < 0.2$, a few particles should be mutated to avoid the local maximum.

The process will be terminated if the objective function reaches the desired value or if the generation reaches the set value.

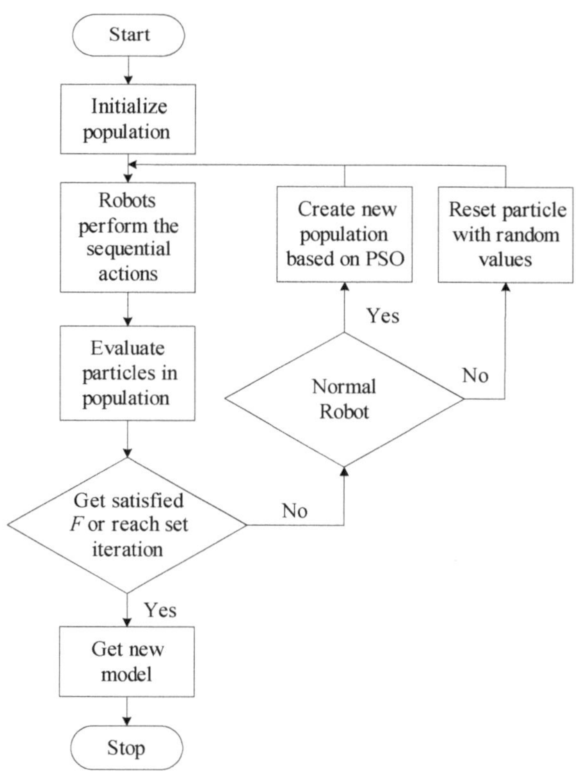

Figure 9. The procedure of an evolutionary based self-recovery method.

3.2. Mudskipper-Inspired Behavior

Because a quadruped robot needs at least three legs to balance its body [19], it is not feasible for the robot to execute movements after it is damaged. In the existing recovery methods, robots cannot walk properly even if one leg has minor damage. As a result, we investigated an alternative method to control the robot based on specific actions. Because such actions can be designed manually based on careful observation, it can be guaranteed that the robot will maintain a good posture at all times. To decide what actions to perform, we consider the motion of a fish called "mudskipper", as shown in Figure 10. Mudskippers are amphibious fish, which means that they can use pectroral fins to "walk" on land. We mimicked crawling behavior of mudskippers because it activates only two fins and the tail to achieve the excellent locomotion over the different types of terrains [21]. Recently, mudskipper-inspired robots have been developed and analyzed. According to McInroe et al. [22], MuddyBot uses its tail and fins to improve its ability to move. This behavior can possibly be used to help a broken quadruped robot move.

Figure 10. Mudskipper laying on a rock.

Mudskipper Behavior

In this study, we focus only on the operation of the two fins of a mudskipper. The simplified model of mudskippers fins is defined as a 2-DOF system, containing two revolute joints with respect to the vertical and the horizontal directions. Figure 11a shows the design model of the mudskipper fins. According to the robot model discussed herein, a quadruped robot consists of 3-DOF per each leg as shown in Figure 11b. Then, the angular rotations of θ_2 and θ_3 are set along a reverse direction. Hence, we can describe the setting of each angle as follows: $\theta_1 = \phi_1$, $\theta_2 = \phi_2$ and $\theta_3 = -\phi_2$. To imitate fin movement, we performed numerical calculation using the sine and cosine functions of time t and one-time-moving period T, as shown below:

$$\phi_1 = w_m \cdot \cos(\frac{2\pi t}{T}), \tag{17}$$

$$\phi_2 = h_m \cdot \sin(\frac{2\pi t}{T}), \tag{18}$$

where w_m and h_m are the width and the height of the moving trajectory, respectively. The moving phase of the leg consists of two different phases, namely, swing and touch. At time $t = 0$ s, the leg will start from the beginning and cover a circular trajectory, and, at $t = T$ s, the leg will end at the starting point. Note that the swing phase is the action in which the leg lifts off the ground, and the touch phase is the action in which the leg touches the ground.

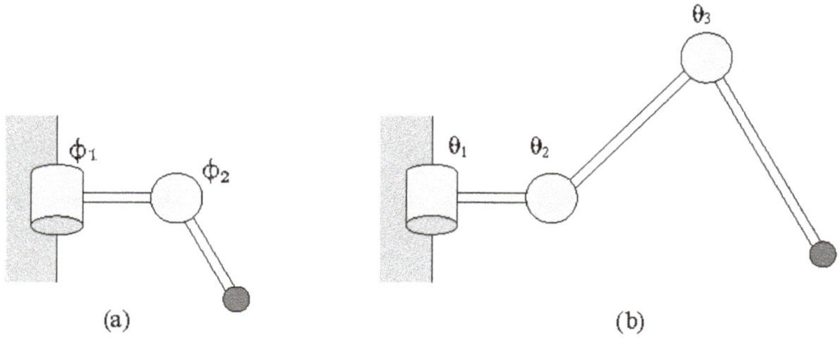

Figure 11. The model of mudskipper's fin and quadruped robot: (**a**) 2-DOFs mudskipper fin; (**b**) robot used in present study.

To control the robot direction, the robot orientation θ_r is used as the control parameter to tune the direction of motion. A simple feedback control scheme is used in this case. The corresponding closed-loop control diagram is shown in Figure 12. In this study, the parameter h_m is set as a constant because a large change in h_m can destabilize the robot's posture, whereas w_m is varied. For example, in case of the loss of two legs, we can adjust the parameter w_m of two legs to increase the size of the trajectory loop along the horizontal direction. If the parameters w_m are set diversely, the robot can be maneuvered to turn left or right. However, it is difficult to tune the parameters. In the next section, we integrate the mudskipper-inspired movement with reinforcement learning to improve the robot with the flexibility to perform in various scenarios.

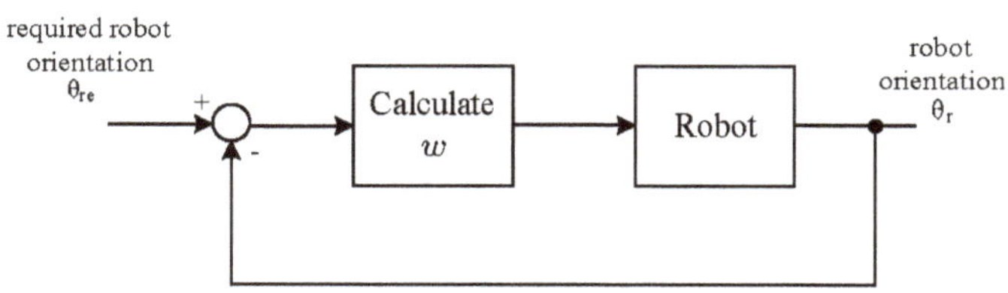

Figure 12. Feedback control diagram for mudskipper-inspired movement.

3.3. *Self-Learning Mudskipper-Inspired Crawling Algorithm (SLMIC)*

To enhance the performance of the robot's mudskipper-inspired movement, the learning approach is integrated to help the robot move in the straight direction. For decades, researchers have made attempts to combine the learning algorithm with robots to improve the flexibility and efficiency of robots. Reinforcement learning have been used in many applications in the field of robotics. In [23], a wheeled robot was programmed to be able to learn online. In addition to high-DOF robots, humanoid robots can learn to walk and crawl successfully by using the methods described in the papers of Lin and Yamaguchi [24,25]. Moreover, the walking speed of quadruped robots can be increased by using the method proposed by Kohl and Stone [26]. Although these reinforcement learning based studies were successful in their own right, a key problem with many of approaches in the literature is the size of the state and action space. Generally, for a multi-joint robot, the state is set as the current position of the joint angles, which means that based on the number of DOFs of the quadruped robot developed, the state would be 12-dimensional. Thus, a new concept of state and action design is proposed in this study.

3.3.1. Q-Learning Algorithm

Reinforcement learning can be simply explained as the process of training pets, such as dogs and cats, to perform certain actions. However, this process can be described as a trial-and-error problem as well. In the beginning of a dog's training process, it seems impossible for a dog to perform a requested task without error. The dog will try to perform some action randomly, such as walking, sitting, or jumping, to get a snack. After it gets the snack, the dog will remember the action that can possibly get it a snack again. By contrast, the dog will not do any action that would lead to an adverse outcome such as a scolding or a beating. Accordingly, the process of reinforcement learning can be thought to be based on reward and punishment. In this study, the quadruped robot is controlled using Q-learning and mudskipper-inspired behavior. The control policy will be adapted with the current state, provided by the surrounding environment, following the Q-learning approach. For example, if the robot is in a state in which it is heading north, but we command it to go to the east, the control policy will be changed to make the robot go to the east. The three main parameters of reinforcement learning are as follows:

- State (S): It is defined as the current scenario of a system, for instance, the position of the robot.
- Action (A): In one system, several actions would be required to be conducted in each state. In wheel-based robots, the actions can be moving forward, turning left, and turning right.
- Reward (R): It depends on the current state and action. It can be positive, negative, or zero for the win, lose, and draw scenarios, respectively. For example, when a robot encounters an obstacle, it needs to avoid the obstacle. If the robot decides to move forward and hit an obstacle, it will get a negative outcome in the form of a punishment. On the contrary, if the robot avoids an obstacle properly, it will receive a positive reward.

Finally, the objective of learning, which is known as policy (π), will be achieved. The best policy is the selection of actions that provide the highest reward, the so-called "Maximum Sum of Expected Rewards".

Q-learning is a reinforcement learning approach with non-model requirements. It employs the concept of Markov decision process (MDP) with finite state arrays to arrive at the optimal policy [27]. The expected reward will be stored in a d-dimensional state-action array. The number of d can be set manually by the user. Q-learning is one algorithm among the various algorithms associated with the temporal-difference method. The Q-values are acquired as follows:

$$V^*(s) = \max_a Q(s,a). \tag{19}$$

Because the Q-function does not need a model for learning and selecting actions, no model is required for state transitions. Q is updated by using the new information obtained after performing an action to correct the old policy. The Q-values are updated numerically based on the temporal-difference concept using Equation (20) [28]:

$$Q(s,a) = Q(s,a) + \alpha(R(s) + \gamma \max_{a'} Q(s',a') - Q(s,a)), \tag{20}$$

where α and γ are the learning rate and the discount factor, respectively. The procedure of the Q-algorithm is summarized in Table 2. The proposed method employs both Q-learning and mudskipper-inspired behavior to control the damaged robot. The controller structure is illustrated in Figure 13.

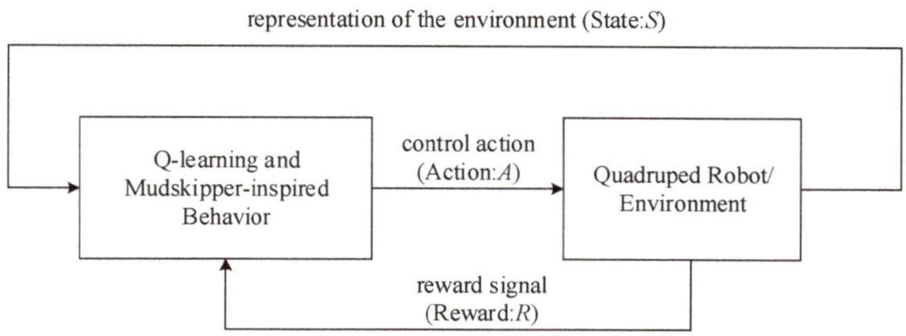

Figure 13. Controller structure for the proposed method (SLMIC).

Table 2. Q-learning procedure for n^{th} episode.

Algorithm: Q-learining Algorithm	
1:	observes its current state s_n.
2:	selects and perform an action a_n.
3:	observes the subsequent state s'_n.
4:	receives an immediate reward r_n.
5:	adjust it Q_{n-1} value using Equation (20)

3.3.2. State

In spite of using joint angles, we employ robot orientation as the state in the Q-learning. The yaw angle of the robot is divided into seven regions, as shown in Figure 14. The regions that separate the state are listed in Table 3. We select the robot orientations as the state to be able to control robot direction properly, and another advantage of doing so is that a robot can learn other actions simultaneously if provided with another Q-value table.

Table 3. The design space region and state rewards.

State	Region	Reward R_s
S0	$\theta_r < -25$	-10
S1	$-25 \leq \theta_r < -15$	-5
S2	$-15 \leq \theta_r < -5$	-1
S3	$-5 \leq \theta_r < 5$	0
S4	$5 \leq \theta_r < 15$	-1
S5	$15 \leq \theta_r < 25$	-5
S6	$25 \leq \theta_r$	-10

3.3.3. Action

Because the mudskipper-inspired movement requires only two legs to realize crawling, the number of robot actions is set to 7 for both legs by tuning the parameter w_m in Equation (17). The parameters w_{m1} and w_{m2} are used to control the legs L_1 and L_2 as shown in Figure 14, and w_{m1} and w_{m2} are calculated using the following equations:

$$w_{m1} = 5 + a_1, \quad (21)$$

$$w_{m2} = 5 + a_2, \quad (22)$$

where a_1 and a_2 are the adjusting parameters, and their values are set as given in Table 4. The numbers used for each action are set unequally to ensure the robot perform different actions, such as turn left and turn right. Therefore, the total number of Q-value is $7 \times 7 = 49$, which is smaller than that in the conventional setting.

Table 4. Design action set.

Action	a_1 (Leg L_1)	a_1 (Leg L_2)
A0	4	1
A1	3	1
A2	2	1
A3	1	1
A4	1	2
A5	1	3
A6	1	4

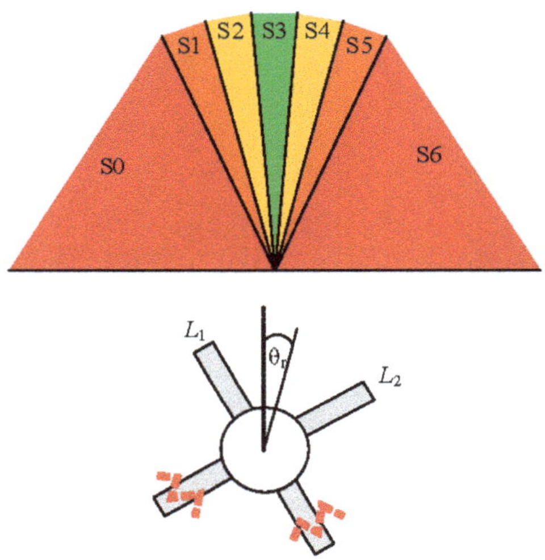

Figure 14. State space of Q-algorithm for robot orientation θ_r.

3.3.4. Reward

Because the aim of the present study is to design a recovery method to ensure that a robot can move after sustaining damage, we set the reward as the summation of two rewards, namely, state reward (R_s) and action reward (R_a):

$$R = R_a + R_s. \tag{23}$$

As a result, we can control not only the robot orientation but also the movability, that is, distance traveled. To facilitate robot motion in straight line, the state reward was designed to be the values listed in Table 3. The robot gets the negative reward (punishment) if it turns left or right. This ensures that, after leaning, the robot moves in the desired direction. Furthermore, the action reward is used to propel the robot to move forward faster, as expressed by Equation (24). If the robot moves backward, it will be punished with -10 reward. By contrast, if the robot can move the longest distance, it will receive +10 reward.

$$R_a = \begin{cases} 10 & \text{if max distance,} \\ d & \\ -10 & \text{if moving backward,} \end{cases} \tag{24}$$

where d is the distance traveled by the robot.

4. Results and Discussion

In this study, two major experiments were conducted in a simulation environment and with a real robot to evaluate the performance and improvement of the proposed robot structural design and the recovery method. In addition, several sub-experiments were conducted to test the proposed robot and recovery method in different scenarios. It starts with the first part of the experiments that pertains to testing the design of the novel quadruped robot structure with caterpillar-inspired prolegs. The performances of the self-learning mudskipper-inspired crawling algorithm (SLMIC) is judged thereafter.

4.1. Results of Caterpillar-Inspired Quadruped Robot (CIQR)

To achieve the designed upright structure, the shape of the legs of CIQR was optimized by numerical simulation as mentioned in the first part of this experiment. Thereafter, the advantages of a new designed structure were analyzed by conducting two sub-experiments, namely, recovering a robot by using conventional recovery methods in a simulation and operating the proposed robot with caterpillar behavior in a real application. Note that, in both sub-experiments, the common structure and the proposed CIQR structure were compared.

4.1.1. Optimization of Robotic Structure

The simulation was run for 20 iterations with μ_1 and μ_2 as 0.5. The weight parameters were set to equal values because the CIQR is required to assign equal weights to both actions. Figure 15 shows the fitness values of optimization with time, and it illustrates that the robot can discover a new structure that can help it walk longer. The evolution of the robot leg can be seen in Table 5. At the beginning, the high prolegs cause the robot to walk slowly in accordance with the fitness values shown in Figure 4. After five iterations, the robot evolves to being suitable for crawling and trotting, such that the robot structure changes, and the fitness value is increased. In iteration = 20, the robot structure changes slightly. A comparison between the proposed model and the conventional model was made in this study. The results show that the optimized model can travel 39.19 cm and 13.34 cm in the trotting and crawling gaits, respectively. By contract, the normal structure can move 36.78 cm and 13.83 cm in the trotting and crawling gaits, respectively. Figure 16 shows the actual CIQR model fabricated using a 3D printer.

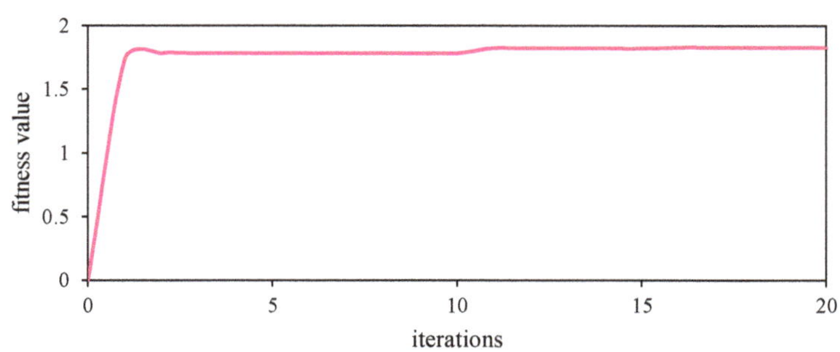

Figure 15. Fitness function of optimization processes with PSO algorithms.

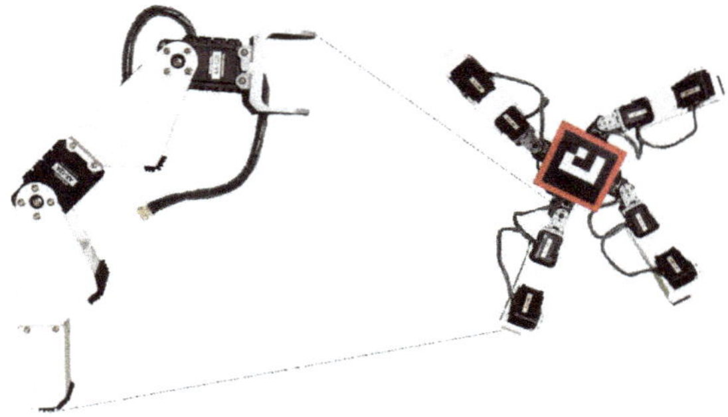

Figure 16. Proposed robot (CIQR).

Table 5. Result of leg structure optimization with different numbers of iterations.

Iteration	l (cm)	h (cm)
0	4.50	3.50
5	3.67	3.38
10	3.67	3.38
15	5.41	3.46
20	5.47	3.33

4.1.2. Simulation Experiments of CIQR with Conventional Recovery Methods

In the experiment, the damage recovery algorithm was employed to compare the performance of the normal design (quadruped robot without prolegs) and the proposed design. The evolutionary adaptive gait, which involves creating sequential actions, was employed in this test; the concept underlying this process was inspired by existing works presented in [8,29]. Three experimental scenarios were tested in this study, namely, one leg loss, two leg loss, and three leg loss, as shown in Figure 17. A simple algorithm was employed to control the damaged robot. For each joint, five rotational angles are controlled in sequences. The robot performed motions step-by-step in round-robin fashion, from the first angle to the last one. Because CIQR has 12 joints, 60 parameters in total are used to control the robot. To determine 60 parameters, PSO was used to execute the discovery process once again [30]. At the beginning, all parameters were set randomly. Next, the robot performed the first action with the first rotational angle of each joint. At $t = T$, the robot performed the second action. This step was iterated until t reached the setting time. The fitness function of this algorithm was set as the distance that the damaged robot could travel. The test results obtained with the damaged robot are given in Table 6. As can be seen, with only one leg lost, CIQR performed better than the normal robot in terms of the total distance traveled. By contrast, the normal robot produced a good result in the second case as it walk about 10 cm more than the proposed robot. The CIQR moved slower in this case possibly because of the additional weight of the prolegs. In case of only one leg, CIQR could travel longer than the normal robot. However, the control method used in this benchmark could not efficiently control the robot because the robot motion was not smooth, and the possibility of the robot flipping over during the evolutionary process prevailed. These problems were tested once again using the proposed recovery algorithm (i.e., SLMIC) described in Section 3.3.

Table 6. Distance traveled by damaged robots

Number of Legs Lost	Normal Robot (cm)	CIQR (cm)
1	48.68	51.67
2	92.42	85.41
3	17.08	131.73

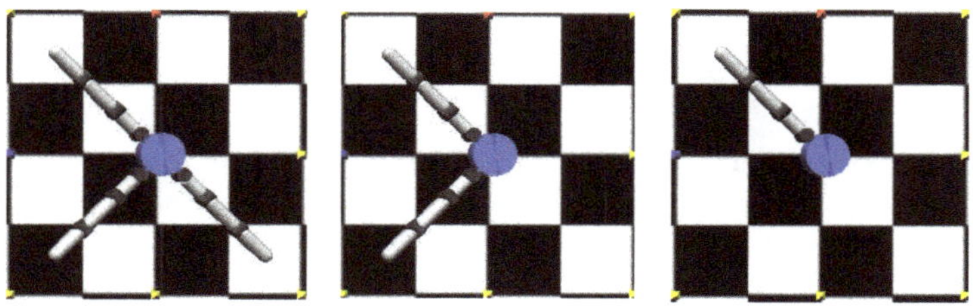

1 leg loss 2 leg loss 3 leg loss

Figure 17. Test cases of damaged robots.

4.1.3. Experimental Results with Caterpillar-Inspired Crawling Behavior

To ensure that the proposed robot could function in real scenarios, an experiment with the actual robot was conducted. In this case, the performance of CIQR was compared with that of the normal robot. The evaluation was conducted with a damaged robot having one functional leg. As in the simulation, both robots were programed to move forward according to Equation (21) with the same set of parameters, that is, $A_2 = A_3 = 40$, $\phi_2 = 0$ and $\phi_3 = 30$, which are the amplitudes for controlling link2 and link3, and the initial offsets of link2 and link3, respectively. Link1 was set to the fixed direction of 0° to ensure the robot traveled straight. The results show that both robots could travel straight but shifted slightly towards the right. However, the novel structure of CIQR traveled longer than the normal robot in the simulation, as shown in Figure 18. The normal robot moved forward and stopped after 20 s, traveling 34.17 cm in the course. In the same period, the CIQR traveled 52.08 cm. Additionally, the positions of both robots in the experiment were measured using an overhead camera and image processing technique.

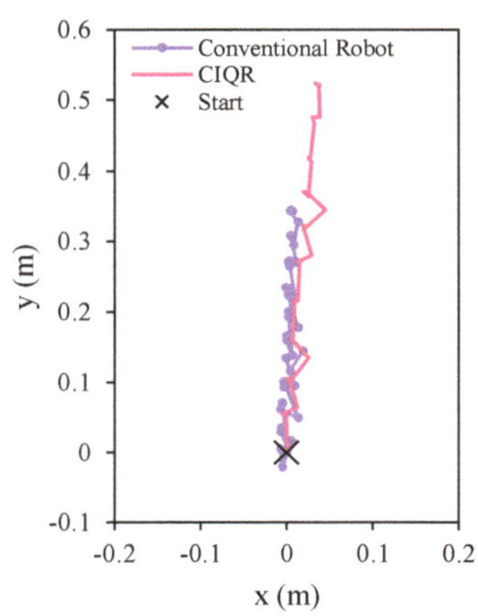

Figure 18. Comparison of trajectories traveled by a conventional robot and CIQR-based robots with three legs lost.

4.2. Evaluation of Self-Learning Mudskipper-Inspired Crawling Method (SLMIC)

In the benchmark, the proposed recovery algorithm was tested in four different scenarios and compared to the conventional evolutionary method without direction (EA), evolutionary method with direction (EAD) and mudskipper-inspired movement (MUD), in the simulation. The test scenarios were as follows:

- case A: one leg lost,
- case B: two adjacent legs lost,
- case C: two diagonal legs lost,
- case D: two adjacent legs and one limb lost.

All the simulation scenarios were run for 20 s with the four methods. After the simulations were executed, only the proposed method, that is, SLMIC, was employed to operate the actual robot because the evolutionary processes were time consuming and SLMIC provided the best performance in terms of recovery distance. The performance of SLMIC was compared with the control method used to control the robot before it was damaged. To achieve the robot position, Inertial Measurement Unit (IMU) was incorporated into the robot and was used to measure the short distance that was required in the recovery process. Image processing was additionally used to obtain the distance traveled by the robot for reporting purposes.

4.2.1. Simulation Results of SLMIC Vis-à-Vis Other Methods

The numerical simulation was conducted separately for each case in the experiments. As shown in Figure 19, the results of four different damage scenarios are as follows:

1. case A: The results show that all methods used in the simulation allowed the robot to be able to move again. With one leg lost, it was easy for the robot to travel with three functional legs. However, not all methods provided the acceptable results. EA helps the robot move the shortest distance compared with other methods, as shown in Figure 19b. EAD helps the robot move longer than EA but it lost out to the MUD and SLMIC methods. By employing the specific actions of mudskippers, both MUD and SLMIC help the robot travel longer distances. However, SLMIC provided the best result in this test because it helped the robot learn to move forward faster.
2. case B: The robot programmed using EA traveled faster than the robot programmed using EAD, as shown in Figure 19c. However, EAD provided the better result in terms of direction. It seemed that, with EAD, the robot optimized multiple objectives, namely, distance traveled and direction of travel. As a result, the robot assigned more importance to direction in optimization, which reduced the distance traveled. MUD and SLMIC provided decent results in terms of distance traveled and direction of travel. Once again, SLMIC provided the best performance.
3. case C: Similar to the two cases in the experiments, with MUD and SLMIC, the robot covered longer distances. However, SLMIC performed better in terms of direction of travel. Opposite to case B, EAD could deal with only the distance traveled. At this time, EAD attempted to optimize the distance traveled by the robot, but it failed to optimize the direction of travel, and thus the robot failed to move straight ahead. With EA, the robot could not perform well because the two diagonal legs affected its balance. The robot flipped over during the recovery process which limited its ability to move. As a result, the robot programmed using EA could travel properly, as shown in Figure 19d.
4. case D: This experiment was the most challenging because of the limited number of functional legs and actuators, as shown in the results in Figure 19e. Given the extremities, the robot programmed using SLMIC could learn to recovery itself with SLMIC and provided the best results in terms of direction and distance. MUD with its specific control method was the second best performer in this experiment. EAD exhibited the worst performance owning to the same reason as in case B, and EA achieved a fair level of performance.

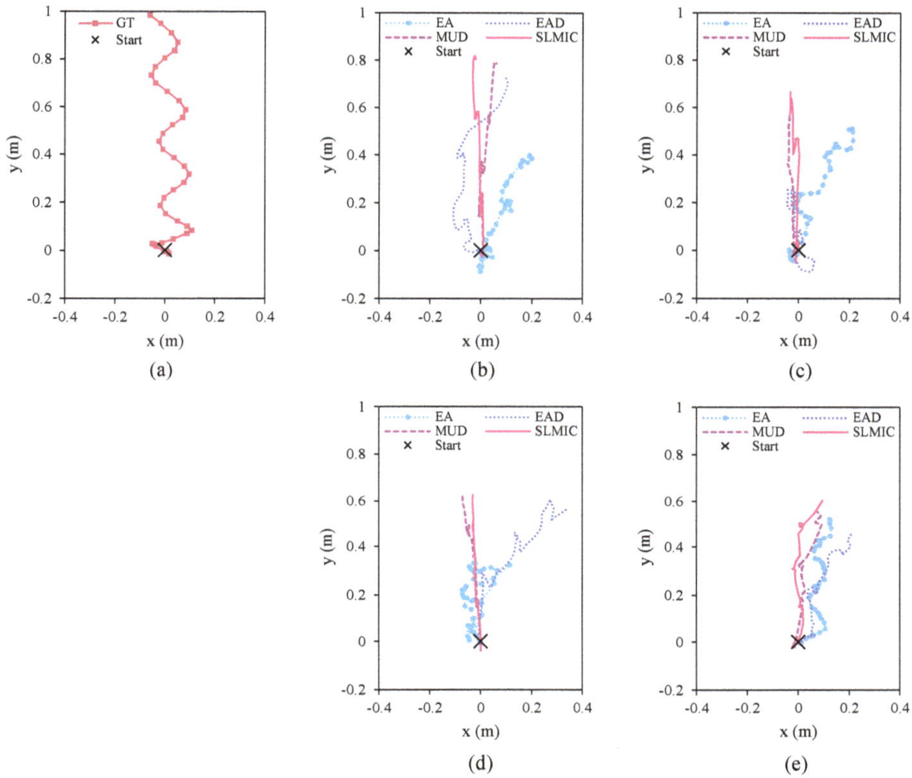

Figure 19. Comparison of simulation results obtained using EA, EAD, MUD and SLMIC: (**a**) ground truth of healthy robot walking with trotting gait; (**b**) one leg lost (case A); (**c**) two adjacent legs lost (case B); (**d**) two diagonal legs lost (case C); (**e**) two adjacent legs and one limb lost (case D).

From the results, it can be concluded that the proposed method (SLMIC) can provide the best results compared to the other methods in terms of direction of travel and the distance traveled. However, a certain learning time is required to achieve the goal. The performance of MUD was inferior to that of SLMIC, but, in most cases, it performed satisfactorily. However, the crucial aspect of MUD is that it employs specific mudskipper-inspired actions with no time required for evolutionary process or learning. The results indicate that it would be difficult to guarantee satisfactory performance in terms of direction and distance traveled with both EA and EAD because differences in robot model can cause task failures, and the methods can get stuck in local minima when performing multi-objectives optimization, which can lead to failure.

4.2.2. Comparison of Experimental Results Obtained Using Previous Control Method and SLMIC

Because SLMIC exhibited the best performance in the simulation, we decided to conduct the experiment involving the actual robot using only with SLMIC. We compared the performance of SLMIC with the previous controller, that is, trotting gait (TG). The test cases were the same as those in the experiments conducted in Section 4.2.1.

1. case A: The results of this test case clearly show that with SLMIC the robot could recover itself to reach the goal, as shown in Figure 20b. Compared to the ground truth (in Figure 20a), the robot programmed with SLMIC almost traveled the same distance as the healthy robot. By contrast,

the robot could not move well when the previous control method (trotting gait controller) was employed. It caused the robot to move back and fourth around a single point in a certain working area.

2. case B: In this case, SLMIC with the damaged robot achieved the same result as the healthy robot. However, the robot moved slightly towards the right part of the working space. By contrast, the robot with two adjacent legs-lost and programmed by the previous method could not perform well, traveling only around the starting point, as shown in Figure 20c.
3. case C: As shown in Figure 20d, with the trotting gait, the broken robot could not function properly and moved backwards during the experiment. This can be one of the reasons why the recovery method is significant for multiple-legged robots. By contrast, the proposed method provided good performance with the learning process. According to the trajectory traveled by the robot programmed with the proposed method, it moved towards the right at the beginning, but it returned to the predetermined direction with the passage of time.
4. case D: Similar to results of the simulation in the previous section, the robot with two adjacent legs and one limb lost found it difficult to achieve the same performance as the healthy robot. However, SLMIC made a big difference compared to the previous controller. Even so, it could not help the robot recover fully, but it did help the damaged robot cover more than half the distance covered by the healthy robot.

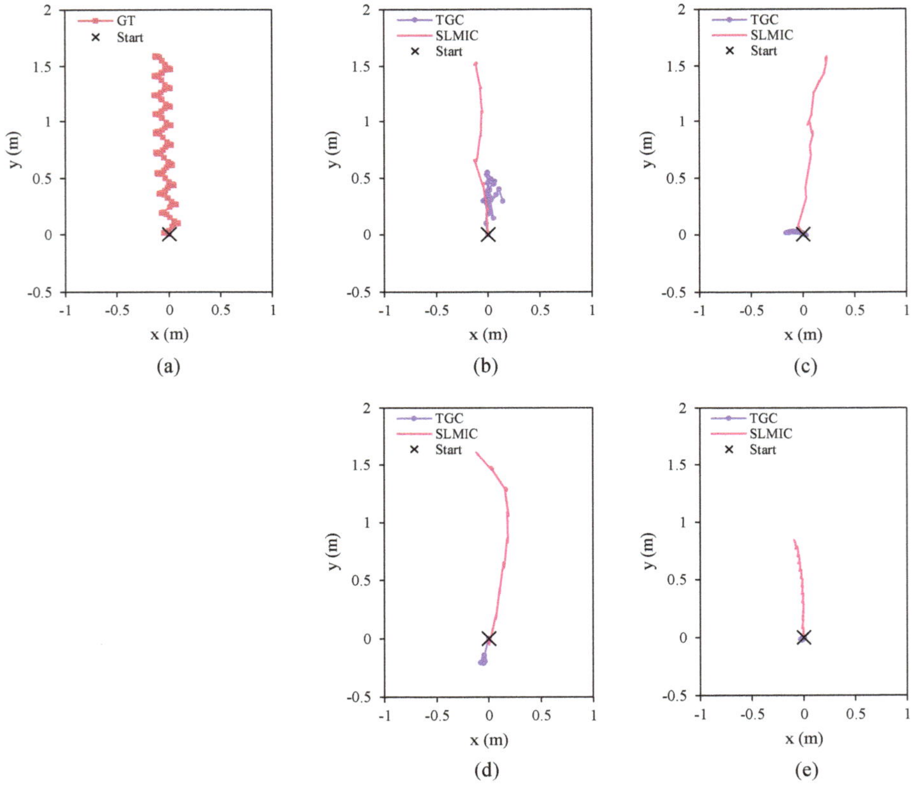

Figure 20. Experimental results of SLMIC compared to previous controller (trotting gait): (**a**) ground truth of healthy robot walking with trotting gait; (**b**) one leg lost (case A); (**c**) two adjacent legs lost (case B); (**d**) two diagonal legs lost (case C); (**e**) two adjacent legs and one limb lost.

After the experiments with actual robots were conducted properly, it was found that the damaged robot could practically not move when the previous controller (trotting gait in this case) was used. As a result, the recovery method that can provide an alternative solution for damaged robots is needed. With SLMIC, the proposed recovery method based on specific actions that guarantee adaptable movement owning to learning could solve most of the problems successfully, and, in one special difficult case, it recovered by approximately 50 percent. The example of robot actions with SLMIC in Figure 21 (case C) shows that the robot turned slightly to the wrong direction at the beginning, but it recovered and moved along the correct direction at the end. Finally, the experiments confirmed that the proposed method (SLMIC) is suitable for quadruped robots with a limited number of functional legs.

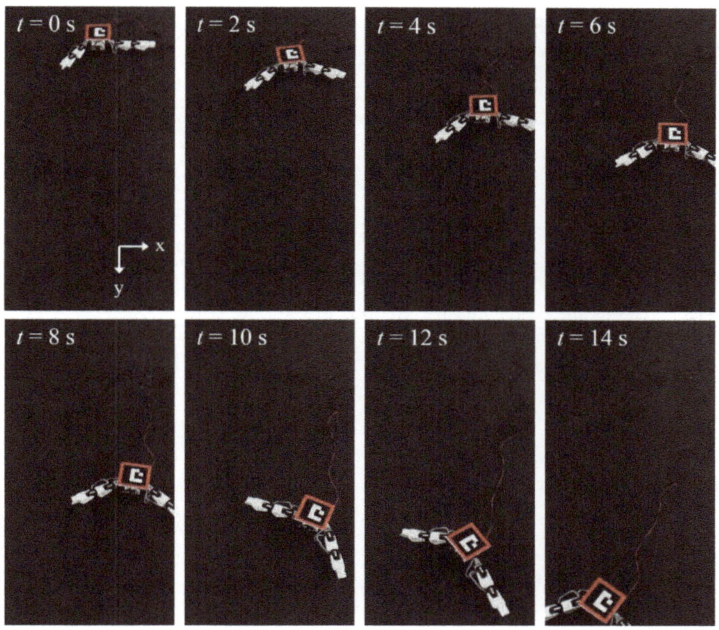

Figure 21. CIQR with two diagonal legs performing recovery action after learning with the SLMIC method.

5. Conclusions

In this paper, a novel structure model of the quadruped robot and self-recovery method were proposed. The CIQR was developed to imitate the crawling locomotion of the caterpillar for the case in which the robot has a small number of active legs. Prolegs similar to those on a caterpillar were added onto the robot limb to improve its ability to move after some parts of the robot were damaged. The proposed bio-inspired quadruped robot structure was optimized to ensure that it can operate in both normal and abnormal scenario, and PSO was used as the optimization method. Moreover, a new bio-inspired locomotion method based on the movement of mudskipper in nature, called SLMIC, was proposed in this paper. The reinforcement learning method, Q-learning, was integrated to improve locomotion adaptability. The specific actions inspired by mudskippers were set as an action for Q-learning processes. The orientation of the robot body was sent to the learning process as a state (current scenario of the robot). The positive and negative rewards were given to the robot when it performed the task. The robot received a positive reward when it moved forward in a straight direction. On the other hand, it received a negative reward when it moved backward.

The results confirm that the proposed structure with prolegs provided better results compared to a normal structure when the robot legs were damaged, especially when the robot had only one leg. With one functional leg, the robot with the proposed structure could travel almost eight times longer than the robot without prolegs. According to the numerical simulation and experiments, the recovery methods are feasible for implementation in a damaged robot that has at least one leg. From the simulation, the robot programmed with SLMIC provided a straighter and longer path than the robot performed with other methods. During the experiment, it was found that the damaged robot controlled with SLMIC could travel in a straight path longer than the robot without recovery methods. The most interesting finding is that, with carefully designed actions, which are based on the simple behaviors of mudskipper, a robot can learn new habits to achieve the same goal as a healthy robot. In the future, it is planned to apply and to develop the proposed structure and self-recovery method with other legged robots to improve the performance.

Author Contributions: S.C. and Y.K. designed the study and conceived of the presented idea; S.C. carried out the simulation and experiments; S.C. took the lead in writing the manuscript. Y.K., T.E., and A.A.R. helped to supervise the study and reviewed the paper.

Acknowledgments: This work is supported by MEXT (the Ministry of Education, Culture, Sports, Science and Technology), Japan.

Conflicts of Interest: The authors declare no conflict of interest.

Abbreviations

The following abbreviations are used in this manuscript:

DOFs	Degrees of Freedom
GA	Genetic Algorithm
PSO	Particle Swarm Optimization
DH	Denavit–Hartenberg
CIQR	Caterpillar-Inspired Quadruped Robot
EA	Evolutionary Algorithm
EAD	Evolutionary Algorithm with Direction
MUD	Mudskipper-Inspired Movement
SLMIC	Self-Learning Mudskipper-Inspired Crawling
TGC	Trotting Gait Controller

References

1. González-de Santos, P.; Garcia, E.; Estremera, J. *Quadrupedal Locomotion: An Introduction to the Control of Four-Legged Robots*; Springer: Berlin/Heidelberg, Germany, 2006.
2. Liu, M.; Li, M.; Pang, J. Fault-tolerant gait implementation of hexapod robot based on finite state automata. In Proceedings of the 2017 29th Chinese Control And Decision Conference (CCDC), Chongqing, China, 28–30 May 2017; pp. 6800–6805.
3. Gao, Z.; Ma, L.; Wang, J. Fault tolerant control method for displacement sensor fault of wheel-legged robot based on deep learning. In Proceedings of the 2018 WRC Symposium on Advanced Robotics and Automation (WRC SARA), Beijing, China, 16 August 2018; pp. 147–152.
4. Dubrova, E. *Fault-Tolerant Design*; Springer Publishing Company: Berlin/Heidelberg, Germany, 2013.
5. Yang, J.M.; Park, Y.K.; Kim, J.G. Fault-Tolerant Gait Planning of Multi-Legged Robots. In *Mobile Robotics, Moving Intelligence*; Buchli, J., Ed.; IntechOpen: Rijeka, Croatia, 2006; Chapter 28.
6. Murakami, M. Task-based Dynamic Fault Tolerance for Humanoid Robots. In Proceedings of the 2006 IEEE International Conference on Systems, Man and Cybernetics, Taipei, Taiwan, 8–11 October 2006; Volume 3, pp. 2197–2202.
7. Bongard, J.; Zykov, V.; Lipson, H. Resilient Machines Through Continuous Self-Modeling. *Science* **2006**, *314*, 1118–1121. [CrossRef] [PubMed]

8. Qiu, G.Y.; Wu, S.H. The Evolutionary Locomotion of Tripedal and Quadrupedal Biomorphic Robots. In Proceedings of the 2011 International Conference on Technologies and Applications of Artificial Intelligence, Las Vegas, NV, USA, 18–21 July 2011; pp. 45–50.
9. Liang, J.; Xue, C. Self identification and control of four-leg robot based on biological evolutionary mechanisms. In Proceedings of the 2010 5th IEEE Conference on Industrial Electronics and Applications, Taichung, Taiwan, 15–17 June 2010; pp. 958–961.
10. Koos, S.; Cully, A.; Mouret, J.B. Fast damage recovery in robotics with the T-resilience algorithm. *Int. J. Robot. Res.* **2013**, *32*, 1700–1723. [CrossRef]
11. Cully, A.; Clune, J.; Tarapore, D.; Mouret, J.B. Robots that can adapt like animals. *Nature* **2015**, *512*, 503–507. [CrossRef] [PubMed]
12. Ren, G.; Chen, W.; Dasgupta, S.; Kolodziejski, C.; Wörgötter, F.; Manoonpong, P. Multiple chaotic central pattern generators with learning for legged locomotion and malfunction compensation. *Inf. Sci.* **2015**, *294*, 666–682. [CrossRef]
13. Zhang, T.; Zhang, W.; Gupta, M.M. Resilient Robots: Concept, Review, and Future Directions. *Robotics* **2017**, *6*. . [CrossRef]
14. Geva, Y.; Shapiro, A. A Novel Design of a Quadruped Robot for Research Purposes. *Int. J. Adv. Robot. Syst.* **2014**, *11*, 95. [CrossRef]
15. Atique, M.M.U.; Ahad, M.A.R. Inverse Kinematics solution for a 3DOF robotic structure using Denavit–Hartenberg Convention. In Proceedings of the 2014 International Conference on Informatics, Electronics Vision (ICIEV), Dhaka, Bangladesh 23–24 May 2014; pp. 1–5.
16. Spong, M.; Hutchinson, S.; Vidyasagar, M. *Robot Modeling and Control*; Wiley: New York, NY, USA 2005.
17. Trimmer, B.A.; Takesian, A.E.; Sweet, B.M.; Rogers, C.B.; Hake, D.C.; Rogers, D.J. Caterpillar locomotion: A new model for soft- bodied climbing and burrowing robots. In Proceedings of the 7th International Symposium on Technology and the Mine Problem, Monterey, CA, USA, 2–5 May 2006.
18. Wei, W.; Dawei, X. A new design and analysis of compliant spine mechanism for caterpillar climbing robot. In Proceedings of the 2012 IEEE International Conference on Robotics and Biomimetics (ROBIO), Guangzhou, China, 11–14 December 2012; pp. 790–795.
19. Darici, O.; Yalcin, M.K.; Temeltas, H. Comparison of gait generation methods in Quadruped walking. In Proceedings of the 2008 IEEE/ASME International Conference on Advanced Intelligent Mechatronics, Xi'an, China, 2–5 July 2008; pp. 1–6.
20. Zhang, H.; Gonzalez-Gomez, J.; Zhang, J. A new application of modular robots on analysis of caterpillar-like locomotion. In Proceedings of the 2009 IEEE International Conference on Mechatronics, Malaga, Spain, 14–17 April 2009; pp. 1–6.
21. Wang, L.; Xu, M.; Liu, B.; Jiang, T.; Zhang, S.; Yang, J. Experimental study on morphology and kinematics of mudskipper in amphibious environments. In Proceedings of the 2013 IEEE International Conference on Robotics and Biomimetics (ROBIO), Shenzhen, China, 12–14 December 2013; pp. 1095–1100.
22. McInroe, B.; Astley, H.C.; Gong, C.; Kawano, S.M.; Schiebel, P.E.; Rieser, J.M.; Choset, H.; Blob, R.W.; Goldman, D.I. Tail use improves performance on soft substrates in models of early vertebrate land locomotors. *Science* **2016**, *353*, 154–158. [CrossRef] [PubMed]
23. Xu, K.; Wu, F.; Zhao, J. Simplified online Q-learning for LEGO EV3 robot. In Proceedings of the 2015 IEEE International Conference on Control System, Computing and Engineering (ICCSCE), Penang, Malaysia, 28–30 November 2015; pp. 77–80.
24. Lin, J.L.; Hwang, K.S.; Jiang, W.C.; Chen, Y.J. Gait Balance and Acceleration of a Biped Robot Based on Q-Learning. *IEEE Access* **2016**, *4*, 2439–2449. [CrossRef]
25. Yamaguchi, A.; Takamatsu, J.; Ogasawara, T. DCOB: Action space for reinforcement learning of high DOF robots. *Autono. Robot.* **2013**, *34*, 327–346. [CrossRef]
26. Kohl, N.; Stone, P. Policy gradient reinforcement learning for fast quadrupedal locomotion. In Proceedings of the 2004 IEEE International Conference on Robotics and Automation (ICRA '04), Barcelona, Spain, 26 April–1 May 2004; Volume 3, pp. 2619–2624.
27. Bellman, R. A Markovian Decision Process. *Indiana Univ. Math. J.* **1957**, *6*, 679–684. [CrossRef]
28. Russell, S.; Norvig, P. *Artificial Intelligence: A Modern Approach*; Prentice Hall Series in Artifi; Prentice Hall: Englewood Cliffs, NJ, USA, 2010.

29. Gao, H.; Wang, T.; Liang, J.; Zhou, Y. Model adaptive gait scheme based on evolutionary algorithm. In Proceedings of the 2013 IEEE 8th Conference on Industrial Electronics and Applications (ICIEA), Melbourne, Australia, 19–21 June 2013; pp. 316–321.
30. Guangyou, Y. A Modified Particle Swarm Optimizer Algorithm. In Proceedings of the 2007 8th International Conference on Electronic Measurement and Instruments, Xi'an, China, 16–18 August 2007; pp. 675–679.

applied sciences

MDPI

Article

Tarantula: Design, Modeling, and Kinematic Identification of a Quadruped Wheeled Robot

Abdullah Aamir Hayat [1,*], Karthikeyan Elangovan [1], Mohan Rajesh Elara [1] and Mullapudi Sai Teja [1,2]

1 Engineering Product Development Pillar, Singapore University of Technology and Design (SUTD), Singapore 487372, Singapore; e_karthikeyan@sutd.edu.sg (K.E.); rajeshelara@sutd.edu.sg (M.R.E.), mullapudisai1812@gmail.com (M.S.T.)

2 Department of Mechanical Engineering, Politecnico di Milano, 20133 Milan, Italy

* Correspondence: abdullahaamir@sutd.edu.sg

Received: 7 November 2018; Accepted: 17 December 2018; Published: 27 December 2018

Abstract: This paper firstly presents the design and modeling of a quadruped wheeled robot named *Tarantula*. It has four legs each having four degrees of freedom with a proximal end attached to the trunk and the wheels for locomotion connected at the distal end. The two legs in the front and two at the back are actuated using two motors which are placed inside the trunk for simultaneous abduction or adduction. It is designed to manually reconfigure its topology as per the cross-sections of the drainage system. The bi-directional suspension system is designed using a single damper to prevent the trunk and inside components from shock. Formulation for kinematics of the wheels that is coupled with the kinematics of each leg is presented. We proposed the cost-effective method which is also an on-site approach to estimate the kinematic parameters and the effective trunk dimension after assembly of the quadruped robot using the monocular camera and ArUco markers instead of high-end devices like a laser tracker or coordinate measurement machine. The measurement technique is evaluated experimentally and the same set up was used for trajectory tracking of the *Tarantula*. The experimental method for the kinematic identification presented here can be easily extended to the other mobile robots with serial architecture designed legs.

Keywords: design and modeling; kinematics; kinematic identification; monocular vision

1. Introduction

Drains are an integral part of every modern city, where drainage systems are entirely subsurface in most countries. Statistics from Asia, Europe, United States show that major cities contain 4000 to 7000 km of drainage lines. The primary purpose of these surface and subsurface sewage systems is to remove excess water in a safe and timely manner, which plays a vital role in controlling water-related diseases or water-borne diseases. Drainage systems have its disadvantages, where these systems give problems to mosquito-borne diseases, clogging, internal damages due to ageing, excessive traffic which causes contamination of groundwater or overflow. To control these problems, serious inspection, monitoring and maintenance of drainage systems is required. At present, this task is labor-intensive as shown in Figure 1a that adds more difficulties in subsurface sewer lines like inaccessible areas with poor lighting, ventilation and safety concerns associated with insect bites. The cross-section of the drainage system with the approximate symmetric design shown in Figure 1b are widely found in Singapore [1]. The width, W, of these drainages typically range from 1.1 to 1.8 m, w from 0.2–0.8 m and the height h between 0.3–1.1 m. Thus, there is a requirement to design the robot to traverse inside this type of drainage systems.

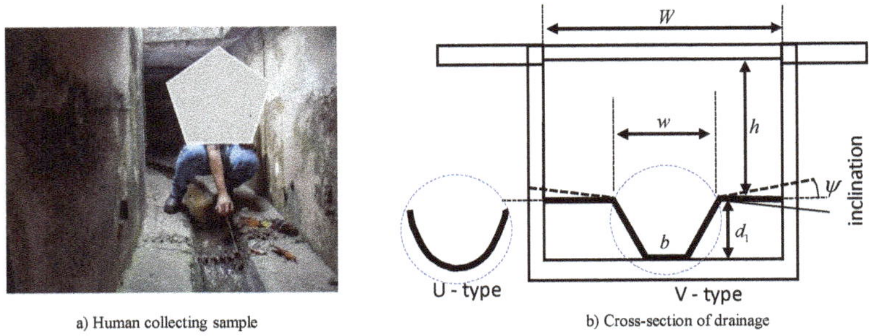

Figure 1. Human collecting water samples inside the drain commonly found in Singapore [1].

The specially designed mechanism with suited locomotion as per the internal geometry of the drainage system is essential. Classification of the inspection robots can be done on the basis of locomotion as tracked, wheeled and legged. PIRAT [2] is a tracked small robot designed for the quantitative assessment of sewer systems surveyed in real time. The development of autonomous body for inspection of liquid filled pipes "pipe rover/pearl rover" has six-legged propulsion [3]. In another work, an autonomous sewer cleaning robot was published that cleans underwater sewers [4]. KARO is a wheeled tethered robot for smart sensor-based sewer inspection equipped with intelligent multi-sensors [5]. KANTARO is a wheeled platform and uses a special mechanism called "naSIR Mechanism" to access straight and even bends pipelines without intelligence of sensors or controllers [6]. KURT is a six-wheeled vehicle that can fit in 600 mm diameter pipelines [7] and MAKRO a worm-shaped wheel, multi-segmented and autonomous bodies for navigation in drain systems [8]. The wheeled robot with fixed morphology finds application in climbing of ropes for inspection task as in [9,10]. Even though a bunch of studies in the literature validates for monitoring or inspection of sewer systems, they mostly suffer from performance issues like modularity and adapting its height as per the geometry of drains that diminish their full potential. One major factor that results in the performance degradation associated with inspection robots design is their fixed morphology. We have proposed the model of quadruped robot for drainage systems that are mainly constructed to carry excess water to reservoirs, unlike the sewage pipes that are used to dispose of solid wastes and water. Tarantula has four-wheel drive and steering locomotion. The drain inspection task can include the identification of the potential mosquito inhabitants and locations that are prone to mosquito-borne diseases as presented in [11] using the images grabbed from the camera mounted on Tarantula in the near future.

Quadruped robots are gaining increased attention among robotics researchers across a wide range of applications with its unique morphology to carry out various kinds of field work. These quadruped robots bring with them the unique advantage of efficiency. Several developments have been made after pioneering research on quadruped robot from MIT [12] and Tokyo University. Since then, a large number of quadruped robots have been developed, such as BISAM [13], which has reptile-like walking and stabilizes itself using a flexible spine. In another work, WARP1 [14] presents a standing posture controller for walking robots, which was successfully tested in simulations and experiments. The pioneering work of Hirose and Fukushima robotics laboratory mainly focused on legged robots for about 40 years. Typical quadruped robots born from this laboratory is TITAN series [15–17] that is the development of a sprawling-type quadruped robot and capable of high velocities and energy efficient walking. Popular among these is TITAN VIII [17]. An introduction to several quadruped robots along with its locomotion and control techniques were presented in [18]. The large dimension quadruped robot equipped with drilling equipment and capable of walking on different terrains by incorporating impedance control for the foot-ground contact was reported in [19]. These quadruped robots were mainly used in the fields like mine detection, walking uneven terrain, etc., but to access the drainage system with varying heights and cross-section, the robot should be designed accordingly to have the ability to reconfigure its morphology. In Table 1, a comparison was made among the existing drainage and sewer inspection and cleaning robots. We have used the word quadruped with *Tarantula* since the kinematics of wheel is coupled with the kinematics of leg. Note that it is not used here in the context of walking, trotting, etc., capabilities of robots.

Table 1. Wheeled and legged robot discussed in this work.

System	Locomotion	n_L, n_W	n_B	R	M	Environment
PCIRs [4]	2-Tracking wheels	–, 2	3	N	N	SP (C)
KARO [5]	4-WID	–, 4	2	N	N	SP (C)
KANTARO [6]	Passively adapted wheels	–,4	2	N	Y	SP (C)
KURT [7]	Wheeled	–, 3	3	N	Y	SP (C)
MAKRO [8]	Wheeled	–, 2n	3	Y	Y	SP (C)
BISAM [13]	Legged	4, –	5	N	Y	RT
Warp1 [14]	Legged	3, –	5	N	Y	RT
TITAN VIII [17]	Legged	1, –	5	N	Y	RT
IPR [20]	Legged	1, –	3	Y	N	SP (C)
Tarantula	Wheeled	4, 4	4	Y	Y	D

R: Reconfigurable, M: Modularity in mechanism, hardware and software, n_L Active degrees of freedom (DOF) in each leg. n_W: number of wheels, n_B: DOF of the moving platform, (C): Circular cross-section. SP: Sewer pipes, RT: Rough Terrain.

An interesting hybrid mode of locomotion robot named PAW used both the wheels and legs to achieve gaits, such as bounding, galloping and jumping, was reported in [21]. In [21], the four legs were having only a single degree of freedom which was used to incline the body and the formulation was presented for inclined turning and the wheel at the distal end to provide the locomotion. Tarantula has four degrees of freedom (DOF) in each leg to provide the change in the height of the body, contact with the inclined surface and for independent steering action. The contribution of this work is the designed mechanism, formulation for the coupled kinematics of legs and wheels along with the identification of the kinematic parameters of each leg.

The mechanical structure and the mechanisms are designed and assembled in CAD. The kinematics of legs is coupled with the wheel steering kinematics for the designed mobile robot Tarantula. The accuracy of these geometric parameters is critical for the control and steering. Hence, it becomes essential to identify the kinematic parameters of the legs after the assembly of the robot. Kinematic identification is a well established area that uses a geometric approach [22] or the optimization based technique [23] to estimate the kinematic parameters. Kinematic calibration of the legged mobile robot is presented in [24] and used the optimization based approach that requires the knowledge of its nominal or theoretical kinematic parameters for its initial guess to find the calibrated parameters and consequently improves the positional accuracy. We have used the geometric approach that needs no prior information of geometric parameters and used the circle point method formulation as presented in [25] to identify the widely used kinematic representation defined by Denavit and Hartenberg [26]. However, Ref. [25] did not account for the robots with prismatic joints. In this work, we have extended the approach proposed in [25] for the prismatic joints as well and demonstrated it with the kinematic identification of each of the four legs of the assembled quadruped robot.

Traditional strategies to recognize kinematic parameters of a robot includes taking the robot to a controlled situation to take pose estimations utilizing a coordinate measurement machine (CMM) [27] or laser tracker [28]. In this work, we have proposed the use of the monocular camera with the AruCo markers to demonstrate it for the identification of Tarantula. Unlike the visual localization which is done using a single marker reported in [29], we have used ArUco markers map (AMM) that resulted in the improved measurement accuracy. The measurement performance of this approach is compared using the standard industrial robot KUKA KR6 R900 robot (KUKA, Augsburg, Germany) [30]. Being cognizant of the above facts, we set the following objectives:

- Design of the robotic platform that can change its height and is holonomic,
- Formulation for kinematics of the wheeled locomotion coupled with the leg kinematics,
- Identification of kinematic parameters after the assembly of the robot, using monocular vision and ArUco markers,
- Trajectory tracking of the robot using the same set-up of monocular vision and ArUco markers.

This paper is divided into five sections. Section 2 lists the design requirements and the mechanical layout, i.e., system architecture of the *Tarantula* is discussed in detail. Section 3 introduces the workspace analysis of the Tarantula along with the kinematics of wheeled locomotion coupled with the leg kinematics. Experiments for identification of the kinematic parameters of the assembled *Tarantula* along with the trajectory tracking in Section 4. Finally, Section 5 concludes the paper.

2. Robot Architecture

In this section, the necessary design requirements for the quadruped robot specifically for the drainage inspection task are discussed first. Then, the mechanical design as per the requirement is discussed. Different components of the robot and the mechanisms developed are explained briefly.

2.1. Design Requirements

The central aspect of the *Tarantula* project is to design a robotic manipulator that can be utilized for the inspection purpose in the hazardous environment inside the drainage system. After surveying the specific drainage geometry and the inspection task to be performed by the robot, the fundamental design considerations are:

- The robotic system should have the capability to move around inside the drain environment. Hence, it must be mobile, unlike fixed industrial robots.
- The mobile platform should reconfigure its height as per the geometry of the drainages (Figure 1)
- The mobile robot should be able to manoeuvre the sharp angular turns inside the drains with minimum turning radius.
- The robot should be modular so that the components can be replaced easily in case of damage.

Considering the above limitations and requirements, and the properties of the cleaning robots reported in the literature, the four-legged, wheeled, and reconfigurable in height robot were conceptualized and developed. Inspired by nature's bilateral body plan of animals and insects, four legs with a reconfigurable structure were used. Taking advantage of the symmetry of terrain as shown in Figure 2, and its variable height, it will be useful to emulate the gate shown by the skater in the designed robot as shown in [31], where the height is changed by maintaining the contact of the wheels with the ground.

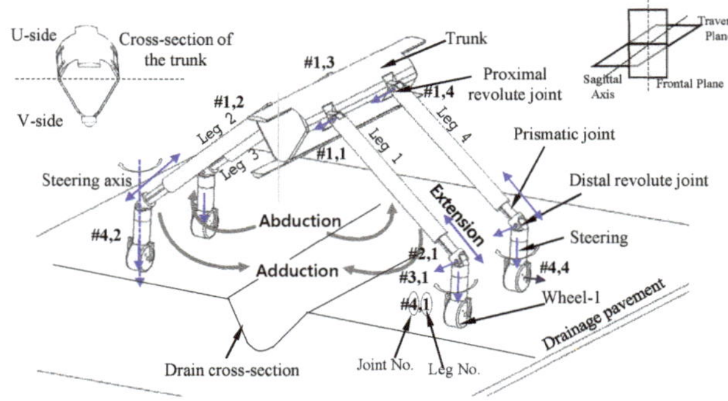

Figure 2. Line diagram of the *Tarantula* on the drainage pavement.

2.2. Mechanical Layout

The *Tarantula* robot is shown in Figure 3. It has four legs each with the four degrees of freedom (DOFs). The four DOF were provided in each leg to change the heights as per the geometry of the drainage systems and for independent steering of each wheel by keeping it in contact with the ground. The four DOFs were constituted by revolute (R), prismatic (P), and two revolute joints as the RPRR (R: revolute and P: prismatic) mechanism. The wheel attached at the end of each leg was considered as the end-effector for each leg. The wheels were used to provide the necessary locomotion. Tarantula is a manually reconfigurable robot (Figure 3) unlike the family of self-reconfigurable cleaning robots [32–36] developed. The mechanisms of the robot are discussed next.

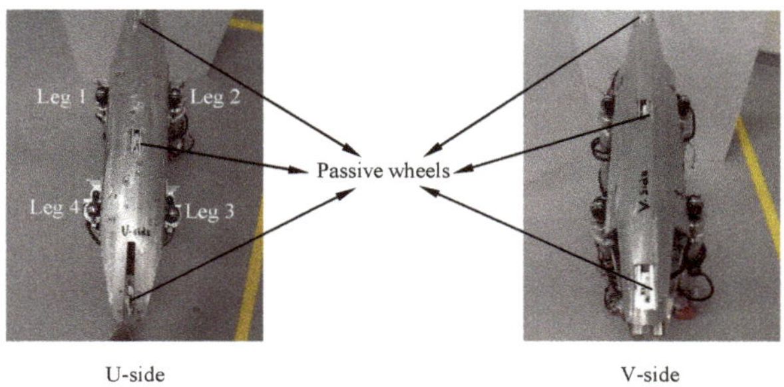

Figure 3. Two manually reconfigurable states and rotating the legs by 180 degrees which turns the body upside down.

2.2.1. Trunk

Figure 3 shows the trunk where the four legs were attached to it. Note that the trunk body has U- and V-cross-sections. This was selected as per the geometry of the drainage cross-section (Figure 2). Figure 3 shows the two manually reconfigurable states in which the robot can be placed. This feature will help to place the trunk of the robot parallel to the drain section. Three passive wheels were provided on the top and the bottom of the trunk respectively to prevent it from rubbing the ground. Inside the trunk, a mechanism for the simultaneous actuation of the two proximal revolute joints in frontal planes, i.e., (#1,1, #1,2) and the two rear legs (#1,3, #1,4) was placed. The trunk contains the necessary electronics, energy source, and sensors. It has the suspension mechanism designed to account for both upside and down configurations to safeguard the robot against jerks transmitted from the ground.

A. Simultaneous Abduction/Adduction Mechanism

The platform is designed specifically for the drain inspection task with its cross-section shown in Figure 1. Figure 4 shows the mechanism assembled inside the trunk to provide the revolute action of each leg. A single motor is used to get the simultaneous abduction or adduction of two adjacent legs in the frontal plane. It is achieved by the transmitting motion from the actuator placed inside the trunk along its length in the sagittal plane. The motor shaft is connected to the gearbox and then to the worm (W) which transfers the motion to worm wheel (WW). The shaft on which the worm wheel is mounted is supported with a boss attached to the chassis, and the shaft ends are placed with the two bevel (B) gears. Bevel gears were used to transmit the rotational motion to the proximal revolute joints of the leg. Here, the two chains (C) and sprocket (S) arrangement were used to actuate each leg. The two chains connected to the leg via sprockets provide the required stability and strength while actuating the legs. This arrangement is suitable for the constrained space, and it also helped in making the legs modular, since it can be easily replaced by removing the pins. The actuation of the proximal joints was limited between zero to 180 degrees.

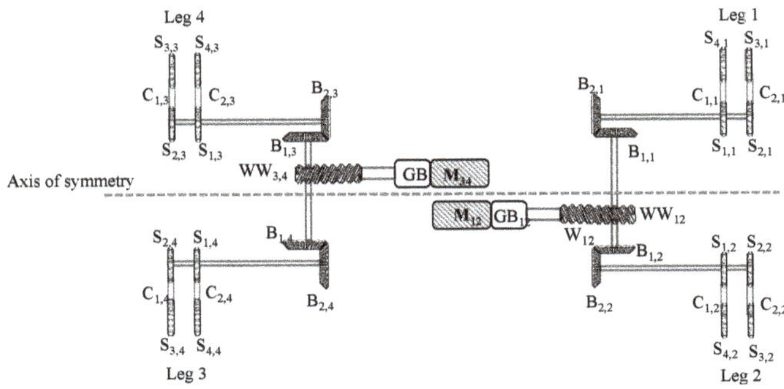

Figure 4. Line diagram of the mechanism for simultaneous abduction or adduction of the links. In the above figure W: Worm, WW: Worm Wheel, B: Bevel Gear, S: Sprocket, C: Chain, GB: Gear Box, M: Motor.

B. Suspension Mechanism

Bidirectional suspension mechanism using a single damper was designed and attached inside the trunk. The adaptive linkages' suspension mechanism was designed to work in two configurations. One is when the U-section is facing towards the ground and the other is with a V-section facing towards the ground as shown in Figure 3. The damper mechanism is attached to the chassis. The damper piston was connected to the cap on which the stabilizer bars rest. Figure 5b shows the upside down position of the mechanism. The Y-swing suspension linkage Y1 is attached to the motor shaft on which the driving sprockets $S_{1,1}$ and $S_{2,1}$ were placed on both sides. These two driving sprockets were connected to the driven sprockets $S_{3,1}$ and $S_{4,1}$ using chains $C_{1,1}$ and $C_{2,1}$, respectively; similarly for the other legs, the Y2 is shown in Figure 5b. The same mechanism is placed for the rear legs. This arrangement provided the suspension that offers a cushion for the actuators and the electronic circuits as well. The compression length was approximately 10 mm. Figure 5c shows the top and front view of the trunk with labeled components.

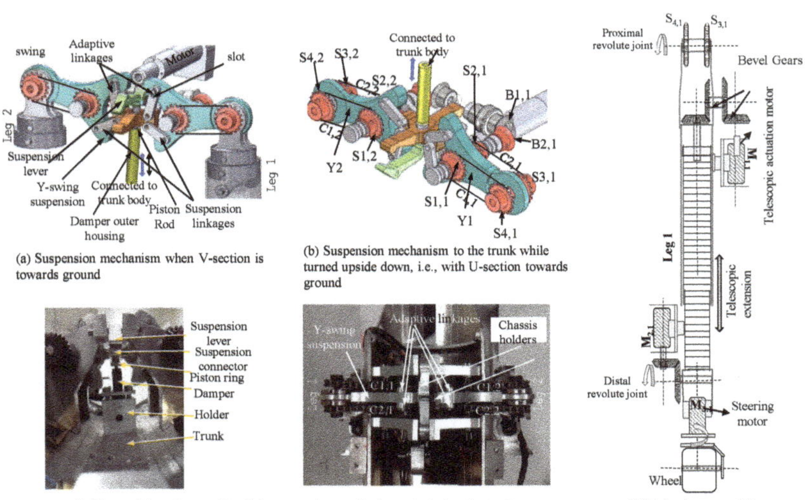

(a) Suspension mechanism when V-section is towards ground

(b) Suspension mechanism to the trunk while turned upside down, i.e., with U-section towards ground

(c) Views of the sub assembly of the suspension mechanism attached to the trunk

d) Mechanism on each leg

Figure 5. Suspension system that provides depression and elevation at the proximal revolute joints.

2.2.2. Telescopic Extension and Distal Revolute Joint

Figure 5d shows the arrangement of two pairs of the bevel gears with shafts used to actuate the screw to get the telescopic action of the leg. Two motors were attached to the body of the telescopic link. The proximal motor, i.e., close to the trunk was used to actuate the telescopic screws, and the distal one was used to provide the revolute action parallel to the abduction/adduction motion. The distal revolute joint was helpful in keeping the wheels in contact with the flat or inclined terrain (Figure 1). The kinematics of the mechanism is discussed in Section 3.

2.2.3. Steering and Wheel Suspension

The wheeled modular mechanism provides the mobility of the robot. Each of the four identical wheel modules has the actuator for steering and the in-wheel motor for the propulsion. The steering action provided to each wheel gave the necessary *four-wheel steering and four-wheel drive* (FWSD). The FWSD is essential as the terrain of the drainage system can have sharp turns and curvatures. The control problem is non-trivial for the FWSD, but the requirement of moving the robot at relatively low speed (1.8 to 2.5 Km h^{-1}) using tethered communication and the simple controller is sufficient. The suspension system provided with the three compressed springs (s1, s2, and s3) attached to each wheel is shown in Figure 6, which provides the required traction to the four wheels. It is also helpful in safeguarding the motor and the micro-controller that is mounted on the wheel hub.

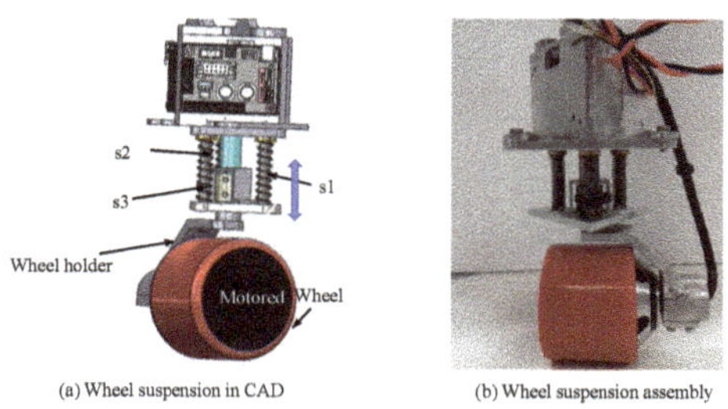

Figure 6. Wheel with suspensions.

2.2.4. Tarantula Electronics

Tarantula is controlled using the simple mechatronics system that uses the mechanical model of the vehicle and steering model to actuate the mechanism. The actuation and locomotion of the wheels were achieved through the coordination between the micro-controller and actuators. An Arduino Atmega2560 16-Bit micro-controller was mounted inside the trunk and was programmed to carry out three major functions. Namely, (a) Control signal generation to the motor driver that controls the motor speed, (b) To receive the feedback of the motor positions, and (c) To obtain the user command from the remote device or the computer. To reduce the number of wires connecting the motors to the controller, the controller area network (CAN) bus interface was used. Thin shielded cables were used to connect the three Maxon motors (DCX22S, M_{11}, M_{21}, M_{31} as shown in Figure 5) with the connectors for controlling the motor modules, i.e., the telescopic action, the distal revolute joint and the steering with the CAN bus interface. The two motors placed inside the trunk for the simultaneous abduction and adduction, namely M_{12} and M_{34} (Figure 4) were connected with the micro-controller separately.

The 24-volt Lithium polymer batteries were kept inside the trunk body cover as the power source. The switching power supply fitted inside the servos allowed for running the servos efficiently at voltages between 8 to 24 volts. These regulators allowed for using thinner wires between the modules to supply sufficient power to the servos. The waterproof skateboard wheels with hub motors were used. These were controlled with the speed and time period for the motors rotations which are defined by pulse-width modulation (PWM) signals from the micro-controller. The system architecture of a single leg is shown in Figure 7. The camera feedback was directly taken to the laptop for the image processing task as discussed in Section 4. A software interface was developed to provide the basic locomotion and reconfiguring the height of the robot. In this reported version of Tarantula, we have not placed any external sensors, i.e., proximity, ultrasonic, infrared (IR) sensors, LiDAR (to map the area), etc. However, the provision for these exists and is part of the future work. The power distribution and management system as done in [36,37] will also be carried out as part of future work.

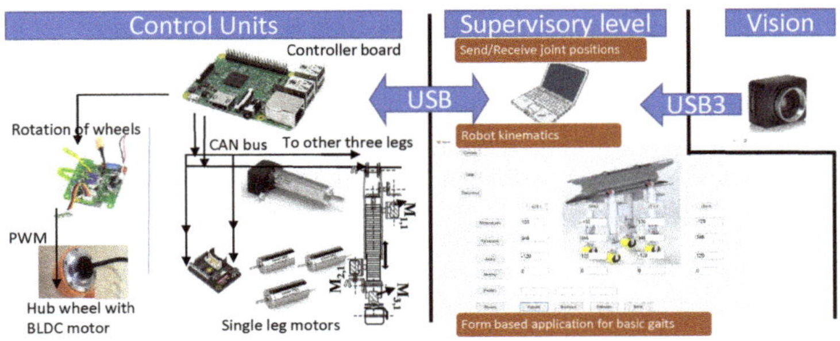

Figure 7. System diagram of *Tarantula*.

3. Modeling and Simulation

This section presents the mathematical modeling for the kinematics of the wheeled robot *Tarantula* coupled with its legged kinematics.

3.1. Kinematic Modeling

The forward kinematics of the robot is modeled using the DH convention [26]. The Denavit and Hartenberg (DH) parameters uses four independent parameters, namely joint offset b_i, joint angle θ_i, link length a_i, and twist angle α_i for the ith link to represent the transformation between two consecutive frames say i and $(i+1)$ in a kinematic chain. The Denavit–Hartenberg (DH) parameters convention used in this paper with the Homogeneous Transformation Matrix (HTM) for a single link are presented in Appendix A. Figure 8 shows the kinematic diagram of the robot platform and the legs. The DH parameters of a single leg for the *Tarantula* robot are listed in Table 2.

Table 2. Denavit and Hartenberg (DH) parameters of the single leg of *Tarantula* robot.

#	α	a	b	θ	Joint Limits Initial	Joint Limits Final	Remarks
1	90	a_1	0	θ_1 (JV)	0 deg.*	180 deg.	Joint near trunk
2	−90	0	b_2 (JV)	0	470 mm	950 mm	To adjust height
3	90	0	0	θ_3 (JV)	0 deg.	180 deg.	$\theta_3 = -\theta_1 \pm \psi$
4	0	0	b_4	θ_4 (JV)	0 deg.	360 deg.	For steering

* deg. is degrees; ψ is the inclination angle of the ground as shown Figure 1.

Figure 8b highlights the second leg and the joint axis vectors attached at each of the joints. The joint axis direction of each joint is denoted by the z-axis and W_2 is the frame attached at the point of contact of the wheel. Assuming all the legs as symmetric, the pose of the wheel W_k w.r.t. the frame F_k on the moving platform attached to the trunk near the proximal revolute joint was calculated using the successive multiplication of the homogeneous transformation matrix (HTM) as:

$$\mathbf{T}_{F,k}^{W,k} = \mathbf{T}_{F,k}^{1}\mathbf{T}_{1}^{2}\mathbf{T}_{2}^{3}\mathbf{T}_{3}^{4}\mathbf{T}_{4}^{W,k}. \quad (1)$$

The subscript k is for the four legs of the quadruped, i.e., $k = 1, \cdots, 4$. Here, it is assumed that the geometry of the legs are identical, and hence the DH parameters remain the same for the four legs. After substituting the values of DH parameters and post multiplying the HTM, the position of the wheels in the frame attached to the proximal revolute joint is:

$$\mathbf{T}_{F,k}^{W,k} = \begin{bmatrix} C\theta_{12,k}C\theta_{4,k} & -C\theta_{12,k}S\theta_{4,k} & S\theta_{13,k} & b_{4,k}S\theta_{13.k} + a_{1,k}C\theta_{1,k} + b_2S\theta_{1,k} \\ S\theta_{12,k}C\theta_{4,k} & -S\theta_{12,k}S\theta_{4,k} & 0 & 0 \\ S\theta_{4,k} & C\theta_{4,k} & 0 & b_{4,k}C\theta_{13.k} - a_{1,k}S\theta_{1,k} + b_2C\theta_{1,k} \\ 0 & 0 & 0 & 1 \end{bmatrix}. \quad (2)$$

The above expression written in the fixed frame attached to the trunk of the robot by prep-multiplying it with the HTM (coordinates and orientation of the frame shown in Figure 8) is:

$$\mathbf{T}_{T}^{W,k} = \mathbf{T}_{T}^{F,k}\mathbf{T}_{F,k}^{W,k}. \quad (3)$$

The first three elements in the fourth column of Equation (2) give the position of the wheel, i.e., $[x_{w,k}, y_{w,k}, z_{w,k}]^T$. The y-coordinates are explicitly shown in Figure 8, and it does not vary w.r.t., the frame attached with the trunk. The two-dimensional graphical representation of the workspace of a single leg is depicted in Figure 9a and, with four legs considering the constraint in Equation (5), is shown in Figure 9b.

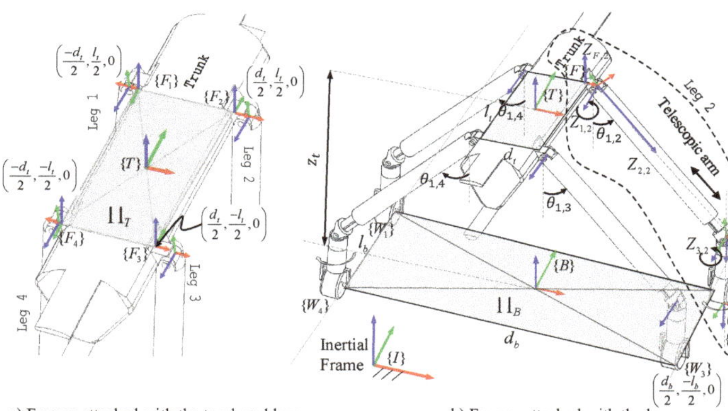

a) Frames attached with the trunk and legs b) Frames attached with the leg

Figure 8. The inertial frame {I}, body fixed frame {T} at the trunk, base frame {B} at the center of the contact point of four wheels with the ground and the DH frames attached with leg 2 having RPRR joints.

The kinematic constraint or dependencies that are utilized for the simultaneous abduction and adduction of all four legs as per the actuation mechanism in the Trunk is written as:

$$\theta_{1,1} = \theta_{1,4} = -\theta_{1,2} = -\theta_{1,3}. \quad (4)$$

Another constraint to maintain the contact of the four wheels with the inclined surface can be written as:

$$\theta_{3,1} = -\theta_{1,1} \pm \psi, \theta_{3,2} = \theta_{1,2} \pm \psi, \theta_{3,3} = -\theta_{1,3} \pm \psi \text{ and } \theta_{3,4} = -\theta_{1,4} \pm \psi, \quad (5)$$

where ψ is the angle of inclination of the pavement as shown in Figure 1. The above constraints for any flat surface perpendicular to the direction of gravity were obtained by substituting $\psi = 0$.

In Section 4.3, the kinematic parameters were identified along with the effective dimension of the trunk in the plane Π_T (Figure 8a).

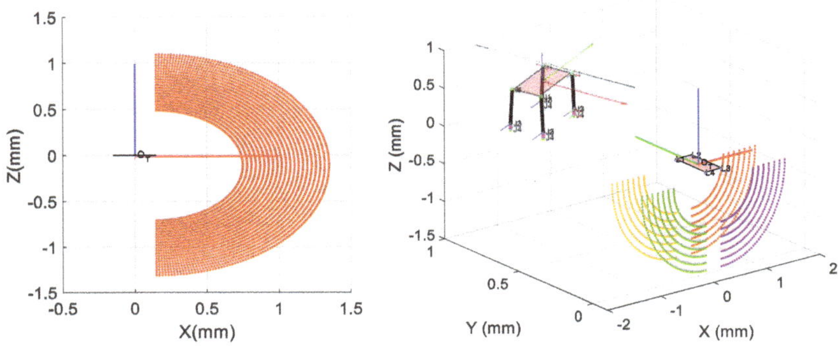

a) Workspace of one robotic arm

b) Workspace of four legs with one section of trunk facing toward ground

Figure 9. Graphical representation of the workspace of Tarantula.

3.2. Kinematics of Wheel

The kinematics of the platform depend upon the arrangement of the wheels and its type selected. They are discussed in detail in [38]. The various architecture of mobile robots rely on the choice of wheel arrangement like differential drive robots, omnidirectional wheels, traction wheels, etc. The four motorized and steered standard wheels were selected, which resulted in greater maneuverability. The formulation is presented for the steering with the varying dimensions of the base plane Π_B (Figure 8b) that is dependent on the leg kinematics.

Figure 10a shows the position of four wheels (W_1, W_2, W_3, W_4) on the base plane. Note that the location of the wheels is subjected to vary as per the change in joint angles θ_{1k} of the legs. It changes the dimension of the rectangle defined by the points of contact of the wheels in the base plane Π_B. The position vector of the wheels in the frame attached with the base is denoted by $\mathbf{l}_{w,k}$. The magnitude and the angle subtended by the position vector are given by:

$$l_{w,k} = \sqrt{x_k^2 + y_k^2} \quad \text{and} \quad \gamma_k = \tanh^{-1}(y_k, x_k), \tag{6}$$

where $\tanh^{-1}$ is the inverse hyperbolic tangent function. The values of x_k were found from Equation (2) and y_k was obtained from the geometry. These point of contacts were experimentally identified in Section 4.3. In this section, the generalized kinematic modeling of the four wheels is presented. The sate vector of the robot's base frame is defined as:

$$\mathbf{v}_B = \begin{bmatrix} \dot{x}_B & \dot{y}_B & \dot{\alpha}_B \end{bmatrix}^T, \tag{7}$$

where $\dot{x}_B, \dot{y}_B$ are the velocity vectors along the x- and y-directions and $\dot{\alpha}_B$ is the angular velocity vector about the z-axis as shown in Figure 10b. Now, taking the component of the velocity vectors at the origin of the base frame, the rolling wheel constraint of the kth wheel was written as:

$$[\sin(\gamma_k + \beta_k) \quad -\cos(\gamma_k + \beta_k) \quad l_{r,k}\cos\beta_k]\mathbf{v}_B - r_w\dot{\varphi} = 0, \tag{8}$$

where $\dot{\varphi} \equiv [\varphi_1\ \varphi_2\ \varphi_3\ \varphi_4]^T$ is the vector of rate of rotation for the wheels. In addition, the no-sliding constraint for the kth wheel is written as:

$$[\cos(\gamma_k + \beta_k) \quad \sin(\gamma_k + \beta_k) \quad l_{w,k} \sin \beta_k]\mathbf{v}_B = 0, \tag{9}$$

where $l_{w,k}$ and γ_k were found as per Equation (6). Equations (8) and (9) were written for all the four wheels as:

$$\mathbf{C}_R \mathbf{v}_B - \mathbf{W}\dot{\phi} = \mathbf{0}, \tag{10}$$

$$\mathbf{C}_S \mathbf{v}_B = \mathbf{0}. \tag{11}$$

The matrix $\mathbf{C}_R$ and $\mathbf{C}_S$ are defined as

$$\mathbf{C}_R \equiv \begin{bmatrix} \sin(\gamma_1+\beta_1) & \cos(\gamma_1+\beta_1) & l_{w,1}\cos\beta_1 \\ \sin(\gamma_2+\beta_2) & \cos(\gamma_2+\beta_2) & l_{w,2}\cos\beta_2 \\ \sin(\gamma_3+\beta_3) & \cos(\gamma_3+\beta_3) & l_{w,3}\cos\beta_3 \\ \sin(\gamma_4+\beta_4) & \cos(\gamma_4+\beta_4) & l_{w,4}\cos\beta_4 \end{bmatrix} \text{ and } \mathbf{C}_S \equiv \begin{bmatrix} \cos(\gamma_1+\beta_1) & \sin(\gamma_1+\beta_1) & l_{r,1}\sin\beta_1 \\ \cos(\gamma_2+\beta_2) & \sin(\gamma_2+\beta_2) & l_{r,2}\sin\beta_2 \\ \cos(\gamma_3+\beta_3) & \sin(\gamma_3+\beta_3) & l_{r,3}\sin\beta_3 \\ \cos(\gamma_4+\beta_4) & \sin(\gamma_4+\beta_4) & l_{r,4}\sin\beta_4 \end{bmatrix}, \tag{12}$$

where the values of γ_k for $k = 1, \cdots, 4$ are obtained from Equation (6). The degree of maneuverability M was found using the sum of the mobility m, i.e., the dimensionality of the null space of the matrix $\mathbf{C}_R$ and steerability s, i.e., the rank of the matrix $\mathbf{C}_S$. In short, the maneuverability is found as 3 ($M = m + s \equiv 0 + 3 = 3$). The state vector in the inertial frame was found by pre multiplying the state vector by the rotation matrix between the inertial and the trunk frames. The experimental results for the trajectory followed by the robot is presented in Section 4.

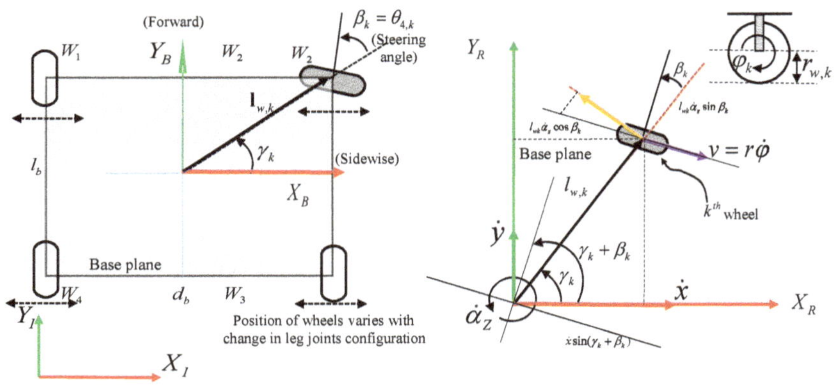

a) Kinematics of the wheel

b) Vector diagram of the kth wheel

Figure 10. Steering kinematics of the wheels with varying base dimensions.

4. Experiments

In the Introduction, it was mentioned that precision and tolerances in fabrication and assembly errors result in the difference between the working model and the CAD. In addition, the robot positioning performance relies on the kinematic model of the robot. This led us to experimentally identify the kinematic parameters, namely the DH parameters by estimating the joint axes' vectors. Here, we have extended the circle point method (CPM) for the prismatic joint present in the RPRR mechanism in each leg of *Tarantula*.

4.1. Setup

Unlike the visual localization done using a single marker reported in [29], we have used ArUco markers map (AMM) for two purposes: one is for the identification of the kinematic parameters of each leg. The other is for the trajectory tracking of the robot. The advantage of using AMM over a single calibration grid or a single marker is that, with a single marker, it is impossible to always keep it in the line of sight of the camera. Figure 11 shows the printed markers from the ArUco library (ARUCO_MIP_36H12) [39,40] on the flat boards.

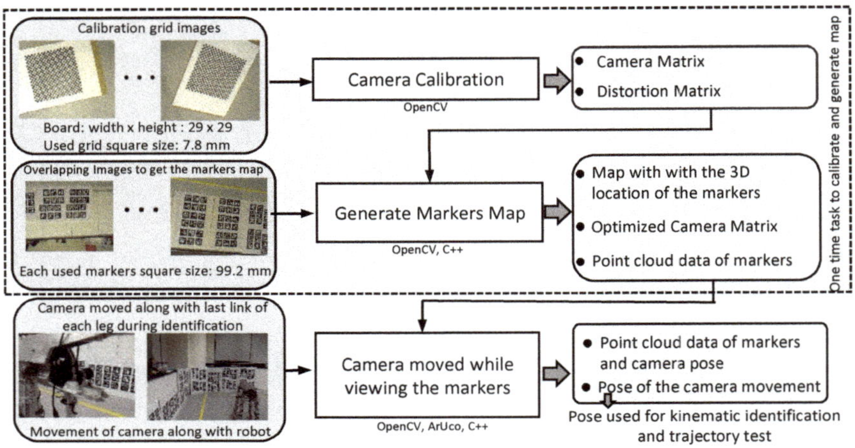

Figure 11. Flow diagram to obtain a pose from the calibrated camera using ArUco markers.

These multiple markers are distinct, and each having a dimension of 99.2 × 99.2 mm was glued on the flat board. These markers were glued without any waviness on the board, which can cause the error in pose reading. The process required building a pairwise markers map database by taking the sequence of images with the common markers appearing in consecutive images using the calibrated camera. This database generated contains the relative pose of each marker concerning one another [29]. We placed the markers such that multiple markers (more than five markers) were visible in a single image or frame. This helped in achieving better pose estimation of the camera using an optimization that involved minimizing the reprojection error of markers in the observed frame [29]. Two approaches for using camera to make position measurements are presented in [41], namely, monocular camera [43,44], and the other is stereo vision [42]. Stereo cameras allows direct pose measurements, but it requires the processing of corresponding images which makes it computational expensive. Whereas, monocular camera usage for pose measurement is an area of research interest because of the following aspects: (i) Synchronizing the captured images taken from multiple cameras is not needed; (ii) Pose readings over a larger workspace can be taken in a region without bothering on the field of view of two or more cameras overlap; (iii) The space required for mount is reduced. Therefore, the use of a monocular camera for pose measurement for kinematic identification and a trajectory followed test is utilized.

Some details of the approach during the experiments are worth noting. The experiments were carried out in ambient lighting conditions. For kinematic identification, the camera was mounted on the last link of a leg, and each joint was actuated one at a time. The frames were grabbed using the Chameleon3 camera from Ptgrey (FLIR Integrated Imaging Solutions Inc., Richmond, BC, Canada) [45] with the 10 mm fixed focal length lens attached to it at 40 frames per second. The position of the markers must not be changed after being placed for a given set of readings. By placing the markers in spread fashion over a larger area, it was possible to actuate the robot's joints in its full workspace as shown in Figure 9b. The frame was grabbed using the camera, and the pose was obtained using the flow diagram presented in Figure 11. The evaluation of the measurement approach proposed here is presented next.

4.2. Measurement Performance

Evaluation of the proposed measurement technique was done by mounting the camera on the KUKA KR6 R900 robot that has the measurement performance of 0.03 mm [30]. The robot's end-effector with the camera mounted was made to move in the x-, y- and z-axis directions of the robots world frame. The initial distance of the camera from the markers was 3.4 m. The initial and final coordinates of the robot were recorded from the teach-pendent (Figure 12a) for movement along each axes. The pose obtained from the camera is plotted in Figure 12b. The measured angle between the given motion about each axis is shown in Figure 12c. Assuming the readings from the robot as the ground truth, the trajectory of the robot end-effector is compared to the one obtained using the monocular camera. Table 3 lists the ideal and the measured distances using the markers. It is observed that variation along the y-direction is highest and in this case, the camera was moving towards the markers, i.e., perpendicular to the plane of the markers. In the rest two of the directions, the variations are within 3 mm. Hence the movement in a plane along the markers is closer to the ideal than the depth one.

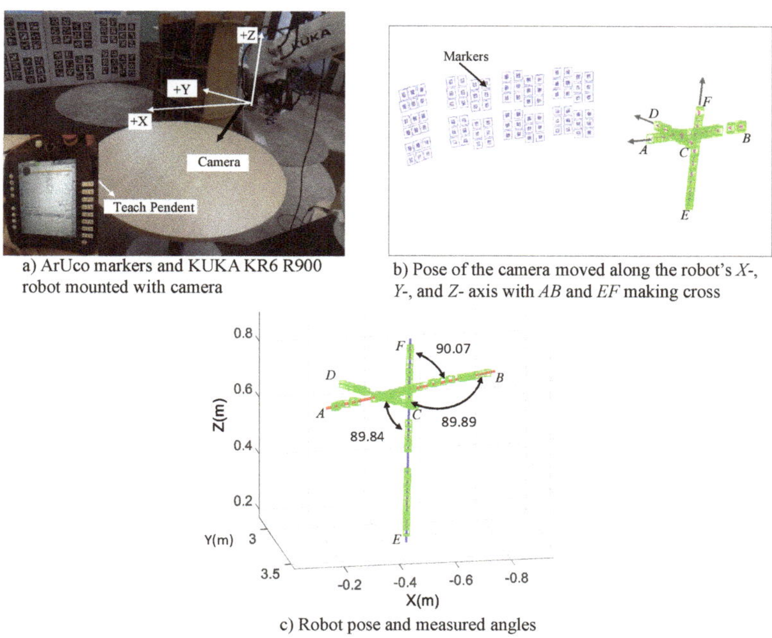

Figure 12. Evaluation of measurements using the ArUco markers and monocular camera.

Table 3. Evaluation of measurement performance using monocular camera and ArUco markers.

	$X:\overline{AB}$ (m)	$Y:\overline{CD}$ (m)	$Z:\overline{EF}$ (m)	$\measuredangle(\overline{AB},\overline{CD})$	$\measuredangle(\overline{CD},\overline{EF})$	$\measuredangle(\overline{EF},\overline{AB})$
Ideal (Robot)	0.400	0.7776	0.7467	90	90	90
Measured (Markers)	0.3987	0.7235	0.7494	89.89	89.84	90.07

X :, Y :, Z : means x-, y- and z-directions in robots world frame, $\measuredangle$: Angles are in degrees.

4.3. Identification of Kinematic Parameters of Tarantula

After the assembly of *Tarantula*, the identification of the proximal revolute joint positions and kinematic parameters of each leg is essential. The mathematical analysis of the measured data to obtain the joint axis vectors (JAVs) presented elsewhere [25] is discussed in short for brevity along with its extension for the prismatic joints. The points traced by the end-effector, i.e., the three-dimensional (3D) coordinates ($\mathbf{x_i} \equiv [x_i, y_i, z_i]^T$) were logged and stacked in a matrix **A** (Equation (13)). The mean of the logged pose data points was subtracted from the elements of **A** which resulted in the transformed set of points in a matrix **B** whose mean is zero. Matrices of three-dimensional data points, **D** and $\bar{\mathbf{D}}$ are shown below:

$$\mathbf{A} = [\mathbf{x}_1 \quad \mathbf{x}_2 \quad \cdots \quad \mathbf{x}_m]^T \; ; \bar{\mathbf{A}} = [(\mathbf{x}_1 - \bar{\mathbf{x}}) \quad (\mathbf{x}_2 - \bar{\mathbf{x}}) \quad \cdots \quad (\mathbf{x}_m - \bar{\mathbf{x}})]^T, \tag{13}$$

where $\bar{\mathbf{x}} \equiv [\bar{x} \; \bar{y} \; \bar{z}]^T = \frac{1}{m}[\sum x_i \; \sum y_i \; \sum z_i]^T$, and m being the number of measurements. Applying singular value decomposition (SVD) [46] on matrix $\bar{\mathbf{A}}$, two square orthogonal matrices **U** and **V** and a rectangular matrix **D** were obtained as:

$$\bar{\mathbf{A}} = \mathbf{V}_{m \times m} \mathbf{D}_{m \times 3} \mathbf{U}_{3 \times m}. \tag{14}$$

The orthogonal columns corresponding to the singular values (SVs) are listed in the column of matrix $\mathbf{U} \equiv [\mathbf{u}_1 \; \mathbf{u}_2 \; \mathbf{u}_3]$. The direction of the joint axis represented by the unit vector **n** in Figure 13a for revolute and Figure 13b for prismatic is given as:

$$\mathbf{n} \equiv \mathbf{u}_3 \quad \texttt{for rotary joints}, \tag{15a}$$

$$\mathbf{n} \equiv \mathbf{u}_1 \quad \texttt{for linearactuating joints}. \tag{15b}$$

The above equations gave the joint axis vector direction. For a point on the JAV, the center of circle **c**, i.e., the center of rotation for revolute joint was obtained using circle fitting method presented in [25], where $\mathbf{c} \equiv c_1\mathbf{u}_1 + c_2\mathbf{u}_2 + \bar{\mathbf{x}}$, c_1 and c_2 are the center of fitted circle in the plane spanned by vectors $\mathbf{u}_1$ and $\mathbf{u}_2$. For prismatic joint the mean of the traced point, i.e., $\mathbf{c} \equiv \bar{\mathbf{x}}$ was taken as the point on the JAV of prismatic joints. The plane of link movement $\mathbf{\Pi}$ can be defined as:

$$\mathbf{\Pi} \equiv \begin{bmatrix} \mathbf{n} \\ -\mathbf{n}^T\bar{\mathbf{x}} \end{bmatrix}. \tag{16}$$

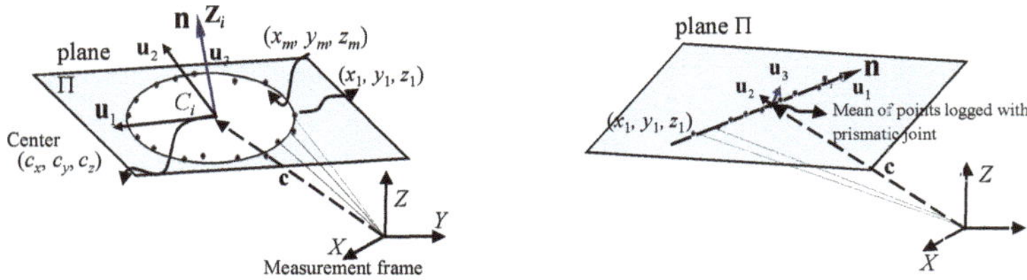

(a) Center and orthogonal directions for revolute joints

(b) Singular values and normal direction for prismatic joints

Figure 13. Position data points on the circle and the line obtained with the revolute and prismatic joint actuation respectively with its singular value direction.

The above information, i.e., the direction of joint axis denoted with **n** and **c** being points on it, were used to extract the DH parameters according to Algorithm 1. The identified parameters are listed in Table 4. The effective length and breadth of the trunk were also identified from the JAVs.

The experimental set-up for the identification experiments is shown in Figure 14b,c. The robot's trunk was rigidly fixed with the frame. Each leg joint was actuated one by one and the frames were grabbed to process it as per the flow diagram shown in Figure 11. The advantage of the circle point method is that the coordinates of the proximal joints were also known. This is helpful in defining the platform plane formed by proximal joints' positions.

Table 4. Identified DH parameters of each leg.

	b_1	a_1	α_1 (deg.)	b_2 (mm)	a_2	α_2 (deg.)	b_3	a_3	α_3 (deg.)	b_4	a_4	α_4 (deg.)
Leg 1	0	0	90.17	485.23 + var	0	−91.13	0	0	89.86	69.21	0	0
Leg 2	0	0	90.01	480.13 + var	0	−91.07	0	0	89.06	70.12	0	0
Leg 3	0	0	89.71	482.72 + var	0	−89.89	0	0	90.32	72.35	0	0
Leg 4	0	0	89.96	482.36 + var	0	−90.14	0	0	89.78	78.16	0	0

var: is the variable length of actuation varying from zero to 0.6 meters and the numeric value is at the compressed state. b_4: these joint offset values are till the position where the camera was placed.

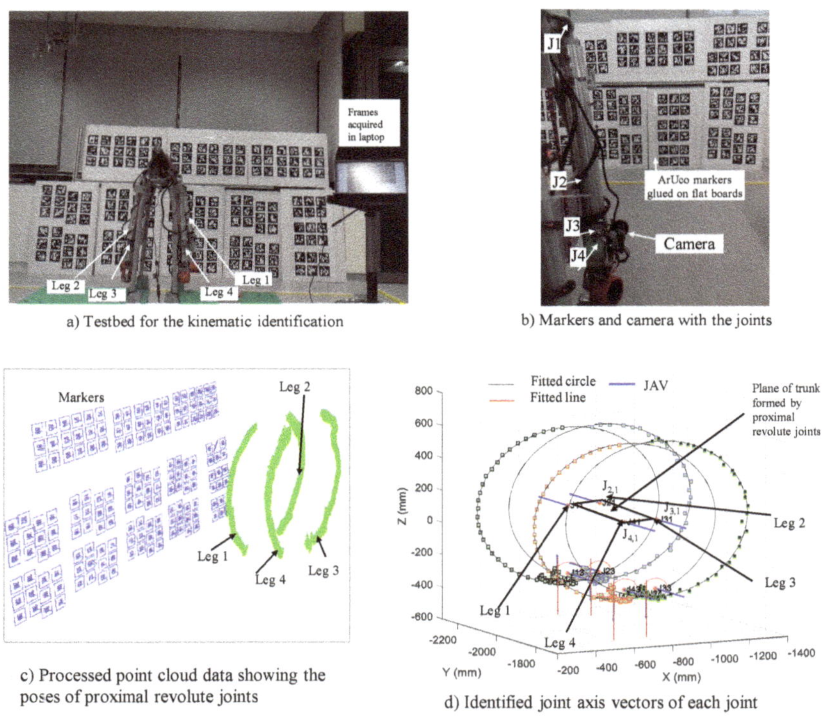

Figure 14. Setup using AruCo markers for identification and localization of robot.

The coordinates of the centre of rotation of the four proximal joints were found that are depicted in Figure 14d as $J_{1,1} \equiv [-0.7038, -1.8386, 0.1208]$, $J_{2,1} \equiv [-0.8728, -1.8400, 0.1219]$, $J_{3,1} \equiv [-0.8693, -2.1265, 0.1205]$ and $J_{4,1} \equiv [-0.6994, -2.1242, 0.1205]$, respectively, in meters, and the dimensions of the rectangle fitted with these four points were found as $l_t = 0.2855$ and $d_t = 0.1696$ m. The identified values were used as y_k in Equation (6), i.e., the magnitude of y_k is half the effective length of the trunk. Next, we have evaluated the measurement technique and performed an initial path followed test.

In order to compare the the identified DH parameters of the four legs using the monocular vision, we selected the parameters from the CAD listed in Table 2. We have used the quintic trajectory using the 3–4–5 interpolating polynomial [47] for the prismatic and revolute joints using l and θ respectively as:

$$\{l,\theta\}(t) = a_0 + a_1 t + a_2 t^2 + a_3 t^3 + a_4 t^4 + a_5 t^5 \equiv \sum_{i=0}^{5} a_{\{l,\theta\}i} t^i, \tag{17}$$

where $a_i (i = 1, \cdots, 5)$ are the coefficients that were derived from the initial and final state of the joints. The velocity and acceleration for linear actuation and rotation, i.e., $\dot{l}, \dot{\theta}$ and $\ddot{l}, \ddot{\theta}$ can be found by taking the first and second order derivatives of Equation (17). The detailed trajectory equations with its derivatives are discussed in [47]. The trajectory as per Equation (17) was used to actuate the joints from its initial to final position, i.e., within the range of 0° to 90° with quintic profile. Figure 15a shows the variation in the position of the wheel hub plotted in in each of the leg frames. The difference in the X-, Y- and Z-positioning of each leg is plotted w.r.t. the CAD parameters in Figure 15a–c, respectively. The variation in x- and y-directions are significant, and working with the identified values in the kinematic model is useful.

Algorithm 1 Algorithm to find the Denavit–Hartenberg (DH) parameters.

- Fix the camera on the last link of the legged robot and keep the trunk fixed, (as shown in Figure 14b).

For i = 1 to n

- Move one joint at a time starting from the first joint while locking the rest. The position of the camera will trace the circular arc and prismatic joint will trace a straight line.

- Log the 3D positions of the camera using ArUco markers map set-up (Section 4). For revolute joint actuate it by ϕ and record the video feed from the camera with aruco markers visible, while for the prismatic joint it was actuated by distance l.

- Find the centre of the circle using 3D circle fitting method for revolute joints and mean of linear points for prismatic joints with the direction of joint axis vector as the normal of the plane of movement.

End for

For $i = n$ to 1

Extract the DH parameters, i.e., b_i, a_i and α_i, using JAVs as inputs to find the perpendicular distances and angles between the two successive joint axis vectors.

End for

- Repeat the above steps for each leg.

- With the coordinates of center of rotation of proximal joints, estimate the effective dimension of the plane of trunk Π_T as shown in Figure 8.

4.4. Trajectory Tracking

The purpose of this section is to test the trajectory followed by the robot. The ArUco markers map were arranged as shown in Figure 16a. The monocular camera was now placed inside the trunk. The same markers arranged in two directions were used to track the pose of the robot. The robot was commanded to move four meters forward and then steer the wheels by 90° and again traverse forward by a meter. Figure 16 shows the point cloud data (pcd) of the pose. Then, it was plotted with the coordinate frame attached at one of the markers. The trajectory followed obtained after fitting the line is shown in Figure 16b with the angle of turn obtained as 87.8 degrees. Note that, in this case, the frames were recorded after every 100 milliseconds for the total trajectory time of 2.3 s. The distance traversed is shown in Figure 16c. The difference in the desired and the obtained trajectory is mainly due to the friction and uncertainties in the model. The limitations of this approach are the measuring variations in the pose that can be ±3 mm at a distance of 4–5 m of the camera from the markers and also the lighting conditions affect the detection of the markers.

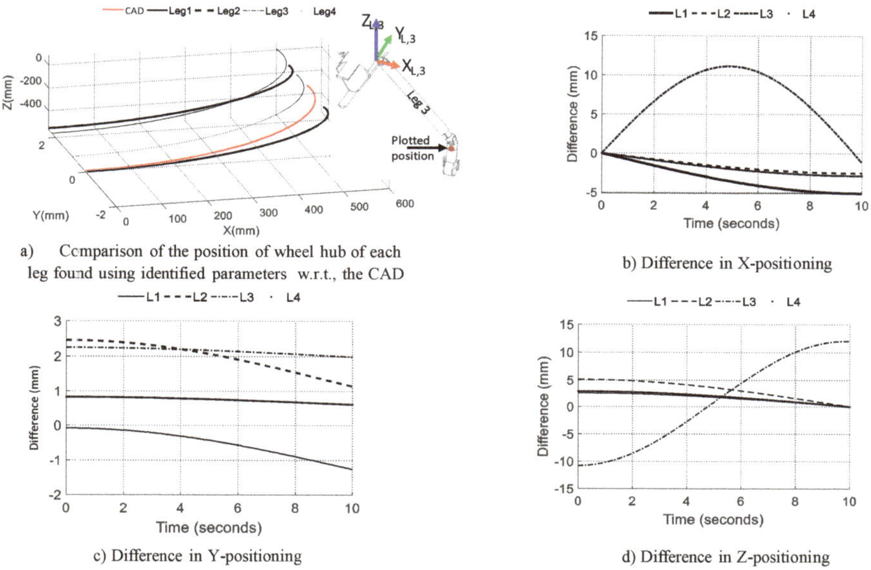

a) Comparison of the position of wheel hub of each leg found using identified parameters w.r.t., the CAD

b) Difference in X-positioning

c) Difference in Y-positioning

d) Difference in Z-positioning

Figure 15. Comparison of the positioning of each leg w.r.t. the CAD.

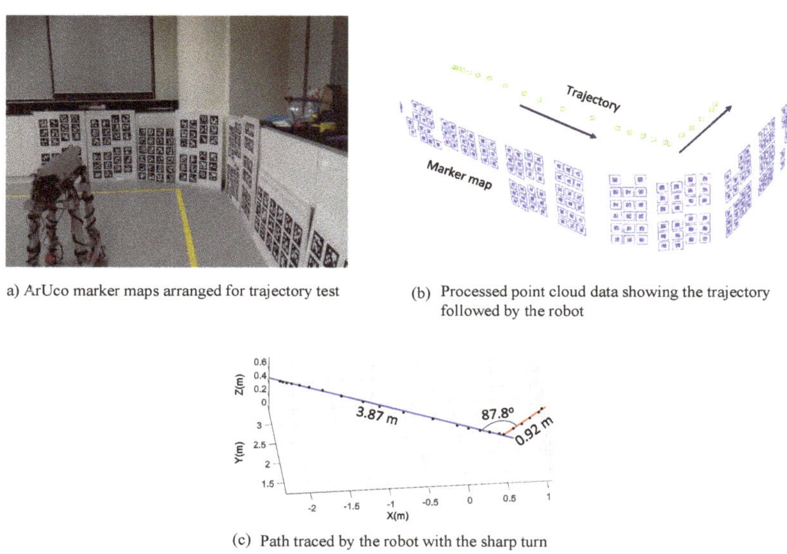

a) ArUco marker maps arranged for trajectory test

(b) Processed point cloud data showing the trajectory followed by the robot

(c) Path traced by the robot with the sharp turn

Figure 16. Experimental setup and the trajectory traced by the robot.

5. Conclusions and Future Works

In this paper, *Tarantula's* design, modeling, and its kinematic parameters identification are presented. The robot was designed for the specific geometry of the drains. The designed manipulator is modular and can adjust its height as per the environment. The simultaneous abduction or adduction of the two legs in a frontal plane was provided using the chain sprocket mechanisms connected using bevel gears arrangement with a single motor. A specially designed bi-directional suspension mechanism was fixed inside the trunk as a shock absorber. The limitation of the chain sprocket is that it imparts the rotational play because of slacking.

To the best of the author's knowledge, this is the first attempt to estimate the kinematic parameters of the assembled robot with four leg using monocular camera mounted on the assembled *Tarantula*. This method also identified the effective dimension of the trunk where the proximal revolute joints were connected. The limitations of the monocular vision and fiducial markers to obtain the pose is that the camera must view the markers in the generated map. We also performed the experiments to evaluate the trajectory tracked by *Tarantula* using a similar set-up. Overall, the contributions of this paper are listed below:

- Design of the modular robot *Tarantula* with the ability to reconfigure its height, mainly for inspecting the drains,
- A mechanism designed for the simultaneous abduction/adduction of legs,
- A methodology to identify the kinematic parameters using ArUco markers and monocular vision for the assembled mobile robot with legs. First, using the calibrated camera, poses of ArUco markers are reconstructed in 3D space. Second, by moving each joint and capturing images, a set of the tracked pose is determined. Then, the DH parameters were evaluated.

The domains set for future research work will mainly focus on: design optimization of the legs to reduce the total weight. Since the design is modular, the newly designed legs can be easily attached to the existing trunk. The second focus will be to mount the Light Detection and Ranging (LiDAR) sensor near the trunk opening of *Tarantula* to map the drains.

Author Contributions: A.A.H. and M.R.E. conceived and designed the experiments;the experiments; A.A.H. and K.E. performed the experiments; A.A.H. and K.E. analyzed the data; K.E., A.A.H. and M.S.T. contributed reagents/materials/analysis tools; A.A.H., K.E., M.S.T. and M.R.E., wrote the paper.

Funding: This work is financially supported by the National Robotics R&D Program Office, Singapore, under the Grant No. RGAST1702 to Singapore University of Technology and Design (SUTD) which are greatly acknowledged to conduct this research.

Conflicts of Interest: The authors declare no conflict of interest.

Appendix A. DH Parameters Notation Used [48]

In this appendix, the definitions of Denavit–Hartenberg (DH) parameters are presented for completeness of the paper. Note that four parameters, namely, b_i, θ_i, a_i and α_i, relate the transformation between two frames i and $i+1$ which are rigidly attached to two consecutive links $\#(i-1)$ and $\#(i)$, respectively, as shown in Figure A1. Their notations and descriptions are summarized in Table A1.

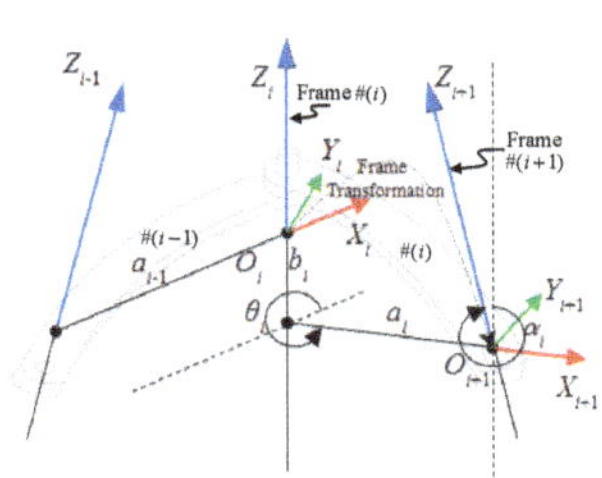

Figure A1. Links and DH parameters.

The resulting coordinate transformation between the frames connected to link i and $i+1$ as:

$$\mathbf{T}_i^{i+1} = \mathbf{T}_{b_i}\mathbf{T}_{\theta_i}\mathbf{T}_{a_i}\mathbf{T}_{\alpha_i} \quad \text{(A1a)}$$

$$= \begin{bmatrix} 1 & 0 & 0 & 0 \\ 0 & 1 & 0 & 0 \\ 0 & 0 & 1 & b_i \\ 0 & 0 & 0 & 1 \end{bmatrix} \begin{bmatrix} C\theta_i & -S\theta_i & 0 & 0 \\ S\theta_i & C\theta_i & 0 & 0 \\ 0 & 0 & 1 & 0 \\ 0 & 0 & 0 & 1 \end{bmatrix} \begin{bmatrix} 1 & 0 & 0 & a_i \\ 0 & 1 & 0 & 0 \\ 0 & 0 & 1 & 0 \\ 0 & 0 & 0 & 1 \end{bmatrix} \begin{bmatrix} 1 & 0 & 0 & 0 \\ 0 & C\alpha_i & -S\alpha_i & 0 \\ 0 & S\alpha_i & C\alpha_i & 0 \\ 0 & 0 & 0 & 1 \end{bmatrix} \quad \text{(A1b)}$$

$$= \begin{bmatrix} C\theta_i & -S\theta_i C\alpha_i & S\theta_i S\alpha_i & a_i C\theta_i \\ S\theta_i & C\theta_i C\alpha_i & -C\theta_i S\alpha_i & a_i S\theta_i \\ 0 & S\alpha_i & C\alpha_i & b_i \\ 0 & 0 & 0 & 1 \end{bmatrix}. \quad \text{(A1c)}$$

Table A1. Notations and descriptions of the DH parameters [26].

Parameters (Name)	Description *
b_i (Joint offset)	$X_i \xrightarrow[@Z_i]{\perp,\text{distance}} X_{i+1}$
θ_i (Joint angle)	$X_i \xrightarrow[@Z_i]{ccw,\text{ rotation}} X_{i+1}$
a_i (Link length)	$Z_i \xrightarrow[@X_{i+1}]{\perp,\text{distance}} Z_{i+1}$
α_i (Twist angle)	$Z_i \xrightarrow[@X_{i+1}]{ccw,\text{ rotation}} Z_{i+1}$

In the table read symbol $\longrightarrow$ as "and", $\perp$ as "perpendicular", @ as "along"and ccw as "counter clockwise".

References

1. Zhang, S.X.; Pramanik, N.; Buurman, J. Exploring an innovative design for sustainable urban water management and energy conservation. *Int. J. Sustain. Dev. World Ecol.* **2013**, *20*, 442–454. [CrossRef]
2. Kirkham, R.; Kearney, P.D.; Rogers, K.J.; Mashford, J. PIRAT—A system for quantitative sewer pipe assessment. *Int. J. Robot. Res.* **2000**, *19*, 1033–1053. [CrossRef]
3. Bradbeer, R. The Pearl Rover Underwater Inspection Robot. In *Mechatronics and Machine Vision;* Billingsley, J., Ed.; Research Studies Press: Baldock, UK, 2000; pp. 255–262.
4. Truong-Thinh, N.; Ngoc-Phuong, N.; Phuoc-Tho, T. A study of pipe-cleaning and inspection robot. In Proceedings of the International Conference on Robotics and Biomimetics (ROBIO), Karon Beach, Phuket, Thailand, 7–11 December 2011; pp. 2593–2598.
5. Kuntze, H.; Schmidt, D.; Haffner, H.; Loh, M. KARO-A flexible robot for smart sensor-based sewer inspection. In Proceedings of the No Dig'95, Dresden, Germany, 19–22 September 1995; pp. 367–374.
6. Nassiraei, A.A.; Kawamura, Y.; Ahrary, A.; Mikuriya, Y.; Ishii, K. Concept and design of a fully autonomous sewer pipe inspection mobile robot KANTARO. In Proceedings of the International Conference on Robotics and Automation, Roma, Italy, 10–14 April 2007; pp. 136–143.
7. Kirchner, F.; Hertzberg, J. A prototype study of an autonomous robot platform for sewerage system maintenance. *Auton. Robots* **1997**, *4*, 319–331. [CrossRef]
8. Streich, H.; Adria, O. Software approach for the autonomous inspection robot MAKRO. In Proceedings of the International Conference on Robotics and Automation, New Orleans, LA, USA, 26 April–1 May 2004; Volume 4, pp. 3411–3416.
9. Baghani, A.; Ahmadabadi, M.N.; Harati, A. Kinematics modeling of a wheel-based pole climbing robot (UT-PCR). In Proceedings of the 2005 IEEE International Conference on Robotics and Automation (ICRA 2005), Barcelona, Spain, 18–22 April 2005; pp. 2099–2104.
10. Ratanghayra, P.R.; Hayat, A.A.; Saha, S.K. Design and Analysis of Spring-Based Rope Climbing Robot. In *Machines, Mechanism and Robotics*; Springer: Berlin, Germany, 2019; pp. 453–462.

11. Fuchida, M.; Pathmakumar, T.; Mohan, R.E.; Tan, N.; Nakamura, A. Vision-based perception and classification of mosquitoes using support vector machine. *Appl. Sci.* **2017**, *7*, 51. [CrossRef]
12. Raibert, M.H. *Legged Robots That Balance*; MIT Press: Cambridge, MA, USA, 1986.
13. Berns, K.; Ilg, W.; Deck, M.; Albiez, J.; Dillmann, R. Mechanical construction and computer architecture of the four-legged walking machine BISAM. *IEEE/ASME Trans. Mechatron.* **1999**, *4*, 32–38. [CrossRef]
14. Ridderstrom, C.; Ingvast, J. Quadruped posture control based on simple force distribution-a notion and a trial. In Proceedings of the International Conference on Intelligent Robots and Systems, Maui, HI, USA, 29 October–3 November 2001; Volume 4, pp. 2326–2331.
15. Hirose, S.; Kato, K. Study on quadruped walking robot in Tokyo Institute of Technology-past, present and future. In Proceedings of the International Conference on Robotics and Automation, San Francisco, CA, USA, 24–28 April 2000; Volume 1, pp. 414–419.
16. Kitano, S.; Hirose, S.; Endo, G.; Fukushima, E.F. Development of lightweight sprawling-type quadruped robot titan-xiii and its dynamic walking. In Proceedings of the International Conference on Intelligent Robots and Systems (IROS), Tokyo, Japan, 3–7 November 2013; pp. 6025–6030.
17. Arikawa, K.; Hirose, S. Development of quadruped walking robot TITAN-VIII. In Proceedings of the IEEE/RSJ International Conference on Intelligent Robots and Systems (IROS 96), Osaka, Japan, 8 November 1996; Volume 1, pp. 208–214.
18. De Santos, P.G.; Garcia, E.; Estremera, J. *Quadrupedal Locomotion: An Introduction to the Control of Four-Legged Robots*; Springer Science & Business Media: Berlin, Germany, 2007.
19. Montes, H.; Armada, M. Force control strategies in hydraulically actuated legged robots. *Int. J. Adv. Robot. Syst.* **2016**, *13*, 50. [CrossRef]
20. Li, T.; Ma, S.; Li, B.; Wang, M.; Li, Z.; Wang, Y. Development of an in-pipe robot with differential screw angles for curved pipes and vertical straight pipes. *J. Mech. Robot.* **2017**, *9*, 051014. [CrossRef]
21. Sharf, I. Dynamic Locomotion with a Wheeled-Legged Quadruped Robot. In *Brain, Body and Machine*; Springer: Berlin, Germany, 2010; pp. 299–310.
22. Barker, L.K. *Vector-Algebra Approach to Extract Denavit–Hartenberg Parameters of Assembled Robot Arms*; NASA Langley Research Center: Hampton, VA, USA, 1983.
23. Hollerbach, J.M.; Wampler, C.W. The calibration index and taxonomy for robot kinematic calibration methods. *Int. J. Robot. Res.* **1996**, *15*, 573–591. [CrossRef]
24. De Santos, P.G.; Jiménez, M.A.; Armada, M.A. Improving the motion of walking machines by autonomous kinematic calibration. *Auton. Robots* **2002**, *12*, 187–199. [CrossRef]
25. Hayat, A.A.; Chittawadigi, R.; Udai, A.; Saha, S.K. Identification of Denavit-Hartenberg parameters of an industrial robot. In Proceedings of the Conference on Advances in Robotics, Pune, India, 4–6 July 2013; pp. 1–6.
26. Denavit, J.; Hartenberg, R.S. A kinematic notation for lower-pair mechanisms based on matrices. *Trans. ASME J. Appl. Mech.* **1955**, *22*, 215–221.
27. Driels, M.R.; Swayze, W.; Potter, S. Full-pose calibration of a robot manipulator using a coordinate-measuring machine. *Int. J. Adv. Manuf. Technol.* **1993**, *8*, 34–41. [CrossRef]
28. Liu, B.; Zhang, F.; Qu, X.; Shi, X. A Rapid coordinate transformation method applied in industrial robot calibration based on characteristic line coincidence. *Sensors* **2016**, *16*, 239. [CrossRef] [PubMed]
29. Babinec, A.; Jurišica, L.; Hubinský, P.; Duchoň, F. Visual localization of mobile robot using artificial markers. *Procedia Eng.* **2014**, *96*, 1–9. [CrossRef]
30. RobotWorx. Available online: https://www.robots.com/robots/kuka-kr-6-r900-fivve (accessed on 27 October 2018).
31. Record Breaking Limbo Skater: 6-Year-Old Skates under 39 Cars—YouTube. Available online: https://www.youtube.com/watch?v=7HEPRZuRWvc (accessed on 1 July 2018).
32. Kee, V.; Rojas, N.; Elara, M.R.; Sosa, R. Hinged-Tetro: A self-reconfigurable module for nested reconfiguration. In Proceedings of the 2014 IEEE/ASME International Conference on Advanced Intelligent Mechatronics (AIM), Besacon, France, 8–11 July 2014; pp. 1539–1546.
33. Prabakaran, V.; Elara, M.R.; Pathmakumar, T.; Nansai, S. Floor cleaning robot with reconfigurable mechanism. *Autom. Constr.* **2018**, *91*, 155–165. [CrossRef]

34. Yuyao, S.; Elara, M.R.; Kalimuthu, M.; Devarassu, M. sTetro: A modular reconfigurable cleaning robot. In Proceedings of the 2018 International Conference on Reconfigurable Mechanisms and Robots (ReMAR), Delft, The Netherlands, 20–22 June 2018; pp. 1–8.
35. Ilyas, M.; Yuyao, S.; Mohan, R.E.; Devarassu, M.; Kalimuthu, M. Design of sTetro: A Modular, Reconfigurable, and Autonomous Staircase Cleaning Robot. *J. Sens.* **2018**, *2018*, 8190802. [CrossRef]
36. Tan, N.; Mohan, R.E.; Elangovan, K. Scorpio: A biomimetic reconfigurable rolling–crawling robot. *Int. J. Adv. Robot. Syst.* **2016**, *13*, 1729881416658180. [CrossRef]
37. Patil, M.; Abukhalil, T.; Patel, S.; Sobh, T. UB robot swarm—Design, implementation, and power management. In Proceedings of the 2016 12th IEEE International Conference on Control and Automation (ICCA), Kathmandu, Nepal, 1–3 June 2016; pp. 577–582.
38. Siegwart, R.; Nourbakhsh, I.R.; Scaramuzza, D. *Introduction to Autonomous Mobile Robots*; MIT Press: Cambridge, MA, USA, 2011.
39. Mapping and Localization from Planar Markers. Available online: http://www.uco.es/investiga/grupos/ava/node/57 (accessed on 1 October 2018).
40. ArUco: A Minimal Library for Augmented Reality Applications Based on OpenCV. Available online: https://www.uco.es/investiga/grupos/ava/node/26 (accessed on 1 November 2018).
41. Hartley, R.I.; Zisserman, A. *Multiple View Geometry in Computer Vision*; Cambridge University Press: Cambridge, MA, USA, 2000; ISBN 0521623049.
42. Bennett, D.J.; Geiger, D.; Hollerbach, J.M. Autonomous robot calibration for hand-eye coordination. *Int. J. Robot. Res.* **1991**, *10*, 550–559. [CrossRef]
43. Rousseau, P.; Desrochers, A.; Krouglicof, N. Machine vision system for the automatic identification of robot kinematic parameters. *IEEE Trans. Robot. Autom.* **2001**, *17*, 972–978. [CrossRef]
44. Meng, Y.; Zhuang, H. Self-Calibration of Camera-Equipped Robot Manipulators. *Int. J. Robot. Res.* **2001**, *20*, 909–921. [CrossRef]
45. Chameleon3 Color USB3 Vision. Available online: https://www.ptgrey.com/chameleon3 (accessed on 4 November 2018).
46. Golub, G.H.; Van Loan, C.F. *Matrix Computations*; JHU Press: Baltimore, MD, USA, 2012; Volume 3.
47. Angeles, J. *Fundamentals of Robotic Mechanical Systems: Theory, Methods, and Algorithms*; Mechanical Engineering Series; Springer International Publishing: Berlin, Germany, 2013.
48. Saha, S.K. *Introduction to Robotics*, 2nd ed; Tata McGraw-Hill Education: New Delhi, India, 2014.

applied sciences

MDPI

Article

Stability Criterion for Dynamic Gaits of Quadruped Robot

Yan Jia [1], Xiao Luo [2,*], Baoling Han [1], Guanhao Liang [3], Jiaheng Zhao [1] and Yuting Zhao [1]

1 School of Mechanical Engineering, Beijing Institute of Technology, No. 5 Zhongguancun South Street, Haidian District, Beijing 100081, China; tictac0324@163.com (Y.J.); hanbl@bit.edu.cn (B.H.); zhaojiaheng91@outlook.com (J.Z.); zytkite@163.com (Y.Z.)

2 School of Computer Science & Technology, Beijing Institute of Technology, No. 5 Zhongguancun South Street, Haidian District, Beijing 100081, China

3 School of Mechatronical Engineering, Beijing Institute of Technology, No. 5 Zhongguancun South Street, Haidian District, Beijing 100081, China; 3120120116@bit.edu.cn

* Correspondence: luox@bit.edu.cn; Tel.: +86-010-6891-8856

Received: 25 October 2018; Accepted: 22 November 2018; Published: 25 November 2018

Featured Application: The new stability-evaluation method can be used to guide the motion planning and posture adjustment of the quadruped robots with dynamic gait.

Abstract: Dynamic-stability criteria are crucial for robot's motion planning and balance recovery. Nevertheless, few studies focus on the motion stability of quadruped robots with dynamic gait, none of which have accurately evaluated the robots' stability. To fill the gaps in this field, this paper presents a new stability criterion for the motion of quadruped robots with dynamic gaits running over irregular terrain. The traditional zero-moment point (*ZMP*) is improved to analyze the motion on irregular terrain precisely for dynamic gaits. A dynamic-stability criterion and measurement are proposed to determine the stability state of the robot and to evaluate its stability. The simulation results show the limitations of the existing stability criteria for dynamic gaits and indicate that the criterion proposed in this paper can accurately and efficiently evaluate the stability of a quadruped robot using such gaits.

Keywords: quadruped robot; stability criterion; dynamic gait

---

1. Introduction

Legged robots that imitate animals have flexible joints and interact with the ground intermittently through their feet, giving them excellent environmental adaptability, which makes them suitable for working in complex environments such as mountains, disaster sites, and warehouses. Compared with other legged robots, quadruped robots have better maneuverability and load capacity. Over the past several decades, there has been an increasing interest in quadruped robots [1–4].

When considering the dynamic motion of legged robots, one must pay attention to their stability. It is very important for robots to complete the assigned task while maintaining balance and moving on an even keel. Researchers have proposed a variety of static- and dynamic-stability-analysis methods to evaluate the stability of legged robot. The static-stability criteria and margins are only applicable for analyzing low-speed static gaits rather than the dynamic motion because they do not consider inertial forces or external impacts [5]. Fukuoka and Kimura proposed a stability criterion named wide stability margin (WSM) [6]. They hypothesized that a robot can maintain its balance if the projection of the center of mass (CoM) of the body is within the support polygon formed by the projection of the current support and swing feet on the horizontal plane; the shortest distance from this projection point to the boundaries of the support polygon is used to evaluate the robot's stability. This stability criterion, in essence, is a static-stability criterion, and it is not suitable for analyzing the stability of high-speed motion. The rationality of the definition of the support polygon has not been explained; there is no experiment to verify that this stability criterion and margin can accurately evaluate the robot's stability. Many dynamic-stability criteria have been presented to address the limitations of static-stability criteria.

The main idea behind the dynamic-stability criterion is that, at a certain state, the motion of the robot is considered to be stable if the torque caused by the ground-reaction force can prevent the robot from tumbling around any support boundary.

The most common dynamic-stability criteria are based on the zero-moment point (*ZMP*) or the center of pressure (CoP). *ZMP* refers to a point on the ground where the net torque in the direction parallel to the ground caused by the gravity and inertial forces of each part of the robot is zero. *ZMP* shows the tendency of the robot to tumble. Researchers believe that the robot is stable if *ZMP* is located inside the support polygon, which is connected to the support feet. The minimum distance between *ZMP* and the support boundaries is used as the stability margin. The larger the minimum distance is, the better the stability that the robot can achieve [7–10]. CoP refers to the point where the net torque caused by the ground-reaction forces is zero; it coincides exactly with *ZMP* when the robot is dynamically stable [11,12]. The traditional *ZMP*-based-stability criteria can only be applied to planar motion, and some researchers have made improvement to analyze the motion on irregular terrain by adjusting the support plane [11,13,14]. The stability criteria based on *ZMP* do not consider the influence of the robot's current speed on its stability, but speed is very important for dynamic stability [15].

Methods using the leg-end-supporting moment (LSM) and tumble stability margin (TSM) [13,16–19], among others, directly consider the resultant moment around each support boundary. The robot will be stable if the resultant moments can counteract the torque which makes the robot tumble. A simpler stability criterion is the force-angle-stability measure (FASM) [20], which considers the angle between the resultant force of gravity, inertial force, and other external forces at the CoM and the perpendicular line from CoM to the support boundary. Other researchers have proposed some energy-based stability criteria [21,22], e.g., the normalized dynamic energy stability margin (NDESM). These methods evaluate the stability of the robot by quantifying the maximum impact energy that the robot can absorb without losing its stability.

The limitation of the above stability criteria is that they are all suitable to the situation in which a support polygon exists when the legged robots are moving. They can be used to analyze the stability of walking gait (when the supporting triangle exists), but are not suitable for dynamic gaits, which only have a support line in most cases, such as trotting, pacing, and bounding.

There have been few studies concerning the stability criteria for quadruped robots with dynamic gaits, all of which focus on trotting. Most researchers believe that a robot with a trot gait is stable only when the current *ZMP* is located on the support line, because there is no tumbling moment around the support line. The distance between *ZMP* and the support line (substantially equivalent to the tumbling moment around the support line) is used to guide the motion control of the robot [14,23,24]. Such analytical methods do not consider the effect of the dynamic process of support-leg alternation on the robot's motion stability. In practice, even if *ZMP* is not located on the current support line, the robot will not necessarily lose its balance and fall. Hence, it is difficult to correctly evaluate the stability of the robot by directly using the distance between *ZMP* and the current support line. Therefore, the dynamic-stability criterion requires further consideration.

Lang et al. [25] proposed a stability analysis method based on Lyapunov theory of switching system. They built the dynamic model of the trotting robot as a nonlinear switching system, and designed a Lyapunov function to prove that the switching system is uniform asymptotic stable. This method is used to choose the landing position of the swing feet to make the position tracking errors of the robot's body converge to zero. The computational complexity of this method is too high, and it cannot evaluate the stability of the robot's motion. Another stability measure is called Landing

Accordance Ratio (LAR) [26]. This method considers the time difference between the moments of the feet in the same phase touch down. If the discordance amount is zero, there is no unexpected moment, and the robot moves stably. There are three major shortcomings of this method: (1) it cannot evaluate the stability of the robot in time, because LAR is updated when the two feet in the same phase both touch the ground; (2) it is not suitable for the situation where the ground-reaction forces at current support feet cannot prevent the robot from tumbling; and (3) it is hard to decide when to update the value of LAR if the robot tips over, because one of the feet will never touch the ground.

The good stability criterion and evaluation method should have the following characteristics: (1) it should determine the stability state of the robot efficiently and correctly; (2) it should evaluate the robot's stability accurately; and (3) it should provide guidance for the robot's motion planning and balance recovery. The purpose of this paper is to present a new stability criterion for the quadruped robots with dynamic gaits over irregular terrain.

When animals experience impacts in nature, they quickly adjust the position of their next set of support feet to maintain stability. This indicates that dynamic stability depends not only on their current motion and support state but also on the positions of the next set of support points. Inspired by quadruped animal moving gesture and referring to the previous dynamic-stability criteria, this paper assumes that a robot is stable if the current or next set of support feet can provide the moment necessary to prevent the robot from tipping in any direction. The less likely the robot is to tip over, the smaller the difference among ground-reaction forces at each support foot are, meaning that the robot has higher capability to stay stable.

Based on the above ideas, by defining and adjusting the virtual-support plane, the traditional *ZMP* can be extended to three-dimensional (3D) space so that it can be used to analyze the motion state on irregular terrain. An extra term containing the velocity of the robot's body is added to the equation for the calculation of *ZMP* to better reflect the robot's motion state. Here, we define the virtual-support quadrilateral and propose a stability criterion for quadruped robots with dynamic gaits. The robot can stay stable if the expanded *ZMP* is located within this quadrilateral, because the current or next set of support feet can provide the moment necessary to prevent the robot from tipping over. The stability is quantified using the distance between *ZMP* and the boundary of the virtual-support quadrilateral, the angle between the vector from CoM to *ZMP* and the normal vector of the support plane, and the distance between *ZMP* and the support line. The simulation results show that the trotting robot can still run stably even if *ZMP* is not on the current support line, and the distance between *ZMP* and the current support line cannot accurately assess the robot's stability. The results also indicate that the proposed stability criterion and measure can efficiently and accurately evaluate the stability of a quadruped robot using dynamic gaits.

This paper is outlined as follows. The method for expanding *ZMP* to 3D space is described in Section 2. The stability criterion for the motion of a quadruped robot with a dynamic gait over irregular terrain is introduced in Section 3. The method of calculating the dynamic-stability measurement is presented in Section 4. The simulation results and discussion are reported in Section 5. Finally, conclusions are presented in Section 6.

2. *ZMP* for Irregular Terrain

In this section, the traditional *ZMP* is modified to deal with the motion on irregular terrain. Because most of the mass is concentrated in the body of the robot, the mass of the legs is ignored to simplify our stability analysis. Traditional *ZMP* refers to a point on the ground at which the net torque caused by gravity, the inertial force, and the inertial moment of each part of the robot is zero; it is only applicable when analyzing the motion on the horizontal plane. In this paper, *ZMP* is extended to 3D space by defining and adjusting the plane of the support feet, and the velocity term is introduced to reflect the motion more efficiently and accurately.

On irregular terrain, *ZMP* should be considered as a point in 3D space.

As shown in Figure 1, the position of the body's CoM (C) in the inertial coordinate system (i) is $P_C^i = \begin{bmatrix} x_C^i & y_C^i & z_C^i \end{bmatrix}^T$. The gravity of the body is $G^i = \begin{bmatrix} G & 0 & 0 \end{bmatrix}^T$, the inertial force at CoM is $F_I^i = \begin{bmatrix} F_{Ix}^i & F_{Iy}^i & F_{Iz}^i \end{bmatrix}^T$, and the inertial moment is $N_I^i = \begin{bmatrix} N_{Ix}^i & N_{Iy}^i & N_{Iz}^i \end{bmatrix}^T$. The position of ZMP in the inertial coordinate system is assumed to be $P_{ZMP}^i = \begin{bmatrix} x_{ZMP}^i & y_{ZMP}^i & z_{ZMP}^i \end{bmatrix}^T$; hence, according to the definition of ZMP,

$$\left(P_C^i - P_{ZMP}^i\right) \times F_I^i + \left(P_C^i - P_{ZMP}^i\right) \times G^i + N_I^i = 0. \tag{1}$$

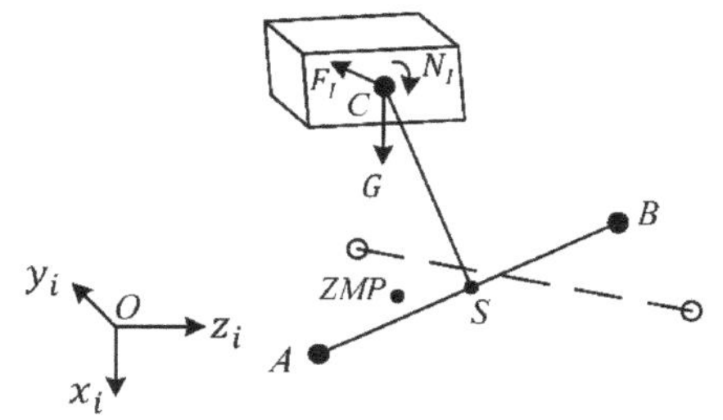

Figure 1. Force-schematic diagram of the robot body.

The following can be obtained from Equation (1):

$$-\left(z_C^i - z_{ZMP}^i\right) F_{Iy}^i + \left(y_C^i - y_{ZMP}^i\right) F_{Iz}^i + N_{Ix}^i = 0; \tag{2}$$

$$\left(z_C^i - z_{ZMP}^i\right)\left(F_{Ix}^i + G_x^i\right) - \left(x_C^i - x_{ZMP}^i\right) F_{Iz}^i + N_{Iy}^i = 0; \tag{3}$$

$$-\left(y_C^i - y_{ZMP}^i\right)\left(F_{Ix}^i + G_x^i\right) + \left(x_C^i - x_{ZMP}^i\right) F_{Iy}^i + N_{Iz}^i = 0. \tag{4}$$

Only two of the three equations are linearly independent [14], but there are three unknowns. An additional condition must be given to solve this problem.

Referring to the assumptions made in traditional ZMP, it is stipulated in this paper that ZMP is always located on the current support plane (which can be nonhorizontal). For the convenience of calculation, the support coordinate system (s) (Figure 2) is defined. The origin of this coordinate system is located at the midpoint S of the current support line $\overline{AB}$, the direction of z_s is from the rear foot to the front foot of the current support line, and x_s is perpendicular to the current support plane and points downward. Another coordinate system used in this paper is the principal axes coordinate system of the body (c), whose origin is located at the body's CoM.

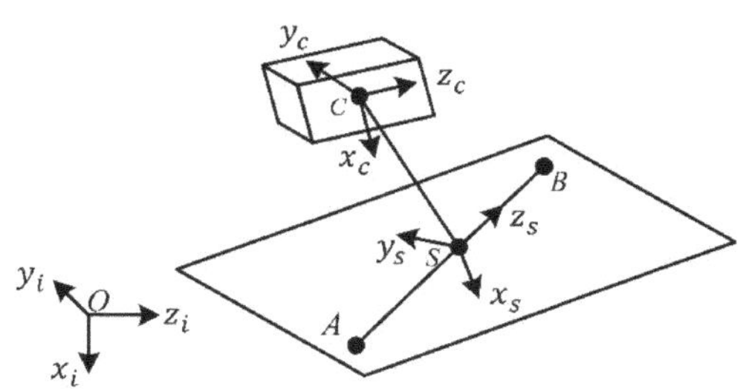

Figure 2. Settings of the coordinate systems.

The position of ZMP in the support coordinate system $P^s_{ZMP} = \begin{bmatrix} 0 & y^s_{ZMP} & z^s_{ZMP} \end{bmatrix}^T$ is calculated by considering the force and moment balance in this system:

$$x^s_{ZMP} = 0; \tag{5}$$

$$y^s_{ZMP} = y^s_C - \frac{x^s_C\left(F^s_{Iy} + G^s_y\right) + N^s_{Iz}}{F^s_{Ix} + G^s_x}; \tag{6}$$

$$z^s_{ZMP} = z^s_C - \frac{x^s_C\left(F^s_{Iz} + G^s_z\right) - N^s_{Iy}}{F^s_{Ix} + G^s_x}. \tag{7}$$

It is difficult to ascertain the actual support plane for some dynamic gaits (trotting, pacing, bounding, etc.) because only two support feet are used most of the time. To solve this problem, the positions of the previous support line, $\overline{A_F B_F}$, and the next support line, $\overline{A_L B_L}$, are comprehensively considered (for gaits with only one support point, such as gallops, more previous and later support points should be used). The definition of the virtual-support plane is proposed.

The following assumptions are made for the current virtual-support plane: (1) this plane always contains the current support line $\overline{AB}$; and (2) from the disappearance of the previous support line to the disappearance of the current support line, the virtual-support plane gradually changes from Σ_1 (determined by the current and previous support lines) to Σ_2 (determined by the current and next support lines). In practice, for uneven terrain, $\overline{AB}$ and $A_F B_F$ (or $\overline{A_L B_L}$) may not lie on the same plane. This paper presents a method to determine the virtual-support plane.

The unit normal of the plane on which A, B, and A_F lie is n_{11} (the angle between this normal and x_i is acute), and that of the plane on which A, B, and B_F lie is n_{12} (it also has an acute angle with x_i). The unit normal vector n_1 of the plane Σ_1 lies on the bisector of the angle formed by n_{11} and n_{12} (Figure 3):

$$n_{11} = \frac{\overrightarrow{AB} \times \overrightarrow{AA_F}}{\|\overrightarrow{AB} \times \overrightarrow{AA_F}\|}; \tag{8}$$

$$n_{12} = \frac{\overrightarrow{AB} \times \overrightarrow{BB_F}}{\|\overrightarrow{AB} \times \overrightarrow{BB_F}\|}. \tag{9}$$

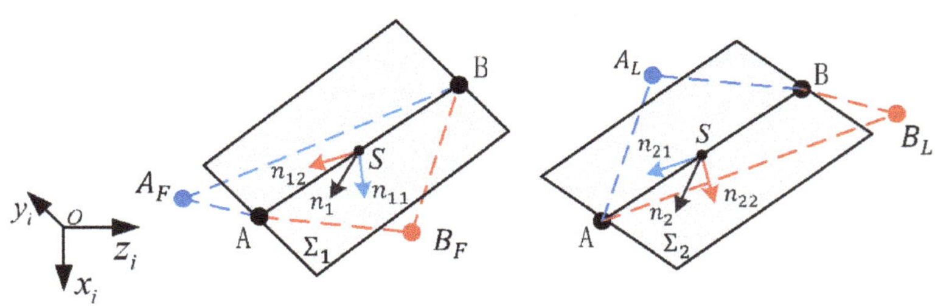

Figure 3. Determination of the virtual-support plane.

Set $n_{11} = n_{12}$, if $\overrightarrow{AB} \times \overrightarrow{AA_F}$ is equal to the zero vector, and $n_{11} = -n_{11}$, if the x_i component of n_{11} is negative. A similar method is used to deal with Equation (9). Hence,

$$n_1 = \frac{n_{11} + n_{12}}{\|n_{11} + n_{12}\|}. \tag{10}$$

n_{21} is the unit normal of the plane on which A, B, and A_L lie and n_{22} is the unit normal of the plane on which A, B, and B_L lie. n_2 is the unit normal vector of Σ_2:

$$n_{21} = \frac{\vec{AB} \times \vec{AA_L}}{||\vec{AB} \times \vec{AA_L}||}; \tag{11}$$

$$n_{22} = \frac{\vec{AB} \times \vec{BB_L}}{||\vec{AB} \times \vec{BB_L}||}; \tag{12}$$

$$n_2 = \frac{n_{21} + n_{22}}{||n_{21} + n_{22}||}. \tag{13}$$

Then, from the disappearance of the previous support line to that of the current support line, the unit normal vector of the virtual-support plane (i.e., x_s in the inertial coordinate system) is

$$x_s^i = \frac{\mu n_1 + (1-\mu) n_2}{||\mu n_1 + (1-\mu) n_2||}, \tag{14}$$

where

$$\mu = -\frac{1}{T_b} t + 1. \tag{15}$$

T_b is the duration between two adjacent steps (half the duration of a gait cycle for trot gait), and t is the time gap between the current running time and the time point when the previous support line disappeared. The z_s axis of the support coordinate system is

$$z_s^i = \frac{\vec{AB}}{||\vec{AB}||}, \tag{16}$$

and the y_s axis is

$$y_s^i = z_s^i \times x_s^i. \tag{17}$$

The rotation matrix from the support to the inertial coordinate system is

$$ {}_s^iR = \begin{bmatrix} x_s^i & y_s^i & z_s^i \end{bmatrix}. \tag{18}$$

The current velocity of the body, which can reflect the robot's current motion state, is not taken into account in the traditional *ZMP*-based stability criteria, however, the velocity is very important for maintaining balance [15], especially under some special condition. For example, when the current ground-reaction force cannot just offset the torque, the robot is made to tumble around a certain direction; however, if the robot's velocity is high enough to allow it to quickly enter a stable area, the robot can maintain its balance. A more accurate assessment of the robot's stability can be proposed by taking the velocity into account. In this paper, the velocity terms are added to the calculation of *ZMP*:

$$x_{ZMP_0}^s = 0; \tag{19}$$

$$y_{ZMP_0}^s = y_{ZMP}^s + \eta \left(v_{yr}^s - v_{yd}^s \right); \tag{20}$$

$$z_{ZMP_0}^s = z_{ZMP}^s + \eta (v_{zr}^s - v_{zd}^s). \tag{21}$$

The modified *ZMP* is denoted by ZMP_0. ZMP_0 still lies on the current virtual-support plane, and the differences between the current actual velocity, v_r^s, and the expected velocity, v_d^s, in the y_s and z_s directions are considered in the calculation of *ZMP*. η is a constant-value coefficient considering the magnitude of the size and velocity of the robot:

$$\eta = \frac{\frac{1}{2}(L+W)}{||v_d||} 0.1 = 0.05 \frac{(L+W)}{||v_d||}, \tag{22}$$

L is the length of the robot's body and W is the width. The expected benefits of doing so are the following: (1) the current stability of the robot can be assessed more accurately and efficiently; and (2) a reference can be provided to eliminate undesired velocity during motion planning.

3. Dynamic-Stability Criterion

The first problem to be solved when considering the stability of the robot is determining whether the robot can maintain balance at the current state. The stability of a robot with a dynamic gait is closely related to two factors: (1) the ground-reaction forces that act on the current support feet; and (2) the ground-reaction forces that will acti on the next support feet. The robot is unstable if neither of these forces can prevent it from tumbling. In addition, the elimination of unexpected speed is also taken into account when evaluating the robot's stability by adding the velocity term into the calculation of *ZMP*.

There are always two legs moving in the same phase for most dynamic gaits; hence, these two legs can be equivalent to one virtual leg from the midpoint of the support line to the robot's CoM. The stability criterion is considered in the current virtual-support plane. $A_L{}'$ and $B_L{}'$ are the projection points on the support plane of the next set of support feet, A_L and B_L, respectively, as shown in Figure 4.

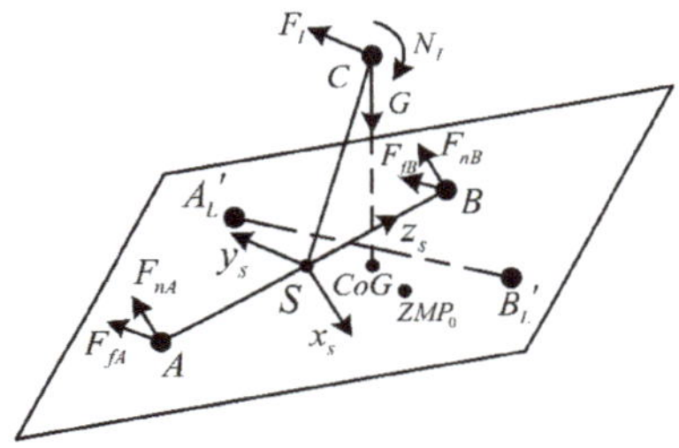

Figure 4. Projection of the next set of support feet and contact force at the current support feet.

It should be noted that the direction of the support force, F_n, is perpendicular to the support plane and points outward, and its main role is to support the robot standing and to maintain the stability of the robot's rotation around the y_s and z_s axes. The frictional force F_f is located on the support plane, which is opposite to the direction of the feet's motion trend with respect to this plane; its function is to push the robot forward and to prevent the robot from rotating around the x_s axis.

The contact forces act on the two support feet, A and B. Hence, the current support line cannot provide the torque around the z_s axis; its main role is to prevent the robot from tumbling around the y_s axis. The torque that prevents the robot from tipping around the z_s axis can only be provided by the next set of support feet. The contact forces at $A_L{}'$ and $B_L{}'$ also play a supplementary role in maintaining rotational stability around y_s. The rotation of the robot around x_s is stopped by the friction at the current support and next support lines, but this is not the main reason the robot loses stability.

ZMP_0 can be considered to be on the extension line of the equivalent force, F_e, acting at the CoM of the body (Figure 5).

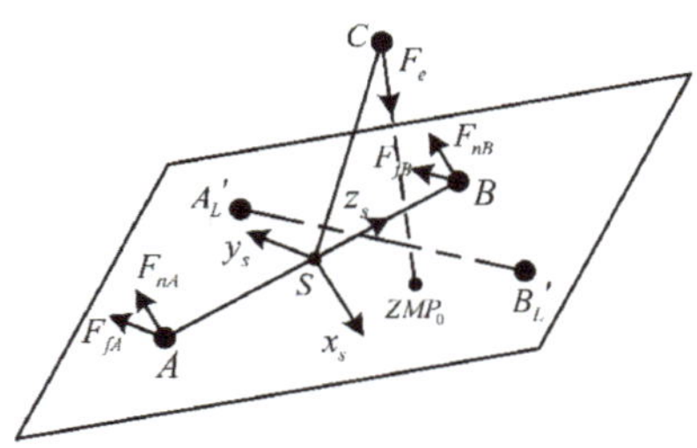

Figure 5. Equivalent force and ZMP_0.

In the x_sSz_s plane, D_1 represents the point with the largest z_s-coordinate among the four support points (A, B, $A_L{}'$, and $B_L{}'$), whereas the z_s-coordinate of D_4 is the smallest one, as shown in Figure 6.

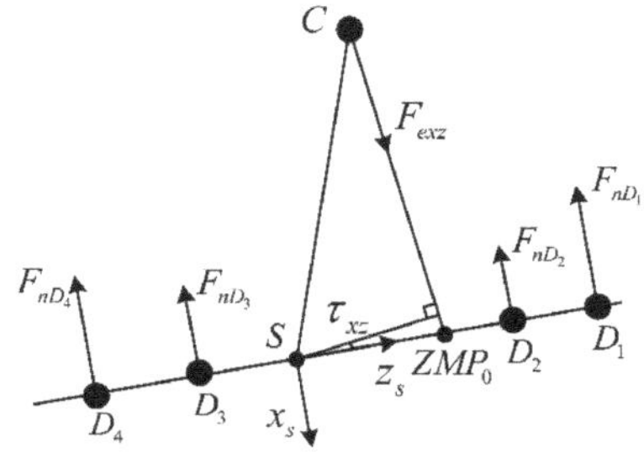

Figure 6. Force diagram in the x_sSz_s plane.

For any moment,

$$\sum_{i=1}^{4} F_{nD_i} = -F_{exz} \cos \tau_{xz}; \tag{23}$$

$$M_{y_s} = \sum_{i=1}^{4} F_{nD_i} z^s_{D_i} + F_{exz} z^s_{ZMP_0} \cos \tau_{xz}. \tag{24}$$

τ_{xz} is the angle between F_{exz} and the support plane; hence,

$$M_{y_s} = \sum_{i=1}^{4} F_{nD_i} \left(z^s_{D_i} - z^s_{ZMP_0} \right). \tag{25}$$

The ground-reaction force at the current or the next set of support feet is unable to maintain the rotational balance of the robot around the y_s axis if $z^s_{ZMP_0}$ is less than the minimum z_s-coordinate of the four support points, or larger than the maximum value, because the direction of the support forces can only be perpendicular to the support plane and point outward. Similarly, the ground-reaction force at the next set of support feet cannot prevent the robot from tumbling around the z_s axis if $y^s_{ZMP_0}$ is less than the minimum or larger than the maximum y_s-coordinate of the four support points.

However, it is insufficient to consider only the rotation around these two directions for stability analysis. The direction of friction is determined, and the magnitude is also limited. Hence, in some specific cases, it is difficult for the ground to provide sufficient friction to push the robot forward and to maintain the rotational balance around the x_s axis. Besides, it is necessary to consider whether the robot can maintain stability easily when the next support line is formed and the current one disappears because the two support lines do not always exist simultaneously. Therefore, referring to the traditional stability criteria based on ZMP, this paper holds that the robot stays stable if ZMP_0 is located within the convex quadrilateral formed by A, B, $A_L{}'$, and $B_L{}'$. This quadrilateral is called the virtual-support quadrilateral. The quadrilaterals $AA_L{}'BB_L{}'$ or $ABB_L{}'A_L{}'$ are formed depending on whether AB intersects $A_L{}'B_L{}'$, as shown in Figure 7.

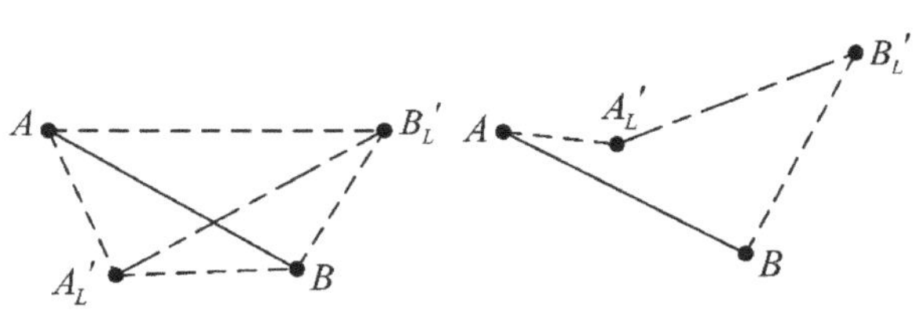

Figure 7. Virtual-support quadrilateral.

4. Dynamic-Stability Measurement

The next problem to be solved is how to evaluate stability if the dynamic motion of the robot is stable. It is clear that the robot rarely tips over if the difference between ground-reaction forces at the support feet is small.

The following three quantities are taken into account to quantify the stability of the robot:

(1) Distances between ZMP_0 and the boundaries of the virtual-support quadrilateral in the virtual-support plane.

These distances not only show the stability of the robot with its current support line but also consider the stability after the next support line appears. The vertical distance between ZMP_0 and each boundary is a comprehensive representation of the distances between ZMP_0 and the two support feet forming the boundary. From Equation (25), it can be seen that, in a certain direction, larger minimum distances between ZMP_0 and the two support feet located at both ends result in smaller differences between the ground-reaction forces at these support feet. The instability of the robot is mainly caused by unbalanced rotation in two mutually perpendicular directions. Therefore, the distances between ZMP_0 and the two sets of opposite boundaries (d_{11} and d_{12}, and d_{21} and d_{22}) should be considered when evaluating the stability of the robot (Figure 8).

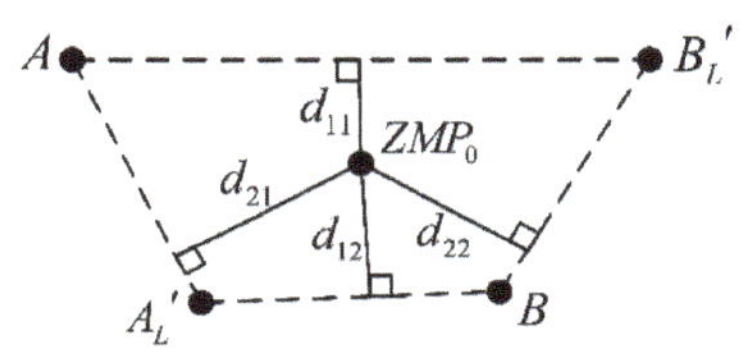

Figure 8. Distances between ZMP_0 and each boundary of the virtual-support quadrilateral.

The two support points forming a certain boundary of the quadrilateral do not necessarily exist at present. Hence, the distances between ZMP_0 and boundaries should be multiplied by their corresponding coefficients ω_i, which are defined as follows:

$$\omega_i = \begin{cases} 1, & two\ support\ feet\ both\ exist \\ 0.5, & only\ one\ support\ foot\ exists \\ 0.2, & otherwise \end{cases} \tag{26}$$

The integrated minimum distance between ZMP_0 and the boundaries is defined as

$$d_{edge} = \min\{\omega_{11}d_{11}, \omega_{12}d_{12}\} + \min\{\omega_{21}d_{21}, \omega_{22}d_{22}\}. \tag{27}$$

d_{edge} is expected to be as large as possible, and it is the most important quantity for evaluating the robot's stability.

(2) Angle between the vector pointing from CoM to ZMP_0 and the normal vector of the virtual-support plane.

The ground-reaction force can be divided into two parts: support and friction forces. The support force is perpendicular to the support plane, and, in theory, it can be provided sufficiently to guarantee the force balance in this direction. The direction of the friction force is opposite to that of the trend of the foot's motion with respect to the support plane if the foot does not slip; the largest static-friction force that the ground can provide is proportional to the magnitude of the support force.

Under a constant equivalent force F_e, larger angles, τ, between the vector pointing from CoM to ZMP_0 (the direction is the same as F_e) and the normal vector of the virtual-support plane (Figure 9) result in lower support forces and greater required friction forces; thus, the feet are more likely to slip, and the stability of the robot becomes worse. $|\sin\tau|$ is used as a key quantity to measure the robot's stability, in which u represents the vector pointing from CoM to ZMP_0:

$$\cos\tau = \frac{u \cdot x_s^i}{\|u\| \|x_s^i\|}; \tag{28}$$

$$|\sin\tau| = \sqrt{1 - \cos\tau^2}. \tag{29}$$

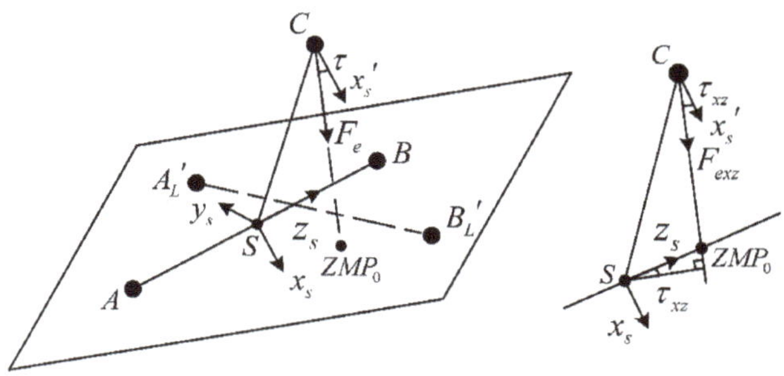

Figure 9. Angle between F_e and x_s.

(3) Distance between ZMP_0 and the support line.

For dynamic gaits, the support lines are alternately formed so that the ground-reaction force can push the robot to move forward and prevent it from falling. What really interacts with the ground are the support legs; therefore, the distance between ZMP_0 and the support lines is also very important for the evaluation of robot stability. The larger is the distance between ZMP_0 and a certain support line, the easier it is for the robot to tip over around this support line.

The stability measurement may change suddenly when the current support line disappears and the next support line forms if only the vertical distance between ZMP_0 and the current support line (denoted by d_C) is considered. The vertical distance between ZMP_0 and the next support line is represented by d_L, as shown in Figure 10. In practice, d_C should be as short as possible during the first half of the support phase, and d_L should be as short as possible during the second half.

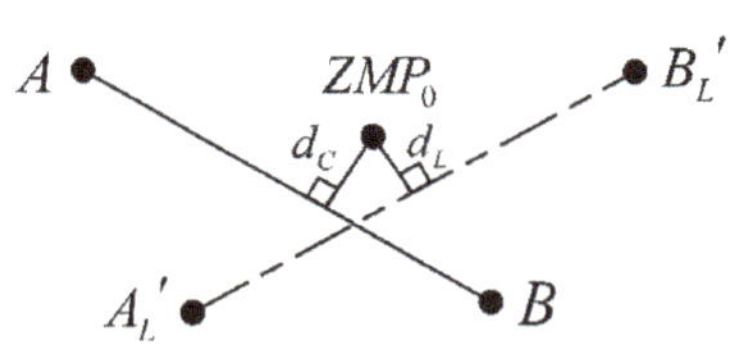

Figure 10. Distances between ZMP_0 and the support lines (d_C and d_L).

The distance between ZMP_0 and the support line is defined as

$$d_{spt} = \omega_c d_c + \omega_L d_L, \tag{30}$$

where ω_C and ω_L are time-varying weights:

$$\omega_C = \begin{cases} 1, & 0 \le t < 0.25T_b \\ -\frac{2}{T_b}t + 1.5, & 0.25T_b \le t < 0.75T_b \\ 0, & 0.75T_b \le t < T_b \end{cases} \tag{31}$$

$$\omega_L = \begin{cases} 0, & 0 \le t < 0.25T_b \\ \frac{2}{T_b}t - 1.5, & 0.25T_b \le t < 0.75T_b \\ 1, & 0.75T_b \le t < T_b \end{cases} \quad (32)$$

The weights are illustrated in Figure 11.

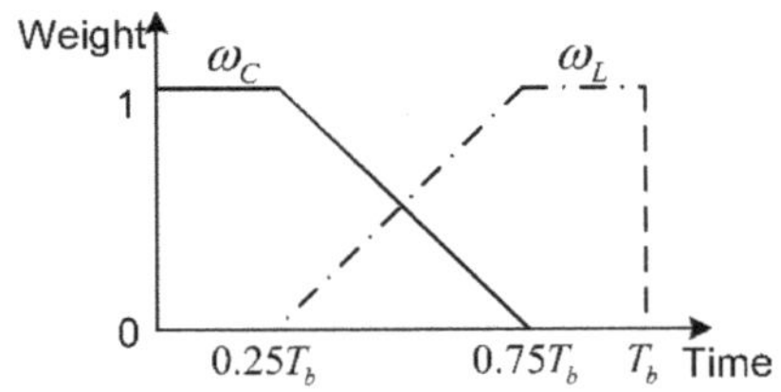

Figure 11. Weights for the distance between ZMP_0 and the two support lines.

Using the above three quantities, this paper proposes the stability measurement $M_{stability}$:

$$M_{stability} = \lambda_1 d_{edge} - \lambda_2|\sin\tau| - \lambda_3 d_{spt}. \quad (33)$$

A larger $M_{stability}$ value indicates better motion stability. The value of the coefficients λ_i should consider: (1) unification of the order of magnitude of the three terms on the right-hand side of Equation (33) (which are related to the size of the robot); and (2) the importance of each parameter. According to the definitions of d_{edge}, $|\sin\tau|$, and d_{spt}, for the dynamically stable gait:

$$0 < d_{edge} \le \frac{1}{2}(L+W) \times 0.5 = \frac{1}{4}(L+W), \quad (34)$$

$$0 \le |\sin\tau| \le 1, \quad (35)$$

$$0 \le d_{spt} \le W, \quad (36)$$

where λ_i can be calculated by:

$$\lambda_1 = 1, \quad (37)$$

$$\lambda_2 = \frac{1}{4}(L+W) \times 0.5, \quad (38)$$

$$\lambda_3 = \frac{\frac{1}{4}(L+W)}{W} \times 0.9. \quad (39)$$

According to the Equations (34)–(39),

$$M_{stability} > \lambda_1 \times 0 - \lambda_2 \times 1 - \lambda_3 \times W = -0.35(L+W), \quad (40)$$

which indicates the possible minimum value of $M_{stability}$, but this does not mean that the motion is stable if $M_{stability} > 0.35(L+W)$.

5. Results and Discussion

The simplified and mechanical models of the quadruped robot used in this paper are shown in Figure 12. Each leg of the robot has three degrees of freedom.

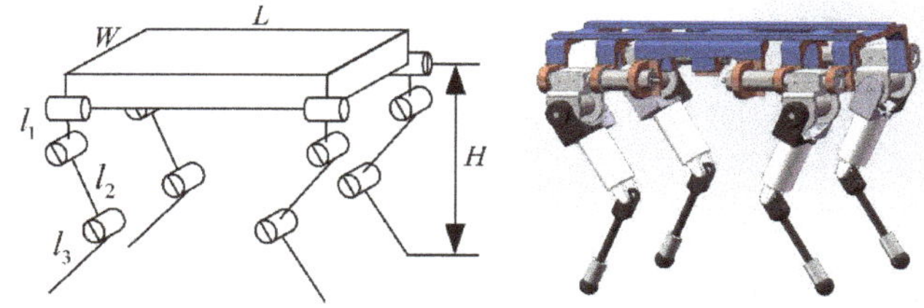

(**a**) (**b**)

Figure 12. (**a**) Simplified; and (**b**) mechanical models of the quadruped robot.

The geometric and weight parameters are shown in Table 1.

Table 1. Geometric and weight parameters of the robot.

Symbol	Value (Unit)	Definition
L	0.5 (m)	Length of the body
W	0.3 (m)	Width of the body
H	0.48 (m)	Height of the body
l_1	0.055 (m)	Length of the leg's first segment
l_2	0.2 (m)	Length of the leg's second segment
l_3	0.2 (m)	Length of the leg's third segment
m_0	15.14 (kg)	Mass of the body
m_1	0.9 (kg)	Mass of the leg's first segment
m_2	1 (kg)	Mass of the leg's second segment
m_3	0.3 (kg)	Mass of the leg's third segment

The moments of inertia of the body around the principal axes are $J_{xx} = 0.74$, $J_{yy} = 0.56$, and $J_{zz} = 0.22$ (kg·m^2).

A simulation model of the quadruped robot was built in the dynamic-simulation software ADAMS. The contact parameters between each foot and ground are shown in Table 2 (referring to the contact parameters between dry rubber and dry asphalt).

Table 2. Contact parameters between the feet and the ground.

Contact Parameters (Unit)	Motion on Horizontal Plane
Stiffness (N/mm)	2855
Damping (N*s/mm)	0.57
Force exponent	1.1
Penetration depth (mm)	0.1
Static coefficient	0.5
Dynamic coefficient	0.3

Trotting is the most commonly used dynamic gait and has been the focus of previous studies; thus, this study used the trot gait as the basic experimental gait. The trot-gait parameters for all simulations are shown in Table 3.

Table 3. Trot-gait parameters.

Gait Parameters (Unit)	Motion on Horizontal Plane
Step length (m)	0.2
Maximum height for the tip of swing legs (m)	0.03
Duration of a gait cycle (s)	1
Velocity (m/s)	0.2
Duty factor	0.6

Three quadruped-robot offline-gait simulations were designed as follow:

- Simulation 1 verified that the quadruped robot can maintain its stable state during dynamic motion when *ZMP* does not lie on the current support line.
- Simulation 2 verified that the stability criterion and measurement proposed in this paper can determine the robot's stability state efficiently and evaluate the stability accurately compared with the stability margins that only consider the distance between *ZMP* and the current support line and the stability analysis method called LAR.
- Simulation 3 showed that the stability-analysis method proposed in this paper is applicable to motion in 3D space.

5.1. Simulation 1

Most existing stability criteria for trotting gaits assume that the robot is in stable motion only when *ZMP* is located on the support line formed by the current support feet. However, this is inconsistent with the movement of animals in nature. To illustrate that these stability criteria cannot accurately determine the stability state of the robot, the following simulation was designed.

The projection of CoM in the direction of gravity onto the virtual-support plane is denoted by CoG. In the trot-gait-motion planning used in this simulation, the CoM of the body moves forward at a constant speed and, at each phase-changing point, CoG is located at the center of the quadrilateral formed by the just-disappearing support line and the newly formed line (see Figure 13).

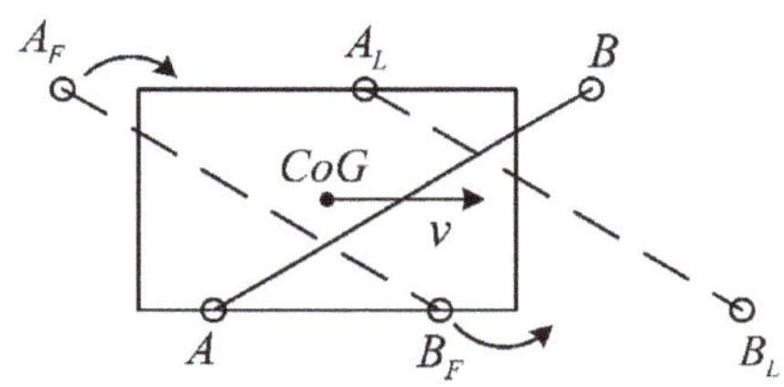

Figure 13. Movement of CoG.

The simulation animation is in Video S1. The position changes of the support lines, CoG, and *ZMP* on the support plane (z_iOy_i plane) during the entire simulation process are shown in Figure 14.

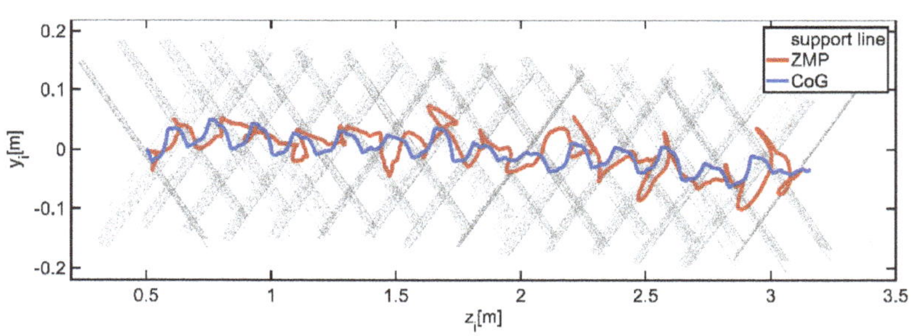

Figure 14. Positions of the support line and *ZMP*.

The gray lines in Figure 14 are the support lines and are shown as several groups of line clusters for two reasons. First, the active or passive compliance was not considered in this simulation process, therefore the feet were bouncing for a short period after landing. Second, the feet were slipping during the support phases. The blue line is the actual motion trajectory of CoG on the horizontal plane. The motion trajectory of CoG was not completely consistent with the planned trajectory under the influence of the unbalanced moment around the current support line, the legs' movement, and the impact force at the feet. The red line is the trajectory of traditional *ZMP*. It can be clearly seen that, during this simulation, *ZMP* was not always or mostly located on the current support line.

As shown in Figure 15, the running direction of the body was slightly skewed (see yaw angle), and the absolute values of the pitch and roll angles were less than 0.11 rad (6°). The curves of these attitude angles show that the robot moved stably and had no obvious tendency to tumble. This indicates that the existing stability criteria for the trot gait cannot accurately determine the stability state of the robot.

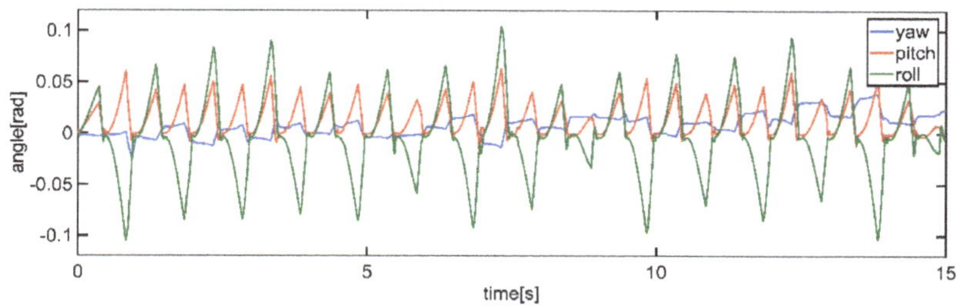

Figure 15. Attitude angles of the body.

5.2. Simulation 2

A second set of comparative simulations was designed to verify the following: (1) it is difficult to correctly evaluate the stability by only considering the distance between *ZMP* and the current support line; (2) the accuracy and effectivity of LAR are worse than the stability-evaluation method proposed in this paper; and (3) the stability-evaluation method proposed in this paper can satisfy the first two requirements of a good-stability criterion and evaluation method presented in Section 1.

In addition to the motion trajectory of the CoG used in Simulation 1 (denoted as Test 1), two other CoG trajectories were used for off-line planning of the quadruped robot's movement (the actual trajectories of CoG did not necessarily coincide with the planned trajectory); refer to the animation (Video S2).

- Test 2: CoG moves around the current support line (the velocity parallel to z_i is 0.2 m/s, and the velocity and acceleration parallel to y_i at the two ends of this direction are zero), as shown in Figure 17.
- Test 3: CoG starts to move from the midpoint of the current support line to that of the next support line at a constant speed when the previous support line has just disappeared, as shown in Figure 16.

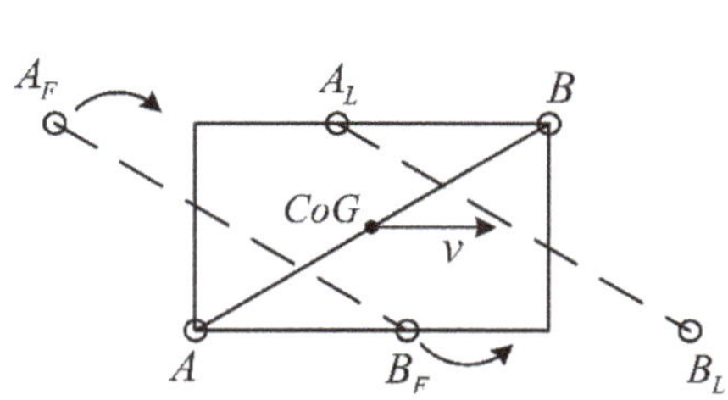

Figure 16. Trajectory of the CoG in Test 3.

The posture angles, positions, and velocities of the body for the three tests are shown in Figure 18a–c, respectively. The driving moments of the three joints of the left hind leg are shown in Figure 18d.

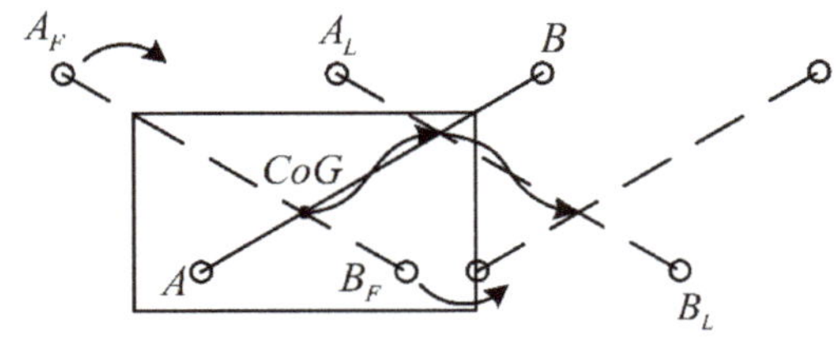

Figure 17. Trajectory of the CoG in Test 2.

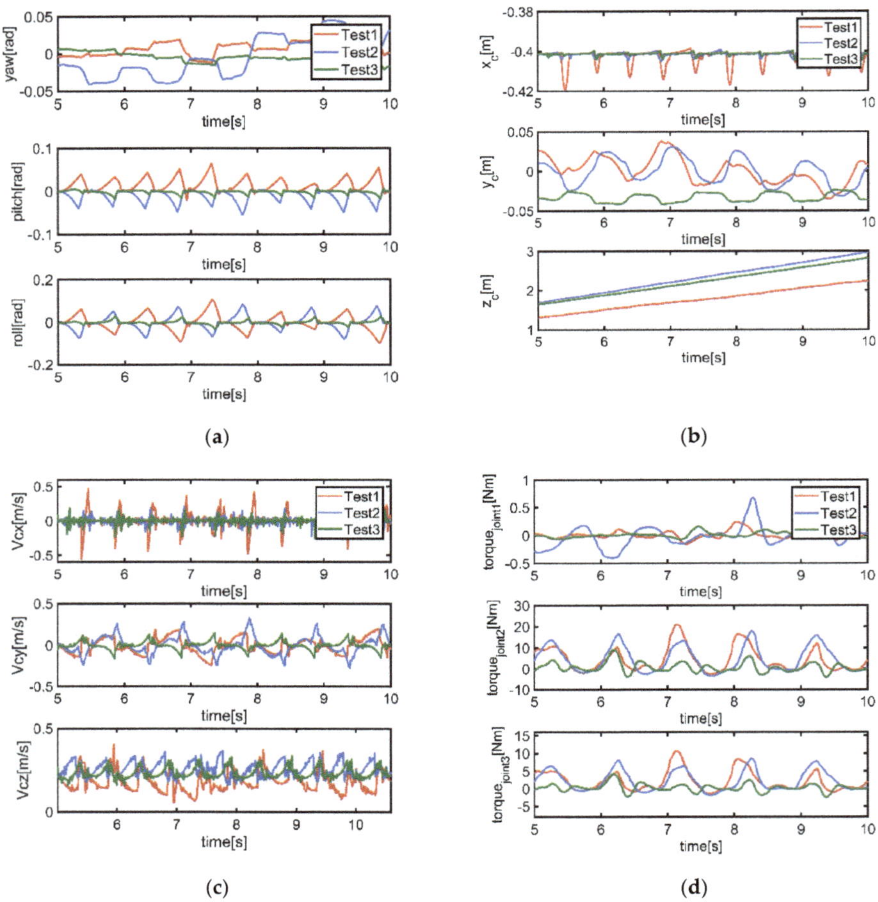

Figure 18. Sensor data of the motion in Simulation 2: (**a**) posture angles; (**b**) positions; (**c**) velocities of the body; and (**d**) driving moments of the left hind leg.

The wave heights of the data curves referring to the difference between adjacent peaks and troughs show the severity of data fluctuations if the fluctuation period is similar. The wave-height-variation range of each dataset and the driving-torque-variation ranges of the three joints of the left hind leg are shown in Figure 19.

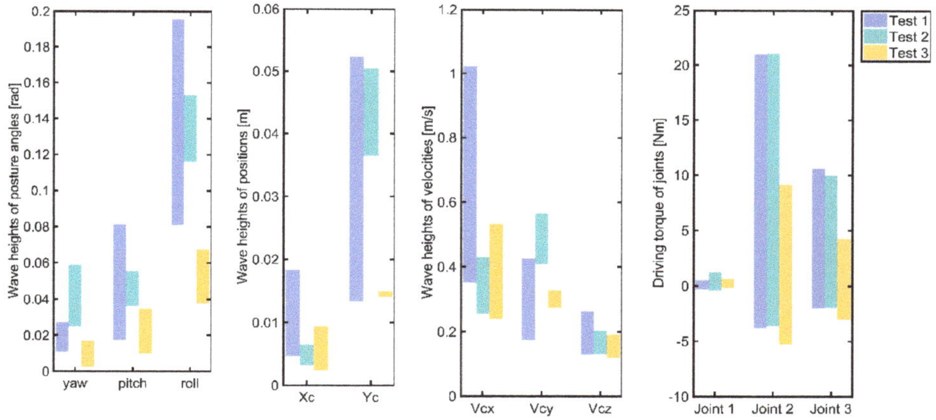

(a) (b) (c) (d)

Figure 19. Variation ranges of: (**a**) wave heights of posture angles; (**b**) wave heights of positions; (**c**) wave heights of velocities; and (**d**) joint torques.

As shown in Figure 19, the motion of Test 3 was the steadiest and required the minimum joint-driving torque, and the stabilities of the other two tests were about the same. In fact, stability cannot be evaluated accurately using individual-sensor data, which is one of the reasons the stability criterion and measurement must be proposed.

The generally accepted stability-evaluation method (denoted as Method 1), i.e., the one that measures the distance between the traditional *ZMP* and the current support line; LAR (denoted as Method 2); and the proposed stability measurement for the trot gait of a quadruped robot (denoted as Method 3) were used to evaluate the stability of the three tests (Figure 20a–c).

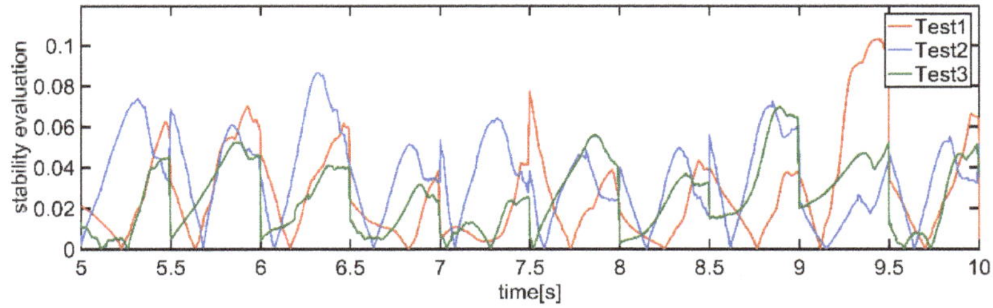

(**a**)

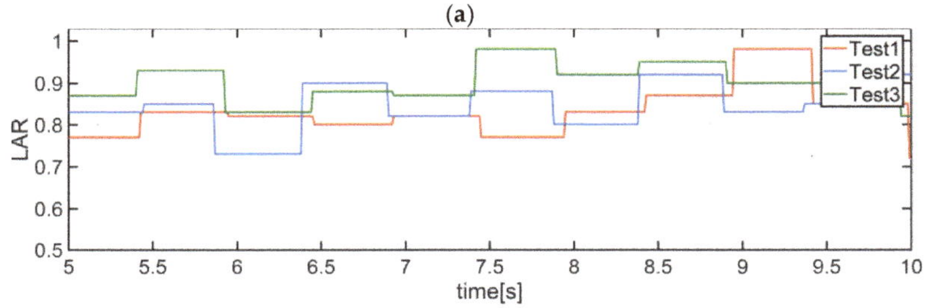

(**b**)

Figure 20. *Cont.*

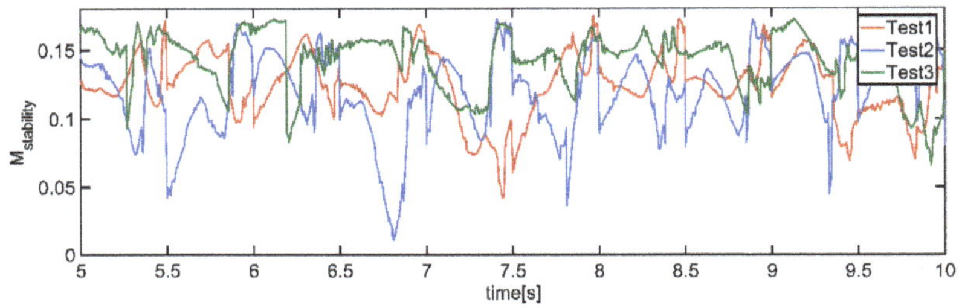

(c)

Figure 20. Stability measure of: (**a**) Method 1; (**b**) Method 2; and (**c**) Method 3.

The average values (over 15 s) of the stability evaluations obtained using these three methods are shown in Table 4.

Table 4. Average values of stability evaluations.

	Method 1	Method 2	Method 3
Test 1	0.0225	0.7899	0.1185
Test 2	0.0365	0.8181	0.1130
Test 3	0.0277	0.8630	0.1402

Test 2 had the lowest stability, followed by Test 3 and then Test 1 according to Method 1. However, this result was not consistent with the actual movement of the robot. Method 2 indicated that the stability of the robot in Test 3 was better than Test 2, followed by Test 1. Using the second method, the stability of Test 1 was found to be slightly better than that of Test 2, with Test 3 being the most stable case. The last two methods could both indicate that the stability of Test 3 was best, but one more simulation should be designed to compare the stability of Test 1 and Test 2.

Two impact forces parallel to y_c in the opposite direction were both applied to the CoM of the body during the trotting in Test 1 and Test 2. The magnitudes, timings, and durations of these forces are shown in Table 5.

Table 5. Impact forces applied in Test 1 and Test 2.

Timing (s)	Magnitude (N)	Duration (s)
5.1	100	0.2
5.1	−80	0.2

The robot is supported by the same set of diagonal legs during the first half of each gait cycle of the trotting gait. The magnitude of the forces that the robot can bear differs in the positive and negative directions of y_c if the positions of ZMP_0 and the next support line are fixed.

For example, in the case shown in Figure 21, the impact forces in the positive and negative directions of y_c can both drive ZMP_0 closer to the boundaries of the virtual-support quadrilateral. However, the force in the positive direction can reduce the distance between ZMP and the support lines, whereas the force in the negative direction increases this distance. Thus, the robot can endure a greater positive impact force along y_c.

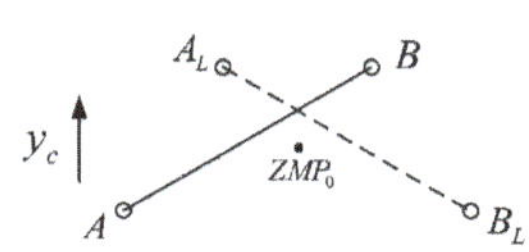

Figure 21. Position of ZMP_0 and the support lines

The changes of the posture angles of these two tests under impacts are shown in Figure 22.

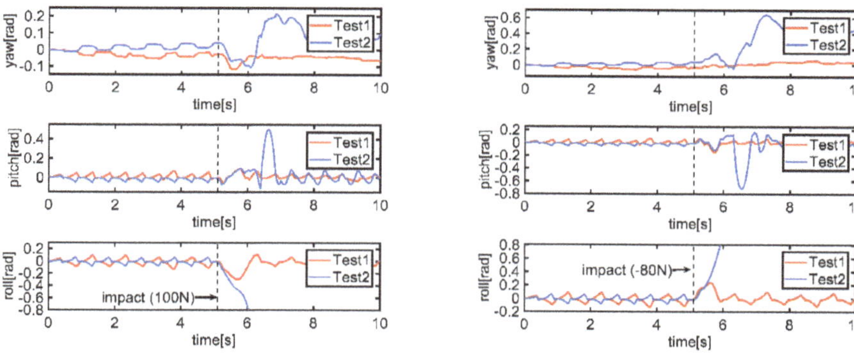

(a) (b)

Figure 22. Posture angles of the two tests with impact forces: (**a**) 100 N; and (**b**) −80 N.

The robot in Test 2 tumbled to the ground after those two impact forces acting in different directions, while the robot in Test 3 could sustain the impacts without losing its stability. This means that the stability of the motion in Test 1 was better than Test 2, thus the method proposed in this paper could evaluate the stability of the quadruped robot with trotting gait more accurately than LAR.

Three impact forces parallel to y_c were applied to the CoM of the body in the three trotting tests, as described above, to further illustrate the accuracy of the proposed stability criterion and measurement. The magnitudes, timings, and durations of these forces are shown in Table 6.

Table 6. Impact forces applied to the body.

Timing (s)	Magnitude (N)	Duration (s)
5	100	0.2
9.2	−82	0.2
14.1	130	0.2

The simulation results of these three tests after impacts are shown in Figure 23.

(a) (b) (c)

Figure 23. Tumbling moments of the three tests: (**a**) Test 1: 9.76 s; (**b**) Test 2: 5.45 s; and (**c**) Test 3: 14.56 s.

Test 2 was the least able to bear the impact force in the positive direction of y_c, since the positions of the next set of support feet differed with respect to ZMP_0 (as can be seen simply from the positions of the next support lines with respect to CoG), tumbling after the first impact (Figure 23b). The robot walked stably in Test 1 until the second impact was applied to the body (Figure 23a), and the robot in Test 3 lost its balance after the last impact (Figure 23c).

The three stability-evaluation methods were also used for the analysis of the three tests with external impacts. To illustrate this intuitively, for Method 2, the stability measurement was set to -0.3 if ZMP_0 was outside the virtual-support quadrilateral and the robot lost its balance according to Equation (40).

As shown in Figure 24, the distance between ZMP and the current support line was not zero for most of the time, making it very difficult to determine when the robot would lose its balance. Using Method 2, it was hard to decide when to update the value of LAR if the robot tips over, because one of the feet would never touch the ground. From the results of Method 3, the robot was considered to lose its balance at 9.753 s (Test 1), 5.109 s (Test 2), and 14.2 s (Test 3), which was consistent with the actual motions of the robot.

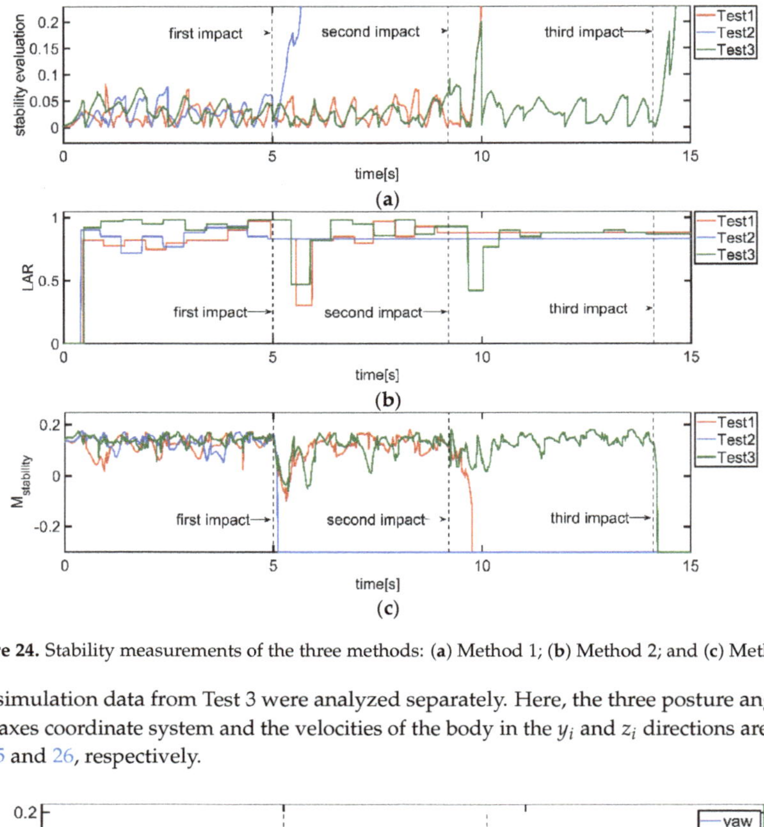

Figure 24. Stability measurements of the three methods: (**a**) Method 1; (**b**) Method 2; and (**c**) Method 3.

The simulation data from Test 3 were analyzed separately. Here, the three posture angles of the principal-axes coordinate system and the velocities of the body in the y_i and z_i directions are shown in Figures 25 and 26, respectively.

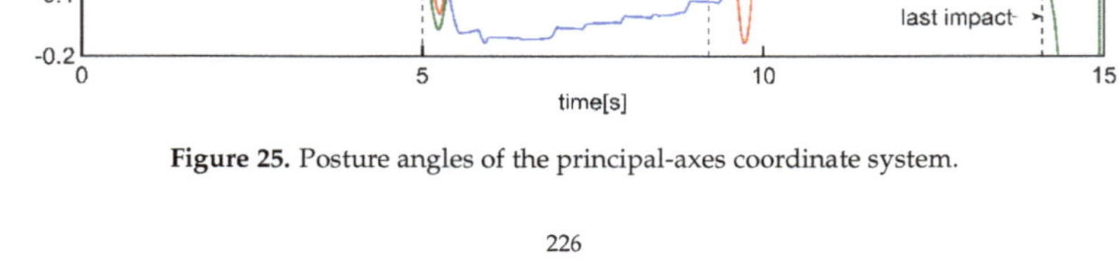

Figure 25. Posture angles of the principal-axes coordinate system.

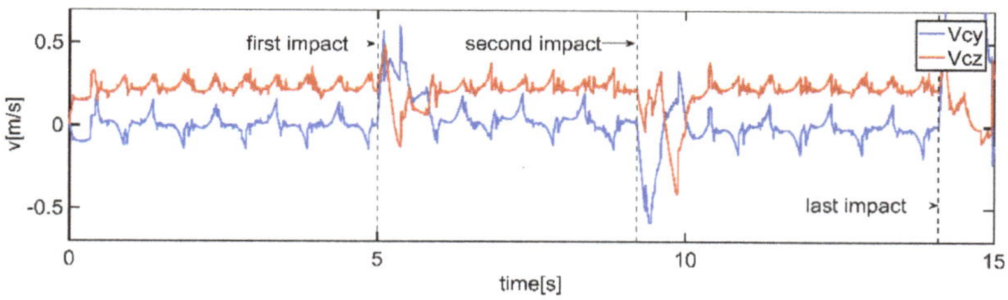

Figure 26. Velocities of the body in the y_i and z_i directions.

The impact forces caused large changes in the velocities along the y_i and z_i directions since the current support lines and these forces were not vertical; thus, the posture angles of the body exceeded the angle-fluctuation range of normal stable motion, reducing the stability of the robot. After the first two impacts, the current ground-reaction force prevented the continued increase of undesired speeds in these two directions to a certain extent since ZMP_0 was located inside the virtual-support quadrilateral, but did not affect the speed in the direction perpendicular to the support line. This part of speed could only be eliminated when the next set of support feet hit the ground, causing a great impact force at the next set of support feet, such that the stability of the robot was still poor and it had to take several steps for the robot to restore normal stable state. Following the last impact, the speed in the y_i direction, the pitch angle, and the roll angle of the body changed sharply, and the ground-reaction force at the support feet could not prevent the robot from tipping over, and the robot lost its balance.

It can be seen in Figure 24 that, after the first impact, the distance between ZMP and the current support line did not exceed the range of the normal stable trotting, and, after the second impact, it was not until 9.8 s that the evaluated stability of Method 1 decreased markedly. Method 2 could show the reduction of the robot's stability correctly after impacts were applied in this test, but it could not evaluate the stability in time, because LAR was updated when the two feet in the same phase both touch the ground. However, from the curve of Method 3, the stability measurement of the robot decreased very quickly and markedly after the first two impacts and then fluctuated significantly within a few steps after impact. After the last impact, the robot lost its balance. This was consistent with the movement previously analyzed.

The positions of ZMP and ZMP_0 were both calculated in this simulation to illustrate that the modified ZMP with a velocity term can reflect the motion state better than the traditional ZMP. Figure 27 shows the curves of ZMP, ZMP_0, and CoG in the y_i direction.

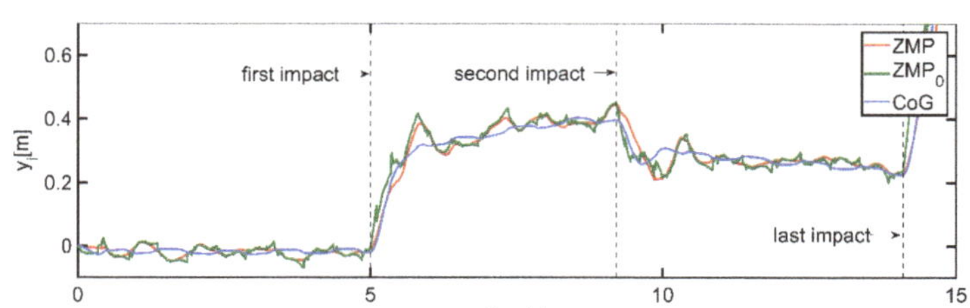

Figure 27. Positions of ZMP, ZMP_0, and CoG in the y_i direction.

After each impact, y_i of ZMP_0 could show the movement trend faster than that of ZMP and CoG, and there was not much difference between the peaks (or troughs) of ZMP and ZMP_0 during the whole movement, meaning that the measurement using ZMP_0 could decrease response time without misjudging the stability state of the robot. As shown in Figure 28, after the final impact, the stability criterion using ZMP_0 determined that the robot lost its balance at 14.2 s, whereas the criterion using

ZMP showed instability at 14.5 s. The time difference between the two adjacent steps of the trot gait used in this study was 0.5 s. Such a time difference is crucial for the robot to adjust the landing positions of the next support line in time to maintain its balance during robot-motion control.

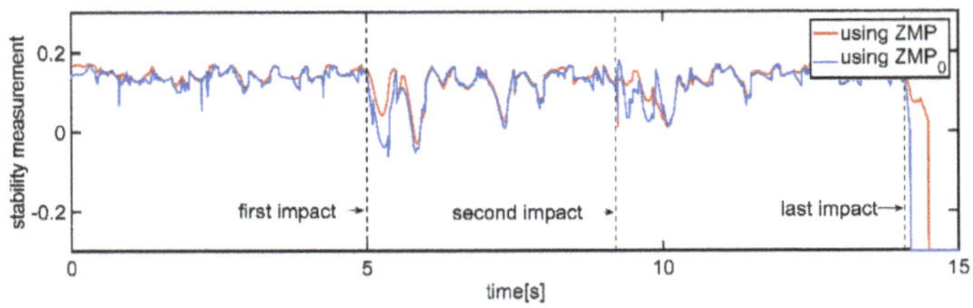

Figure 28. Stability measurements using *ZMP* and ZMP_0.

5.3. Simulation 3

Simulation 3 was designed to prove that the stability-evaluation method proposed in this paper is also applicable to the motion of a robot in 3D space (Figure 29). The robot moved on a 12° slope as shown in the animation (Video S3), and the body's trajectory was similar to that used in Test 3 of the previous simulation: CoG started to move from the midpoint of the current support line to that of the next support line at a constant speed when the previous support line had just disappeared. The contact parameters between each foot and the ground and the trot-gait parameters of this simulation were set to be identical to Test 3.

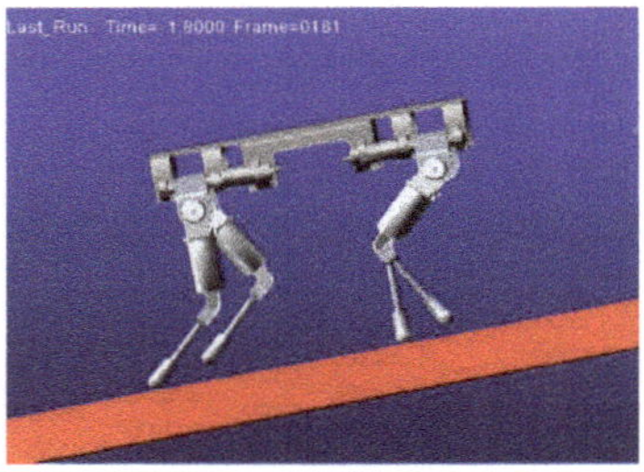

Figure 29. Simulation of Simulation 3.

The attitude angles of the principal-axes coordinate system, and the variation of the stability measurements calculated using above three methods with time are shown in Figure 30a,b.

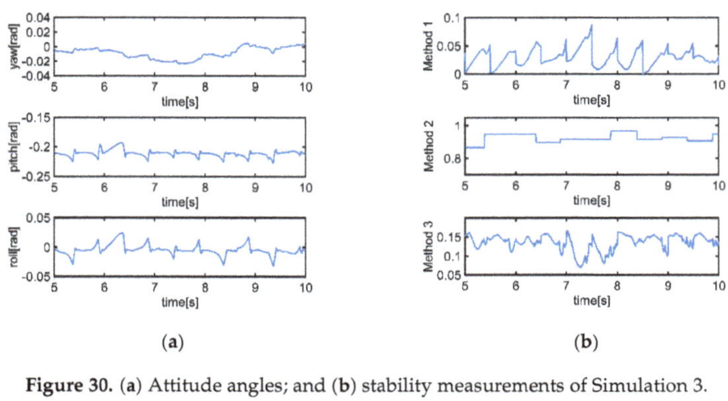

Figure 30. (**a**) Attitude angles; and (**b**) stability measurements of Simulation 3.

The movement of the robot trotting on the slope was steady. The average values (over 15 s) of the stability evaluations obtained using these three methods of Simulation 3 and Test 3 are listed in Table 7.

Table 7. Average values of stability evaluations when the robot trotting on the 12° slope and horizontal plane.

	Method 1	Method 2	Method 3
On slope (Simulation 3)	0.0278	0.9007	0.1354
On horizontal plane (Test 3)	0.0277	0.8630	0.1402

Method 1 showed that those two motions had almost the same stability. Method 2 indicated that the stability of the motion on slope was much higher than that on horizontal plane, which was contrary to the result of Method 3. Three same impact forces (as shown in Table 6) were applied to the CoM of the body during Simulation 3. The posture angles of the principal-axes coordinate system, the positions and velocities of the body are shown in Figure 31a–c, respectively.

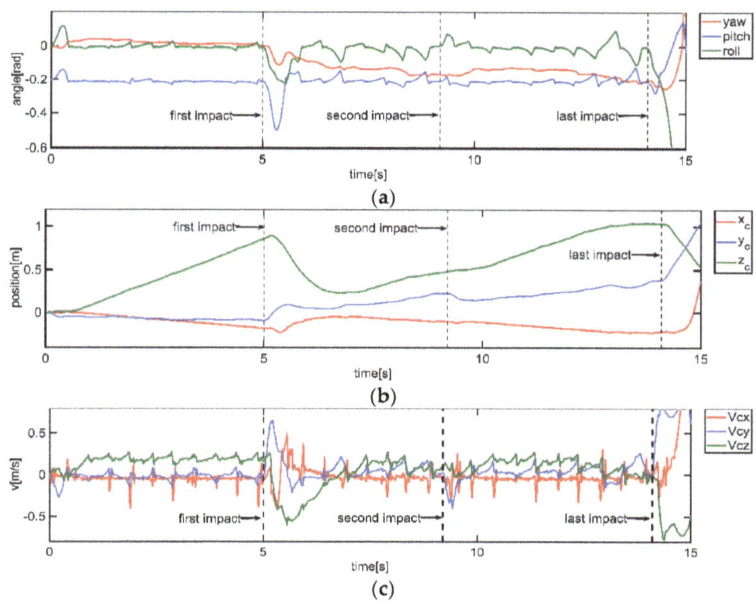

Figure 31. Sensor data of the motion with impact in Simulation 3: (**a**) posture angles of the principal-axes coordinate system; (**b**) positions; and (**c**) velocities of the body.

The robot did not lose its balance before the last impact, but its feet skidded on the slope after the first impact because it was hard for the ground to provide enough friction to push the robot upward. The wave heights of the posture angles were not able to recover to the states before impact. Therefore, the robot was more likely to lose its balance on the slope, and the stability of this motion was worse than in Test 3. Three Methods were used to analyze the motion on slope with impacts (Figure 32). The stability-evaluation method proposed in this paper could measure the stability of motion in 3D space more accurately and efficiently than Methods 1 and 2.

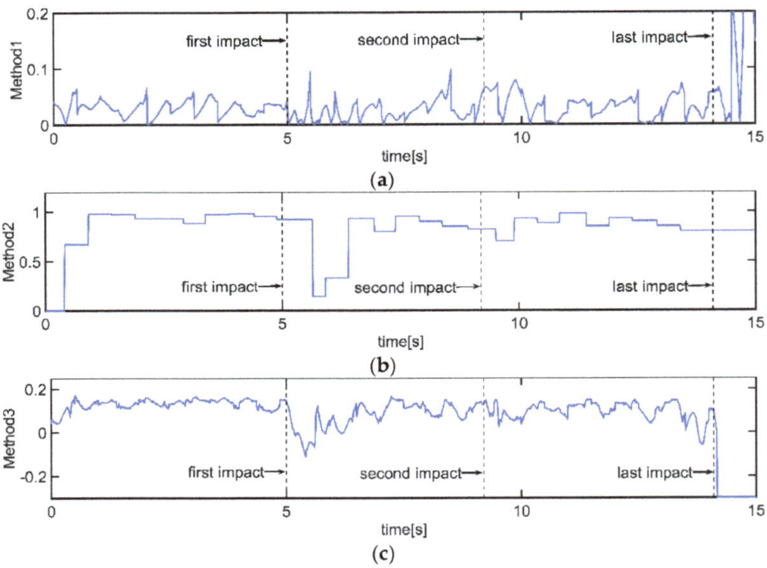

Figure 32. Stability measurements of the motion on slope under the three methods: (**a**) Method 1; (**b**) Method 2; and (**3**) Method 3.

6. Conclusions

Owing to the limitations in the previous studies based on the stability criteria of the dynamic gaits of a quadruped robot, a new stability-evaluation method is proposed in this paper. First, the traditional *ZMP* is improved so that it can be used to analyze the robot's 3D motion, and the influence of velocity on the motion state is considered. Then, the stability criterion for a dynamic gait of the quadruped robot is presented to determine whether the robot can maintain its balance during movement. Finally, the stability is evaluated using d_{edge}, $|\sin\tau|$ and d_{spt}.

The following three conclusions can be obtained through simulation experiments: (1) the robot may still maintain its stability even if *ZMP* is not located on the current support line; (2) the dynamic-stability criterion and measurement proposed in this paper can determine the robot's stability state and evaluate the stability efficiently and accurately; and (3) the stability-evaluation method proposed in this paper is applicable for measuring the stability of motion in 3D space. The new stability-evaluation method is simple for calculations and has a wide range of applications. The future work includes using this method to guide the motion planning and attitude adjustment of the robot and extending it to the stability evaluations of other legged robots.

Supplementary Materials: The following are available online at https://zenodo.org/record/1525073#.W_tT8naYOUk, Video S1: Animation of Simulation 1, Video S2: Animation of Simulation 2, and Video S3: Animation of Simulation 3.

Author Contributions: Conceptualization, Y.J.; Methodology, Y.J. and G.L.; Supervision, B.H.; Writing—original draft, Y.J.; and Writing—review and editing, Y.J., X.L., J.Z. and Y.Z.

Funding: This research was funded by National Natural Science Foundation for Young Scientists of China, grant number 61501034.

Acknowledgments: Thanks to Jianan Zhao for providing language help, Yi Yao for editing videos and all our colleagues for providing all types of help during the preparation of this manuscript.

Conflicts of Interest: The authors declare no conflict of interest.

References

1. Raibert, M.H.; Tello, E.R. Legged Robots That Balance. *IEEE Expert* **1986**, *1*, 89. [CrossRef]
2. Hutter, M.; Gehring, C.; Bloesch, M.; Hoepflinger, M.A.; Remy, C.D.; Siegwart, R. Starleth: A compliant quadrupedal robot for fast, efficient, and versatile locomotion. In *Adaptive Mobile Robotics*; World Scientific: Singapore, 2012.
3. Seok, S.; Wang, A.; Chuah, M.Y.; Hyun, D.J.; Lee, J.; Otten, D.M.; Lang, J.H.; Kim, S. Design Principles for Energy-Efficient Legged Locomotion and Implementation on the MIT Cheetah Robot. *IEEE/ASME Trans. Mechatron.* **2015**, *20*, 1117–1129. [CrossRef]
4. Hutter, M.; Gehring, C.; Jud, D.; Lauber, A.; Bellicoso, C.D.; Tsounis, V.; Hwangbo, J.; Bodie, K.; Fankhauser, P.; Bloesch, M.; et al. ANYmal—A highly mobile and dynamic quadrupedal robot. In Proceedings of the 2016 IEEE/RSJ International Conference on Intelligent Robots and Systems (IROS), Daejeon, Korea, 9–14 October 2016; pp. 38–44.
5. Garcia, E.; Estremera, J.; Santos, P.G.D. A comparative study of stability margins for walking machines. *Robotica* **2002**, *20*, 595–606. [CrossRef]
6. Kimura, H. Adaptive dynamic walking of a quadruped robot on natural ground based on biological concepts. *Int. J. Robot. Res.* **2007**, *26*, 475–490. [CrossRef]
7. Kang, D.-O.; Lee, Y.-J.; Lee, S.-H.; Hong, Y.-S.; Bien, Z. A study on an adaptive gait for a quadruped walking robot under external forces. In Proceedings of the International Conference on Robotics and Automation, Albuquerque, NM, USA, 20–25 April 1997; Volume 2774, pp. 2777–2782.
8. Erbatur, K.; Okazaki, A.; Obiya, K.; Takahashi, T.; Kawamura, A. A study on the zero moment point measurement for biped walking robots. In Proceedings of the 7th International Workshop on Advanced Motion Control, Proceedings (Cat. No. 02TH8623), Maribor, Slovenia, 3–5 July 2002; pp. 431–436.
9. Hirukawa, H.; Hattori, S.; Harada, K.; Kajita, S.; Kaneko, K.; Kanehiro, F.; Fujiwara, K.; Morisawa, M. A universal stability criterion of the foot contact of legged robots—Adios ZMP. In Proceedings of the 2006 IEEE International Conference on Robotics and Automation (ICRA 2006), Orlando, FL, USA, 15–19 May 2006; pp. 1976–1983.
10. Vukobratović, M.; Herr, H.M.; Borovac, B.; Raković, M.; Popovic, M.; Hofmann, A.; Jovanović, M.; Potkonjak, V. Biological principles of control selection for a humanoid robot's dynamic balance preservation. *Int. J. Humanoid Robot.* **2008**, *05*, 639–678. [CrossRef]
11. Sardain, P.; Bessonnet, G. Forces acting on a biped robot. Center of pressure-zero moment point. *IEEE Trans. Syst. Man Cybern. Part A Syst. Hum.* **2004**, *34*, 630–637. [CrossRef]
12. Gonzalez de Santos, P.; Jimenez, M.A.; Armada, M.A. Dynamic Effects in Statically Stable Walking Machines. *J. Intell. Robot. Syst.* **1998**, *23*, 71–85. [CrossRef]
13. Yoneda, K.; Hirose, S. Tumble stability criterion of integrated locomotion and manipulation. In Proceedings of the IEEE/RSJ International Conference on Intelligent Robots and Systems (IROS'96), Osaka, Japan, 4–8 November 1996; Volume 872, pp. 870–876.
14. Khorram, M.; Moosavian, S.A.A. A 3D stable trot of a quadruped robot over uneven terrains. *Proc. Inst. Mech. Eng. Part C J. Mech. Eng. Sci.* **2015**, *231*, 555–573. [CrossRef]
15. Karčnik, T. Stability in legged locomotion. *Boil. Cybern.* **2004**, *90*, 51–58. [CrossRef] [PubMed]
16. Lee, W.; Raibert, M. Control of hoof rolling in an articulated leg. In Proceedings of the 1991 IEEE International Conference on Robotics and Automation, Sacramento, CA, USA, 9–11 April 1991; Volume 1382, pp. 1386–1391.
17. Goswami, A. Foot rotation indicator (FRI) point: A new gait planning tool to evaluate postural stability of biped robots. In Proceedings of the 1999 IEEE International Conference on Robotics and Automation (Cat. No. 99CH36288C), Detroit, MI, USA, 10–15 May 1999; Volume 41, pp. 47–52.
18. Zhou, D.; Low, K.H.; Zielinska, T. A stability analysis of walking robots based on leg-end supporting moments. In Proceedings of the 2000 ICRA—Millennium Conference, IEEE International Conference on Robotics and Automation, Symposia Proceedings (Cat. No. 00CH37065), San Francisco, CA, USA, 24–28 April 2000; Volume 2833, pp. 2834–2839.
19. Lin, B.; Song, S. Dynamic modeling, stability and energy efficiency of a quadrupedal walking machine. In Proceedings of the IEEE International Conference on Robotics and Automation, Atlanta, GA, USA, 2–6 May 1993; Volume 363, pp. 367–373.

20. Papadopoulos, E.; Rey, D. The force-angle measure of tipover stability margin for mobile manipulators. *Veh. Syst. Dyn.* **2000**, *33*, 29–48. [CrossRef]
21. Gonzalez-de-Santos, P. An improved energy stability margin for walking machines subject to dynamic effects. *Robotica* **2005**, *23*, 13–20.
22. Garcia, E.; de Santos, P.G. On the Improvement of Walking Performance in Natural Environments by a Compliant Adaptive Gait. *IEEE Trans. Robot.* **2006**, *22*, 1240–1253. [CrossRef]
23. Yoneda, K.; Iiyama, H.; Hirose, S. Intermittent trot gait of a quadruped walking machine dynamic stability control of an omnidirectional walk. In Proceedings of the IEEE International Conference on Robotics and Automation, Minneapolis, MN, USA, 22–28 April 1996; Volume 3004, pp. 3002–3007.
24. Zhang, S.; Gao, J.; Duan, X.; Li, H.; Yu, Z.; Chen, X.; Li, J.; Liu, H.; Li, X.; Liu, Y.; et al. Trot pattern generation for quadruped robot based on the ZMP stability margin. In Proceedings of the 2013 ICME International Conference on Complex Medical Engineering, Beijing, China, 25–28 May 2013; pp. 608–613.
25. Lin, L.; Jian, W.; Hongxu, M.; Qing, W. Dynamic stability analysis of a trotting quadruped robot on unknown rough terrains. In Proceedings of the 2015 Chinese Automation Congress (CAC), Wuhan, China, 27–29 November 2015; pp. 320–325.
26. Won, M.; Kang, T.H.; Chung, W.K. Gait planning for quadruped robot based on dynamic stability: Landing accordance ratio. *Intell. Serv. Robot.* **2009**, *2*, 105–112. [CrossRef]

applied sciences

MDPI

Article

Learning an Efficient Gait Cycle of a Biped Robot Based on Reinforcement Learning and Artificial Neural Networks

Cristyan R. Gil [1], Hiram Calvo [1,*] and Humberto Sossa [1,2]

1 Centro de Investigación en Computación, Instituto Politécnico Nacional, Ciudad de México 07738, Mexico; b151159@sagitario.cic.ipn.mx (C.R.G.); hsossa@cic.ipn.mx (H.S.)
2 Tecnológico de Monterrey, Campus Guadalajara, Zapopan 45138, Mexico
* Correspondence: hcalvo@cic.ipn.mx; Tel.: +52-(55)-5729-6000 (ext. 56516)

Received: 2 January 2019; Accepted: 26 January 2019; Published: 1 February 2019

Featured Application: The final product is an algorithm that allows a simulated robot to learn sequences of poses (single configurations of its joints) in order to learn to walk (and extendable to other tasks) by finding the combination of these poses.

Abstract: Programming robots for performing different activities requires calculating sequences of values of their joints by taking into account many factors, such as stability and efficiency, at the same time. Particularly for walking, state of the art techniques to approximate these sequences are based on reinforcement learning (RL). In this work we propose a multi-level system, where the same RL method is used first to learn the configuration of robot joints (poses) that allow it to stand with stability, and then in the second level, we find the sequence of poses that let it reach the furthest distance in the shortest time, while avoiding falling down and keeping a straight path. In order to evaluate this, we focus on measuring the time it takes for the robot to travel a certain distance. To our knowledge, this is the first work focusing both on speed and precision of the trajectory at the same time. We implement our model in a simulated environment using q-learning. We compare with the built-in walking modes of an NAO robot by improving normal-speed and enhancing robustness in fast-speed. The proposed model can be extended to other tasks and is independent of a particular robot model.

Keywords: q-learning; Q-networks; reinforcement learning; gait cycle; biped robots

1. Introduction

Biped robots are designed with a physical structure that tries to emulate a human being with the purpose of providing versatility in terms of movement in such a way that these robots can move through irregular terrains and are better adapted in comparison with robots with wheels [1]. This kind of robots is easily adapted to the human environment due their locomotion, and humans can adapt easily to their interaction because of their similarity in terms of physical structure. Given their kinematic complexity (more than twenty degrees of freedom), the methods to control them are highly complex, with this level of complexity and their extensive potential a recurrent theme of study for many researchers, both in the robotics area and the artificial intelligence area. Each field proposes new control methods that offer better velocity of displacement, better stability, or the capacity to walk in different terrains. In this work, we propose a suitable neural architecture in order to reach a fixed distance in the shortest time. Our main goal is, given a set of zero moment point (ZMP) regions, to enable a biped robot to find a convenient gait cycle. We aim to generalize the definitions of the actions, states, and rewards to generate walks showing a stable behavior, i.e., the robot should not fall down. Additionally, the gait cycle must be as fast as possible.

There is a large number of articles related to the control of biped robots, many of which propose highly complex techniques achieving good performance [2–4]. Recently, several works use machine learning techniques, such as reinforcement learning (RL), to achieve control of the gait cycle of a biped robot [5,6]. Because of this, the goal of this work is to develop a learning system that allows a biped robot to perform the walking cycle with the aim to improve its performance.

Conventional control methods of the biped robots involve many parameters such as the length of the robot, its weight, the speed at which the engines respond, its center of mass or its center of gravity, among others. That is, the design of a control strategy is specific for a robot, with particular parameters for that robot, whereby a strategy that is originally designed for robot X will work in robot Y only if the parameters of robot Y are the same or at least similar to the ones in X; however, if the parameters of robot Y are different, the control strategy that was used in robot X will not be applicable to robot Y and a new control has to be designed, this time considering only robot Y. We are talking about strategies that even when they work, they are not flexible, and attempting to generalize them becomes a very complicated task.

Having said this, the effort to develop a control strategy that is versatile enough to be applicable in different robots without any substantial modification of its structure makes sense and is precisely in that niche that our proposal is born: the idea of using self-learning algorithms to overcome such limitations. In this work we opt for the reinforced learning method because the way it works is very similar to the way humans learn, that is, through trial and error. Due to the number of experiments required to obtain a practical set of movements, the ideal environment for implementing such a method is a simulated environment, following with the idea since the inception of simulators, that they should allow a straightforward transfer to real robots [7]. The use of simulators to avoid costly testing of new ideas and improve efficiency is a common practice amongst participants of the DARPA (Defense Advanced Research Projects Agency) Robotics challenge [8,9].

The structure of the rest of this paper is as follows. In Section 2, we present a compilation of some recent works about the control of a biped robot; it is divided in three subchapters, each one corresponding to a different approach: classical, ZMP, and alternative approaches. Section 3 describes our proposal in detail; we explain our vision of solving the problem of the gait cycle of a robot (but not restricted only for that problem), how the different knowledge modules are designed and the leaning framework, and most importantly, how these modules will lead the robot to achieve a complete gait cycle. Later on, in Section 4, we show the results of our method applied to a virtual robot and compare results to measure the performance of our algorithm. Finally, in Section 5 we summarize our results and present a general prospect of the paths that can be followed from this work.

2. State of the Art

There are many ways to walk, and even subjects can be identified by analyzing their gait [10]. Thus, finding an optimal gait cycle for a biped robot has been a recurrent task in the literature. In this section we present a brief summary of the latest works on the biped robot gait cycle research in order to have a better understanding of the problem and to emphasize the fact that the problem has been addressed in many different ways.

2.1. Classical Approaches

Early biped walking of robots involved static walking with a very low walking speed [6]. The step time was over 10 seconds per step and the balance control strategy was performed through the use of COG (center of gravity). Here the projected point of COG onto the ground always falls within the supporting polygon that is made by two feet. During the static walking, the robot can stop the walking motion any time without falling down. The disadvantage of static walking is that the motion is too slow and wide for shifting the COG. Below, there are some recent works solving the problem of biped walking based on classical approaches.

Jung-Yup, Ill-Woo, and Jun-Ho [11] describe a walking control algorithm for biped humanoid robots that considers an uneven and inclined floor. Some online controllers worked well for a slightly uneven and inclined floor, but the robot immediately fell down when the floor inclinations exceeded a certain threshold. Hence, six online controllers (upright pose controller, landing angular momentum controller, landing shock absorber, landing timing controller, landing position controller, and vibration reduction controller) were developed, designed through simple mathematical models and experiments, and then suitable activation periods were planned in a walking cycle. Each online controller has a clear objective and the controllers are decoupled from each other. To validate the performance of the online controllers, walking experiments on an uneven and inclined aluminum plate were performed.

The authors in [12] describe a proposed method for planning walking patterns, which includes the ground conditions, dynamic stability constraint, and relationship between walking patterns and actuator specifications. They generate hip and foot trajectories, and based on the computation of the zero moment point, they select the most stable trajectory. As a result of this approach, the system can adjust to the ground condition by adjusting the values of foot parameters. The system was validated using a dynamic simulator and a real robot. Recently, deviation control was implemented in the planning stage with a trajectory correction during the gait cycle [13].

In [2], an online gait trajectory generation method is proposed. The gait trajectory has continuity, and smoothness in variable period and stride to realize the various bipedal walking gaits. The method is tested using simulation and experiment. Recently, control techniques have been applied to balance a biped robot while walking in slopes [14], along with the use of central pattern generation [15,16].

2.2. Alternative Approaches

In this section we present some recent works solving the problem of biped walking based on alternative approaches.

In [1], given a sample trajectory, they use several coupled generic central pattern generators (CPG) to control a bipedal robot. They use one CPG for each degree of freedom. They show that starting from a sample trajectory a controller can be build that modulates the speed of locomotion and the step length of a robot.

In [17], the authors introduce the concept of simplified walking to describe the complete biped motions with the unit of a walking step. The robot follows a predefined trajectory using several movements including forward, sideways walking, turning, and so on.

The authors in Reference [18] use a central pattern generator to model the gait of the robot. To achieve it, a quality function combines diverse parameters in order to allow the robot to search for an optimal gait. At the end, the robot could walk quickly and with stability.

In [19], Meriçli and Veloso record a complete gait cycle from a given walk algorithm. Once it is recorded, it is reproduced in a cycle loop, making real-time corrections with human feedback.

In [20], a robot learns how to walk without prior knowledge of an explicit dynamics model. Initially, a large collection of poses is defined, but with the use of reinforcement learning, the authors reduced the number of poses available so the robot could walk. More than 50 poses were manually defined.

Numerical optimization methods have been used to adjust optimal gaits at runtime for a robot walking on a treadmill with a speed change [21]. As in most works, in this work, we neglected the friction between the robot's feet and floor. Currently, there are few works that deal with this problem [22].

2.3. Zero Moment Point Approaches

In general, the walking control strategies using the ZMP can be divided in two approaches. First, the robot can be modeled by considering several point masses, the locations of the point masses, and the mass moments of inertia of the linkages. The walking pattern is then calculated by solving

ZMP dynamics derived from the robot model with a desired ZMP trajectory [23]. During walking, sensory feedback is used to control the robot.

In the second approach, the robot is modeled using a simple mathematical model such as an inverted pendulum system, and then the walking pattern is designed based on the limited information of a simple model and experimental hand tuning. During walking, diverse online controllers are activated to compensate the walking motion through the use of various sensory feedback data including the ZMP. Figure 1 illustrates stable and unstable poses.

The first approach can derive a precise walking pattern that satisfies the desired ZMP trajectory, but it is hard to generate the walking pattern in real-time due to the large calculation burden. Further, if the mathematical model is different from the real robot, the performance is diminished. On the contrary, the second approach can easily generate the walking pattern online. However, several kinds of online controllers are needed to compensate the walking pattern in real-time because the prescribed walking pattern cannot satisfy the desired ZMP trajectory. In addition, this method depends strongly on the sensory feedback, and hence the walking ability is limited to the sensor's performance and requires considerable experimental hand tuning [11].

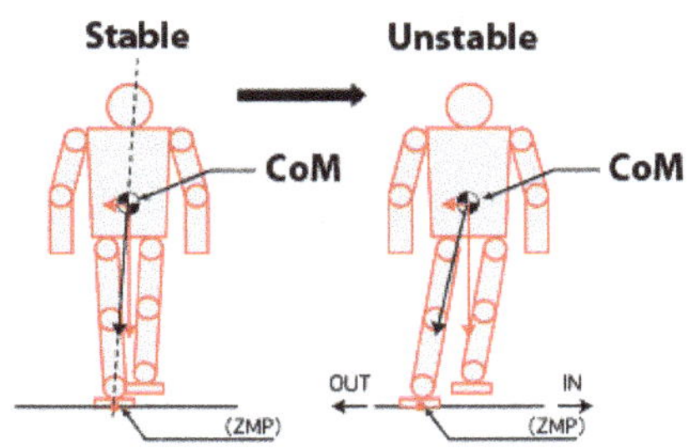

Figure 1. Zero Moment Point (ZMP) concept (CoM is the center of mass).

Below, some recent works solving the problem of biped walking based on ZMP approaches are described.

In Reference [24], the authors design an omnidirectional ZMP-based engine to control an NAO robot, they use an inverted pendulum model, to generate, with a preview controller, dynamically balanced centers of mass trajectories. The importance of this work lies in the implementation on a NAO robot, where the authors explain that even when omnidirectional walking and preview control have been explored extensively, in many articles, they do not provide detailed working of the walk engine and the results are often based on simulated experiments. Balance on different road surfaces (for example soil, grass, ceramic) was studied in Reference [25]. The authors used fuzzy control theory based on ZMP information from various sensors. In Reference [26], the authors compute the zero moment point on a robot walking with a human-like gait obtained from a bio-inspired controller. They design a 2D and a 3D walking gait showing better performance in 2D. The computation time reached in this work is around 0.0014 ms. Other works take further the idea of comparing with human locomotion by restricting certain joints and observing the effects and torque changes in the robot [27]. Symmetricity of gait cycle patterns is analyzed in detail in Reference [28]. A thorough survey on stability criteria in biped robots is presented in Reference [28] as well.

In Reference [29], the authors present a method for biped walking using reinforcement learning; this method uses the motion of the robot arms and legs to shift the ZMP on the soles of the robot. The algorithm is implemented in both a simulated robot and a real robot. The work presented in [6] is an improvement of the previous work, with the difference that it converges faster because the action space is smaller; they reduce from 24 actions to 16 actions. Their results are very similar to the previous work.

3. Proposed Method

This section presents our proposed solution, starting with a general description of the framework. Even when this work is focused on solving the problem of the walking of a biped robot, it is thought of as part of a general solution that allows for solving other problems following a similar approach. A section delimiting our approach is presented. Later on, we will show the definition of the base level modules, which are the modules that interact with the robot by increasing or decreasing a specific joint by a small amount, and how these are learned by the robot.

Afterwards, the definition of the second level modules will be presented, i.e., the *poses*, and the way they are learned by the robot. In both cases, namely the first and second level modules, there is a proposal of the architecture of a Q-network along with the activation function for each of its layers.

3.1. General Framework

As we have presented in the state of the art section, there are several works related to the biped robot control problem. It is very important to find a better strategy to overcome this problem since it is a hard task to try to manage complex kinematics such as having more than 20 degrees of freedom in this kind of robot.

In some areas of computer science such as structured programming or structures analysis, it is necessary to use what is called *decomposition*. Decomposition means breaking a complex problem or system into parts that are easier to conceive, understand, program, and maintain. In other words, divide a large problem into smaller ones.

Thinking in *decomposition* and the complexity of the biped robots, we would like to state our vision regarding how a robot can learn complex behavior, which involves learning in a multiple-level system using a framework made of different levels: base level, first level, second level, and third level.

At the lowest level of this framework, the base level, lies the direct interaction with the angles of all the joints of the robot, and in the first level, we have specific configuration of the joints to conform a pose. The second level consists of simple activities, and finally, at the upmost, third level, more complex activities (tasks) are defined. In this framework each level uses the knowledge of the previous level to build a new knowledge level. Figure 2 illustrates a general schema for robots learning activities.

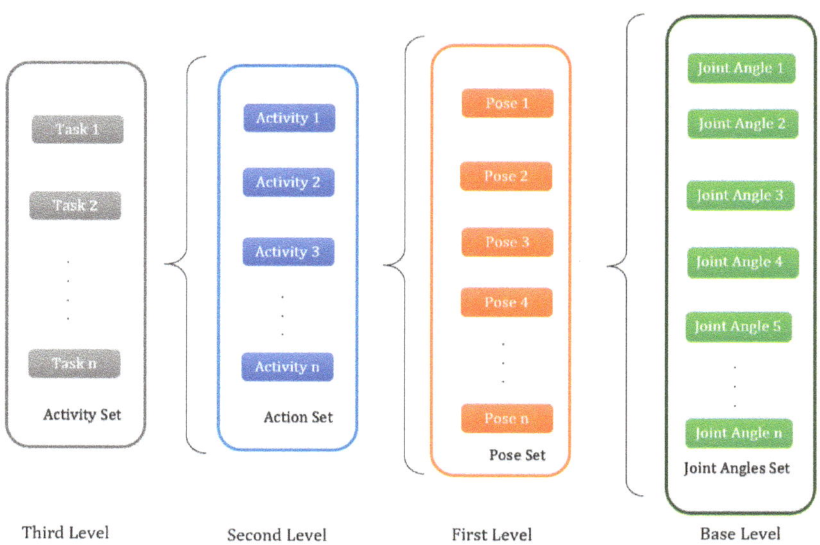

Figure 2. Learning activities for robots using a decomposition framework.

In the following subsections, we provide details on each level of the decomposition framework.

3.1.1. Joint Angle (Base Level)

A *joint angle* is a change in the angle of a particular joint; for instance, if the robot's elbow joint is 30°, calling an elbow extension will change it to 25°, for example. In other words, calling a joint angle module will interact directly with the increase or decrease of the angle of a joint.

This level is the base of the decomposition framework and its modules (extension or flexion) are directly defined by us.

3.1.2. Pose (First Level)

A *pose* is a specific configuration of the robot joints in an instant t, which is a list that contains the information about the position of the whole body of the robot. For instance, a pose can be written as:

Pose = [Left-Shoulder, Left-Elbow, Right-Shoulder, Right-Elbow, Left-Hip, Left-Knee, Left-Ankle, Right-Hip, Right-Knee, Right-Ankle]

This pose definition has ten different values to be filled; for example, we can have a pose with the next values:

$$pose1 = [10^\circ, 30^\circ, 20^\circ, 50^\circ, 0^\circ, 25^\circ, 10^\circ, 0^\circ, 5^\circ, 2^\circ]$$

As can be seen in *pose1*, the left elbow of the robot has a value of 30°, its left hip a value of 0°, and so on. This level of the decomposition framework is learned using the information that the previous level provides. For example, imagine that we want to reach the values of *pose1*, but we have the follow configuration:

$$[6^\circ, 30^\circ, 20^\circ, 50^\circ, 0^\circ, 25^\circ, 10^\circ, 0^\circ, 5^\circ, 2^\circ]$$

For this example, the only joint with a different value is the left shoulder, that has 6° instead of 10°, so we need to change it. To do so, we call a module from the lower level, e.g., *left-shoulder extension*, which increases the value of the left shoulder by 1° every time it is called. Then, we should call it four times to reach the desired position.

3.1.3. Activity (Second Level)

An *activity* is a combination of poses that correspond to a specific action. For instance an activity can be a movement of a robot, such as the NAO's hello gesture. Therefore, let us suppose that our system has already learned four different poses; placing them in a determined order will let the robot achieve the "hello" gesture.

That is the essence of this level, the idea (as in all the levels) is to use the previous knowledge—the poses—to complete an activity. To accomplish it, the system must combine the available poses, in different orders, until it is able to reach its goal.

Some examples of activities that can be completed with poses are *walking forward*, *walking backward*, *sitting down*, *standing up*, *turning left*, *turning right*, *taking an object*, etc. For example, Figure 3 shows the decomposition of the hello gesture of NAO.

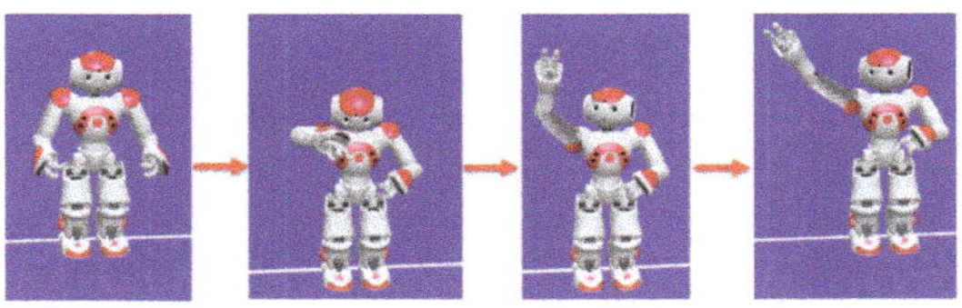

Figure 3. NAO's hello gesture decomposition.

3.1.4. Task (Third Level)

A *task* is a combination of activities to achieve a specific goal; the idea is to use the previous knowledge to achieve more complex tasks. The previous level is related to learning activities, so now we need to learn how to combine these activities to complete more difficult tasks.

An example of a task can be "serve a drink." If in the previous level, the system learned the activities "rotate hand," "take an object," "open a bottle," and "pour the content," we expect that at this level, the robot is able to perform tasks combining activities in a specific way. This level is mentioned only as part of a general proposal of a decomposition scheme. For the scope of this work, tasks will not be covered. The scope of this work will reside in learning an action using previously learned poses.

The chosen activity is *walking* (see Figure 4). It was selected because walking is a very complex action, as we have seen in the state of the art, and there are a lot of written works aiming to solve this problem. Achieving a good performance of this activity using the proposed learning framework would hint toward evidence supporting our decomposition-based proposal.

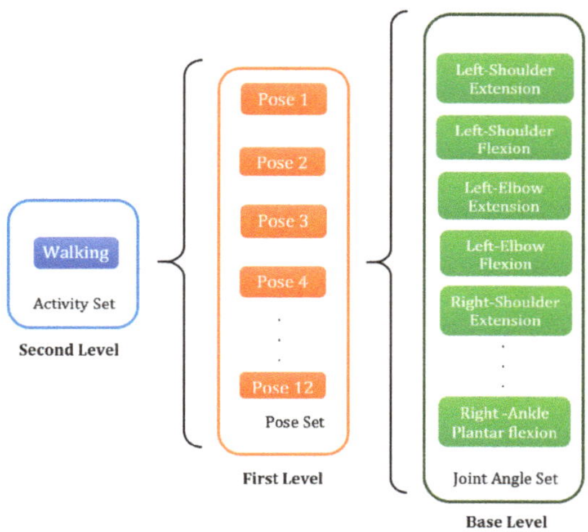

Figure 4. *Walking* learning framework.

We will begin by establishing the base level, i.e., the joint angle modules. Then, we will find some poses that should finally allow the robot to walk. Before we describe our procedure in detail, we need to refer to the specifications of the robot that will be used, i.e., the NAO Robot (https://www.softbankrobotics.com/emea/en/nao).

3.2. Base Level Modules Definition

In the base level, we have the joint angle modules. These modules have to be defined by us because this is the basis for learning in the next levels.

Human movements are described in three dimensions based on a series of planes and axes. There are three planes of motion that pass through the human body: the sagittal plane, the frontal plane, and the transverse (horizontal) plane. The sagittal plane lies vertically and divides the body into right and left parts. The frontal plane also lies vertically and divides the body into anterior and posterior parts. The transverse plane lies horizontally and divides the body into superior and inferior parts [30]. For this work we considered the joints that are utilized when walking on the sagittal plane, that is, 10 of the 25 joints the NAO has, as shown in Section 3.1.2.

Each of the 10 joints considered has 2 movements: extension and flexion. Taking this into account, we have a total of 20 modules defined for the base level. A brief description of the 20 modules in the base level is listed in Table 1, two modules per each movement (see http://doc.aldebaran.com/1-14/family/nao_h25/joints_h25.html for a detailed description of NAO's joints.).

Table 1. Base-level movements, range of movement, and limitations selected in this work.

No.	Movement	NAO Joint (in °)		Limit (in °)	
		From	To	From	To
1	Left shoulder pitch	−119.50	119.50	70.00	110.00
2	Left elbow roll	−88.50	−2.00	−65.00	−25.00
3	Left, hip pitch	−88.00	27.73	−35.00	−15.00
4	Left knee pitch	−5.29	121.04	45.00	65.00
5	Left ankle pitch	−68.15	52.86	−45.00	−25.00
6	Right shoulder pitch	−119.50	119.50	70.00	110.00
7	Right elbow roll	−88.50	−2.00	−65.00	−25.00
8	Right hip pitch	−88.00	27.73	−35.00	−15.00
9	Right knee pitch	−5.29	121.04	45.00	65.00
10	Right ankle pitch	−68.15	52.86	−45.00	−25.00

3.3. *First Level Modules Definition*

Once we have set the base methods, we are able to define the poses, that is, the modules of the first level in the decomposition framework.

The success of an upper level of the framework resides in the knowledge of the previous level. That is, in the upper level we want to complete the activity of *walking*; thus, in this level we need to find poses that help to achieve it.

Just like we have seen in the state of the art, many of the works on biped walking use ZMP trajectories. Basically, what they do is to set the ZMP on one foot of the robot, and with that coordinate, compute the position of the angles of the joints using inverse kinematics. Hence, the modules of this level will be defined as configurations of the joints that lead the robot to a specific position of the ZMP.

First, we have to discuss how we compute the ZMP. According to Reference [6], the ZMP coordinates can be calculated as follows:

$$x = \frac{W\cdot((f_2 + f_4) - (f_1 + f_3))}{2\cdot(f_1 + f_2 + f_3 + f_4)} \tag{1}$$

$$y = \frac{L\cdot((f_1 + f_2) - (f_3 + f_4))}{2\cdot(f_1 + f_2 + f_3 + f_4)} \tag{2}$$

where W is the width and L the length of the foot sole and $f_1 + f_2 + f_3$ and f_4 are the four sensors located in the sole of a robot's foot; the NAO robot already comes with these sensors built-in.

3.3.1. Zero Moment Point (ZMP)-Based Poses

In order to learn poses, we select them based on the ZMP criterion because, as can be found in Reference [6], following a ZMP trajectory enables the robot to walk. We define 12 modules in the first level of knowledge, where each of them will be calculated by considering the ZMP criterion, i.e., following a specific position of it. The *x*,*y*-coordinates of the ZMP of each pose from 1 to 6 located in the sole of the right foot (shown in Figures 5 and 6) and are defined as:

Pose 1: $-0.25 \leq Y \leq 0.25$ and $-0.25 \leq X \leq 0.25$
Pose 2: $1.25 \leq Y \leq 1.75$ and $-0.25 \leq X \leq 0.25$
Pose 3: $2.75 \leq Y \leq 3.25$ and $-0.25 \leq X \leq 0.25$
Pose 4: $4.25 \leq Y \leq 4.75$ and $-0.25 \leq X \leq 0.25$
Pose 5: $1.75 \leq Y \leq 2.25$ and $0.75 \leq X \leq 1.25$
Pose 6: $1.75 \leq Y \leq 2.25$ and $-1.25 \leq X \leq -0.75$

Poses from 7 to 12 are exactly the same but the ZMP coordinate is computed using the force sensors in the left foot instead of the right foot.

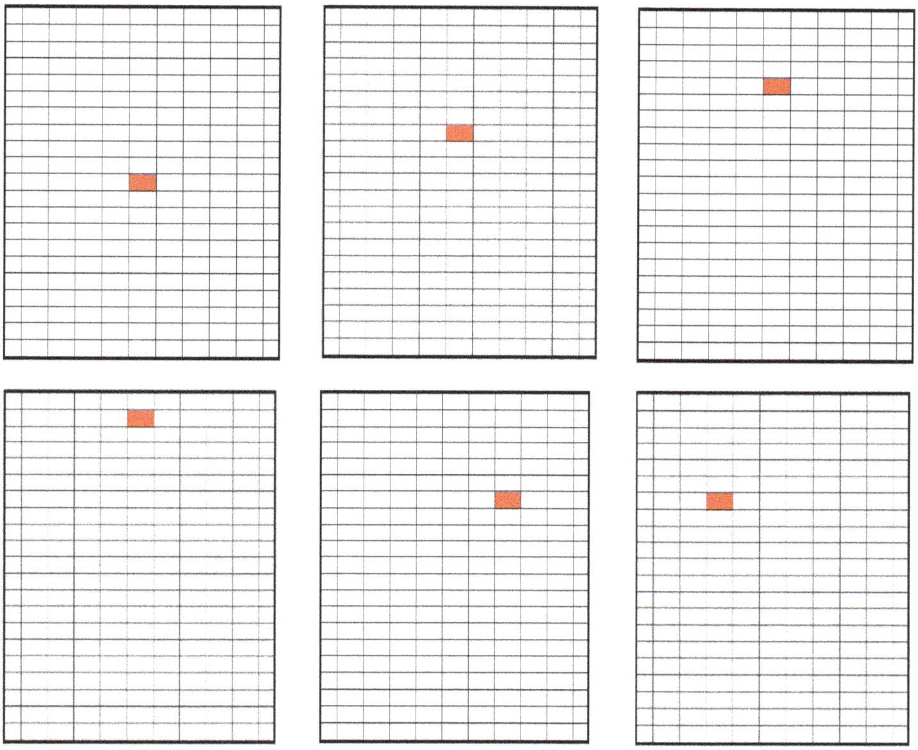

Figure 5. Poses 1 to 6.

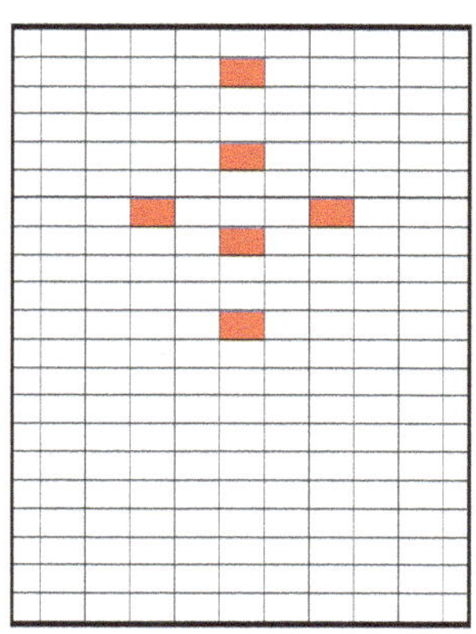

Figure 6. Poses in one-foot sole.

3.3.2. Learning the Poses

Once we have described the elements that our method requires to automatically learn the parameters at the first level, we will describe how poses are learned using q-learning and artificial neural networks.

In order to use reinforcement learning, we need to define states, actions, and rewards. The use of artificial neural networks allows us to generalize the states. That is, instead of stating that a specific state corresponds to certain action, we generalize groups of states that correspond to that particular action. Following Reference [31], we will refer to artificial neural networks combined with q-learning as Q-networks.

The *actions* of the Q-network in this level of knowledge are the methods of the previous level, in this case, the 20 joint angle methods defined in the previous level; for instance, an action in this level is the *left-knee extension* or the *right-hip flexion*.

The *state* has a length of four values and it is conformed with the combination of the ZMP coordinates of both feet, which seems pretty obvious because we want the robot to have a precise coordinate of the ZMP.

The *reward* proposed was 10 if the robot reached the goal coordinates, and -10 if the robot falls down. It is 0 in any other case. The purpose was to avoid falling down by giving a negative reward; remember that for q-learning, the goal is to maximize reward.

3.3.3. First Level Q-Network

The Q-network used in this work is based on Reference [31], where the input of the network is the state and the output is a layer of several neurons, one neuron for each available action. The output yields, as a result, a q-value for each action. For instance, if we have four available actions, output might be something like (1.52, 0.3, -0.5, 2.8). When the training is finished, the maximum value is selected. In this example, we would choose to perform the fourth action because its value is the largest of all four. To update the weights of the Q-network, we use backpropagation with the difference that we do not have a static target y vector, so we need to use the following equation to compute our target (based on http://outlace.com/rlpart3.html) for every state-action pair, except when we reach a terminal state where the reward update is simply r_{t+1}.

$$Q(S_t, A_t) \leftarrow r_{t+1} + \gamma \cdot maxQ(S_{t+1}, A_{t+1}) \quad (3)$$

3.3.4. First Level Q-Network Size

To finish this section, the only thing left is the size of the Q-network. We propose a Multi-Layer Perceptron (MLP) model with four neurons in the input layer, two hidden layers with hyperbolic tangent (tanh) activation function (one with 165 neurons and the other with 120 neurons), and finally an output layer with 20 neurons, each of them corresponding to one available action.

The input layer and the output layer are defined for the problem. We have chosen hyperbolic tangent for the activation functions of the hidden layers because we expect, in some neurons, negative values. This architecture was proposed experimentally ensuring convergence of the system (see Figure 7 for the architecture of this network).

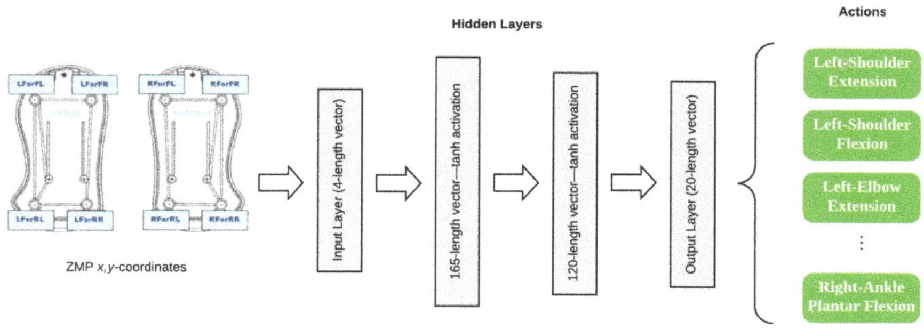

Figure 7. First Level Q-Network.

3.4. Second Level Module Definition

Once we have the poses and the modules of the first level, we are able to obtain the next level method (action). Remember that for this work, the action to learn is *walking*. In the previous level we used Q-networks to learn the poses, so in this level we will use the same technique in order to have a uniform way of learning in each level. The goal is to find a combination of poses that allows the robot to walk without falling down, as fast as possible.

The Q-network in this level is slightly different from the one defined before; below, there is a redefinition of the action, state, and reward.

3.4.1. Actions of the Second Level

The *actions* available for this network are the modules learned in the previous level, the poses, where each of the poses learned before are options to accomplish our goal in this level.

In several works in the state of the art, the authors set a ZMP trajectory, and according to that, they compute the angles of each joint [6]. In our case, the system has already learned the angles that led to that position, so the problem is reduced to finding which poses and what order of these poses make the robot walk.

3.4.2. State of the Second Level

In reinforcement learning, it is fundamental to properly define states. If the defined states do not provide enough information about the environment, the algorithm will not converge to a solution. That is why we need to include all relevant information in the description of a state.

States at this level are conformed according to the real value of the joints in both legs of the robot by considering the hip, the knee, and the ankle. These values correspond to the current position of the robot.

Additionally, we need the information about the distance. This is important because our system needs the notion of displacement. To obtain this information we use the ultrasonic sensor of the NAO robot (see http://doc.aldebaran.com/2-1/family/robots/sonar_robot.html for the location of these sensors).

Finally, in order to reduce the time of convergence of the algorithm, we restrict the walking to only walking in a straight line; that is, if the robot deviates to the left or to the right by more than 10°, the algorithm will be penalized, with the expectation that it avoids the repetition of the movement that led to that angle in future.

3.4.3. Reward of the Second Level

The *reward* is set in such a way that the robot is able to walk. It seems obvious that the distance covered is a very important factor if we want the system to have sense of displacement. For that reason, the positive reward will be obtained by considering the information of the ultrasonic sensor of the NAO.

Even when there are other techniques to measure the displacement of a robot, such as using a GPS or recording a video from the initial position to the end position, we decided to use the ultrasonic sensor for simplicity; using these sensors meant that there was no need to add any other hardware to the robot, making it easier to focus in the algorithm instead of the hardware. However, taking this decision brought about some complications. For example, the robot needed an obstacle in front of it to be able to measure the distance. Additionally, the obstacle must be at the correct angle, otherwise the measure of the sensor will be wrong. In future work, other options to measure the goal distance will be considered to improve our system.

The reward was set in a way that, if the robot advanced 8 cm, it will receive a reward of 10. If the robot fell or if the robot deviated from the straight path it will receive a −10 reward. The q-learning framework aimed to maximize the reward. With a negative reward, the system would try to avoid any path that led to this condition. In any other case, the reward was 0.

3.4.4. Second Level Q-Network

The proposed Q-network in this level is composed of one input layer with 11 neurons, where 8 of them are taken from the joints of the robot: (1) left hip pitch, (2) left knee pitch, (3) left ankle pitch, (4) left ankle roll, (5) right hip pitch, (6) right knee pitch, (7) right ankle pitch, and (8) right ankle roll.

Notice that in this state, we are only considering both legs of the robot, four values for each one; this is again, to reduce the complexity. After some experiments, we found that this information is enough to make the algorithm converge while accomplishing the goal task.

The remaining neurons correspond to the ultrasonic sensors (two neurons) and the indicator of the deviation of the robot. If the robot deviates in an angle of more than 10 degrees to the left or to the right, this indicator value is −1, otherwise the indicator gives a value of +1.

The proposed net has two hidden layers, the first one has 150 neurons with a tanh activation function, and in the second hidden layer, there are 120 neurons with the same tanh activation function.

Finally, an output layer of 12 neurons with the activation function ReLU (rectifier linear unit) is created, with each of the neurons corresponding to one action. In this level, the actions available are the poses that were learned before; remember in the first layer the system learned 12 poses, so these are the available actions for this level. This architecture is shown in Figure 8.

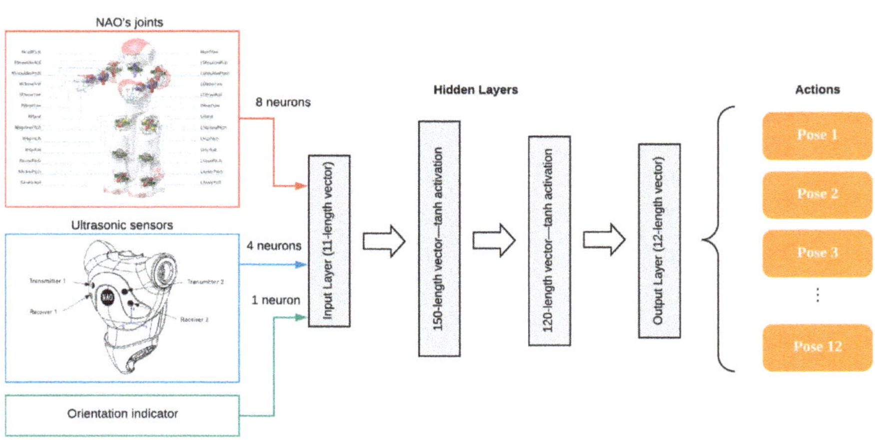

Figure 8. Second Level Q-network.

4. Experiment and Results

This section begins with a brief overview of the software used for simulating and training the networks used in this method. Afterwards, details of our fall module to detect if a robot fell down or not are given. To finish this section, we show the results of training the robot to walk, we compare the speed reached with the speed of the NAO built-in walking controller, and we show the results of training changing a parameter called *action time* and its effects in the performance of the system (the implementation of our neural network, as well as interface with the simulator environment, can be found at http://idic.likufanele.com/~{}calvo/gait/).

4.1. Simulation Software

Following the recent trend of finding optimal parameters in simulated environments, to avoid costly experiments in terms of time and physical wear [32,33], we use a simulator software called Webots (http://www.cyberbotics.com). Webots is a development environment used to model, program, and simulate mobile robots [34]. With Webots, the user can design complex robotic setups, with one or several similar or different robots in a shared environment. The properties of each object, such as shape, color, texture, mass, friction, etc., are chosen by the user. A large choice of simulated sensors and actuators is available to equip each robot.

Webots allows for the launch of a simulated NAO moving in a virtual world (for details, see http://doc.aldebaran.com/1-14/software/webots/webots_index.html). This simulator is a convenient choice because in this simulator we can read all the sensors of the robot, the force sensors to compute the ZMP, the joints sensors that give us the real value of the joints, the ultrasonic sensor we need to measure the distance covered, etc. (refer to Figure 9). The simulator allows the user to interface with C, C++, and Python. In the environment, we can add multiple obstacles and we can even access the cameras of the NAO.

Figure 9. Webots for NAO.

4.2. Q-Network Algorithm

To communicate with the NAO robot, we use the NAOqi Framework (for details, see http://doc.aldebaran.com/2-1/ref/index.html). NAOqi is the name of the main software that runs on the robot and controls it. The NAOqi Framework is the programming framework used to program NAO. It answers to common robotics needs including: parallelism, resources, synchronization, and events. This framework allows homogeneous communication between different modules (motion, audio, and video), homogeneous programming, and homogeneous information sharing (see http://doc.aldebaran.com/1-14/dev/naoqi/index.html#naoqi-framework-overview).

4.2.1. Programming Environment

The whole program is written in Python 2.7 with a Python Application Programming Interface (API) developed by Aldebaran robotics to communicate with the NAOqi framework.

We designed and trained the artificial neural networks with the Python library Keras; this is a high-level neural network API, written in Python, and is capable of running on top of either TensorFlow or Theano. It was developed with a focus on enabling fast experimentation (retrieved from https://keras.io/).

The library contains numerous implementations of commonly used neural network building blocks such as layers, objectives, activation functions, optimizers, and a lot of tools to make working with images and text data easier. From these, we selected and loaded only the required modules in order to have an efficient use of memory.

```
Training time: 1.944000 seconds
Reward:  0 i: 0
Game #: 0
Epoch 1/1

40/40 [==============================] - 1s - loss: 0.1258
Training time: 1.861000 seconds
Reward:  0 i:  0
Game #: 0
Epoch 1/1

40/40 [==============================] - 2s - loss: 0.1256
Training time: 2.175000 seconds
Reward:  0 i:  0
Game #: 0
```

Figure 10. Training Time using a Central Processing Unit (CPU).

Each time a new action is taken, the reward that this action yields must be evaluated and the model is trained again based on this information. Because of this, Graphical Processing Unit (GPU)-optimization is needed. We use the Keras backend with Theano to take advantage of the GPU of the NVIDIA Geforce GTX 580; otherwise, we would not have been able to compute the results fast enough. The motors of the NAO take 0.1 seconds to complete a movement instruction, while computing the update of a Q-network using the Central Processing Unit (CPU) took approximately 2 seconds (this was implemented on an Intel Core i5 computer with 8 GB of RAM; see Figure 10). Evidently, this is not fast enough. Fortunately, using the GPU, we could compute the network update in around 0.002 seconds (see Figure 11).

```
Training time: 0.002000 seconds
Reward:  0 i: 0
Game #: 0
Epoch 1/1

40/40 [==============================] - 0s - loss: 1.0335e-04
Training time: 0.002000 seconds
Reward:  0 i:  0
Game #: 0
Epoch 1/1

40/40 [==============================] - 0s - loss: 1.0304e-04
Training time: 0.002000 seconds
Reward:  0 i:  0
Game #: 0
```

Figure 11. Training Time using a Graphical Processing Unit (GPU).

4.2.2. Fall Module

A key element in the algorithm is the capacity to detect the falling of a robot with the *fall module*. This module is a pre-trained MLP that allows one to identify whether the robot has fallen or not. Even when the NAO robot comes with a built-in module of this type, the answer is very slow, as it provides an answer in 3 seconds, while we needed it in less than 0.2 seconds. This is why the development of a new fall module was needed.

In order to implement this module, we acquired data from the simulation and store in a dataset of 790 values, and we used that data in a previously labeled format to train a Multi-Layer Perceptron (MLP) with an input layer of ten neurons; from these ten neurons, eight belonged to the force sensors in the soles of the feet, and the other two belonged to the inertial unit of the NAO robot. This inertial unit told us the torso angle of the robot with respect to the ground; it was computed using the accelerometer and the gyrometer of the NAO (see Figure 12).

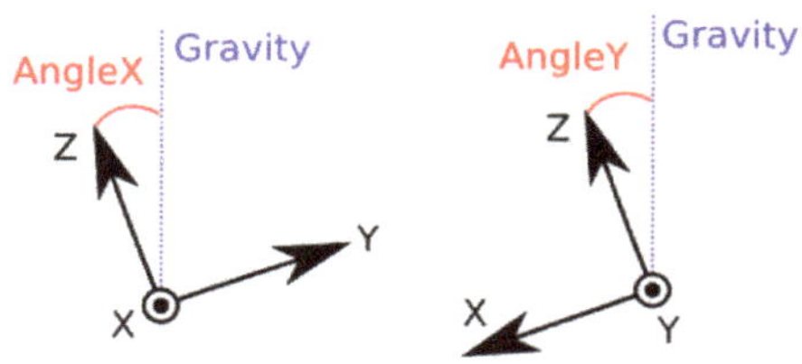

Figure 12. NAO Inertial Unit.

The fall-module MLP has two hidden layers, one with twelve neurons with the ReLU activation function, and another with eight neurons with a sigmoid activation function. Lastly, there was an output layer with only one neuron, which outputs 0 if the robot fell down or 1 if it did not. This architecture is shown in Figure 13.

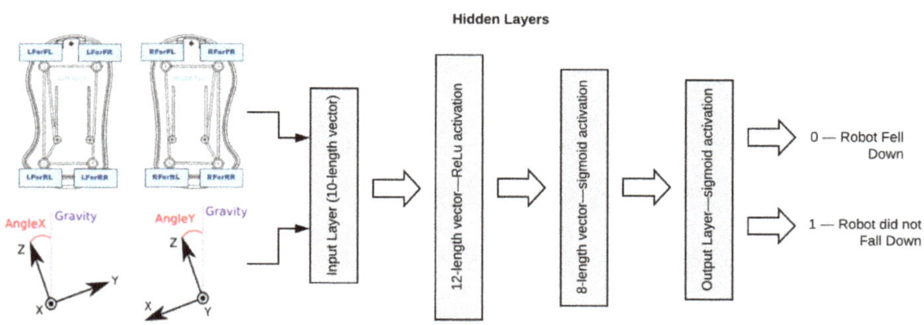

Figure 13. Fall Module Multi-Layer Perceptron (MLP).

4.2.3. Algorithm Description

First, we need to load the fall module to detect if the robot has fallen (see Figure 14). Every time the simulated robot fell down, we needed to restore the simulation manually because the simulated robot could not rise by itself. For this reason, we could not train the Q-network in one run; however, we needed to keep training the network just after the last run. For this reason, we needed to save the network model at the end of each run and to load it at the beginning of each new run.

We needed to initialize the NAO proxies, the motion proxy that allows to send signals to the NAO motors; the posture proxy that allows one to set the robot in a "home position"; the sonar proxy, which tells us the information of the ultrasonic sensors; and the memory proxy to access the data recorded, v. gr. the angle of the joints. After that, the robot required activation. We could do this by setting the stiffness of the body to 1—this was achieved using the instruction motion.setStiffnesses ("Body", 1.0).

We initialized the number of *epochs* to 50. Notice that we need many more iterations due to the fact that when the robot falls down, it cannot always stand up by itself, and the simulation has to be restored manually. For instance, imagine we set the number of epochs to 2000 and the robot fell down in the second iteration, then we would have to wait until the 2000th iteration to restore the simulation, which would take considerably more time.

Figure 14. NAO trying to stand up and failing.

Afterwards, we initialized gamma to 0.9, which means experience is considered, the closer gamma to zero, the less we take into consideration the experience of previous trials.

In Reference [31], the authors used a technique to let the algorithm converge via *experience replay*. It works in the following way: During the run of the algorithm, all the experiences $< s_t$, a, r, $s_{t+1} >$ are stored in a *replay memory*. When training the network, random mini batches from the replay memory are used instead of the most recent transition. Before training we needed to define a variable for this purpose with *replay* as an empty list; however, as we were running the algorithm multiple times, in our case it is not an empty list, but it was the list previously filled in a previous run, i.e., *replay* = *lastreplay*.

We opened a loop from $i = 0$ until $i =$ epochs, set the robot in the "home position," which was the NAO posture "StandInit," then we read sonars with the memory proxy and stored it in *sonar*. We then read the *state*; remember that the state in the first level will be a vector of 4 values while in the second level is a vector of 11 values.

Then, we opened a while loop with the condition that the NAO had not reached a terminal state. A terminal state was when the robot fell or when it reached the goal; that is, in the first level a ZMP was located in a specific interval, and in the second level, a distance of 8 cm was covered.

Later, we ran the Q-network forward and stored the result in *qval* (a vector). Suppose that we have the vector (1.2, −0.36, 2.2, 0.0). The $action_{t+1}$ will be the action 3 because the third value is the greatest; however, we wanted to explore more options in order to not fall in a local minimum, so with a probability of *epsilon*, we chose between the max action or a random action. This epsilon gradually decreased in such a way that after many iterations it became 0.

Then, we applied the $action_{t+1}$ and observed the *reward* and the *new state*, where the values of the reward were described in the previous section. At this point, we had [*state*, *action*, *reward*, *new state*]. We needed to store this tuple in *replay* and repeat the process until the length of replay was the same as *buffer*; we set buffer as 60. Once the buffer was filled, we got a mini batch, which was a random sample of length 30 from the replay array.

Thereafter, we looped over each element of the mini batch. Each element was a list of four values that we used to set (*old state, action, $reward_2$, new $state_2$*), then we ran forward our Q-network using as input the *old state* and stored the result in *old qval*, where we selected the greatest value of *old qval* and saved the index of it in *maxQ*. Later, we defined a vector X with the same values of *old state*, and a vector Y with the same values of *old qval*.

At this point, we needed to check whether the $reward_2$ belonged to a terminal state, i.e., if the variable *update* was equal to the value of the $reward_2$; if not, $update = (reward_2 + (gamma \times maxQ))$. This variable *update* was the rule used to compute the update of the neural network.

The value of the variable *update* needed to replace the value in the Y vector in the position of *index*. For instance, suppose we have a $Y = [1.2, -0.36, 2.2, 0.0]$, $index = 3$, and $update = 10$, such that the result after replacing is $Y = [1.2, -0.36, 10, 0.0]$. This will be the target for the Q-network, which should be done for all the elements in the minibatch. Once we finished the loop, we used the entire Y and X vector to train the Q-network using backpropagation. The Q-network algorithm is shown in Figure 15.

Q-network Algorithm

```
Load fall module
Initialize NAO proxies
Define the size of the Q-network
Set NAO stiffness on
Initialize epochs
Set gamma
Set batchSize
Set buffer
replay ← empty list
for i = 1, epochs:
    read state s_t
    while not terminal state:
        Run Q-network forward and store result in qval
        With probability ε, select random action a
            Otherwise, select a_t ← max (qval)
        Execute action a
        Read new state s_t+1
        Read reward r
        replay ← (st, a, s_t+1, r)
        if replay length is equal to buffer length:
            Set minibatch as a random sample of replay of length batchSize
            for memory, minibatch:
                Extract from memory (old_state, action, reward2, new_state2)
                Run Q-network forward. Store result in newQ
                maxQ ← max(newQ)
                if reward2 is not -10 or 10:
                        update ← reward2 + gamma·maxQ
                else:
                        update ← reward2
                end if
                Y[action] ← update
                X_train ← old_state
                Y_train ← Y
            end for
            Use X_train and Y_train to train the model
    end if
    end while
end for
```

Figure 15. Q-network Algorithm.

4.3. Learning the Poses

During the first level of training, the results we got were the configurations of the joints that we had previously defined as poses. The results are presented in Table 2, showing the value of the degrees of each joint. These values rendered the pose shown in Figure 16. Note that some angles remain fixed, such as the head pitch and yaw.

Table 2. Angles of Pose 1.

Head	
Head Pitch	0
Head Yaw	0

Left Arm		**Right Arm**	
L Shoulder Pitch	105.20	**R Shoulder Pitch**	105.08
L Shoulder Roll	15.69	**R Shoulder Roll**	−15.31
L Elbow Yaw	−85.87	**R Elbow Yaw**	85.85
L Elbow Roll	−29.59	**R Elbow Roll**	29.72
L Wrist Yaw	0.00	**R Wrist Yaw**	0.00

Left Leg		**Right Leg**	
L Hip Yaw Pitch	0.00	**R Hip Yaw Pitch**	0.00
L Hip Roll	2.87	**R Hip Roll**	2.87
L Hip Pitch	−18.51	**R Hip Pitch**	−18.51
L Knee Pitch	48.04	**R Knee Pitch**	48.04
L Ankle Pitch	−29.53	**R Ankle Pitch**	−29.53
L Ankle Roll	−6.33	**R Ankle Roll**	−6.33

Figure 16. Pose 1.

4.4. Learning to Walk

It was then time to learn the activity of *walking*. This was done using the previously learned knowledge, i.e., the poses. In the previous sections, we described the use of Q-networks to learn to walk, the state as defined by the combination of some selected robot joints, the value of the ultrasonic sensor, and an additional value that identified whether the robot was deviating or not.

The actions for the network are the poses, and by this moment, the system had already deduced the angle of each joint of the robot to achieve the 12 proposed poses and the reward was given by the distance advanced (the goal was to reach 8 centimeters).

The Q-network proposed in the methodology remained the same: a 4-layer Multi-Layer Perceptron (MLP), with one input layer of 11 neurons; 2 hidden layers, both with tanh activation function; one with 150 neurons and other with 120 neurons; and finally, an output layer with 12 neurons, one for each pose.

The algorithm was tested in a simulated NAO robot on the Webots simulator, having in mind that the motors response was approximately 0.08 seconds, the actions performed by the algorithm were set to 0.1s. In other words, we chose a pose of the 12 available, then we sent the signal to the motor, but the result would not be reflected until 0.08 s later; we had to wait this time so that we could read the new state and the reward. This waiting time between actions is called the *action time*.

Even when it was first thought the action time would always be 0.1 s, we realized that modifying this action time would have an effect in the performance of the algorithm. For this reason, the algorithm was tested using different values of the action time (0.1 s, 0.15 s, 0.2 s, and 0.25 s).

The four tests were done in the same conditions, with the goal distance being 8 cm, and the available actions for the Q-network, the 12 poses, and the size of the Q-network were also the same.

4.4.1. Results for an Action Time of 0.1 s

The first test was using an action time of 0.10 s. It took around 2500 iterations to converge. The poses chosen for the algorithm in this test were pose 10, pose 1, pose 3, pose 4, pose 7, pose 7, pose 8, and pose 12, as illustrated in Figure 17. The robot, following these poses, took on average 1.621 s to reach the 8 centimeters marked as the goal.

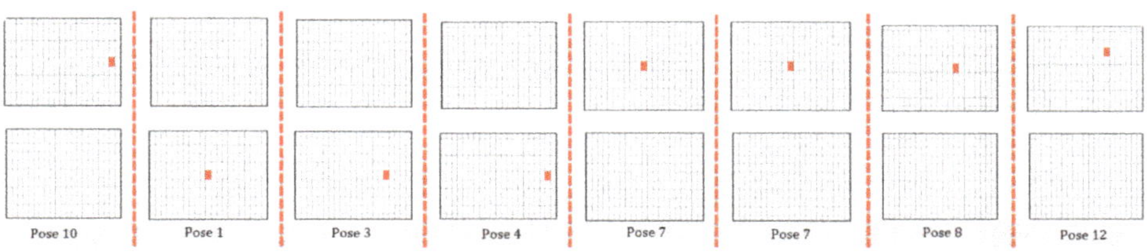

Figure 17. Selected poses for an Action Time of 0.1 s.

4.4.2. Results with an Action Time of 0.15 s

The next test used an action time of 0.15 s. It took around 2800 iterations to converge, where the poses chosen for the algorithm in this test were pose 1, pose 3, pose 4, pose 7, and pose 12, as illustrated in Figure 18. The robot, following these poses, took an average of 1.520s to reach the 8 centimeters marked as the goal.

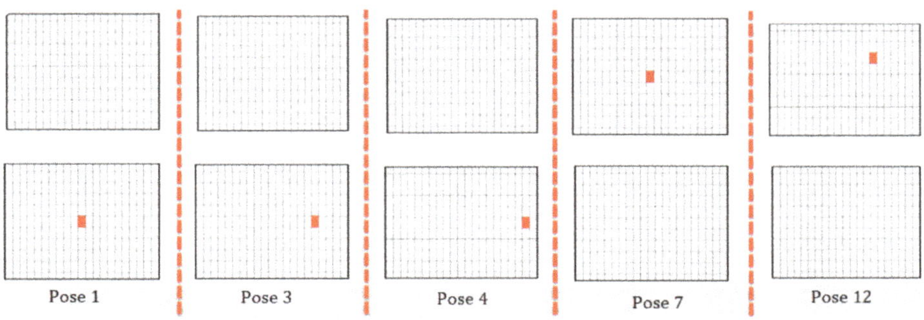

Figure 18. Selected Poses for an Action Time of 0.15 s.

4.4.3. Testing with an Action Time of 0.2 s

The next test used an action time of 0.20 s. It took around 2500 iterations to converge. The poses chosen for the algorithm in this test were the same that the test before: pose 1, pose 3, pose 4, pose 7, and pose 12; however, in this case, the robot took 2.2 s on average to reach the 8 centimeters marked as the goal.

4.4.4. Testing with an Action Time of 0.25 s

The last test used an action time of 0.25 s. It took around 2600 iterations to converge. The poses chosen for the algorithm in this test were pose 1, pose 4, pose 7, and pose 12. The robot, using these poses, took 2.015 s on average to reach the 8 centimeters marked as the goal. See Figure 19.

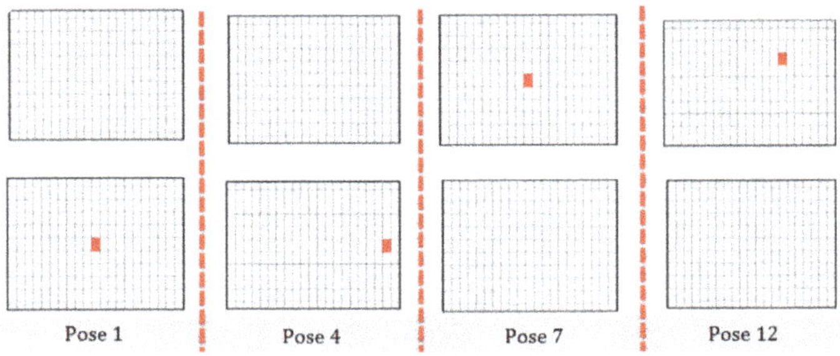

Figure 19. Selected poses for an Action Time of 0.2 s.

Comparing the performance of the four tests we have, the fastest one was the test done with the action time of 0.15 s, and the slowest was the one with the action time of 0.2 s.

A comparison table of the four tests is shown in Table 3.

Table 3. Q-network Results.

Q-Network Results					
Action Time	**Number of Poses**	**List of Poses**	**Distance**	**Time**	**Speed**
0.10 s	8	10, 1, 3, 4, 7, 7, 8, 12	8 cm	1.621 s	4.93 cm/s
0.15 s	5	1, 3, 4, 7, 12	8 cm	1.520 s	5.26 cm/s
0.20 s	5	1, 3, 4, 7, 12	8 cm	2.200 s	3.63 cm/s
0.25 s	4	1, 4, 7, 12	8 cm	2.015 s	3.97 cm/s

4.4.5. Comparing Results

The NAO robot came with a built-in controller based on an inverted pendulum to control the gait cycle, and we could configure the speed of the NAO with this controller. Normally, the NAO reached a speed of 3.92 cm/s; however, in fast walking mode, it could reach a speed of 6.153 cm/s (see Table 4).

Table 4. NAO's built-in walking controller.

NAO Built-In Walking Controller			
Mode	**Distance**	**Time**	**Speed**
Normal Walking	8 cm	2.04 s	3.92 cm/s
Fast Walking	8 cm	1.30 s	6.153 cm/s

When comparing the result of the four tests with the *normal speed* of the NAO, we can see that tests with action times of 0.15 s and 0.1 s were faster than the normal speed of the robot. Furthermore, the test with an action time of 0.25 s was still slightly better. These results are encouraging because the performance was enough to be considered as a good performance.

Nevertheless, when comparing with the *fast walking* mode of the NAO, none of the four tests was faster than the NAO's built-in fast walking, the closest one being the test with the action time of 0.15 s.

However, when testing whether the robot fell down or not, when we tested the resultant Q-networks during 10 tests of 5 s each, the robot never fell down; however, when testing the NAO's built-in fast walking, in 3 out of 10 tests done with that speed, the robot fell down (the comparison of stability and deviation between the normal walking and fast walking of the NAO, as well as the four obtained results, can be analyzed in detail in Video S1; see Figure 20).

Figure 20. NAO in fast walking mode that had fallen down.

5. Conclusions and Future Work

In this section, we summarize the results and conclusions obtained in this work, as well as the analysis about stability and the ability to not fall down. Later, there is an explanation of how we believe the definitions of the action, state, and reward can be applicable to other robots. Afterwards we expose why the learning framework can be considered successful. Finally, future work directions are discussed.

5.1. Stability and Not Falling Down

After training the whole system, the robot was able to walk while displaying a stable behavior, which was to walk with balance. In addition, the robot did walk in straight line, i.e., it did not deviate. On the other hand, the algorithm provided for the robot allowed us to find: (i) a collection of twelve poses based on a ZMP criterion, and (ii) a combination of these poses in different action times that allow the robot to reach the goal distance without falling down and without deviating. Allowing the robot to avoid falling down is a very important part of this work and during the experimentation we found that our approach accomplished this objective.

5.2. Actions, States, and Rewards Applicable to Other Robots

The state of the first level of the learning framework was obtained from the four force-sensors in each foot in such a way that this definition of state will work for every robot that has these kinds of sensors.

The state of the second level was formed with the information of the current angle of the joints and the value of the ultrasonic sensors of the robot. If a robot had the opportunity to read the angle of its joints and there was a readable value of the sensors that measured the distance, then the definition of the state in this level was applicable for that robot.

The reward of the first level was computed depending on the region where the ZMP was. Therefore, for any robot that had the force sensors on its feet, we would be able to use the same reward definition for this level.

The reward of the second level was computed using the values of the ultrasonic sensors, that is, the measure of the distance. Because of this, in any robot in which we could measure the distance, we could apply the same definition of reward.

On the first level of the Q-network, the available actions proposed here were modules that interacted directly with the current position of a joint, and they increased or decreased the value of a joint. For that, if we could send a command to a robot that modified the position of a joint of that robot, then we could use this definition of actions for that robot.

On the second level of the Q-network, the actions are the poses that the learning framework previously learned; therefore, if we could apply the Q-network of the first level on a robot and it succeeded, then that robot may be able to learn and to use the poses of the first level. The actions were the same proposed here; of course, the poses were not expected to be the same but the way they were learned and how they are used can be the same as in this work.

5.3. Learning Framework

At the end of this work, we verified that the approach proposed here works using the same machine learning technique (q-learning) such that we could induce the poses that conformed to the first level of our hypothetical learning framework, and we could make the robot walk just by combining these poses.

We showed that with this algorithm, it was not necessary to consider the kinematics of the robot, and it was unnecessary to compute the inverse kinematics of the joints since we just have to know the information of some sensors to deduce some modules that we called poses to further used them in a more complex activity. The convergence time of the algorithm in each pose and in each walking test was large, but the results were satisfying.

It is clear that the approach is far from perfect, even when comparing the results with the built-in build of the robot showing that performance was good, there were many scenarios where this algorithm failed because in this work we have only considered a specific setting for walking, that is, walking in straight line on a flat floor. If we tried the algorithm as it is on a ramp, this will lead to the robot falling down.

Additional settings for walking and the possibility of expanding the number of activities at level two to achieve a task of level three in the learning framework is what leads us to describe the future work.

5.4. Future Work

On one hand, we can explore different stages of walking, that is, walking on an irregular floor, walking on ramps, or omnidirectional walking, i.e., to walk in any direction, not only in a straight line. On the other hand, we can keep adding activities, i.e., modules of level two of the learning framework. Until now, we have proposed and tested only one activity, i.e., walking, but the idea is to learn more activities. Of course, more poses are needed to achieve such new activities. For instance, we can add "sit down," "turn around," or "take an object." Lastly, to transfer the simulated algorithm to physical robots is left as future work.

Once our system has more activities available, we can ask the system to learn a module of level three; remember that we defined these modules as tasks, so the next step will be to learn a task just by using the previous information, i.e., the activities.

Supplementary Materials: The following are available online at http://www.mdpi.com/2076-3417/9/3/502/s1, Video S1: Comparing NAO walking modes. , Video S1: Comparing NAO walking modes.

Author Contributions: Conceptualization: C.R.G. and H.C.; Methodology: C.R.G., H.C. and H.S.; Software: C.R.G.; Investigation: H.C.; Writing: H.C. and C.R.G.; Supervision: H.S.

Funding: This research was funded by CONACyT-SNI, Fronteras de la Ciencia 65, and Instituto Politecnico Nacional (IPN), particularly through grants SIP20182114, SIP20180730, SIP20190007, 20195886, EDI, and COFAA-SIBE.

Conflicts of Interest: The authors declare no conflict of interest.

References

1. Shirashaka, S.; Machida, T.; Igarashi, H. Leg Selectable Interface for Walking Robots on Irregular Terrain. In Proceedings of the 2006 SICE-ICASE International Joint Conference, Busan, Korea, 18–21 October 2006.
2. Ill-Woo, P.; Jung-Yup, K.; Jungho, L. Online Free Walking Trajectory Generation for Biped. In Proceedings of the IEEE Internactional Conference on Robotics and Automation (ICRA 2006), Orlando, FL, USA, 15–19 May 2006; pp. 1231–1236.
3. Huang, Q.; Kajita, S.; Koyachi, N.; Kaneko, K.; Yokoi, K.; Kotoku, T.; Arai, H.; Komoriya, K.; Tanie, K. Walking patterns and actuator specifications for a biped robot. In Proceedings of the 1999 IEEE/RSJ International Conference on Intelligent Robots and Systems. Human and Environment Friendly Robots with High Intelligence and Emotional Quotients (Cat. No.99CH36289), Kyongju, Korea, 17–21 October 1999.
4. Tay, A.B. Walking Nao Omnidirectional Bipedal Locomotion. Bachelor's Thesis, Bachelor of Science, School of Computer Science and Engineering, The University of New South Wales, Kensington, UK, 2009; pp. 1–49.
5. Morimoto, J.; Cheng, G.; Atkeson, C.G.; Zeglin, G. A simple reinforcement learning algorithm for biped walking. In Proceedings of the IEEE International Conference on Robotics and Automation, 2004. Proceedings. ICRA '04. 2004, New Orleans, LA, USA, 26 April–1 May 2004.
6. Jin-Ling, L.; Kao-Shing, H.; Wei-Cheng, J.; Yu-Jen, C. Gait Balance and Acceleration of a Biped Robot. *IEEE Access* **2016**, *4*, 2439–2449.
7. Michel, O. Webots: Symbiosis between virtual and real mobile robots. In *International Conference on Virtual Worlds*; Lecture Notes in Computer Science 1434; Springer: Berlin/Heidelberg, Germany, 1998; pp. 254–263.
8. Hubicki, C.; Abate, A.; Clary, P.; Rezazadeh, S.; Jones, M.; Peekema, A.; Van Why, J.; Domres, R.; Wu, A.; Martin, W.; et al. Walking and running with passive compliance: Lessons from engineering a live demonstration of the ATRIAS biped. *IEEE Robot. Autom. Mag.* **2018**, *25*, 23–39. [CrossRef]
9. Martin, W.C.; Wu, A.; Geyer, H. Robust spring mass model running for a physical bipedal robot. In Proceedings of the 2015 IEEE International Conference on Robotics and Automation (ICRA), Seattle, WA, USA, 26–30 May 2015; pp. 6307–6312.
10. Hong, S.; Kim, E. A New Automatic Gait Cycle Partitioning Method and Its Application to Human Identification. *Int. J. Fuzzy Log. Intell. Syst.* **2007**, *17*, 51–57. [CrossRef]
11. Jung-Yup, K.; Ill-Woo, P.; Jun-Ho, O. Walking Control Algorithm of Biped Humanoid Robot on Uneven and Inclined Floor. *J. Intell. Robot. Syst.* **2007**, *48*, 457–484.
12. Huang, Q.; Yokoi, K.; Kajita, S.; Kaneko, K.; Arai, H.; Koyachi, N.; Tanie, K. Planning Walking Patterns for a Biped Robot. *IEEE Trans. Robot. Autom.* **2001**, *17*, 280–289. [CrossRef]
13. Lu, Q.; Zheng, Y.; Lu, X. Deviation Correction Control of Biped Robot Walking Path Planning. In *IOP Conference Series: Materials Science and Engineering*; IOP Publishing: Bristol, UK, 2018; Volume 394, p. 032029.
14. Ito, S.; Nishio, S.; Ino, M.; Morita, R.; Matsushita, K.; Sasaki, M. Design and adaptive balance control of a biped robot with fewer actuators for slope walking. *Mechatronics* **2018**, *49*, 56–66. [CrossRef]
15. Liu, C.; Yang, J.; An, K.; Chen, Q. Rhythmic-Reflex Hybrid Adaptive Walking Control of Biped Robot. *J. Intell. Robot. Syst.* **2018**, *7*, 1–17. [CrossRef]
16. Righetti, L.; Jan Ijspeert, A. Programmable Central Pattern Generators: An application to biped locomotion control. In Proceedings of the IEEE International Conference on Robotics and Automation, Orlando, FL, USA, 15–19 May 2006; pp. 1585–1590.
17. Liu, J.; Chen, X.; Veloso, M. Simplified walking: A new way to generate flexible biped patterns. In *Mobile Robotics: Solutions and Challenges*; World Scientific: Singapore, 2010; pp. 583–590.
18. Rivas, F.M.; Canas, J.M.; González, J. Automatic learning of walking modes for a humanoid robot (in Spanish). In Proceedings of the Robot2011 III Robotics Workshop, Experimental robotics, Seville, Spain, 28–29 November 2011; pp. 120–127.
19. Meriçli, Ç.; Veloso, M. Biped walk learning on NAO through playback and real-time corrective demonstration. Workshop on Agents Learning Interactively from Human Teachers. In Proceedings of the 9th International Conference on Autonomous Agents and Multiagent Systems (AAMAS 2010), Toronto, ON, Canada, 9–14 May 2010.
20. Kao-Shing, H.; Keng-Hao, Y.; Jia-Yan, L. The Study on the Learning of Walking Gaits for Biped Robots. *Int. Fed. Autom. Control* **2013**, *46*, 589–593.

21. Gasparri, G.M.; Manara, S.; Caporale, D.; Averta, G.; Bonilla, M.; Marino, H.; Catalano, M.; Grioli, G.; Bianchi, M.; Bicchi, A.; et al. Efficient Walking Gait Generation via Principal Component Representation of Optimal Trajectories: Application to a Planar Biped Robot with Elastic Joints. *IEEE Robot. Autom. Lett.* **2018**, *3*, 2299–2306. [CrossRef]
22. Corral, E.; Marques, F.; García, M.J.G.; Flores, P.; García-Prada, J.C. Passive walking biped model with dissipative contact and friction forces. In *European Conference on Mechanism Science*; Springer: Cham, Switzerland, 2019; pp. 35–42.
23. Dekker, M. *Zero-Moment Point Method for Stable Biped Walking*; University of Technology: Eindhoven, The Netherlands, 2009.
24. Strom, J.; Slavov, G.; Chown, E. Omnidirectional Walking Using ZMP and Preview Control for the NAO Humanoid Robot. *Lect. Notes Comput. Sci.* **2010**, *5949*, 378–389.
25. Lee, H.W.; Yang, J.L.; Zhang, S.Q.; Chen, Q. Research on the Stability of Biped Robot Walking on Different Road Surfaces. In Proceedings of the 2018 1st IEEE International Conference on Knowledge Innovation and Invention (ICKII), Jeju, Korea, 23–27 July 2018; pp. 54–57.
26. Van der Noot, N.; Barrera, A. Zero-Moment Point on a Bipedal Robot. In Proceedings of the 17th IEEE Mediterranean Electrotechnical Conference, Beirut, Lebanon, 13–16 April 2014.
27. Liu, Y.; Zang, X.; Zhang, N.; Liu, Y.; Wu, M. Effects of unilateral restriction of the metatarsophalangeal joints on biped robot walking. In Proceedings of the 2018 Eighth International Conference on Information Science and Technology (ICIST), Cordoba, Spain, 30 June–6 July 2018; pp. 395–400.
28. Duysens, J.; Forner-Cordero, A. Walking with perturbations: A guide for biped humans and robots. *Bioinspir. Biomim.* **2018**, *13*, 061001. [CrossRef] [PubMed]
29. Kao-Shing, H.; Jin-Ling, L.; Jhe-Syun, L. Biped Balance Control by Reinforcement Learning. *J. Inf. Sci. Eng.* **2016**, *32*, 1041–1060.
30. McGinnis, P.M. *Biomechanics of Sport and Exercise*; Human Kinetics Publishers: Champaign, IL, USA, 2015.
31. Mnih, V.; Kavukcuoglu, K.; Silver, D.; Graves, A. Playing Atari with Deep Reinforcement Learning. *arXiv*, 2013; arXiv:1312.5602.
32. Singla, A.; Bhattacharya, S.; Dholakiya, D.; Bhatnagar, S.; Ghosal, A.; Amrutur, B.; Kolathaya, S. Realizing Learned Quadruped Locomotion Behaviors through Kinematic Motion Primitives. *arXiv*, 2018; arXiv:1810.03842.
33. Rai, A.; Antonova, R.; Meier, F.; Atkeson, C.G. Using Simulation to Improve Sample-Efficiency of Bayesian Optimization for Bipedal Robots. *arXiv*, 2018; arXiv:1805.02732.
34. Michel, O. Webots: Professional Mobile Robot Simulation. *Int. J. Adv. Robot. Syst.* **2004**, *1*, 39–42. [CrossRef]

applied sciences

MDPI

Article

Turning Gait Planning Method for Humanoid Robots

Tianqi Yang [1], Weimin Zhang [1,2,*], Xuechao Chen [1,3], Zhangguo Yu [1,2] and Libo Meng [1,2] and Qiang Huang [1,3]

1. Intelligent Robotics Institute, School of Mechatronical Engineering, Beijing Institute of Technology, Beijing 100081, China; Yangtianq930208@163.com (T.Y.); chenxuechao@bit.edu.cn (X.C.); yuzg@bit.edu.cn (Z.Y.); menglibo@bit.edu.cn (L.M.); qhuang@bit.edu.cn (Q.H.)
2. Beijing Advanced Innovation Center for Intelligent Robots and Systems,Beijing 100081, China
3. Key Laboratory of Biomimetic Robots and Systems, Ministry of Education, Beijing 100081, China

* Correspondence: zhwm@bit.edu.cn; Tel.: +86-10-6891-3122

Received: 11 June 2018 ; Accepted: 24 July 2018 ; Published: 30 July 2018

Abstract: The most important feature of this paper is to transform the complex motion of robot turning into a simple translational motion, thus simplifying the dynamic model. Compared with the method that generates a center of mass (COM) trajectory directly by the inverted pendulum model, this method is more precise. The non-inertial reference is introduced in the turning walk. This method can translate the turning walk into a straight-line walk when the inertial forces act on the robot. The dynamics of the robot model, called linear inverted pendulum (LIP), are changed and improved dynamics are derived to make them apply to the turning walk model. Then, we expend the new LIP model and control the zero moment point (ZMP) to guarantee the stability of the unstable parts of this model in order to generate a stable COM trajectory. We present simulation results for the improved LIP dynamics and verify the stability of the robot turning.

Keywords: non-inertial reference frame; centrifugal force; turning model LIP; trajectory planning

1. Introduction

The basic functionality required for humanoid robots is the ability to achieve various human movements. There are several existing methods for the planning and control of walking without revolving around the axis perpendicular to the horizontal plane [1–5]. The more challenging problem inherent to controlling a biped robot is maintaining its stability when it is moving. A widely used method for determining the stability of the robots is whether the zero moment point (ZMP) is in the supporting area [6,7]. In the trajectory planning method, the ZMP is seen as a linear inverted pendulum (LIP) which is a simplified model. The robot is regarded as a point, and the entire mass is concentrated at the center of the mass. Another trajectory planning method considers the humanoid robot as a seven-link model. The position of the ZMP can be calculated by the state of the COM of every link rod including the position, the velocity and the acceleration [8]. By comparing the ZMP trajectories, the trajectory with the highest stability margin is selected as the off-line trajectory. These methods have unique advantages and disadvantages. Recently, newer gait-planning methods [9–13] have been developed that do not rely on the ZMP and have indeed produced marked improvements in humanoid robot walking.

However, it is also very important for humanoid robots to be able to turn while walking at a high speed. For the robot to achieve various types of locomotion and efficiently complete any given task, it must be capable of turning while walking at a high speed. When the robot turns at a very low speed, the trajectory of the center of mass can be generated by the traditional LIP. However, the robot is likely to tip over during fast turns, but only if the trajectory is only generated by the LIP. Unlike straight-line walk (or "linear walk"), the so-called "turning walk" requires highly complex dynamical systems. During turning walk, the COM moves as it does during linear walk but while simultaneously performing circular motions, so in the world reference, the robot can not be seen as a particle. There are two choices that can be chosen to deal with the turning walk. One is to establish the humanoid robot whole body dynamics model in the world reference. The other is to translate the turning walk into the straight-line walk by choosing an appropriate non-inertial reference, and, in the non-inertial reference, the LIP model can also be used to generate the trajectory while the dynamics of the LIP are changed. There is an extended model of the LIP that is applied to the trajectory planning of the robot. The most popular model is the spring loaded inverted pendulum (SLIP) [14–16] which is used in robot running. In robot running, the LIP model causes big collisions to the ground, so in order to reduce the force that is generated by the collisions, the SLIP is proposed and has produced a good effect.

Previous researchers have indeed explored dynamic turning. The most famous example is the ASIMO robot, which can turn while walking and even while running. Soichiro Suzuki [17] achieved a quasi-passive turn walk by utilizing a mechanical oscillator. There has also been research into slip turns; Kanehiro presented a novel hierarchical controller for walking torque-controlled humanoid robots [18,19] which is capable of executing quick slip-turns on a HRP-4C on its toes. Koeda at al. achieved slip-turns in the HOAP-2 robot [20–23]. Their method minimizes the turning angle based on variations in friction across the floor. Despite these valuable contributions, there has been relatively little research on turning while the robot is walking.

Most research has applied straight-line walking models directly to the turning walks [24]. This method does not take into account the effect of the robot's own rotation on the actual ZMP so in the lateral direction, the actual ZMP will be different form the planned ZMP. Due to the large model error, the robot will be easily unstable. So, the proposed method can reduce the model error to guarantee the stability of the robot. The main contribution of the paper is that our method takes into account the rotation factors in the turning process without increasing the complexity of the dynamics. So, the model is more accurate and the trajectory can be generated very fast.

The remainder of the paper is organized as follows. In Section 2, we analyze the scope of application of LIP and the limitations of the LIP during turning walk. Then, we introduce a non-inertial reference that can convert the turning walk to the straight-line walk. In Section 3, we establish the improved LIP model in the non-inertial reference and extend its dynamics. Then, we analyze the stable component and unstable component of the LIP and get the trajectory of the center of mass (COM) by ensuring boundedness of COM trajectories for a given reference ZMP trajectory. In Section 4, we give the trajectory planning of the foot and convert the COM trajectory under non-inertial system to the trajectory in a world coordinate system which can be used in the control of the robot based on the current state. In Section 5, results from the simulations and experiments are presented to verify the feasibility of the proposed method.

2. Non-Inertial Reference in Turning Walk

In the translational movement, the robot can be seen as a mass point because when it is treated as a rigid, the motion of the COM can represent the motion of the whole body. However, in the turning walk, there is not only translational movement but also rotation around the axis perpendicular to the horizontal plane. So, the motion of the COM cannot represent the motion of the whole robot which will make the robot fall down. In this paper, we introduce the non-inertial reference in which the motion of the turning walk is the straight-line walk.

The motion of the non-inertial reference in the turning walking is shown in Figure 1, where the green curve represents the robot trajectory and the red curve represents the non-inertial reference. The non-inertial trajectory is defined as involute. The reference frame not only moves along the curve but also rotates which makes the Y-axis always point to the robot. In this paper, we call this the "involute reference frame" (IRF). In the IRF, the motion of the robot can be described similarly to straight-line walk.

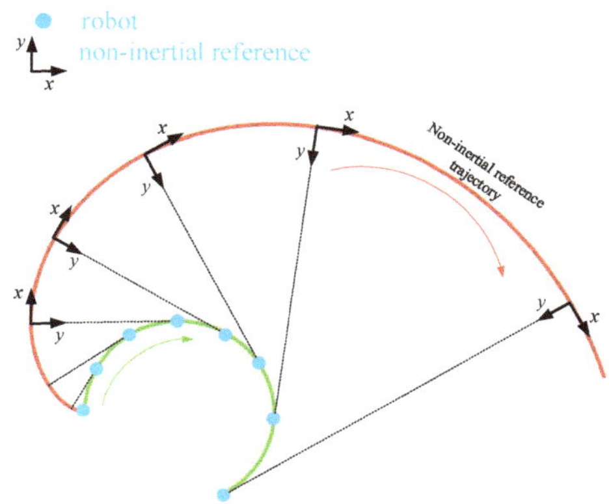

Figure 1. Non-inertial Reference in turning walking.

For the IRF, its motion has tangential acceleration, normal acceleration, and centripetal acceleration that rotates synchronously with circular motion. The motion of the object moving in the IRF is affected by inertial forces. The motion of the IRF and the inertial forces acting on the robot are shown in Figure 2. a_t is the tangential acceleration; a_n is the normal acceleration; and ω is the angular velocity. f_{a_n} is generated due to normal acceleration, f_{a_t} is generated due to tangential acceleration and f_{rin} is generated due to centripetal acceleration. f_{CoF} is the coriolis force. f_ω is the force generated due to the changing rate of angular velocity. All the inertial force expressions can be obtained as shown in Formula (1). l is the distance that the robot walks in the forward direction and is also the length between the origin of the IRF and the tangent point of the circle.

$$\begin{cases} f_{\mathrm{a_n}} = m\omega^2 r \\ f_{\mathrm{a_t}} = m(\dot{\omega} l + \omega \dot{r}) \\ f_{\mathrm{rin}} = m\omega^2 r \\ f_{\mathrm{CoF}} = 2m\omega v \\ f_\omega = m\dot{\omega} r. \end{cases} \tag{1}$$

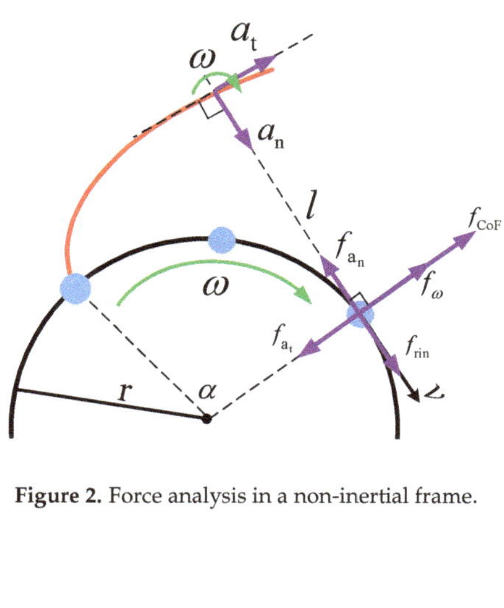

Figure 2. Force analysis in a non-inertial frame.

where m is the mass of the object moving in the non-inertial frame, and r is the turning radius. The directions of the inertial centrifugal force (f_{rin}) and normal force (f_{a_n}) are opposite, so the effect of the two forces cancels out. It is only necessary to calculate the f_ω, f_{rin} and f_{CoF} in the IRF. Through the relationship between the physical quantities shown in Formula (2), we can simplify this force based on the involute properties in Formula (3).

$$l = \alpha r, v = \omega r, \dot{\alpha} = \omega \tag{2}$$

$$\begin{cases} f_{\text{CoF}} = 2m\omega^2 r \\ f_\omega = m\dot{v}\alpha \\ f_{\text{a}_\text{t}} = m(\dot{v}\alpha + \frac{v^2}{r}). \end{cases} \tag{3}$$

where v is the velocity in the forward direction related to the IRF and it is also the tangential velocity of circular motion related to the global reference. α is the angle that the robot turns. f_{a_t} is a changing force which is determined by the acceleration and velocity of the object.

In Formula (4), through the the vector superposition of these inertial forces, f_{sum} is the resultant of inertial force applied to an object moving in the IRF, and its direction is the same as that of f_{a_t}

$$\begin{gathered} \mathbf{f}_{\text{sum}} = \mathbf{f}_{\text{CoF}} + \mathbf{f}_{\text{a}_\text{t}} + \mathbf{f}_{\text{a}_\text{n}} + \mathbf{f}_{\text{rin}} + \mathbf{f}_\omega \\ f_{\text{sum}} = m(\frac{v^2}{r}). \end{gathered} \tag{4}$$

The discussion above centers around a scenario in which the turning radius is constant. During turning walk, however, the radius must change so that the robot can reach its target destination. Turning walk can be divided into several movements across a constant radius, where the expression of the resultant force does not change. As shown in Figure 3, if the radius is r_1, the force is $f'_{\text{sum}} = m\frac{v^2}{r_1}$. Similarly, r_2, r_3, and so on.

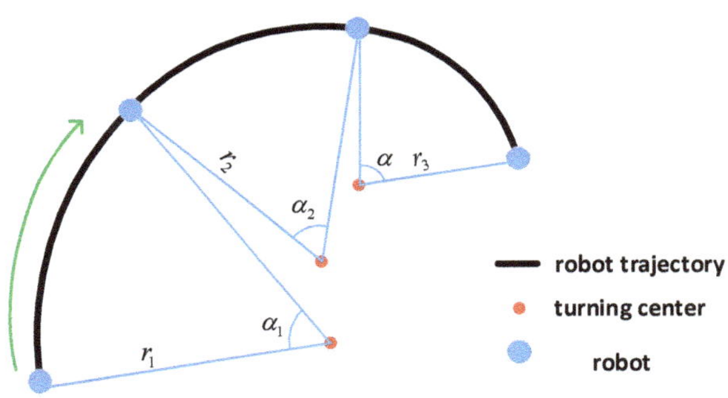

Figure 3. Different radii of the robot turning walk.

3. Turning Gait Planning in the Non-Inertial Reference (IRF)

3.1. Planning of the COM Trajectory

In order to make the description more intutive, we first define three directions related to the robot. The forward direction is called the y direction. The leftward and rightward direction is called the x direction. The vertical direction is called the z direction.

As the basic model of the robot walking, the LIP model simplifies the complex dynamic model of the robot. As is shown in Figure 4, the trajectory of the COM can represent the robot motion when the robot only moves in translation.

$$\ddot{x}_{\text{com}} = \omega^2(x_{\text{com}} - x_{\text{zmp}}). \tag{5}$$

Here, $\omega = \sqrt{\frac{g}{h_{\text{com}}}}$. h_{com} is the height of the COM. x_{zmp} is the position of the ZMP which can be planned ahead. x_{com} is the position of the COM. The input of the LIP model is the x_{zmp} and the output is the x_{com}. This model is applicable when the robot does not rotate around the z-axis.

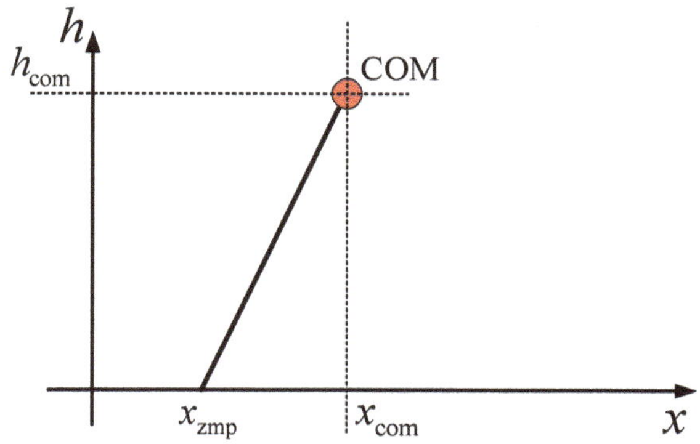

Figure 4. Linear inverted pendulum of the robot.

When compared with the straight-line walk, the system for the turning walk is more complex. For the turning walk, because of the existence of f_{a_n}, f_{a_t}, f_{rin}, and f_{ω}. we must improve the traditional LIP model. As we can know from Section II, for the y direction in the IRF, the resultant of the inertial force is zero, so the traditional LIP is still applicable. However, for the x direction in the IRF, the resultant force is f_{sum}, so we can get the model of the turning walk from Figure 5. In this section, we call the f_{sum} force another name, f_{ct}, in order to simplify the subsequent derivation.

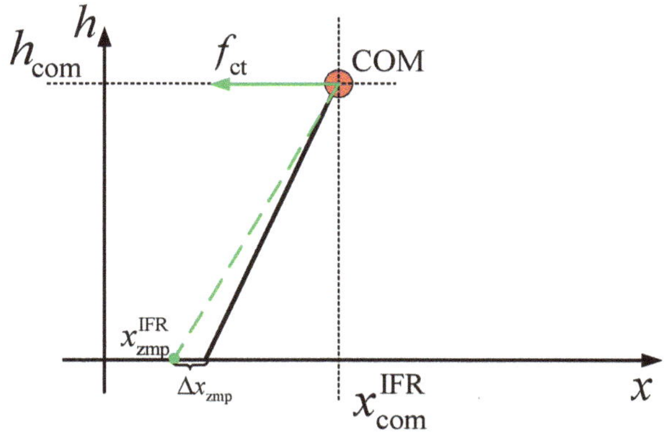

Figure 5. Model of the linear inverted pendulum (LIP) of the robot's turning walk in the x direction.

Because of force f_{ct}, the ZMP moves by a short distance. Based on this new model, we can then obtain the new dynamics.

$$\ddot{x}^{\text{IRF}}_{\text{com}} = \omega^2[x^{\text{IRF}}_{\text{com}} - (x_{\text{zmp}} + \frac{\dot{y}^{\text{IRF}\,2}_{\text{com}}}{\omega^2 r_{\text{com}}})], \tag{6}$$

where r_c is the turn radius of the COM. $x^{\text{IRF}}_{\text{zmp}}$ is the ZMP position in the IRF, and $x^{\text{IRF}}_{\text{com}}$ is the COM position in the IRF. The $x^{\text{IRF}}_{\text{zmp}}$ position of the new ZMP has changed compared to that in the straight-line walk, as shown in Formula (7):

$$x^{\mathrm{IRF}}_{\mathrm{zmp}} = x_{\mathrm{zmp}} + \frac{\dot{y}^{\mathrm{IRF}\;2}_{\mathrm{com}}}{\omega^2 r_{\mathrm{com}}}. \tag{7}$$

So, the new LIP model can be expressed as follows:

$$\ddot{x}^{\mathrm{IRF}}_{\mathrm{com}} = \omega^2 (x^{\mathrm{IRF}}_{\mathrm{com}} - x^{\mathrm{IRF}}_{\mathrm{zmp}}), \tag{8}$$

We can, therefore, treat turning walk as straight walk when we add the ZMP deviation: The new state space of the LIP system can then be expressed as $x^{\mathrm{IRF}} = [x^{\mathrm{IRF}}_{\mathrm{com}}, \dot{x}^{\mathrm{IRF}}_{\mathrm{com}}, x^{\mathrm{IRF}}_{\mathrm{zmp}}]^T$. The input of the system is the velocity of the ZMP:

$$\dot{x}^{\mathrm{IRF}} = \begin{bmatrix} 0 & 1 & 0 \\ \omega^2 & 0 & -\omega^2 \\ 0 & 0 & 0 \end{bmatrix} x^{\mathrm{IRF}} + \begin{bmatrix} 0 \\ 0 \\ 1 \end{bmatrix} \dot{x}^{\mathrm{IRF}}_{\mathrm{zmp}}, \tag{9}$$

After doing the above simplification, turning walk can be converted into straight-line walk in the IRF when the straight-line walk is applied to a force (f_{ct}), as shown in Figure 6. Because the velocity of the COM in the y direction is not constant, the force $f_{ct}(\mathrm{t})$ will also change all the time. According to Formula (7), we know that the $\Delta x_{\mathrm{zmp}}(\mathrm{t})$ is also not constant:

$$f_{ct}(\mathrm{t}) = \mathrm{m}\omega^2 (x^{\mathrm{IRF}}_{\mathrm{zmp}}(\mathrm{t}) - x_{\mathrm{zmp}}(\mathrm{t})) = \mathrm{m}\omega^2 \Delta x_{\mathrm{zmp}}(t), \tag{10}$$

where m is the mass of the robot.

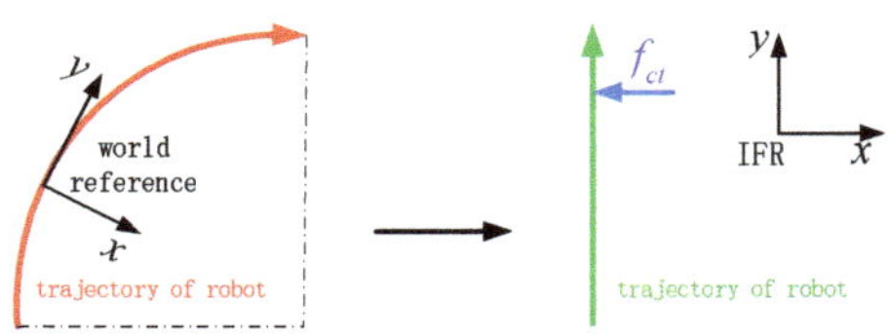

Figure 6. Transformation between straight and turning walk.

Through changes in the coordinates, this model can be divided into two components [7,25]. One is the stable mode and the other is the unstable mode, as is shown in Formula (11):

$$\begin{aligned} x_{\mathrm{un}} &= \frac{\omega x^{\mathrm{IRF}}_{\mathrm{com}} - \dot{x}^{\mathrm{IRF}}_{\mathrm{com}}}{\omega} \\ x_{\mathrm{st}} &= \frac{\omega x^{\mathrm{IRF}}_{\mathrm{com}} + \dot{x}^{\mathrm{IRF}}_{\mathrm{com}}}{\omega}. \end{aligned} \tag{11}$$

We can then obtain an expression for x_{un}, as shown in Formula (12). Formula (12) is a first-order differential equation, and its eigenvalue is an integrity, so this mode is a divergent system. The input of the subsystem is the position of the ZMP because the value of ω is constant:

$$\dot{x}_{\mathrm{un}}(t) - \omega x_{\mathrm{un}}(t) = -\omega x^{\mathrm{IRF}}_{\mathrm{zmp}}(t). \tag{12}$$

In order to match the state space shown in Formula (9), the input of the system is the velocity of the ZMP. We can rewrite the expression in the Formula (13) when the radius of the COM is approximately constant:

$$\dot{x}_{\mathrm{un}}(t) - \omega x_{\mathrm{un}}(t) = -\omega \int_0^t \dot{x}^{\mathrm{IRF}}_{\mathrm{zmp}}(t) dt \approx -\omega \int_0^t (\dot{x}_{\mathrm{zmp}}(t) + \frac{2\dot{y}^{\mathrm{IRF}}_{\mathrm{zmp}}(t)\ddot{y}^{\mathrm{IRF}}_{\mathrm{zmp}}(t)}{\omega^2 r_{\mathrm{com}}}) dt. \tag{13}$$

In the IRF, there is no inertial force in the forward direction y so the dynamics are the same as in the straight-line walk. We can calculate the $\dot{y}^{\mathrm{IRF}}_{\mathrm{zmp}}$ and $\ddot{y}^{\mathrm{IRF}}_{\mathrm{zmp}}$ based on the traditional LIP model ahead of time. Although x_{un} is divergent, we can find a ZMP trajectory whose initial condition satisfies $x_{\mathrm{un}}(t_{initial}) = \omega \int_{t_{initial}}^{\infty} e^{-\omega(t-t_{\mathrm{initial}})} x^{\mathrm{IRF}}_{\mathrm{zmp}}(t)dt$ to make the trajectory of the COM stable. According to the definition of x_{un}, each control cycle $t_{initial}$=$k\delta$ can be used as the initial time for future trajectory planning. k can be $1, 2 \cdots n$, and δ is the control cycle. We can obtain the relationship between the state space and the velocity of ZMP to find this initial condition and the ZMP trajectory which is shown in Formula (14):

$$\frac{\omega x^{\mathrm{IRF}}_{\mathrm{com}} - \dot{x}^{\mathrm{IRF}}_{\mathrm{com}}}{\omega} = -\omega \int_{t_{\mathrm{initial}}}^{\infty} e^{-\omega(t-t_{\mathrm{initial}})} \int_{t_{\mathrm{initial}}}^{\infty} \dot{x}^{\mathrm{IRF}}_{\mathrm{com}} dtdt. \tag{14}$$

Here, we can obtain the relationship between the current state space of the COM, including the position and velocity and the future velocity of the ZMP. As is mentioned above, ZMP should be always in the supporting area to guarantee the stability of the robot, so the constraints of the velocity of ZMP are shown as Formula (15):

$$x^{\mathrm{min}}_{\mathrm{zmp}}(t) \leq x^{\mathrm{IRF}}_{\mathrm{zmp}}(t_{initial}) + \int_{t_{\mathrm{initial}}}^{t} \dot{x}^{\mathrm{IRF}}_{\mathrm{zmp}}(\tau)d\tau \leq x^{\mathrm{max}}_{\mathrm{zmp}}(t). \tag{15}$$

There are a lot of solutions to the COM trajectories that can satisfy the stable conditions for Formula (14). Through the optimal control theory, the input of the system is the velocity of ZMP, and we can reduce the value of the COM velocity deviating from the average speed, thus reducing the change in the COM velocity. We define the cost function (f) as the Formula (16):

$$f(\dot{x}^{\mathrm{IRF}}_{\mathrm{com}}(t), t) = \int_{t_{\mathrm{initial}}}^{\infty} (\dot{x}^{\mathrm{IRF}}_{\mathrm{com}}(t) - x^{\mathrm{average}}_{\mathrm{com}})^2 dt. \tag{16}$$

We discretize Formula (16) and obtain Formula (17):

$$f(\dot{x}^{\mathrm{IRF}}_{\mathrm{com}}(k+j), k) = \sum_{j=0}^{j=N} (\dot{x}^{\mathrm{IRF}}_{\mathrm{com}}(k+j) - x^{\mathrm{average}}_{\mathrm{com}})^2 \delta, \tag{17}$$

where N is the number of sampling points, and $\dot{x}^{\mathrm{IRF}}_{\mathrm{com}}((k+j)$ is the predicted COM velocity at the jth sampling point. The control variable is the ZMP velocity. We should thus derive the relationship between the velocities of the COM and the ZMP.

We can use the recursive method to predict the COM velocity after the velocities within the predicted time are known. These expressions are shown as Formula (18):

$$\begin{aligned}
&\dot{x}^{\mathrm{IRF}}_{\mathrm{com}}(k+1) \\
&= \dot{x}^{\mathrm{IRF}}_{\mathrm{com}}(k) + \ddot{x}^{\mathrm{IRF}}_{\mathrm{com}}(k)\delta = \dot{x}^{\mathrm{IRF}}_{\mathrm{com}}(k) + \omega^2(x^{\mathrm{IRF}}_{\mathrm{com}}(k) - x^{\mathrm{IRF}}_{\mathrm{zmp}}(k))\delta, \\
&\dot{x}^{\mathrm{IRF}}_{\mathrm{com}}(k+2) \\
&= \dot{x}^{\mathrm{IRF}}_{\mathrm{com}}(k+1) + \ddot{x}^{\mathrm{IRF}}_{\mathrm{com}}(k+1)\delta \\
&= \dot{x}^{\mathrm{IRF}}_{\mathrm{com}}(k+1) + \omega^2(x^{\mathrm{IRF}}_{\mathrm{com}}(k+1) - x^{\mathrm{IRF}}_{\mathrm{zmp}}(k) - \dot{x}^{\mathrm{IRF}}_{\mathrm{zmp}}(k)\delta)\delta, \\
&\vdots \\
&\dot{x}^{\mathrm{IRF}}_{\mathrm{com}}(k+N) \\
&= \dot{x}^{\mathrm{IRF}}_{\mathrm{com}}(k+N-1) + \ddot{x}^{\mathrm{IRF}}_{\mathrm{com}}(k+N-1)\delta \\
&= \dot{x}^{\mathrm{IRF}}_{\mathrm{com}}(k+N-1) + \omega^2(x^{\mathrm{IRF}}_{\mathrm{com}}(k+N-1) - x^{\mathrm{IRF}}_{\mathrm{zmp}}(k) \\
&- \dot{x}^{\mathrm{IRF}}_{\mathrm{zmp}}(k)\delta - \cdots - \dot{x}^{\mathrm{IRF}}_{\mathrm{zmp}}(k+N-2)\delta)\delta.
\end{aligned} \tag{18}$$

We therefore have the cost function, the stability constraint, and the ZMP constraint. We can obtain the ZMP velocity as the input for the new LIP model. We can, therefore, plan the ZMP and COM trajectories.

3.2. Foot Position and Allowable ZMP Region Planning

The foot position determines the area supporting the robot, so it must be determined first. For the turning walk, we use a circular curve interpolation to generate the foot trajectory, which is shown in Figure 7. In the single support period, the foot position of the swinging leg will follow the curve of the circle.

At this stage, we know the foot step, the walking cycle period, and the turning angle. In turning walk, the foot steps of the two legs are not the same. Therefore, we must first calculate the right foot radius ($r_{st,r}$). The foot step is approximately equal to the arc length.

The radius of the left foot should be calculated as shown in Formula (19) in order to calculate the right foot radius:

$$r_{st,l} = l_{st,l}/\theta. \tag{19}$$

Suppose now that the robot turns right while walking. Here, $l_{st,l}$ is the right foot step, and θ is the turning angle. The right foot radius is $r_{st,r} = r_{st,l} + w_{hip}$. We can also obtain the right foot step as $l_{st,r} = \theta(r_{st,l} + w_{hip})$, and w_{hip} is the distance between two feet. The foot trajectory is then generated by spline interpolation. The foot position can be calculated when the turning angle ($\theta(t)$) is known. The turning angle ($\theta(t)$) is obtained by cubic spline interpolation. This can guarantee that the foot speed at the beginning and end of the single support period is zero.

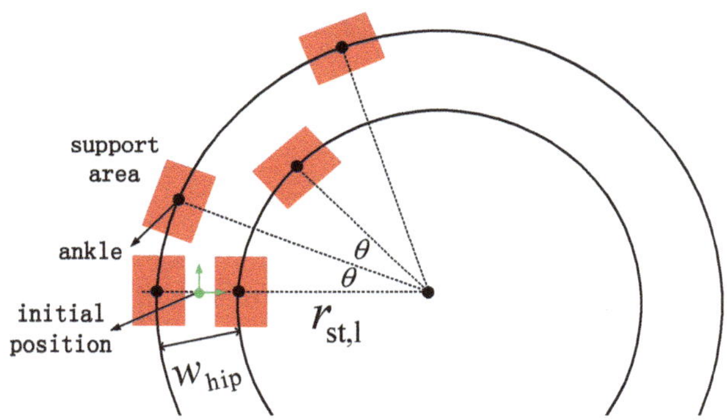

Figure 7. Foot position planning.

In straight-line walking, the foot coordinate only has translation related to the hip coordinate, so the hip yaw joint does not turn, but in turning walk, besides the translation, the foot coordinate also has rotation related to the hip coordinate, so the hip yaw joint will play a role in coordinate rotation shown in Figure 8.

After the foot positions have been determined, the next step is planning the allowable ZMP region (AZR). This is a prerequisite for planning the ZMP trajectory. The AZR is the polygon that is surrounded by the supporting feet. However, for turning walk, the front foot direction differs from that of the rear foot. Therefore, in the double support period, the AZR is as shown in Figure 9.

The directions of the x and y coordinates are the same as those in the IRF. We suppose that the robot is at the position shown in Figure 9, and the position of AZR at the world coordinate reference is also known. So, we should express the AZR in the IRF. After we have determined the α and the position of y_{com}^{IRF}, we can easily obtain the expression of the AZR in the IRF through the positive kinematics.

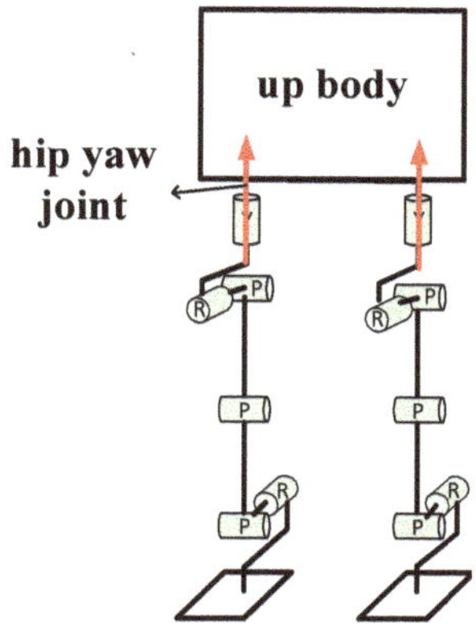

Figure 8. Hip yaw joint.

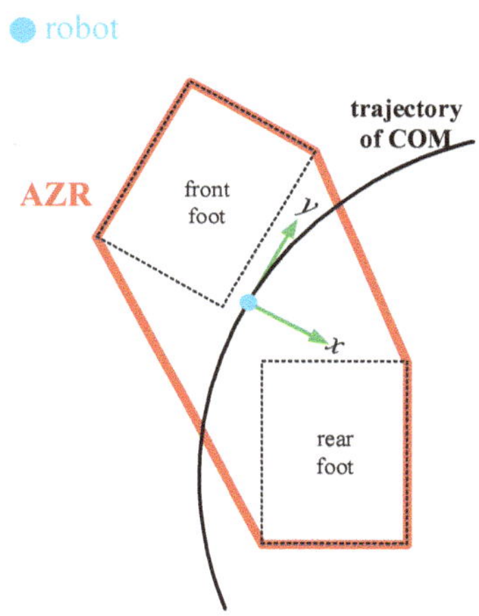

Figure 9. AZR of the double support period.

4. Transformation from the IRF to the World Coordinate Reference

For the humanoid robot walk, what we need is the trajectory in the world coordinate reference so we need to convert the trajectory in the IRF into the trajectory under the world coordinate reference.

In the turning walk, the forward direction of the COM is always the tangential direction of the COM circle, so the COM direction is always changing. Once we have planned the foot position and the allowable ZMP region (AZR) in the world reference, we need to express them in the world coordinate reference. In the IRF, the distance passed through in the forward direction is the arc length of the turning walk in the world coordinate reference. First, we know that the radius of the COM differs from the radius of the foot. The radius of the average COM trajectory (r_c) is in the range between the radii of the two feet. We can then obtain r_c from Formula (20). w_{hip} is the width of the robot hip:

$$r_c = r + w_{hip}/2. \tag{20}$$

In one control cycle, the change in the angle of the COM direction ($\Delta\alpha$) can also be calculated as shown in Figure 10.

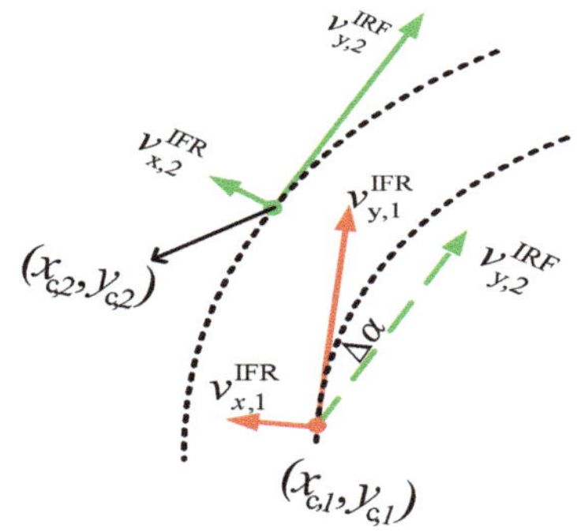

Figure 10. Rotation angle of the center of mass (COM) coordinate system.

$$\Delta\alpha = \frac{y_{c,1} + v_{y,1} \cdot \delta}{r_c + x_{c,1} + v_{x,1} \cdot \delta} - \frac{y_{c,1}}{r_c + x_{c,1}}, \tag{21}$$

where, $v_{x,1}^{IRF}$, $v_{y,1}^{IRF}$ are the velocities of the COM in the IRF at the current moment. $x_{c,1}$, $y_{c,1}$ is the position of the COM in the world coordinate reference at the current moment. $v_{x,2}^{IRF}$, $v_{y,2}^{IRF}$, $x_{c,2}$ and $y_{c,2}$ have the same meaning in the next control period. δ is the control cycle which depends on the controller's operating speed. However, we can establish the relationship between the position in the world coordinate reference and the velocities in the IRF. The expression is shown in Formula (22):

$$\begin{bmatrix} x_{c,2} \\ y_{c,2} \end{bmatrix} = \begin{bmatrix} x_{c,1} \\ y_{c,1} \end{bmatrix} + \delta \begin{bmatrix} \cos(\alpha) & \sin(\alpha) \\ -\sin(\alpha) & \cos(\alpha) \end{bmatrix} \begin{bmatrix} v_{x,1}^{IRF} \\ v_{y,1}^{IRF} \end{bmatrix}, \tag{22}$$

where

$$\alpha = \sum \Delta\alpha_i.$$

Turning Walk Planning Method

We can now provide an outline of the turning walk algorithm. First, we require the input data, including the foot steps and the turning angle. This information is then used to calculate the turning radius, the trajectory of the foot. and the AZR over the entire time span. What we need to know is the position and velocity of the COM and the ZMP position. In the initialization stage, these values are determined by hand, while future values are calculated by the model of the new LIP model.

The general $k-$th iteration proceeds as follows:

- Use the new cost function with stability and ZMP constraints to compute the ZMP velocity based on the new LIP model in the preceding section;
- Based on the planned ZMP velocity, we then obtain the ZMP position and can also get the position and velocity of COM in the IRF;
- Compute $\Delta\alpha$ based on the COM position and velocity;
- Based on $\Delta\alpha$, we can calculate the COM velocity and the ZMP position in the world coordinate reference. This is the turning walk data. We then return to the first step.

5. Simulations and Experiments

We now present some simulations of the BHR-6 humanoid robot of Beijing Institute of Technology to illustrate the performance of the turning gait planning method. The robot parameters are shown in

Table 1. We also present some comparisons between two different walking speeds. Through comparing different walking speeds, we can determine the function of the proposed gait planning method.

Table 1. Parameters of robot BHR-6.

Parameters	Value
Degrees of freedom	23
Weight	50.0 kg
Height	1.65 m
Foot length	24 cm
Foot width	14 cm
Hip width	15 cm
Contact type	Point contact

The planned ZMP as-obtained by the cost function (f) is shown in Figure 11. We set the walking step to T_{step}= 1.2 s and sampling cycle to T_{sample}= 0.004 s. The walking step was $D_{\text{step}} = 0.33$ m, and the turning radius was $r_c = 1.06$ m; these are the planning results in the global reference frame. The ZMP was in the boundary of the supporting area and continued moving forward whether in the double supporting period or single supporting period. This planning result is similar to that of linear walking.

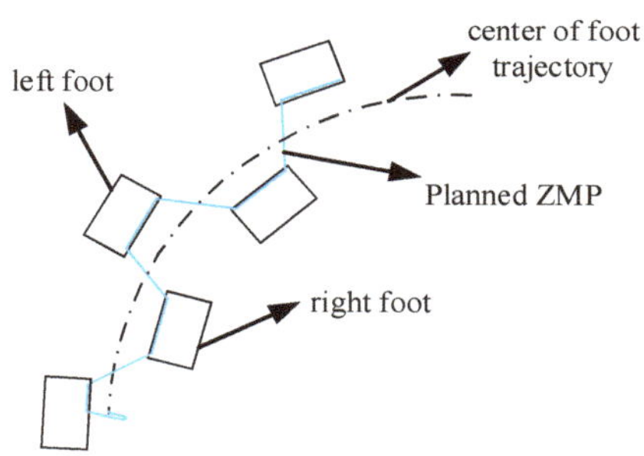

Figure 11. ZMP planning for the robot turning walk.

Once the ZMP trajectory was planned, the new trajectory was generated based on the new LIP model expressed in Formula (6). The formula is easy to be discretized. We simulated two walking speed trajectories with different radii and walking speeds to test the practicability of this method. As shown in Figure 12, the trajectory was generated as the robot turned to the right. As opposed to straight-line walk, turning walk creates a trajectory offset to the direction in which the robot is moving. When the robot turns right, the trajectory of COM will shift to the right and for left turns, it will shift to the left. As shown in Figure 13, the faster the robot moves, the more the trajectory is offset. The foot width of our BHR-6 was 14 cm but when the robot walking speed was 2 km per hour, the offset achieved 4.5 cm with the radius $r_c = 0.67$ m. When the robot walks as slowly 1 km/h, the offset fell to 0.8 cm with the same turn radius. For the same walking speed, the offsets caused by different radii were also different. For the walking speed 2 km/h per hour, the offset was 2.5 cm with the radius $r_c = 1.06$ m. When the robot walks slowly, the centroid shift caused by centripetal force can be neglected and the trajectory can be planned in the same manner as for straight-line walking.

We reached the results shown in Figure 14 by calculating the Δzmp. Thus we can see the influence of robot rotation on the robot's ZMP position. Both walking speed and turning radius had significant impacts on the offset of the ZMP. A decrease in the radius and increase in walking speed caused Δzmp to increase. As shown in Figure 14d, when the radius was 0.67 m and the walking speed was 2 km per hour, the greatest offset of ZMP was 6.2 cm; this offset markedly affects the planning of the reference ZMP trajectory.

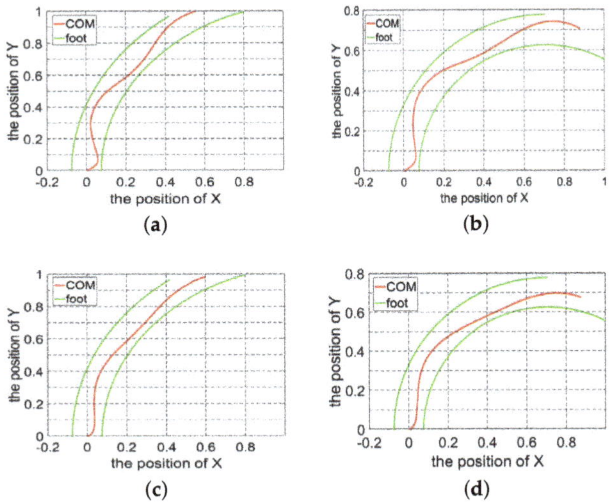

Figure 12. COM trajectories: (**a**) $r_c = 1.06$ m, $v_{ave} = 1$ km/h, (**b**) $r_c = 0.67$ m, $v_{ave} = 1$ km/h, (**c**) $r_c = 1.06$ m, $v_{ave} = 2$ km/h, (**d**) $r_c = 0.67$ m, $v_{ave} = 2$ km/h.

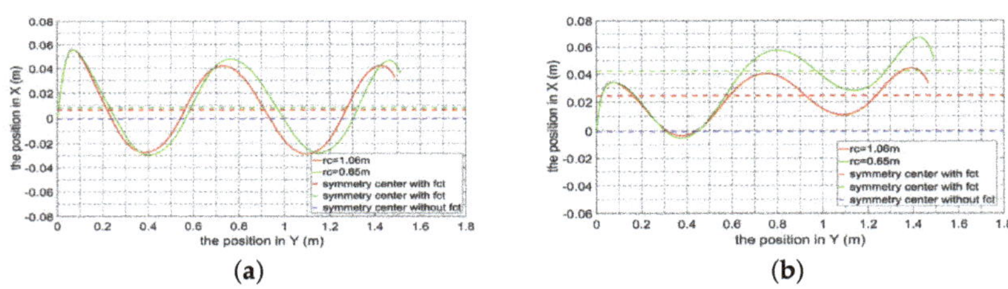

Figure 13. COM trajectories in the involute reference frame (IRF) (**a**) $v_{ave} = 1$ km/h, (**b**) $v_{ave} = 2$ km/h.

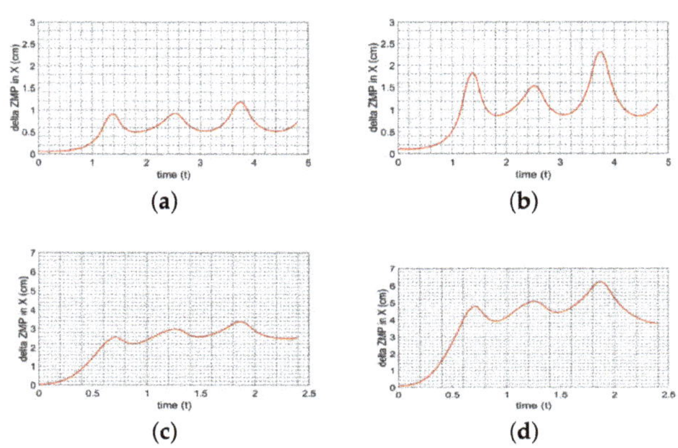

Figure 14. the value of the Δx_{zmp}: (**a**) $r_c = 1.06$ m, $v_{ave} = 1$ km/h, (**b**) $r_c = 0.67$ m, $v_{ave} = 1$ km/h, (**c**) $r_c = 1.06$ m, $v_{ave} = 2$ km/h, (**d**) $r_c = 0.67$ m, $v_{ave} = 2$ km/h.

As opposed to straight-line walk, during turning walk, the hip yaw joints underwent major rotation at the angle shown in Figure 15. During a single support period, the supporting leg does not turn, and only the swinging leg turns. This means that the body does not turn. During a double support period, both hip yaw joints turn, and the body turns through an angle. Because the support area in the double support period is much larger than that in the single support period, the robot is less likely to slide on the ground when it turns fast. These phenomena correspond with the foot position planning results.

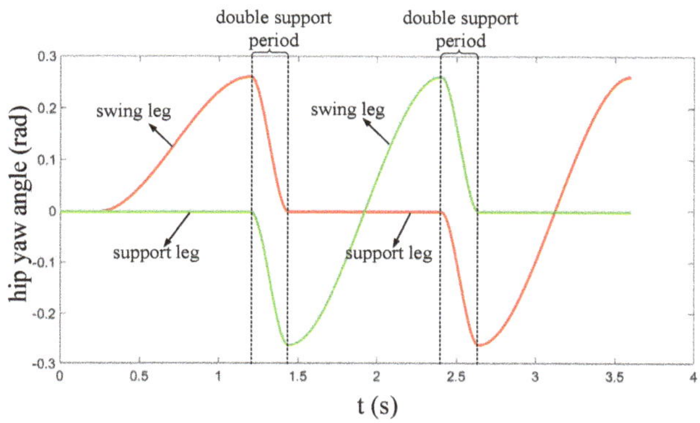

Figure 15. Hip yaw angle characteristics.

There are a few angles that must be defined properly for the purpose of turning walk. In Figure 16, the green line marks the COM angle (tangent direction of the turning motion) which equals the value of the upbody angle. It was obtained based on Formula (21) and Formula (22). The blue line is the foot angle, and the foot angle is planned before plotting the COM trajectory. In the single support period, the foot of the swing leg turns while the support leg maintains its previous state and in the double support period, the foot maintains its previous state. All these angles increase at the same average speed in one walk cycle so that at the end of every walk cycle, the direction of the feet and upbody is the same.

We next applied the proposed method to an actual humanoid robot identical to the one modeled in the simulation platform and under the same parameters as the simulation. As shown in Figure 17, $D_{step} = 0.33$ m, $T_{step} = 1.2$ s, and the robot achieved turning walk at a speed of 1 km/h.

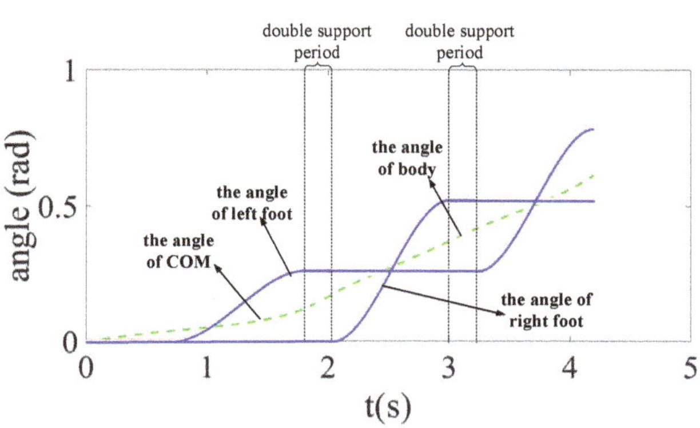

Figure 16. Angle data for gait planning.

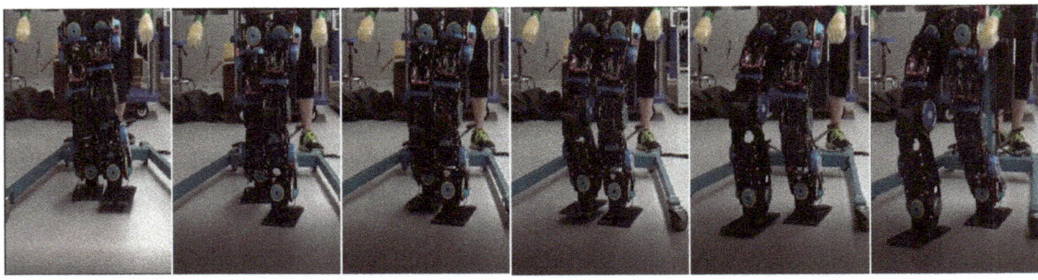

Figure 17. Experiments in the real humanoid robot.

Then, we measured the actual ZMP of the robot calculated by the force sensors and converted it to the IRF, as shown in Figure 18. Another experiment was also done which generated the robot's COM trajectory directly by the LIP as is shown in Figure 19. By using the proposed method, The actual ZMP trajectory of the robot was able to better follow the planned ZMP both in the lateral direction and the longitudinal direction. This agrees with the results of the simulation results. In the simulation shown in Figure 13, the COM trajectory deviated to one side in order to make the actual ZMP trajectory follow the planned ZMP trajectory. However, the traditional LIP does not take into account the effect of the robot's own rotation on the actual ZMP, so in the lateral direction, the actual ZMP is offset to one side by a distance form the planned ZMP. In addition, in the longitudinal direction, there are no changes in dynamics, so the actual ZMP can also follow the planned ZMP well.

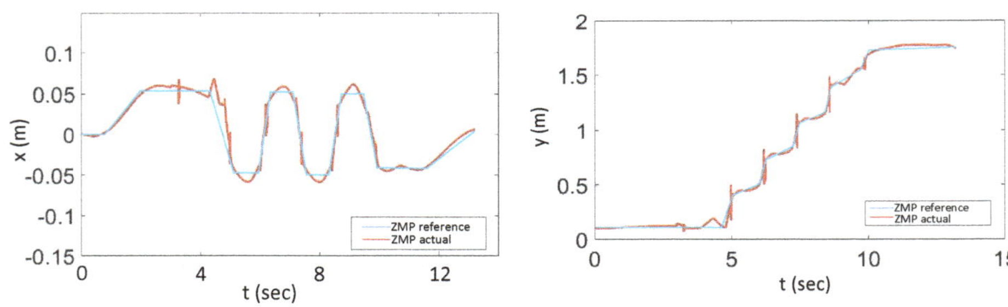

(**a**) (**b**)

Figure 18. ZMP reference vs. actual with proposed method (**a**): Lateral ZMP, (**b**): Longitudinal zero moment point (ZMP).

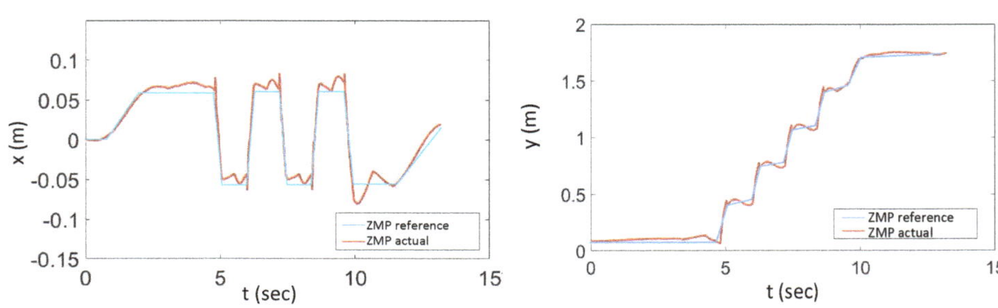

(**a**) (**b**)

Figure 19. ZMP reference vs. actual with previous LIP (**a**): Lateral ZMP, (**b**): Longitudinal ZMP.

6. Conclusions

In view of the limitations of the current LIP model in the gait planning of robot turning, we presented a turning walk method for humanoid robots that can reduce the error between the model and the actual robot. This method not only makes the model accurate, but also avoids complex dynamic analysis. By introducing the IRF, the dynamics of the turning walk model were also simplified using the LIP model with centripetal forces acting in the left and right directions. We performed some simulations and experiments to verify the correctness of the proposed method. The results proved that this method can achieve a stable turning walk.

The proposed method involves coordinate system transformations between world coordinate references and the IRF. So when the robot's COM coordinate and the world coordinate are relatively rotated horizontally, this means the robot will turn and the proposed method will play a part.

Working from another perspective, the straight-line walk can be seen as a special case of the turn walk. For the straight-line walk, the turn radius can be seen as infinity $r_c = \infty$. We can obtain Formula (5) from Formula (6), and $\Delta\alpha$ in Formula (21) has no value, so all the formulas are the same as those in the straight-line walk.

The results show that our method is more effective during rapid turning of a robot which can greatly improve the efficiency of the robot movement. However, for turns with small radii, the calculated turning angle is large, so the robot will easily slip when the robot is also carrying out a fast turn. For the robot slip turn, there has been some research, as is shown in the first section, so the robot can make a slip turn in order to improve efficiency of the turn.

Author Contributions: Tianqi Yang conceived of the presented idea. Tianqi Yang and Weimin Zhang developed the theoretical formalism. Tianqi Yang, Xuechao Chen and Qiang Huang designed the simulation experiments and analyzed the data. Tianqi Yang, Libo Meng and Zhangguo Yu discussed the results and wrote the paper.

Funding: This work was supported in part by the Pre-research Project under Grant 41412040101 and in part by the National Natural Science Foundation of China under Grant 61533004, Grant 91748202 and Grant 61703043.

Conflicts of Interest: The authors declare no conflict of interest.

References

1. Wieber, P.B. Trajectory free linear model predictive control for stable walking in the presence of strong perturbations. In Proceedings of the IEEE-RAS International Conference on Humanoid Robots, Genova, Italy, 4–6 December 2006; pp. 137–142.
2. Pratt, J.; Carff, J.; Drakunov, S.; Goswami, A. Capture point: A step toward humanoid push recovery. In Proceedings of the IEEE-RAS International Conference on Humanoid Robots, Genova, Italy, 4–6 December 2006; pp. 200–207.
3. Herdt, A.; Perrin, N.; Wieber, P.B. Walking without thinking about it. In Proceedings of the IEEE/RSJ International Conference on Intelligent Robots and Systems, Taipei, Taiwan, 18–22 October 2010; pp. 190–195.
4. Morisawa, M.; Kajita, S.; Kanehiro, F.; Kaneko, K.; Miura, K.; Yokoi, K. Balance control based on capture point error compensation for biped walking on uneven terrain. In Proceedings of the IEEE-RAS International Conference on Humanoid Robots (Humanoids), Osaka, Japan, 29 November–1 December 2012; pp. 734–740.
5. Deits, R.; Tedrake, R. Footstep planning on uneven terrain with mixed-integer convex optimization. In Proceedings of the IEEE-RAS International Conference on Humanoid Robots, Madrid, Spain, 18–20 November 2014; pp. 279–286.
6. Kajita, S.; Kanehiro, F.; Kaneko, K.; Fujiwara, K.; Harada, K.; Yokoi, K.; Hirukawa, H. Biped walking pattern generation by using preview control of zero-moment point. In Proceedings of the International Conference on Robotics and Automation, Taipei, Taiwan, 14–19 September 2003; Volume 2, pp. 1620–1626.
7. Lanari, L.; Hutchinson, S.; Marchionni, L. Boundedness issues in planning of locomotion trajectories for biped robots. In Proceedings of the IEEE-RAS International Conference on Humanoid Robots, Madrid, Spain, 18–20 November 2014; pp. 951–958.
8. Huang, Q.; Yokoi, K.; Kajita, S.; Kaneko, K.; Arai, H.; Koyachi, N.; Tanie, K. Planning walking patterns for a biped robot. *IEEE Trans. Robot. Autom.* **2001**, *17*, 280–289. [CrossRef]

9. Koolen, T.; Boer, T.D.; Rebula, J.; Goswami, A.; Pratt, J. Capturability-based analysis and control of legged locomotion, Part 1: Theory and application to three simple gait models. *Int. J. Robot. Res.* **2012**, *31*, 1094–1113. [CrossRef]
10. Mombaur, K.; Truong, A.; Laumond, J.P. *From Human to Humanoid Locomotion—An Inverse Optimal Control Approach*; Kluwer Academic Publishers: Norwell, MA, USA, 2010.
11. Faraji, S.; Pouya, S.; Ijspeert, A. Robust and Agile 3D Biped Walking With Steering Capability Using a Footstep Predictive Approach. In Proceedings of the Robotics Science and Systems (RSS), Berkeley, CA, USA, 12–16 July 2014.
12. Sulistijono, I.A.; Setiaji, O.; Salfikar, I.; Kubota, N. Fuzzy walking and turning tap movement for humanoid soccer robot EFuRIO. In Proceedings of the IEEE International Conference on Fuzzy Systems, Barcelona, Spain, 18–23 July 2010; pp. 1–6.
13. Hu, Y.; Mombaur, K. Bio-Inspired Optimal Control Framework to Generate Walking Motions for the Humanoid Robot iCub Using Whole Body Models. *Appl. Sci.* **2018**, *8*, 278. [CrossRef]
14. Han, B.; Luo, X.; Liu, Q.; Zhou, B.; Cheng, X. Hybrid control for SLIP-based robots running on unknown rough terrain. *Robotica* **2014**, *32*, 1065–1080. [CrossRef]
15. Shahbazi, M.; Babuska, R.; Lopes, G.A.D. Unified Modeling and Control of Walking and Running on the Spring-Loaded Inverted Pendulum. *IEEE Trans. Robot.* **2016**, *32*, 1178–1195. [CrossRef]
16. Poulakakis, I. Spring Loaded Inverted Pendulum embedding: Extensions toward the control of compliant running robots. *Robot. Autom.* **2010**, *58*, 5219–5224.
17. Cao, Y.; Suzuki, S.; Hoshino, Y. Turn Control of a Three-Dimensional Quasi-Passive Walking Robot by Utilizing a Mechanical Oscillator. *Engineering* **2014**, *6*, 93–99. [CrossRef]
18. Miura, K.; Kanehiro, F.; Kaneko, K.; Kajita, S.; Yokoi, K. Slip-Turn for Biped Robots. *IEEE Trans. Robot.* **2013**, *29*, 875–887. [CrossRef]
19. Miura, K.; Kanehiro, F.; Harada, K.; Kaneko, K.; Yokoi, K.; Kajita, S. Slip turn generatioin of humanoid robots based on an analysis of friction model. In *Emerging Trends in Mobile Robotics*; World Scientific: Sigapore, 2015; pp. 761–768.
20. Miura, K.; Kanehiro, F.; Kaneko, K.; Kajita, S.; Yokoi, K. Quick slip-turn of HRP-4C on its toes. In Proceedings of the IEEE International Conference on Robotics and Automation, Saint Paul, MN, USA, 14–18 May 2012; pp. 3527–3528.
21. Koeda, M.; Ueno, M.; Serizawa, T. Shuffle Parallel Translation and Pivot Turn of Humanoid Robot in Dynamics Simulator. In Proceedings of the International Conference on Artificial Intelligence, Modelling and Simulation (AIMS), Kota Kinabalu, Malaysia, 3–5 December 2013; pp. 217–220.
22. Hashimoto, K.; Yoshimura, Y.; Kondo, H.; Lim, H.-O.; Takanishiet, A. Realization of quick turn of biped humanoid robot by using slipping motion with both feet. In Proceedings of the IEEE International Conference on Robotics and Automation, Shanghai, China, 9–13 May 2011; pp. 2041–2046.
23. Peng, S.; Shui, H.; Ma, H. A simulation and experiment research on turning gait planning of blackmann-II humanoid robot. In Proceedings of the IEEE International Conference on Control and Automation, Xiamen, China, 9–11 June 2010; pp. 719–724.
24. Kajita, S.; Hirukawa, H.; Harada, K.; Yokoi, K. *Introduction to Humanoid Robotics*; Springer Publishing Company: Berlin, Germany, 2014.
25. Scianca, N.; Cognetti, M.; de Simone, D.; Lanari, L.; Oriolo, G. Intrinsically stable MPC for humanoid gait generation. In Proceedings of the IEEE-RAS 16th International Conference on Humanoid Robots, Cancun, Mexico, 15–17 November 2016; pp. 601–606.

applied sciences

MDPI

Article

Design and Experimental Evaluation of a Single-Actuator Continuous Hopping Robot Using the Geared Symmetric Multi-Bar Mechanism

Long Bai [1,*], Fan Zheng [1], Xiaohong Chen [1], Yuanxi Sun [1] and Junzhan Hou [2]

1 State Key Laboratory of Mechanical Transmission, Chongqing University, Chongqing 400044, China; 20160702066@cqu.edu.cn (F.Z.); xhchen@cqu.edu.cn (X.C.); sunyuanxi@cqu.edu.cn (Y.S.)
2 Xi'an Institute of Applied Optics, Xi'an 710065, China; PG29146111@e.ntu.edu.sg
* Correspondence: bailong@cqu.edu.cn

Received: 3 December 2018; Accepted: 18 December 2018; Published: 20 December 2018

Abstract: This paper proposes the design and performance evaluation of a miniaturized continuous hopping robot RHop for unstructured terrain. The hopping mechanism of RHop is realized by an optimized geared symmetric closed-chain multi-bar mechanism that is transformed from the eight-bar mechanism, and the actuator of RHop is realized by a servo motor and the clockwork spring, thereby enabling RHop to realize continuous hopping while its motor rotates continuously only in one direction. Comparative simulations and experiments are conducted for RHop. The results show that RHop can realize better continuous hopping performance, as well as the improvement of energy conversion efficiency from 70.98% to 76.29% when the clockwork spring is applied in the actuator. In addition, comparisons with some state-of-the-art hopping robots are conducted, and the normalized results show that RHop has a better energy storage speed.

Keywords: mobile robot; jumping robot; hopping robot; continuous hopping; single actuator

1. Introduction

With the development of science and technology, robots have become indispensable automation equipment in human society by assisting or replacing humans in various situations. Among mobile robots, hopping robots show a stronger ability to overcome obstacles in the fields of planetary surface exploration, unstructured terrain search, etc. [1–4] when compared with wheeled robots [5], legged robots [6], and tracked robots [7]. The hopping robot was first proposed by Oberth and Seifert [8] in 1967. Then the one-leg jumping model proposed by Raibert [9] became the theoretical research basis for later single-leg jumping robots. As time goes, the NASA three-generation hopping robot [10] and the MIT MICROBOTS [11] significantly progressed the hopping robot research.

According to hopping characteristics of the hopping robot, hopping robots can be classified into two categories: continuous hopping robot and intermittent hopping robot. A continuous hopping robot can recharge its hopping energy in the flying/landing phase, and immediately hop again after contacting the ground, while the intermittent hopping robot needs to realize its energy recharge and hopping attitude adjustment in another ground phase. Currently, continuous hopping robots usually adopt hydraulic [12,13] and pneumatic [14] actuators to realize continuous actuation, which can enable the continuous hopping robot to have high power density and fast drive-response. However, it is still difficult to realize miniaturization and multi-drive coordinated control for them. The intermittent hopping robots usually use their motor as their actuator and use springs as an elastic components [15,16]. Then, by slowly recharging and locking the energy of elastic components by the motor and the hopping mechanism, the intermittent hopping robot can realize hopping once the elastic energy is released. Due to the limitation of the motor and the complexity of additional energy locking and releasing mechanism in the intermittent hopping robot [15–17], the preparation time that is required to complete a hop will become relatively longer than continuous hopping robots.

The energy locking and releasing mechanism of the hopping robot can be divided into two categories:

(1) Take advantage of the characteristics of the actuator, either by increasing the number of actuators or by using the forward and reverse rotation of the actuator. For instance, NASA's second-generation frog-inspired hopping robot uses an additionally actuated latching mechanism [10], while the NPU kangaroo-inspired hopping robot additionally uses an actuated ratchet–pawl mechanism [15,16] to realize energy locking and releasing. In addition, the locust-inspired hopping robot designed by Zaitsev et al. [18] and the integrated jumping–crawling robot designed by Jung et al. [17] realizes energy release by reversing the actuating motor. The energy storage and release of these aforementioned robots are all distributed; therefore, it is difficult to absorb the landing impact energy for the next hop.
(2) Using special devices and irregular component contours such as eccentric cam [19–21], incomplete gear [22–27], etc. to achieve the conversion between energy storage and release. For example, a miniature jumping robot proposed by Zhao et al. [28] uses a quick release detent ball mechanism to hold the spring, and it uses the lever as a strike mechanism to strike the push shaft to release the stored energy. Faraji et al. used two eyebolts located at opposite ends of the rear leg and the fishing line to hold energy for the spider-inspired hopping robot [29], which can release the energy stored in the spring by cut the fishing line. In addition, quadruped robot with jumping ability uses the ratchet and pawl [30], while the first-generation MSU Jumper [31] and the second-generation micro-robots uses one-way bearings. Moreover, the cylinder-shaped robot [32] uses the latch and the hook to realize the conversion of energy state.

Therefore, existing electrically actuated hopping robots have the following disadvantages: (1) Increasing the number of actuators in the hopping robot may result in an increase in robot weight as well as the control difficulty. In the meantime, high-intensity forward and reverse rotation of the motor will reduce the electric energy efficiency of the robot, due to energy loss caused by the moment of inertia of the transmission system. (2) The design of the special devices increases the difficulty of robot control and reduces the reaction speed of the hopping robot. (3) Hopping robots use irregular contours may also suffer from reduced machining accuracy and the wear problem of the contour. The parameters of the incomplete gear and the trajectory of the cam contour, the movement tracks of the hopping mechanism and the limit position of the spring deformation are all required complex coupling design calculations.

Therefore, in this paper, we aim to design a fast-response continuous hopping robot with a single low-power motor that rotates unidirectionally, and the energy storage and release should be simple and reliable without an additional locking and releasing mechanism. (1) Using mechanical evolution, we realize the design of the hopping mechanism while retaining the advantages of the pure linkage mechanism, using the method of 'higher pairs replacing lower pairs' and isomeric mechanisms with same kinematics to meet design objectives. Furthermore, taking into account the transformation characteristics of the compound closed-chain multi-bar mechanism, a symmetrical double-gear-pair 10-bar based continuous hopping robot (named RHop) is proposed (the robotic prototype is shown in Figure 1). (2) Using the characteristics of the periodic motion and the acceleration characteristics during the collinearity of the crank–rocker mechanism, rapid energy storage and release are achieved by the continuous unidirectional rotation of the motor. (3) A clockwork spring is added to the actuation design to amplify the torque of the motor as well as to absorb the landing impact, thereby enabling its continuous and steady hopping.

The rest of this paper is organized as follows: Section 2 introduced the structural design of the RHop; Section 3 proposed the structural optimization and the actuation design; Section 4 performed the prototype experiment and discussion, and Section 5 concluded the paper.

Figure 1. Prototype of the RHop.

2. Structural Design

2.1. Hopping Mechanism Design

Based on the design goals specified in Section 1, the design of the hopping mechanism will be carried out using the method of 'higher pairs replacing lower pairs' and isomeric mechanisms with the same kinematics. The preliminary design used pure linkage mechanisms to retain the advantages of the traditional mechanism, and the new mechanism was evolved to meet design objectives when the pure linkage mechanism was unable to meet demand. Finally, the mechanism was improved in series to meet the design goals better.

2.1.1. Design Objectives

To meet the requirements of continuous hopping with fast response, simple and reliable energy storage/release, the first research objectives of this paper is established as:

(1) Using only one motor that is as small as possible.
(2) The motor should only rotate unidirectionally.

Due to the inherent characteristic in kinematics of the hopping mechanism, it will add unnecessary angular momentum to the robot system if there is a horizontal component of ground reaction force (GRF) in the direction of take-off, thereby endangering the stability of hopping posture. For example, the direction of the GRF should be close to the vertical direction when the robot's take-off angle is 90°. Based on this requirement, the second research objective of this paper is proposed as: achieving non-rotation or minimal rotation of the robot (sagittal plane) to take-off, which means the robot take-off with near-zero angular velocity and the system has no additional angular momentum. To materialize this requirement, we present the following two design objectives:

(3) The foot trajectory curve of the hopping mechanism should (a) be a straight line; (b) pass through the center of mass (CM) of the robot; (c) be consistent with the take-off direction of the hopping robot.
(4) Minimize the moment acting on the robot's CM when the robot component moves.

At the same time, for the miniaturization of the robot, the following design objectives are proposed:

(5) The foot trajectory curve of the hopping mechanism should be located within its mechanism.
(6) The structure of the hopping mechanism is compact.

2.1.2. Preliminary Configuration

In order to design the configuration of the hopping mechanism by evolving the closed-chain multi-bar isomeric mechanisms with same kinematics, as well as satisfying the aforementioned design requirements (DR.), the degree of freedom (DOF) of the planar mechanism is first studied.

The equation for calculating the DOF of a planar mechanism is:

$$F = 3n - 2p_{\mathrm{L}} - p_{\mathrm{H}} \tag{1}$$

where N represents the number of components in the planar mechanism, where $n = N - 1$ is the number of moving components, p_{L} is the number of lower pairs and p_{H} is the number of higher pairs.

To satisfy DR.1, the DOF of the robot needs to be set to one. When using the pure linkage mechanism to design the hopping mechanism, the number of higher pairs ($p_{\mathrm{H}} = 0$) can be omitted, i.e.,

$$F = 3n - 2p_{\mathrm{L}} = 1 \tag{2}$$

Table 1 lists the desirable permutations of Equation (2).

Table 1. The desirable permutations of Equation (2).

Config.	config.1	config.2	config.3	config.4	config.5	...
n	1	3	5	7	9	...
p_{L}	1	4	7	10	13	...

Config.1 does not exist; config.2 is a common single closed-chain four-bar mechanism; config.3 is a closed-chain six-bar linkage (Watt type and Stephenson type), as shown in Figure 2. Config.4 is a closed-chain eight-bar linkage, in which 16 main configurations of kinematical chain are shown in Figure 3 [33]. Config.5 is a closed-chain 10-bar linkage, which includes 230 specific configurations [34].

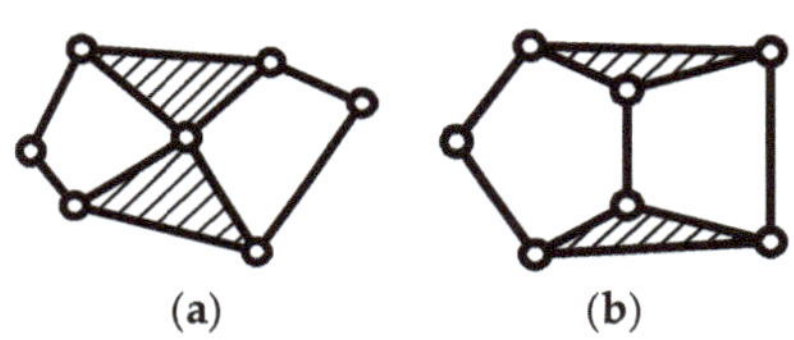

Figure 2. Kinematical chain of the closed-chain six-bar mechanism with a single degree of freedom (DOF). (**a**) Watt type. (**b**) Stephenson type.

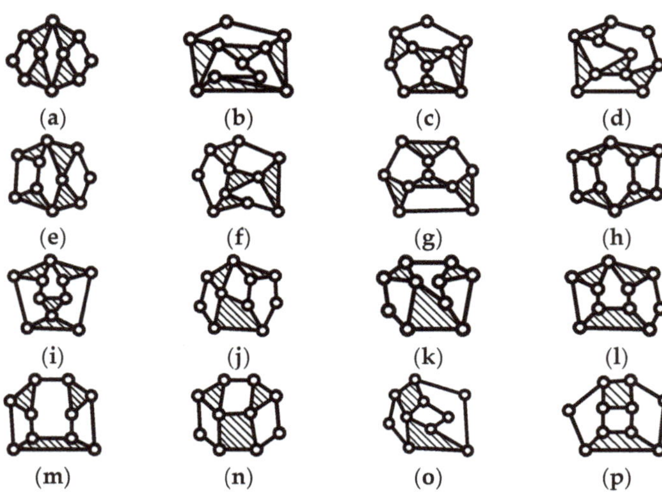

Figure 3. Kinematical chain of closed-chain eight-bar mechanism with a single DOF.

The four-bar mechanism is difficult to meet the requirements of complex motion performance and rich trajectory due to its few optimization parameters. The configuration of the 10-bar linkage is too complicated and changeable. Therefore, the single DOF six-bar linkage and eight-bar linkage are preselected for the mechanical design of RHop.

DR.4 requires that the horizontal acceleration of the CM caused by the inertia of the moving component needs to be eliminated. Therefore, the bilateral symmetric kinematical chain configuration is chosen as the basic structural configuration of the hopping robot. Watt and Stephenson types of the six-bar linkage and the type (a), (b), (e), (f), (g), (h), (i), (m), (n), (o), and (p) of the eight-bar linkage shown in Figure 3 can all meet this requirement. According to the drive mode indicated in DR.2, the input bar needs the function of circumferential rotation to form the crank-rocker mechanism. The kinematical chain configurations in the eight-bar linkage satisfy the condition are (m), (n), and (p), and the six-bar linkage does not satisfy the condition of existing the crank-rocker mechanism.

2.1.3. Configuration Adjustment

To use a common rotary motor as the power source, the two cranks need to rotate at the same speed with an opposite direction. Therefore, the gear pair with a confirmed transmission ratio is introduced into the eight-bar mechanism. Since the revolute pairs in the desired eight-bar mechanism are all lower pairs, it is necessary to perform 'higher pairs replacing lower pairs' to convert the two cranks and the related bars into a pair of geared crank–rocker mechanisms.

Any planar mechanism with lower pairs can be considered as consisting of several connected Assur groups [35]. To find a suitable mechanism, in which revolute pairs can be successfully replaced by gear pairs [36,37], attempts of transforming configuration (m), (p), and (n) via Assur groups are depicted in Figure 4.

As can be obtained from Figure 4, only the configurations (n) can be totally transformed, in which the crank mechanism consists of the bar 4 and 6.

2.1.4. Structural Improvement

P_7 and P_8 (Figure 4c) in the bilateral symmetric gear–crank–rocker mechanism are set as the output. The motion trajectories of points P_7 and P_8 are circular arcs with points P_5 and P_6 as their arc center, respectively. However, this output trajectory cannot meet DR.3. Therefore, additional motion components should be added to meet the required foot motion trajectory. Since the required hopping mechanism in this paper is a bilateral symmetric mechanism, and it does not contradict with DR.1 and DR.4, then the additional number of moving components n', the number of lower pairs p'_{L} and the number of higher pairs p'_{H} must satisfy the following Equation (3):

$$3n' = 2p'_{\mathrm{L}} + p'_{\mathrm{H}} \tag{3}$$

After full investigation of common mechanism, the gear-linkage mechanism was finally selected as the additional moving component for the hopping mechanism. To ensure that the structure of the hopping mechanism is compact without changing its symmetry properties, a symmetrical gear-linkage mechanism with the least number of components, that is, the symmetrical gear–three-bars mechanism shown in Figure 4d, was selected. The output bars P_5P_7 and P_6P_8 were extended to bars with three revolute pairs, the newly added revolute pairs in which are articulated to the non-gear ends of the symmetrical gear–three-bar mechanism. Therefore, the whole bilateral symmetric gear–crank–rocker mechanism finally evolved into a symmetrical double-gear-pair ten-bar mechanism with a single DOF, as shown in Figure 4d, which was consistent with DR.1 to DR.4.

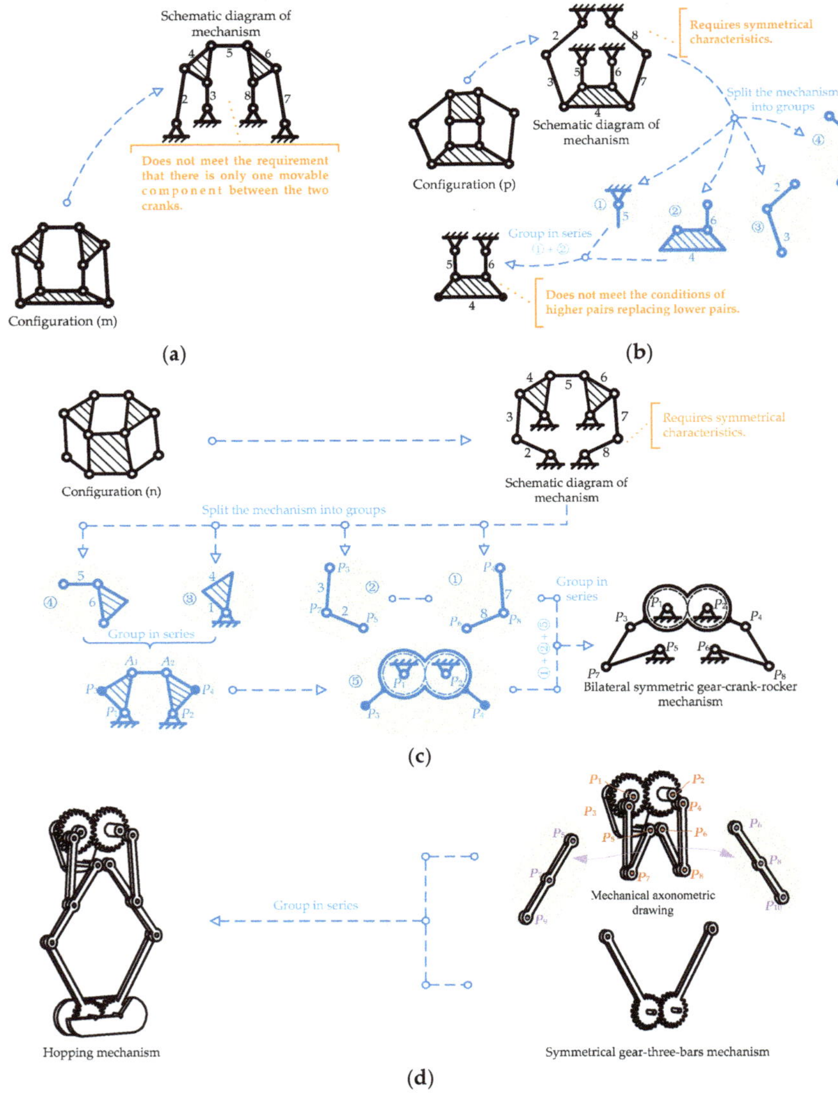

Figure 4. Configuration evolution of the eight-bar mechanism. (**a**) The evolution of the configuration (m). (**b**) The evolution of the configuration (p). (**c**) The evolution of the configuration (n). (**d**) The process of structural improvement.

2.2. Energy Mechanism Design

The energy mechanism should be designed to realize the storage/release of elastic energy for hopping. Most existing electrically actuated hopping robots use springs as the energy storage/release elements, which have the advantages of small mass and high controllability. Therefore, in this paper, we will utilize the tension spring to design the energy mechanism for the RHop.

When a crank works as the driving linkage, the rocker will be the driven linkage with a reciprocating swing in variable speed. The crank will have two chances of being collinear, with the connecting rods in one locomotion cycle. When the crank and the connecting rod are collinear for the second time, the swing speed of the rod will reach its tipping point, i.e., its acceleration will reach the largest. Based on these characteristics, we finally placed the tension spring between two points P_9 and P_{10}, as shown in Figure 5. This design has the following benefits:

Assume that crank P_1P_3 is the active component that rotates one turn in one motion cycle, and that the horizontal axis represents the rotation angle α of the crank.

(1) When the rotation angle is $\alpha_1 = 0$, the crank and the connecting rods will be collinear for the first time. This will be the crucial moment where the distance between points P_9 and P_{10} is the smallest. This also will be the initial position of the energy storage phase, as shown in Figure 5a.
(2) Within the range of α_1–α_2, the distance between P_9 and P_{10} will gradually increase, and the length of the spring will gradually grow.
(3) When the rotation angle reaches α_3, the crank and the connecting rods are collinear for the second time. At this time, the distance between point P_9 and P_{10} will be the largest, which will be the end position of the energy storage phase, as shown in Figure 5c. It can also be noticed that the feature of the crank–rocker mechanism is properly compatible with the required hopping burst characteristics.
(4) Within the range of α_3–α_4, the distance between the point P_9 and P_{10} will gradually decrease. Figure 5d shows the general position of this energy release phase.
(5) When the rotation angle is $\alpha_5 = 360°$, the mechanism will return to its initial position of the energy storage phase, thus preparing for the next hopping cycle.

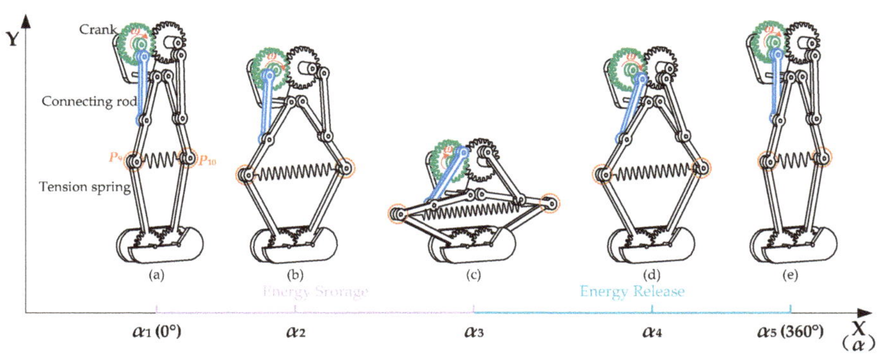

Figure 5. The energy storage and release phases. α in the horizontal axis represents the rotation angle of the crank. (**a**) The initial position of the energy storage phase, which means that the tension spring is in a natural state. (**b**) The general position of the energy storage phase. (**c**) The end position of the energy storage phase, which means that the energy storage of the tension spring is maximum. (**d**) The general position of the energy release phase. (**e**) The initial position of the energy storage phase.

3. Structural Optimization and Actuation Design

3.1. Structural Optimization

3.1.1. Kinematics Analysis

According to the selected configuration of the hopping mechanism in Figure 4d, the kinematics model of the hopping mechanism is depicted in Figure 6. Because the selected configuration of the hopping mechanism is symmetrical (the branched chain P_2–P_4–P_8–P_{10}–P_{12}–P_6 is symmetric with the branched chain P_1–P_3–P_7–P_9–P_{11}–P_5), and only one of the branched chains is selected for kinematic analysis (the branched chain P_2–P_4–P_8–P_{10}–P_{12}–P_6 is selected).

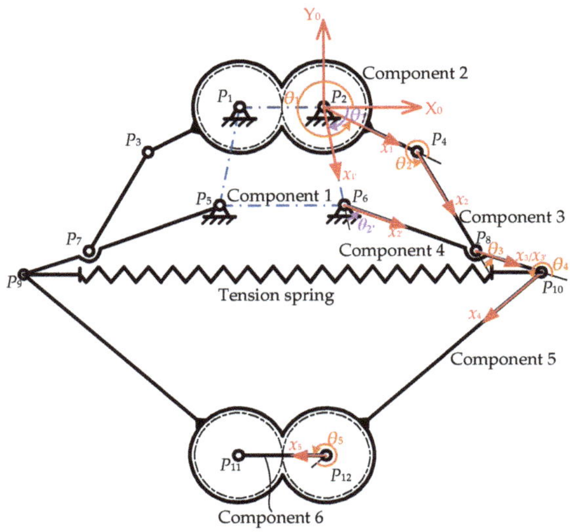

Figure 6. The kinematics model of the hopping mechanism. Where component 1 is the frame $P_1P_5P_6P_2$; component 2 is the crank–gear–bar P_2P_4; component 3 is the connecting rod P_4P_8; component 4 is the rocker $P_6P_8P_{10}$; component 5 is the gear-bar $P_{10}P_{12}$; component 6 is the connecting rod $P_{12}P_{11}$ that is connected with the rotation center of the gear–bar.

According to the Denavit–Hartenberg Method [38], the homogeneous transformation matrix from the *i*-1 coordinate system to the *i* coordinate system is:

$$ {}^{i-1}T_i = \begin{bmatrix} C\theta_i & -S\theta_i & 0 & a_{i-1} \\ C\alpha_{i-1}S\theta_i & C\alpha_{i-1}C\theta_i & -S\alpha_{i-1} & -d_iS\alpha_{i-1} \\ S\alpha_{i-1}S\theta_i & S\alpha_{i-1}C\theta_i & C\alpha_{i-1} & d_iC\alpha_{i-1} \\ 0 & 0 & 0 & 1 \end{bmatrix} \tag{4} $$

where $C\theta_i \triangleq C(\theta_i) \triangleq \cos(\theta_i)$, $S\theta_i \triangleq S(\theta_i) \triangleq \sin(\theta_i)$.

In Figure 6, the branched chain P_2–P_4–P_8–P_{10}–P_{12}–P_6 can be further subdivided into two branched chains, the main chain P_2–P_4–P_8–P_{10}–P_{12} and the secondary chain P_2–P_6–P_8–P_{10}. The specific D-H parameters of the main and secondary chain are shown in Table A1 in Appendix A.

According to Table A1 and Equation (4), the position vectors of each hinge in the main chain and the secondary chain in the fixed coordinate system are:

$$ \begin{cases} {r_{P_4}}^{\mathrm{T}} = l_{P_2P_4}\begin{bmatrix} C\theta_1 & S\theta_1 & 0 \end{bmatrix} \\ {r_{P_8}}^{\mathrm{T}} = {r_{P_4}}^{\mathrm{T}} + l_{P_4P_8}\begin{bmatrix} C(\theta_{1+2}) & S(\theta_{1+2}) & 0 \end{bmatrix} \\ {r_{P_{10}}}^{\mathrm{T}} = {r_{P_8}}^{\mathrm{T}} + l_{P_8P_{10}}\begin{bmatrix} C(\theta_{1+2+3}) & S(\theta_{1+2+3}) & 0 \end{bmatrix} \\ {r_{P_{12}}}^{\mathrm{T}} = {r_{P_{10}}}^{\mathrm{T}} + l_{P_{10}P_{12}}\begin{bmatrix} C(\theta_{1+2+3+4}) & S(\theta_{1+2+3+4}) & 0 \end{bmatrix} \\ {r'_{P_6}}^{\mathrm{T}} = l_{P_2P_6}\begin{bmatrix} C\theta_{1'} & -S\theta_{1'} & 0 \end{bmatrix} \\ {r'_{P_8}}^{\mathrm{T}} = {r'_{P_6}}^{\mathrm{T}} + l_{P_6P_8}\begin{bmatrix} C(\theta_{1'-2'}) & -S(\theta_{1'-2'}) & 0 \end{bmatrix} \end{cases} \tag{5} $$

where r_{A} is the position vector of the hinge A in the main chain; $r_{\mathrm{A}}= [\mathrm{A}_x\ \mathrm{A}_y\ \mathrm{A}_z]^{\mathrm{T}}$; r'_{A} is the position vector of the hinge A in the secondary chain; $r'_{\mathrm{A}} = [\mathrm{A}_{x'}\ \mathrm{A}_{y'}\ \mathrm{A}_{z'}]^{\mathrm{T}}$; $\theta_{i+j} = \theta_i + \theta_j$, $\theta_{i-j} = \theta_i - \theta_j$.

The hopping mechanism has the characteristics of symmetry and is combined with Equation (5); the expression for the position vector of each joint point in the branched chain P_1–P_3–P_7–P_9–P_{11}–P_5 is:

$$\begin{cases} r_{P_3}{}^{\mathrm{T}} = \begin{bmatrix} -P_{4x} - CL & P_{4y} & 0 \end{bmatrix} \\ r_{P_7}{}^{\mathrm{T}} = \begin{bmatrix} -P_{8x} - CL & P_{8y} & 0 \end{bmatrix} \\ r_{P_9}{}^{\mathrm{T}} = \begin{bmatrix} -P_{10x} - CL & P_{10y} & 0 \end{bmatrix} \\ r_{P_{11}}{}^{\mathrm{T}} = \begin{bmatrix} -P_{12x} - CL & P_{12y} & 0 \end{bmatrix} \\ r'_{P_5}{}^{\mathrm{T}} = \begin{bmatrix} -P_{6x'} - CL & P_{6y'} & 0 \end{bmatrix} \end{cases} \tag{6}$$

where $CL = l_{P_1P_2}$.

3.1.2. Establishment of Optimization

It is important and necessary to optimize the dimension of the closed-chain mechanism [39]. Because of the dual-objective optimization, the special method of TOPSIS (Technique for Order Preference by Similarity to an Ideal Solution) was chosen for dimension optimization. The same trend of evaluation indicators facilitates a reasonable compromise between the two optimization objectives. Initially proposed by Hwang and Yoon in 1981 [40–42], two vectors are constructed:

$$F^* = [f_m^*, f_n^*]^{\mathrm{T}} \tag{7}$$

$$F(\boldsymbol{X}) = [f_m(\boldsymbol{X}), f_n(\boldsymbol{X})]^{\mathrm{T}} \tag{8}$$

The vector represented by Equation (7) is regarded as an ideal point of the vector function represented by Equation (8). Then the smaller the deviation between $F(\boldsymbol{X})$ and F^*, the closer the objective function will be to the ideal point, i.e., the closer the design variable will be to the optimal solution.

The distance equation of the TOPSIS Method is:

$$F_{\mathrm{d}} = \|F(\boldsymbol{X}) - F^*\| = \left\{ \sum_{i=1}^{n} \lambda_j [f_j(x_i) - f_j{}^*]^p \right\}^{\frac{1}{p}} \tag{9}$$

where $p \subset [1, +\infty)$. When $p = 1$, it is the Hamming distance or absolute distance; when $p = 2$, it is the Euclidean distance; when $p = \infty$, it is the Chebyshev distance. In this paper, $p = 2$ was selected due to the optimization function dimension and distance characteristics.

(1) Objective function

First, to maximize the energy that is stored in the hopping mechanism, the tension spring is required to have the greatest amount of tensile deformation. Second, to make the structure more compact, the minimum longitudinal dimension of the mechanism is also required. Assuming that the sampling position in one hopping cycle is n, then the dual-objective optimization objective function can be expressed as:

$$\begin{cases} f_1(\boldsymbol{X}) = \min\left[\min\limits_{i=1}^{n}(P_{10x_i} - P_{9x_i}) - \max\limits_{i=1}^{n}(P_{10x_i} - P_{9x_i})\right] \\ f_2(\boldsymbol{X}) = \min\left[\max\limits_{i=1}^{n}(P_{2y_i} - P_{12y_i})\right] \end{cases} \tag{10}$$

where $\boldsymbol{X}$ represents the design variable matrix; P_{10x_i} is the horizontal position of P_{10} at the ith sampling position; P_{9x_i} is the horizontal position of P_9 at the ith sampling position; P_{2y_i} is the vertical position of P_2 at the ith sampling position; P_{12y_i} is the vertical position of P_{12} at the ith sampling position.

(2) Design variables

It can be seen from Equation (10) that the optimization objective is related to the coordinates of points P_9, P_{10}, and P_{12}. As can be also noticed from Section 3.1.1, the coordinates of these three points are determined by (1) the length of each linkage, (2) the angle between the line connecting the hinge points P_2 & P_6 and the X_0-axis direction, and (3) the position of the crank–gear–bar. Therefore, the optimization variables matrix $\boldsymbol{X}$ can be concluded as being:

$$\boldsymbol{X} = [x_1, x_2, x_3, x_4, x_5, x_6, x_7, x_8, x_9, x_{10}]^{\mathrm{T}} = [l_1, l_2, l_3, l_4, l_5, l_6, l_7, l_8, \beta_1, \beta_2]^{\mathrm{T}} \tag{11}$$

where $l_1 = l_{P_1P_2}$; $l_2 = l_{P_2P_4}$; $l_3 = l_{P_2P_6}$; $l_4 = l_{P_6P_8}$; $l_5 = l_{P_4P_8}$; $l_6 = l_{P_8P_{10}}$; $l_7 = l_{P_{10}P_{12}}$; $l_8 = l_{P_{12}P_{11}}$; $\beta_2 = \theta_{1'}$. β_1 is the angle between the initial position of the crank–gear–bar P_2P_4 and the positive direction of the X_0-axis.

(3) Constraint conditions

According to the conditions for the existence of a crank–rocker mechanism in the planar four-bar linkage $P_2P_4P_8P_6$, the bars l_2, l_3, l_4, and l_5 have the following linear constraints:

$$\boldsymbol{A}_1\boldsymbol{X} \leq \boldsymbol{b}_1 \tag{12}$$

Considering the limitation of the processing technology of the component and the requirement for the non-interference of installation position, the linear constraints of geometric dimensions are given as:

$$\boldsymbol{A}_2\boldsymbol{X} \leq \boldsymbol{b}_2 \tag{13}$$

According to the requirements of the overall size of the robot, design variables should be restrained, that is:

$$\begin{cases} \boldsymbol{X} \geq \boldsymbol{d}_{min} \\ \boldsymbol{X} \leq \boldsymbol{d}_{max} \end{cases} \tag{14}$$

Based on the actual processing requirements of the frame, the limitations of the original length and tensile deformation of the tension spring, the following nonlinear constraints exist:

$$\boldsymbol{\Phi}(\boldsymbol{X}) \leq G \tag{15}$$

The main chain and the secondary chain are coupled at the joint P_8. The slopes of the link P_6P_8 and P_8P_{10} are equal. This is because the relative distance between point P_{11} and P_{12} along the X_0-axis is constant, i.e., there is a constraint relationship $P_{12x} - P_{11x} = l_8$. Therefore, the closed-chain vector constraint of the hopping mechanism given is:

$$\boldsymbol{H}(\boldsymbol{X}) = 0 \tag{16}$$

The specific process of establishing the constraint conditions of the optimization model is detailed in Appendix A.

(4) Optimization Model

According to Equation (7)–(16), the dimension optimization mathematical model of the robot hopping mechanism can be obtained as follows:

$$\begin{cases} \boldsymbol{X} = [x_1, x_2, \ldots, x_{10}]^{\mathrm{T}} \\ \min\ f_1(\boldsymbol{X}); f_2(\boldsymbol{X}) \\ s.t.\ \ \boldsymbol{A}_1\boldsymbol{X} \leq \boldsymbol{b}_1;\ \ \boldsymbol{A}_2\boldsymbol{X} \leq \boldsymbol{b}_2 \\ \qquad \boldsymbol{X} \leq \boldsymbol{d}_{max};\ \ \boldsymbol{X} \geq \boldsymbol{d}_{min} \\ \qquad \boldsymbol{\Phi}(\boldsymbol{X}) \leq G;\ \ \boldsymbol{H}(\boldsymbol{X}) = 0 \end{cases} \tag{17}$$

3.1.3. Optimization Process and Results

The flow chart of the optimization process is shown in Figure 7. The program is implemented in Matlab, and the optimization results are listed in Table 2.

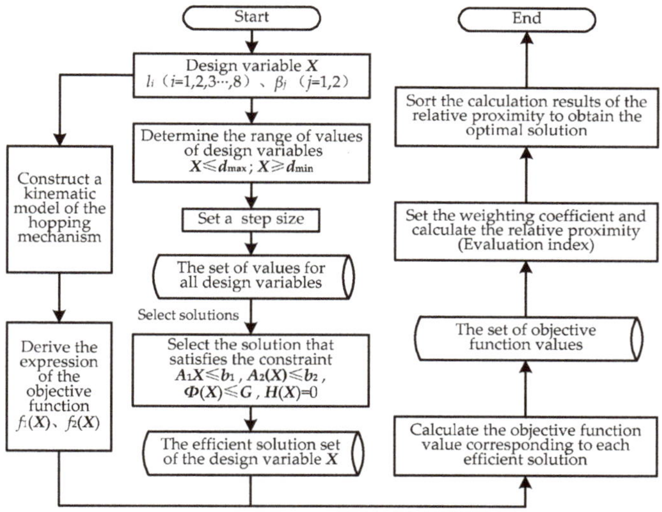

Figure 7. The flow chart of the optimization.

Table 2. Optimization results of the hopping mechanism dimensions.

Optimization Parameters	l_1/mm	l_2/mm	l_3/mm	l_4/mm	l_5/mm
Optimization Results	30.000	15.729	32.000	50.939	52.851
Optimization Parameters	l_6/mm	l_7/mm	l_8/mm	β_1/(°)	β_2/(°)
Optimization Results	34.061	75.000	25.000	276.180°	134.768°

By analyzing the data shown in Table 2, it can be proven that: (1) The optimization result satisfies the range of values of the design variables; (2) The bars l_2, l_3, l_4, and l_5 satisfy the constraints of the crank–rocker mechanism; (3) The dimension of the bars l_1, l_2, l_3 avoid interference during assembly; (4) Optimized data meets the constraints on the kinematical chain. The optimized method is used to gain a set of ideal data.

The graphical optimization result is shown in Figure 8a. To avoid interference during hopping, changes are made to the optimization result, and the solid hopping mechanism model is shown in Figure 8b.

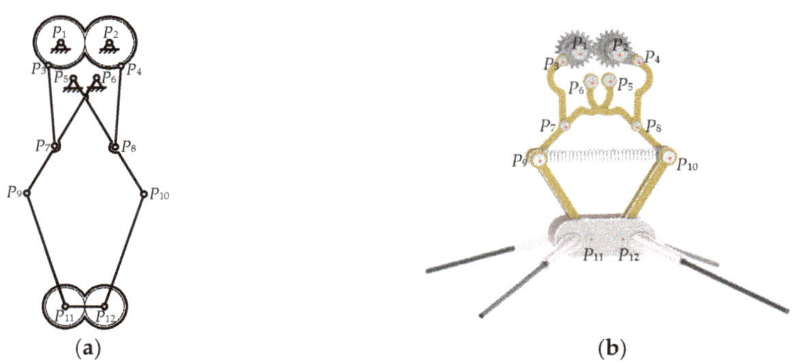

Figure 8. Hopping mechanism. (**a**) Optimization result. (**b**) Solid model.

3.2. Actuation Design

3.2.1. Theoretical Analysis

The actuator of the hopping robot has the following requirements.

(1) In the process of energy storage, the actuator needs to provide the crank–gear–bar great driving torque to realize the relative movement between point P_9 and P_{10}, thereby realizing the elongation of the tension spring.
(2) When the energy is released, the angular velocity of the crank will change dramatically within a very short period. Therefore, it is necessary to reduce the resistance from the motor.
(3) When the hopping robot lands, it is necessary to absorb the landing shock to reduce the damage to the motor.

Therefore, a clockwork spring that is capable of storing and releasing angular energy is introduced between the output shaft of the motor and the crank–gear–bar as an auxiliary energy storage element, as shown in Figure 9. There are three advantages of the proposed special actuator.

(1) The input torque of the drive motor is amplified by the clockwork spring, thereby reducing the requirement for the motor output power. At the same time, the great torque provided by the actuator can be released in a very short time.
(2) The great torque releasing speed mentioned in (1) can drive the crank–gear–bar to cross the limit position of the hopping mechanism quickly, thereby realizing the explosive hop of the robot.
(3) The clockwork spring also absorbs the impact of hopping on the motor.

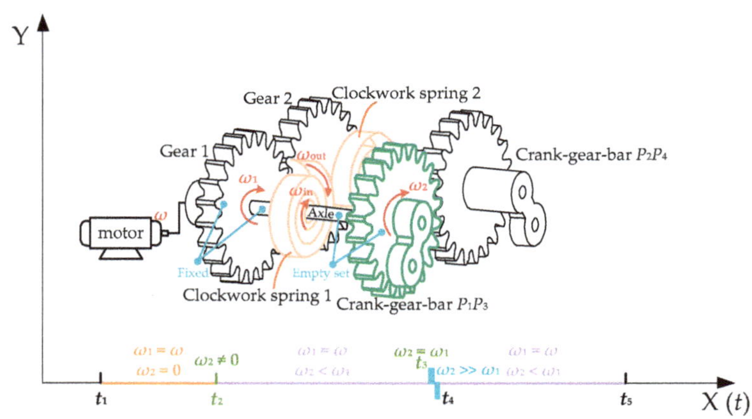

Figure 9. Energy-stored type actuator. The clockwork spring is introduced between the motor output shaft and the crank-gear-bar as an auxiliary energy storage element.

The crank–gear–bar P_1P_3 and P_2P_4 are an empty set in the shaft, the gear 1 and the gear 2 are fixed at the other end of the two shafts respectively, and mesh with each other. Gear 1 is driven by the motor, and the inner ring of the clockwork spring 1 and 2 are fixed on gear 1 and 2 respectively, which means $\omega_{in} = \omega_1$. The outer rings are fixed on crank–gear–bar P_1P_3 and P_2P_4 respectively, which means $\omega_{out} = \omega_2$. When the motor continuously rotates in the clockwise direction, the angular velocity of the motor is ω. The relationship between the actuator and the active component is shown in Figure 9.

(1) During the period t_1–t_2, $\omega_1 = \omega$ and $\omega_2 = 0$. At this phase, the inner and outer ring of the clockwork spring generates a difference in speed, and the clockwork spring begins to store energy.

(2) At the moment t_2, the angular velocity of the crank–gear–bar changes, i.e., $\omega_2 \neq 0$. During the period $t_2 - t_3$, $\omega_2 \neq 0$ but $\omega_2 < \omega_1$. There is still a difference in the speed between the inner and outer rings of the clockwork spring, which means that the spring can still store energy, but the energy storage speed slows down.
(3) At the moment t_3, the clockwork spring rotates to its limit position, the angular velocity is $\omega_2 = \omega_1$. The inner and outer rings of the clockwork spring rotate at the same speed, which means that the process of energy storage of the clockwork spring is over.
(4) In a very short time from $t_3 - t_4$, the energy stored in the clockwork spring is released, and the angular velocity is $\omega_2 >> \omega_1$.
(5) During the period $t_4 - t_5$, the angular velocity is $\omega_2 < \omega_1$. When the time reaches t_4, the next hopping cycle begins.

3.2.2. Actuator Verification

To explore the influence of the clockwork spring on the hopping performance of the robot, as well as the above theoretical analysis, dynamics simulations were performed for the hopping robot HR-A (with special actuator) and hopping robot HR-B (only with the rigid motor).

Figure 10 shows the deformation velocity of the tension spring and the angular velocity of the crank–gear–bar P_1P_3 of the HR-A. The X-axis is the simulation time. The blue dashed line represents the deformation velocity of the tension spring (corresponding to the right Y-axis), and the red solid line represents the angular velocity of the P_1P_3 (corresponding to the left Y-axis).

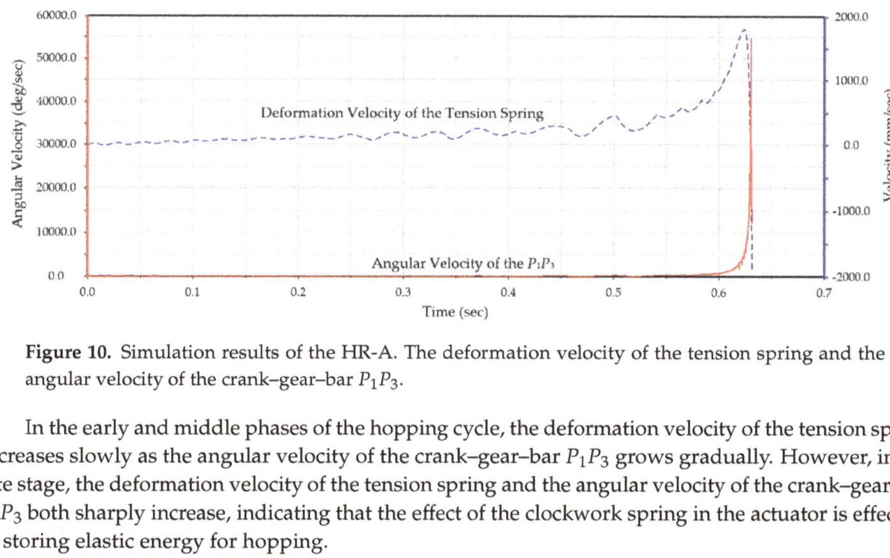

Figure 10. Simulation results of the HR-A. The deformation velocity of the tension spring and the angular velocity of the crank–gear–bar P_1P_3.

In the early and middle phases of the hopping cycle, the deformation velocity of the tension spring increases slowly as the angular velocity of the crank–gear–bar P_1P_3 grows gradually. However, in the late stage, the deformation velocity of the tension spring and the angular velocity of the crank–gear–bar P_1P_3 both sharply increase, indicating that the effect of the clockwork spring in the actuator is effective in storing elastic energy for hopping.

To further investigate the performance of the proposed actuator, the deformation velocities of the tension springs in HR-A and HR-B are compared, as shown in Figure 11a. Where the blue dashed line represents the deformation velocity of the tension spring in HR-A, and the red solid line represents the deformation velocity of the tension spring in HR-B. For HR-A, the deformation velocity of its tension spring increases slowly before t = 0.521 s. After this time, the energy of the clockwork spring is released, then the deformation speed of the tension spring increases rapidly, reducing the energy storage time to 0.63 s. However, the deformation velocity of tension spring in HR-B only grows and decreases steadily in its whole energy storage phase, which reaches the length of 0.704 s. Key positions in the simulation of HR-A and HR-B are compared in Figure 11b.

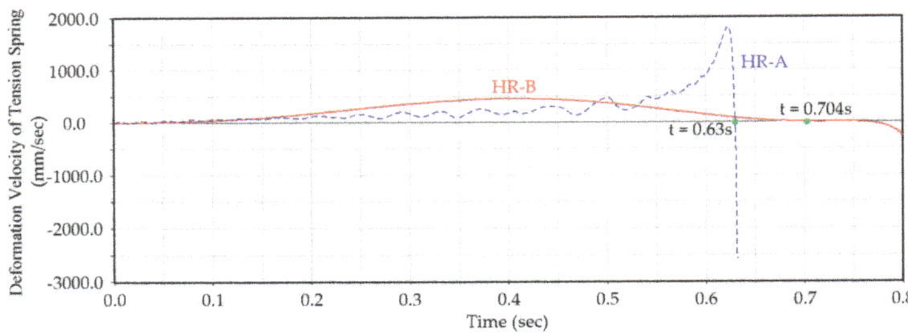

(**a**)

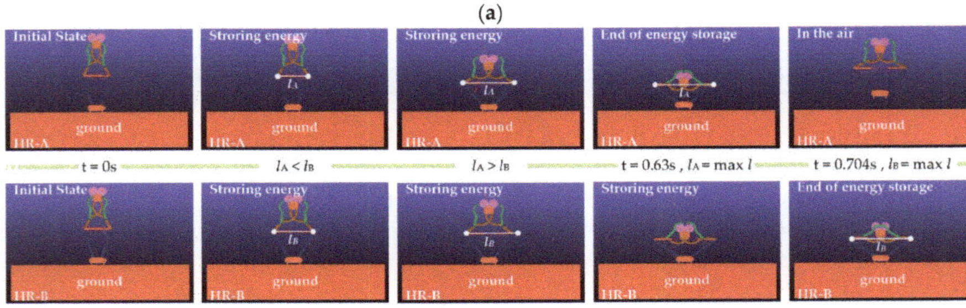

(**b**)

Figure 11. Comparison of simulation results for HR-A and HR-B. (**a**) The deformation velocity of the tension spring. (**b**) Key positions in the simulation. l_i means the length of the tension spring.

4. Prototype Implementation and Experiment

4.1. Prototype Design

The prototype of RHop is shown in Figure 12. The main structural parts are manufactured by 3D printing, and the axles are made of carbon fiber rods to reduce the overall weight of the RHop. The driving motor adopts a servo that can rotate around the whole circumference, with an output driving torque of 0.22 N·m. The prototype size is 252 mm × 155 mm × 85 mm, and its weight is about 560 g (including batteries, motor, etc.).

(**a**)

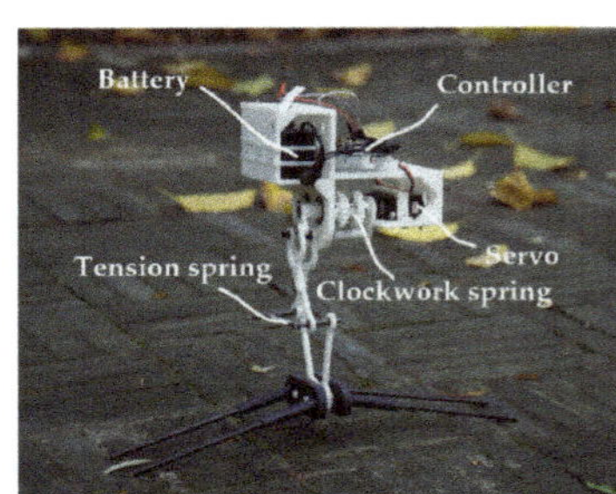

(**b**)

Figure 12. Robotic prototype. (**a**) Front view. (**b**) Side view.

4.2. Hopping Performance Analysis

4.2.1. Foot Trajectory Verification

By substituting the optimization results of the hopping mechanism into Equations (5) and (6), the movement posture of the hopping mechanism in the energy storage phase can be obtained, as shown in Figure 13a. Point *A* of the connecting link $P_{11}P_{12}$ was selected as the reference point to obtain the foot trajectory curve.

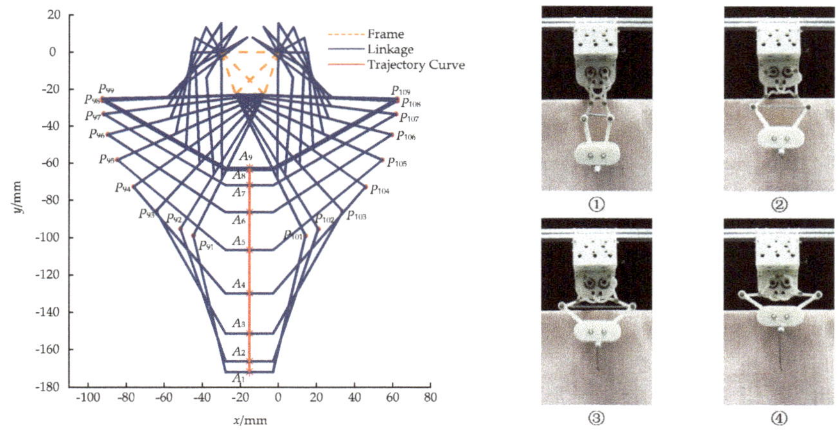

(a) (b)

Figure 13. (**a**) Theoretical movement posture of the hopping mechanism at the energy storage phase. (**b**) Experimental process diagram of trajectory reproduction at energy storage phase. ① represents the initial position; ② and ③ represents the general position; ④ represents the end position.

As shown in Figure 13a, the theoretical foot trajectory is a straight line perpendicular to the X-axis, and meets the requirements of DR.3–DR.6. In addition, the stretching deformation of the optimized hopping mechanism during the energy storage phase can reach Δl = 96 mm, which is longer than the height of the whole hopping robot RHop.

To obtain the foot trajectory of the prototype, the body of the hopping robot was fixed. A paint pen was installed at the center of the foot of the robot, thereby enabling the trajectory to be drawn on the scale paper. In the scale paper, the horizontal and vertical distance of each grid was 1 mm. Figure 13b shows the experimental process.

The comparison of the theoretical and the experimental foot trajectories are depicted in Figure 14. Specific comparison results are listed in Table 3.

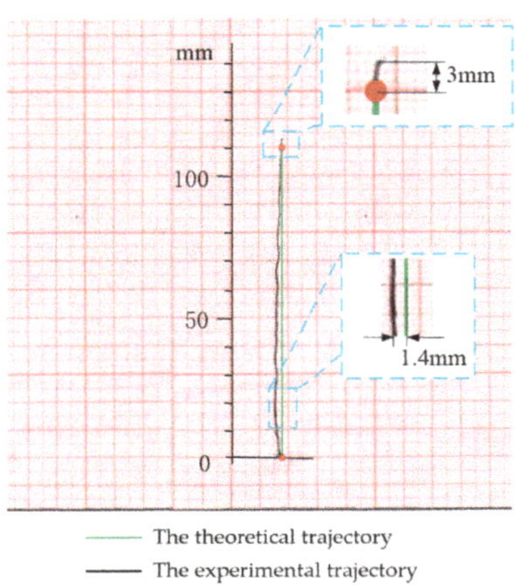

Figure 14. Comparison of the foot trajectory between theory and experiment.

Table 3. Trajectory sampling analysis results.

Horizontal Deviation	Vertical Deviation	Average Deviation	Root Mean Squared Error
1.4 mm	3 mm	0.7345 mm	0.8543 mm

It can be seen that the foot trajectory obtained from the experiment is an approximate straight line perpendicular to the x-axis, which differs by about 3 mm from the theoretical trajectory in the vertical direction, and the maximum horizontal deviation of the trajectory is about 1.4 mm.

The main reasons for the deviation are concluded as follows:

(1) Insufficient accuracy of 3D printing manufacture, which makes it hard to achieve a completely symmetrical structure during assembly.
(2) Vibration in the experiment.

4.2.2. Hopping Performance Verification

To verify the correctness of the theory of the mechanism and the actuator design, as well as the feasibility of using a single motor with continuous circumferential rotation in one direction to achieve continuous hop, the prototype was mounted on a vertical slide rail to constrain its lateral deflection for the avoidance of the negative effects of unsteady landing and taking-off in continuous hopping (The video file titled "Experimental Video-RHop.mp4", which can be found in the Supplementary Materials).

Two prototypes were tested in the experiment, i.e., (a) the prototype equipped with the special actuator, and (b) the prototype, just equipped with a rigid motor.

Tension springs with different stiffness were tested for whether they could achieve energy storage. When the stiffness of the tension spring was less than or equal to 398.3 N/m, both prototypes could achieve the energy storage and hopping. When the stiffness of the tension spring was greater than 455.2 N/m, neither type of prototype could achieve energy storage and hop; When the stiffness was within the range of aforementioned limit values, only the first prototype could achieve the energy storage and hopping. This proves that the robot that equipped with the special actuator could amplify its output torque with the same motor, thereby improving the hopping performance. Figure 15 shows the experimental process of the two prototypes using k = 398.3 N/m tension springs. The energy storage time required for the first prototype was 0.871 s, and it could hop up to 129 mm. However, the second prototype needed 1.236 s and could just hop up to 120 mm. Let the ratio γ between the gravitational potential energy at the highest point of the robot E_p and the elastic potential energy of the tension spring E_{ps} be defined as the energy conversion efficiency of the robotic hopping mechanism. Then, the energy equation of two prototype can be measured as E_{ps} = 1.392 J, E_{p1} = 1.062 J and E_{p2} = 0.988 J, i.e., the energy conversion efficiency of the first prototype is γ_1 = 76.29%, and the second one is γ_2 = 70.98%, which proves that the first prototype can greatly reduce its energy storage time and improves its energy efficiency by 5.31%. The main reasons for the loss of elastic potential energy are the friction at the hinged joint and the friction of the slide rail.

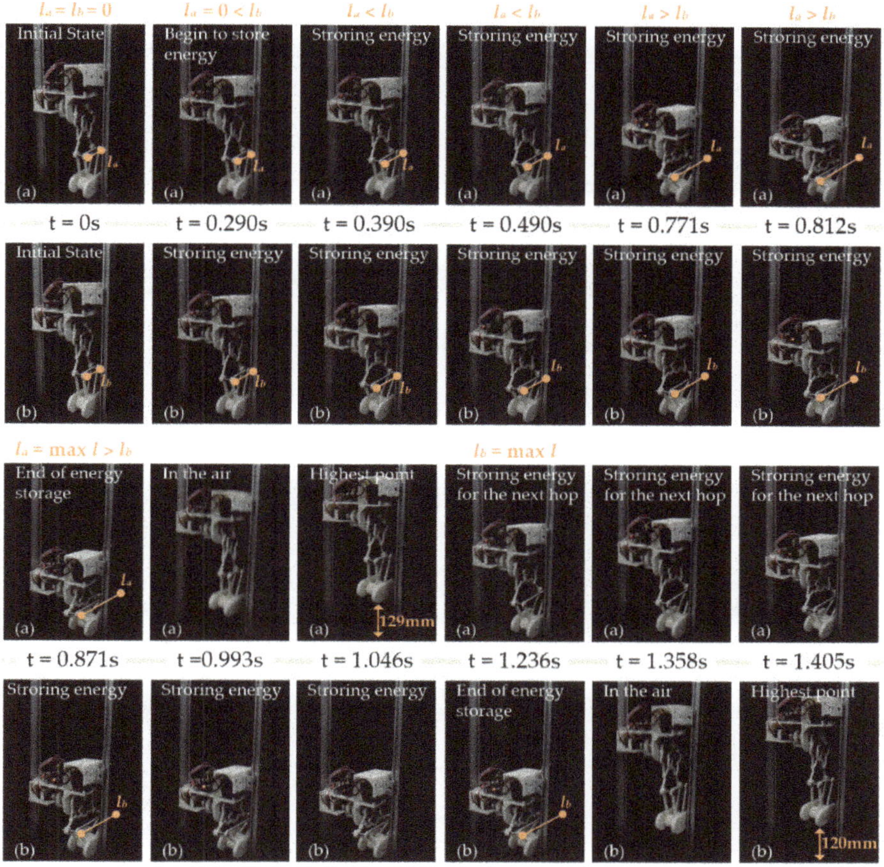

Figure 15. Experimental results comparison for different driving modes. (**a**) The first prototype equipped with a special actuator; (**b**) The second prototype equipped with only a rigid motor. l_i means the length of the tension spring.

4.2.3. Continuous Hopping Verification

Figure 16 shows the experimental process of first five continuous hopping cycles within a one hundred hop continuous hopping test. During the experiment, the driving motor rotated continuously in one direction with a constant speed, and the robot hopped repeatedly in the vertical direction along the slide rail until the motor stopped. The time required for the hopping mechanism to complete the first energy storage was about 0.663 s, and the time required for the second and subsequent energy storage was about 0.373 s. The main reason for this is that: after the robot completed the first hopping, each subsequent hopping motion could absorb the partial landing impact energy of the previous hopping, thereby shortening the energy storage time of the next hopping and improving the hopping efficiency and energy utilization.

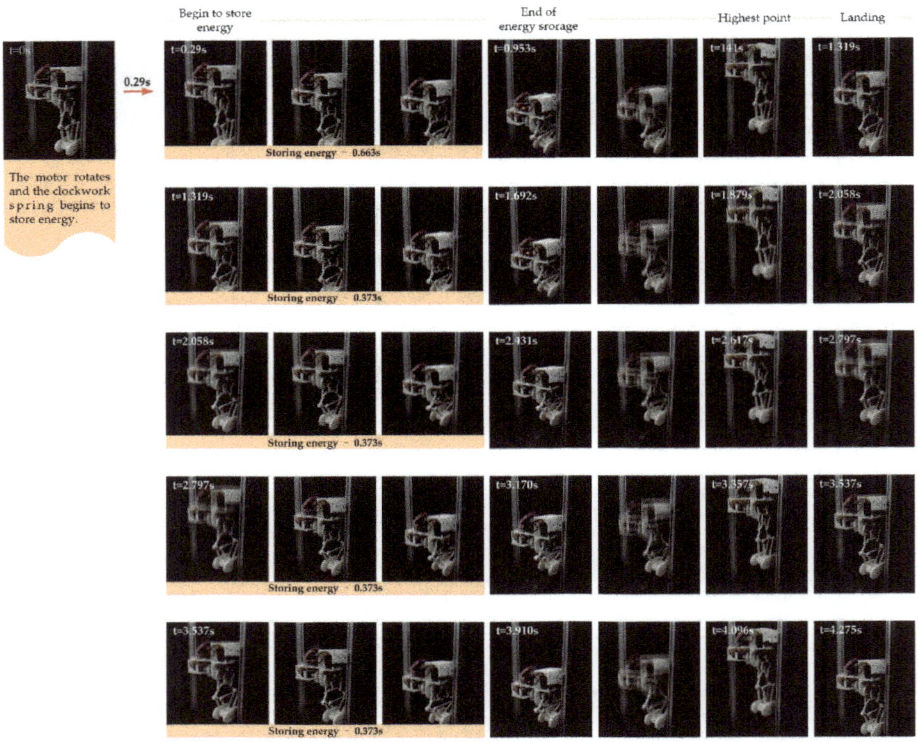

Figure 16. Experimental process of hopping performance test.

4.3. Comparison With Other Robots

Since the hopping height and the hopping distance differs when the take-off angle of the hopping robot changes, the hopping performance of some state-of-the-art robots were normalized, along with their hopping heights and hopping distances. The specific derivation process of the normalization method is detailed in Appendix B, and the comparison results are shown in Table 4.

As shown in Table 4, RHop performed best in the energy storage speed, which means that when the hopping height is consistent (indicating that the robot is flying in the air for the same time), RHop will complete a hop with less time than other hopping robots (standing on the ground for a shorter time), thereby enhancing its mobility.

Among the existing electrically driven hopping robots, one hopping robot is very special, it is called Salto [3,43,44]. It is driven in the same way as this paper, using more than just a rigid motor. The difference is that the elastic element used in this paper is a clockwork spring, and it uses a solid section of latex, which is also its energy storage component. Using an eight-bar mechanism as a limb with an advanced drive strategy (SEA+MA), Salto has good vertical jumping agility, and can achieve extremely high hops with extremely short standing times. The Salto-1P also adds an attitude control scheme [45,46]. In contrast, the hopping ability of RHop is not as good as that of Salto-1P, but meets both of the requirements mentioned in the paper for the foot trajectory curve and the minimization of the moment acting on the robot's CM when the robot component moves. Salto-1P uses design exploration, combined with kinematic tuning to obtain. Due to the symmetrical nature of its mechanism, RHop does not require any calculations to achieve these requirements, i.e., the exploration process of Salto-1P is more complicated. Additionally, although the control strategy of Salto-1P is advanced, it is still difficult to control complexity and mechanism coupling. RHop adopts the control mode of continuous circumferential rotation in one direction with constant speed, which is simpler and more operable.

Table 4. Comparison of hopping performance of existing electrically driven robots.

Robot Name	Normalized Hopping Height (m)	Energy Storage Time (s)	Energy Storage Speed (m/s)
Flea-inspired Jumping Robot [47]	0.688	19	0.036
A Jumping Robot [48]	1.071	60	0.018
TAUB [17]	3.335	18.4	0.181
Grillo III [22]	0.125	12.5	0.010
MSU Jumper [31]	0.930	10	0.093
A Bio-inspired Jumping Robot [12]	1.026	60	0.017
A Surveillance Robot [15]	0.410	7.2	0.057
An Integrated Jumping-Crawling Robot [10]	2.900	28	0.104
RHop	**0.145**	**0.373**	**0.389**

5. Conclusion

This paper describes the design and experiment of RHop, a miniaturized continuous hopping robot. Using mechanical evolution, the hopping mechanism in RHop is realized by a geared multi-bar mechanism without an additional locking and releasing mechanism. It is driven by a single motor with continuous circumferential rotation in one direction. It satisfies the proposed DR.1–DR.4, and the optimization results also fulfill the DR.5 and DR.6. The special actuator designed in this paper can amplify the motor torque and reduce the energy storage time, thereby enhancing the overall hopping performance of RHop. Comparative simulations and experiments are conducted for RHop. As the experiments show, the theoretical model, the simulation model, and the prototype are in approximate agreement with each other. In a single hopping, the energy conversion efficiency of RHop reaches a high value of 76.29%. In the continuous hopping experiment, it is proven that a single motor with continuous circumferential rotation in one direction to achieve continuous hopping is feasible. Furthermore, RHop has a good energy storage speed when compared with other state-of-the-art hopping robots.

Future efforts to improve the performance of RHop will include the power matching between the motor and the elastic element, and the coupling and modulating between the special actuator and the tension spring. At the same time, the hopping mechanism that satisfies the design requirements presented in this paper is not unique. The advantages of other combination mechanisms and different design methods can be used to obtain new mechanisms. The design thoughts presented in this paper are still applicable to the exploration of other mechanisms, which will be referenced by other researchers. Additionally, it is also possible to apply the prototype that is developed in this paper to other mobile robots to study multi-mobile robots; particularly, combining this with wheeled movement to improve the obstacle performance could be insightful in improving the environmental adaptability and mobility of the robot.

Supplementary Materials: The following are available online at http://www.mdpi.com/2076-3417/9/1/13/s1.

Author Contributions: Conceptualization, L.B., F.Z. and X.C.; Methodology, L.B. and F.Z.; Software, F.Z.; Validation, F.Z. and X.C.; Formal Analysis, L.B. and F.Z.; Investigation, X.C. and J.H.; Resources, L.B. and Y.S.; Data Curation, F.Z. and X.C.; Writing—Original Draft Preparation, L.B.; Writing—Review & Editing, L.B., F.Z., X.C., and Y.S.; Supervision, L.B..; Project Administration, L.B.

Funding: This research was funded by the Foundation for the Sci & Tech Research Project of Chongqing Science & Technology Commission (grant no. cstc2016jcyjA0472, cstc2017zdcy-zdzxX0007, and cstc2015jcyjA70002), the Special Cooperation Program for Higher Education Institutions Collaborative Innovation (grant no. KH2016006), the National Natural Science Foundation of China (grant no. 51505044, 51705050, and 51709023), and the Fundamental Research Funds for the Central Universities (Grant No. 2018CDGFJX0022, 2018CDQYHK0029).

Conflicts of Interest: The authors declare no conflict of interest.

Appendix A

Appendix A.1 D-H Parameter Table

The D-H parameter table of the main and secondary chain mentioned in Section 3.1.1 is shown in the following table.

Table A1. D-H parameter table of the main and secondary chain.

Coordinate System (i)	θ_i	d_i	a_i	α_i
0	-	-	0	0
1	θ_1	0	$l_{P_2P_4}$	0
2	θ_2	0	$l_{P_4P_8}$	0
3	θ_3	0	$l_{P_8P_{10}}$	0
4	θ_4	0	$l_{P_{10}P_{12}}$	0
5	θ_5	0	-	-
$1'$	$2\pi - \theta_{1'}$	0	$l_{P_2P_6}$	0
$2'$	$\theta_{2'}$	0	$l_{P_6P_8}$	0
$3'$	0	0	-	-

Appendix A.2 Constraint Conditions

1. Constraint conditions of the crank–rocker mechanism

According to the conditions for the existence of a crank–rocker mechanism in the planar four-bar linkage $P_2P_4P_8P_6$, the bars l_2, l_3, l_4, and l_5 have the following six linear constraints:

$$\begin{cases} l_2 - l_i \le 0 \ (i = 3,4,5) \\ l_2 + l_3 - l_4 - l_5 \le 0 \\ l_2 + l_4 - l_3 - l_5 \le 0 \\ l_2 + l_5 - l_3 - l_4 \le 0 \end{cases} \quad \text{(A1)}$$

$\boldsymbol{X}$ represents the 10 dimensions of the optimization design variable, and the 6×10 dimensions matrix $\boldsymbol{A_1}$ represents the coefficient matrix of the constraint Equation (A1). The six dimensions of vector $\boldsymbol{b_1}$ represent the constraint vector of Equation (A1), and so Equation (A1) can be changed into:

$$\boldsymbol{A}_1\boldsymbol{X} \le \boldsymbol{b}_1 \quad \text{(A2)}$$

2. Linear constraints of geometric dimensions

According to the requirements of the overall size of the robot, design variables should be restrained, that is:

$$\begin{cases} l_{i\min} \le l_i \le l_{i\max} \ (i = 1,2,3,\ldots,8) \\ \beta_{j\min} \le \beta_j \le \beta_{j\max} \ (j = 1,2) \end{cases} \quad \text{(A3)}$$

The 10 dimensions of vector $\boldsymbol{d_{\min}}$ represent the lower-bound matrix of the design variable; the 10 dimensions of vector $\boldsymbol{d_{\max}}$ represent the upper-bound matrix of the design variable, then Equation (A3) can be changed into:

$$\begin{cases} \boldsymbol{X} \ge \boldsymbol{d_{min}} \\ \boldsymbol{X} \le \boldsymbol{d_{max}} \end{cases} \quad \text{(A4)}$$

Considering the limitations of processing technology of the component, and the requirement for non-interference of the installation position, the linear constraints of the geometric dimensions are given as:

$$\begin{cases} l_2 - l_1 \le d_1 \\ l_2 - l_3 \le d_2 \end{cases} \quad \text{(A5)}$$

The 2 × 10 dimensional matrix $\boldsymbol{A_2}$ represents the coefficient matrix of the constraint Equation (A5). The two dimensional vector $\boldsymbol{b_2}$ represents the constraint vector of Equation (A5), so that Equation (A5) can be changed into:

$$A_2 X \leq b_2 \tag{A6}$$

3. Non-linear constraints of geometric dimensions

Based on the actual processing requirements of the frame, the limitations of the original length, and the tensile deformation of the tension spring, the following nonlinear constraints exist:

$$\boldsymbol{\Phi}(\boldsymbol{X}) \leq \boldsymbol{G} \tag{A7}$$

where $\boldsymbol{\Phi}(\boldsymbol{X})$ is a 3-dimensional function vector and $\boldsymbol{G}$ is a 3-dimensional constant vector.

$$\boldsymbol{\Phi}(\boldsymbol{X}) = \begin{pmatrix} \varphi_1(\boldsymbol{X}) \\ \varphi_2(\boldsymbol{X}) \\ \varphi_3(\boldsymbol{X}) \end{pmatrix} = \begin{pmatrix} P_{6x'} \\ P_{9x} - P_{10x} \\ P_{6y'} - P_{9y} \end{pmatrix} \quad \boldsymbol{G} = \begin{pmatrix} g_1 \\ g_2 \\ g_3 \end{pmatrix} = \begin{pmatrix} d_3 \\ d_4 \\ d_5 \end{pmatrix}$$

4. Constraints on the kinematical chain

The main chain and the secondary chain are coupled at the joint P_8, which means:

$$\begin{cases} h_1(\boldsymbol{X}) = l_2 C\theta_1 + l_5 C(\theta_{1+2}) - l_3 C\theta_{1'} - l_4 C(\theta_{1'-2'}) = 0 \\ h_2(\boldsymbol{X}) = l_2 S\theta_1 + l_5 S(\theta_{1+2}) + l_3 S\theta_{1'} + l_4 S(\theta_{1'-2'}) = 0 \end{cases} \tag{A8}$$

Point P_6, P_8, and P_{10} form a link with three elements, i.e., the slope of the link P_6P_8 and P_8P_{10} are equal:

$$h_3(\boldsymbol{X}) = \frac{S(\theta_{1+2+3})}{C(\theta_{1+2+3})} + \frac{S(\theta_{1'-2'})}{C(\theta_{1'-2'})} = 0 \tag{A9}$$

Since the relative distance between point P_{11} and P_{12} along the X_0-axis is constant, i.e., there is a constraint relationship $P_{12x} - P_{11x} = l_8$, this means:

$$h_4(\boldsymbol{X}) = 2[l_2 C_1 + l_5 C_2 + l_6 C_3 + l_7 C_4] + \frac{1}{2} l_1 - l_8 \tag{A10}$$

where $C_1 = C(\theta_1)$, $C_2 = C(\theta_{1+2})$, $C_3 = C(\theta_{1+2+3})$, $C_4 = C(\theta_{1+2+3+4})$.

Integrating Equation (A8)–(A10), the closed-chain vector constraint of the hopping mechanism is:

$$\boldsymbol{H}(\boldsymbol{X}) = \begin{pmatrix} h_1(\boldsymbol{X}) \\ h_2(\boldsymbol{X}) \\ h_3(\boldsymbol{X}) \\ h_4(\boldsymbol{X}) \end{pmatrix} = 0 \tag{A11}$$

Appendix B

When the take-off angle of a robot is not 90°, let the origin of the coordinate system be at the take-off point, and let the x-axis be along the horizontal direction and the y-axis along the vertical direction. Then, the robot's hopping trajectory can be described as:

$$\begin{cases} x = v_0 t \cos\theta \\ y = v_0 t \sin\theta - \frac{1}{2} g t^2 \end{cases} \tag{A12}$$

where v_0 is the take-off speed, t is the hopping time, and θ is the take-off angle.

The hopping height h and hopping distance d of the robot can be obtained from Equation (A12):

$$\begin{cases} h = \frac{v_0^2 \sin^2 \theta}{2g} \\ d = \frac{v_0^2 \sin 2\theta}{g} \end{cases} \tag{A13}$$

The normalized hopping height with a take-off angle of 90° is:

$$\begin{cases} \theta = \arctan\left(\frac{4h}{d}\right)(\theta \neq 90^\circ) \\ H = \frac{h}{\sin^2 \theta} \end{cases} \tag{A14}$$

Thus the energy storage speed is:

$$v_s = \frac{H}{T} \tag{A15}$$

where T is the time of the energy storage phase when the robot completes a hop.

According to Equations (A14) and (A15), combined with the hopping height, hopping distance, and energy storage time of the compared robot, the normalized hopping height and energy storage speed of the robot can be calculated, as shown in Table A2.

Table A2. Specific calculation results.

Robot Name	Hopping Height (m)	Hopping Distance (m)	Normalized Hopping Height (m)	Energy Storage Time (s)	Energy Storage Speed (m/s)
Flea-inspired Jumping Robot [47]	0.640	0.700	0.688	19	0.036
A Jumping Robot [48]	1.050	0.600	1.071	60	0.018
TAUB [17]	3.130	3.200	3.335	18.4	0.181
Grillo III [22]	0.100	0.200	0.125	12.5	0.010
MSU Jumper [31]	0.872	0.898	0.930	10	0.093
A Bio-inspired Jumping Robot [12]	1.000	0.650	1.026	60	0.017
A Surveillance Robot [15]	0.410	-	0.410	7.2	0.057
An Integrated Jumping–Crawling Robot [10]	2.900	-	2.900	28	0.104
RHop	**0.145**	**-**	**0.145**	**0.373**	**0.389**

References

1. Yoshikawa, K.; Otsuki, M.; Kubota, T.; Maeda, T.; Ushijima, M.; Watanabe, S.; Sakamoto, K.; Kunii, Y.; Umeda, K. A new mechanism of smart jumping robot for lunar or planetary satellites exploration. In Proceedings of the 2017 IEEE on Aerospace Conference, Big Sky, MT, USA, 4–11 March 2017; pp. 1–9.
2. Zhang, Z.; Zhao, J.; Chen, H.; Chen, D. A survey of bioinspired jumping robot: Takeoff, air posture adjustment, and landing buffer. *Appl. Bionics Biomech.* **2017**, *2017*, 1–22. [CrossRef]
3. Plecnik, M.M.; Haldane, D.W.; Yim, J.K.; Fearing, R.S. Design exploration and kinematic tuning of a power modulating jumping monopod. *J. Mech. Robot.* **2017**, *9*, 011009. [CrossRef]
4. Zhu, Y.; Chen, L.; Liu, Q.; Qin, R.; Jin, B. omnidirectional jump of a legged robot based on the behavior mechanism of a jumping spider. *Appl. Sci.* **2018**, *8*, 51. [CrossRef]
5. Ikeda, H.; Kawabe, T.; Wada, R.; Sato, K. Step-climbing tactics using a mobile robot pushing a hand cart. *Appl. Sci.* **2018**, *8*, 2114. [CrossRef]
6. Gart, S.W.; Li, C. Body-terrain interaction affects large bump traversal of insects and legged robots. *Bioinspir. Biomim.* **2018**, *13*, 026005. [CrossRef]
7. Ducros, C.; Hauser, G.; Mahjoubi, N.; Girones, P.; Boisset, L.; Sorin, A.; Jonquet, E.; Falciola, J.M.; Benhamou, A. RICA: A tracked robot for sampling and radiological characterization in the nuclear field. *J. Field Robot.* **2017**, *34*, 583–599. [CrossRef]
8. Seifert, H.S. The lunar pogo stick. *J. SPACECRAFT* **1967**, *4*, 941–943. [CrossRef]
9. Raibert, M.H.; Tello, E.R. *Legged Robots That Balance*; MIT Press: Cambridge, MA, USA, 1986; p. 89.
10. Burdick, J.; Fiorini, P. Minimalist jumping robots for celestial exploration. *Int. J. Robot. Res.* **2003**, *22*, 653–674. [CrossRef]

11. Dubowsky, S.; Kesner, S.; Plante, J.-S.; Boston, P. Hopping mobility concept for search and rescue robots. *Ind. Robot.* **2008**, *35*, 238–245. [CrossRef]
12. Wiedebach, G.; Bertrand, S.; Wu, T.; Fiorio, L.; McCrory, S.; Griffin, R.; Nori, F.; Pratt, J. Walking on partial footholds including line contacts with the humanoid robot atlas. In Proceedings of the 2016 IEEE-RAS 16th International Conference on Humanoid Robots (Humanoids), Cancun, Mexico, 15–17 November 2016; pp. 1312–1319.
13. Yan, H.; Li, H.; Zhou, S. Study on hopping height control and detection for the pneumatic actuator. In Proceedings of the 5th International Conference on Electrical Engineering and Automatic Control, Harbin Inst Technol, Weihai, China, 16–18 October 2015; pp. 1121–1128.
14. Graichen, K.; Hentzelt, S.; Hildebrandt, A.; Kärcher, N.; Gaißert, N.; Knubben, E. Control design for a bionic kangaroo. *Control Eng. Practice* **2015**, *42*, 106–117. [CrossRef]
15. Long, B.; Wenjie, G.; Xiaohong, C.; Qian, T.; Rong, X. Landing impact analysis of a bioinspired intermittent hopping robot with consideration of friction. *Math. Probl. Eng.* **2015**, *2015*.
16. Long, B.; Wenjie, G.; Xiaohong, C.; Meng, X.-y. Hopping capabilities of a bio-inspired and mininally actuated hopping robot. In Proceedings of the 2011 International Conference on Electronics, Communications and Control (ICECC), Ningbo, China, 9–11 September 2011; pp. 1485–1489.
17. Jung, G.-P.; Casarez, C.S.; Jung, S.-P.; Fearing, R.S.; Cho, K.-J. An integrated Jumping-crawling robot using height-adjustable jumping module. In Proceedings of the 2016 IEEE International Conference on Robotics and Automation (ICRA), Stockholm, Sweden, 16–21 May 2016; pp. 4680–4685.
18. Zaitsev, V.; Gvirsman, O.; Hanan, U.B.; Weiss, A.; Ayali, A.; Kosa, G. A locust-inspired miniature jumping robot. *Bioinspir. Biomim.* **2015**, *10*, 066012. [CrossRef]
19. Jun, B.; Kim, Y.; Jung, S. Design and control of jumping mechanism for a kangaroo-inspired robot. In Proceedings of the 2016 6th IEEE International Conference on Biomedical Robotics and Biomechatronics (BioRob), UTown, Singapore, 26–29 June 2016; pp. 436–440.
20. Nguyen, Q.-V.; Park, H.C. Design and demonstration of a locust-like jumping mechanism for small-scale robots. *J. Bionic Eng.* **2012**, *9*, 271–281. [CrossRef]
21. Zhang, J.; Song, G.; Li, Y.; Qiao, G.; Song, A.; Wang, A. A bio-inspired jumping robot: Modeling, simulation, design, and experimental results. *Mechatronics* **2013**, *23*, 1123–1140. [CrossRef]
22. Li, F.; Liu, W.; Fu, X.; Bonsignori, G.; Scarfogliero, U.; Stefanini, C.; Dario, P. Jumping like an insect: Design and dynamic optimization of a jumping mini robot based on bio-mimetic inspiration. *Mechatronics* **2012**, *22*, 167–176. [CrossRef]
23. Song, G.; Yin, K.; Zhou, Y.; Cheng, X. A surveillance robot with hopping capabilities for home security. *IEEE Trans. Consum. Electron.* **2009**, *55*, 2034–2039. [CrossRef]
24. Wang, H.; Song, G.; Zhang, J.; Meng, T. A Bio-inspired Jumping Robot for Mobile Sensor Networks over Rough Terrain. In *Future Communication, Computing, Control and Management*; Springer: New York, NY, USA, 2012; pp. 57–62.
25. Lambrecht, B.G.A.; Horchler, A.D.; Quinn, R.D. A small, insect-inspired robot that runs and jumps. In Proceedings of the 2005 IEEE International Conference on Robotics and Automation, Barcelona, Spain, 18–22 April 2005; pp. 1240–1245.
26. Sun, Y.; Ge, W.; Zheng, J.; Dong, D. Design and evaluation of a prosthetic knee joint using the geared five-bar mechanism. *IEEE Trans. Neural Syst. Rehabil. Eng.* **2015**, *23*, 1031–1038. [CrossRef]
27. Sun, Y.; Ge, W.; Zheng, J.; Xia, F.; Dong, D. Optimization of actuating torques in multi-bar prosthetic joints with springs. *Eng. Optimiz.* **2016**, *49*, 1183–1196. [CrossRef]
28. Zhao, J.; Xi, N.; Gao, B.; Mutka, M.W.; Xiao, L. Design and testing of a controllable miniature jumping robot. In Proceedings of the 2010 IEEE/RSJ International Conference on Intelligent Robots and Systems, Taipei, Taiwan, 18–22 October 2010; pp. 3346–3351.
29. Faraji, H.; Tachella, R.; Hatton, R.L. Aiming and vaulting: Spider inspired leaping for jumping robots. In Proceedings of the 2016 IEEE International Conference on Robotics and Automation (ICRA), Stockholm, Sweden, 16–21 May 2016; pp. 2082–2087.
30. Chan, C.Y.; Liu, Y.C. Towards a walking, turning, and jumping quadruped robot with compliant mechanisms. In Proceedings of the 2016 IEEE International Conference on Advanced Intelligent Mechatronics (AIM), Banff, AB, Canada, 12–15 July 2016; pp. 614–620.

31. Zhao, J.; Jing, X.; Bingtuan, G.; Ning, X.; Fernando, J.C.; Mutka, M.W.; Li, X. MSU Jumper: A Single-Motor-Actuated Miniature Steerable Jumping Robot. *IEEE Trans. Robot.* **2013**, *29*, 602–614. [CrossRef]
32. Zhao, J.; Yan, W.; Xi, N.; Mutka, M.W.; Xiao, L. A miniature 25 grams running and jumping robot. In Proceedings of the 2014 IEEE International Conference on Robotics and Automation (ICRA), Hong Kong, China, 31 May–7 June 2014; pp. 5115–5120.
33. Freudenstein, F.; Dobrjanskyj, L. On a theory for the type synthesis of mechanisms. In *Applied Mechanics*; Springer: New York, NY, USA, 1966; pp. 420–428.
34. Woo, L.S. Type synthesis of plane linkages. *J. Bionic Eng.* **1967**, *89*, 159.
35. Sun, Y.; Ge, W.; Zheng, J.; Dong, D. Solving the kinematics of the planar mechanism using data structures of assur groups. *J. Mech. Robot.* **2016**, *8*, 061002. [CrossRef]
36. Buchsbaum, F.; Freudenstein, F. Synthesis of kinematic structure of geared kinematic chains and other mechanisms. *J. Mech. Robot.* **1970**, *5*, 357–392. [CrossRef]
37. Yokoyama, Y. Studies on the geared linkage mechanisms: 1st report, classification and analysis of geared four-bar linkage. *Bull. JSME* **1974**, *17*, 1332–1339. [CrossRef]
38. Rocha, C.R.; Tonetto, C.P.; Dias, A. A comparison between the Denavit–Hartenberg and the screw-based methods used in kinematic modeling of robot manipulators. *Comput.-Integr. Manuf.* **2011**, *27*, 723–728. [CrossRef]
39. Nansai, S.; Rojas, N.; Elara, M.R.; Sosa, R.; Iwase, M. On a Jansen leg with multiple gait patterns for reconfigurable walking platforms. *Adv. Mech. Eng.* **2015**, *7*, 1–18. [CrossRef]
40. Zelany, M. A concept of compromise solutions and the method of the displaced ideal. *Comput. Oper. Res.* **1974**, *1*, 479–496. [CrossRef]
41. Hwang, C.L.; Yoon, K. *Multiple Attribute Decision Making*; Springer: New York, NY, USA, 1981; pp. 1–531.
42. Shih, H.S.; Shyur, H.J.; Lee, E.S. An extension of TOPSIS for group decision making. *Math. Comput. Model.* **2007**, *45*, 801–813. [CrossRef]
43. Haldane, D.W.; Plecnik, M.M.; Yim, J.K.; Fearing, R.S. Robotic vertical jumping agility via series-elastic power modulation. *Sci. Rob.* **2016**, *1*, eaag2048. [CrossRef]
44. Haldane, D.W.; Plecnik, M.; Yim, J.K.; Fearing, R.S. A power modulating leg mechanism for monopedal hopping. In Proceedings of the 2016 IEEE/RSJ International Conference on Intelligent Robots and Systems (IROS), Daejeon, Korea, 9–14 October 2016; pp. 4757–4764.
45. Haldane, D.W.; Yim, J.K.; Fearing, R.S. Repetitive extreme-acceleration (14-g) spatial jumping with Salto-1P. In Proceedings of the 2017 IEEE/RSJ International Conference on Intelligent Robots and Systems (IROS), Vancouver, BC, Canada, 24–28 September 2017; pp. 3345–3351.
46. Yim, J.K.; Fearing, R.S. Precision Jumping Limits from Flight-phase Control in Salto-1P. In Proceedings of the 2018 IEEE/RSJ International Conference on Intelligent Robots (IROS), Madrid, Spain, 1–5 October 2018.
47. Noh, M.; Kim, S.-W.; An, S.; Koh, J.-S.; Cho, K.-J. Flea-Inspired Catapult Mechanism for Miniature Jumping Robots. *IEEE Trans. Robot.* **2012**, *28*, 1007–1018.
48. Zhang, J.; Song, G.; Qiao, G.; Meng, T.; Sun, H. An Indoor Security System with a Jumping Robot as the Surveillance Terminal. *IEEE Trans. Consum. Electron.* **2011**, *57*, 1774–1781. [CrossRef]

applied sciences

MDPI

Article

Optimal Collision-Free Grip Planning for Biped Climbing Robots in Complex Truss Environment

Shichao Gu [1], Haifei Zhu [1,*], Hui Li [2], Yisheng Guan [1] and Hong Zhang [1,3]

1 School of Electromechanical Engineering, Guangdong University of Technology, Guangzhou 510006, China; xuguqian9@163.com (S.G.); ysguan@gdut.edu.cn (Y.G.); hzhang@ualberta.ca (H.Z.)

2 Key Laboratory of Biomimetic Robots and Systems, Ministry of Education, Beijing Institute of Technology, Beijing 100081, China; lihui2011@bit.edu.cn

3 Department of Computing Science, University of Alberta, Edmonton, AB T6G2H1, Canada

* Correspondence: hfzhu@gdut.edu.cn; Tel.: +86-20-3932-2212

Received: 2 November 2018; Accepted: 4 December 2018; Published: 7 December 2018

Abstract: Biped climbing robots (BiCRs) can overcome obstacles and perform transition easily thanks to their superior flexibility. However, to move in a complex truss environment, grips from the original point to the destination, as a sequence of anchor points along the route, are indispensable. In this paper, a grip planning method is presented for BiCRs generating optimal collision-free grip sequences, as a continuation of our previous work on global path planning. A mathematic model is firstly built up for computing the operational regions for negotiating obstacle members. Then a grip optimization model is proposed to determine the grips within each operational region for transition or for obstacle negotiation. This model ensures the total number of required climbing steps is minimized and the transition grips are with good manipulability. Lastly, the entire grip sequence satisfying the robot kinematic constraint is generated by a gait interpreter. Simulations are conducted with our self-developed biped climbing robot (Climbot), to verify the effectiveness and efficiency of the proposed methodology.

Keywords: grip planning; biped climbing robots; collision avoidance; grip optimization

1. Introduction

Spatial trusses consisting of members are widely used in the construction of roofs, towers, bridges, and the like. However, so far truss-associated routine tasks such as construction, painting, inspection, maintenance, and so on rely highly on manual labor. These routine tasks are usually high-rise and high-intensity, signifying a great risk to workers' safety. Thus, a kind of biped climbing robot has been designed as an ideal assistant or substitute for human workers carrying out these tasks. Typical representatives of BiCRs include SM2 [1], ROMA [2], Shady3D [3], 3DCLIMBER [4,5], PoleClimbingRobot [6] and Treebot [7]. These BiCRs generally comprise of an arm-like serial body for locomotion and grippers at both ends for attachment. Thanks to their biped climbing patterns, BiCRs can agilely move in complex 3D truss environments. Motivated by these characteristics, we also developed a biped climbing robot [8], named Climbot as shown in Figure 1. For the system implementation details and the climbing performance of Climbot, refer to [9].

To complete a given task, for example inspecting the connection reliability of truss joints, BiCRs must be capable of motion planning. Basically, the motion planning of BiCRs consists of grip planning and single-step motion planning [10]. In the grip planning procedure, a list of discrete grip locations, following which robots can navigate from the starting point to the destination, is determined. While in the single-step motion planning procedure, the shifting motion between adjacent grips is generated. It should be noted that not only the grips but also the single-step shifting motion must be free of collision. In this paper, we focus on the study of collision-free grip planning, assuming the truss environment is known or captured with integrated sensors such as in [11].

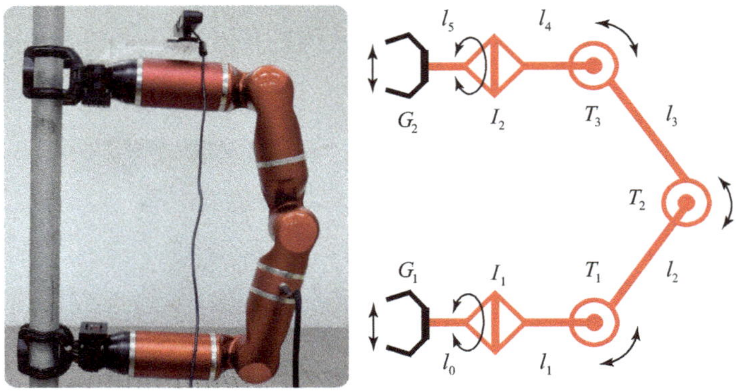

Figure 1. The 5-DoF Climbot and its kinematic diagram.

Figuring out the grip sequences in a 3D truss environment is challenging for BiCRs. Besides the demand for collision avoidance, grips must satisfy the robot kinematic constraints for continuous cycles of climbing from the point of view of reachability. Furthermore, grips should contribute to the formation of a reasonable combination of various gaits with corresponding step lengths, to save the climbing time and energy.

The grip sequences to BiCRs are like the footprint sequences to humanoid robots. Therefore, we can learn from the foot placement problem of humanoid robots that has been addressed in the literature. There are two families of approaches for solving the humanoid robot foot placement problem, which is based on discrete search and continuous optimization, respectively. The discrete search methods normally require a pre-generated set of potential footprints before planning. Then classical methods, such as A^* and RRT, are commonly used for searching. Based on the terrain map and a discrete set of footstep placement positions, Kuffner [12,13] presented a global dynamic programming approach using greedy heuristics to plan safe navigation strategies for biped robots moving in obstacle-cluttered environments. Chestnutt et al. [14,15] used an A^* search algorithm to generate a sequence of collision-free footstep locations to reach a given goal state. A tiered planning strategy was introduced in [16] that split the planner into three layers to traverse different terrain types. Ayaz et al. [17,18] presented a global reactive footstep planning strategy based upon a humanistic approach, in which a heuristic cost based on the complexity of stepping motion was used to assign foot placements and an exhaustive search was employed to identify the best path. These approaches can easily handle obstacle avoidance but introduce the trade-off between computational efficiency and solution precision. The continuous optimization approaches operate directly on the poses of the footsteps as continuous decision variables. Thus, it can make up for the precision deficiency of the discrete search methods. In [19,20], the authors presented a novel footstep optimization method with mixed-integer convex constraints that could solve the problem to its global optimum. However, the generation of each reachable region deeply relied on the position of the previous step, increasing the time consumption. In addition, adjusting the parameters to approximate the reachable regions was always difficult and thus reduced the planning accuracy. Guan et al. [21,22] built global optimization models with non-linear constraints to find the maximum heights of the obstacles that can be overcome. The results were then used as a priori knowledge and a database for surmounting obstacles. However, they focused on how to stride across one obstacle only, but not generating the nearby footprint sequences. Please note that these footstep planning methods for humanoid robots are always applied in 2D or 2.5D environments, which differs from the grip planning for BiCRs in complex 3D truss environments.

To the best of our knowledge, the collision-free grip planning problem has rarely been studied in the literature. Balaguer et al. [2,23] treated the BiCRs' climbing path planning as a TSP-like problem. They modeled the environment as a graph with two categories of primitives. They then proposed a heuristic algorithm to solve the climbing path, along which the robot visited all beam faces without repetition. Detweiler et al. [24] presented a path planning algorithm to optimize the locomotion sequences for the Shady3D robot. A set of potential gripping points was first dispersed in the truss environment up to a certain density. Then the shortest locomotion sequence represented with gripping points was computed with the Dijkstra's distance. Based on the framework of conventional genetic algorithm, Chung et al. [25] adopted the concept of genetic modification to design a new genetic operator. They used this method to solve the climbing path with the minimum energy demand. However, these methods all suffer limitations from spending long planning time, ignoring specific transition movement, and not thinking about obstacle avoidance. Lam et al. [7] discretized the tree surface into finite grasp points, then used a dynamic programming algorithm to identify the global path and adopted a motion planning algorithm to generate the single-step climbing motion. Therefore, this method was only applicable to BiCRs climbing on the object surface with non-enclosure grippers. It can only obtain a near-optimal solution. Zhu et al. [26] presented two optimal strategies to select a collision-free grip from its potential region based on three criteria step by step. However, this method lacks global guidance in searching, and hence, is inefficient most of the time.

For BiCRs rapidly generating optimal collision-free grips, we have proposed a novel framework, which further subdivides the grip generation procedure into three steps:

(1) quick determination of all feasible climbing routes in global, outputting member sequence and corresponding grip orientation and operational regions for transition;
(2) optimal arrangement of collision-free grips on the operational regions on each member along each feasible climbing route;
(3) generation of the entire grip sequence with a gait interpreter.

For Step (1), we have presented a high-efficiency global path planning method in [27]. Further to our previous work, we present an optimal collision-free grip planning method to minimize the number of climbing steps in this paper.

The novelty of this paper is the first systematic presentation of an optimal collision-free grip planner for the biped climbing robots generating grip sequences in a complex truss environment. This grip planner not only handles well with the collision between the robot and the truss, but also guarantees the robot kinematics, the minimum number of climbing steps, the good manipulability of transition grips. It should be noted that this grip planner is able to solve the grip planning problem of more than 30 grips in a scene of 25 members within 0.65 s. Another novelty of this paper is the mathematical model for computing the operational regions for negotiating obstacle members.

The remainder of this paper is organized as follows. We briefly review the global path planning and its output, followed by introducing the idea of collision-free grip planning in Section 2. We then create a mathematical model for computing the operational regions to negotiate obstacles in Section 3. We construct a mathematical optimization model to determine the collision-free grips in the operational regions in Section 4. A gait interpreter and its implementation are described in Section 5. In Section 6, we conduct simulations with Climbot to verify the proposed analysis and algorithms. Finally, we conclude our work in Section 7.

2. Problem Statement

2.1. Global Path Planning and Feasible Routes

In our framework, the crucial point of global path planning is to provide global guidance for the subsequent processes, i.e., grip planning and single-step motion planning. It concentrates on the fast determination of all feasible routes, Γ. The distinguishing merit is that it largely narrows down the searching space to a limited number of members, and thus largely increases the searching efficiency.

As a global guidance for BiCRs, a feasible route, Γ_i, should indicate the entire climbing path with the member sequence $\mathfrak{M}$, the gripping orientations ${}^{W}\boldsymbol{R}$ on the members, and the operational regions $\mathfrak{R}_t$ for transitions. Figure 2 shows us an illustration of one feasible route in a scene consisting of 7 members, taken from [27]. In the scene, each member is described in the world frame $\{W\}$ as

$$ {}^{W}\boldsymbol{P} = {}^{W}\boldsymbol{P}_0 + t \cdot {}^{W}\boldsymbol{P}_{dir}, \ \ 0 \leq t \leq L_{mem}, \tag{1} $$

where ${}^{W}\boldsymbol{P}_0$, ${}^{W}\boldsymbol{P}_{dir}$ and L_{mem} are the reference point, the direction unit vector and the length of the member, respectively. While t is a scale, which can be used to specify the gripping position on the member. For the convenience of expression and understanding, ${}^{j}x$ is denoted as a grip position on the j-th member from where a transition begins (the takeoff segment) and ${}^{j}y$ on the same member to where a transition ends (the landing segment). That is to say, ${}^{j}x$ and ${}^{j}y$ are also scales the same as t. Based on this notation, the operational regions for transiting from j-th member to (j + 1)-th member are the set of ${}^{j}x$ and ${}^{j+1}y$, i.e.,

$$ \begin{cases} {}^{j}x \in [{}^{j}\underline{x},\ {}^{j}\overline{x}] \\ {}^{j+1}y \in [{}^{j+1}\underline{y},\ {}^{j+1}\overline{y}] \end{cases}, \tag{2} $$

where the boundaries are computed by the transition analyzer during global path planning. We proved that ${}^{j+1}y$ and ${}^{j}x$ are linear for BiCRs like Climbot with planar configurations in [27]. Therefore, their relationship can be written as

$$ {}^{j+1}y = \sigma\, {}^{j}x + \delta, \tag{3} $$

where σ and δ are constants. In other words, the gripping points within the operational regions for transitions are one-to-one mapping.

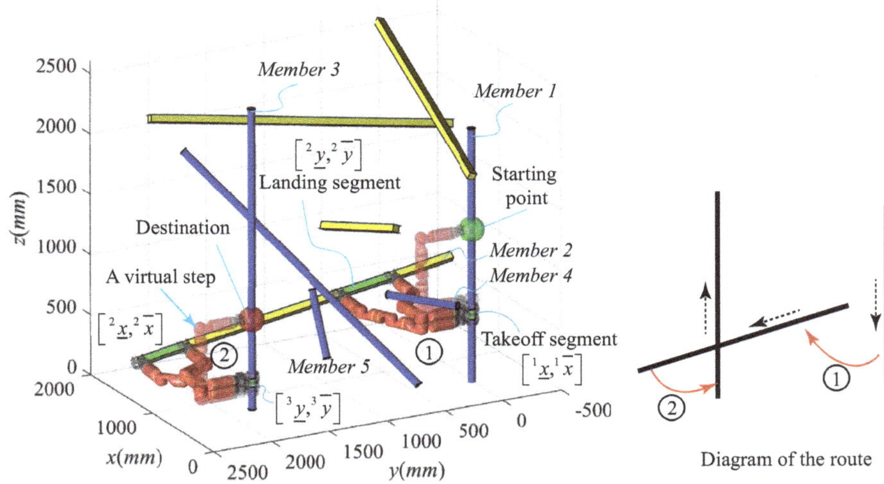

Figure 2. An illustration of a feasible route. Green and red spheres in the figure stand for the initial position and the destination, respectively. Nearby numbers indicate the sequence of transitions to be performed along the route.

2.2. The Problem of Optimal Collision-Free Grip Planning

From Figure 2, we can see that BiCRs may collide with the truss members during climbing, no matter when it moves forward on a member or transits between two adjacent members. Without loss of generality, the collision-free grip planning problem in a truss environment can be simplified as the model shown in Figure 3. For ease of understanding, we define three categories of operational regions. They are (1) the operational region $\mathfrak{R}_{init}$ representing the starting point t_{init} and the destination t_{goal}, which are given beforehand, (2) the operational region $\mathfrak{R}_t$ for performing transitions between adjacent members, which is output from the global path planner, and (3) the operational region $\mathfrak{R}_o$ for negotiating obstacles when moving on a member. The operational regions for transitions are highlighted with green, while that for negotiating obstacles are with red in Figure 3. Accordingly, the optimal collision-free grip planning problem is to compute $\mathfrak{R}_o$, then optimally determine grips within $\mathfrak{R}_t$ and $\mathfrak{R}_o$, and finally arrange grips between $\mathfrak{R}_t$ and $\mathfrak{R}_o$ so that adjacent grips satisfy the robot kinematics.

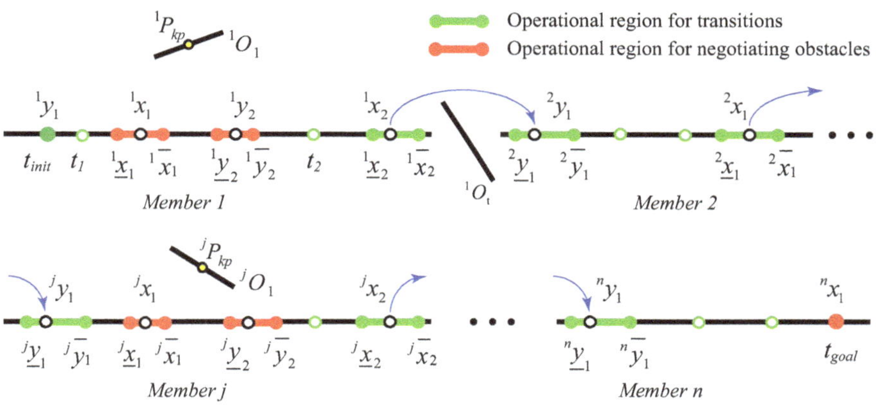

Figure 3. A general model of collision-free grip planning for BiCRs.

Figure 4 shows us the flowchart of our proposed optimal collision-free grip planning algorithm. Each feasible route $\mathbf{\Gamma}_i = \{\mathfrak{M}, {}^W\boldsymbol{R}, \mathfrak{R}_t\}$ is extracted for grip planning successively. For each member $\mathfrak{M}_j$ of $\mathbf{\Gamma}_i$, the operational region ${}^j\mathfrak{R}_o$ for negotiating obstacle members will be computed with a mathematical model (which will be presented in Section 3). After getting the operational regions for negotiating obstacles on all via members, an optimization model is used to determine the grips $\mathbb{S}_{or}(G_{or}, C_{or})$ within the operational regions. The objective of this optimization model is to achieve the minimum number of climbing steps and the good manipulability of the transition grips. The optimization model also takes the relationship between step length and climbing gaits into account. Finally, a gait interpreter is designed to arrange grips outside of the operational regions, to obtain the entire grip sequence $\mathbb{S}(G, C)$. Guiding by the output of the grip optimization model, the gait interpreter can ensure a shifting configuration exists for each pair of adjacent grips. This shifting configuration will be the input of single-step motion planning. During the grip planning procedure, the route will be determined to be blocked if solving of the operational region ${}^j\mathfrak{R}_o$ for negotiating obstacles fails or no solution for the grip optimization model.

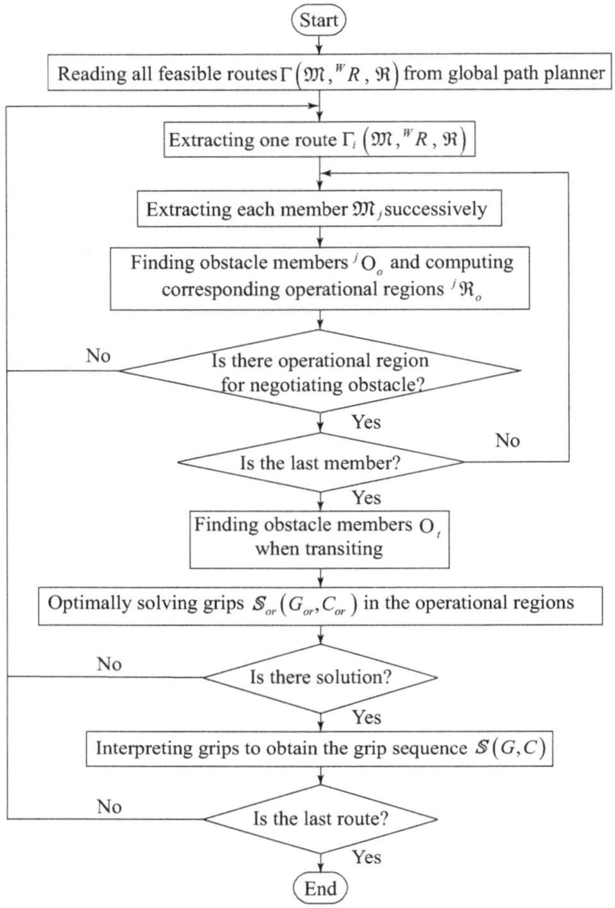

Figure 4. The flowchart of collision-free grip planning.

3. Computation of Operational Regions for Negotiating Obstacles

In this section, we study how BiCRs negotiate obstacles to move from the transition landing segment to the next takeoff segment on a specific member.

3.1. The Key Point for Negotiating Obstacles

As mentioned earlier, for the j-th member, the applicable gripping orientation ${}^W\boldsymbol{R}_j$, the transition landing segment $[{}^j\underline{y}, {}^j\overline{y}]$ and next takeoff segment $[{}^j\underline{x}, {}^j\overline{x}]$ are determined by the global path planner. BiCRs with planar configuration can only move within a plane, i.e., the robot plane η as shown in Figure 5, along the via member. Therefore, the possible range of collision can be segmented by a cuboid Δ. Denote l_{cub}, w_{cub} and h_{cub} as the length, width and height of this cuboid, respectively. The dimension of the cuboid can be set as,

$$\begin{cases} l_{cub} = max(||{}^j x - {}^j y||) \\ w_{cub} = 2\,(r_{link} + r_{mem}) \\ h_{max} = l_2 \sin\left(\arcsin\left(l_{safe}/l_2\right) + |\theta_{max}|/2\right) + l_1 + r_{link} + r_{mem} \end{cases}, \tag{4}$$

where r_{link} and r_{mem} are the radii of the robot links and the truss members; l_1 and l_2 refer to the robot link lengths indicated in Figure 1, while θ_{max} is the maximum rotation angle of the T-type joint of the robot; l_{safe} is a pre-defined safe distance to facilitate the gripping and grip-releasing operations. In this paper, only the truss members are considered as potential obstacles during climbing. Thus after defining this cuboid, the robot can only collide with those members that intersect the cuboid during the movement from the transition landing segment to the next takeoff segment. Conversely, the potential obstacle members can be extracted out by an intersection check between each member and the cuboid.

Next, we define a *Key Point* which plays a key role to determine the reasonable obstacle-negotiating configuration. As shown in Figure 5, the obstacle-negotiating configuration is a state that the robot grips on the *j*-th member with both grippers, surrounding the obstacle with its body (Figure 6a) or avoiding collision with the obstacle by lowering its body (Figure 6b). The *Key Point* is actually the point that most likely to collide with the robot in the obstacle-negotiating configuration.

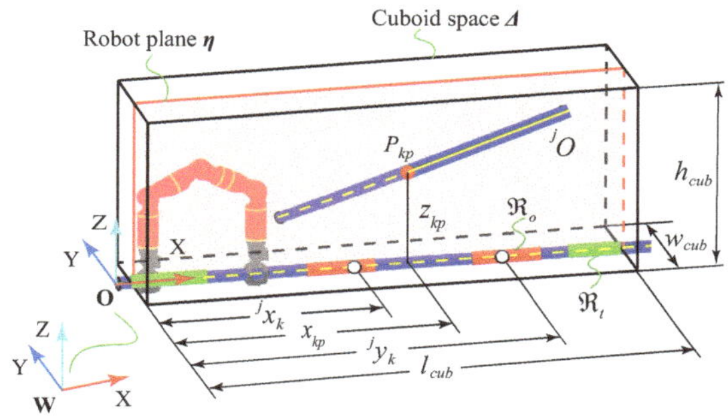

Figure 5. The definition of collision avoidance space and *Key Point*.

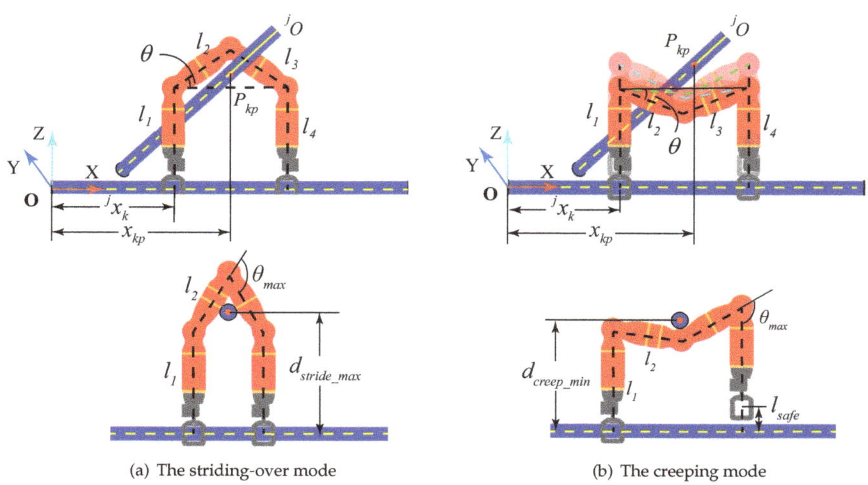

(a) The striding-over mode

(b) The creeping mode

Figure 6. Negotiating obstacles with the striding-over mode and the creeping mode.

Suppose the angle between the obstacle member axis and the robot plane is λ. If $\lambda < \lambda_{threshold}$, where $\lambda_{threshold}$ is a small angle, the obstacle member is approximately parallel to the robot plane. In this case, the *Key Point* is one of the end points of the obstacle member axis, according to the striding-over mode or the creeping mode selected for negotiating obstacles. If $\lambda > \lambda_{threshold}$, where the obstacle member intersects the robot plane, the *Key Point* is selected as,

- the intersection point between the obstacle member axis and η, if this intersection point locates within the cuboid, or
- the intersection point between the obstacle member axis and the cuboid, if the intersection point between the obstacle member axis and η is outside the cuboid. The *Key Point*, in this case, is closest to η.

3.2. The Mode to Negotiate Obstacles

Basically, BiCRs have two modes of surmounting obstacles, which are the striding-over mode and the creeping mode, respectively, illustrated with Climbot in Figure 6. Denote d_o as the distance between the *Key Point* and the gripping member. The maximum distance d_{stride_max} for striding-over and the minimum distance d_{creep_min} for creeping, can be calculated with the robot's parameters. Obviously, the robot can surmount the obstacle member only with the striding-over mode when d_o is smaller than d_{creep_min} as shown in Figure 6a, and with the creeping mode when d_o is larger than d_{stride_max} as shown in Figure 6b. In the case of $(d_{creep\_min} < d_o < d_{stride\_max})$, the robot can overcome the obstacle with either of these two modes. Further reflecting in the mathematical model (Equation (5)) to compute the operational regions for negotiating obstacles, the applicable mode determines the boundaries of the rotation angle of the T_1 or T_3 joint of the robot.

3.3. The Mathematical Model of Operational Regions for Negotiating Obstacles

Assume n_j periods of obstacle crossings are required to move from ${}^jy$ to ${}^jx$ on the *j-th Member*. For the k-th period of obstacle crossing, referring to Figures 5 and 6, its operational region $\mathfrak{R}_o$ is dependent on two variables, ${}^jx_k$ and the rotation angle θ of the T_1 or T_3 joint of the robot. Moreover, the minimum and maximum of ${}^jx_k$ are exactly the boundaries of the takeoff segment, while ${}^jx_k$ and θ determine the boundaries of the landing segment. Therefore, the following mathematical model is built up to calculate the maximum operational region $\mathfrak{R}_o = \{[f_1, f_2], [f_3, f_4]\}$:

$$\begin{aligned}
&\underset{{}^jx_k}{\text{minimize}} && f_1 = {}^jx_k \\
&\underset{{}^jx_k}{\text{maximize}} && f_2 = {}^jx_k \\
&\underset{{}^jx_k,\,\theta}{\text{minimize}} && f_3 = \left({}^jx_k + 2l_2\cos(\theta)\right) \\
&\underset{{}^jx_k,\,\theta}{\text{maximize}} && f_4 = \left({}^jx_k + 2l_2\cos(\theta)\right) \\
&\text{subject to:} && \\
& && d_i \geq (r_{link} + r_{mem}), \\
& && \theta_{min} \leq \theta \leq \theta_{max}, \\
& && \left(x_{kp} - 2l_2\right) \leq {}^jx_k \leq x_{kp} - r_{link} - r_{mem}, \\
& && x_{kp} + r_{link} + r_{mem} \leq {}^jx_k + d_{grips},
\end{aligned} \tag{5}$$

where d_i is the distance between the robot link and the obstacle member; $\left[\theta_{min} \quad \theta_{max}\right]$ is the rotation limitation of the T-type joint of the robot; d_{grips} is the distance between two grips distributed in the takeoff and landing segments, respectively. It should be noted that d_i in the pre-gripping and gripper-releasing procedures must be also considered.

4. Optimal Collision-Free Grip Planning

In this section, we discuss the optimal determination of grips within each operational region. Thus, hereafter the three categories of operational regions will be treated equally if not specified.

4.1. Objective: Minimum Climbing Steps and Good Manipulability

4.1.1. Minimum Climbing Steps

To form a firm grip, BiCRs must accomplish several procedures, i.e., detecting the target member, adjusting the gripper's orientation accordingly, and then approaching the member and gripping it. According to our experiments, the grip operation normally consumes two-thirds of the time in each climbing cycle [28,29]. Therefore, minimizing the total number of climbing steps is a promising solution to reduce the time consumption. Suppose n members are involved in the feasible route, where $(n-1)$ periods of transitions are required. The objective function to minimize the total number of climbing steps can be written as:

$$\underset{^jx_k,\,^jy_k}{\text{minimize}} \sum_{j=1}^{n}\sum_{k=1}^{n_j+1}\left(g(^jx_k,\,^jy_k)+1\right), \tag{6}$$

where ^jy_k is generally a grip within the landing segment of an operational region, and ^jx_k is another one within the takeoff segment, while $g(^jx_k,\,^jy_k)$ is the function calculating the minimum number of steps required to move from ^jy_k to ^jx_k. This function will be discussed in detail in Section 4.2.1. It should be noted that the landing grip $^jy_{k+1}$ is corresponding to the takeoff grip ^jx_k in a transiting gait.

In Equation (6), one step should be counted for performing transitions or negotiating obstacles. For the last member, a virtual step after reaching the destination (as shown in Figure 2) is added to maintain the consistency of the form.

4.1.2. Good Manipulability

The configurations corresponding to the boundaries of operational regions for transitions always suffer from singularity or joint rotation limitations. So optimal transition grips should keep a distance from their boundaries. To achieve this goal, we construct a potential field, i.e., Equation (7), to adjust the transition grips as close to the midpoint of the corresponding operational regions as possible. Owing to the function characteristics, closer to the midpoint, the value of the function is smaller and the robot has better manipulability when moves nearby the grips.

$$\underset{^jx_{n_j+1}}{\text{minimize}} \sum_{j=1}^{n}(^jx_{n_j+1}-{^j\underline{x}}_{n_j+1})(^jx_{n_j+1}-{^j\overline{x}}_{n_j+1}), \tag{7}$$

where $^jx_{n_j+1}$ is actually the last grip on the j-th *Member*. That is to say, in fact, $^jx_{n_j+1}$ is ^jx in Equation (2).

4.1.3. Combination

To unify the above two objectives in a function, Equations (6) and (7) are normalized. Equation (6) can be rewritten as,

$$\underset{^jx_k,\,^jy_k}{\text{minimize}} \sum_{j=1}^{n}\sum_{k=1}^{n_j+1}\frac{g(^jx_k,\,^jy_k)+1}{^jN_{min}+1}, \tag{8}$$

where $^jN_{min}$ is the ideal minimum number of climbing steps going through the j-th member without consideration of collision, which is computed as,

$$^jN_{min} = \text{minimize}\ g(^jx_{n_j+1},\,^jy_1). \tag{9}$$

Equation (7) can be written as,

$$\underset{^jx_{n_j+1}}{\text{minimize}} \sum_{j=1}^{n}\frac{(^jx_{n_j+1}-{^j\underline{x}}_{n_j+1})(^jx_{n_j+1}-{^j\overline{x}}_{n_j+1})}{({^j\overline{x}}_{n_j+1}-{^j\underline{x}}_{n_j+1})^2/4}. \tag{10}$$

For the destination, this component is set to −1 to place the last grip exactly in the destination. Combining Equations (8) and (10) yields the final objective function in Equation (12).

4.2. Constraints

4.2.1. Moving Distance and the Gaits

Benefiting from the bipedal climbing pattern inspired by arboreal primates, BiCRs normally have several climbing gaits. Climbot, for example, has at least three basic gaits, namely the inchworm-like gait, the swinging-around gait, and the flipping-over gait [8]. Each gait has its own characteristics, among which the step length plays a key role in the determination of the required number of climbing steps given a moving distance. Table 1 summarizes the minimum and maximum step lengths of each gait. In the table, the so-called hybrid gait refers to a three-step climbing pattern derived from the basic gaits, as shown in Figure 7.

Table 1. Minimum and maximum step lengths with different gaits.

Gaits	Step Lengths	
	Minimum	**Maximum**
Inchworm-like gait	$S_0 = 0$	$S_1 = S_3 - S_2$
Hybrid gait	S_1	$S_2 = 2l_2(1 - cos(\theta_{max}/2))$
Swinging-around or Flipping-over gait	S_2	$S_3 = \sqrt{(2l_2)^2 - l_{safe}^2}$
Hybrid gait	S_3	$S_4 = 2l_2$

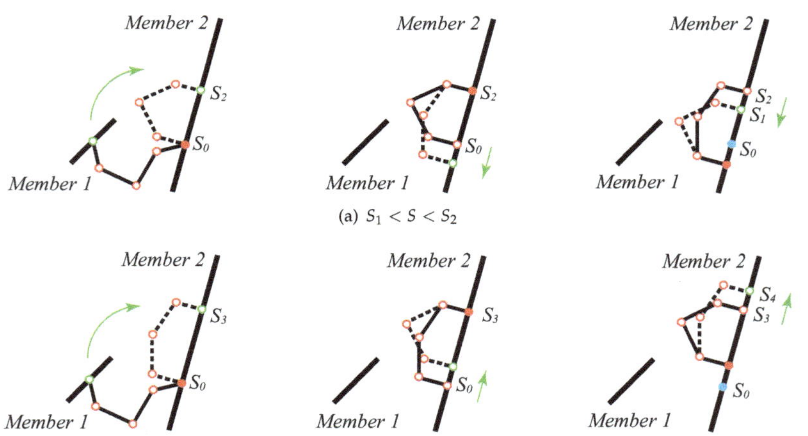

Figure 7. The three-step hybrid gait. The solid lines represent the current configuration, while the dashed ones stand for the following configuration one step forward.

To go through a given distance, i.e., from a landing segment to a takeoff segment on a certain member, Climbot always can climb with a combination of different gaits. Based on Table 1, we analyze the relationship between the moving distance and the minimum number of required climbing steps as follows.

- DiMov (direct movement) gait: the moving distance is equal to S_0. In this case, the landing segment partially (or completely) overlaps with the takeoff segment. Climbot can pass through just with one grip performing two climbing cycles continuously. As a result, the required number of climbing steps in this case is 0.

- Inch gait: the moving distance is within (S_0, S_1). Only the inchworm-like gait is applied to this condition. At least two steps (stretching and shrinking) are required when climbing with the Inch gait.
- SaFo gait: the moving distance is within (S_2, S_3). One step climbing with the swinging-around gait or flipping-over gait meets the distance well.
- Hyb gait: the moving distance is within (S_1, S_2) or (S_3, S_4). Under such circumstances, a mixture of the inchworm-like gait and swinging-around or flipping-over gait should be applied (hybrid gait). Figure 7 illustrates the climbing patterns with Climbot. It moves a step with the swinging-around gait or the flipping-over gait, and then climbs two steps with the inchworm-like gait. Thus, at least three steps are required.
- SaFo gait: the moving distance is larger than S_4. The minimum number of climbing steps can be calculated as $\lceil S_k/S_3 \rceil$, where the symbol $\lceil\ \rceil$ represents the ceiling operation. Accordingly, the robot moves with the swinging-around gait or the flipping-over gait.

In summary, the mathematical relationship between the minimum number of climbing steps N_k and the corresponding moving distance S_k on a member can be expressed as a piecewise function,

$$g({}^j x_k, {}^j y_k) = N_k = \begin{cases} 0, & S_k = 0, \\ 2, & S_k \in (S_0\ S_1], \\ 3, & S_k \in (S_1\ S_2), \\ 1, & S_k \in [S_2\ S_3], \\ 3, & S_k \in (S_3\ S_4), \\ \lceil S_k/S_3 \rceil, & S_k \in [S_4\ \infty), \end{cases} \tag{11}$$

where $S_k = |{}^j x_k - {}^j y_k|$. This piecewise function is plotted in Figure 8. Given the moving distance, the required minimum climbing steps can be computed with Equation (11).

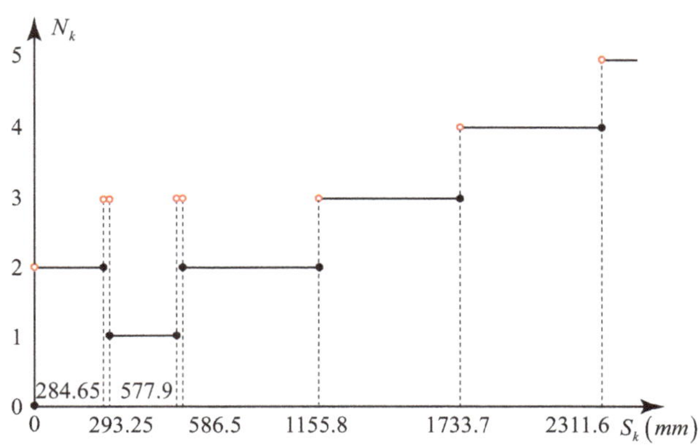

Figure 8. The relationship between moving distance (on a member) and minimum climbing steps.

4.2.2. Collision Avoidance during Transitions

Potential collision must be also taken into account when BiCRs perform transitions. As a result, it is necessary to extract the potential obstacle members before solving the grip optimization model, in order to guarantee safe transitions.

Figure 9 shows our strategy to find all the potential obstacles ${}^j\mathcal{O}_t$. Supposing BiCRs transit from *Member 1* to *Member 2*, a conservative *Sphere* Ω is constructed based on the operational region $\mathfrak{R}_t$ for transiting. The center P_C of *Sphere* Ω coincides with the center point of a set of T_1 and T_3 joint center points corresponding to the boundary configurations. While the radius r_Ω of *Sphere* Ω is set as the furthest distance between its center and the gripping points. Thus, this sphere space covers the range where potential collision may happen during transition. Obstacle members are then found out by carrying out intersection checks between the sphere and each truss member. Taking the case in Figure 9 for example, *Member* 4, *Member* 5 and *Member* 6 will be extracted as the potential obstacles.

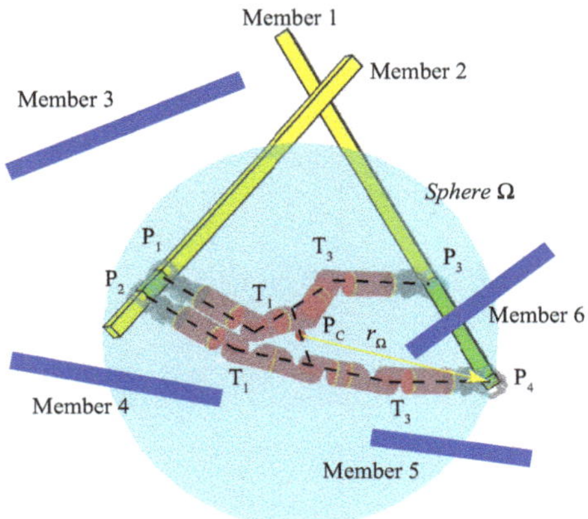

Figure 9. The method to extract potential obstacles when transiting between adjacent members.

4.3. The Optimization Model

Putting all the pieces together gives us the entire optimization model for grip planning, as

$$
\begin{aligned}
&\underset{{}^j x_k,\ {}^j y_k}{\text{minimize}} \quad f = \sum_{j=1}^{n} \left(\sum_{k=1}^{n_j+1} \frac{g({}^j x_k,\ {}^j y_k) + 1}{{}^j N_{min} + 1} + \frac{({}^j x_{n_j+1} - {}^j \underline{x}_{n_j+1})({}^j x_{n_j+1} - {}^j \overline{x}_{n_j+1})}{({}^j \overline{x}_{n_j+1} - {}^j \underline{x}_{n_j+1})^2/4} \right) \\
&\text{subject to:} \\
&\qquad g({}^j x_k, {}^j y_k) \ \ s.t.\ Equation\ (11), \\
&\qquad {}^j N_{min} \ \ s.t.\ Equation\ (9), \\
&\qquad S_k = |{}^j x_k - {}^j y_k|, \\
&\qquad {}^{j+1} y_1 = \sigma\, {}^j x_{n_j+1} + \delta, \\
&\qquad {}^j x_k \in [{}^j \underline{x}_k,\ {}^j \overline{x}_k], \\
&\qquad {}^j y_k \in [{}^j \underline{y}_k,\ {}^j \overline{y}_k], \\
&\qquad ({}^j b_k - {}^j x_k)({}^j x_k - {}^j c_k) \leq 0,\ \ if\ {}^j x_k \in [{}^j \underline{x}_k,\ {}^j b_k] \cap [{}^j c_k,\ {}^j \overline{x}_k], \\
&\qquad \mathcal{R} \bigcap {}^j \mathcal{O}_t = \varnothing, \\
&\qquad {}^1 y_1 = t_{init}, \\
&\qquad {}^n x_{n_j+1} = t_{goal},
\end{aligned}
\tag{12}
$$

where $\mathcal{R}$ represents the robot.

Please note that some operational regions have two segments [27], whose boundaries increase in turn. Then the constraint, $({}^j b_k - {}^j x_k)({}^j x_k - {}^j c_k) \leq 0$, is added to ensure that ${}^j x_k$ is within the valid operational region.

5. Gait Interpreter

After determining the grips within each operational region, it is time to generate grips connecting the operational regions. That is to say, at this time, the grips are generated outside of operational regions. These remaining grips must be arranged in such a proper way that finally any pair of adjacent grips satisfies the robot kinematic constraints. A dedicated gait interpreter is proposed to implement this function in this section.

Algorithm 1 gives out the basic structure of the proposed gait interpreter. The outputs of the optimal collision-free grip planner are $GripsInfo$, inclusive of $\mathbb{S}_{or}$ (G_{or}, C_{or}) and $GaitInfo$. $GaitsInfo$ actually packages the parameters calculated in Section 4.2.1, including the moving distance ${}^{j}S_k$, number of climbing steps ${}^{j}N_k$ and corresponding *gait* type. These parameters are important reference for the gait interpreter arranging the remaining grips. Therefore, the inputs to the gait interpreter are the truss environment and $GripsInfo$. During processing, the gait interpreter takes the transition, the obstacle negotiation, or the movement between them as a unit. For each unit, the gait interpreter simply arranges grips by calling an appropriate sub-function, according to the corresponding *gait* type. Sub-functions InchwormGait, HybGait and SaFoGait are designed to generate grips and configurations for the movement between operational regions. While PerformTransition and NegotiateObstacle are designed to pack the transition and obstacle-negotiating grips into the grip sequence.

Algorithm 1: The gait interpreter

```
Input  : WTruss: The truss environment;
         GripsInfo: Output of the optimal collision-free grip planner.
Output: Grips: The entire grip sequence.
MemN ⇐ MembersNumber(GripsInfo);
for i = 1 to MemN do
    cMem ⇐ GetMember(WTruss, GripsInfo(i));
    GaitsInfo ⇐ GripsInfo(i);
    ObsN ⇐ ObstaclesNumber(GaitsInfo);
    for k = 1 to 2(ObsN + 1) do
        Gait ⇐ GaitType(GaitsInfo(k));
        switch Gait do
            case Inch do
                Grips.AddGrip(InchwormGait(cMem, GaitsInfo(k))) ;   // Algorithm 2
            case Hyb do
                Grips.AddGrip(HybGait(cMem, GaitsInfo(k))) ;        // Algorithm 3
            case SaFo do
                Grips.AddGrip(SaFoGait(cMem, GaitsInfo(k)));
            case NegotObs do
                Grips.AddGrip(NegotiateObstacle(cMem, GaitsInfo(k)));
            case Transit do
                Grips.AddGrip(PerformTransition(cMem, GaitsInfo(k)));
        end
    end
end
end
```

Algorithm 2 is the implementation of the sub-function InchwormGait. InchwormGait takes charge of generating grips for the inchworm-like gait. According to Section 4.2.1, one extra grip t_{insert} should be inserted for connecting the operational regions. Based on $GaitsInfo$, the function

CALCULATESTEPLENGTH is called to calculate the proper step length $StepLen$ with respect to the gripping position y. Then t_{insert} can be obtained by an offset from the gripping position y. After that, the adjacent grips of y, t_{insert} and x will be sent for kinematic check. If y and t_{insert} or t_{insert} and x do not satisfy the kinematic constraints, t_{insert} will be updated to the opposite direction on the member. Reflecting on the movement of the robot, it is moving forward or backward to adjust the gripping position. If the kinematic check is passed, the function GENERATEGRIP is then used to generate $Grips$ of an inchworm-like gait. Otherwise, the robot cannot continue to climb along this route.

Algorithm 2: INCHWORMGAIT: generating grips for the inchworm-like gait

```
Input  : cMem: The current member where the robot climbing on;
         GaitsInfo: The information of the inchworm gait.
Output: Grips: The grip sequence of the inchworm gait.
cR ⇐ GaitsInfo.cR ;                          // The orientation for gripping cMem
y ⇐ GaitsInfo.y ;                    // The gripping position in the landing segment
x ⇐ GaitsInfo.x ;                    // The gripping position in the takeoff segment
StepLen ⇐ CALCULATESTEPLENGTH(GaitsInfo, y);
t_insert ⇐ y + StepLen;
if CHECKKINEMATICS(y, x, t_insert) ≠ true then
    t_insert ⇐ y − StepLen;
    if CHECKKINEMATICS(y, x, t_insert) ≠ true then
        return Grips ⇐ ∅;
    else
        return Grips ⇐ GENERATEGRIP(cR, cMem, y, t_insert, x);
    end
else
    return Grips ⇐ GENERATEGRIP(cR, cMem, y, t_insert, x);
end
```

Algorithm 3 is the implementation of the sub-function HYBGAIT, in charge of generating grips for the hybrid gait. To climb with the hybrid gait, two extra grips need to be inserted. According to Section 4.2.1, every hybrid gait contains an inchworm-like gait. Therefore, one of these two grips will be solved firstly, then the function INCHWORMGAIT is called to generate another. The direction $InsDir$ of inserting first grip is firstly initialized. Based on the distance between y and x, two different strategies are used by the algorithm. If this distance is in the range of (S_1, S_2), an offset $Offset$ is calculated with the function CALCULATEOFFSET according to $GaitsInfo$ and $InsDir$. Then t_{insert} can be solved easily. The other grip $Grips_{new}$ can be obtained by calling the function INCHWORMGAIT. If $Grips_{new}$ is empty, $Offset$ will be updated by changing to another side to insert the grip. Then INCHWORMGAIT is called again. If this operation successes, $Grips$ between t_{insert} and x are generated by the function GENERATEGRIP. Otherwise, the robot cannot reach the destination via this route. If the moving distance is in (S_3, S_4), two grips always can be inserted between y and x successfully since there is enough adjustment space.

In the case of $SaFo$ gait, $(\lceil S_i/S_2 \rceil - 1)$ grips are distributed uniformly between the initial and finished gripping position. Verifications of kinematics and collision avoidance should be applied to each step.

Algorithm 3: HybGait: generating grips for the hybrid gait

```
Input  : cMem: The current member where the robot climbing on;
         GaitsInfo: The information of the hybrid gait.
Output: Grips: The grip sequence of the hybrid gait.
cR ⇐ GaitsInfo.cR ;                          // The orientation for gripping cMem
y ⇐ GaitsInfo.y;  x ⇐ GaitsInfo.x ;
MovDir ⇐ InitializeMoveDirction(y, x) ;      // Initialize the direction
if |x − y| ∈ (S1, S2) then
    GaitsInfo_new ⇐ GaitsInfo;
    OffSet ⇐ CalculateOffSet(GaitsInfo, MovDir, y);
    t_insert ⇐ y + OffSet;
    GaitsInfo_new.y ⇐ t_insert;
    Grips_new ⇐ InchwormGait(cMem, GaitsInfo_new) ;        // Algorithm 2
    if Grips_new = ∅ then
        GaitsInfo_new ⇐ GaitsInfo;
        OffSet ⇐ CalculateOffSet(GaitsInfo, −MovDir, y);
        t_insert ⇐ y + OffSet;
        GaitsInfo_new.x ⇐ t_insert;
        Grips_new ⇐ InchwormGait(cMem, GaitsInfo_new) ;    // Algorithm 2
        if Grips_new = ∅ then
            return Grips ⇐ ∅
        else
            Grips ⇐ Grips_new;
            return Grips.AddGrip(GenerateGrip(cR, cMem, t_insert, x));
        end
    else
        Grips ⇐ GenerateGrip(cR, cMem, y, t_insert);
        return Grips.AddGrip(Grips_new);
    end
else
    GaitsInfo_new ⇐ GaitsInfo;
    OffSet ⇐ CalculateOffSet(GaitsInfo, MovDir);
    t_insert ⇐ y + OffSet;  GaitsInfo_new.x ⇐ t_insert;
    Grips ⇐ InchwormGait(cMem, GaitsInfo_new);
    return Grips.AddGrip(GenerateGrip(cR, cMem, t_insert, x));
end
```

6. Simulation

To verify the proposed analysis and algorithms, simulations are conducted with Climbot. A simulation environment is developed, and algorithms are implemented on the platform of MATLAB R2015b. All the simulations are launched on a desktop with Intel Core i7-7700K CPU and 16GB RAM, running with the 64-bit operating system Windows 10 Pro. In the simulations, the starting point and the destination are specified manually but arbitrarily, and are highlighted with a green and red sphere, respectively.

6.1. The Result of Operational Region of Negotiating Obstacle

This part of simulations is to verify the effectiveness and efficiency of the computation of operational regions for negotiating obstacles. In the first scene, Climbot is assumed to move on a member, while another member acts as an obstacle. The pose of the obstacle member is randomly generated. Four snapshots of results are shown in Figure 10. In the figure, the segments highlighted in green are the operational regions obtained for surmounting the obstacle member. A collision-free obstacle-negotiating configuration is provided as an illustration of the effectiveness. We changed the pose of the obstacle member 588 times, 305 of which got operational region. The average time consumed in the computation procedure is 0.13 s, and the maximum one is 0.44 s.

We conduct another simulation with a scene consisting of multiple obstacles, to test the applicability of our model to multi-obstacle cases. The simulation result is shown in Figure 11. In this scene, the robot is requested to move from the left end to the right end on *Member 8*, overcoming three groups of obstacles successively. In most cases, a group of obstacles, for example {*Member 1, 2, 3*} or {*Member 4, 5*} in the figure, should be considered at one shot. Because these obstacles must be overcome once and for all. In other cases, for instance, the robot overcome *Member 6* and *Member 7* with two adjacent steps, the operational region can be simply obtained by an intersection operation of the results for negotiating each individual obstacle. However, how to group the obstacle members reasonably is pending for further study.

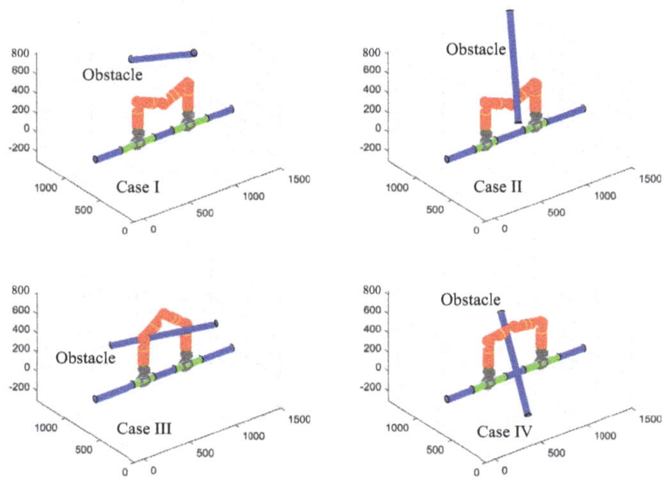

Figure 10. The results of computing operational regions for negotiating an obstacle in various pose.

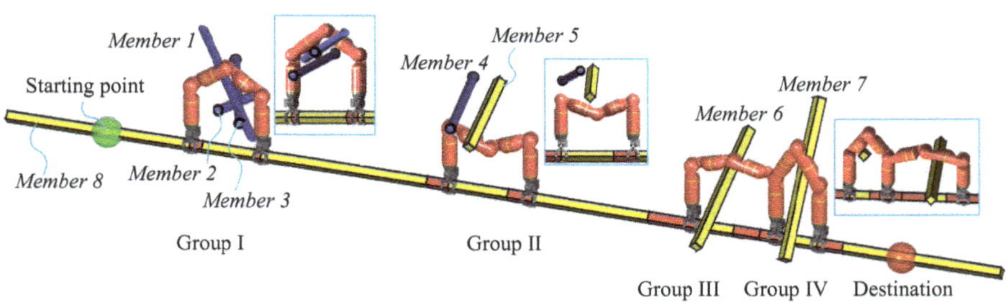

Figure 11. The results of computing operational regions for negotiating multiple obstacles.

6.2. The Result of Good Manipulability

This part of simulation is conducted to verify the necessity of considering the manipulability of transition grips. In the simple scenario in Figure 12, three members comprise the climbing environment. The robot is requested to start from *Member 1*, pass *Member 2* and finally reach *Member 3*. The operational regions for transitions, obtained in the global path planning procedure, are highlighted in green.

The grip planning results with and without consideration of the grip manipulability are shown in the figure, illustrated with transition configurations. The two slightly transparent configurations, corresponding grips on the boundaries of operational regions, are the results without consideration of the grip manipulability. They are very close to the robot's singularity (the T_2 joint is close to 0°). Further in the gripping procedure, it will affect the ability of the robot to adjust its gripper to align with the target member. As a comparison, the other two configurations represent the results considering the grip manipulability but keeping the number of climbing steps the same. From the figure, we can see that the grips are close to the midpoints of the operational regions.

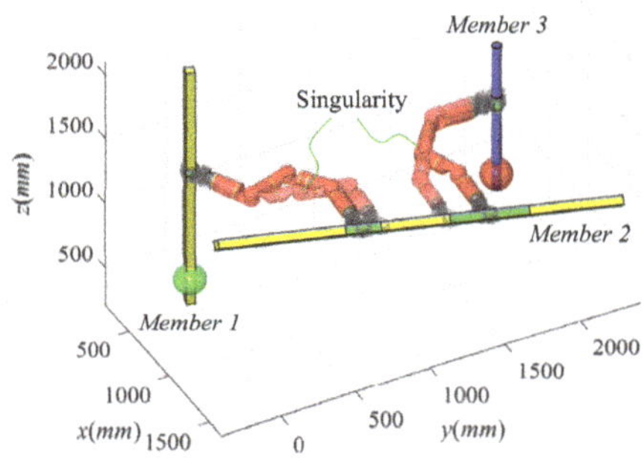

Figure 12. The comparison of determination of transition grips with and without consideration of the grip manipulability.

6.3. The Results of Collision-Free Grip Planning

The simulations in this part are conducted to verify the proposed overall grip planning algorithm. The two scenes used in our previous work, consisting of 9 and 25 members, respectively, are again deployed for simulations. Figures 13 and 14 show us the results.

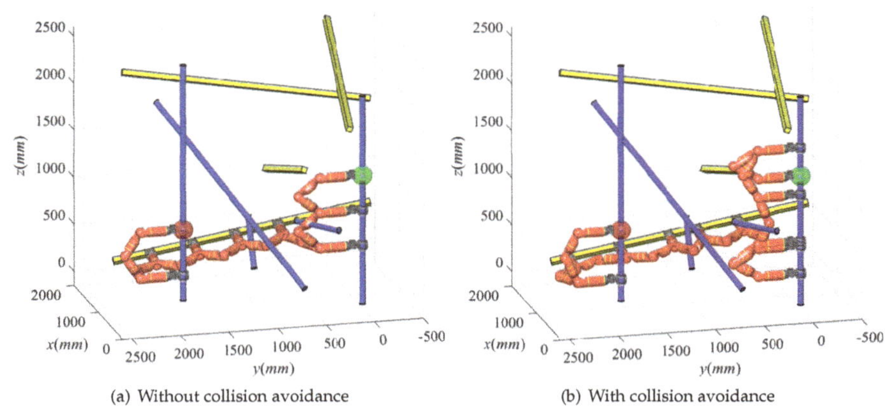

Figure 13. Comparison of results of grip planning with and without consideration of collision avoidance in a scene of 9 members.

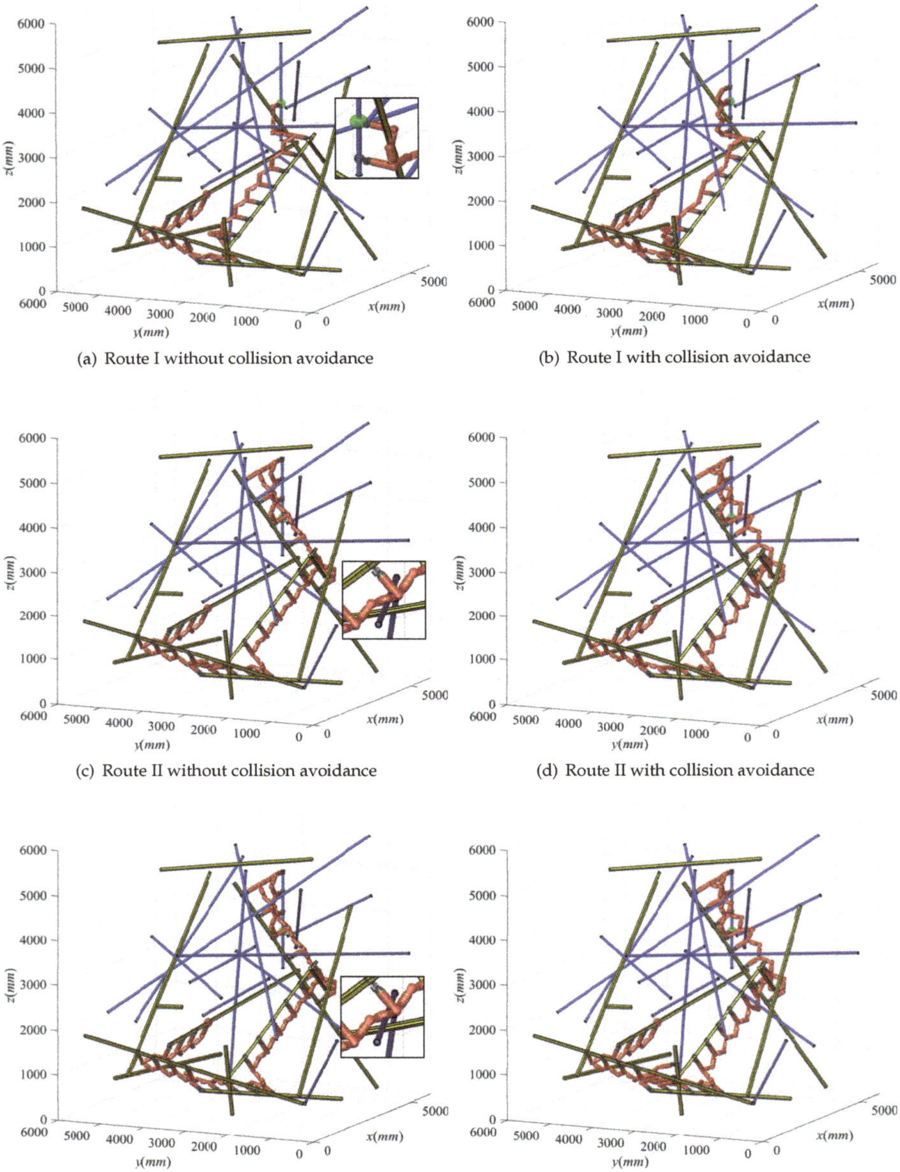

(a) Route I without collision avoidance

(b) Route I with collision avoidance

(c) Route II without collision avoidance

(d) Route II with collision avoidance

(e) Route III without collision avoidance

(f) Route III with collision avoidance

Figure 14. Comparison of results of grip planning with and without consideration of collision avoidance in a scene of 25 members.

If collision avoidance is not taken into account, we do not need to compute the operational regions for negotiating obstacles prior to grip planning. Only the operational regions for transitions are considered in Equation (12). The optimization objectives are to achieve the minimum number of climbing steps, and good manipulability of transition grips as well. Figure 13a shows us the result planned in this way. Eight climbing steps are required from the starting point to the destination. In the figure, we can see that the robot collides with a member when it moves downwards on the first member. Therefore, in this case, the robot actually cannot execute the climbing movement. If collision avoidance is considered, the planner can generate a collision-free solution, where twelve climbing steps are required. As shown in Figure 13b, four extra adjustment steps are added to the route, to change the grip locations. The robot uses two inchworm-like gaits to adjust the gripping positions in climbing. One time is to get ready to surmount the obstacle member with the creeping mode, while another is to prepare for transition since the operational region is short. The planning procedure considering collision avoidance costs 0.78 s.

Figure 14 shows the comparison of results of grip planning, with and without consideration of collision avoidance, in a complex truss environment with 25 members. According to the outputs of the global path planner, there are three feasible routes in total. On the left column are the results obtained without collision avoidance. As a result, the collision between the robot and the truss members can be found in these sub-figures. The collision can be seen more clearly from the zoom-in views. On the right column are the results with collision avoidance. Collision-free grip sequences are successfully solved for each feasible route. Table 2 summaries the comparison of the results. From the table, we know that obstacle members are detected for each route. The time consumption increases four times if collision avoidance is considered. However, the entire planning procedure is still very quick, which costs approximately 0.65 s only. Moreover, the number of grips increases slightly, as the cost of avoiding collision.

Table 2. Comparison of results of grip planning with and without the consideration of collision avoidance.

Items	Collision Avoidance	Route I	Route II	Route III
Number of via members	\	6	6	7
Number of detected obstacle members	without	\	\	\
	with	13	7	9
Number of grips	without	21	28	29
	with	28	31	34
Time consumption (s)	without	0.12	0.11	0.13
	with	0.65	0.62	0.63

7. Conclusions and Future Work

Biped climbing robots have bright and broad application prospects in the field of performing high-rise truss-related routine tasks. To perform such a task, BiCRs must have the ability to plan their grips prior to climbing.

In this paper, we presented a novel optimal collision-free grip planner for BiCRs generating grip sequence in complex truss environments. The planner essentially consists of three components: (1) a mathematical model for computing the operational regions for surmounting obstacles; (2) a grip placement optimizer for determining the grips within operational regions; and (3) a gait interpreter for generating grips between adjacent operational regions. The idea behind this novel scheme is to use the priority to determine the grips at key places, i.e., operational regions for surmounting obstacles and that for performing transitions, then move to other places. Because normally there is less room to choose and adjust grips for these key places. Simulation results verified that the planner was able to plan approximately 30 grips within 0.65 s, taking collision avoidance, grip manipulability and number of climbing steps into account. However, the current planner only considers the forward and backward movement on a member, which limits the applicable objects as BiCRs with, or less than, five degrees of freedom. In addition, the total time consumed in grip planning is related to the initial values for optimization.

In the near future, we will study the optimal climbing gait type and the collision-free motion for the robot shifting between adjacent grips. Extensive climbing experiments on various trusses will be conducted to further verify our planning algorithms.

Author Contributions: S.G. and H.Z. (Haifei Zhu) co-organized the work, conceived and designed the planner and performed the simulation work. S.G. wrote the manuscript. H.L. and Y.G. co-worked to prepare the final manuscript. Y.G. and H.Z. (Hong Zhang) co-supervised the research.

Funding: This research was funded in part by the National Natural Science Foundation of China (Grant Nos. 51605096, 51705086), the Frontier and Key Technology Innovation Special Funds of Guangdong Province (Grant Nos. 2015B090922003, 2017B050506008), the Program of Foshan Innovation Team of Science and Technology (Grant No. 2015IT100072) and the Open Funds of Beijing Advanced Innovation Centre for Intelligent Robots and Systems (Grant No. 2016IRS16).

Conflicts of Interest: The authors declare no conflict of interest.

Nomenclature

$\boldsymbol{\Gamma}$	set of all feasible routes
$\boldsymbol{\Gamma}_i$	i-th feasible route, $\boldsymbol{\Gamma}_i = \{\mathfrak{M}, {}^W\boldsymbol{R}, \mathfrak{R}_t\}$
$\mathfrak{M}$	member sequence along the route
$\mathfrak{M}_j$	j-th member
$\mathfrak{R}$	operational region $\mathfrak{R} = \{\mathfrak{R}_o, \mathfrak{R}_t, \mathfrak{R}_{init}\}$, each inclusive of the takeoff segement and the landing segement
$\mathfrak{R}_t$	operational region for transiting between adjacent members
$\mathfrak{R}_o$	operational region for negotiating obstacle members
$\mathfrak{R}_{init}$	operational region representing starting point and destination
${}^W\boldsymbol{R}$	set of gripping orientations corresponding to member sequence $\mathfrak{M}$
${}^W\boldsymbol{R}_j$	gripping orientation on the j-th member
${}^W\boldsymbol{P}_0, {}^W\boldsymbol{P}_{dir}, L_{mem}, r_{mem}$	reference point, direction unit vector, length and radius of the j-th member
t	a scale to specify the gripping position on the j-th member
${}^W\boldsymbol{Truss}$	truss environment
${}^jx_k$	k-th gripping position in the takeoff segment $[{}^j\underline{x}_k, {}^i\overline{x}_k]$ on the j-th member
${}^jy_k$	k-th gripping position in the landing segment $[{}^j\underline{y}_k, {}^i\overline{y}_k]$ on the j-th member
${}^j\mathcal{O}$	obstacles when moving on the j-th member
$\mathbb{S}_{or}\,(G_{or}, C_{or})$	grips in the operational regions
G_{or}	a grip in the operational region, inclusive of gripping position and orientation
C_{or}	a configuration corresponding to G_{or}
$\mathbb{S}\,(G, C)$	entire grip sequence
$\boldsymbol{P}_{kp}$	key point for negotiating an obstacle member, $\boldsymbol{P}_{kp} = [x_{kp}, y_{kp}, z_{kp}]$
η	robot plane
Δ	the cuboid space where collision may happen, for moving on a member
Ω	the sphere space where collision may happen, for transiting
$l_{cub}, w_{cub}, h_{cub}$	length, width and height of the cuboid space
l_{safe}	a pre-defined safe distance to facilitate the gripping and grip-releasing operations
${}^jN_{min}$	minimum number of climbing steps when passing the j-th member
${}^jN_k$	minimum number of climbing steps from the k-th landing segment to the $(k+1)$-th takeoff segment
S_0, S_1	minimum and maximum step lengths of the inchworm-like gait
S_2, S_3	minimum and maximum step lengths of the flipping-over gait and swinging-around gait
S_4	maximum step length of the hybrid gait
$\mathcal{R}$	robot

References

1. Xu, Y.; Brown, B.; Aoki, S.; Kanade, T. Mobility and Manipulation of a Light-weight Space Robot. In Proceedings of the IEEE/RSJ International Conference on Intelligent Robots and Systems, Munich, Germany, 12–16 September 1994; pp. 11–19.
2. Balaguer, C.; Gimenez, A.; Pastor, J.M.; Padron, V.M.; Abderrahim, M. A Climbing Autonomous Robot for Inspection Applications in 3D Complex Environments. *Robotica* **2000**, *18*, 287–297. [CrossRef]
3. Detweiler, C.; Vona, M.; Yoon, Y.; Yun, S.K.; Rus, D. Self-assembling Mobile Linkages. *IEEE Robot. Autom. Mag.* **2007**, *14*, 45–55. [CrossRef]
4. Tavakoli, M.; Marjovi, A.; Marques, L.; Almeida, A.T.D. 3DCLIMBER: A Climbing Robot for Inspection of 3D Human Made Structures. In Proceedings of the IEEE/RSJ International Conference on Intelligent Robots and Systems, Nice, France, 22–26 September 2008; pp. 4130–4135.
5. Tavakoli, M.; Marques, L. 3DCLIMBER: Climbing and Manipulation over 3D Structures. *Mechatronics* **2011**, *21*, 48–62. [CrossRef]
6. Tavakoli, M.; Zakerzadeh, M.R.; Vossoughi, G.; Bagheri, S. A Hybrid Pole Climbing and Manipulating Robot with Minimum DOFs for Construction and Service Applications. *Ind. Robot. Int. J.* **2005**, *32*, 171–178. [CrossRef]
7. Lam, T.L.; Xu, Y. A Flexible Tree Climbing Robot: Treebot—Design and Implementation. In Proceedings of the IEEE International Conference on Robotics and Automation, Shanghai, China, 9–13 May 2011; pp. 5849–5854.
8. Guan, Y.; Jiang, L.; Zhang, X.; Zhang, H. Climbing Gaits of a Modular Biped Climbing Robot. In Proceedings of the IEEE/ASME International Conference on Advanced Intelligent Mechatronics, Singapore, 14–17 July 2009; pp. 532–537.
9. Guan, Y.; Jiang, L.; Zhu, H.; Wu, W.; Zhou, X.; Zhang, H.; Zhang, X. Climbot: A Bio-inspired Modular Biped Climbing Robot—System Development, Climbing Gaits, and Experiments. *J. Mech. Robot.* **2016**, *8*, 021026. [CrossRef]
10. Chen, S.; Zhu, H.; Guan, Y.; Wu, P. Collision-free Single-step Motion Planning of Biped Pole-climbing Robots in Spatial Trusses. In Proceedings of the IEEE International Conference on Robotics and Biomimetics, Shenzhen, China, 12–14 December 2013; pp. 280–285.
11. Chen, W.; Gu, S.; Zhu, L.; Zhang, H.; Zhu, H.; Guan, Y. Representation of Truss-style Structures for Autonomous Climbing of Biped Pole-climbing Robots. *Robot. Auton. Syst.* **2018**, *101*, 126–137. [CrossRef]
12. Kuffner, J.; Nishiwaki, K.; Kagami, S.; Inaba, M. Footstep Planning among Obstacles for Biped Robots. In Proceedings of the IEEE/RSJ International Conference on Intelligent Robots and Systems, Maui, HI, USA, 29 October–3 November 2001; pp. 500–505.
13. Kuffner, J.; Nishiwaki, K.; Kagami, S.; Inaba, M.; Inoue, H. Motion Planning for Humanoid Robots. *Auton. Robot.* **2003**, *12*, 692–698.
14. Chestnutt, J.; Lau, M.; Cheung, G.; Kuffner, J.; Hodgins, J.; Kanade, T. Footstep Planning for the Honda ASIMO Humanoid. In Proceedings of the IEEE International Conference on Robotics and Automation, Orlando, FL, USA, 15–19 May 2006; pp. 629–634.
15. Chestnutt, J.; Kuffner, J.; Nishiwaki, K.; Kagami, S. Planning Biped Navigation Strategies in Complex Environments. In Proceedings of the IEEE International Conference on Humanoid Robotics, Karlsruhe, Germany, 29–30 September 2003.
16. Chestnutt, J.; Kuffner, J. A Tiered Planning Strategy for Biped Navigation. In Proceedings of the IEEE/RSJ International Conference on Humanoid Robots, Santa Monica, CA, USA, 10–12 November 2004; pp. 422–436.
17. Ayaz, Y.; Munawar, K.; Malik, M.B.; Konno, A.; Uchiyama, M. Human-Like Approach to Footstep Planning Among Obstacles for Humanoid Robots. In Proceedings of the IEEE/RSJ International Conference on Intelligent Robots and Systems, San Diego, CA, USA, 29 October– 2 November 2007; pp. 5490–5495.
18. Ayaz, Y.; Owa, T.; Tsujita, T.; Konno, A. Footstep planning for Humanoid Robots among Obstacles of Various Types. In Proceedings of the IEEE International Conference on Humanoid Robots, Paris, France, 7–10 December 2009; pp. 361–366.
19. Deits, R.; Tedrake, R. Footstep Planning on Uneven Terrain with Mixed-integer Convex Optimization. In Proceedings of the IEEE/RSJ International Conference on Humanoid Robots, Madrid, Spain, 18–20 November 2014; pp. 279–286.

20. Kuindersma, S.; Deits, R.; Fallon, M.; Valenzuela, A.; Dai, H.; Permenter, F.; Koolen, T.; Marion, P.; Tedrake, R. Optimization-based Locomotion Planning, Estimation, and Control Design for the ATLAS Humanoid Robot. *Auton. Robot.* **2016**, *40*, 429–455. [CrossRef]
21. Guan, Y.; Yokoi, K.; Tanie, K. Feasibility: Can Humanoid Robots Overcome Given Obstacles? In Proceedings of the IEEE International Conference on Robotics and Automation, Orlando, FL, USA, 15–19 May 2006; pp. 1054–1059.
22. Guan, Y.; Neo, E.S.; Yokoi, K.; Tanie, K. Stepping over Obstacles with Humanoid Robots. *IEEE Trans. Robot.* **2006**, *22*, 958–973. [CrossRef]
23. Balaguer, C.; Padron, V.M.; Gimenez, A.; Pastor, J.M.; Abderrahim, M. Path Planning Strategy of Autonomous Climbing Robot for Inspection Applications in Construction. In Proceedings of the International Symposium on Automation and Robotics in Construction, Madrid, Spain, 22–24 September 1999.
24. Detweiler, C.; Vona, M.; Kotay, K.; Rus, D. Hierarchical Control for Self-assembling Mobile Trusses with Passive and Active Links. In Proceedings of the IEEE International Conference on Robotics and Automation, Orlando, FL, USA, 15–19 May 2006; pp. 1483–1490.
25. Chung, W.K.; Xu, Y. Minimum energy demand locomotion on space station. *J. Robot.* **2013**, *2013*, 723535. [CrossRef]
26. Zhu, H.; Guan, Y.; Su, M.; Cai, C. Evaluation of Graspable Region and Selection of Footholds for Biped Pole-climbing Robots. In Proceedings of the IEEE International Conference on Robotics and Biomimetics, Bali, Indonesia, 5–10 December 2014; pp. 1757–1762.
27. Zhu, H.; Gu, S.; He, L.; Guan, Y.; Zhang, H. Transition Analysis and Its Application to Global Path Determination for a Biped Climbing Robot. *Appl. Sci.* **2018**, *8*, 122. [CrossRef]
28. Xiao, Z.; Wu, W.; Wu, J.; Zhu, H. Gripper Self-alignment for Autonomous Pole-grasping with a Biped Climbing Robot. In Proceedings of the IEEE International Conference on Robotics and Biomimetics, Guangzhou, China, 11–14 December 2012; pp. 181–186.
29. Gu, S.; Su, M.; Zhu, H.; Guan, Y.; Rojas, J.; Zhang, H. Efficient Pole Detection and Grasping for Autonomous Biped Climbing Robots. In Proceedings of the IEEE International Conference on Robotics and Biomimetics, Macau, China, 5–8 December 2017; pp. 246–251.

applied sciences

MDPI

Article

Design and Experiment of a Novel Façade Cleaning Robot with a Biped Mechanism

Shunsuke Nansai [1,*], Keichi Onodera [2], Prabakaran Veerajagadheswar [3], Mohan Rajesh Elara [4] and Masami Iwase [2]

1 Department of Advanced Machinery Engineering, School of Engineering, Tokyo Denki University, Tokyo 120-8551, Japan
2 Department of Advanced Multidisciplinary Engineering, Graduate School of Advanced Science and Technology, Tokyo Denki University, Tokyo 120-8551, Japan; onodera@ctrl.fr.dendai.ac.jp (K.O.); iwase@fr.dendai.ac.jp (M.I.)
3 SUTD-JTC I[3] Center, Singapore University of Technology and Design, Singapore 487372, Singapore; prabakaran@sutd.edu.sg
4 Engineering Product Development Pillar, Singapore University of Technology and Design, Singapore 487372, Singapore; rajeshelara@sutd.edu.sg
* Correspondence: nansai@ctrl.fr.dendai.ac.jp; Tel.: +81-3-5284-5120

Received: 10 October 2018; Accepted: 19 November 2018; Published: 26 November 2018

Abstract: Façade cleaning in high-rise buildings has always been considered a hazardous task when carried out by labor forces. Even though numerous studies have focused on the development of glass façade cleaning systems, the available technologies in this domain are limited and their performances are broadly affected by the frames that connect the glass panels. These frames generally act as a barrier for the glass façade cleaning robots to cross over from one glass panel to another, which leads to a performance degradation in terms of area coverage. We present a new class of façade cleaning robot with a biped mechanism that is able overcome these obstacles to maximize its area coverage. The developed robot uses active suction cups to adhere to glass walls and adopts mechanical linkage to navigate the glass surface to perform cleaning. This research addresses the design challenges in realizing the developed robot. Its control system consists of inverse kinematics, a fifth polynomial interpolation, and sequential control. Experiments were conducted in a real scenario, and the results indicate that the developed robot achieves significantly higher coverage performance by overcoming both negative and positive obstacles in a glass panel.

Keywords: glass façade cleaning robot; wall climbing robot; biped mechanism

1. Introduction

Robots have been advancing exponentially over the last three decades, moving beyond the traditional bounds of industrial applications into service missions and sharing social spaces with humans. Frey and Osborne have estimated that 47% of total US employment will be replaced by robots and/or artificial intelligence (AI) in the near future [1]. Early evidence points to significant improvements in productivity and safety over a number of service tasks that are dull, dirty, and/or dangerous [2]. Façade cleaning of high-rise buildings and skyscrapers offers enormous opportunities for the use of robots. In recent decades, skyscrapers such as Burj Khalifa [3–6] in Dubai and the Shanghai Tower [7–9] in Shanghai have been built as a result of improvements in construction technologies and processes. Even in such newfangled skyscrapers, the façades are generally cleaned by humans. In the case of some skyscrapers, including Burj Khalifa, equipment, such as gondolas, is not installed, and humans are employed to clean façades by hanging from a rope. The use of a manual workforce poses a high risk of fall, accidents, and even fatalities. Numerous incidences of accidents have been reported, even with the use of a gondola, in façade cleaning jobs. One such situation involved a gondola that lost control caused by a gust of wind at the Shanghai World Financial Center [10]. In another instance, a gondola became suspended in mid-air at a height of 240 m at the One World Trade Center in New York [11]. Robotic solutions offer enormous potential in significantly minimizing the risk to humans and improving productivity in façade cleaning jobs.

To overcome such issues, research on glass façade cleaning robots has been reported. For example, a series of Skycleaners, which are entirely driven by pneumatic actuators, has been developed to move and clean on the glass wall by utilizing vacuum suction cups [12,13]. The Skycleaner is installed with suction cups on both ends of the actuators in an X–Y stage, and these suction cups are connected to a vacuum pump [14]. Sun et al. developed a wall climbing robot with an X–Y stage mechanism as well as a rotational joint and demonstrated the effectiveness of the developed robot and the proposed strategies through field trials at both the City University of Hong Kong and the British Broadcasting Corporation (BBC) [15]. The efficacy and validity of the developed system has been demonstrated through a range of experiments [16]. Serbot AG, an industrial company in Switzerland, developed a wall and solar panel cleaning robot called GEKKO [17]. GEKKO is equipped with a vacuum suction cup crawler which rotates horizontally. The crawler has a semi-circular shape and has a straight line on one side and an arc line on the other side, through the use of which both linear and rotational motions are realized. By changing the number of brushes and crawlers, the developed robot can be manually reconfigured to adapt to different wall areas. Seo et al. developed ROPE RIDE [18]. ROPE RIDE enables one to climb up a vertical surface by utilizing a rope dropped down from the top of the building, two additional propeller thrusters are installed to securely adhere to the wall. A cleaning unit for a cleaning wall has been installed [19]. A new impedance control system has been proposed, and the effectiveness of the force control system to press brushes to the wall with a constant force has been shown through an experiment [20]. SIRIUS is a wall cleaning robot developed by Fraunhofer IFF, which is an industrial company in Germany [21–24]. SIRIUS realizes up-and-down motion by utilizing a crane installed on buildings for façade maintenance. It adheres to the wall by a suction system which locomotes like a linear actuator [25–27]. Ceplina et al. developed a wall climbing robot capable of spraying different cleaning and disinfecting liquids in hospital and clinic nursing environments assuming a full flat wall surface [28]. Akinfiev et al. developed a climbing cleaning robot hanging down from a crane. They prototyped the robot and demonstrated its effectiveness via experiments in real conditions [29]. The methods of the literature are effective and valid, as shown by the experimental results. Even though numerous studies in the literature have demonstrated the use of glass façade cleaning robots, conventional platforms experience a system failure with respect to area coverage, which degrades their full efficiency. One major factor in glass façade structures that affects the performances of façade cleaning systems is the frames that connect the two glass panels. These frames generally act as a barrier for glass façade cleaning robots to cross over from one glass panel to another, which leads to a performance degradation in terms of area coverage. One viable approach to overcome these difficulties in façade structures is to develop a façade cleaning system with a biped mechanism that can overcome these obstacles.

In the field of robotics, there have been several biped robots that have been reported in the literature for different application scenarios. Christian Ott et al. [30] proposed a novel biped walking machine for domestic service purpose. The proposed robot was designed as an experimental system for studying biped locomotion based on torque-controlled joints, and the experimental results were discussed by utilizing the scheme. In another work, Asiya M. Al-Busaidi developed a low-cost educational system to study and control a biped robot in real time [31]. The proposed system used MATLAB for visualization control. Arduino was used as a low-level controller to manage the walking gaits of the biped robot. Chen-Yuan Liu and Jhen-Cheng Wang developed a novel biped robot to obtain patient data from a Japanese patient office [32]. This particular study applied a three S-Curve model to analyze the acquired data. In addition, the analyzed information was used to foresee the development trend in the field of biped robot walking. Biped robots have been proposed for wall climbing and inspection purposes. The authors in Reference [33] proposed Climbot, which was developed based on the climbing patterns of leech worms. The presented robot was developed under a biped concept with five degrees of freedom and two special grippers as end effectors. Similarly, an urban reconnaissance climbing biped robot is presented in Reference [34]. The developed robot was under-actuated and made smaller in order to access confined spaces. The study discusses the evaluation process of the robot's performance while navigating on inclined surfaces. In spite of the fact that numerous studies in the literature address biped robotics, they are largely limited to inspections and rescue applications. In addition, none of the previous work in biped robotics targets glass façade cleaning, which presents significant opportunities for research and development.

Our ultimate goal is to develop a glass façade cleaning robot using a biped mechanism capable of crossover from one glass panel to another without any difficulties, with an objective of maximizing its coverage area. In order to attain a working robotic system that can achieve a superior area coverage performance, we adopted a parallel linkage mechanism for the developed platform. The major challenges encountered during the development of our façade cleaning robot were its parallel linkage mechanism design, the assimilation of cleaning modules, and the translation of the theoretical design into a physical prototype. This paper summarizes all these aspects and concludes with experimental results that validate its ability to move from one glass pane to another.

This paper is organized as follows. The design and development of the biped façade cleaning robot satisfying the basis strategy is described in Section 2. Section 3 describes the design of the control system to locomote the developed biped robot on the glass surface. It is verified in Section 4 that the developed robot is capable of moving on the glass surface by utilizing the designed trajectory generator. Section 5 concludes this paper.

2. Development of Biped Robot

This section discusses the design challenges of the glass façade cleaning robot in terms of the glass cleaning process and area coverage. In particular, the abilities and tasks involved in glass façade cleaning are discussed, and ways in which to realize these abilities and tasks are proposed. Additionally, the hardware/software design challenges are described.

2.1. Design Challenges

In order to design an efficient façade cleaning robot, it is critical to (1) clean the window from top to bottom and (2) move the squeegee to track the scrubber, as shown in Figure 1. The design challenges involved in these key points are discussed in this section. The basic strategy is to resolve functions to achieve each key point as an individual design/control issue. Figure 2 shows a solution for window size and window shape for terrain adaptability. The solution is a biped system and consists of individual legs. The developed system extends its cleaning area and is controlled in a coordinated way. If the window shape is changed, it is able to adapt by extending the cleaning area of each individual leg. Additionally, Figure 2 shows the technique of the scrubber and the squeegee. The solution consists of a swarm control system to control the robots systematically. The swarm control system has high adaptability, as shown in Reference [35]. Patil et al. designed and implemented a UB swarm robot system consisting of five robots that are heterogeneous in sensory units, microcontroller, functionality, and size [36]. Our proposed swarm robot system consists of multiple biped robots equipped with a cleaning unit on each robot's foot so that each robot can move from top to bottom and that the squeegee is installed next to the scrubber. Keeping the squeegee in contact with the glass surface during cleaning is a control issue of the robot. Additionally, cleaning in one direction is a transition plan of the developed robotic system as well. Thus, the challenges regarding the hardware and software are listed as follows:

Hardware Design Challenges

- Development of the biped robot
- Development of a cleaning unit with the squeegee installed next to the scrubber

Software Design Challenges

- Design of a control system capable of keeping both the scrubber and the squeegee in contact with the glass surface.
- Design of a transition planner in order to clean in one direction

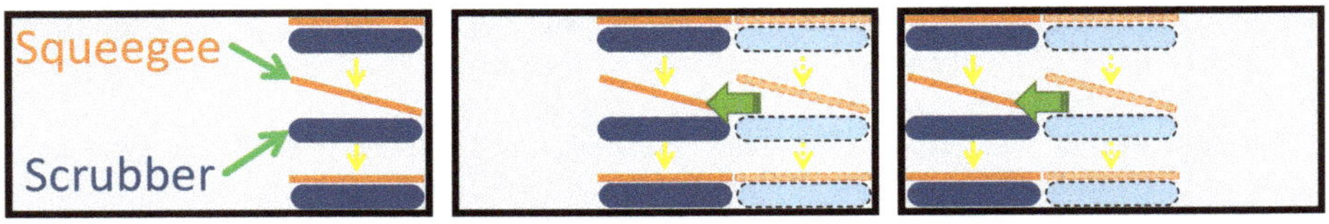

Figure 1. Two key points that need to be addressed.

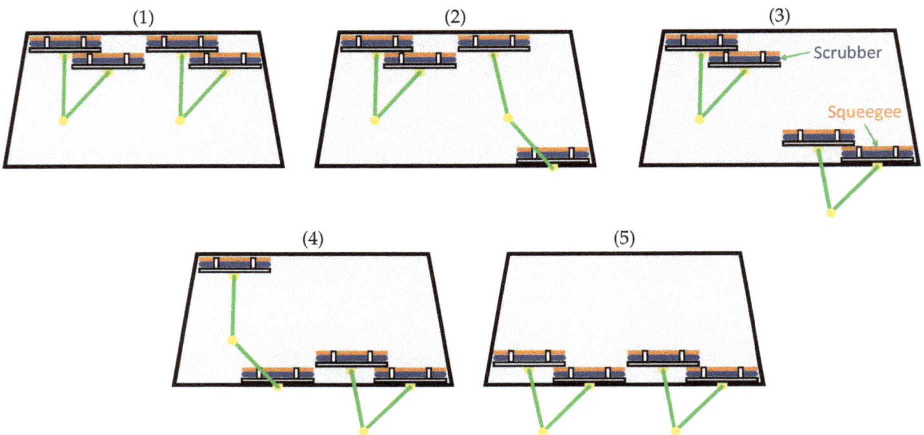

Figure 2. A biped robot adapts to distinct window frames. The process of our strategy is illustrated from (**1**) to (**5**).

2.2. Biped Robot Design

Our strategy consisting of the biped mechanism and its control system has been described. In this section, specifications of the biped robot are analyzed quantitatively, and the development of the biped robot is shown.

Figure 3 is a CAD design of the biped robot. Each foot unit includes a cleaning unit, a suction cup, a vacuum pump, a seal mechanism, a valve to break vacuum, and servo motors for joint actuation. The morphology adopts a parallel linkage mechanism that keeps the foot unit of the robot and the glass surface parallel without any control system. That is, since the degree of freedom of the system is able to be decreased, the number of actuators for locomotion can also be decreased. Hence, adopting a parallel linkage mechanism provides an advantage in terms of weight saving. Weight saving is important because it allows one to decrease the amount of energy needed to maintain the robot's weight. The area of both the adhesion/sealing mechanism and the driving mechanism are separated. This separation provides two advantages: the driving range in the rotational direction is extended, and maintenance is easy. Since only one attachment is required between each part, hardware updates in the future will be easy. Support parts, such as legs, can be installed onto the adhesion component. The support parts are pushed to the glass surface when the suction cup is in vacuum mode. This makes the robot robust against glass surfaces.

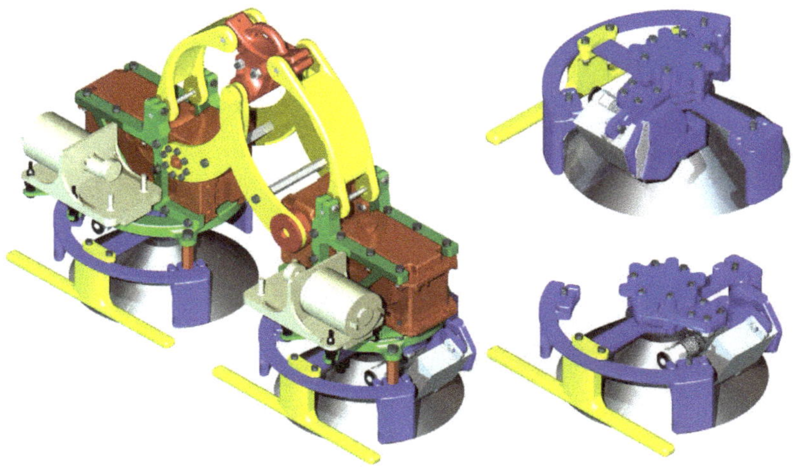

Figure 3. CAD design of the robot. Each foot unit includes a cleaning unit, a suction cup, a vacuum pump, a seal mechanism, a valve to break vacuum, and servo motors for joint actuation. The robot installs a parallel linkage mechanism to keep the foot unit of the system and the glass surface parallel without any control system.

Figure 4 shows a schematic figure of the designed robot. The designed robotic system can be supposed as a simple two-link manipulator in a vertical plane on the arm. The origin is set on the center of the supporting leg, and the center of the idling leg (x_h, y_h) is

$$\begin{cases} x_h = l(\cos\phi_1 + \cos\phi_2) + 2d, \\ y_h = l(\sin\phi_1 - \sin\phi_2) \end{cases} \tag{1}$$

where l [cm] and d [cm] represent the length of the link and the distance between the center of the foot to the joint, respectively. The angles of each link $\phi_{\{1,2\}}$ are limited within Equation (2) as a result of the installation of the parallel linkage mechanism:

$$\frac{\pi}{4} \leq \phi_{\{1,2\}} \leq \frac{3\pi}{4}. \tag{2}$$

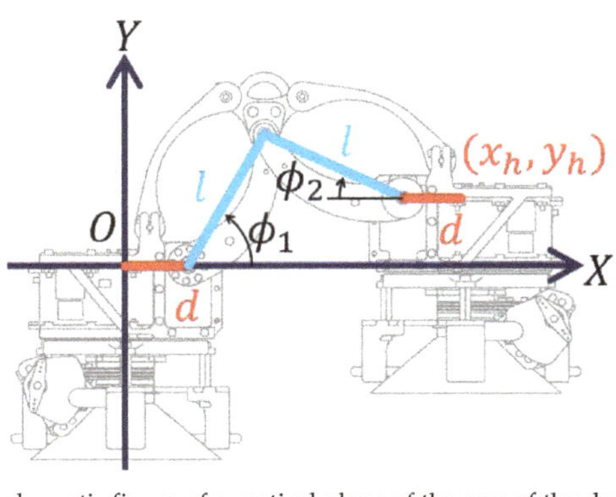

Figure 4. A schematic figure of a vertical plane of the arm of the designed robot.

From Equation (1) within Equation (2), the theoretical movement range of the model is shown in Figure 5, where the actual movement range is smaller than Figure 5 due to collisions between the foot units. On the basis of Figure 5 and Equation (1), the robot is capable of moving ±13.66 cm along the z axis and 22.54 cm along the x axis. In the case of the X–Y stage type, its movement range depends on one of its actuators. On the other hand, biped type robots, like the designed robot, can obtain enough movement range, even though one of the actuators is limited, since it mainly depends on its mechanism and link lengths.

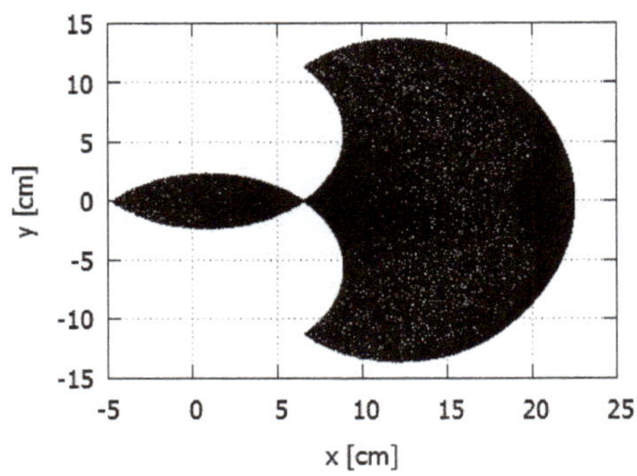

Figure 5. The theoretical movement range of the vertical plane of the arm of the designed robot. The system is capable of moving ±13.66 cm along the z axis and 22.54 cm along the x axis.

The MX-106T and the RX-28 Dynamixel Robot Servo Actuator were adopted for driving actuators and were utilized as a base actuator and an arm actuator, respectively. Table 1 shows their specifications. Regarding the base actuator, the required torque $\tau_{b\{1,2\}}$ [Nm] for driving is obtained by Equation (3), based on a simplified model shown in Figure 6.

$$\tau_{b\{1,2\}} \geq m_{\{2,1\}} g(2d + l_1 \cos\phi_1 + l_2 \cos\phi_2) \cos\theta_{b\{1,2\}}. \tag{3}$$

From Equation (3), if $\theta_{b\{1,2\}} = \phi_{\{1,2\}} = 0$ rad, the required torque becomes maximum. Since the stall torque of the servo motor is 8.4 Nm from Table 1, the driving torque is 2.8 Nm with a safety factor of 3. The relationship between link length and the weight of the foot unit is shown in Figure 7. From Figure 7, "×" highlighted in red represents the actual design parameter, and it satisfies Equation (3). Thus, MX-106T is able to generate enough driving torque as a base actuator. Regarding the base actuator, if $\theta_{b1} = -\pi/2$, the required torque for driving $\tau_{\{1,2\}}$ [Nm] is obtained by Equation (4), based on a simplified model shown in Figure 8.

$$\tau_{\{1,2\}} \geq m_{\{1,2\}} g l_{\{1,2\}} \sin\phi_{\{1,2\}}. \tag{4}$$

From Equation (4), if $\phi_{\{1,2\}} = \pi/2$ rad, the required torque becomes maximum. Since the stall torque of the servo motor is 3.7 Nm from Table 1, the driving torque is 1.23 Nm with a safety factor of 3. The relationship between link length and the weight of the foot unit is shown in Figure 9. From Figure 9, "×" highlighted in red represents the actual design parameter, and it satisfies Equation (4). Thus, RX-28 is able to generate enough driving torque as a base actuator.

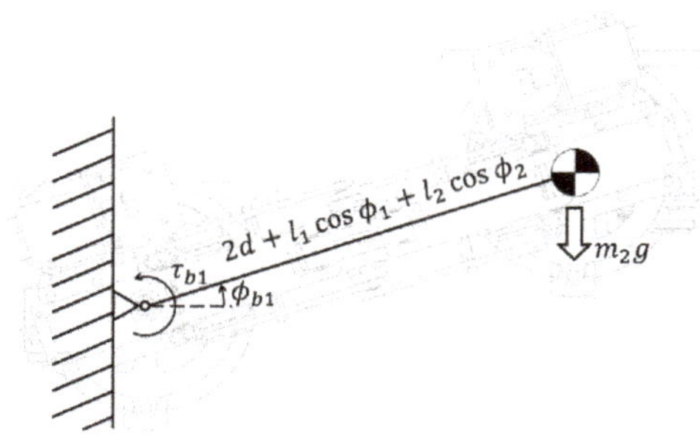

Figure 6. A schematic figure of the simplified model regarding the base actuator.

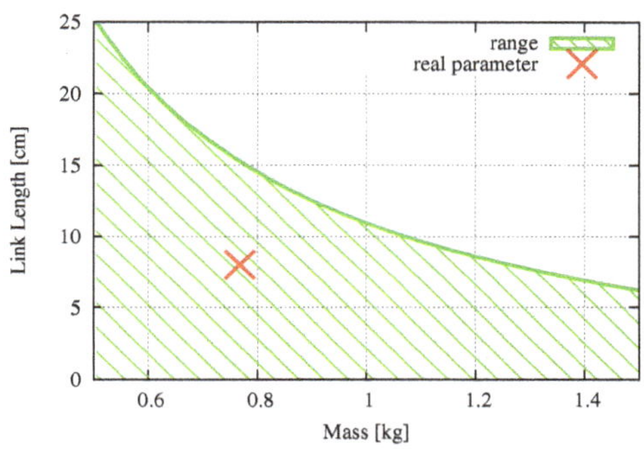

Figure 7. Designable range satisfying Equation (3). "×" highlighted in red represents the actual design parameter, and it satisfies Equation (3). Thus, MX-106T is able to generate enough driving torque as a base actuator.

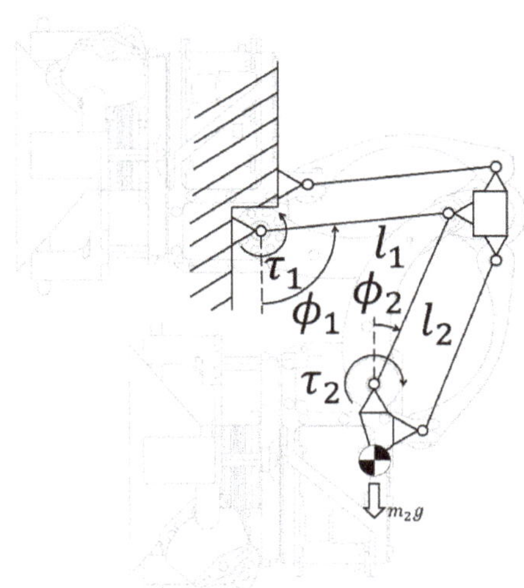

Figure 8. A schematic figure of the simplified model regarding the arm actuator.

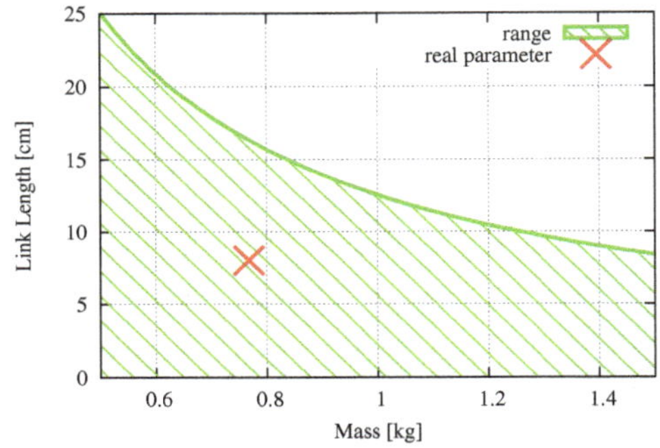

Figure 9. Designable range satisfying Equation (4). "×" highlighted in red represents the actual design parameter, and it satisfies Equation (4). Thus, RX-28 is able to generate enough driving torque as a base actuator.

Table 1. Specifications of the servo motors.

Model Number	MX-106T	RX-28	HS-5085MG
Maker	ROBOTIS	ROBOTIS	HITEC
Stall Torque	8.4 Nm	3.7 Nm	4.3 kg·cm
No Load Speed	45 rpm	85 rpm	0.13 rec/60deg
Weight	153 g	72 g	21.9 g
Voltage	12 V	12 V	6 V

DP0125 is a vacuum pump, and its specification is shown in Table 2. A mechanical valve was installed on the switch vacuum and was controlled by an RC servo motor (shown in Table 1). A diagram of the vacuum system is shown in Figure 10. Since the diameter of the installed suction cup is 4 in. (approximately 10.16 cm), the suction force in the vertical situation W [N] is derived as follows:

$$\begin{aligned} W &= CP0.1f, \\ &= 10.16^2\pi \times 33.3 \times 0.1 \times \frac{1}{8} \\ &= 134.987, \end{aligned}$$

where C [cm^2], P [-kPa], and f represent the area of the suction cup, the vacuum pressure, and the safety factor, respectively, and the safety factor is generally set as 1/8 for the vertical situation. The weight of the developed robot is 1.483 kg and ensures that the designed vacuum system can secure itself to the vertical surface. Finally, the actual developed robot from CAD is shown in Figure 11.

Table 2. Specification of the vacuum pump.

Model Number	DP0125
Maker	Nitto Kohki
Attainable Vacuum	−33.3 kPa
Free Air Displacement	2.5 L/min
Voltage	12 V
Weight	80 g

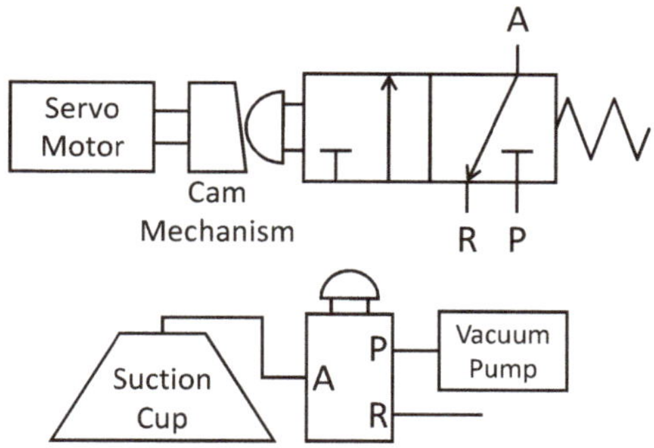

Figure 10. A diagram of the vacuum system. **Top**: The switching valve is controlled by the RC servo motor via a cam mechanism. **Bottom**: Chamber air in the suction cup is sucked once the toggle is pushed in.

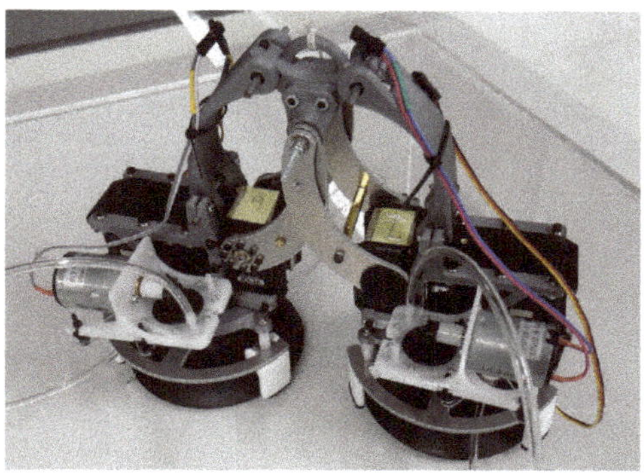

Figure 11. Photograph of the actual robot.

3. Trajectory Generation

This section shows how the trajectories to locomote the developed robot individually were generated. Generally, biped robots alternately moves their left foot and then and right foot forward. However, such locomotion is not effective for glass façade cleaning, as discussed. In order to satisfy our basic strategy, the system locomotes one step at a time. We developed a gait to locomote a nest target step position indicated by an operator as a first step to full automation. A straight line basis gait should be generated (rather than a smooth gait as animals have) because when engaging/disengaging, the suction cup should approach the glass surface vertically to prevent involving the edge of the suction cup. Straight line basis foot trajectories were generated by utilizing a fifth polynomial interpolation and inverse kinematics. We assumed that the foot position would vary sequentially, and generated it utilizing sequential control.

3.1. Inverse Kinematics

Firstly, target angles of each actuator, ϕ_{b1}, ϕ_{b2}, ϕ_1, and ϕ_2 [rad], to realize target positions of the foot unit x_r, y_r, z_r [cm] were formulated by utilizing inverse kinematics. Figure 12 and Table 3 show a schematic figure and parameters of the robot.

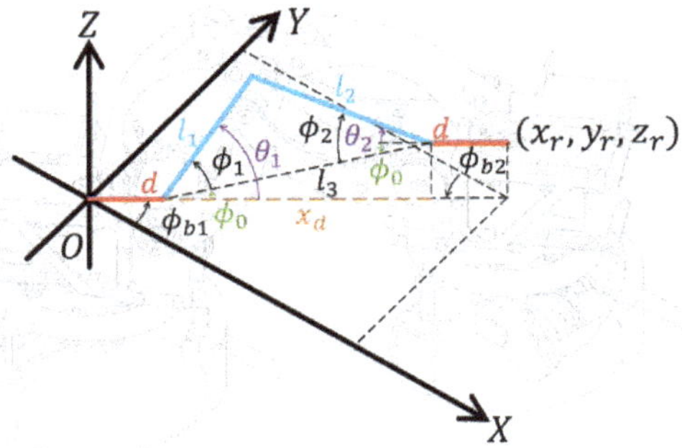

Figure 12. The schematic figure of the biped robot in three-dimensional space.

Table 3. Physical Parameters. ($i = 1, 2$).

Notation	Parameters
l_i	Link length
m_i	Mass
d	Length from center of the foot unit to the joint
l_3	Length between the joint of each foot unit
x_d	Length of l_3 on the X–Y plane

The target angles of the base actuators are

$$\phi_{b\{1,2\}} = \tan^{-1}\frac{y_r}{x_r}. \tag{5}$$

Moreover,

$$x_d = \sqrt{x_r^2 + y_r^2} - 2d,$$
$$l_3 = \sqrt{x_d^2 + z_r^2},$$
$$\phi_0 = \tan^{-1}\frac{x_d}{z_r}.$$

By cosine theorem,

$$\phi_1 = \cos^{-1}\frac{l_1^2 + l_3^2 - l_2^2}{2l_1l_3},$$
$$\phi_2 = \cos^{-1}\frac{l_2^2 + l_3^2 - l_1^2}{2l_2l_3}.$$

Thus, the target angles of each link $\theta_{\{1,2\}}$ are

$$\theta_1 = \phi_0 + \phi_1, \tag{6}$$
$$\theta_2 = -\phi_0 + \phi_2. \tag{7}$$

The trajectories of each joint to move the system were derived by utilizing inverse kinematics formulated in Equations (5)–(7).

3.2. Fifth-Degree Polynomial Interpolation

A fifth-degree polynomial interpolation generates a smooth trajectory from a starting point of a gait to an ending point as the speed and the acceleration are represented as fourth and third degrees, respectively. Since all trajectory parameters (position, speed, and acceleration) in this approach are represented as a continuous function, load forces to the actuators can be decreased. Given these advantages, we used this technique for our generation of trajectories, an approach that has been successfully used previously in the community [37–39].

The trajectories of position $x_r(t)$, $y_r(t)$, $z_r(t)$, speed $\dot{x}_r(t)$, $\dot{y}_r(t)$, $\dot{z}_r(t)$, and acceleration $\ddot{x}_r(t)$, $\ddot{y}_r(t)$, $\ddot{z}_r(t)$ in terms of time t are defined as follows:

$$\{x, y, z\}_r(t) = \sum_{i=0}^{5} a_{\{x,y,z\}i} t^i \tag{8}$$

$$\{\dot{x}, \dot{y}, \dot{z}\}_r(t) = \sum_{i=1}^{5} i a_{\{x,y,z\}i} t^{i-1} \tag{9}$$

$$\{\ddot{x}, \ddot{y}, \ddot{z}\}_r(t) = \sum_{i=2}^{5} i(i-1) a_{\{x,y,z\}i} t^{i-2} \tag{10}$$

where $a_i(i = 0, \cdots, 5)$ represents coefficients and these are derived from initial states and final states of a given gait. The initial and final states are defined as $\{x, y, z\}_s$ and $\{x, y, z\}_f$, respectively. Let transformation time T [s] be constant. Using Equations (8)–(10), the initial states at the starting times, a_0, a_1, and a_2, are derived as follows:

$$\{x, y, z\}_r(0) = a_{x,y,z0} = \{x, y, z\}_s \tag{11}$$

$$\{\dot{x}, \dot{y}, \dot{z}\}_r(0) = a_{x,y,z1} = \{\dot{x}, \dot{y}, \dot{z}\}_s = 0 \tag{12}$$

$$\{\ddot{x}, \ddot{y}, \ddot{z}\}_r(0) = 2a_{x,y,z2} = \{\ddot{x}, \ddot{y}, \ddot{z}\}_s = 0. \tag{13}$$

Additionally, from the relation of the final states at the ending time, we have

$$\{x, y, z\}_r(T) = \sum_{i=0}^{5} a_{\{x,y,z\}i} T^i = \{x, y, z\}_f \tag{14}$$

$$\{\dot{x}, \dot{y}, \dot{z}\}_r(T) = \sum_{i=1}^{5} i a_{\{x,y,z\}i} T^{i-1} = \{\dot{x}, \dot{y}, \dot{z}\}_f \tag{15}$$

$$\{\ddot{x}, \ddot{y}, \ddot{z}\}_r(t) = \sum_{i=2}^{5} i(i-1) a_{\{x,y,z\}i} t^{i-2} = \{\ddot{x}, \ddot{y}, \ddot{z}\}_f. \tag{16}$$

By representing Equations (14)–(16) as a matrix form, a_3, a_4, and a_5 are derived as follows:

$$\begin{bmatrix} a_{\{x,y,z\}5} \\ a_{\{x,y,z\}4} \\ a_{\{x,y,z\}3} \end{bmatrix} = A^{-1} \begin{bmatrix} \{x, y, z\}_f - \{x, y, z\}_s \\ 0 \\ 0 \end{bmatrix}, \tag{17}$$

where

$$A = \begin{bmatrix} T^5 & T^4 & T^3 \\ 5T^4 & 4T^3 & 3T^2 \\ 20T^3 & 12T^2 & 6T \end{bmatrix}.$$

As the fifth-degree polynomial interpolation can be calculated uniquely depending upon the initial conditions $\{x, y, z\}_s$, the final conditions $\{x, y, z\}_f$, and the transformation time T, the gait generation strategy is therefore designed by defining these states.

3.3. Gait Generation Based on Sequential Control

Hereafter, the sequence of the target foot position is discussed. As discussed above, in order to generate the straight basis gait, the target foot position is set as follows:

- Lift up the idling leg vertically
- Move the idling leg to the target step position horizontally against the glass surface
- Bring down the idling leg to the glass surface
- Switch the idling leg and the supporting leg, and repeat 1∼3

This paper starts from a simple algorithm to avoid complexity as a first step to realize automation control. A strategy for this is to utilize only one kinematics model to generate the trajectories. A related gait from one foot unit is generated by utilizing the single kinematics model. By setting a reference point in the center of the left foot unit, the relative position of the right foot unit with respect to each sequence is set as shown in Table 4. In Table 4, x_t, y_t, and z_t represent the target position of the biped robot, and $S_x = \mathrm{sgn}(x_t)$. The sequence shown in Table 4 generates trajectories that start the step from the right foot unit when moving the robot to the right and from the left foot unit when moving the robot to the left. The trajectory between each target position in Table 4 is generated via the fifth polynomial interpolation. Finally, the target angles of each actuator to track the generated trajectory are derived by inverse kinematics.

Table 4. Target positions in each sequence.

Sequence	x [cm]	y [cm]	z [cm]
1	x_0	y_0	$S_x z_t$
2	$S_x x_t$	$S_x y_t$	$S_x z_t$
3	$S_x x_t$	$S_x y_t$	z_0
4	$S_x x_t$	$S_x y_t$	$-S_x z_t$
5	x_0	y_0	$-S_x z_t$
6	x_0	y_0	z_0

4. Experiments

This section verifies that the developed biped robot is capable of locomoting on the glass surface with the designed trajectory generator. The purpose of these experiments is to show that the locomotion abilities of the developed biped robot satisfies the requirements of accomplishing our proposing strategy. Thus, in these experiments, it was shown that the developed biped robot is able to locomote on the vertical glass wall robustly and to overcome window frame obstacles. More advanced controls including obstacle avoidance control and servo control will be designed in future work.

4.1. System Architecture and Experiments

Figure 13 shows a system architecture for the experiments. From Figure 13, the system mainly consists of a laptop computer for the target position indication, the biped robot, and a control board. Since the robotic system is suspended on a rope for safety purposes, the control board and power supplies such as the safety rope can be installed externally. CM-700 produced by Robotis is the main computer and performs the communication between the components and trajectory calculations. The RC servo motor and the vacuum pump for the vacuum system are controlled by Arduino MEGA 2560 via the receiving of control signals from CM-700. A power supply (6 V) for driving the RC servo motor is installed on the control board, and the 12 V supply is delivered from outside via cable. In this paper, two experiments were performed to verify the effectiveness of both its normal locomotion and its ability to overcome obstacles.

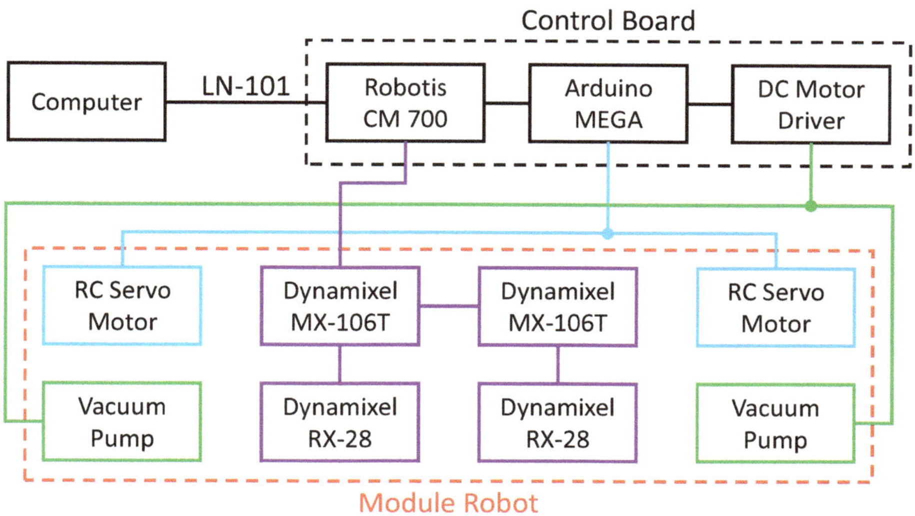

Figure 13. A system architecture for the experiments. The system mainly consists of a laptop computer for the target position indication, the biped robot, and a control board. Since the biped robot is suspended on a rope for safety purposes, the control board and power supplies such as the safety rope can be installed externally.

Regarding the experiment of normal locomotion, x_t [cm] and y_t [cm] were provided by the laptop computer. z_t [cm] was set as 2.5 cm on the glass surface. The transformation time of the fifth polynomial interpolation was set $T = 1.5$ s.

The experiment of normal locomotion aimed to demonstrate the motion performance of the biped robot. The key points are as follows: clean the window from top to bottom, move the squeegee to track the scrubber, keep the squeegee in contact with the glass surface while cleaning a column, and clean in one direction. Snapshots and graphs of the experiment of normal locomotion with time variation are shown in Figures 14 and 15.

Regarding the experiment of overcoming obstacles, x_t, y_t, and z_t [cm] were set as $x_t = 15\cos(-85\pi/180)$, $y_t = 15\sin(-85\pi/180)$ and $z_t = 4$. The transformation time of the fifth polynomial interpolation was set as $T = 1.5$ s. In addition, an obstacle supposed to represent the window frame (a 6 mm height and a 20 mm width) was attached on the glass surface. This experiment demonstrated that the biped robot is capable of overcoming a window frame obstacle. Snapshots of the experiment of overcoming obstacles with time variation are shown in Figure 16.

4.2. Discussions

Figure 14 shows that the motion performance without any human aid met the first and second key criteria: cleaning the window from top to bottom and moving the squeegee to track the scrubber. In Figure 15a indicates the trajectory of the biped robot on the $x - y$ plane, and (b) and (c) indicate the time series data of the biped robot on the x and y axis. The origin was set at the initial position of the marker equipped on the left foot unit. Figure 15 shows that the robot achieved the second key point and the fourth key point: cleaning in one direction. However, two issues were found: First, the trajectory strayed to positive in the x direction, as shown in Figure 15. This is because the controller is currently feedforward only. The experiment in this paper focused only on motion performance, rather than motion accuracy. A feedback control and state estimation system, such as SLAM, will be designed in our future work. Second, as Figure 15c shows, the support leg moved while the idling leg was moving, even though it was fixed by the vacuum suction cup, as shown in Figure 14. This is due to marker locations that are not equipped on the rotational axis of the base actuator. Since the markers are located outside the base actuator, the marker position moves according to the rotation of the base actuator. Hence, this does not represent a serious technical issue. On the basis of the above discussion, the left foot unit is always located behind the right foot unit. It is thus demonstrated that the developed biped robot is capable of locomoting on the glass surface smoothly utilizing the designed system and trajectory generator.

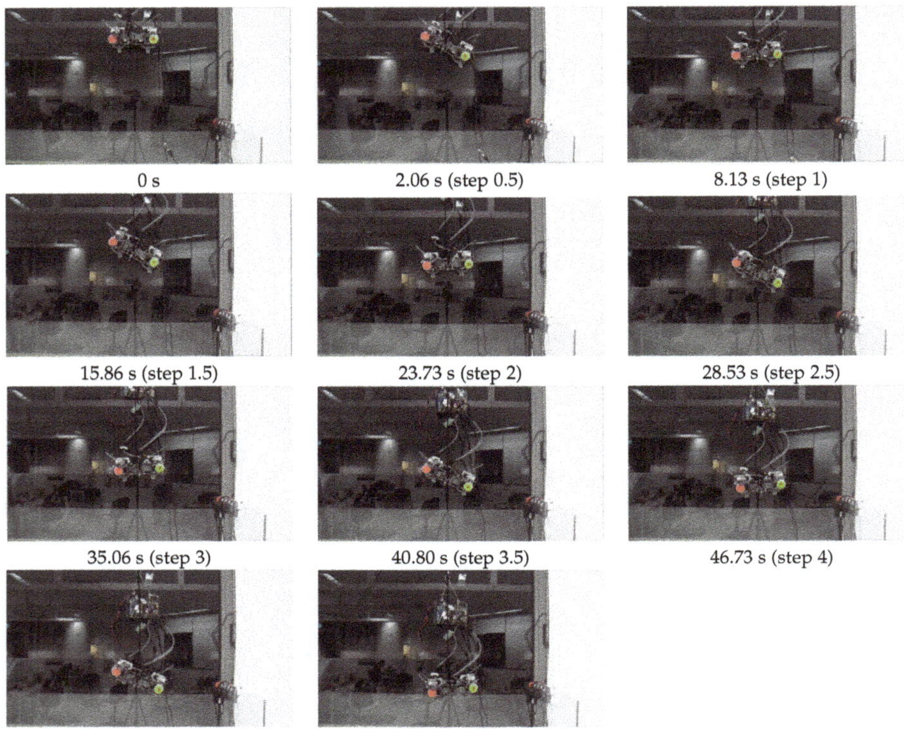

0 s | 2.06 s (step 0.5) | 8.13 s (step 1)

15.86 s (step 1.5) | 23.73 s (step 2) | 28.53 s (step 2.5)

35.06 s (step 3) | 40.80 s (step 3.5) | 46.73 s (step 4)

56.46 s (step 4.5) | 61.65 s (step 5)

Figure 14. The snapshots of the normal-locomotion experiment with time variation. Five normal locomotions within 61.65 s were realized without any human aid.

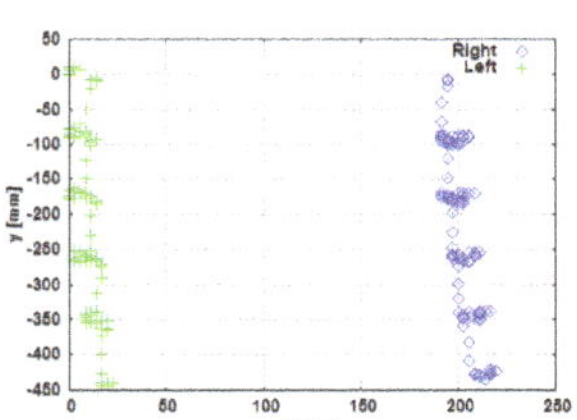

(**a**) The trajectory of each foot unit on the $x - y$ plane

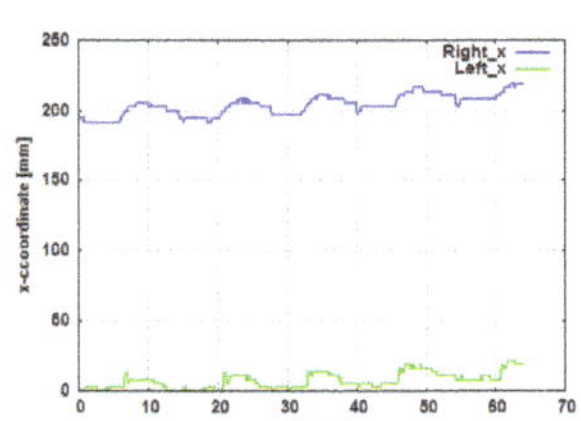

(**b**) The time series data of each foot unit on the x axis

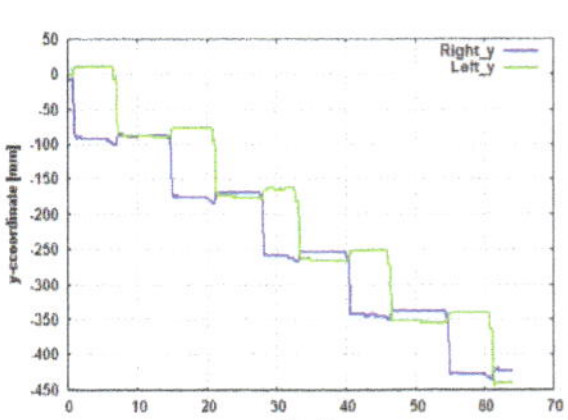

(**c**) The time series data of each foot unit on the y axis

Figure 15. The graph of the normal locomotion experiment. (**a**–**c**) are the graphs of numerical data related to the normal locomotion between five steps. The origin was set on the initial left foot unit marker.

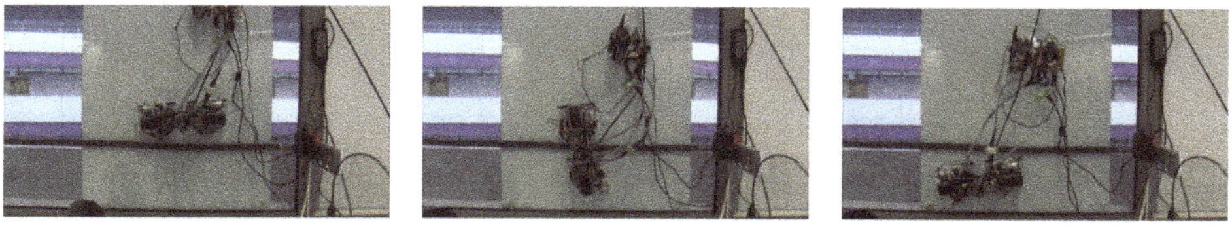

Figure 16. Snapshots of the overcoming obstacles experiment with time variation. Obstacles were overcome within 40 s without any human aid.

From Figure 16, obstacles were overcome within 40 s without any human aid. It is thus demonstrated that the developed robot can overcome window frames safely using the designed application.

5. Conclusions

This paper proposes a new class of façade cleaning robot with a biped mechanism that can overcome glass frames to maximize its area coverage. The developed robot uses active suction cups to adhere to glass walls and adopts mechanical linkage to navigate the glass surface to perform cleaning. This paper also discusses the control system of the developed robot, which consists of inverse kinematics, a fifth polynomial interpolation, and sequential control. Experiments were conducted in a real-time scenarios in order to validate the robot's ability to cross over from one glass panel to another without any difficulties, with the objective of maximizing the coverage area. The result indicates that the developed robot achieves significantly higher performance in terms of maximum area by overcoming both negative and positive obstacles on a glass panel. Future research will focus on (1) the integration of range and inertial sensors for autonomous navigation, (2) the integration of cleaning modules in the developed system, (3) exhaustive experiments in highly diverse and complex environments to quantify the complexity of the developed platform, and (4) the development of additional features to improve power management issues.

Author Contributions: S.N. and M.R.E. conceived and designed the experiments; S.N., K.O. and P.V. performed the experiments; S.N. and P.V. analyzed the data; S.N., K.O., M.R.E. and M.I. wrote the paper.

Funding: This research is supported by the SUTD-JTC Industrial Infrastructure Innovation Center under grants IPJTCT 31501.

Conflicts of Interest: The authors declare no conflict of interest.

References

1. Frey, C.B.; Osborne, M.A. The future of employment: How susceptible are jobs to computerisation. *Technol. Forecast. Soc. Chang.* **2013**, *7*. [CrossRef]
2. International Fedration of Robotics. Service Robot Statistics. Available online: http://www.ifr.org/service-robots/statistics/ (accessed on 21 November 2018).
3. Aldred, J. Burj Khalifa—A new high for high-performance concrete. In *Proceedings of the Institution of Civil Engineers-Civil Engineering*; Thomas Telford Ltd.: London, UK, 2010; Volume 163, pp. 66–73.
4. Baker, W.F. The World's tallest building. *STRUCTURE*, June 2011.
5. Baker, W.F.; Korista, D.S.; Novak, L.C. Burj Dubai: engineering the world's tallest building. *Struct. Des. Tall Spec. Build.* **2007**, *16*, 361–375. [CrossRef]
6. Weismantle, P.A.; Smith, G.L.; Sheriff, M. Burj Dubai: an architectural technical design case study. *Struct. Des. Tall Spec. Build.* **2007**, *16*, 335–360. [CrossRef]
7. Zeljic, A.S. Shanghai Tower Façade Design Process. In Proceedings of the International Conference of Building Envelope Systems, Vancouver, BC, Canada, 27–30 June 2010. Available online: http://www.gensler.com/uploads/documents/Shanghai_Tower_Facade_Design_Process_11_10_2011.pdf (accessed on 21 November 2018).

8. Xia, J.; Poon, D.; Mass, D. Case study: Shanghai Tower. *CTBUH J.* **2010**, *2010*, 12–18.
9. Zhaoa, X.; Ding, J.; Suna, H. Structural design of shanghai tower for wind loads. *Procedia Eng.* **2011**, *14*, 1759–1767. [CrossRef]
10. BBC. Shanghai Window Cleaning Cradle Swings out of Control. Available online: http://www.bbc.com/news/world-asia-china-32176401 (accessed on 21 November 2018).
11. BBC. Window Washers Rescued from High up World Trade Center. Available online: http://www.bbc.com/news/world-us-canada-30028969 (accessed on 21 November 2018).
12. Zhang, H.; Zhang, J.; Zong, G.; Wang, W.; Liu, R. Sky cleaner 3: A real pneumatic climbing robot for glass-wall cleaning. *IEEE Robot. Autom. Mag.* **2006**, *13*, 32–41. [CrossRef]
13. Zhang, H.; Zhang, J.; Zong, G. Effective nonlinear control algorithms for a series of pneumatic climbing robots. In Proceedings of the 2006 IEEE International Conference on Robotics and Biomimetics, Kunming, China, 17–20 December 2006; IEEE: Piscataway, NJ, USA, 2006; pp. 994–999.
14. Zhang, H.; Zhang, J.; Zong, G. Requirements of glass cleaning and development of climbing robot systems. In Proceedings of the 2004 International Conference on Intelligent Mechatronics and Automation, Chengdu, China, 26–31 August 2004; IEEE: Piscataway, NJ, USA, 2004; pp. 101–106.
15. Sun, D.; Zhu, J.; Lai, C.; Tso, S. A visual sensing application to a climbing cleaning robot on the glass surface. *Mechatronics* **2004**, *14*, 1089–1104. [CrossRef]
16. Sun, D.; Zhu, J.; Tso, S.K. *A Climbing Robot for Cleaning Glass Surface with Motion Planning and Visual Sensing*; INTECH Open Access Publisher: London, UK, 2007.
17. Serbot-AG. Available online: http://www.serbot.ch/en/ (accessed on 21 November 2018).
18. Seo, K.; Cho, S.; Kim, T.; Kim, H.S.; Kim, J. Design and stability analysis of a novel wall-climbing robotic platform (ROPE RIDE). *Mech. Mach. Theory* **2013**, *70*, 189–208. [CrossRef]
19. Kim, T.Y.; Kim, J.H.; Seo, K.C.; Kim, H.M.; Lee, G.U.; Kim, J.W.; Kim, H.S. Design and Control of a Cleaning Unit for a Novel Wall-Climbing Robot. *Appl. Mech. Mater.* **2014**, *541–542*, 1092–1096. [CrossRef]
20. Kim, T.; Seo, K.; Kim, J.H.; Kim, H.S. Adaptive impedance control of a cleaning unit for a novel wall-climbing mobile robotic platform (ROPE RIDE). In Proceedings of the 2014 IEEE/ASME International Conference on Advanced Intelligent Mechatronics (AIM), Besacon, France, 8–11 July 2014; IEEE: Piscataway, NJ, USA, 2014; pp. 994–999.
21. Fraunhofer-IFF. Available online: http://www.iff.fraunhofer.de/en.html (accessed on 21 November 2018).
22. Böhme, T.; Schmucker, U.; Elkmann, N.; Sack, M. Service robots for facade cleaning. In Proceedings of the 24th Annual Conference of the IEEE Industrial Electronics Society, Aachen, Germany, 31 August–4 September 1998; IEEE: Piscataway, NJ, USA, 1998; Volume 2, pp. 1204–1207.
23. Elkmann, N.; Schmucker, U.; Scharfe, H.; Schoop, C.; Kubbe, I. Drive Device for Moving a Robot or Vehicle on Flat, Inclined or Curved Surfaces, Particularly of a Glass Construction and Robot with Drive Device. U.S. Patent 5,959,424, 28 September 1999.
24. Elkmann, N.; Hortig, J.; Fritzsche, M. Cleaning automation. In *Springer Handbook of Automation*; Springer: Berlin, Germany, 2009; pp. 1253–1264.
25. Elkmann, N.; Felsch, T.; Sack, M.; Saenz, J.; Hortig, J. Innovative service robot systems for facade cleaning of difficult-to-access areas. In Proceedings of the 2002 IEEE/RSJ International Conference on Intelligent Robots and Systems, Lausanne, Switzerland, 30 September–4 October 2002; IEEE: Piscataway, NJ, USA, 2002; Volume 1, pp. 756–762.
26. Elkmann, N.; Kunst, D.; Krueger, T.; Lucke, M.; Böhme, T.; Felsch, T.; Stürze, T. SIRIUSc–Façade cleaning robot for a high-rise building in munich, germany. In *Climbing and Walking Robots*; Springer: Berlin, Germnay, 2005; pp. 1033–1040.
27. Bridge, B.; Elkmann, N.; Lucke, M.; Krüger, T.; Kunst, D.; Stürze, T.; Hortig, J. Kinematics, sensors and control of the fully automated façade-cleaning robot SIRIUSc for the Fraunhofer headquarters building, Munich. *Ind. Robot Int. J.* **2008**, *35*, 224–227.
28. Cepolina, F.E.; Muscolo, G.G. Design of a robot for hygienization of walls in hospital environments. In Proceedings of the ISR/Robotik 2014; 41st International Symposium on Robotics, Munich, Germany, 2–3 June 2014; pp. 1–7.
29. Akinfiev, T.; Armada, M.; Nabulsi, S. Climbing cleaning robot for vertical surfaces. *Ind. Robot Int. J.* **2009**, *36*, 352–357. [CrossRef]

30. Ott, C.; Baumgärtner, C.; Mayr, J.; Fuchs, M.; Burger, R.; Lee, D.; Eiberger, O.; Albu-Schäffer, A.; Grebenstein, M.; Hirzinger, G. Development of a biped robot with torque controlled joints. In Proceedings of the 2010 10th IEEE-RAS International Conference on Humanoid Robots (Humanoids), Nashville, TN, USA, 6–8 December 2010; IEEE: Piscataway, NJ, USA, 2010; pp. 167–173.
31. Al-Busaidi, A.M. Development of an educational environment for online control of a biped robot using MATLAB and Arduino. In Proceedings of the Mechatronics (MECATRONICS), 2012 9th France-Japan & 7th Europe-Asia Congress on and 2012 13th Int'l Workshop on Research and Education in Mechatronics (REM), Paris, France, 21–23 November 2012; IEEE: Piscataway, NJ, USA, 2012; pp. 337–344.
32. Liu, C.Y.; Wang, J.C. Forecasting the development of the biped robot walking technique in Japan through S-curve model analysis. *Scientometrics* **2009**, *82*, 21–36. [CrossRef]
33. Guan, Y.; Jiang, L.; Zhu, H.; Zhou, X.; Cai, C.; Wu, W.; Li, Z.; Zhang, H.; Zhang, X. Climbot: A modular bio-inspired biped climbing robot. In Proceedings of the 2011 IEEE/RSJ International Conference on Intelligent Robots and Systems (IROS), San Francisco, CA, USA, 25–30 September 2011; IEEE: Piscataway, NJ, USA, 2011; pp. 1473–1478.
34. Minor, M.; Dulimarta, H.; Danghi, G.; Mukherjee, R.; Tummala, R.L.; Aslam, D. Design, implementation, and evaluation of an under-actuated miniature biped climbing robot. In Proceedings of the 2000 IEEE/RSJ International Conference on Intelligent Robots and Systems, Takamatsu, Japan, 31 October–5 November 2000; IEEE: Piscataway, NJ, USA, 2000; Volume 3, pp. 1999–2005.
35. AbuKhalil, T.; Sobh, T.; Patil, M. Survey on decentralized modular robots and control platforms. In *Innovations and Advances in Computing, Informatics, Systems Sciences, Networking and Engineering*; Springer: Berlin, Germany, 2015; pp. 165–175.
36. Patil, M.; Abukhalil, T.; Patel, S.; Sobh, T. UB robot swarm—Design, implementation, and power management. In Proceedings of the 2016 12th IEEE International Conference on Control and Automation (ICCA), Kathmandu, Nepal, 1–3 June 2016; IEEE: Piscataway, NJ, USA, 2016; pp. 577–582.
37. Okazaki, Y.; Yamamoto, M.; Komatsu, M.; Tsusaka, Y.; Adachi, Y. Development of a Human Safe, Multi-Degree-of-Feedom Robot Arm Technology using Pneumatic Muscles (in Japanese). *J. Robot. Soc. Jpn.* **2010**, *28*, 62–70. [CrossRef]
38. Noguchi, Y.; Iwase, M.; Hatakeyama, S.; Izutsu, M. A Yoyo Trick Realized by Parallel-Link Manipulator. In Proceedings of the 39th Annual Conference of the IEEE Industrial Electronics Society, Vienna, Austria, 10–13 November 2013; pp. 4049–4054.
39. Takeda, Y.; Okada, M. Development of a Jumping Robot with a Non-Circular Gear. In Proceedings of the 2012 JSME Conference on Robotics and Mechatronics, Hamamatsu, Japan, 27–29 May 2012; Volume 3, pp. 2A2–V04.

applied sciences

MDPI

Article

An Adaptive Neural Non-Singular Fast-Terminal Sliding-Mode Control for Industrial Robotic Manipulators

Anh Tuan Vo and Hee-Jun Kang *

School of Electrical Engineering, University of Ulsan, 93 Daehak-ro, Nam-gu, Ulsan 680-749, Korea; voanhtuan2204@gmail.com

* Correspondence: hjkang@ulsan.ac.kr; Tel.: +82-52-259-2207

Received: 20 November 2018; Accepted: 7 December 2018; Published: 10 December 2018

Featured Application: The proposed control methodology could be applied to not only the joint position tracking control for industrial robotic manipulators such as serial, parallel robots, and an electrohydraulic series elastic manipulator, but also other mechanical systems that belong to the class of general nonlinear second-order system. For example, it could be applied for the stabilization or trajectory tracking of mechanical systems as a piezo positioning stage, magnetic levitation systems, or for the chaos control, synchronization, and anti-synchronization of chaotic complex systems.

Abstract: In this study, a robust control strategy is suggested for industrial robotic manipulators. First, to minimize the effects of disturbances and dynamic uncertainties, while achieving faster response times and removing the singularity problem, a nonsingular fast terminal sliding function is proposed. Second, to achieve the proposed tracking trajectory and chattering phenomenon elimination, a robust control strategy is designed for the robotic manipulator based on the proposed sliding function and a continuous adaptive control law. Furthermore, the dynamical model of the robotic system is estimated by applying a radial basis function neural network. Thanks to those techniques, the proposed system can operate free of an exact robotic model. The suggested system provides high tracking accuracy, robustness, and fast response with minimal positional errors compared to other control strategies. Proof of the robustness and stability of the suggested system has been verified by the Lyapunov theory. In simulation analyses, the simulated results present the effectiveness of the suggested strategy for the joint position tracking control of a 3-degree of freedom (3-DOF) PUMA560 robot.

Keywords: non-singular fast-terminal sliding-mode control; industrial robotic manipulator; external disturbance; dynamic uncertainty; adaptive control law

---

1. Introduction

Literature regarding robotic manipulators has introduced many control systems focused on achieving high performance against various uncertainties, including external noise. These control methods were derived to fundamentally control the motion of robot manipulators, and include the proportional-derivative (PD) controller [1], nonlinear PD controller [2], and the proportional-integral-derivative (PID) controller [3,4]. The advantages of the cited control systems were to provide a simple and basic approach to implementation, as they do not require an exact dynamic model. However, these systems could not obtain the desired performance in the presence of disturbances and dynamic uncertainties. Several advanced control approaches have been proposed to advance system performance, such as the fuzzy controller [5–7] and neural network controller [8–10], but they demand complicated calculations, and the effectiveness of each solution still has several limitations. The control scheme design strategy is based on the robot dynamic model, where the whole dynamic model is computed and compensated explicitly to achieve the desired performance. Therefore, other enhanced methods were suggested to improve the motion tracking for robot manipulators, including a computed torque controller (CTC) [11,12], adaptive controller [13–15], and sliding mode controller (SMC) [16–20]. Among those controllers, SMC has been confirmed to offer high robustness against uncertainties and disturbances for nonlinear systems. Therefore, the SMC has been widely applied in real applications [16–20]. However, the traditional SMC still possesses drawbacks such as requiring an exact dynamic model, singularity problems, a chattering phenomenon, and finite-time convergence. Some research efforts have focused on overcoming these disadvantages. For the system states to approach the sliding variable within a finite-time, the use of the terminal sliding mode control (TSMC), based on the nonlinear sliding surface, has been reported in the literature [13,21–23]. Nonetheless, the TSMC convergence time is slow when compared to the conventional SMC, and still contains a singularity glitch. To solve convergence time and singularity issues, several fast TSMC (FTSMC) [24–26] and nonsingular TSMC (NTSMC) [27–29] approaches have been proposed. Practically, private algorithms, such as FTSMC or NTSMC, have only treated an individual weakness or failed to solve the other disadvantages of the classical SMC. Consequently, the nonsingular fast TSMC (NFTSMC) has been introduced [30–34]. Here, NFTSMC can solve many disadvantages of the classical SMC or other control algorithms based on TSMC. However, chattering behavior has not been removed by applying a high-frequency switching control law to the control input of the above methods, which include TSMC, FTSMC, NTSMC, and NFTSMC. Therefore, some effective techniques have been introduced to handle this topic by application of the saturation function (refer to [35]), full-order sliding mode control (FOSMC) [36,37], or high-order sliding mode control (HOSMC) [35,38].

One of the main tasks in the design of a control method based on SMC or TSMC is to develop an exact dynamic model of the robot manipulator, which one does not readily know in advance for real robot systems. To estimate this unknown dynamic model, several computing approaches have been proposed such as neural networks [39–41] and fuzzy logic systems [42–44] due to their universal approximation capabilities.

While each disadvantage of the classical SMC and TSMC has been treated individually, this report focuses on simultaneous resolution of the disadvantages of SMC and TSMC, including the requirement for an exact dynamic model, as well as the presence of a singularity problem, chattering phenomenon, and finite-time convergence.

Consequently, the goal of this research is to develop a robust control strategy for robotic manipulators based on an adaptive neural non-singular fast-terminal sliding-mode control (ANNFTSMC) scheme. The main advantages of the suggested control strategy include:

- The inheritance of NFTSMC advantages in terms of non-singularity, finite-time convergence, fast transient response, low steady-state errors, and high position tracking accuracy.
- The achievement of smooth control inputs with chattering behavior elimination.
- The removal of demand for an exact dynamic model by applying an adaptive radial basis function neural network to approximate an unknown robot function.
- Better tracking performance and less impact by disturbances and uncertainties compared to classic SMC and other control methods based on TSMC.
- Improved robustness and stability of the robot system, as demonstrated by Lyapunov theory.

The remainder of the report is structured as follows. Following the introduction, the problem statements are presented, succeeded by the design approach for the proposed control strategy, where the proposed strategy is utilized to allow joint position tracking control simulation for a 3-degree of freedom (3-DOF) robot manipulator. Here, its tracking performance is compared with SMC and TSMC to analyze the effectiveness of the proposed control strategy. Finally, conclusions are presented.

2. Problem Statements

2.1. Radial Basis Function Neural Network

Previous research on the universal approximation theory proved that any nonlinear function over a compact set with arbitrary accuracy can be approximated by the radial basis function neural network (RBFNN). Here, RBFNNs have several advantages, including ease of design, good generalization, strong tolerance to input noise, and online learning ability. Compared with a multiplayer neural network, an RBFNN is simpler and converges faster. An RBFNN includes three layers: the input layer, hidden layer, and output layer, all of which are expressed in Figure 1.

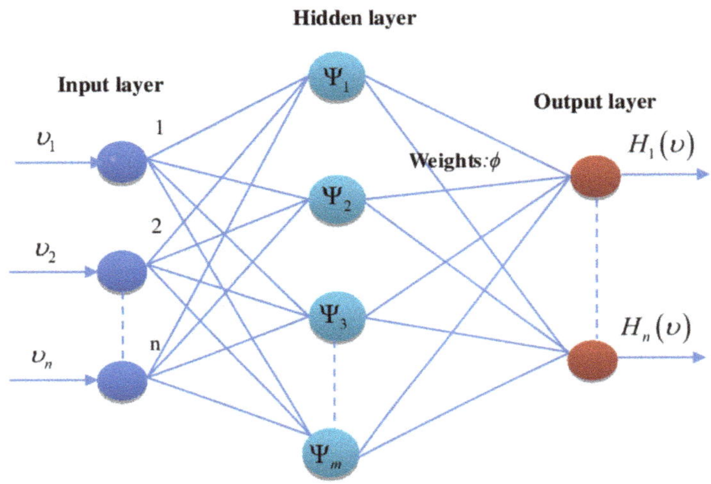

Figure 1. Structure of radial basis function neural network.

The output of the RBFNN can be computed as

$$H(v) = \phi^T \Psi(v) + \xi(v) \tag{1}$$

where $v \in \mathbb{R}^n$ and $H(v)$ are the neural network input and output, respectively. Here, $\phi^T \in \mathbb{R}^{n \times m}$ is the weight matrix connecting the hidden layer and the output layer, $\Psi(v)$ is the nonlinear function of the hidden nodes, and $\xi(v) \in \mathbb{R}^n$ is an approximation error of the neural network (NN).

A Gaussian fit is selected for the nonlinear function as follows:

$$\Psi(v) = \exp\left(\frac{-(v-\mu_l)^T(v-\mu_l)}{\delta_l^2}\right), l = 1, 2, \ldots, m, \tag{2}$$

where δ and μ correspond to the width and center of the Gaussian function, respectively.

2.2. Dynamic Model of the Robot Manipulator

For an n-link rigid robotic manipulator, the dynamic model can be described as (refer to [45,46])

$$M(q)\ddot{q} + C_m(q, \dot{q})\dot{q} + G(q) + F_r(\dot{q}) + \tau_d(t) = \tau(t), \tag{3}$$

where q, $\dot{q}$ and $\ddot{q} \in \mathbb{R}^n$ correspond to the position, velocity, and acceleration of the robot manipulator, respectively. Additionally, $M(q) \in \mathbb{R}^{n \times n}$ is the invertible inertia matrix, $C_m(q, \dot{q}) \in \mathbb{R}^{n \times 1}$ is the matrix from the centrifugal force and Coriolis, $G(q) \in \mathbb{R}^{n \times 1}$ is the gravitational force matrix, $F_r(\dot{q}) \in \mathbb{R}^{n \times 1}$

denotes the friction matrix, $\tau(t) \in \mathbb{R}^{n\times 1}$ is the designed actuation input of actuators, and $\tau_d(t) \in \mathbb{R}^{n\times 1}$ is a load disturbance matrix.

To simplify the approach and analysis, Equation (3) is given as

$$\ddot{q} = \Xi(q,\dot{q}) + B(q)\tau(t) + \Delta_u(q,\dot{q},t). \tag{4}$$

where $\Xi(q,\dot{q}) = M^{-1}(q)[-C_m(q,\dot{q})\dot{q} - G(q)]$ is the nominal dynamic model of the robot manipulator without perturbations and uncertainties, $\Delta_u(q,\dot{q},t) = M^{-1}(q)[-F_r(\dot{q}) - \tau_d(t)]$ represents the unknown perturbation and uncertainty terms, and $B(q) = M^{-1}(q)$.

The hypothesis here is that the control variables will follow the desired trajectory, with high performance, in finite-time under a robust control strategy. In this case, the proposed system does not need an exact robotic model.

The following assumptions are crucial for the design approach.

Assumption 1. *The inertia matrix $M(q)$, is an invertible, positive definite, and symmetric matrix that adheres to the bounded condition,*

$$\theta_1 \le M(q) \le \theta_2, \tag{5}$$

where θ_1 and θ_2 represent positive constants.

Assumption 2. *The unknown perturbations, uncertainties, and approximation errors of NN have an upper-bound satisfying the following relation,*

$$|\Delta_u(q,\dot{q},t)| \le \Omega, \tag{6}$$

where Ω is an unknown positive constant.

3. Design Procedure for a Control Strategy

In this section, a new control strategy is suggested for a robot manipulator using Equation (3), which is described by the two following main tasks.

3.1. Design Non-Singular Fast-Terminal Sliding Variable

Based on the TSMC design approach, a state variable termed as the NFTSM variable was previously designed, where the novel NFTSM variables are proposed from the tracking positional error as

$$s_i = \dot{\varsigma}_i + h_{1i} sign[\varsigma_i] + h_{2i}\varsigma_i^{[\alpha_i]}, \tag{7}$$

where h_{1i}, h_{2i} are positive values, $\alpha_i > 1$, and the variable ς_i is selected as

$$\varsigma_i = e_i + \int_0^t \left(\Gamma_{1i}{e_i}^{[2-\vartheta_i]} + \Gamma_{2i}e_i + \Gamma_{3i}{e_i}^{[\vartheta_i]}\right)d\sigma, \tag{8}$$

where $e_i = q_i - q_{ir}$ $(i = 1, 2, \ldots, n)$ is the tracking positional error, q_{ir} is described as the desired path value, ς_i is the sliding surface variable, Γ_{1i}, Γ_{2i}, Γ_{3i} are positive coefficients satisfying the relation $4\Gamma_{1i}\Gamma_{3i} > \Gamma_{2i}^2$, $0 < \vartheta_i < 1$ $(i = 1, 2, \ldots, n)$ and ${e_i}^{[\vartheta_i]}$ is as described in [47]

$${e_i}^{[\vartheta_i]} = |e_i|^{\vartheta_i} sign[e_i]. \tag{9}$$

Remark 1. *Once the tracking positional error $|e_i|$ is much greater than 1, $\Gamma_{1i}{e_i}^{[2-\vartheta_i]} + \Gamma_{2i}e_i$ contributes to the task by offering a fast convergence. While the tracking positional error $|e_i|$ is much smaller than 1, $\Gamma_{3i}{e_i}^{[\vartheta_i]}$ contributes by producing finite time convergence.*

According to the SMC manner, once the state variable proceeds in sliding mode, the following constraints are imposed (refer to [16–20]):

$$s_i = 0 \text{ and } \dot{\varsigma}_i = 0, \tag{10}$$

$$\varsigma_i = 0 \text{ and } \dot{\varsigma}_i = 0. \tag{11}$$

Combining Equation (10) constraints with Equation (7) yields

$$\dot{\varsigma}_i = -h_{1i} sign[\varsigma_i] - h_{2i} \varsigma_i^{[\alpha_i]}, \tag{12}$$

and combining Equation (11) constraints with Equation (8) gives

$$\dot{e}_i = -\Gamma_{1i} e_i{}^{[2-\vartheta_i]} - \Gamma_{2i} e_i - \Gamma_{3i} e_i{}^{[\vartheta_i]}. \tag{13}$$

It must be proved that once the second-order sliding motion takes place, i.e., $s_i = 0$, the first-order sliding motion takes place in finite-time, i.e., $\varsigma_i = 0$, and the state variable system of Equation (13) reaches zero in finite-time. The following theorems have been established for this proof.

Theorem 1. *Consider the dynamic system shown in Equation (12). The original point $\varsigma_i = 0$ is globally balanced in finite-time and the state variable of the system (10) converges to zero in finite-time $T_{s_i} \leq \varsigma_i^2(0)/\sqrt{2}h_{1i}$.*

Proof. The positive-definite Lyapunov functional is investigated as

$$V_1 = \frac{\varsigma_i^2}{2}. \tag{14}$$

With Equation (12), the time derivative of Equation (14) is computed as

$$\begin{aligned} \dot{V}_1 &= \varsigma_i\left(-h_{1i} sign[\varsigma_i] - h_{2i}\varsigma_i^{[\alpha_i]}\right) \\ &= -h_{1i}|s_i| - h_{2i} s_i^{[\alpha_i+1]} \\ &\leq -h_{1i}|s_i| \\ &= -\sqrt{2}h_{1i} V_1^{1/2} \end{aligned} . \tag{15}$$

It can be seen that (15) has the form $\dot{V}_1 + \sqrt{2}h_{1i}V_1^{1/2} \leq 0$. Therefore, the defined finite-time is given by [48]:

$$T_{s_i} \leq \frac{\varsigma_i^2(0)}{\sqrt{2}h_{1i}}. \tag{16}$$

This completes the proof. □

Theorem 2. *Consider the dynamic system (13). The original point $e_i = 0$ consists of globally balanced points in finite-time and the state variable of the system (13) as it converges to zero in finite-time $T_{e_i} \leq T_{e_i}^f$. $T_{e_i}^f$ is defined as*

$$T_{e_i}^f = \frac{2}{(1-\vartheta_i)}\left(\frac{\pi}{2} - \tan^{-1}\frac{\Gamma_{2i}}{\sqrt{4\Gamma_{1i}\Gamma_{3i} - \Gamma_{2i}^2}}\right)\frac{1}{\sqrt{4\Gamma_{1i}\Gamma_{3i} - \Gamma_{2i}^2}}. \tag{17}$$

Proof. The Lyapunov function candidate is investigated as

$$V_2 = e_i{}^2. \tag{18}$$

With Equation (13), the time derivative of Equation (18) is calculated as

$$\begin{aligned} \dot{V}_2 &= 2e_i\dot{e}_i \\ &= 2e_i\left(-\Gamma_{1i}e_i{}^{[2-\vartheta_i]} - \Gamma_{2i}e_i - \Gamma_{3i}e_i{}^{[\vartheta_i]}\right) \\ &= 2\left(-\Gamma_{1i}e_i{}^{[3-\vartheta_i]} - \Gamma_{2i}e_i{}^2 - \Gamma_{3i}e_i{}^{[1+\vartheta_i]}\right) \\ &= 2\left(-\Gamma_{1i}V_2^{(3-\vartheta_i)/2} - \Gamma_{2i}V_2 - \Gamma_{3i}V_2^{(\vartheta_i+1)/2}\right) \end{aligned} . \tag{19}$$

□

To arrive at a conclusion from Equation (19), the following Lemma is used.

Lemma 1. [49]: *For any real numbers $z_1 > 0$, $z_2 > 0$, and $0 < \varphi < 1$, an extended Lyapunov function condition of finite-time stability can be given in the form of a fast-terminal sliding mode as $\dot{L}(x) + z_1L(x) + z_2L^{\varphi}(x) \leq 0$, where the settling time can be estimated by*

$$T \leq \frac{1}{z_1(1-\varphi)} \ln \frac{z_1L^{1-\varphi}(x(0)) + z_2}{z_2}. \tag{20}$$

From Equation (19), $\vartheta_i + 1/2 < 1$ indicates that $\dot{V}_2 \leq 0$. Based on Lemma 1, the original point $e_i = 0$ is a globally balanced point in finite-time. In the next step, proof that the error state variable of the system (13) converges to zero in finite-time will be given.

Equation (19) can be shown as

$$\dot{V}_2 = 2V_2^{(\vartheta_i+1)/2}\left(-\Gamma_{1i}V_2^{1-\vartheta_i} - \Gamma_{2i}V_2^{(1-\vartheta_i)/2} - \Gamma_{3i}\right). \tag{21}$$

Equation (21) can be expressed as

$$\begin{aligned} dV_2 &= 2V_2^{(\vartheta_i+1)/2}\left(-\Gamma_{1i}V_2^{1-\vartheta_i} - \Gamma_{2i}V_2^{(1-\vartheta_i)/2} - \Gamma_{3i}\right)dt \\ \Rightarrow dt &= -\frac{dV_2}{2V_2^{(\vartheta_i+1)/2}\left(\Gamma_{1i}V_2^{1-\vartheta_i} + \Gamma_{2i}V_2^{(1-\vartheta_i)/2} + \Gamma_{3i}\right)} \\ &= -\frac{1}{1-\vartheta_i}\frac{dV_2^{(1-\vartheta_i)/2}}{\left(\Gamma_{1i}V_2^{1-\vartheta_i} + \Gamma_{2i}V_2^{(1-\vartheta_i)/2} + \Gamma_{3i}\right)} \end{aligned} . \tag{22}$$

Setting $V_2(T_{e_i}) = 0$ and taking the integral of Equation (22) during the time period where $0 \to T_{e_i}$ gives

$$T_{e_i} = \frac{2}{(1-\vartheta_i)}\frac{1}{\sqrt{4\Gamma_{1i}\Gamma_{3i} - \Gamma_{2i}^2}}\left(\tan^{-1}\frac{2\Gamma_{1i}V_2^{(1-\vartheta_i)/2}(e_i(0))}{\sqrt{4\Gamma_{1i}\Gamma_{3i} - \Gamma_{2i}^2}} - \tan^{-1}\frac{\Gamma_{2i}}{\sqrt{4\Gamma_{1i}\Gamma_{3i} - \Gamma_{2i}^2}}\right). \tag{23}$$

It can be seen that T_{e_i} is limited by $T_{e_i}^f = \frac{2}{(1-\vartheta_i)}\left(\frac{\pi}{2} - \tan^{-1}\frac{\Gamma_{2i}}{\sqrt{4\Gamma_{1i}\Gamma_{3i}-\Gamma_{2i}^2}}\right)\frac{1}{\sqrt{4\Gamma_{1i}\Gamma_{3i}-\Gamma_{2i}^2}}$. In fact, $V_2(T_{e_i}) = 0$ means $e_i(T_{e_i}) = 0$. In addition, it can be seen that the upper-bound of $T_{e_i}^f$ is only dependent on the design constants, as Γ_{1i}, Γ_{2i}, Γ_{3i}, ϑ_i and the tracking positional error in Equation (13) approach zero in finite-time. Therefore, the proof of Theorem 2 is complete.

The proposed control strategy forces the error state variables to reach sliding variables in finite time, as will be presented next.

3.2. Design an Adaptive Neural Non-Singular Fast-Terminal Sliding-Mode Control for Robotic Manipulators

To achieve the desired control performance for the system in Equation (3), the control method is performed as follow:

Substituting Equation (8) into Equation (7) provides

$$s = \dot{e} + \Gamma_1 e^{[2I_n - \vartheta]} + \Gamma_2 e + \Gamma_3 e^{[\vartheta]} + h_1 sign[\varsigma] + h_2 \varsigma^{[\alpha]}, \tag{24}$$

where $s = \begin{bmatrix} s_1, & \ldots, & s_n \end{bmatrix}^T$, I_n is the unit matrix, $\vartheta = diag(\vartheta_1, \cdots, \vartheta_n)$, $\alpha = diag(\alpha_1, \cdots, \alpha_n)$, $\Gamma_1 = diag(\Gamma_{11}, \cdots, \Gamma_{1n})$, $\Gamma_2 = diag(\Gamma_{21}, \cdots, \Gamma_{2n})$, $\Gamma_3 = diag(\Gamma_{31}, \Gamma_{32}, \cdots, \Gamma_{3n})$, $h_1 = diag(h_{11}, \cdots, h_{3n})$, $h_2 = diag(h_{21}, \cdots, h_{2n})$, $sign[\varsigma] = \begin{bmatrix} sign[\varsigma_1], & \ldots, & sign[\varsigma_n] \end{bmatrix}^T$, $e = [e_1, \ldots, e_n]^T$. $e^{[2I_n - \vartheta]}$, $e^{[\vartheta]}$, and $\varsigma^{[\alpha]}$ are vectors defined as

$$e^{[\vartheta]} = diag(sign[e]) \cdot |e|^{\vartheta} = \begin{bmatrix} e_1^{[\vartheta_1]}, & e_2^{[\vartheta_2]}, & \cdots, & e_n^{[\vartheta_n]} \end{bmatrix}^T. \tag{25}$$

To simplify the analysis, the following notion is applied

$$\frac{de^{[\vartheta]}}{dt} = \vartheta diag\left(|e|^{\vartheta - I_n}\right) \cdot \dot{e}. \tag{26}$$

Using Equation (26), the time derivative of Equation (24) is derived as

$$\dot{s} = \ddot{e} + \Gamma_1 (2I_n - \vartheta) diag\left(|e|^{I_n - \vartheta}\right)\dot{e} + \Gamma_2 \dot{e} + \Gamma_3 \vartheta diag\left(|e|^{\vartheta - I_n}\right)\dot{e} + h_2 \alpha diag\left(|\varsigma|^{\alpha - I_n}\right)\dot{\varsigma}. \tag{27}$$

From Equation (4), $\ddot{e}$ is presented as

$$\begin{aligned} \ddot{e} &= \ddot{q} - \ddot{q}_d \\ &= \Xi(q, \dot{q}) + B(q)\tau(t) + \Delta_u(q, \dot{q}, t) - \ddot{q}_d \end{aligned}. \tag{28}$$

Substituting Equation (28) into Equation (27) gives

$$\dot{s} = \Xi(q, \dot{q}) + B(q)\tau(t) + \Delta_u(q, \dot{q}, t) - \ddot{q}_d + \Pi(e, \varsigma), \tag{29}$$

where $\Pi(e, \varsigma) = \Gamma_1 (2I_n - \vartheta) diag\left(|e|^{I_n - \vartheta}\right)\dot{e} + \Gamma_2 \dot{e} + \Gamma_3 \vartheta diag\left(|e|^{\vartheta - I_n}\right)\dot{e} + h_2 \alpha diag\left(|\varsigma|^{\alpha - I_n}\right)\dot{\varsigma}$.

To obtain the desired performance, the proposed control algorithm is designed for system (3) as

$$\tau(t) = B^{+}(q)\left(\tau_{eq}(t) + \tau_s(t)\right), \tag{30}$$

where $B^{+}(q) = B^T(q)\left[B(q)B^T(q)\right]^{-1}$, the equivalent control law is constructed as

$$\tau_{eq}(t) = -\left(\Xi(q, \dot{q}) + \Pi(e, \varsigma) - \ddot{q}_d\right), \tag{31}$$

and the switching control term is designed as

$$\tau_s = -(\Omega + \rho_1) sign(s) \tag{32}$$

in which Ω and ρ_1 are positive constants.

Substituting control laws (30)–(32) into Equation (29) provides

$$\dot{s} = -(\Omega + \rho_1) sign(s) + \Delta_u(q, \dot{q}, t). \tag{33}$$

The positive-definite Lyapunov functional is selected as

$$V_3 = \frac{1}{2}s^T s. \tag{34}$$

With Equation (33), the time derivative of Equation (34) is derived as

$$\begin{aligned} \dot{V}_3 &= s^T \dot{s} \\ &= s^T(-(\Omega + \rho_1)sign(s) + \Delta_u(q,\dot{q},t)) \\ &= -\Omega|s| - \rho_1|s| + \Delta_u(q,\dot{q},t)s \le -\rho_1|s| \end{aligned} . \tag{35}$$

Accordingly, based on the Lyapunov criterion [47], it can be verified that the stability of the tracking error is secured under control laws (30)–(32) despite the presence of external disturbances and system uncertainties.

Unfortunately, robot manipulators have complicated dynamic models with many parametric uncertainties (e.g., friction, sensor noise, payload, perturbations). Therefore, it is not trivial to precisely calculate the uncertainty upper-bounds and provide an exact robot dynamic function in the equivalent control law. To overcome these difficulties, a robust control strategy will be constructed for robotic manipulators based on an adaptive neural non-singular fast terminal sliding mode control (ANNFTSMC) scheme. Here, an adaptive radial basis function neural network will be utilized to approximate an unknown robot function, while an adaptive law will be used to estimate the uncertainty upper bounds and estimated error of the NN. In this report, RBFNN is used to approximate the dynamic robot model as follows:

$$\mathrm{f}(x) = \Xi(q,\dot{q}), \tag{36}$$

where $x = [x_1, x_2]^T$, assign $x_1 = q$, and $x_2 = \dot{q}$.

Define $\hat{\mathrm{f}}(x)$ as an approximated function of $\mathrm{f}(x)$, $\hat{\mathrm{f}}(x)$ can be described by an NN, as follows

$$\hat{\mathrm{f}}(x) = \hat{\phi}^T \Psi(x). \tag{37}$$

Here, $\hat{\phi}$ is the adaptable parameter vector.

The optimal parameter ϕ^* can be described, as follows:

$$\phi_H^* = \text{argmin}\left\{ \sup_{x\in\Theta_x} \left| \mathrm{f}(x) - \hat{\mathrm{f}}(x,\hat{\phi}) \right| \right\}. \tag{38}$$

Accordingly, RBFNN (37) can exactly approximate the arbitrary value of $\mathrm{f}(x)$ which is given by the following Lemma.

Lemma 2. *For any given real continuous function* $\mathrm{f}(X)$ *on the compact set* $\Theta_X \in R^n$ *and arbitrary positive coefficient* $\xi > 0$*, there is a neural approximator existence* $\hat{\mathrm{f}}(X)$ *that possesses a similar form as Equation (37), such that*

$$\sup_{X\in\Theta_X} \left| \mathrm{f}(X) - \hat{\mathrm{f}}(X,\hat{\phi}) \right| < \xi. \tag{39}$$

Therefore, the robot dynamic model can be described as

$$\ddot{q} = \phi^{*T}\Psi(x) + B(q)\tau(t) + W, \tag{40}$$

where $W = \Delta_u(q,\dot{q},t) + \xi$ is the lumper uncertainty, including disturbances, dynamic uncertainties, and NN approximation error. In this step, the lumper uncertainty is assumed to be bounded by an unknown positive constant, $|W| \le \Phi$.

The proposed control law as depicted in Figure 2 is designed as follows:

$$\tau(t) = B^+(q)(\tau_{eq}(t) + \tau_{as}(t)). \tag{41}$$

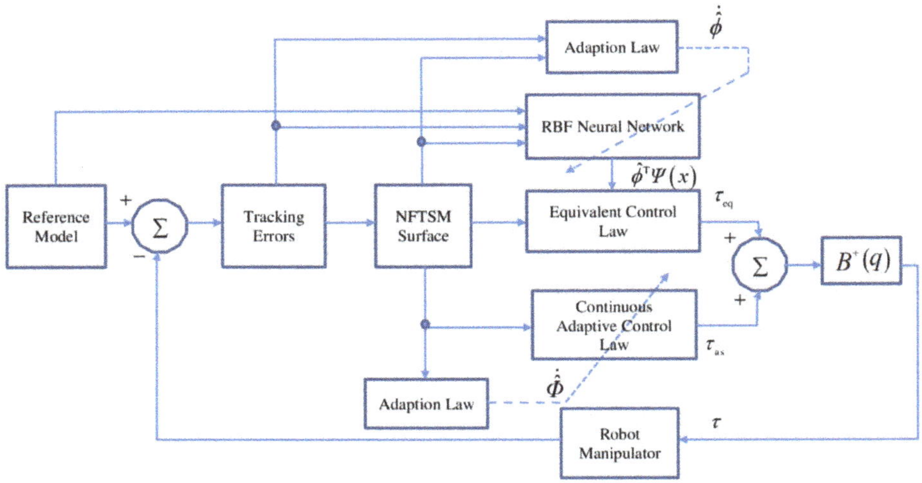

Figure 2. Block diagram of the proposed control method. RBR = radial basis function; NFTSM = nonsingular fast terminal sliding mode control.

Here, the equivalent control law is constructed as

$$\tau_{eq}(t) = -\left(\hat{\phi}^T\Psi(x) + \Pi(e,\varsigma) - \ddot{q}_d\right), \tag{42}$$

and $\tau_{as}(t)$ is an adaptive control term for replacing the control law $\tau_s(t)$ in Equation (32), describing $\tau_{as}(t)$ as

$$\tau_{as} = -(\hat{\Phi} + \rho_1)sign(s), \tag{43}$$

and the adaptive updating rules are given as

$$\dot{\hat{\Phi}} = \frac{1}{\gamma}|s|, \tag{44}$$

$$\dot{\hat{\phi}} = \frac{1}{\omega}s\Psi(x), \tag{45}$$

where $\hat{\Phi}$ is the estimated value of the design parameter Φ, ρ_1 is a positive constant, and γ, ω indicate the adaptive gains.

The control design approach for the robot system is summarized in Theorem 3 below.

Theorem 3. *For the system (3), if the suitable NFTSM variables have been selected as (7) and (8) and the control input signal is constructed as (41)–(43) with its parameter updating rules designed as (44) and (45), then the sliding variable motion is a certainty, and the tracking error variables converge to zero.*

Proof. Define the adaptive estimation error and NN weight approximation error, respectively, as follows

$$\tilde{\Phi} = \hat{\Phi} - \Phi, \tag{46}$$

$$\tilde{\phi} = \phi^* - \hat{\phi}. \tag{47}$$

□

The time derivative of the sliding surface in Equation (29) is rewritten as

$$\dot{s} = \phi^* \Psi(x) + B(q)\tau(t) + W - \ddot{q}_d + \Pi(e,\varsigma). \tag{48}$$

Substituting control laws (41)–(43) into Equation (48) provides

$$\dot{s} = \widetilde{\phi}^T \Psi(x) - (\hat{\Phi} + \rho_1) sign(s) + W. \tag{49}$$

The positive-definite Lyapunov functional is selected as

$$V_4 = \frac{s^T s}{2} + \frac{\gamma \widetilde{\Phi}^T \widetilde{\Phi}}{2} + \frac{\omega \widetilde{\phi}^T \widetilde{\phi}}{2}. \tag{50}$$

With the result of Equation (49), the time derivative of Equation (50) is derived as

$$\begin{aligned} \dot{V}_4 &= s^T \dot{s} + \gamma \widetilde{\Phi}^T \dot{\widetilde{\Phi}} - \omega \widetilde{\phi}^T \dot{\hat{\phi}} \\ &= s^T (\widetilde{\phi}^T \Psi(x) - (\hat{\Phi} + \rho_1) sign(s) + W) + \gamma(\hat{\Phi} - \Phi)\dot{\hat{\Phi}} - \omega \widetilde{\phi}^T \dot{\hat{\phi}} \cdot \\ &= s^T \widetilde{\phi}^T \Psi(x) - \hat{\Phi}|s| - \rho_1 |s| + Ws + \gamma(\hat{\Phi} - \Phi)\dot{\hat{\Phi}} - \omega \widetilde{\phi}^T \dot{\hat{\phi}} \end{aligned} \tag{51}$$

Applying the updating laws (41)–(43) to (51) yields

$$\begin{aligned} \dot{V}_4 &= -\hat{\Phi}|s| - \rho_1 |s| + \Phi s + (\hat{\Phi} - \Phi)|s| \\ &= -\rho_1 |s| + Ws - \Phi |s| \\ &\le -\rho_1 |s| \end{aligned} . \tag{52}$$

If the parameter ρ_1 is selected to be greater than zero, $\dot{V}_4$ will be negative-definite. Based on the Lyapunov principle [47], $\dot{V}_4$ becoming negative-definite indicates that s and $\widetilde{\Phi}$ reach zero. Therefore, the tracking error variables converge to the sliding variables. Therefore, Theorem 3 is proven.

Remark 2. *In practical systems, the parameter drift problem typically occurs under the adaptive control rule (44). Consequently, the bounded approach is implemented to set up the adaptive estimator as*

$$\dot{\hat{\Phi}} = \begin{cases} 0 & \text{if} \quad |s| \le \varpi \\ \frac{1}{\gamma}|s| & \text{if} \quad |s| > \varpi \end{cases}, \tag{53}$$

in which $\varpi > 0$ is an arbitrary positive value.

Remark 3. *[35]: The chattering phenomenon can be significantly alleviated by replacing the $sign(\cdot)$ function with a saturation function in the control input signal, such as*

$$\text{sat}\left(\frac{s}{\varepsilon^*}\right) = \begin{cases} sign(s) & \text{if } |s| \ge (\varepsilon^*)^2 \\ \frac{s}{\varepsilon^*} & \text{if } |s| < \varepsilon^* \end{cases} \tag{54}$$

in which $0 < \varepsilon^* < 1$ is a minor positive coefficient called boundary layer thickness, and $\varepsilon^* = 0.1$.

4. Simulation Analyses

To demonstrate the effectiveness of the proposed control strategy, the strategy was applied to a pathway tracking control for the first three joints of a PUMA560 manipulator, and its tracking performance was compared with those of a classical SMC [16,17] and NFTSMC [49,50]. The dynamic model with the crucial parameters found in a 3-DOF PUMA560 robot manipulator was explained by Armstrong et al. [51]. We utilized the MATLAB/Simulink environment for all simulation analysis with the sampling rate set to 10^{-3} s. In this work, only the first three joints of a robot manipulator were investigated (the last three joints were blocked). The simulations were implemented to compare the controllers in terms of their positional accuracy, response speed, and the resulting chattering phenomenon in their control inputs.

To ascertain the robustness of all control methods, we evaluated the system performance in three operation stages, where disturbances and uncertainties were modeled as follows:

$$F_r(\dot{q}) + \tau_d(t) = \begin{bmatrix} 0.9\dot{q}_1 + 1.0\sin(3q_1) + 1.7\sin(\dot{q}_1) \\ 1.8\dot{q}_2 + 1.85\sin(2q_2) + 1.65\sin(\dot{q}_2) \\ -2.1\dot{q}_3 + 2.5\sin(2q_3) + 0.57\sin(\dot{q}_3) \end{bmatrix}. \quad (55)$$

Stage 1: Robot system was assumed to run under normal operation from time 0 s to 15 s.
Stage 2: Robot system was assumed to run under operation condition, but there was an external disturbance impacting the first joint between 15 s and 50 s. This external disturbance had a value defined as $(15\sin(q_1q_2) + 1.5\cos(\dot{q}_1q_2) + 5.5\cos(\dot{q}_1\dot{q}_2))$.
Stage 3: Robot system was assumed to run under operation condition, but there was a partial loss (75%) of control input effectiveness at the second joint between 25 s and 50 s.

The desired joint pathways for the position tracking were

$$q_r = \begin{bmatrix} \cos(\frac{t}{5\pi}) - 1, & \sin(\frac{t}{5\pi} + \frac{\pi}{2}), & \sin(\frac{t}{5\pi} + \frac{\pi}{2}) - 1 \end{bmatrix}^T. \quad (56)$$

The RBFNN architecture consisted of seven nodes, the initial weight matrix of the network was selected as 0, the width and center of the Gaussian function was set as $\delta = 0.2$, and the center of the Gaussian function μ was selected in range $(-1.5 \div 1.5)$ with $\mu_l = 0.5$. The matrix used in an adaptive law of RBFNN was selected as $\omega = 15I_7$, and the NN input was selected as $v = \begin{bmatrix} e & \dot{e} & q_r & \dot{q}_r & \ddot{q}_r \end{bmatrix}$.

The SMC control input was set as

$$\tau(t) = -B^{-1}(q)(\Xi(q,\dot{q}) + \eta(\dot{q} - \dot{q}_r) - \ddot{q}_r + (\Phi_2 + \rho_2)sign(s)). \quad (57)$$

Here, η, Φ_2, ρ_2 are positive constants, s is a linear sliding function, and q_r is defined as a desired trajectory value.

The NFTSMC control input was set as

$$\tau(t) = -B^{-1}(q)\left(\Xi(q,\dot{q}) - \ddot{q}_r + \beta\frac{h}{d}(\dot{e})^{2-\frac{d}{h}} + (\Phi_3 + \rho_3)\frac{s}{\|s\| + \nu}\right). \quad (58)$$

Here, β, Φ_3, ρ_3 are positive constants, s is a nonlinear sliding function, ν is a small positive scalar, q_r is defined as a desired trajectory value, and $d,\ h$ are positive odd integers satisfying the condition $1 < d/h < 2$.

The control parameter selection for the varying control strategies, including classical SMC, NTSMC, and the proposed control strategy is shown in Table 1.

The averaged tracking errors were calculated according to the following equation $E_i^{av} = \sqrt{\frac{1}{n}\sum_{k=1}^{n}\left(\|e_i\|^2\right)}$ $i = 1,\ 2,\ 3$ in which n is the number of simulation steps.

The trajectory tracking performances, including tracking positions and tracking errors at each of the first three joints with three controllers, are illustrated in Figures 3 and 4. In Stage 1 (from 0 s to 15 s), three of the control systems give similar good path tracking performance. In Stage 2 (from time greater than 15 s) and in Stage 3 (from time greater than 25 s), it is clear that the classical SMC provides the poorest path tracking performance, where robot operation becomes unstable when a large disturbance or uncertainty is applied. From Table 2 and Figure 4, it is observed that NFTSMC provides less path tracking error and faster transient response than classical SMC. However, tracking performance is also diminished upon application of a large disturbance. It is noteworthy that the proposed sliding surface is designed based on the sliding function integral in Equation (8), and this integral portion has a significant role in providing fast transient response and robustness against uncertainty and disturbances. Therefore. the proposed control strategy gives the best path tracking performance and fastest transient response among the compared control strategies, due to the role of the proposed surfaces, an adaptive compensator, and a main contribution of the proposed controller.

Table 1. The control parameter selection for the varying control strategies. SMC = sliding mode controller; ANNFTSMC = adaptive neural non-singular fast-terminal sliding-mode control.

Control Strategy	Control Parameters	Parameter Value
Classical SMC	η, Φ_2, ρ_2	2, 9.9, 1
NFTSMC	d, h, β	5, 3, 2
	Φ_3, ρ_3, ν	9.9, 1, 0.1
Proposed Control Strategy (ANNFTSMC)	h_1, h_2	$diag(10, 10, 10), diag(6, 6, 6)$
	$\Gamma_1, \Gamma_2, \Gamma_3$	$diag(3, 3, 3), diag(3, 3, 3), diag(2, 2, 2)$
	ϑ, α	$diag(0.4, 0.4, 0.4), diag(1.2, 1.2, 1.2)$
	$\gamma, \rho_1, \varpi, \varepsilon^*$	0.5, 0.1, 0.01, 0.1

Table 2. The averaged tracking errors under control input signals of the control strategy.

Error Control Strategy	E_1^{av}	E_2^{av}	E_3^{av}
SMC	0.1943	0.8708	0.0060
NFTSMC	0.1542	0.1218	0.0038
ANNFTSMC	0.0031	0.0031	0.0029

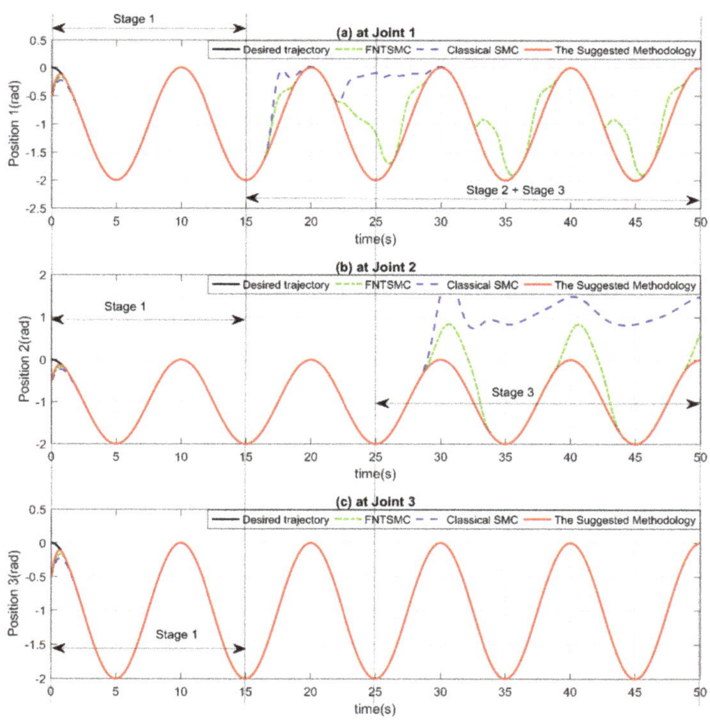

Figure 3. Trajectory tracking positions: (**a**) at Joint 1, (**b**) at Joint 2, and (**c**) at Joint 3.

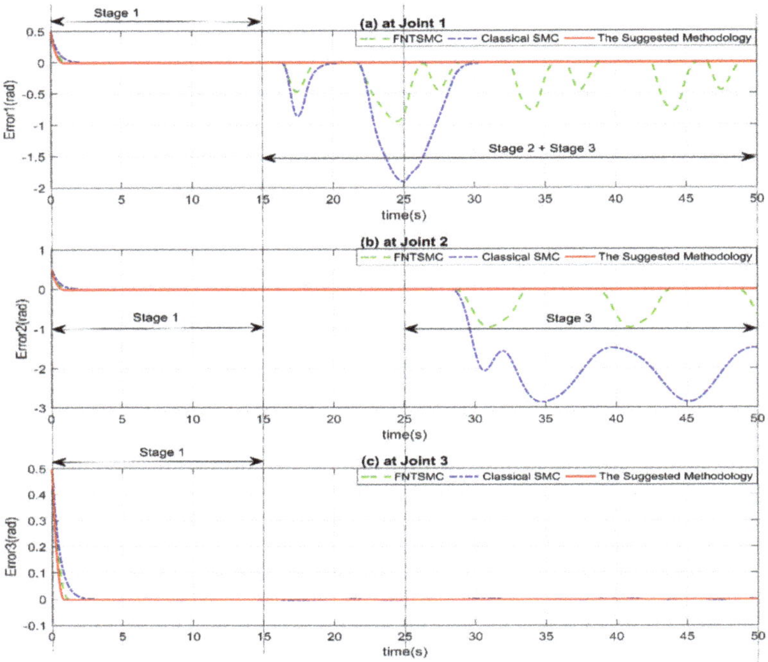

Figure 4. Trajectory tracking errors: (**a**) at Joint 1, (**b**) at Joint 2, and (**c**) at Joint 3.

The control input signals for all control types, including classical SMC, NFTSMC, and the suggested system are shown in Figure 5. In Figure 5a, it is clear that the NFTSMC offers a continuous control signal by using a boundary technique [35]. However, the weakness of this technique is that a choice must be made between chattering phenomenon removal and path tracking precision. Consequently, this technique decreases the robustness of the system while also increasing the tracking error. In Figure 5b, the SMC offers a discontinuous control signal with serious chattering behavior. On the contrary, the suggested system offers a continuous control signal for the robot manipulator without the loss of its effectiveness, as shown in Figure 5c.

The adaptations of the estimated parameters are shown in Figure 6. These adaptive gains are estimated according to the variation of the influences of disturbances and uncertainties, and they will attain a constant value once the error variables converge to the sliding surface in a stable phase.

From the simulation performance, we conclude that the proposed controller gives the best performance compared to a classical SMC and NFTSMC in terms of tracking precision, transient response, chattering deletion, and small steady state error.

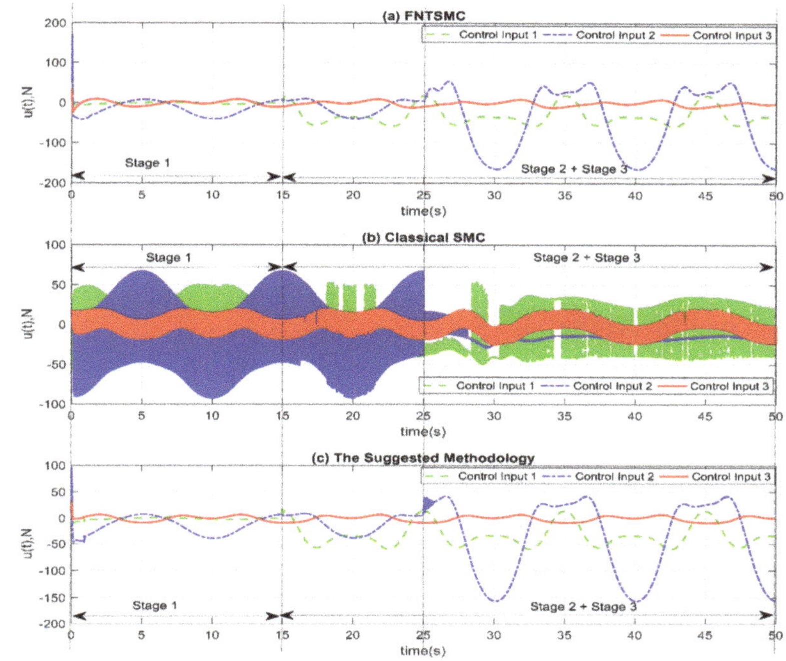

Figure 5. Control input signals: (**a**) FNTSMC, (**b**) classical SMC, and (**c**) the suggested control methodology. FNTSMC = fast nonsingular terminal sliding mode control.

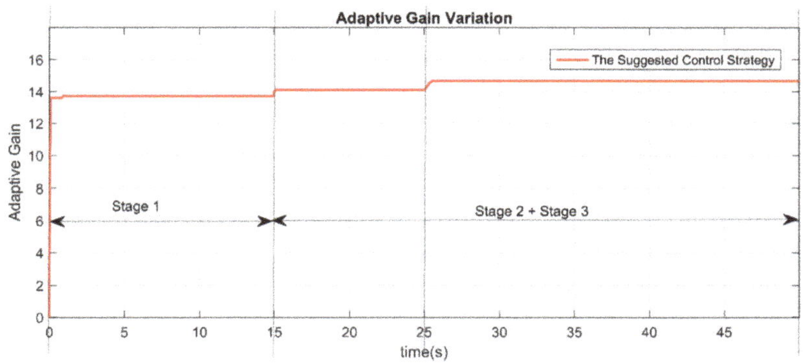

Figure 6. Time history of adaptive gain.

5. Conclusions

In this report, a robust trajectory tracking control strategy was developed for robot manipulators. From the simulation results and performance comparison with two other control strategies for a 3-DOF PUMA560 robot manipulator, our control strategy offered the best performance in terms of tracking positional accuracy, small steady-state errors, fast convergence, and chattering phenomenon rejection. The suggested control solution has the following benefits: (1) inherits the advantages of the NFTSMC, including non-singularity, finite-time convergence, fast transient response, low steady-state errors, and high position tracking accuracy; (2) achieves smoothness with elimination of chattering behavior; (3) does not demand an exact dynamic model for the robot manipulator by applying an adaptive radial basis function neural network to approximate an unknown robot function; (4) compared to the classical SMC and another control methods based on TSMC, the proposed control strategy offers better tracking performance and stronger resistance against disturbances and uncertainties; (5) robustness and stability of the robot system was demonstrated fully by Lyapunov theory.

Author Contributions: All authors contributed equally to this article and accepted the final report.

Funding: This research was funded by the Ministry of Education, grant number (NRF-2016R1D1A3B03930496).

Acknowledgments: This work was supported by the Basic Science Research Program through the National Research Foundation of Korea (NRF) funded by the Ministry of Education (NRF-2016R1D1A3B03930496).

Conflicts of Interest: All authors announce that they have no conflict of interest in relation to the publication of this article.

References

1. Yang, C.; Huang, Q.; Jiang, H.; Peter, O.O.; Han, J. PD control with gravity compensation for hydraulic 6-DOF parallel manipulator. *Mech. Mach. Theory* **2010**, *45*, 666–677. [CrossRef]
2. Ouyang, P.R.; Zhang, W.-J.; Wu, F.-X. Nonlinear PD control for trajectory tracking with consideration of the design for control methodology. In Proceedings of the IEEE International Conference on Robotics and Automation, ICRA'02, Washington, DC, USA, 10–17 May 2002; Volume 4, pp. 4126–4131.
3. Arimoto, S. Stability and robustness of PID feedback control for robot manipulators of sensory capability. In Proceedings of the Robotics Research: The First International Symposium, Bretton Woods, NH, USA, 25 August–2 September 1983; pp. 783–799.
4. Su, Y.; Müller, P.C.; Zheng, C. Global Asymptotic Saturated PID Control for Robot Manipulators. *IEEE Trans. Control Syst. Technol.* **2010**, *18*, 1280–1288. [CrossRef]
5. Guo, Y.; Woo, P.-Y. An adaptive fuzzy sliding mode controller for robotic manipulators. *IEEE Tran. Syst. Man Cybern.-Part A Syst. Hum.* **2003**, *33*, 149–159.
6. Tran, X.-T.; Kang, H.-J. TS fuzzy model-based robust finite time control for uncertain nonlinear systems. *Proc. Inst. Mech. Eng. Part C J. Mech. Eng. Sci.* **2015**, *229*, 2174–2186. [CrossRef]
7. Vo, A.T.; Kang, H.-J.; Le, T.D. An Adaptive Fuzzy Terminal Sliding Mode Control Methodology for Uncertain Nonlinear Second-Order Systems. In Proceedings of the International Conference on Intelligent Computing, Wuhan, China, 15–18 August 2018; pp. 123–135.
8. Sanger, T.D. Neural network learning control of robot manipulators using gradually increasing task difficulty. *IEEE Trans. Robot. Autom.* **1994**, *10*, 323–333. [CrossRef]
9. Hsia, T.C.; Jung, S. A simple alternative to neural network control scheme for robot manipulators. *IEEE Trans. Ind. Electron.* **1995**, *42*, 414–416. [CrossRef]
10. Vo, A.T.; Kang, H.-J.; Nguyen, V.-C. An output feedback tracking control based on neural sliding mode and high order sliding mode observer. In Proceedings of the 2017 10th International Conference on Human System Interactions (HSI), Ulsan, Korea, 17–19 July 2017; pp. 161–165.
11. Van, M.; Kang, H.-J.; Suh, Y.-S.; Shin, K.-S. A robust fault diagnosis and accommodation scheme for robot manipulators. *Int. J. Control Autom. Syst.* **2013**, *11*, 377–388. [CrossRef]
12. Shang, W.; Cong, S. Nonlinear computed torque control for a high-speed planar parallel manipulator. *Mechatronics* **2009**, *19*, 987–992. [CrossRef]
13. Zhihong, M.; O'day, M.; Yu, X. A robust adaptive terminal sliding mode control for rigid robotic manipulators. *J. Intell. Robot. Syst.* **1999**, *24*, 23–41. [CrossRef]
14. Moreno, J.A.; Negrete, D.Y.; Torres-González, V.; Fridman, L. Adaptive continuous twisting algorithm. *Int. J. Control* **2016**. [CrossRef]
15. Lim, K.; Eslami, M. Robust adaptive controller designs for robot manipulator systems. *IEEE J. Robot. Autom.* **1987**, *3*, 54–66. [CrossRef]
16. Utkin, V.I. *Sliding Modes in Control and Optimization*; Springer Science & Business Media: Berlin, Germany, 2013.
17. Edwards, C.; Spurgeon, S. *Sliding Mode Control: Theory and Applications*; Crc Press: Boca Raton, FL, USA, 1998.
18. Eker, I. Sliding mode control with PID sliding surface and experimental application to an electromechanical plant. *ISA Trans.* **2006**, *45*, 109–118. [CrossRef]

19. Kamal, S.; Moreno, J.A.; Chalanga, A.; Bandyopadhyay, B.; Fridman, L.M. Continuous terminal sliding-mode controller. *Automatica* **2016**. [CrossRef]
20. Xiao, B.; Hu, Q.; Zhang, Y. Adaptive sliding mode fault tolerant attitude tracking control for flexible spacecraft under actuator saturation. *IEEE Trans. Control Syst. Technol.* **2012**, *20*, 1605–1612. [CrossRef]
21. Man, Z.; Paplinski, A.P.; Wu, H.R. A robust MIMO terminal sliding mode control scheme for rigid robotic manipulators. *IEEE Trans. Autom. Control* **1994**, *39*, 2464–2469. [CrossRef]
22. Wu, Y.; Yu, X.; Man, Z. Terminal sliding mode control design for uncertain dynamic systems. *Syst. Control Lett.* **1998**, *34*, 281–287. [CrossRef]
23. Tang, Y. Terminal sliding mode control for rigid robots. *Automatica* **1998**, *34*, 51–56. [CrossRef]
24. Yu, X.; Zhihong, M. Fast terminal sliding-mode control design for nonlinear dynamical systems. *Circuits Syst. I Fundam. Theory* **2002**, *49*, 261–264. [CrossRef]
25. Mobayen, S. Fast terminal sliding mode controller design for nonlinear second-order systems with time-varying uncertainties. *Complexity* **2015**, *21*, 239–244. [CrossRef]
26. Madani, T.; Daachi, B.; Djouani, K. Modular-controller-design-based fast terminal sliding mode for articulated exoskeleton systems. *IEEE Trans. Control Syst. Technol.* **2017**, *25*, 1133–1140. [CrossRef]
27. Eshghi, S.; Varatharajoo, R. Nonsingular terminal sliding mode control technique for attitude tracking problem of a small satellite with combined energy and attitude control system (CEACS). *Aerosp. Sci. Technol.* **2018**, *76*, 14–26. [CrossRef]
28. Safa, A.; Abdolmalaki, R.Y.; Shafiee, S.; Sadeghi, B. Adaptive nonsingular terminal sliding mode controller for micro/nanopositioning systems driven by linear piezoelectric ceramic motors. *ISA Trans.* **2018**. [CrossRef] [PubMed]
29. Lin, C.-K. Nonsingular terminal sliding mode control of robot manipulators using fuzzy wavelet networks. *IEEE Trans. Fuzzy Syst.* **2006**, *14*, 849–859. [CrossRef]
30. Su, Y.; Zheng, C.; Mercorelli, P. Global finite-time stabilization of planar linear systems with actuator saturation. *IEEE Trans. Circuits Syst. II Express Briefs* **2017**, *64*, 947–951. [CrossRef]
31. Su, Y.; Zheng, C. Robust finite-time output feedback control of perturbed double integrator. *Automatica* **2015**, *60*, 86–91. [CrossRef]
32. Xu, S.S.-D.; Chen, C.-C.; Wu, Z.-L. Study of nonsingular fast terminal sliding-mode fault-tolerant control. *IEEE Trans. Ind. Electron.* **2015**, *62*, 3906–3913. [CrossRef]
33. Yang, L.; Yang, J. Nonsingular fast terminal sliding-mode control for nonlinear dynamical systems. *Int. J. Robust Nonlinear Control* **2011**, *21*, 1865–1879. [CrossRef]
34. Chen, G.; Jin, B.; Chen, Y. Nonsingular fast terminal sliding mode posture control for six-legged walking robots with redundant actuation. *Mechatronics* **2018**, *50*, 1–15. [CrossRef]
35. Utkin, V. Discussion aspects of high-order sliding mode control. *IEEE Trans. Autom. Control* **2016**, *61*, 829–833. [CrossRef]
36. Feng, Y.; Zhou, M.; Zheng, X.; Han, F.; Yu, X. Full-order terminal sliding-mode control of MIMO systems with unmatched uncertainties. *J. Frankl. Inst.* **2018**, *355*, 653–674. [CrossRef]
37. Feng, Y.; Han, F.; Yu, X. Chattering free full-order sliding-mode control. *Automatica* **2014**, *50*, 1310–1314. [CrossRef]
38. Rubio-Astorga, G.; Sánchez-Torres, J.D.; Cañedo, J.; Loukianov, A.G. High-order sliding mode block control of single-phase induction motor. *IEEE Trans. Control Syst. Technol.* **2014**, *22*, 1828–1836. [CrossRef]
39. Nentwig, M.; Mercorelli, P. Throttle valve control using an inverse local linear model tree based on a fuzzy neural network. In Proceedings of the 7th IEEE International Conference on Cybernetic Intelligent Systems, London, UK, 9–10 September 2008; pp. 1–6.
40. Li, X.-J.; Yang, G.-H. Neural-network-based adaptive decentralized fault-tolerant control for a class of interconnected nonlinear systems. *IEEE Trans. Neural Netw. Learn. Syst.* **2018**, *29*, 144–155. [CrossRef] [PubMed]
41. Sun, R.; Wang, J.; Zhang, D.; Shao, X. Neural network-based sliding mode control for atmospheric-actuated spacecraft formation using switching strategy. *Adv. Space Res.* **2018**, *61*, 914–926. [CrossRef]
42. Shen, Q.; Jiang, B.; Cocquempot, V. Adaptive fuzzy observer-based active fault-tolerant dynamic surface control for a class of nonlinear systems with actuator faults. *IEEE Trans. Fuzzy Syst.* **2014**, *22*, 338–349. [CrossRef]

43. Huo, B.; Xia, Y.; Yin, L.; Fu, M. Fuzzy adaptive fault-tolerant output feedback attitude-tracking control of rigid spacecraft. *IEEE Trans. Syst. Man Cybern. Syst.* **2017**, *47*, 1898–1908. [CrossRef]
44. Wang, S.-Y.; Liu, F.-Y.; Chou, J.-H. Adaptive TSK fuzzy sliding mode control design for switched reluctance motor DTC drive systems with torque sensorless strategy. *Appl. Soft Comput.* **2018**, *66*, 278–291. [CrossRef]
45. Spong, M.W.; Vidyasagar, M. *Robot Dynamics and Control*; John Willey & Sons: New York, NY, USA, 1989.
46. Islam, S.; Liu, X.P. Robust sliding mode control for robot manipulators. *IEEE Trans. Ind. Electron.* **2011**, *58*, 2444–2453. [CrossRef]
47. Polyakov, A.; Fridman, L. Stability notions and Lyapunov functions for sliding mode control systems. *J. Frankl. Inst.* **2014**, *351*, 1831–1865. [CrossRef]
48. Bhat, S.P.; Bernstein, D.S. Finite-time stability of continuous autonomous systems. *SIAM J. Control Opt.* **2000**, *38*, 751–766. [CrossRef]
49. Yu, S.; Yu, X.; Shirinzadeh, B.; Man, Z. Continuous finite-time control for robotic manipulators with terminal sliding mode. *Automatica* **2005**, *41*, 1957–1964. [CrossRef]
50. Feng, Y.; Yu, X.; Man, Z. Non-singular terminal sliding mode control of rigid manipulators. *Automatica* **2002**, *38*, 2159–2167. [CrossRef]
51. Armstrong, B.; Khatib, O.; Burdick, J. The explicit dynamic model and inertial parameters of the PUMA 560 arm. In Proceedings of the 1986 IEEE International Conference on Robotics and Automation, San Francisco, CA, USA, 7–10 April 1986; Volume 3, pp. 510–518.

applied sciences

MDPI

Article

A Novel Approach for a Inverse Kinematics Solution of a Redundant Manipulator

Michal Kelemen [1], Ivan Virgala [1,*], Tomáš Lipták [1], Ľubica Miková [1], Filip Filakovský [1] and Vladimír Bulej [2]

1 Faculty of Mechanical Engineering, Technical University of Košice, 04200 Košice, Slovakia; michal.kelemen@tuke.sk (M.K.); tomas.liptak@tuke.sk (T.L.); lubica.mikova@tuke.sk (Ľ.M.); filip.filakovsky@tuke.sk (F.F.)

2 Faculty of Mechanical Engineering, University of Žilina, 01026 Žilina, Slovakia; vladimir.bulej@fstroj.uniza.sk

* Correspondence: ivan.virgala@tuke.sk; Tel.: +421-55-602-2455

Received: 18 September 2018; Accepted: 8 November 2018; Published: 12 November 2018

Abstract: Kinematically-redundant manipulators present considerable difficulties, especially from the view of control. A high number of degrees of freedom are used to control so-called secondary tasks in order to optimize manipulator motion. This paper introduces a new algorithm for the control of kinematically-redundant manipulator considering three secondary tasks, namely a joint limit avoidance task, a kinematic singularities avoidance task, and an obstacle avoidance task. For path planning of end-effector from start to goal point, the potential field method is used. The final inverse kinematic model is designed by a Jacobian-based method considering weight matrices in order to prioritize particular tasks. Our approach is based on the flexible behavior of priority value due to the acceleration of numerical simulation. The results of the simulations show the advantage of our approach, which results in a significant decrease of computing time.

Keywords: computing time; inverse kinematics; joint limit avoidance; kinematic singularity; manipulator; obstacle avoidance; potential field

1. Introduction

Kinematically-redundant manipulators are mechanisms which have more degrees of freedom (DOF) than is required for the execution of a given task. The advantage of kinematically-redundant manipulators in comparison with conventional manipulators is in the utilization of redundant manipulator joints for optimization tasks [1,2]. These optimization tasks are secondary tasks of the inverse kinematic or dynamic model. Manipulator redundancy is used for tasks such as avoidance of collision with obstacles, avoidance of kinematic singularities, maintenance of the admissible joint ranges, increasing of manipulability in specified directions, optimization of execution time, minimizing energy consumption, etc. [3,4]. On the other hand, kinematic redundancy causes disadvantages, such as the requirements of greater structural complexity of manipulator construction (higher number of actuators, sensors, costs, etc.). It is additionally important to note that control algorithms for inverse kinematic and dynamic model are considerably more complicated [5].

This study investigates kinematically-redundant mechanisms moving in an environment with obstacles. The investigated mechanisms additionally have to deal with a joint limit avoidance task and a kinematic singularity avoidance task. There are several methods to solve the mentioned problems, namely Jacobian-based methods, null-space methods, and task augmentation methods [6,7]. Many approaches have been used for the kinematic control of manipulators with secondary tasks. The gradient projection method (GPM) is one of them. It was first used in [8] to deal with joint limit avoidance. A later study [9] additionally introduced an iterative approach for joint limit avoidance.

These approaches used null space or enlarged space. A problem occurs when the number of all tasks exceeds the number of DOF of manipulators. The weighted least-norm (WLN) method is a method which deals with constraints all the time. Considering the joint limit avoidance task, WLN uses self-motion only when it is necessary in comparison with GPM [10]. An approach with consideration of WLN solution was suggested by Whitney [11], and has been used for minimizing energy by using inertia matrix as the weighting matrix. This approach was also used in [12] for minimizing joint torques and in [13] for minimizing joint velocities. Whitney and Chan [14,15] describe the role of weight matrices and the priority of its choice for emphasizing or de-emphasizing the role of some components in the computing process. Another algorithm, namely the clamping loop algorithm, ensures the avoidance of joint limits, however is fairly time consuming [16]. Earlier research attempted to assign the priority of the particular tasks by weight matrices [17,18]. However, in some cases the task requirements cannot be achieved [19,20]. In [21], the authors assigned a lower priority to obstacle avoidance task. Problems occurred when the secondary task was not compatible with the main task and the numerical simulations failed. Some works allow the activation or deactivation of secondary tasks by continuous inverse of Jacobian multiplied by the activation matrix [22]. In [23], the authors proposed a new task-regulation framework based on a hierarchy of quadratic program. Within their framework it is possible to forward the constraint task separation across priority levels, eliminating the need for converting inequality constraints into equalities.

Our developed approach deals with Jacobian-based method using weight matrices to set the priority of primary as well as all secondary tasks. Our approach is based on changing value of task priority during numerical simulation. The behavior of priority changing is based on numerical computing smoothness. During computing, all secondary tasks are active.

This paper is organized as follows. First, the paper deals with path-planning for end-effector of a manipulator moving in an environment with obstacles. For this purpose, the potential field method is introduced and an environment with obstacles is modeled. Next, the inverse kinematic model is derived. Consequently, the low-level control of the experimental model is introduced. The paper describes algorithms for all mentioned areas. Then, the simulations and experiment are performed, and the results are compared and discussed in the conclusion.

2. Path Planning Task for End-Effector

The control of robot motion is a very complex and difficult task. The control system has to deal with many circumstances and changes of conditions in robot environment, while the computing algorithms are often excessively difficult from the view of computing power [24].

The aim of this section is to introduce a means of path planning for the manipulator end-effector. As was mentioned earlier, the investigated manipulator will move in an environment with obstacles. The manipulator has to move its end-effector from start point to goal point and the control algorithm has to ensure the avoidance of any collision between manipulator links and obstacles. The path from start to goal point of the end-effector is planned by means of potential field method, as described in the following.

Potential Field Method

Our aim is to move the manipulator end-effector from its start position $s_{start} = [x_S, y_S]^T$ to its goal position $s_{goal} = [x_G, y_G]^T$, while the control system has to ensure the avoidance of collision with obstacles. In this research we use potential field method for the purposes of path planning task. The generated path is the shortest path from start point to goal point. This research assumes planar motion of the manipulator.

The main idea of the potential field method is very simple, and at the same time the method is very powerful for robot navigation. The potential field method deals with two kinds of fields, namely an attractive field and a repulsive field [25].

In general, the attractive potential field represents the relation between each point of the robot workspace and the goal point. The workspace of the robot can be divided into a defined number of points according to the chosen grid. The softer the workspace grid is, the more precise the planned path will be. There are several ways to mathematically model the attractive field. The commonly used relation of attractive potential function is [26,27]:

$$U_{att}(s) = \frac{1}{2}\xi \parallel s_{goal} - s_{grid} \parallel^{k} \quad (1)$$

where ξ is a positive scalar variable, s_{grid} is the position of every point from the workspace, and k is a number higher than zero. For k = 1 the potential field has a conic shape and for k = 2 the potential field has a parabolic shape. In this research we use k = 1. The corresponding force function can be expressed as:

$$F_{att}(s) = -\nabla U_{att}(s) = -\frac{\partial U_{att}(s)}{\partial s} \quad (2)$$

By modeling the attractive field one obtains the function which has a local extreme at the goal point. We can imagine the attractive field as, for example, a ball falling down a hill, which stops at the lowest point. The example of attractive field is shown in Figure 1.

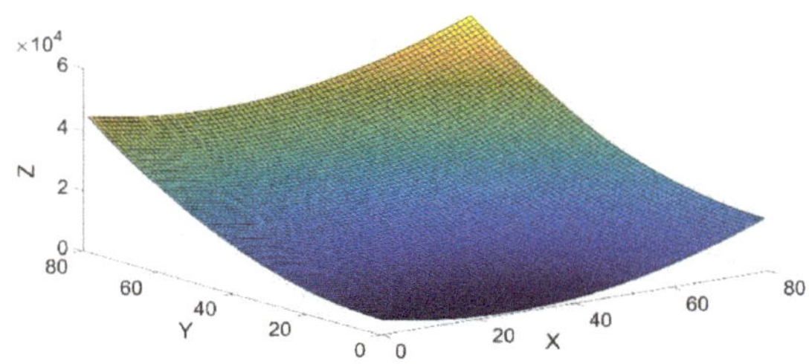

Figure 1. Example of attractive field.

In Figure 1, the dark blue color in the position (0, 30, 0) represents the local minimum of the potential function, that is, the goal point.

Besides the attractive field there is also the repulsive field. The repulsive field represents the environment, with which the manipulator cannot collide. In our study, the repulsive field is represented by obstacles of circular shape. The repulsive potential function usually takes the following form:

$$U_{rep}(s) = \begin{cases} \eta\left(\frac{1}{d} - \frac{1}{d_0}\right)^2, d \leq d_0 \\ 0,\ d > d_0 \end{cases} \quad (3)$$

where $d = \parallel s_{obstacle} - s_{robot} \parallel$ is the distance between the obstacle and manipulator link, d_0 represents the influence of the obstacle, and η is a scalar parameter. The gradient corresponding to this function is :

$$F_{rep}(s) = \nabla U_{rep}(s) = \begin{cases} \eta\left(\frac{1}{d} - \frac{1}{d_0}\right)^2 \frac{(s-s_0)}{d^3},\ d \leq d_0 \\ 0, d > d_0 \end{cases} \quad (4)$$

where s is the actual position of the manipulator link and s_0 is the distance from the manipulator link to the obstacle. The aim of the repulsive forces is to affect out from the obstacle. The example of the repulsive field applied in this research can be seen in Figure 2.

The final potential field is given by the sum of the attractive and repulsive functions:

$$F(q) = -\nabla U_{att} + \nabla U_{rep} \quad (5)$$

The set of all obstacles in the workspace of the investigated manipulator can be represented by the matrix $O \in R^{m \times h}$, where h is the number of obstacles and m represents the dimension of the performed task. Since our task is planar, the parameter m equals 2. By summing the attractive field $U_{att}(s)$ and repulsive field $U_{rep}(s)$ we obtain the matrix F(q), which includes obstacles of the manipulator environment as well as course of the attractive function with goal point (local extreme). Based on this matrix the shortest way from the start point to the goal point can be obtained. The principle is as follows. Each point of the workspace is represented by a numerical value. The goal point has the lowest numerical value from each point of the workspace. Starting at the start point, the next step is to move to the adjoining point, which has a lower numerical value than the start point. After this step, an adjoining point has to be found with a lower numerical value than the previous point. Thus, the path from the start point to the goal point can be generated (see Figure 3). The repulsive field, which is included in the aforementioned final matrix, ensures that the obstacles have high numerical values and, therefore, the path never goes through these obstacles.

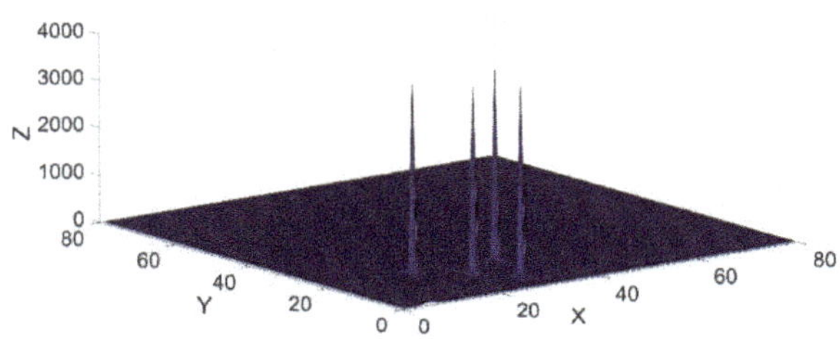

Figure 2. Example of the repulsive field.

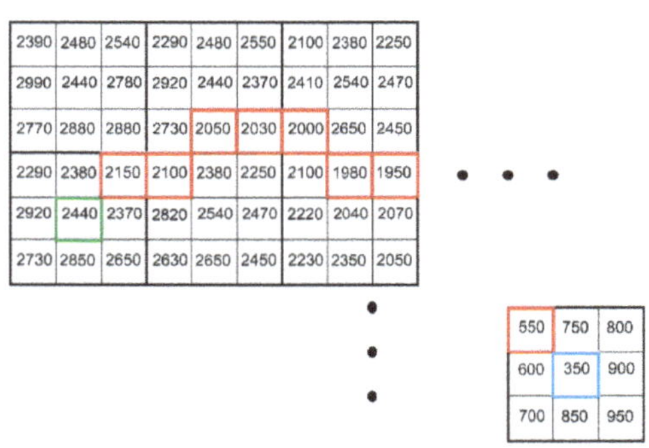

Figure 3. Final potential field matrix, which is used to find the shortest path from the start point to the goal point.

In Figure 3, the square outlined in green represents the start point of the manipulator end-effector and the square outlined in blue represents the goal point of the end-effector. The squares outlined in red show the planned path according to the algorithm introduced above. As can be seen in Figure 3, the minimum value of all values in the matrix is 350. This represents the goal point of the manipulator end-effector.

Next, the algorithm for the path planning is introduced.

In the first step of the algorithm, the obstacles are determined. In this research it is irrelevant whether the obstacles are scanned by camera, sensor, or whether they are defined by the user. The generation of the arrays U_{att} and U_{rep} are performed in Steps 2–3. In the FOR cycles (Steps 4–5), the relation between each point of the manipulator workspace in regard to the particular obstacles is investigated. The output of Algorithm 1 (Generation of Attractive and Repulsive Fields) is an attractive field and a repulsive field in the form of matrices. The constants x_{max} and y_{max} characterize the workspace in which the manipulator works. Following the generation of the final function, it is then necessary to find the shortest path from the start point of the manipulator end-effector to its goal point in the workspace, according to Figure 3. This can be achieved by the following algorithm.

Algorithm 1 Generation of Attractive and Repulsive Fields

```
Determination (scan) of the obstacles
U_att - > zero matrix, U_att ∈ R^(x_max × y_max)
U_rep - > unit matrix, U_rep ∈ R^(x_max × y_max)
FOR x = x_min: x_max
  FOR y = y_min: y_max
      Computation of −∇U_att(x, y)
      FOR obstacle = 1: number_of_obstacles
        IF || s_obstacle − s_robot ||≤ d_0
           ∇U_rep(x, y) = ∇U_rep + η(1/d − 1/d_0)^2
         ELSE
           ∇U_rep(x, y) = ∇U_rep(x, y) + 0
         END IF
       END FOR
     END FOR
END FOR
F(q) = −∇U_att + ∇U_rep
```

Algorithm 2 (Path Planning) works with the output of Algorithm 1 (Generation of Attractive and Repulsive Fields). The aim of this algorithm is to determine the shortest path from the start point to the goal point. The output of this algorithm is the matrix $P \in R^{r \times 2}$, where r is the number of workspace positions between the start point and the goal point. The matrix P is then used as the input to the inverse kinematic model to control the manipulator links.

The designed environment in our study, including the planned path from the start to the goal point, can be seen in the Figure 4.

Figure 4 shows the goal position in (2, 140, 0), which is the local extreme of the attractive field. This environment with five obstacles and exactly defined start and goal points is used for all of the case studies in this paper. A different view of the generated potential field can be seen in Figure 5. The field surrounding the obstacles affects out from the obstacles and each point of the workspace tends to the goal point.

The planned path of the manipulator end-effector is subsequently used as input to the inverse kinematic model of manipulator.

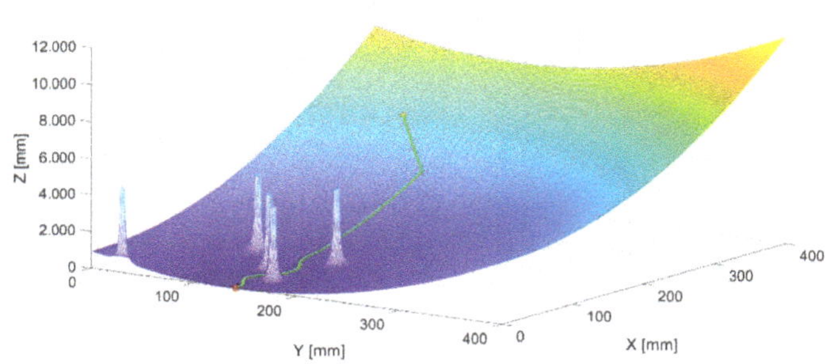

Figure 4. The final potential field with the obstacles and planned path.

Algorithm 2 Path Planning

```
x_start = x_position_of_end-effector,
  y_start = y_position_of_end-effector
flag = non zero value
WHILE flag ≠ 0
    i = i + 1
    P[i,1] = x_start, P[i,2] = y_start
    flag = F(q) [x_start, y_start]
    U_temp[1] = F(q) [x_start-1, y_start]
    U_temp[2] = F(q) [x_start+1, y_start]
    U_temp[3] = F(q) [x_start, y_start-1]
    U_temp[4] = F(q) [x_start, y_start+1]
    U_temp[5] = F(q) [x_start-1, y_start-1]
    U_temp[6] = F(q) [x_start-1, y_start+1]
    U_temp[7] = F(q) [x_start+1, y_start-1]
    U_temp[8] = F(q) [x_start+1, y_start+1]
    k = position_of_min_value_of_U_temp
    SWITCH (k)
        1:x_start = x_start-1, y_start = y_start
        2:x_start = x_start+1, y_start = y_start
        3:x_start = x_start, y_start = y_start-1
        4:x_start = x_start, y_start = y_start+1
        5:x_start = x_start-1, y_start = y_start-1
        6:x_start = x_start-1, y_start = y_start+1
        7:x_start = x_start+1, y_start = y_start-1
        8:x_start = x_start+1, y_start = y_start+1
    END SWITCH
END WHILE
```

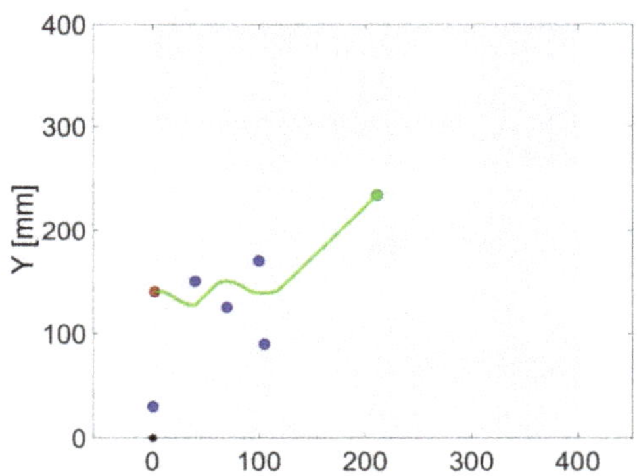

Figure 5. Planned path using potential field method.

3. Inverse Kinematic Model and Computing Algorithm

The inverse kinematic model serves to find such vector of generalized variables $q \in R^n$, (n—number of DOF) in the joint space, by which the end-effector of the manipulator reaches the required position in the task space $x \in R^m$. The vector of generalized variables is defined as $q = [q_1, q_2, \ldots, q_n]^T$. The solution of the inverse kinematic model is significantly more difficult to obtain than that of the direct kinematic model. In many cases there are no analytical solutions. This especially holds in cases of kinematically-redundant manipulators [28]. In such cases, a numerical solution of inverse kinematics has to be applied. The solution arises from the following Equation (6), which represents the relation between joint space and task space:

$$\dot{x} = J\dot{q} \tag{6}$$

where $\dot{x}$ is vector of end-effector velocity, q is vector of generalized variables–joint velocities, and $J \in R^{m \times n}$ is the Jacobian matrix. The indices m and n represent the dimension of task space and the dimension of joint space, respectively. The inverse kinematics are usually based on the numerical solution of:

$$\dot{q} = J^{-1}\dot{x} \tag{7}$$

Equation (7) can be solved when the Jacobian matrix is symmetric. For non-symmetric Jacobian matrices, any method developed for the purpose of Equation (7) solving has to be applied, such as the pseudo-inverse of the Jacobian matrix, or its transposition. In this study, the damped least squares method is used. The Equation (7) includes primary solution-finding, such as q, by which the end-effector reaches the required position in the task space. Subsequently, the advantage of kinematically-redundant manipulators can be applied. This advantage relates to the use of the redundant degrees of freedom for optimization tasks. From the viewpoint of mathematics, kinematically-redundant manipulators can reach the desired position of end-effector in an infinite amount of ways. In other words, the required position x can be reached by an infinite amount of generalized variable configurations.

The optimization tasks solved in this study are a joint limit avoidance task, a kinematic singularity avoidance task, and an obstacle avoidance task. These tasks are so-called secondary tasks which can be completed while the primary task is also performed. There are several methods to solve these optimization tasks. This study uses a method which is part of the class of Jacobian-based methods class. This method considers weight matrices in order to prioritize particular tasks. Very often, some constraints cannot be satisfied simultaneously, although they can be satisfied separately [29]; accordingly, some compromise has to be made.

3.1. Kinematic Singularities Avoidance Task

The kinematic singularities avoidance task plays a significant role during numerical computing. Around the singular positions, the manipulator loses its manipulability and the numerical computation slows down until it fails [30]. Kinematic singularities represent the problem with the mapping of task space to joint space. This problem occurs when the determinate of the Jacobian matrix equals zero, that is, $\det J(q) = 0$. There are many methods dealing with these computing problems. In this study, the damped least squares method arises from the objective function H. The damped least squares method was used for the first time in 1986 by Nakamura [31] and also independently by Wampler [32]. H is given as:

$$H = \|J\dot{q} - \dot{x}\|^2 + \|\rho\dot{q}\|^2 \tag{8}$$

where the first term provides primary task solution and the second term deals with kinematic singularities by suitable choice of non-zero positive parameter ρ. The vector of joint velocity $\dot{q}$ is derived by $\frac{dH}{d\dot{q}^T} = 0$:

$$\dot{q} = J^T\left(JJ^T + \rho^2 I\right)^{-1}\dot{x} \tag{9}$$

where $I \in R^{m \times m}$ is a square diagonal unit matrix with the dimension of end-effector task space.

3.2. Joint Limit Avoidance Task

The joint limit avoidance task deals with the range of motion of particular manipulator links. In the case of revolute joints, the construction of real manipulators usually does not allow full joint revolution (360°). During motion control of the manipulator, this limitation of link motion has to be considered in order to prevent the destruction of manipulator construction.

For the joint limit avoidance task, we use an approach with changing of value of weight variable W_{li} based on joint position. If the joint is in admissible range, the value of the weight variable is set to be zero. When the joint reaches the boundary of its range motion, the value of the weight variable increases. When the joint reaches a value out of its admissible range, the value of the weight variable increases to its maximum. This approach can be expressed by Equation (10) [33]:

$$W_{li} = \begin{cases} W_W \leftarrow q_i < q_{imin} \\ \frac{W_W}{2}\left\{1 + \cos\left[\pi\left(\frac{q_i - q_{imin}}{\rho_i}\right)\right]\right\} \leftarrow q_{imin} \leq q_i \leq q_{imin} + \rho_i \\ 0 \leftarrow q_{imin} + \rho_i \leftarrow q_i \leftarrow q_{imax} - \rho_i \\ \frac{W_W}{2}\left\{1 + \cos\left[\pi\left(\frac{q_{imax} - q_i}{\rho_i}\right)\right]\right\} \leftarrow q_{imax} - \rho_i \leq q_i \leq q_{imax} \\ W_W \leftarrow q_i > q_{imax} \end{cases} \tag{10}$$

The value of the weight variable has to be set for every joint of the manipulator which needs to be limited in the range of motion. Particular weight variables W_{li} are parts of the final weight matrix of the joint limit avoidance task $W_l \in R^{n \times n}$. The final weight matrix is the diagonal matrix:

$$W_l = \begin{bmatrix} W_{l1} & & & & \\ & W_{l2} & & & \\ & & W_{l3} & & \\ & & & \cdots & \\ & & & & W_{ln} \end{bmatrix} \tag{11}$$

The weight matrix W_l is used with the corresponding Jacobian matrix $J_l \in R^{n \times n}$. The Jacobian matrix for the joint limit avoidance task is $J_l = \frac{\partial e}{\partial q}$. If a particular joint does not consider the joint limit avoidance task, the value of J_l is set to be zero; otherwise it is set to be one. The limit of all links of the manipulator investigated in this study is $\pm$ 100°.

3.3. Obstacle Avoidance Task

During the obstacle avoidance task, the control system investigates the relation between manipulator links and obstacles in their environment. The aim of this secondary task is to prevent the collision between any part of the manipulator and potential obstacles, regardless of whether the shape of the obstacle is regular or irregular. Every obstacle of irregular shape can be geometrically modeled as a cylinder, with the obstacle being situated in the center of the cylinder; the diameter of the cylinder determines the distance of influence of this obstacle.

The coordinate of an obstacle in the end-effector task space is s_o. The projection of the line from the i-th joint of the manipulator link to the center of a particular cylinder on the i-th link is [33]:

$$p_i = e_i^T (s_0 - s_i) \tag{12}$$

The coordinate of the potential link point which could collide with the obstacle is:

$$s_{ci} = s_i + p_i e_i \tag{13}$$

The distance between the potential point of collision on the link and the center of the cylinder is expressed as:

$$d_{ci} = \| s_{ci} - s_0 \| \tag{14}$$

The unit vector of the potential point of collision to the center of the obstacle is:

$$u_i = \frac{s_{ai} - s_0}{d_{ci}} \tag{15}$$

Analogous to the joint limit avoidance task, the Jacobian matrix also has to be determined for the obstacle avoidance task. The i-th row of the Jacobian matrix can be written as:

$$J_{ci} = -u_i^T J_{s_{ci}} \tag{16}$$

The matrix $J_{s_{ci}}$ is:

$$J_{s_{ci}} = \frac{\partial s_{ci}}{\partial q} \tag{17}$$

The Jacobian matrix J_c consists of submatrices J_{ci}. The dimension of the Jacobian matrix is $J_c \in R^{c\times c}$, where c represents the number of manipulator links which could collide with the obstacles.

3.4. Final Inverse Kinematic Model

For the final inverse kinematic model, a Jacobian-based method is used. This method is based on the minimization of the objective function, which deals with the primary task as well as secondary tasks. The advantage of this method is that the number of secondary tasks is not limited, as it is in task augmentation methods [33].

In this study, we investigate the algorithms for a five-link and 20-link manipulator moving in the plane. The redundancy of the investigated manipulator is used for the abovementioned optimization tasks, namely the obstacle avoidance task, the joint limit avoidance task, and the kinematic singularities avoidance task. The redundancy problem can be expressed by finding a vector q which approximately satisfies Equation (7) by minimizing the objective function H. The final inverse kinematic model can be derived based on the same idea as mentioned in Section 3.1:

$$H = \| J\dot{q} - \dot{x} \|^2 + \| J_c\dot{q} - \dot{x}_c \|^2 + \| J_L\dot{q} - \dot{x}_L \|^2 + \| \rho\dot{q} \|^2 \tag{18}$$

where $J_c \in R^{c\times c}$ is the Jacobian matrix for the obstacle avoidance task, c is the number of links which can collide with an obstacle, $J_L \in R^{l\times l}$ is the Jacobian matrix for the joint limit avoidance task, l is the number of joints in which its motion limit is considered, and ρ is a scalar constant which overcomes computing problems around kinematic singularities. Equation (18) considers the primary task represented by the term $\| J\dot{q} - \dot{x} \|^2$ and other secondary tasks. The weight matrix is assigned to each task in order to set the priority of particular tasks. This can be achieved by using weight matrices, as follows:

$$H = \left(J\dot{q} - \dot{x}\right)^T W\left(J\dot{q} - \dot{x}\right) + \left(J_c\dot{q} - \dot{x}_c\right)^T W_c\left(J_c\dot{q} - \dot{x}_c\right) + \left(J_L\dot{q} - \dot{x}_L\right)^T W_L\left(J_L\dot{q} - \dot{x}_L\right) + W_s\dot{q}^T\dot{q} \tag{19}$$

where $W \in R^{m\times m}$ is the weight matrix of the primary task, $W_c \in R^{c\times c}$ is the weight matrix of the obstacle avoidance task, $W_L \in R^{l\times l}$ is the weight matrix of the joint limit avoidance task, and $W_s \in R^{n\times n}$ is the weight matrix of the kinematic singularities avoidance task. These weight matrices are diagonal matrices multiplied by coefficients to set the level of priority of the given task. The choice of these coefficients is subjective. The solution of Equation (19) is given by $\frac{dH}{d\dot{q}^T}$:

$$\frac{dH}{d\dot{q}^T} = 2\left(J^T W_e J + J_c^T W_c J_c + J_l^T W_l J_l + W_s\right)\dot{q} - 2\left(J^T W_e \dot{x} + J_c^T W_c \dot{x}_c + J_l^T W_l \dot{x}_l\right) \tag{20}$$

The vector of joint velocities $\dot{q}$ is expressed by $\frac{dH}{d\dot{q}^T} = 0$:

$$\dot{q} = \left(J^{-1}WJ + J_c^T W_c J_c + J_l^{-1} W_l J_l + W_s\right)^{-1} \left(J^T W\dot{x} + J_c^T W_c \dot{x}_c + J_l^T W_l \dot{x}_l\right) \tag{21}$$

Considering Equation (21), the joint velocities $\dot{x}_c$ and $\dot{x}_L$ have to be set to zero in order to avoid the joint limits and collisions with the obstacles. In other words, to prevent these secondary tasks, their velocities have to be zero. In this study, the dimensions of matrices W_c, W_L, and W_s are 5×5 for the five-link manipulator and 20×20 for the 20-link manipulator. Consequently, the final inverse kinematic model is:

$$\dot{q} = \left(J^{-1}WJ + J_c^T W_c J_c + J_l^{-1} W_l J_l + W_s\right)^{-1} \left(J^T W\dot{x}\right) \tag{22}$$

Next, the algorithm for the inverse kinematics solution will be presented. The aim of Algorithm 3 (Inverse kinematic model) is the positioning of the end-effector of the manipulator through the points of the planned path from Section 2 while manipulator links hold all secondary tasks.

Algorithm 3 Inverse kinematic model

1: CYCLE WHILE 1
2: Determination of new required vector $x_d \in R^m$ from the matrix of planned path $P \in R^{r \times 2}$
3: CYCLE WHILE 2
4: Computation of Jacobian matrix J (damped least squares method)
5: Determination of actual end-effector position in the task space $x \in R^m$ with actual generalized variables $q \in R^n$
6: Computation of general equation
$\dot{q} = \left(J^{-1}WJ + J_c^{-1} W_c J_c + J_l^{-1} W_l J_l + W_s\right)^{-1} \left(J^T W\dot{x}\right)$
7: $q = q_{previous} + \dot{q}dt$
8: $q_{previous} = q$
9: IF $x_d = x$ THEN
END CYCLE WHILE 2
ELSE
CYCLE WHILE 2 continues
END IF
10: END CYCLE WHILE 2
11: END CYCLE WHILE 1

The output of Algorithm 2 (Path Planning) is used as the input to Algorithm 3 (Inverse kinematic model). The "WHILE 1" cycle ensures the positioning of the end-effector through each point of the planned path given by $P \in R^{r \times 2}$. The "WHILE 2" cycle finds the solution for x_d by means of the final inverse kinematic model given by Equation (22). This cycle ends when the end-effector position in the task space equals the required position x_d—the point from the planned path. The solution also assumes a certain tolerance, which is given by the user.

One of the challenges of this method using weight matrices is the setting of the values of particular weight matrices. Since this choice is subjective, the incorrect choice of weight matrices can cause the computation to slow down or fail. Our contribution to this field is the modification of the computing algorithm in using flexible tasks priority.

3.5. A New Algorithm of Inverse Kinematic Model—Acceleration of Computing

The problem in numerical modeling occurs in the case of inappropriate choice of the priority for all tasks solved in the inverse kinematic model. Let us consider the inverse kinematic model, including joint limit and obstacle and kinematic singularities avoidance tasks. By setting the priority for all tasks, the control system could work according to our requirements. For example, by adding obstacles to the manipulator workspace, while the priority of weight matrices remains the same, the numerical computing could not work as we expect. In many cases, the computing process not only slows down but the process even fails. For this reason, we have improved this method in order to deal with these problems during the simulation. Our approach is based on changing the priority of a particular task of the inverse kinematic model during the simulation, according to simulation behavior.

During the performance of the "WHILE 2" cycle, the variable *counter* increases in each cycle while the actual position of the end-effector x does not equal the required position x_d. If the value of the variable *counter* is greater than *max. admissible value*, which means that the calculation time is too long, the priority of the chosen task decreases in order to accelerate the computation. When the solution for the required position x_d has been found and at the same time if the value of the variable *counter* is lower than *min. admissible value*, the priority of the chosen task increases. Increasing, as well as decreasing, the priority of the chosen task is in certain boundaries defined function (linear, quadratic, etc.).

Our new approach, namely flexible priority solution (FPS), significantly accelerates the computation of the inverse kinematic model with secondary tasks that will be shown in the simulation results.

4. Low Level Control

For experimental purposes, we used a fuve-link manipulator with five Dynamixel AX-12 servomechanisms (ROBOTIS, Seoul, Korea) with a torque of 1.49 Nm and a speed of 0.169/60°. The servomechanisms were connected in series, as shown in Figure 6.

The servomechanisms communicate together through UART (Universal asynchronous receiver-transmitter) communication protocol using circuit SN74LS241N. Every transmitted and received packet has the following form:

0xFF–0xFF–Id–Length–Instruction–Parameter 1– . . . –Parameter N–Check sum

The first two bytes indicate the start of the received or transmitted packet. By other bytes we set the required operation from all available functions of Dynamixel AX12. The servomechanisms of the experimental manipulator were controlled by an ATmega162 microcontroller running at 16 MHz.

The ATmega162 microcontroller has RISC (Reduced instruction set computer) architecture allowing up to 16 MIPS (Million Instruction Per Second) throughput at 16 MHz. The simplified model of algorithm running on ATmega162 describes Algorithm 4 (Low level control):

Algorithm 4 Low level control

```
CYCLE WHILE 1
  Find out positions of servomechanisms (UART 1)
  Send positions to PC (UART 2)
  CYCLE WHILE 2
    wait for all required positions from PC
  END CYCLE WHILE 2
  Move servomechanisms to required positions
END CYCLE WHILE 1
```

The "WHILE 1" cycle is an infinite cycle running on microcontroller ATmega162. The microcontroller uses two independent UART communications, the first one for communication with the servomechanism inner controller and the second one for communication with a PC. Both of these communications run at a speed of 200 kB/s.

The scheme of information flow is shown in Figure 7.

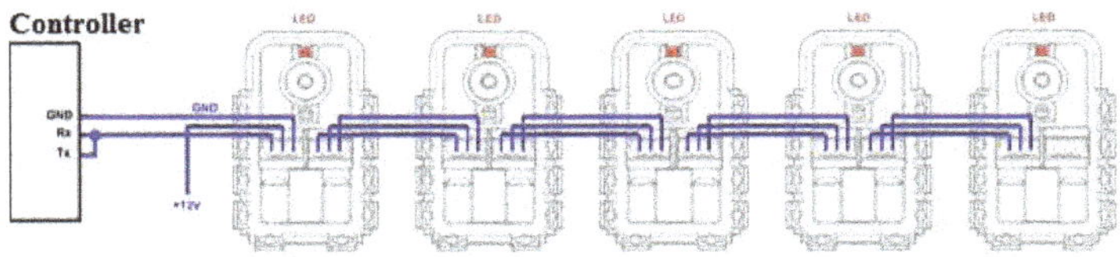

Figure 6. Dynamixel AX12 connection.

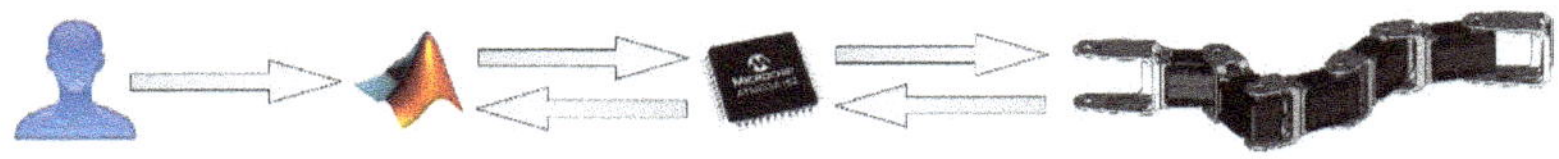

Figure 7. Scheme of information flow.

The user determines the input parameters, such as the priority of primary and secondary tasks. The control systems are run using MATLAB (MathWorks, Natick, MA, USA) software, which transmits the requirements of servomechanisms positions to the microcontroller and microcontroller to the inner control system of the servomechanisms. The actual positions of the servomechanisms are transmitted to the microcontroller by the inner control system of servomechanisms. The microcontroller processes these data and transmits them to MATLAB. Algorithm 5 (Modified algorithm for inverse kinematic model) is run in MATLAB, while Algorithm 4 (Low level control) runs in the ATmega162 microcontroller.

Algorithm 5 New algorithm for inverse kinematic model

CYCLE WHILE 1

  Determination of new required vector $x_d \in R^m$ from the matrix of planned path $P \in R^{r \times 2}$

  CYCLE WHILE 2

    increase counter

    IF counter > max. admissible value

      decrease priority of chosen task by chosen function

      counter = 0

    END IF

    Computation of Jacobian matrix J (damped least squares method)

    Determination of actual end-effector position in the task space $x \in R^m$ with actual generalized variables $q \in R^n$

    Computation of general equation

$$\dot{q} = \left(J^{-1}WJ + J_O{}^{-1}W_cJ_c + J_l{}^{-1}W_lJ_l + W_s\right)^{-1}\left(J^TW\dot{x}\right)$$

    $q = q_{previous} + \dot{q}dt$

    $q_{previous} = q$

    IF $x_d = x$ THEN

      END CYCLE WHILE 2

    ELSE

      CYCLE WHILE 2 continues

    END IF

  END CYCLE WHILE 2

  IF counter < min. admissible value

    increase priority of chosen task by chosen function

    counter = 0

  END IF

  counter = 0

END CYCLE WHILE 1

5. Numerical Computing and Results

In this section, the testing of several case studies is presented. The first one assumed a 20-link manipulator and the second one a 5-link manipulator. These case studies used the same initial conditions as number and positions of obstacles, the same start and goal point of end-effector, and range of links motion of $\pm 100°$. The link length of the five-link manipulator was 67 mm and the link length of the 20-link manipulator was 16.75 mm. Both manipulators had a link length of 335 mm. The admissible tolerance of end-effector positioning was set to be 5 mm.

All of the abovementioned algorithms were subsequently applied by numerical computing. All case studies used the same scenario according to Figure 4. Case studies were run using an Intel Core™ i7-3770 3.40 GHz CPU. The resulting computing time for particular case studies was an average value based on 10 repeated simulations.

5.1. Case Study 1

The first case study assumes a 20-link manipulator with a link length of 16.75 mm. The values of the weight matrices are as follows: the weight matrix of the primary task, $W = 0.1I$, where **I** is a unit matrix with dimension 2×2; the weight matrix of the obstacle avoidance task, $W_c = 20I$, where **I** is a unit matrix with dimension 20×20; the weight matrix of the joint limit avoidance task, $W_l = 50I$, where **I** is a unit matrix with dimension 20×20; and the weight matrix of the kinematic singularities avoidance task, $W_s = 50I$, where **I** is a unit matrix with dimension 20×20.

In the case of the introduced method working with Algorithm 3 with constant weight matrices, the simulation time was 46.3663 s. Using FPS, the simulation time was 34.9354 s. Our approach speeds up the simulation by about 24.65%. Graphical representations of the simulation of the 20-link manipulator are shown in Figure 8.

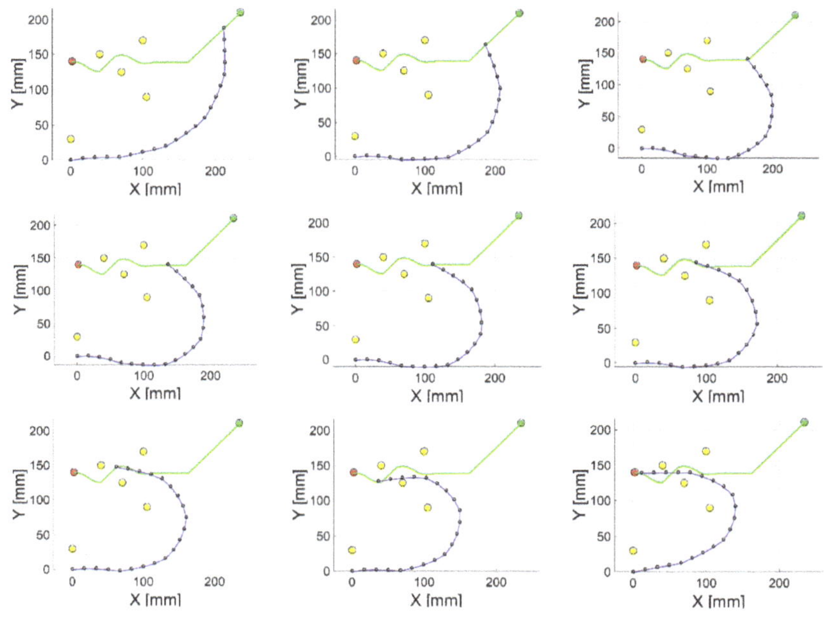

Figure 8. Graphical representations of the simulation of the 20-link manipulator.

5.2. Case Study 2

The second case study assumes a 5-link manipulator with a link length of 67 mm. The values of all weight matrices are the same as in Case Study 1. The only difference is in the dimension of weight matrices.

In the case with constant weight matrices, the simulation time was 8.8106 s. Using our algorithm, the simulation time decreased to 5.6325 s. Using FPS speeds up the simulation by about 36.07%. Graphical representations of the simulation of Case Study 2 are shown in Figure 9.

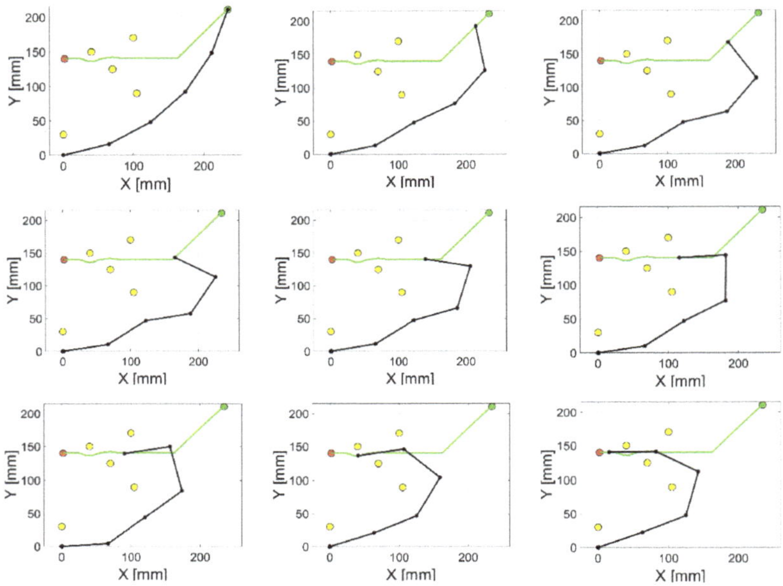

Figure 9. Graphical representations of the simulation of the five-link manipulator.

Algorithm 5 (New algorithm for inverse kinematic model) appeared to be significantly accelerative. Using Algorithm 3, in some cases the computing fails, whereas using our approach the computing finished successfully. The effectiveness of the improved approach is due to the suitable choice of the parameters max. admissible value and min. admissible value from Algorithm 5. Despite the fact that our algorithm consists of more computing instructions in comparison with the original algorithm, it is significantly faster; in the case of the five-link manipulator, it is 36.07% times faster and in the case of the 20-link manipulator it is 24.65% faster.

For all case studies, all weight matrices were constant besides the weight matrix of the primary task. This means that the joint limit avoidance task, kinematic singularities avoidance task, and obstacle avoidance task have more priority than the precise positioning of the end-effector through each point of the planned path. In other words, it is better to move the end-effector slightly less precisely than for the manipulator to collide with the obstacles, since this could result in destructive consequences for the manipulator or its environment in real applications.

The second case study, considering the 5-link manipulator, was also tested by an experimental model composed of five Dynamixel AX12 servomechanisms. The following figures compare generalized variables $q = [q_1, q_2, \ldots, q_n]^T$ from the simulation model to those of the experimental model.

Figure 10 shows that the real manipulator, except for small deviations, almost exactly copies the simulated values of the generalized variables. The small deviations are caused by the control system of the servomechanisms not being properly tuned [34]. Figure 11 shows the end-effector positioning error. The error of simulation is roughly 5 mm. This error was caused by the predetermined tolerance of positioning, which is 5 mm.

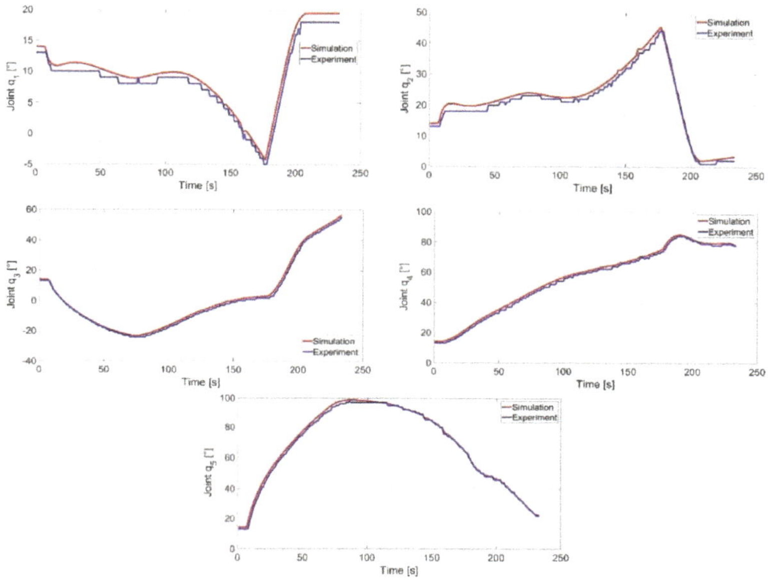

Figure 10. The time behavior of generalized variables.

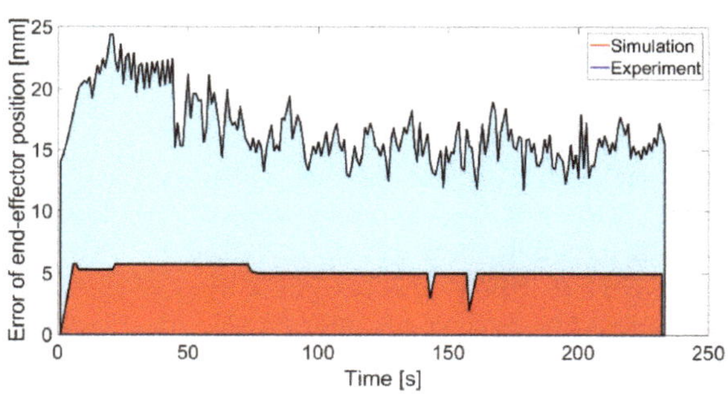

Figure 11. End-effector positioning error.

Figure 12 shows images of the motion of the experiment using the five-link manipulator according to the simulation from Figure 9. The video sequence was recorded using MATLAB image equipment [35].

Figure 12 corresponds to the time behavior of the experimentally-given generalized variables shown in Figure 10.

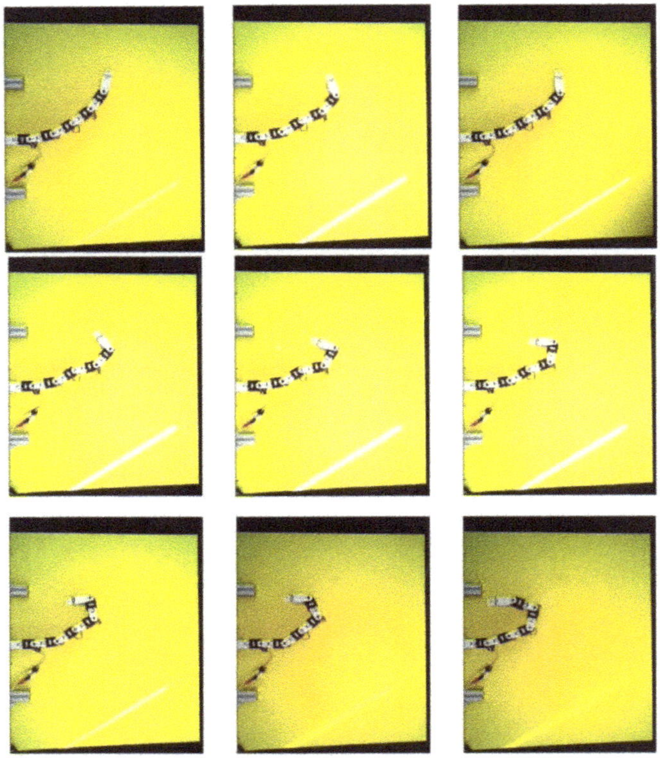

Figure 12. Images from the experiment using the five-link manipulator.

5.3. Comparison with Other Methods

In the previous section, the WLN original approach working with task priorities was compared with our developed FPS approach. In this section, analyses considering the next methods are presented. Different simulation conditions will also be used.

The first method analyzed is from the gradient projection methods (GPM) class. GPM are probably the most frequently discussed and used method for inverse kinematics of kinematically-redundant manipulators. GPM was firstly introduced by Liegeois [8] to utilize kinematic redundancy to avoid joint limits. By extension of the basic inverse kinematic model, a new model was derived [8]:

$$\dot{q} = J^{+}\dot{x} + \alpha\left(J^{+}J - I_n\right)\nabla z \tag{23}$$

where J^{+} is the pseudoinverse matrix of J, I_n is the identity matrix $I_n \in R^{n\times n}$, ∇z is the vector of objective function, and α is a weighting parameter larger than zero. The term $J^{+}\dot{x}$ is a minimum-norm solution and the term $\left(J^{+}J - I_n\right)$ is a null-space projection matrix. The homogeneous term $\left(J^{+}J - I_n\right)\nabla z$ is orthogonal to $J^{+}\dot{x}$ and is referred to as the self-motion of the mechanism (manipulator) in joint space with any influence on end-effector motion in task space. An arbitrary secondary task can be applied for the objective function z according to requirements. For following case study, the same secondary tasks that were introduced in previous sections were used. Equation (23) is presented in the following tables as generalized GPM (GGPM).

Another widely used method for inverse kinematics is the closed-loop inverse kinematics (CLIK) algorithm, which was developed to overcome the joint drift for open-chain robot manipulators [36]. The CLIK algorithm at velocity level can be formulated as follows. The expression of location error and its derivative is:

$$e = x_d - x \tag{24}$$

$$\dot{e} = \dot{x}_d - \dot{x} \tag{25}$$

The vector of joint velocity has to be set so the error tends to zero. Considering the pseudoinverse solution, the generalized CLIK algorithm can be expressed as:

$$\dot{q} = J^{\dagger}\left[\dot{x}_d + K_P(x_d - x)\right] \tag{26}$$

Equation (26) combined with Equation (6) gives:

$$\dot{e} = K_P e \tag{27}$$

where K_P is a symmetric positive definite matrix. The final CLIK solution can be expressed as:

$$\dot{q} = J^{\dagger}\left[\dot{x}_d + K_P(x_d - x)\right] + \alpha\left(J^{\dagger}J - I_n\right)\nabla z \tag{28}$$

Subsequently, all of the mentioned methods were applied in the simulations in order to compare them in terms of computation time. The scenario was the same as in the first two case studies (Figure 4). For secondary tasks expression, the same approach as that introduced in previous sections is used. In the CLIK algorithm, matrix K_P is set to be a diagonal matrix with values 10. The priority of the obstacle avoidance task was then decreased from 20 to 2. In the following simulations, obstacle influence was also changed. The obstacle influence is the parameter which causes increasing sense of obstacle avoidance task. This parameter arises from Section 3.3. The higher this parameter is, the more difficult the passage through the obstacles is. In other words, in the case of a high obstacle influence parameter, the passage for manipulator motion is narrower. Thus, high obstacle influence represents highly rugged terrain.

Based on numerical simulations, computing times are determined for all the described methods in the following Tables 1–4. The simulations again consider a five-link and a 20-link manipulator.

Table 1. Simulation results with an obstacle influence of 10 mm.

Method	Computation Time (s)—5 Links	Computation Time (s)—20 Links
FPS	0.99	45.51
WLN	1.11	49.38
GGPM	19.15	92.62
CLIK	18.21	52.14

Table 2. Simulation results with an obstacle influence of 20 mm.

Method	Computation Time (s)—5 Links	Computation Time (s)—20 Links
FPS	1.24	44.54
WLN	1.57	49.61
GGPM	19.94	Failure
CLIK	18.39	Failure

Table 3. Simulation results with an obstacle influence of 30 mm.

Method	Computation Time (s)—5 Links	Computation Time (s)—20 Links
FPS	1.29	45.52
WLN	1.82	69.96
GGPM	28.80	Failure
CLIK	26.79	Failure

Table 4. Simulation results with an obstacle influence of 40 mm.

Method	Computation Time (s)—5 Links	Computation Time (s)—20 Links
FPS	1.51	50.61
WLN	2.41	140.79
GGPM	31.09	Failure
CLIK	28.72	Failure

The simulation results show that our FPS method has significant utility. The computing time is considerably lower, especially for the cases with a large number of DOF. The importance of FPS also increases in cases with highly rugged terrain (i.e., those with large obstacle influence). The mark "Failure" in the tables means that simulations lasted too long or the self-motion of particular joints collided with obstacles or with other links.

6. Conclusions and Future Work

This study investigates the algorithms for investigating the positioning of manipulator end-effector while secondary tasks are considered, namely a joint limit avoidance task, an obstacle avoidance task, and a kinematic singularities avoidance task. The paper deals with path planning for end-effector using potential field method. The output of path planning is used as the input to the inverse kinematic model, which is designed by a Jacobian-based method. This approach includes the use of weight matrices for primary and secondary tasks in order to set the priority of each task. Using this method, during numerical simulations the computation significantly slowed down in some cases. In this paper, a new approach is introduced which uses flexible choice of priority for task of inverse kinematic model due to acceleration of computation. The choice of task priority was performed based on simulation behavior. The simulations were performed using a 5-link and a 20-link manipulator. If the computing power slowed down for a certain required position, the priority of the main task decreased. Consequently, if the computing power during the simulation increased, the priority of the main tasks also increased in order to increase the motion precision. In the case of the 20-link manipulator, our approach decreased computation time by 24.65%; and in the case of the 5-link manipulator it decreased computation time by 36.07%. Our algorithm consists of more computing instructions than the original algorithm, yet it is also faster. This paper also presents analysis including the comparison of four methods for inverse kinematics. The analysis expresses the utility of FPS, especially for cases when the manipulator has a large number of DOF and there is highly rugged terrain. The results show the effectiveness of our approach.

In the future, we hope to continue in improving the new approach, that is, FPS, which is introduced in this study. The aim of our future work will be to improve our approach by testing different functions for increasing and decreasing level of task priority. Subsequently, we hope to utilize our approach in a dynamically changing manipulator environment and to test it on an industry manipulator.

Author Contributions: I.V. and T.L.—mathematical model of redundant manipulator; Ľ.M.—simulations; M.K. and F.F.—programming and experiments; and V.B.—data analysis.

Funding: This research was funded by Slovak Grant Agency VEGA1/0872/16 "Research of synthetic and biological inspired locomotion of mechatronic systems in rugged terrain" and by project Slovak Grant Agency VEGA 1/0389/18 "Research on kinematically redundant mechanisms".

Conflicts of Interest: The authors declare no conflict of interest. The founding sponsors had no role in the design of the study; in the collection, analyses, or interpretation of data; in the writing of the manuscript; or in the decision to publish the results.

References

1. Chiaverini, S.; Oriolo, G.; Walker, I.D. *Kinematically Redundant Manipulators*; Springer: Berlin/Heidelberg, Germany, 2008; pp. 245–268.
2. Wei, Q.; Yang, C.; Fan, W.; Zhao, Y. Design of Demonstration-Driven Assembling Manipulator. *Appl. Sci.* **2018**, *8*. [CrossRef]
3. Kilin, A.; Bozek, P.; Karavaev, Y.; Klekovkin, A.; Shestakov, V. Experimental investigations of a highly maneuverable mobile omniwheel robot. *Int. J. Adv. Robot. Syst.* **2017**, *14*. [CrossRef]
4. Siciliano, B. Kinematic Control of Redundant Robot Manipulators: A Tutorial. *J. Intell. Robot. Syst.* **1990**, *3*, 201–212. [CrossRef]
5. Wang, J.; Li, Y.; Zhao, X. Inverse Kinematics and Control of a 7-DOF Redundant Manipulator Based on the Closed-Loop Algorithm. *Int. J. Adv. Robot. Syst.* **2010**, *7*, 1–10. [CrossRef]
6. Flacco, F.; De Luca, A.; Khatib, O. Motion control of redundant robots under joint constraints: Saturation in the null space. In Proceedings of the IEEE International Conference on Robotics and Automation, Saint Paul, MN, USA, 14–18 May 2012; pp. 285–292.
7. Flacco, F.; De Luca, A. Discrete-time redundancy resolution at the velocity level with acceleration/torque optimization properties. *Robot. Auton. Syst.* **2015**, *70*, 191–201. [CrossRef]
8. Liegeois, A. Automatic supervisory and control of the configuration and behavior of multibody and mechanisms. *IEEE Trans. Syst. Man Cybern.* **1977**, *12*, 868–871. [CrossRef]
9. Chaumette, F.; Marchand, R. A redundancy-based iterative approach for avoiding joint limits: Application to visual servoing. *IEEE Trans. Robot. Autom.* **2001**, *17*, 719–730. [CrossRef]
10. Konstantinov, M.S.; Markov, M.D.; Nencheu, D.N. Kinematic control of redundant manipulators. In Proceedings of the 11th Znt. Symposium on Industrial Robots, Tokyo, Japan, 7–9 October 1981; pp. 561–568. [CrossRef]
11. Whitney, D.E. The mathematics of coordinated control of prosthetic arms and manipulators. *ASME J. Dyn. Syst. Meas. Cont.* **1972**, *94*, 303–309. [CrossRef]
12. Hollerbach, J.M.; Suh, K.C. Redundancy resolution of manipulators through torque optimization. *IEEE J. Robot. Autom.* **1987**, *3*, 1016–1021. [CrossRef]
13. RunBin, C.; YangZheng, C.; Lin, L.; Jian, W.; Xu, M.H. Inverse Kinematics of a New Quadruped Robot Control Method. *Int. J. Adv. Robot. Syst.* **2013**, *10*. [CrossRef]
14. Whitney, D.E. Resolved Motion Rate Control of Manipulators and Human Prostheses. *IEEE Trans. Man-Mach. Syst.* **1969**, *10*, 47–53. [CrossRef]
15. Chan, T.F.; Dubey, R.V. A Weighted Least-Norm Solution Based Scheme for Avoiding Joint Limits for Redundant Joint Manipulators. *IEEE Trans. Robot. Autom.* **1995**. [CrossRef]
16. Huang, S.; Peng, Y.; Wei, W.; Xiang, J. Clamping weighted least-norm method for the manipulator kinematic control with constraint. *Int. J. Cont.* **2016**, *89*, 2240–2249. [CrossRef]
17. Chiaverini, S. Singularity-Robust Task-Priority Redundancy Resolution for Real-Time Kinematic Control of Robot Manipulators. *IEEE Trans. Robot. Autom.* **1997**, *13*, 398–410. [CrossRef]
18. Park, J.; Choi, Y.; Chung, W.K.; Youm, Y. Multiple Tasks Kinematics Using Weighted Pseudo-Inverse for Kinematically Redundant Manipulators. In Proceedings of the IEEE International Conference on Robotics and Automation, Seoul, Korea, 21–26 May 2001.
19. Lee, J.; Mansard, N.; Park, J. Intermediate Desired Value Approach for Task Transition of Robots in Kinematic Control. *IEEE Trans. Robot.* **2012**, *28*, 1260–1277. [CrossRef]
20. Zlajpah, L.; Nemec, B. Kinematic control algorithms for on-line obstacle avoidance for redundant manipulators. In Proceedings of the IEEE/RSJ International Conference on Intelligent Robots and Systems, Lausanne, Switzerland, 30 September–4 October 2002; pp. 1898–1903. [CrossRef]
21. Maciejewski, A.A.; Klein, C.A. Obstacle Avoidance for Kinematically Redundant Manipulators in Dynamically Varying Environments. *Int. J. Robot. Res.* **1985**, *4*, 109–117. [CrossRef]

22. Mansard, N.; Khatib, O.; Kheddar, A. A Unified Approach to Integrate Unilateral Constraints in the Stack of Tasks. *IEEE Trans. Robot.* **2009**, *25*, 670–685. [CrossRef]
23. Kanoun, O.; Lamiraux, F.; Wieber, P.B. Kinematic control of redundant manipulators: Generalizing the task priority framework to inequality tasks. *IEEE Trans. Robot.* **2011**, *27*, 785–792. [CrossRef]
24. Duchoň, F.; Babinec, A.; Kajan, M.; Beňo, P.; Florek, M.; Fico, T.; Jurišica, L. Path planning with modified a star algorithm for a mobile robot. *Procedia Eng.* **2014**, *96*, 59–69. [CrossRef]
25. Montiel, O.; Sepúlveda, R.; Orozco-Rosas, U. Optimal Path Planning Generation for Mobile Robots using Parallel Evolutionary Artificial Potential Field. *J. Intell. Robot. Syst.* **2015**, *79*. [CrossRef]
26. Cosfo, F.A.; Padilla Castaneda, M.A. Autonomous Robot Navigation using Adaptive Potential Fields. *Math. Comp. Model.* **2004**, *40*. [CrossRef]
27. Silva-Ortigoza, R.; Márquez-Sánchez, C.; Carrizosa-Corral, F.; Hernández-Guzmán, V.M.; García-Sánchez, J.R.; Taud, H.; Marciano-Melchor, M.; Álvarez-Cedillo, J.A. Obstacle Avoidance Task for a Wheeled Mobile Robot—A Matlab Simulink Based Didactic Application. *Intech* **2014**. [CrossRef]
28. Žlajpah, L.; Petrič, T. Obstacle Avoidance for Redundant Manipulators as Control Problem, Serial and Parallel Robot Manipulators—Kinematics, Dynamics, Control and Optimization. *Intech* **2012**. [CrossRef]
29. Baerlocher, P.; Boulic, R. An inverse kinematics architecture enforcing an arbitrary number of strict priority levels. *Vis. Comput.* **2004**, *20*, 402–417. [CrossRef]
30. Buss, S.R. Introduction to Inverse Kinematics with Jacobian Transpose, Pseudoinverse and Damped Least Squares methods. *IEEE Trans. Robot. Autom.* **2004**, *17*, 1–19.
31. Nakamura, Y.; Hanafusa, H. Inverse kinematics solutions with singularity robustness for robot manipulator control. *J. Dyn. Syst. Meas. Cont.* **1986**, *108*, 163–171. [CrossRef]
32. Wampler, C.W. Manipulator inverse kinematic solutions based on vector formulations and damped least squares methods. *IEEE Trans. Syst. Man Cybern.* **1986**, *16*, 93–101. [CrossRef]
33. Fahimi, F. *Autonomous Robots: Modeling, Path Planning, and Control*; Springer: Berlin/Heidelberg, Germany, 2009; ISBN 978-0-387-09537-0.
34. Buscarino, A.; Famoso, C.; Fortuna, L.; Frasca, M. Passive and active vibrations allow self-organization in large-scale electromechanical systems. *Int. J. Bifurc. Chaos* **2016**, *26*. [CrossRef]
35. Koniar, D.; Stofan, S.; Hargas, L.; Hrianka, M.; Simonova, A. Hardware conditioning in process of high speed imaging. *Adv. Electr. Electron. Eng.* **2012**, *13*, 567–574. [CrossRef]
36. Liegeois, A. Automatic Supervisory Control of Configuration and Behavior of Multibody Mechanism. *IEEE Trans. Syst. Man Cybern.* **1977**, *7*, 868–871. [CrossRef]

applied sciences

MDPI

Article

Solving the Time-Varying Inverse Kinematics Problem for the Da Vinci Surgical Robot

Long Bai *, Jianxing Yang, Xiaohong Chen, Pei Jiang, Fuqiang Liu, Fan Zheng and Yuanxi Sun

State Key Laboratory of Mechanical Transmission, Chongqing University, Chongqing 400044, China; 20132295@cqu.edu.cn (J.Y.); chenxh@cqu.edu.cn (X.C.); peijiang@cqu.edu.cn (P.J.); liufq@cqu.edu.cn (F.L.); 20160702066@cqu.edu.cn (F.Z.); sunyuanxi@cqu.edu.cn (Y.S.)

* Correspondence: bailong@cqu.edu.cn

Received: 15 November 2018; Accepted: 3 February 2019; Published: 6 February 2019

Featured Application: The work of this paper is mainly applied for the inverse kinematics solving of Da Vinci surgical robot and its similar robotic systems that do not satisfy the Pieper principle.

Abstract: A dialytic-elimination and Newton-iteration based quasi-analytic inverse kinematics approach is proposed for the 6 degree of freedom (DOF) active slave manipulator in the Da Vinci surgical robot and other similar systems. First, the transformation matrix-based inverse kinematics model is derived; then, its high-dimensional nonlinear equations are transformed to a high-order nonlinear equation with only one unknown variable by using the dialytic elimination with a unitary matrix. Finally, the quasi-analytic solution is eventually obtained by the Newton iteration method. Simulations are conducted, and the result show that the proposed quasi-analytic approach has advantages in terms of accuracy (error < 0.00004 degree (or mm)), solution speed (<20 ms) and is barely affected by the singularity during intermediate calculations, which proves that the approach meets the real-time and high-accuracy requirements of master–slave mapping control for the Da Vinci surgical robots and other similar systems. In addition, the proposed approach can also serve as a design reference for other types of robotic arms that do not satisfy the Pieper principle.

Keywords: minimally invasive surgery robot; inverse kinematics; dialytic elimination; Newton iteration

1. Introduction

In recent years, medical robots have increasingly been used as fundamental surgical instrument/equipment [1], among which the Da Vinci system developed by Intuitive Surgical Company is the most successful [2]. The Da Vinci robotic system consists of three components: the surgeon console (master manipulators), the patient trolley (slave manipulators), and the imaging system [3]. The surgeon manipulates two master handles at the master remote console, which can acquire high-resolution, binocular, three-dimensional, magnified views of the operative field as compared with open surgery [4]. However, this kind of master–slave operation mode poses a great challenge to the real-time and accurate performance of the master–slave mapping control, among which the inverse kinematics problem is the first obstacle. The Da Vinci system and similar surgical robot systems satisfy the following two features:

(1) contain mechanical structures to generate Remote Center of Motion [5] (RCM, a fixed point on the active arm, there are two methods to generate RCM based motion: (a) virtual RCM, (b) physically constrained RCM [6]), and the structure can be simplified;
(2) do not satisfy the Pieper principle.

At present, the main methods of robotic inverse kinematics include the analytic method and the numerical method. The analytic approach uses vector, spinor or Lie algebra to obtain results [7–9], which can only obtain satisfied solutions in limited robotic configurations that should satisfy the Pieper principle [10,11], etc. However, to imitate the surgeon operation, the manipulators of most thoracic and abdominal surgery robots, including the Da Vinci system, contain rotational/translational kinematic pairs, tilt angle joints, and offset joints at the tip (artificial wrist part) that are designed for additional flexibility, which is beyond the solution capability of the analytic approach.

One feasible strategy is to simplify the kinematic model to acquire analytical results. For example, Ai et al. and Podsedkowski et al. [12,13] obtained an approximate solution by separating the position and orientation of the slave manipulators, the solution deviation of which needs to be manually eliminated by experienced surgeons via visual inspection. Zhang et al. [14] solved the inverse kinematics of a 4-DOF surgical robot using the elimination method, and Mezouar et al. [15] used unit dual quaternions to model the kinematics, however, outcomes of their studies benefited strongly from the simplicity of the arm configuration. In addition, Fu et al. [16,17] obtained an approximate solution based on the differential transformation that increases error feedbacks to compensate the cumulative error, the final precision of which can reach 0.18 mm.

Another way to solve the inverse kinematics problem of the surgical robot that does not satisfy the Pieper principle is the Jacobian matrix-based numerical method [18]. However, this method involves a Jacobian matrix, which may lose versatility due to its singularity problem as well as the low convergence speed of the iterative solutions. To improve the numerical solution performance, a numerical method based on the position increment of robotic joints was proposed [18]. Specifically, to make this method suitable for the inverse kinematics analysis of a robotic manipulator with an offset wrist, Bu et al. [19] used the method of decoupling DOFs based on a cut-off point that is limited by the manipulator configuration. In addition, modern methods such as the neural network method [20], the genetic algorithm [21], the electromagnetism-like mechanism [22,23], the hybrid electromagnetism-like mechanism [24], etc., have been gradually explored for possible solutions of the inverse kinematics of surgical robots. However, these methods either require a large amount of sample data or suffer from poor convergence speed; therefore, these methods are not suitable for the rapid development of a master–slave manipulator in surgical robots.

To address the lack of efficient and accurate inverse kinematics solutions for the Da Vinci system, we propose a quasi-analytic solution method that involves dialytic elimination and the Newton iteration method. The dialytic elimination method [25] is used to transform the high-dimensional equations into a nonlinear equation containing only one unknown variable, which is then solved by the Newton iteration method. In applying this method to the inverse kinematics solution of slave manipulator, we show dual superiorities: (a) barely affected by the singularity problem and (b) maintain rapidity.

The steps of our proposed method as follows:

(1) simplify the structure of RCM, using an equivalent mechanism to take the place of the RCM mechanism.
(2) establish the coordinate system, this step can be realized by the D–H (Denavit–Hartenberg) method or combine the characteristics of the structure to make sure there are more origins of the coordinate system can be set up at the same point.
(3) obtain the mathematical model of forward and inverse kinematics via the matrix transformation method. Then, transform the high-dimensional equations into a nonlinear equation containing only one unknown parameter using the dialytic elimination method, and solve it by the Newton iteration method.

The rest of this paper is organized as follows: Section 2 presented the kinematics analysis of the Da Vinci system; Section 3 solves the inverse kinematics by the Newton iteration method; Section 4 analyzes our solution and compared it with other methods via a simulation; Section 5 concludes the paper.

2. Kinematics Analysis

2.1. Forward Kinematics Analysis

During surgery using the Da Vinci medical robot, dual articulated arms (slave manipulators) work as the right and left hand of the surgeon, respectively. The coordinate frame of the master/slave manipulator is set the same as the coordinate frame of human sight, i.e., the world coordinate frame. Therefore, in Figure 1, the origin of the base coordinate frame of the slave manipulator is located at the bottom of the robot frame. The Z_0-axis is upward, along the robot frame.

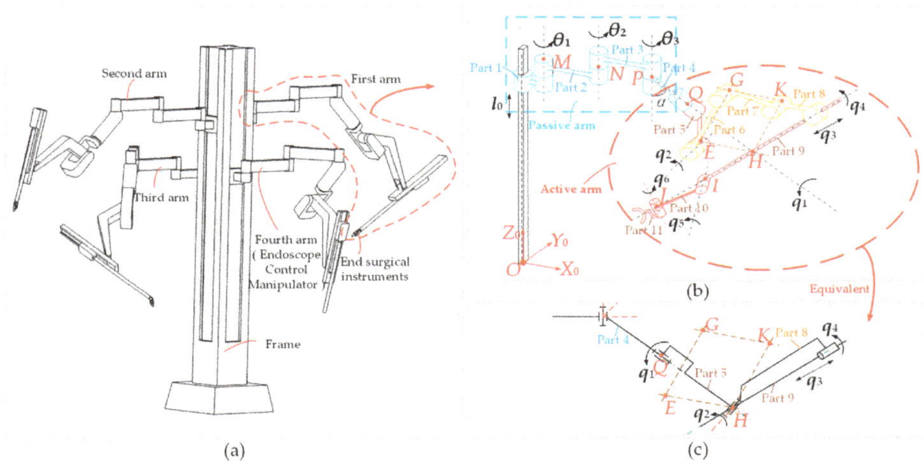

Figure 1. Structure schematic of the Da Vinci patient trolley. (**a**) Composition of the patient trolley, including four arms (first arm, second arm, third arm and fourth arm) and a frame. (**b**) Structure schematic of the first arm (slave manipulator), including the passive arm (dashed rectangle area) and active arm (dashed ellipse area), and point H, which is the RCM. (**c**) Equivalent diagram of the parallelogram mechanism (yellow part).

Before surgery, the angles of passive joints M, N, and P and the position of the M joint on the frame are adjusted manually and then remain fixed during surgery. The joints that mainly assist the surgeon are the active joints of the slave manipulator: Q, E, G, K, H, I, and J in Figure 1b.

The DOFs of the slave manipulator includes four DOFs: l_0, θ_1, θ_2 and θ_3. The parts 6, 7, and 8 in the active arm belong to the parallelogram mechanism (i.e., these three parts cannot move freely but under the motion constraint of the parallelogram mechanism in Figure 1c. Then only parts 4, 5, 8 and 9 can move freely in the slave manipulator, i.e., part 5 rotates around part 4, there is rotation between part 6 and part 5, and there is rotation between part 9 and part 8. Therefore, the entire slave manipulator will have 6 DOFs when the additional 3 DOFs of the H joint are considered.

2.2. Mathematical Model of Forward Kinematics

According to the previous analysis, an active arm coordinate system is established in Figure 2. The origin of the base coordinate $x_0y_0z_0$ (0-coordinate frame) of the active arm is established at the joint Q between the passive and active arms, the direction of which is the same as that of the frame coordinate $X_0Y_0Z_0$.

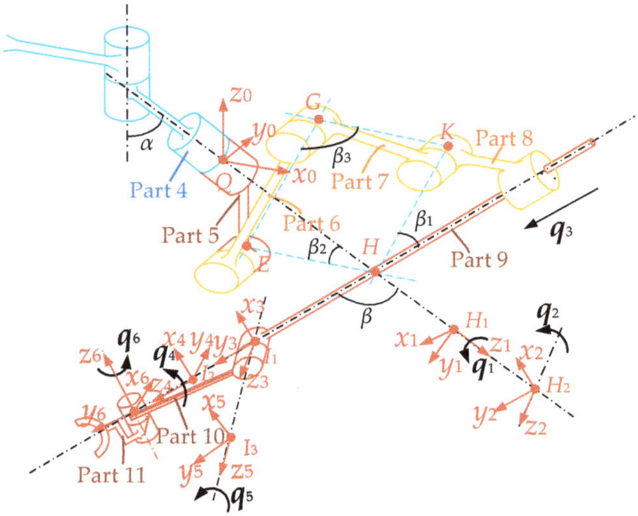

Figure 2. Structure diagram of the active arm of the slave manipulator. To more clearly, we draw H and I as H, H_1, H_2 and I_1, I_2, I_3 respectively. 1-coordinate frame fixed on part 5; 2-coordinate frame fixed on part 8; 3-coordinate frame fixed on part 9; 4-coordinate frame fixed on part 9; 5-coordinate frame fixed on part 10; 6-coordinate frame fixed on part 11.

Figure 2 shows the six DOFs of the active arm in the slave manipulator, namely five rotating joints: q_1, q_2, q_4, q_5, q_6, and a translational joint: q_3.

Six transform matrixes frame from $x_1y_1z_1$ (1-coordinate frame) to $x_6y_6z_6$ (6-coordinate frame) can be obtained as follows:

$$\begin{cases} T = \{-(\pi-\alpha)\}_x\{q_1\}_z\{l_{QH}\}^z \\ {}_2^1T = \{-(\beta-\frac{\pi}{2})\}_x\{\frac{\pi}{2}\}_y\{q_2\}_z \\ {}_3^2T = \{q_3\}^y \\ {}_4^3T = \{-\frac{\pi}{2}\}_x\{q_4\}_z \\ {}_5^4T = \{\frac{\pi}{2}\}_x\{q_5\}_z \\ {}_6^5T = \{l_{IJ}\}^y\{\frac{\pi}{2}\}_y\{q_6\}_z \end{cases}, \tag{1}$$

where ${}^i{}_jT$ means transform matrix from the i-coordinate frame to the j-coordinate frame; $\{\ \}_x$, $\{\ \}_y$ and $\{\ \}_z$ represent the rotation angle around the x-axis, y-axis and z-axis, respectively; $\{\ \}^x$, $\{\ \}^y$ and $\{\ \}^z$ denotes the translational movement along the x-axis, y-axis and z-axis, respectively.

Then, the pose matrix of the active arm at the endpoint of the slave manipulator can be obtained as follows:

$${}_6^0T = {}_1^0T{}_2^1T{}_3^2T{}_4^3T{}_5^4T{}_6^5T = \begin{bmatrix} n_x & o_x & a_x & p_x \\ n_y & o_y & a_y & p_y \\ n_z & o_z & a_z & p_z \\ 0 & 0 & 0 & 1 \end{bmatrix} = \begin{bmatrix} \boldsymbol{R} & \boldsymbol{P} \\ \boldsymbol{0} & 1 \end{bmatrix}, \tag{2}$$

where $\boldsymbol{R}$ and $\boldsymbol{P}$ are the orientation and position of the slave manipulator endpoint, respectively. Substituting Equation (1) into Equation (2) yields Equation (3):

$${}_6^0T = \{-(\pi-\alpha)\}_x\{q_1\}_z\{l_{QH}\}^z\{-(\beta-\frac{\pi}{2})\}_x\{\frac{\pi}{2}\}_y\{q_2\}_z\{q_3\}^y\{-\frac{\pi}{2}\}_x\{q_4\}_z\{\frac{\pi}{2}\}_x\{q_5\}_z\{l_{IJ}\}^y\{\frac{\pi}{2}\}_y\{q_6\}_z. \tag{3}$$

2.3. Mathematical Model of Inverse Kinematics

In the case of inverse kinematics, the known pose matrix is ${}^0{}_6\boldsymbol{T}$ in Equation (3), and the unknowns are q_1, q_2, q_3, q_4, q_5 and q_6.

Left-multiplying the matrices $\boldsymbol{T} = \{q_3\}^{y^{-1}}\{q_2\}_z^{-1}\{\frac{\pi}{2}\}_y^{-1}\{-(\beta-\frac{\pi}{2})\}_x^{-1}\{l_{QH}\}^{z^{-1}}\{q_1\}_z^{-1}\{-(\pi-\alpha)\}_x^{-1}$ on both sides of Equation (3) yields Equation (4):

$$T_6^0T = \left\{-\frac{\pi}{2}\right\}_x\{q_4\}_z\left\{\frac{\pi}{2}\right\}_x\{q_5\}_z\{l_{IJ}\}^y\left\{\frac{\pi}{2}\right\}_y\{q_6\}_z. \tag{4}$$

To simplify the complexity of the formula, we construct a unitary matrix [26] $\boldsymbol{U}$ as follows. After introducing the matrix $\boldsymbol{U}$, the variables in the ${}^0{}_6\boldsymbol{T}$ can be replaced by the Euler formula.

$$\boldsymbol{U} = \begin{bmatrix} \frac{1}{\sqrt{2}}i & -\frac{1}{\sqrt{2}}i & 0 & 0 \\ \frac{1}{\sqrt{2}} & \frac{1}{\sqrt{2}} & 0 & 0 \\ 0 & 0 & 1 & 0 \\ 0 & 0 & 0 & 1 \end{bmatrix} \tag{5}$$

In addition, we know that $\boldsymbol{E} = \boldsymbol{U}^{-1}\boldsymbol{U}$. Left-multiply the inverse of unitary matrix $\boldsymbol{U}^{-1}$ and right-multiply the matrix $\boldsymbol{U}$ on both sides of Equation (4). Then, multiply the $\boldsymbol{E}$ matrix between these two matrices to transform Equation (4) into the following:

$$\boldsymbol{Y_3X_2C_1X_1C_2C_3} = \boldsymbol{C_4X_4C_5X_5C_6X_6}. \tag{6}$$

Here, $\boldsymbol{C}_1 = \boldsymbol{U}^{-1}[\{l_{QH}\}^z\{\pi/2 - \beta\}_z]^{-1}\boldsymbol{U}$, $\boldsymbol{C}_2 = \boldsymbol{U}^{-1}[\{\alpha - \pi\}_x]^{-1}\boldsymbol{U}$, $\boldsymbol{C}_3 = \boldsymbol{U}^{-1}{}^0{}_6\boldsymbol{TU}$, $\boldsymbol{C}_4 = \boldsymbol{U}^{-1}\{-\pi/2\}_z\boldsymbol{U}$, $\boldsymbol{C}_5 = \boldsymbol{U}^{-1}\{\pi/2\}_z\boldsymbol{U}$, and $\boldsymbol{C}_6 = \boldsymbol{U}^{-1}\{l_{IJ}\}^y\{\pi/2\}_y\boldsymbol{U}$. Clearly, $\boldsymbol{C}_1$, $\boldsymbol{C}_2$, $\boldsymbol{C}_3$, $\boldsymbol{C}_4$, $\boldsymbol{C}_5$ and $\boldsymbol{C}_6$ are constant matrices or parameter matrices related to the structure of the robot. In addition, $\boldsymbol{X}_j$ (j = 1, 2, 4, 5, 6) is as follows:

$$\boldsymbol{X}_j = \begin{cases} T_{Xj}^{-1} & j=1,2 \\ T_{Xj} & j=4,5,6 \end{cases}, \tag{7}$$

where $\boldsymbol{T}_{\mathrm{Xj}}$ (j = 1, 2, 4, 5, 6) represents the results of the transform matrix of q_{j} rotating around the z-axis, which undergoes the left-multiplication of $\boldsymbol{U}^{-1}$ and right-multiplication of $\boldsymbol{U}$. Let $x_{\mathrm{j}} = \cos(q_{\mathrm{j}}) + i\sin(q_{\mathrm{j}})$, $x_{\mathrm{j}}^{-1} = \cos(q_{\mathrm{j}}) - i\sin(q_{\mathrm{j}})$; thus, $\boldsymbol{T}_{\mathrm{Xj}}$ can be expressed as follows:

$$\boldsymbol{T}_{Xj} = \boldsymbol{U}^{-1}\{q_i\}_z\boldsymbol{U} = \begin{bmatrix} x_j & 0 & 0 & 0 \\ 0 & x_j^{-1} & 0 & 0 \\ 0 & 0 & 1 & 0 \\ 0 & 0 & 0 & 1 \end{bmatrix}. \tag{8}$$

Let $\boldsymbol{Y}_{\mathrm{j}}$ represents the result of the transform matrix q_{j} along the y-axis, which undergoes the left-multiplication of $\boldsymbol{U}^{-1}$ and right-multiplication of $\boldsymbol{U}$, and $x_{\mathrm{j}} = q_{\mathrm{j}}$; then, $\boldsymbol{Y}_{\mathrm{j}}$ can be expressed as follows:

$$Y_j = \boldsymbol{U}^{-1}\{q_i\}^y\boldsymbol{U} = \begin{bmatrix} 1 & 0 & 0 & -\frac{\sqrt{2}}{2}x_j \\ 0 & 1 & 0 & -\frac{\sqrt{2}}{2}x_j \\ 0 & 0 & 1 & 0 \\ 0 & 0 & 0 & 1 \end{bmatrix}. \tag{9}$$

Take $\boldsymbol{A} = \boldsymbol{Y_3X_2C_1X_1C_2C_3}$, $\boldsymbol{B} = \boldsymbol{C_4X_4C_5X_5C_6X_6}$. Then the dimensions of matrix $\boldsymbol{A}$ and matrix $\boldsymbol{B}$ are both 4 × 4. Each element in matrix $\boldsymbol{A}$ is a function containing x_1, x_2, and x_3, while the elements in matrix $\boldsymbol{B}$ are functions containing x_4, x_5, and x_6. In addition, corresponding elements in matrix $\boldsymbol{A}$ and matrix $\boldsymbol{B}$ are equal, i.e., $\boldsymbol{A}ij = \boldsymbol{B}ij$. The specific contents of the elements are detailed in Appendix A.

To realize element elimination, we evaluated elements in matrix $\boldsymbol{A}$ and $\boldsymbol{B}$ that do not contain x_6 and $x_6{}^{-1}$: $\boldsymbol{B}_{13}$, $\boldsymbol{B}_{14}$, $\boldsymbol{B}_{23}$, $\boldsymbol{B}_{24}$, $\boldsymbol{B}_{33}$ and $\boldsymbol{B}_{34}$ to establish Equation (10):

$$\left.\begin{array}{l} A_{13} = B_{13} \\ A_{14} = B_{14} \\ A_{23} = B_{23} \\ A_{24} = B_{24} \\ A_{33} = B_{33} \\ A_{34} = B_{34} \end{array}\right\}. \tag{10}$$

To further eliminate x_4 and x_5, the linear combination of x_4 and x_5, and their reciprocals are treated as new variables, i.e., $\boldsymbol{X} = [x_4x_5, x_4x_5{}^{-1}, x_4{}^{-1}x_5, x_4{}^{-1}x_5{}^{-1}, x_4{}^{-1}, x_5{}^{-1}, x_4, x_5, x_4{}^0x_5{}^0]^{\mathrm{T}}$, which contains nine variables. Because there are only six equations in Equation (10), three more equations are required, which are attainable by conducting nonlinear computations on the elements in Equation (10), as shown in Equation (11).

$$\left.\begin{array}{l} A^{L1} = -A_{34}A_{23} + A_{24}A_{33} = -B_{34}B_{23} + B_{24}B_{33} = B^{L1} \\ A^{L2} = -A_{24}A_{13} + A_{14}A_{23} = -B_{24}B_{13} + B_{14}B_{23} = B^{L2} \\ A^{L3} = -A_{13}A_{34} + A_{14}A_{33} = -B_{13}B_{34} + B_{14}B_{33} = B^{L3} \end{array}\right\} \tag{11}$$

Then, by combining Equations (10) and (11):

$$[\boldsymbol{D}(x_1, x_2, x_3)]_{9\times 9}\boldsymbol{X} = 0. \tag{12}$$

The matrix $[\boldsymbol{D}(x_1, x_2, x_3)]_{9\times9}$ contains only the variables x_1, x_2, and x_3 since $\boldsymbol{X}$ is nonzero. The condition that Equation (12) has a solution is that the determinant of $[\boldsymbol{D}(x_1, x_2, x_3)]_{9\times9}$ equals zero. The following equation, which contains x_1, x_2, and x_3, can be obtained provided that $|\boldsymbol{D}(x_1, x_2, x_3)| = 0$:

$$\left(C_1'x_1 + C_2' + C_3'x_1^{-1}\right)x_2 = -\left(C_4'x_1 + C_5' + C_6'x_1^{-1}\right)x_2^{-1}, \tag{13}$$

where $\boldsymbol{C}'_{\mathrm{j}}$ is a constant related to the robot's parameters. Considering that x_1, x_2, x_4, x_5, and x_6 can be defined as $x_{\mathrm{j}} = \cos(q_{\mathrm{j}}) + i\sin(q_{\mathrm{j}})$ and $x_{\mathrm{j}}{}^{-1} = \cos(q_{\mathrm{j}}) - i\sin(q_{\mathrm{j}})$, then x_1, x_2, x_4, x_5, and x_6 are all non-zero. Therefore, x_2 can be expressed by x_1 according to Equation (13):

$$x_2^2 = \frac{-\left(X_1'x_1 + C_2' + C_3'x_1^{-1}\right)}{\left[\left(C_4'x_1 + C_5' + C_6'x_1^{-1}\right)\right]} = f_1(x_1). \tag{14}$$

To represent x_3 by x_1 and x_2, we construct the equations $A^{\mathrm{L4}} = A_{23}A_{14} + A_{13}A_{24} + A_{33}A_{34}$ and $B^{\mathrm{L4}} = B_{23}B_{14} + B_{13}B_{24} + B_{33}B_{34}$. Let $A^{\mathrm{L4}} = B^{\mathrm{L4}}$, then

$$\left(C_7'x_1 + C_9' + C_{12}'x_1^{-1}\right)x_3x_2 + \left(C_8'x_1 + C_{11}' + C_{13}'x_1^{-1}\right)x_3x_2^{-1} + C_{10}' = 0. \tag{15}$$

Equation (15) contains only three variables, x_1, x_2, and x_3, therefore, x_3 can be expressed by x_1 and x_2 as:

$$x_3 = f_2(x_1, x_2). \tag{16}$$

Because both A^{L2} and A^{L3} are functions of x_1, x_2 and x_3, by combining $A^{\mathrm{L2}} = B^{\mathrm{L2}}$ with $A^{\mathrm{L3}} = B^{\mathrm{L3}}$, the following can be obtained:

$$\begin{bmatrix} x_4 \\ x_4^{-1} \end{bmatrix} = \begin{bmatrix} -\frac{i}{l_{IJ}} & -\frac{\sqrt{2}i}{l_{IJ}} \\ -\frac{i}{l_{IJ}} & \frac{\sqrt{2}i}{l_{IJ}} \end{bmatrix} \begin{bmatrix} A^{L2} \\ A^{L3} \end{bmatrix}. \tag{17}$$

Therefore,

$$x_4 = -\frac{i}{l_{IJ}}A^{L2} - \frac{\sqrt{2}i}{l_{IJ}}A^{L3} = f_3(x_1, x_2, x_3). \tag{18}$$

Similarly, by combining $A_{13} = B_{13}$ and $A_{14} = B_{14}$ with Equation (16), the following can be obtained:

$$\begin{bmatrix} x_5 \\ x_5^{-1} \end{bmatrix} = \begin{bmatrix} -\frac{\sqrt{2}}{(A^{L2}i - l_{IJ})} & -\frac{\sqrt{2}il_{IJ}}{(A^{L2}i - l_{IJ})} \\ -\frac{\sqrt{2}i}{(A^{L2} - l_{IJ}i)} & -\frac{\sqrt{2}l_{IJ}}{(A^{L2} - l_{IJ}i)} \end{bmatrix} \begin{bmatrix} A_{14} \\ A_{13} \end{bmatrix}. \tag{19}$$

Therefore, x_5 can be represented by x_1, x_2, and x_3:

$$x_5 = -\frac{\sqrt{2}A_{14}}{(A^{L2}i - l_{IJ})} - \frac{\sqrt{2}il_{IJ}A_{13}}{(A^{L2}i - l_{IJ})} = f_4(x_1, x_2, x_3). \tag{20}$$

Combined with Equations (16) to (20) and $A_{11} = B_{11}$, x_6 can finally be represented by x_1:

$$x_6 = \frac{(b_1 i + b_2)x_1 x_2^{-1} + (b_3 i + b_4)x_2^{-1} + (b_5 i + b_6)x_1^{-1}x_2^{-1}}{\frac{1}{8}x_5 x_4 + \frac{i}{4}x_4 - \frac{1}{8}x_5^{-1}x_4 + \frac{i}{4}x_5 + \frac{1}{4}x_5^{-1} + \frac{1}{8}x_5 x_4^{-1} - \frac{i}{4}x_4^{-1} - \frac{1}{8}x_5^{-1}x_4^{-1}} = f_5(x_1), \tag{21}$$

where b_j is a constant related to the robot's parameters.

From Equations (16) to (21), the variables x_2, x_3, x_4, x_5, and x_6 can be represented with the variable x_1. Let A_{24} and B_{24} be equal, then:

$$F(x_1) = A_{24} - B_{24} = 0. \tag{22}$$

By solving Equation (22), x_1 can be solved, then x_2, x_3, x_4, x_5, and x_6 can be subsequently solved by Equation (16) to (21). Specifically, we know from Equation (14) that $x_2 = \pm\sqrt{f_1(x_1)}$, therefore, $F(x_1)$ will have two expressions:

$$F(\xi_1) = \begin{cases} F_1(x_1)x_2 = \sqrt{f_1(x_1)} \\ F_2(x_1)x_2 = -\sqrt{f_1(x_1)} \end{cases}. \tag{23}$$

Equation (23) will be electively solved according to the minimum norm of the distance the joint moved.

3. Inverse Kinematics Solution

From the derivation of the forward and inverse kinematics of the Da Vinci slave manipulator, it is known that the key to solving inverse kinematics is to solve the nonlinear Equation (22).

3.1. Inverse Kinematics Solution Method

The Newton–Raphson method is an efficient numerical iterative approach to solve nonlinear equations in real and complex domains. In this paper, we construct the following iterative formula:

$$x_{n+1} = x_n - \frac{F(x_n)}{F'(x_n)}. \tag{24}$$

3.2. Initial Value of the Inverse Kinematics Solution

The motion ranges of each joint in the Da Vinci slave manipulator are as follows: θ_1 from $-90°$ to $90°$, θ_2 from $-135°$ to $0°$, l_{IH} from 0 to 315 mm, θ_3 from $-180°$ to $180°$, θ_4 from $-90°$ to $90°$, and θ_5 from $-90°$ to $90°$. In the previously defined $x_1 = \cos(q_1) + i\sin(q_1)$, q_1 is the rotation angle of the first joint with a range of $[-\pi/2, \pi/2]$. A finite number of values are obtained by extracting values within these available ranges at a certain angle interval (our subsequent calculation results show that a less than $5°$ angle interval is acceptable) to solve the forward kinematics, the results of which are stored as the initial value database for the inverse kinematics solution. By selecting the data from the initial value database that is the least squares approximate to the required inverse kinematics as the initial values, the Newton iteration method will successfully converge.

3.3. Inverse Kinematics Solution Process

According to the inverse kinematics of the Da Vinci slave manipulator and the derived solution process, a detailed solution flow chart is constructed in Figure 3.

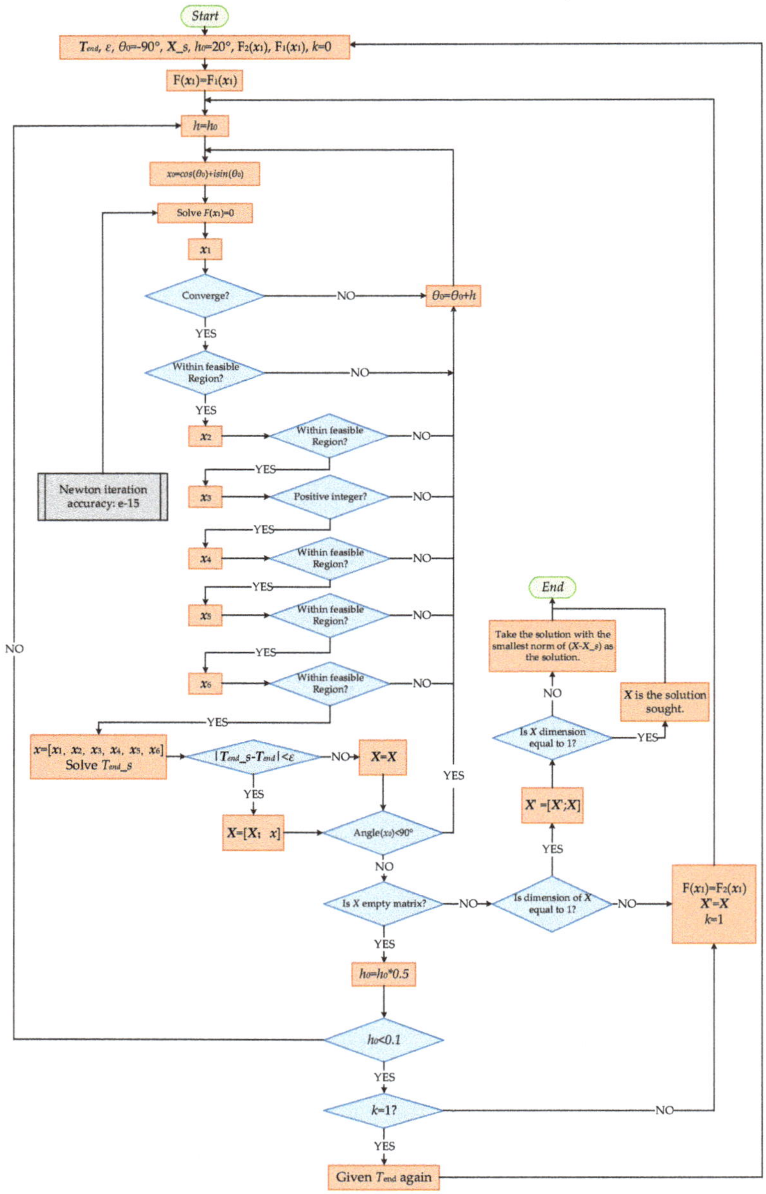

Figure 3. Inverse kinematics solution process of the Da Vinci slave manipulator.

In Figure 3, T_{end} is the pose matrix at the endpoint of the robot (i.e., ${}^{0}{}_{6}T$), ε is the precision of the solution, θ_0 is the initial value, h is the initial step value, and X_s represents the values of each joint at previous positions. The value k is used to determine whether both forms of $F(x_1)$ have been solved.

4. Example Analysis and Discussion

4.1. Parameters of the Da Vinci Slave Manipulator

According to Figure 4, the size parameters of the Da Vinci slave manipulator are shown in Table 1.

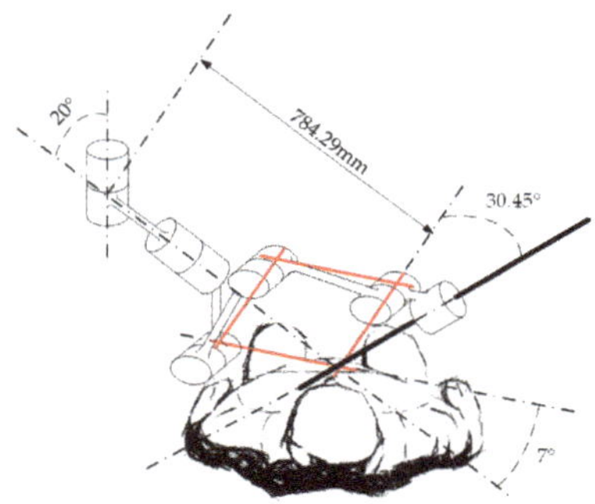

Figure 4. Range of motion of the slave manipulator.

Table 1. Structure parameters of the Da Vinci slave manipulator.

β_1 (°)	β_2 (°)	l_{QH} (mm)	β_3 (°)	α (°)	l_{IJ} (mm)
30.45	7	784.29	0	20	10

4.2. Solution of Proposed Method

As shown in Table 2 (the form is referenced by the book: An Introduction to Error Analysis [27]), the input values are a set of arbitrarily selected joint parameters within the effective range of joint space. The outputs in Table 2 are the calculated values by our proposed method. h represents the step size of the solution, and the error is calculated as follows:

$$\varepsilon = |\text{Output} - \text{Input}|. \tag{25}$$

In this paper, the solution precision of the Newton–Raphson method is 1E–10 ε = 1E-7 for T_{end}, θ_0 takes the minimum value of $-90°$ within the feasible region, and the initial step size is h_0 = 20°. The test carried on CodeBlocks 16.01 based on the hardware that a PC platform running with an Intel (R) Core (TM) i5-4460 CPU, clocked at 3.2 GHz (memory: 4 GB).

We also use the same input to run the procedure on the "R12CCPU-V" MELSEC iQ-R series C language controller from Mitsubishi Electric and VxWorks 6.9 operating system, the results are list in Table 3.

Table 2. Inverse kinematics solution of the Da Vinci slave manipulator.

Joint \ Step (°)	Output h = 20	Error	Input	Joint \ Step (°)	Output h = 20	Error	Input
θ_1 (°)	10.00000000000004	0.00000000000004	10	θ_1 (°)	15.39999999999996	0.00000000000003	15.4
θ_2 (°)	−124.999999995	0.000000005	−125	θ_2 (°)	−116.499999995	0.000000005	−116.5
l_{IH} (mm)	100.0000000000001	0.0000000000001	100	l_{IH} (mm)	101.3000000000001	0.0000000000001	101.3
θ_3 (°)	99.999999995	0.000000005	100	θ_3 (°)	86.3999999999996	0.0000000000004	86.4
θ_4 (°)	−14.0000000000008	0.0000000000008	−14	θ_4 (°)	13.6999999999998	0.0000000000002	13.7
θ_5 (°)	18.000000002	0.000000002	18	θ_5 (°)	−15.899999996	0.000000004	−15.9
t (s)	0.421			t (s)	0.406		
Joint \ Step (°)	**Output h = 10**	**Error**	**Input**	**Joint \ Step (°)**	**Output h = 20**	**Error**	**Input**
θ_1 (°)	−50.240000000000009	0.000000000000009	−50.24	θ_1 (°)	48.3689999999998	0.0000000000002	48.369
θ_2 (°)	−60.759999999999998	0.000000000000002	−60.76	θ_2 (°)	−80.59700000000002	0.00000000000002	−80.597
l_{IH} (mm)	126.94000000000001	0.00000000000001	126.94	l_{IH} (mm)	125.4680000000002	0.0000000000002	125.468
θ_3 (°)	101.379999995	0.000000005	101.38	θ_3 (°)	−80.593999999998	0.000000000002	−80.594
θ_4 (°)	−15.9400000000002	0.0000000000002	−15.94	θ_4 (°)	−68.5920000000003	0.0000000000003	−68.592
θ_5 (°)	24.78999999	0.00000001	24.79	θ_5 (°)	15.237999998966513		15.238
t (s)	0.484			t (s)	0.390		
Joint \ Step (°)	**Output h = 20**	**Error**	**Input**	**Joint \ Step (°)**	**Output h = 20**	**Error**	**Input**
θ_1 (°)	−56.54835000000004	0.00000000000004	−56.54835	θ_1 (°)	43.36587495200008	0.00000000000008	43.365874952
θ_2 (°)	−53.89575999999999	0.00000000000001	−53.89576	θ_2 (°)	−125.687459795	0.000000005	−125.6874598
l_{IH} (mm)	168.85324	0.00004	168.8532	l_{IH} (mm)	136.8452159699998	0.0000000000002	136.84521597
θ_3 (°)	−96.358469995	0.000000005	−96.35847	θ_3 (°)	75.312548960004	0.000000000004	75.31254896
θ_4 (°)	−50.268430000001	0.000000000001	−50.26843	θ_4 (°)	25.984572680002	0.000000000002	25.98457268
θ_5 (°)	50.468509999	0.000000001	50.46851	θ_5 (°)	−42.58794151	0.00000001	−42.587941523
t (s)	0.437			t (s)	0.437		

Table 3. Inverse kinematics solution of the Da Vinci slave manipulator (based on better hardware and software).

Joint \ Step (°)	Output h = 20	Error	Input	Joint \ Step (°)	Output h = 20	Error	Input
θ_1 (°)	9.9999999999999998	0.000000000000002	10	θ_1 (°)	15.39999999999995	0.00000000000005	15.4
θ_2 (°)	−124.999999995	0.000000005	−125	θ_2 (°)	−116.499999995	0.000000005	−116.5
l_{IH} (mm)	99.9999999999999	0.0000000000001	100	l_{IH} (mm)	101.3000000000002	0.0000000000002	101.3
θ_3 (°)	99.999999995	0.000000005	100	θ_3 (°)	86.3999999999998	0.0000000000002	86.4
θ_4 (°)	−13.9999999999995	0.0000000000005	−14	θ_4 (°)	13.69999999999991	0.00000000000009	13.7
θ_5 (°)	18.000000002	0.000000002	18	θ_5 (°)	−15.899999996	0.000000004	−15.9
t (s)	0.016			t (s)	0.016		
Joint \ Step (°)	**Output h = 20**	**Error**	**Input**	**Joint \ Step (°)**	**Output h = 20**	**Error**	**Input**
θ_1 (°)	−50.240000000000002	0.000000000000002	−50.24	θ_1 (°)	48.3689999999998	0.0000000000002	48.369
θ_2 (°)	−60.759999999999998	0.000000000000002	−60.76	θ_2 (°)	−80.59699999999998	0.00000000000002	−80.597
l_{IH} (mm)	126.9400000000001	0.0000000000001	126.94	l_{IH} (mm)	125.4680000000002	0.0000000000002	125.468
θ_3 (°)	101.379999995	0.000000005	101.38	θ_3 (°)	−80.593999999998	0.0000000000002	−80.594
θ_4 (°)	−15.9400000000008	0.0000000000008	−15.94	θ_4 (°)	−68.5920000000003	0.0000000000003	−68.592
θ_5 (°)	24.78999999	0.00000001	24.79	θ_5 (°)	15.237999999	0.000000001	15.238
t (s)	0.016			t (s)	0.016		
Joint \ Step (°)	**Output h = 20**	**Error**	**Input**	**Joint \ Step (°)**	**Output h = 20**	**Error**	**Input**
θ_1 (°)	−56.54835000000004	0.00000000000004	−56.54835	θ_1 (°)	43.36587495200008	0.00000000000008	43.365874952
θ_2 (°)	−53.89575999999999	0.00000000000001	−53.89576	θ_2 (°)	−125.687459795	0.000000005	−125.6874598
l_{IH} (mm)	168.85320000000007	0.00000000000007	168.8532	l_{IH} (mm)	136.8452159699999	0.0000000000001	136.84521597
θ_3 (°)	−96.358469995	0.000000005	−96.35847	θ_3 (°)	75.3125489599998	0.0000000000002	75.31254896
θ_4 (°)	−50.268430000001	0.000000000001	−50.26843	θ_4 (°)	25.9845726799996	0.0000000000004	25.98457268
θ_5 (°)	50.468509999	0.000000001	50.46851	θ_5 (°)	−42.58794151	0.00000001	−42.587941523
t (s)	0.016			t (s)	0.016		

In order to verify the proposed method, we build a simulation model in Adams. Firstly, we gave a random movement for the simulation model, then we obtained the matrix of the end pose from the simulation result (Figure 5a), which included the position and the orientation. Secondly, the matrixes were imported to the algorithm we proposed as input. Thirdly, we used our solution to drive the robot model. In the final results, depicted in Figure 5b, we see that the curve obtained by our algorithm is very close to the target curve. This means the proposed method is of high accuracy.

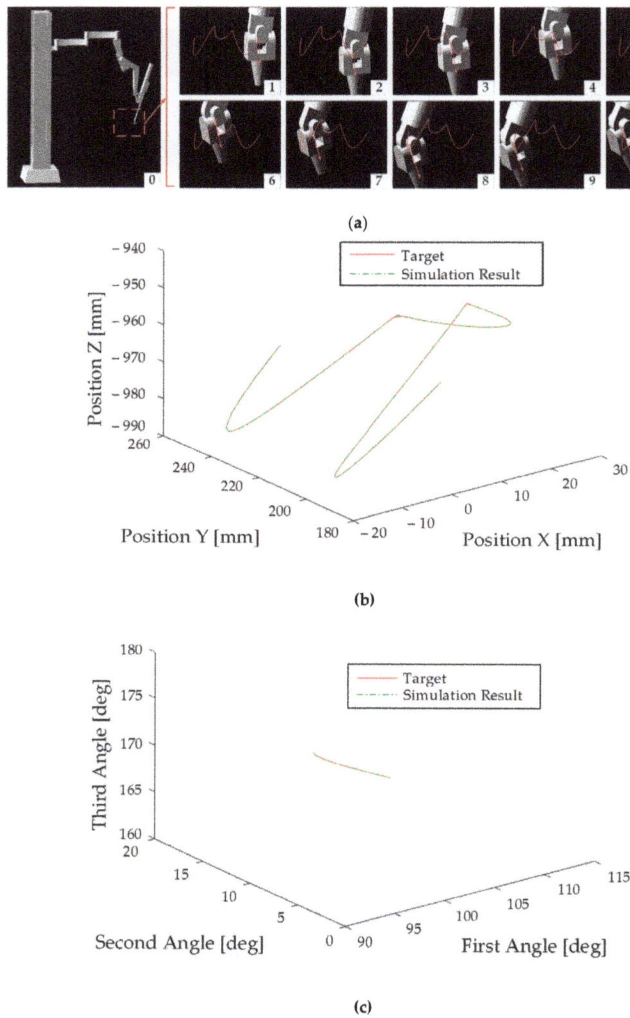

Figure 5. Simulation results: (**a**) simulation in Adams, (**b**) end position comparison between target and our result, (**c**) the end orientation (Euler angles) comparison between target and our result.

4.3. Discussion of the Solution

As noted in the previous section, the main focus of the solution includes the solution accuracy, real-time performance (solution speed) and the singularity problem (based on the joint position solution or not).

(1) Solution accuracy

Table 2 shows that for any input joint parameter, whether an integer or a decimal value accurate to nine decimal places, the error between the calculated values (output) and the given values (input) of six joint parameters of the slave manipulator is within 0.00004 degree (or mm), which demonstrates that the proposed approach for the Da Vinci surgical robot inverse kinematics is capable of finding effective solutions with high accuracy.

(2) Solution speed

The number of calculation steps of the algorithm/program is one of the influential factors that affect the solution speed. As schematically shown in Figure 3, the number of steps of the inverse kinematics approach proposed in this paper depends mainly on two factors:

① The number of Newton iterations in the best-case scenario, the numerical solution can be obtained by one iteration when the initial value is perfectly selected. In general, if the given initial value ensures convergence, then the solution of $F(x_1) = 0$ is expected to take approximate 20 iterations; in cases where the convergence fails, the program is designed to iterate at most 50 times and then exit to avoid a dead loop.

② The step size. when the step size h is large, the iteration times are reduced, thereby shorten the calculation time. However, the selection of initial values will be relatively imprecise, resulting in higher possibility of solution failure due to unreasonable initial values. In cases where the step size is small, the initial values tend to be more precise, resulting in a greater possibility of obtaining a solution. However, the resulting shortcoming is the demand for more calculation steps and longer time.

An improved calculation process is proposed by taking both factors into account: first, start the procedure by following the flow chart shown in Figure 3 with a relatively large step size; second, reduce the step size gradually for more precise initial values if the equation has no solution. This approach boosts the speed of calculation but also reduces the possibility of solution failure.

(3) Singularity influences

Our method abandons the use of Jacobian matrix inversion. Therefore, there will be barely no singular solution in the condition of a good initial value from a mathematical perspective. On the other hand, because our algorithm is based on the position increment of the manipulator's joints, Then, the motion joint position trajectory is generated by interpolation based on the initial joint position and the end joint position. The joint position interpolation is not affected by the singularity.

4.4. Comparison with Other Methods

Table 4 compared our method with another two typical methods in applicability, complexity of calculation, solution precision, solution speed and singularity influences.

Table 4. Comparison between the proposed method and others.

Method	Applicability	Complexity of Calculation	Solution Precision	Solution Speed	Singularity Influences
separates the position and orientation [10,11]	especially for the structure satisfied Pieper principle	low	high	quick	need to consider about
method based on the Jacobian matrix [16,28]	all	high	middle	not quick	need to consider about
The proposed method	especially for the structure not satisfied Pieper principle	middle	high	quick	barely considering about

Particularly, we choose separates the position and orientation method (name as "separate P&R method" below) to applied on the Da Vinci slave manipulator, and the comparison with the proposed method within the same simulation model is shown in Figure 6. The position curve of "separate P&R method" is following the target with some fluctuation, while the orientation result is far away from the correct solution, which means that the Separate P&R method is not suitable for the slave manipulator that dissatisfied the Pieper principle. Even though increasing the error feedback to compensate for the results may increase the solution precision, the compensation operation will take a large amount of time, which proves the superiority of our proposed method in terms of solution precision and speed.

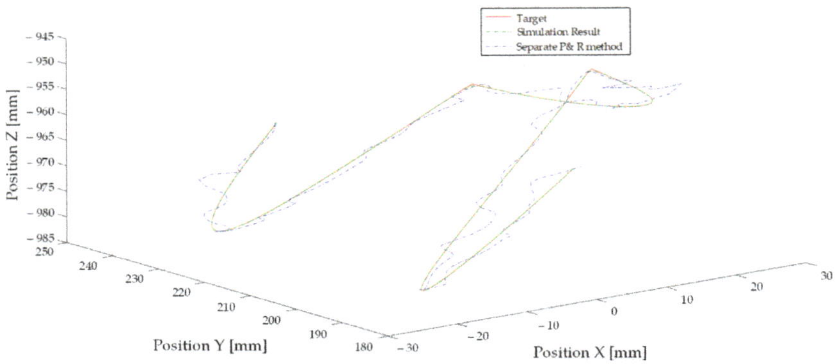

(a)

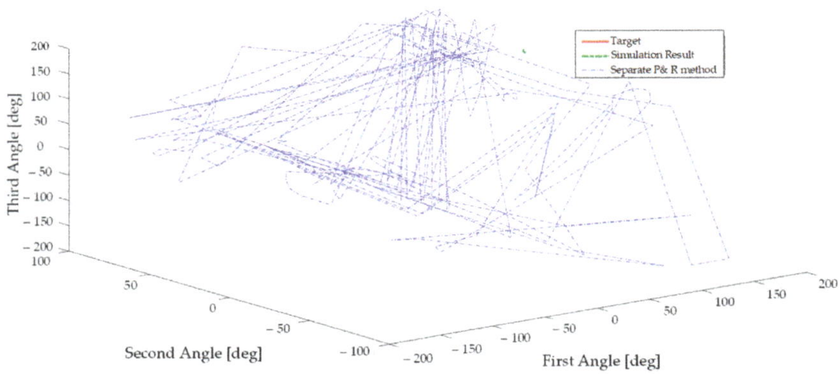

(b)

Figure 6. The endpoint comparison between results of our proposed method and the separate P&R method. (**a**) The position result comparison; (**b**) the orientation result comparison.

5. Conclusions

(1) In this paper, an inverse kinematics mathematical model of the 6-DOF active arm of the Da Vinci surgical robot established by using a coordinate transformation matrix method according to its mechanical motion diagram and working characteristics. By introducing a unitary matrix, the mathematical model on the real plane is transformed to a complex plane, consequently avoiding the relatively complex trigonometric calculations associated with the conventional solution procedure. Following the steps proposed, this approach can apply to other similar robot kinematic problems.

(2) The dialytic elimination method serves to individually eliminate the joint variables and ultimately obtains a high-order nonlinear equation with only one unknown variable of the first joint. Subsequently, the Newton iteration method is introduced to solve the nonlinear equation on the basis of selecting multiple initial values reasonably in the feasible region. The test shows that carefully changing the step size and selecting initial values in the feasible region for the iteration solution of the nonlinear equation will significantly reduce the possibility of solution failure.

(3) In theory, the accuracy of the approach can reach an error of < 0.00004 degree (or mm) and achieve a solution time of 0.5 s (Intel (R) Core i5-4460 CPU, clocked at 3.2 GHz (memory: 4 GB)). Running the procedure on better hardware and software (based on the "R12CCPU-V" MELSEC iQ-R series C language controller from Mitsubishi Electric and VxWorks 6.9 operating system) decreases the solution time to 20 ms, which fully satisfies the real-time performance requirement of surgical robots.

(4) A method that combines dialytic elimination and the Newton iteration method is applied to solve the inverse kinematics problem of the slave manipulators of the master–slave surgical robot. The proposed approach achieves higher accuracy than other solutions based on position and orientation separation while the real-time performance is satisfied. Moreover, in contrast to other numerical solutions, the solution method is based on the position of the joint and is therefore barely not affected by any singularity. Further, the application of our approach removes the restriction of the design of the master manipulator configuration.

Author Contributions: Conceptualization, L.B. and J.Y.; methodology, X.C.; software, F.Z.; validation, J.Y., L.B., Y.S. and X.C.; formal analysis, J.Y.; investigation, P.J.; resources, F.L.; data curation, L.B.; writing—original draft preparation, J.Y.; writing—review and editing, L.B., X.C. and Y.S.; visualization, X.C.; supervision, L.B.; project administration, L.B.; funding acquisition, X.C.

Funding: This research was funded by the Special Cooperation Program for Higher Education Institutions Collaborative Innovation (grant no. KH2016006), the Foundation for Sci & Tech Research Project of Chongqing Science & Technology Commission (grant nos. cstc2016jcyjA0472, and cstc2017zdcy-zdzxX0007), the National Natural Science Foundation of China (grant nos. 51705050, and 51709023), and Fundamental Research Funds for the Central Universities (grant no. 2018CDGFJX0022, 2018CDQYHK0029).

Acknowledgments: The authors gratefully acknowledge Zixiang Zhang for his valuable support with language editing.

Conflicts of Interest: The authors declare no conflict of interest.

Appendix A

$$\boldsymbol{A}_{11} = (c_1 i + c_2)\boldsymbol{x}_1\boldsymbol{x}_2^{-1} + (c_3 i + c_4)\boldsymbol{x}_2^{-1} + (c_5 i + c_6)\boldsymbol{x}_1^{-1}\boldsymbol{x}_2^{-1};$$
$$\boldsymbol{A}_{12} = (c_7 i + c_8)\boldsymbol{x}_1\boldsymbol{x}_2^{-1} + (c_9 i + c_{10})_5\boldsymbol{x}_2^{-1} + (c_{11} i + c_{12})\boldsymbol{x}_1^{-1}\boldsymbol{x}_2^{-1};$$
$$\boldsymbol{A}_{13} = (c_{13} i + c_{14})\boldsymbol{x}_1\boldsymbol{x}_2^{-1} + (c_{15} i + c_{16})\boldsymbol{x}_2^{-1} + (c_{17} i + c_{18})\boldsymbol{x}_1^{-1}\boldsymbol{x}_2^{-1};$$
$$\boldsymbol{A}_{14} = (c_{19} i + c_{20})\boldsymbol{x}_1\boldsymbol{x}_2^{-1} + (c_{21} i + c_{22})\boldsymbol{x}_2^{-1} - \tfrac{\sqrt{2}}{2}\boldsymbol{x}_3 + (c_{23} i + c_{24})\boldsymbol{x}_1^{-1}\boldsymbol{x}_2^{-1};$$
$$\boldsymbol{A}_{21} = (c_{25} i + c_{26})\boldsymbol{x}_1\boldsymbol{x}_2 + (c_{27} i + c_{28})\boldsymbol{x}_2 + (c_{29} i + c_{30})\boldsymbol{x}_2\boldsymbol{x}_1^{-1};$$
$$\boldsymbol{A}_{22} = (c_{31} i + c_{32})\boldsymbol{x}_1\boldsymbol{x}_2 + (c_{33} i + c_{34})\boldsymbol{x}_2 + (c_{35} i + c_{36})\boldsymbol{x}_2\boldsymbol{x}_1^{-1};$$
$$\boldsymbol{A}_{23} = (c_{37} i + c_{38})\boldsymbol{x}_1\boldsymbol{x}_2 + (c_{39} i + c_{40})\boldsymbol{x}_2 + (c_{41} i + c_{42})\boldsymbol{x}_2\boldsymbol{x}_1^{-1};$$
$$\boldsymbol{A}_{24} = (c_{43} i + c_{44})\boldsymbol{x}_1\boldsymbol{x}_2 + (c_{45} i + c_{46})\boldsymbol{x}_2 - \tfrac{\sqrt{2}}{2}\boldsymbol{x}_3 + (c_{47} i + c_{48})\boldsymbol{x}_2\boldsymbol{x}_1^{-1};$$
$$\boldsymbol{A}_{31} = (c_{49} i + c_{50})\boldsymbol{x}_1 + (c_{51} i + c_{52})\boldsymbol{x}_1^{-1};$$
$$\boldsymbol{A}_{32} = (c_{53} i + c_{54})\boldsymbol{x}_1 + (c_{55} i + c_{56})\boldsymbol{x}_1^{-1};$$
$$\boldsymbol{A}_{33} = (c_{57} i + c_{58})\boldsymbol{x}_1 + (c_{58} - c_{57} i)\boldsymbol{x}_1^{-1};$$
$$\boldsymbol{A}_{34} = (c_{59} i + c_{60})\boldsymbol{x}_1 + (c_{60} - c_{59} i)\boldsymbol{x}_1^{-1};$$
$$\boldsymbol{B}_{11} = \tfrac{1}{8}\boldsymbol{x}_6\boldsymbol{x}_5\boldsymbol{x}_4 + \tfrac{i}{4}\boldsymbol{x}_6\boldsymbol{x}_4 - \tfrac{1}{8}\boldsymbol{x}_6\boldsymbol{x}_5^{-1}\boldsymbol{x}_4 + \tfrac{1}{4}\boldsymbol{x}_6\boldsymbol{x}_5 + \tfrac{1}{4}\boldsymbol{x}_6\boldsymbol{x}_5^{-1} + \tfrac{1}{8}\boldsymbol{x}_6\boldsymbol{x}_5\boldsymbol{x}_4^{-1} - \tfrac{i}{4}\boldsymbol{x}_6\boldsymbol{x}_4^{-1} - \tfrac{1}{8}\boldsymbol{x}_6\boldsymbol{x}_5^{-1}\boldsymbol{x}_4^{-1};$$
$$\boldsymbol{B}_{12} = \tfrac{1}{8}\boldsymbol{x}_5\boldsymbol{x}_4\boldsymbol{x}_6^{-1} - \tfrac{i}{4}\boldsymbol{x}_4\boldsymbol{x}_6^{-1} - \tfrac{1}{8}\boldsymbol{x}_5^{-1}\boldsymbol{x}_4\boldsymbol{x}_6^{-1} + \tfrac{1}{4}\boldsymbol{x}_5\boldsymbol{x}_6^{-1} + \tfrac{1}{4}\boldsymbol{x}_5^{-1}\boldsymbol{x}_6^{-1} + \tfrac{1}{8}\boldsymbol{x}_5\boldsymbol{x}_4^{-1}\boldsymbol{x}_6^{-1} + \tfrac{i}{4}\boldsymbol{x}_4^{-1}\boldsymbol{x}_6^{-1} - \tfrac{1}{8}\boldsymbol{x}_5^{-1}\boldsymbol{x}_4^{-1}\boldsymbol{x}_6^{-1};$$
$$\boldsymbol{B}_{13} = \tfrac{-i\sqrt{2}}{8}\boldsymbol{x}_5\boldsymbol{x}_4 - \tfrac{i\sqrt{2}}{8}\boldsymbol{x}_5^{-1}\boldsymbol{x}_4 - \tfrac{i\sqrt{2}}{4}\boldsymbol{x}_5 + \tfrac{i\sqrt{2}}{4}\boldsymbol{x}_5^{-1} - \tfrac{i\sqrt{2}}{8}\boldsymbol{x}_5\boldsymbol{x}_4^{-1} - \tfrac{i\sqrt{2}}{8}\boldsymbol{x}_5^{-1}\boldsymbol{x}_4^{-1};$$
$$\boldsymbol{B}_{14} = \left(\tfrac{\sqrt{2}}{8}\boldsymbol{x}_5\boldsymbol{x}_4 - \tfrac{\sqrt{2}}{8}\boldsymbol{x}_5^{-1}\boldsymbol{x}_4 + \tfrac{\sqrt{2}}{4}\boldsymbol{x}_5 + \tfrac{\sqrt{2}}{4}\boldsymbol{x}_5^{-1} + \tfrac{\sqrt{2}}{8}\boldsymbol{x}_5\boldsymbol{x}_4^{-1} - \tfrac{\sqrt{2}}{8}\boldsymbol{x}_5^{-1}\boldsymbol{x}_4^{-1}\right) l_{IJ};$$
$$\boldsymbol{B}_{21} = -\tfrac{1}{8}\boldsymbol{x}_6\boldsymbol{x}_5\boldsymbol{x}_4 - \tfrac{i}{4}\boldsymbol{x}_6\boldsymbol{x}_4 + \tfrac{1}{8}\boldsymbol{x}_6\boldsymbol{x}_5^{-1}\boldsymbol{x}_4 + \tfrac{1}{4}\boldsymbol{x}_6\boldsymbol{x}_5 + \tfrac{1}{4}\boldsymbol{x}_6\boldsymbol{x}_5^{-1} - \tfrac{1}{8}\boldsymbol{x}_6\boldsymbol{x}_5\boldsymbol{x}_4^{-1} + \tfrac{i}{4}\boldsymbol{x}_6\boldsymbol{x}_4^{-1} + \tfrac{1}{8}\boldsymbol{x}_6\boldsymbol{x}_5^{-1}\boldsymbol{x}_4^{-1};$$
$$\boldsymbol{B}_{22} = -\tfrac{1}{8}\boldsymbol{x}_6^{-1}\boldsymbol{x}_5\boldsymbol{x}_4 + \tfrac{i}{4}\boldsymbol{x}_6^{-1}\boldsymbol{x}_4 + \tfrac{1}{8}\boldsymbol{x}_6^{-1}\boldsymbol{x}_5^{-1}\boldsymbol{x}_4 + \tfrac{1}{4}\boldsymbol{x}_6^{-1}\boldsymbol{x}_5 + \tfrac{1}{4}\boldsymbol{x}_6^{-1}\boldsymbol{x}_5^{-1} - \tfrac{1}{8}\boldsymbol{x}_6^{-1}\boldsymbol{x}_5\boldsymbol{x}_4^{-1} + \tfrac{i}{4}\boldsymbol{x}_6^{-1}\boldsymbol{x}_4^{-1} + \tfrac{1}{8}\boldsymbol{x}_6^{-1}\boldsymbol{x}_5^{-1}\boldsymbol{x}_4^{-1};$$
$$\boldsymbol{B}_{23} = \tfrac{i\sqrt{2}}{8}\boldsymbol{x}_5\boldsymbol{x}_4 + \tfrac{i\sqrt{2}}{8}\boldsymbol{x}_5^{-1}\boldsymbol{x}_4 - \tfrac{i\sqrt{2}}{4}\boldsymbol{x}_5 + \tfrac{i\sqrt{2}}{4}\boldsymbol{x}_5^{-1} + \tfrac{i\sqrt{2}}{8}\boldsymbol{x}_5\boldsymbol{x}_4^{-1} + \tfrac{i\sqrt{2}}{8}\boldsymbol{x}_5^{-1}\boldsymbol{x}_4^{-1};$$
$$\boldsymbol{B}_{24} = \left(-\tfrac{\sqrt{2}}{8}\boldsymbol{x}_5\boldsymbol{x}_4 + \tfrac{\sqrt{2}}{8}\boldsymbol{x}_5^{-1}\boldsymbol{x}_4 + \tfrac{\sqrt{2}}{4}\boldsymbol{x}_5 + \tfrac{\sqrt{2}}{4}\boldsymbol{x}_5^{-1} - \tfrac{\sqrt{2}}{8}\boldsymbol{x}_5\boldsymbol{x}_4^{-1} + \tfrac{\sqrt{2}}{8}\boldsymbol{x}_5^{-1}\boldsymbol{x}_4^{-1}\right) l_{IJ};$$
$$\boldsymbol{B}_{31} = -\tfrac{\sqrt{2}}{8}\boldsymbol{x}_6\boldsymbol{x}_5\boldsymbol{x}_4 - \tfrac{i\sqrt{2}}{4}\boldsymbol{x}_6\boldsymbol{x}_4 + \tfrac{\sqrt{2}}{8}\boldsymbol{x}_6\boldsymbol{x}_5^{-1}\boldsymbol{x}_4 + \tfrac{\sqrt{2}}{8}\boldsymbol{x}_6\boldsymbol{x}_5\boldsymbol{x}_4^{-1} - \tfrac{i\sqrt{2}}{4}\boldsymbol{x}_6\boldsymbol{x}_4^{-1} - \tfrac{\sqrt{2}}{8}\boldsymbol{x}_6\boldsymbol{x}_5^{-1}\boldsymbol{x}_4^{-1};$$
$$\boldsymbol{B}_{32} = -\tfrac{\sqrt{2}}{8}\boldsymbol{x}_6^{-1}\boldsymbol{x}_5\boldsymbol{x}_4 + \tfrac{i\sqrt{2}}{4}\boldsymbol{x}_6^{-1}\boldsymbol{x}_4 + \tfrac{\sqrt{2}}{8}\boldsymbol{x}_6^{-1}\boldsymbol{x}_5^{-1}\boldsymbol{x}_4 + \tfrac{\sqrt{2}}{8}\boldsymbol{x}_6^{-1}\boldsymbol{x}_5\boldsymbol{x}_4^{-1} + \tfrac{i\sqrt{2}}{4}\boldsymbol{x}_6^{-1}\boldsymbol{x}_4^{-1} - \tfrac{\sqrt{2}}{8}\boldsymbol{x}_6^{-1}\boldsymbol{x}_5^{-1}\boldsymbol{x}_4^{-1};$$
$$\boldsymbol{B}_{33} = \tfrac{i}{4}\boldsymbol{x}_5\boldsymbol{x}_4 + \tfrac{i}{4}\boldsymbol{x}_5^{-1}x_4 - \tfrac{i}{4}\boldsymbol{x}_5\boldsymbol{x}_4^{-1} - \tfrac{i}{4}\boldsymbol{x}_5^{-1}\boldsymbol{x}_4^{-1};$$

$B_{34} = \left(-\frac{1}{4}x_5x_4 + \frac{1}{4}x_5^{-1}x_4 + \frac{1}{4}x_5x_4^{-1} - \frac{1}{4}x_5^{-1}x_4^{-1}\right)l_{IJ}$,
where c_i is a function related to the structural parameters of the robot, which is known and can be regarded as a constant.

References

1. Taylor, R.H. A Perspective on Medical Robotics. *Proc. IEEE* **2006**, *94*, 1652–1664. [CrossRef]
2. Turchetti, G.; Palla, I.; Pierotti, F.; Cuschieri, A. Economic evaluation of Da Vinci-assisted robotic surgery: A systematic review. *Surg. Endosc.* **2012**, *26*, 598. [CrossRef] [PubMed]
3. Pugin, F.; Bucher, P.; Morel, P.; Bucher, P.; Morel, P. History of robotic surgery: From AESOP and ZEUS to Da Vinci. *J. Visc. Surg.* **2011**, *148*, e3–e8. [CrossRef] [PubMed]
4. Ishikawa, N.; Watanabe, G. Robot-assisted cardiac surgery. *Ann. Thorac. Cardiovasc. Surg.* **2015**, *21*, 322. [CrossRef] [PubMed]
5. Taylor, R.H.; Funda, J.; Grossman, D.D.; Karidis, J.P.; LaRose, D.A. Remote center-of-motion robot for surgery. U.S. Patent 5397323 A, 14 March 1995.
6. Zhou, X.; Zhang, H.; Feng, M.; Zhao, J.; Fu, Y. New remote centre of motion mechanism for robot-assisted minimally invasive surgery. *Biomed. Eng. Online* **2018**, *17*, 170. [CrossRef] [PubMed]
7. Ozgoren, M.K. Topological analysis of 6-joint serial manipulators and their inverse kinematic solutions. *Mech. Mach. Theory* **2002**, *37*, 511–547. [CrossRef]
8. Balkan, T.; Ozgoren, M.K.; Arikan, M.A.S.; Baykurt, H.M. A kinematic structure-based classification and compact kinematic equations for six-dof industrial robotic manipulators. *Mech. Mach. Theory* **2001**, *36*, 817–832. [CrossRef]
9. Balkan, T.; Ozgoren, M.K.; Arikan, M.A.S.; Baykurt, H.M. A method of inverse kinematics solution including singular and multiple configurations for a class of robotic manipulators. *Mech. Mach. Theory* **2000**, *35*, 1221–1237. [CrossRef]
10. Vol, N. Structural kinematics of 6-revolute-axis robot manipulators. *Mech. Mach. Theory* **1996**, *31*, 647–658.
11. Pieper, D.L. *The Kinematics of Manipulators under Computer Control*; Stanford University: Stanford, CA, USA, 1968.
12. Ai, Y.; Pan, B.; Niu, G.; Fu, Y.; Wang, S. Master-slave control technology of isomeric surgical robot for minimally invasive surgery. In Proceedings of the IEEE International Conference on Robotics and Biomimetics, Qingdao, China, 3–7 December 2017; pp. 2134–2139.
13. Podsędkowski, L. Forward and Inverse Kinematics of the Cardio-Surgical Robot with Non-Coincident Axis of the Wrist. *IFAC Proc. Vol.* **2003**, *36*, 437–442. [CrossRef]
14. Zhang, X.; Nelson, C.A. Kinematic Analysis and Optimization of a Novel Robot for Surgical Tool Manipulation. *J. Med. Devices* **2008**, *2*, 737–745. [CrossRef]
15. Mezouar, Y. Kinematic modeling and control of a robot arm using unit dual quaternions. *Robot. Auton. Syst.* **2016**, *77*, 66–73.
16. Xie, Q.; Pan, B.; Fu, Y.; Wang, S. Master-slave control technology research based on abdominal minimally invasive surgery robot. *Robot* **2011**, *33*, 53–58. [CrossRef]
17. Fu, Y.; Li, H.; Xie, Q. Master-slave control technology research for abdominal minimally invasive surgery robot. In *ASME 2010 International Mechanical Engineering Congress and Exposition*; American Society of Mechanical Engineers: New York, NY, USA, 2010; pp. 53–58.
18. Kajita, S. *Humanoid Robots*; Tsinghua University Press: Beijing, China, 2007.
19. Bu, W.; Liu, Z.; Tan, J. Inverse displacement analysis of 6R robots with offset wrists based on decoupling degrees of freedom at the cutoff points. *J. Mech. Eng.* **2010**, *46*, 1–5. [CrossRef]
20. Koker, R.; Oz, C.; Cakar, T.; Ekiz, H. A study of neural network based inverse kinematics solution for a three-joint robot. *Robot. Auton. Syst.* **2004**, *49*, 227–234. [CrossRef]
21. Lin, Y.; Zhao, H.; Ding, H. Solution of inverse kinematics for general robot manipulators based on multiple population genetic algorithm. *J. Mech. Eng.* **2017**, *53*, 1. [CrossRef]
22. Birbil, S.I.; Fang, S.C. An Electromagnetism-like Mechanism for Global Optimization. *J. Glob. Optim.* **2003**, *25*, 263–282. [CrossRef]
23. Birbil, S.I.; Fang, S.C.; Sheu, R.L. On the Convergence of a Population-Based Global Optimization Algorithm. *J. Glob. Optim.* **2004**, *30*, 301–318. [CrossRef]

24. Ren, Z.; Wang, Z.; Li, J.; Sun, L. Inverse Kinematics Solution for Robot Manipulator Based on Hybrid Electromagnetism-like Mechanism Algorithm. *J. Mech. Eng.* **2012**, *48*, 21–28. [CrossRef]
25. Yin, B.; Yan, B.; Shuai, J.; Ren, W.-B. Solution for direct kinematics of 3-PRS parallel manipulator using sylvester dialytic elimination method. In Proceedings of the International Conference on Mechanics and Civil Engineering, Orlando, FL, USA, 23–25 June 2014.
26. Roger, A.; Horn, R. *Matrix Analysis: English*; People Post Press: Beijing, China, 2015.
27. Taylor, R.J. *An Introduction to Error Analysis*; University Science Books: New York, NY, USA, 1997.
28. Hayashibe, M.; Suzuki, N.; Hashizume, M.; Konishi, K.; Hattori, A. Robotic surgery setup simulation with the integration of inverse-kinematics computation and medical imaging. *Comput. Methods Programs Biomed.* **2006**, *83*, 63–72. [CrossRef] [PubMed]

applied sciences

MDPI

Article

Serpens: A Highly Compliant Low-Cost ROS-Based Snake Robot with Series Elastic Actuators, Stereoscopic Vision and a Screw-Less Assembly Mechanism

Filippo Sanfilippo *, Erlend Helgerud, Per Anders Stadheim and Sondre Lieblein Aronsen

Department of Science and Industry Systems, University of South-Eastern Norway (USN), Post Box 235, 3603 Kongsberg, Norway; erlend.helgerud@usn.no (E.H.); peranders.stadheim@gmail.com (P.A.S.); sondre.lieblein@gmail.com (S.L.A.)

* Correspondence: filippo.sanfilippo@usn.no; Tel.: +47-942-58-929

Received: 28 December 2018; Accepted: 21 January 2019; Published: 24 January 2019

Abstract: Snake robot locomotion in a cluttered environment where the snake robot utilises a sensory-perceptual system to perceive the surrounding operational environment for means of propulsion is defined as perception-driven obstacle-aided locomotion (POAL). From a control point of view, achieving POAL with traditional rigidly-actuated robots is challenging because of the complex interaction between the snake robot and the immediate environment. To simplify the control complexity, compliant motion and fine torque control on each joint is essential. Accordingly, intrinsically elastic joints have become progressively prominent over the last years for a variety robotic applications. Commonly, elastic joints are considered to outperform rigid actuation in terms of peak dynamics, robustness, and energy efficiency. Even though a few examples of elastic snake robots exist, they are generally expensive to manufacture and tailored to custom-made hardware/software components that are not openly available off-the-shelf. In this work, *Serpens*, a newly-designed low-cost, open-source and highly-compliant multi-purpose modular snake robot with series elastic actuator (SEA) is presented. *Serpens* features precision torque control and stereoscopic vision. Only low-cost commercial-off-the-shelf (COTS) components are adopted. The robot modules can be 3D-printed by using Fused Deposition Modelling (FDM) manufacturing technology, thus making the rapid-prototyping process very economical and fast. A screw-less assembly mechanism allows for connecting the modules and reconfigure the robot in a very reliable and robust manner. The concept of modularity is also applied to the system architecture on both the software and hardware sides. Each module is independent, being controlled by a self-reliant controller board. The software architecture is based on the Robot Operating System (ROS). This paper describes the design of *Serpens* and presents preliminary simulation and experimental results, which illustrate its performance.

Keywords: snake robot; series elastic actuator; SEA; Robot Operating System; ROS

1. Introduction

In nature, limbless organisms such as snakes may exploit rocks, stones, branches, obstacles, or other irregularities in the terrain as a means of propulsion to achieve locomotion [1]. This remarkable ability allows biological snakes to be exceptionally adaptable to various types of environments. Snake robots that can replicate this range of behaviour could enable a variety of possible applications for use in challenging real-life operations and hazardous or confined areas that conventional robots (i.e., wheeled, tracked and legged) and humans are unable to access, such as explorations of earthquake-hit areas, pipe inspections for the oil and gas industry, fire-fighting operations, and search-and-rescue activities (SAR) [2,3]. Snake robot locomotion in a cluttered environment where the snake robot utilises a sensory-perceptual system to exploit the surrounding operational space and identifies walls, obstacles, or other external objects for means of propulsion can be defined as *perception-driven obstacle-aided locomotion* (POAL) [4,5]. The development of POAL is known to be challenging because of the complex interaction between the snake robot and the adjacent cluttered environment [6]. From a control point of view, achieving POAL requires precisely identifying potential push-points and to accurately determine achievable contact reaction forces. Accomplishing this with traditional rigidly-actuated robots is extremely demanding because of the absence of compliance.

Traditional gear-motor-driven actuators are designed and built for industrial automation. These conventional motors commonly actuate stiff linkages while encoders are employed to make position control possible. The control and overall performance of rigidly actuated robots is heavily impacted by actuator dynamics (i.e., masses, reflected inertias, and stiffness) [7]. Rigid actuators transmit high (virtually infinite) mechanical impedance, thus forcing the robot to resist motion when subject to a force. Rigidly actuated robots are also characterised by having a high bandwidth, which forces them to promptly move to commanded positions regardless of what external forces act on their joints. This characteristic is adequate for industrial automation because it allows robots for tracking trajectories in static or mapped environments, i.e., pick-and-place applications, but it is not suitable for robots that interact with unmapped and dynamic environments or need to navigate terrains cluttered with obstacles, such as snake robots. Indeed, conventional rigid actuators are unable to store and release energy or exploit natural dynamics, making them too inefficient or undesired for mobile locomotion applications. Moreover, when considering POAL, the high reflected inertia of traditional gear-motor-driven actuators can cause potential collisions that may damage both the robot and the environment.

To facilitate the control complexity for robots that interact with unmapped and dynamic environments or need to navigate rough terrains cluttered with obstacles, compliant motion and fine torque control on each joint is desirable. Consequently, intrinsically elastic joints have become progressively prominent over the last years for a variety of robotic applications. Commonly, elastic joints are considered to outperform rigid actuation in terms of peak dynamics, robustness, and energy efficiency [8]. Even though a few examples of elastic snake robots exist [9,10], they are generally costly to produce and tailored to custom-made hardware/software components that are not openly available off-the-shelf.

To give researchers a novel snake robot that is inexpensive to manufacture, easily customisable, and fast to fabricate, a newly-designed low-cost, open-source, and highly-compliant multi-purpose modular snake robot with series elastic actuators (SEA) is introduced in this work. The presented snake robot is named *Serpens* ("the Serpent", Greek Ὄφις) after the homonym constellation of the northern hemisphere. *Serpens* is shown in Figure 1. *Serpens* features compliant torque-controlled actuators and stereoscopic vision. Only low-cost commercial-off-the-shelf (COTS) components are adopted to achieve a sustainable prototyping process. The robot modules can be 3D-printed by using Fused Deposition Modelling (FDM) manufacturing technology [11], thus making the rapid-prototyping process very economical and quick. A screw-less assembly mechanism allows for connecting the modules and for reconfiguring the robot in a very reliable and robust manner. By combining the rapid-prototyping approach with the modular concept, different configurations can be achieved. By using a low-cost sensing approach, functions for torque sensing at the joint level, sensitive collision detection and joint compliant control are possible. The concept of modularity is also applied to the system architecture on both the software and hardware sides. Each module is independent, being controlled by a self-reliant controller board. The software architecture is based on the Robot Operating System (ROS) [12]. The authors intend this work to be the first in a series of open-source designs to be released, and through the contributions of the open-source user community, result in a large number of design modifications and variations available to researchers.

Figure 1. *Serpens*: A low-cost ROS-based snake robot with series elastic actuator (SEA), precision torque control and a screw-less assembly mechanism.

The paper is organised as follows. A review of the related research work is given in Section 2. In Section 3, we focus on the description of the mechanical overview. A software/hardware overview is described in Section 4.6. In Section 5, some preliminary simulation and experimental results are outlined. Finally, conclusions and future works are discussed in Section 6.

2. Related Research Work

To achieve locomotion in a cluttered and irregular terrain, a snake robot must be able to adapt its body motion to the environment. This requires that the robot can sense environment contact forces acting along its body [13]. To the best of our knowledge, the works in [13–18] present snake robot designs featuring contact sensing capabilities. However, the vast majority of snake robots that have been designed thus far adopt traditional gear-motor-driven actuators. The fact that these robots employ rigid actuators requires a very high degree of awareness of their surroundings to achieve POAL. When adopting traditional gear-motor-driven actuators, this implies that a very precise mathematical model that includes the interaction between the snake robot and the surrounding operational environment is needed. Furthermore, when considering POAL, the high reflected inertia of rigidly actuated robots can cause possible collisions that may damage both the robot and the environment.

To avoid the risk of rigid collisions, an alternative approach is inspired by the ability of biological mechanisms to accurately achieve compliance (passively and/or by precisely control torque). Based on this idea, series elastic actuators (SEA) were introduced in [19] as a means of achieving compliant motion and force control with traditional gear-motor-driven actuators. Thereafter, the design and control of SEA has been widely exploited in the fields of legged locomotion [20,21], humanoid robots [22] and manipulators [23]. Regarding snake robots, different methods of achieving compliant motion by controlling the torques exerted by the joints of the robot were presented by the Robotics Institute at the Carnegie Mellon University [10,24]. These control strategies are implemented on a snake robot that includes SEA and torque sensing at each joint, and demonstrate compliant locomotion that adapts naturally to the robot's surrounding terrain. This work is very pragmatic and has shown some success. However, the underlying idea is based on a relatively simplistic oscillation and adaptation of the torque to the surrounding obstacles. We hypothesise that exploiting full knowledge of the robot's configuration and surrounding environment can be more beneficial and can produce more reliable results with hopefully better performance. Moreover, the proposed robot design adopts financially demanding components and the software is not completely open-source.

To the best of our knowledge, a 3D-printable highly compliant multi-purpose modular robot that features SEA, precise torque control, open software/hardware and a screw-less assembly mechanism has not been released yet.

3. Mechanical Overview

In this section, the mechanical overview of *Serpens* is depicted by highlighting the selected design principles, the mechanical design, the proposed screw-less assembly mechanism and the adopted series elastic actuators.

3.1. Design Principles

The module-design of *Serpens* is inspired by the following principles:

- *Principle of minimalism*. To make the robot inexpensive, easily customisable, and fast to fabricate, each module is equipped with the simplest mechanical structure, the minimum number of actuators (only a single SEA per module with one degree of freedom (DOF) and the simplest set of sensors. During the design process, the main focus is to keep the amount of parts as low as possible, and at the same time minimise the number of assembly operations.
- *Principle of symmetry*. To facilitate the interaction with the environment, a symmetric design is adopted for each module with a flat profile for the interaction with the terrain. The symmetric design is also selected to store and release energy in a balanced manner.
- *Screw-less assembly mechanism*. To ease the connection and reconfiguration of the robot modules, a screw-less assembly mechanism is designed in a very reliable and robust manner.

3.2. Mechanical Design

The construction of *Serpens* consists of similarly designed modules that are shown in Figure 2 and include a head module, a varying number of joint modules, and a tail module. The head module contains an *Intel RealSense D435* stereoscopic camera [25]. The tail module only contains an anchorage mechanism for the external power supply cable. The joint module of *Serpens* is characterised by the parameters summarised in Table 1.

Table 1. Parameters of *Serpens*'s joint module.

Parameter	Value
Weight	$\sim$500 g
Width/height	75 mm
Length between joint axes	200 mm
Degrees of freedom	1
Max joint travel	$\pm 90°$
Max continuous joint torque	3.0 Nm (at 12 V)
Max joint speed with no load	77 RPM (at 12 V)
Operating Temperature (actuators)	-5 °C$\sim$80 °C

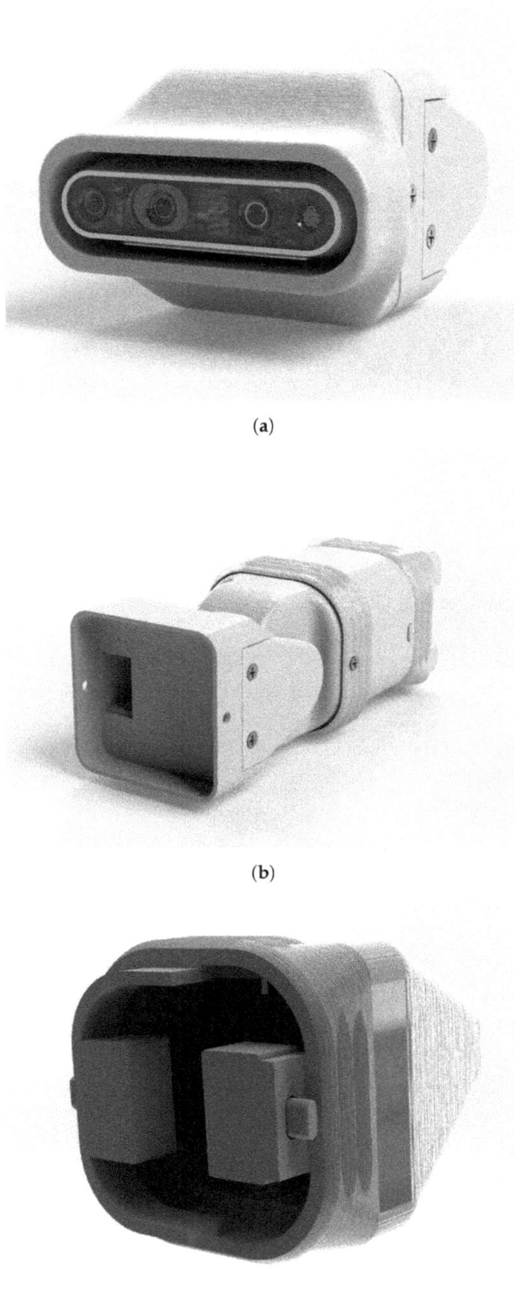

(**a**)

(**b**)

(**c**)

Figure 2. The head, joint, and tail modules of *Serpens*: (**a**) the head module of *Serpens* with the *Intel RealSense D435* stereoscopic camera; (**b**) one of the joint modules of *Serpens*; and (**c**) the tail module of *Serpens*.

3.3. Estimated Production Cost

The estimated production cost for the joint module is 250 USD, including the following elements:

- 3D-printing cost;
- cost of COTS mechanical parts (e.g., springs, nuts, bolts, bearings); and
- electrical components (e.g., micro-controller, sensors, and actuator).

As shown in Figure 3, *Serpens* allows for realising different connections, such as pitch connection, yaw connection and pitch–yaw connection.

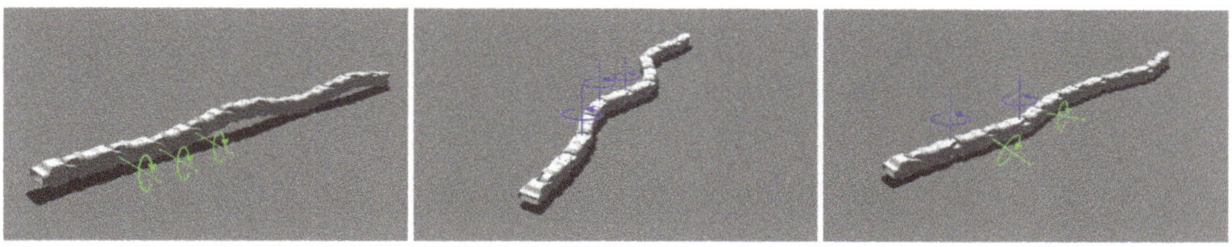

(a) (b) (c)

Figure 3. Two different connections can be achieved with *Serpens*: (**a**) pitch connection; (**b**) yaw connection; and (**c**) pitch–yaw connection.

The pitch connection allows *Serpens* to move only in 1D, forward or backward. The yaw-connecting configuration makes it possible to move *Serpens* similar to real snakes with all the joints rotate around the yaw axis. The pitch–yaw-connecting configuration enables *Serpens* to have some modules that rotate around the pitch axis and others around the yaw axis, respectively. This makes it possible to achieve new locomotion capabilities, such as sidewinding, rotating and rolling [26].

An exploded view of the joint module design is shown in Figure 4. To strengthen the modular attributes of *Serpens* through a generic and reusable module-design, each joint module is fitted with a screw-less assembly mechanism, a micro-controller, an actuator, an elastic gear, a rotary encoder, a bearing mechanism and a battery-pack. The proposed design enables the assembly process to be performed in an uncritical manner with respect to ordering and rotation of joints, while contributing to a better weight distribution throughout the body.

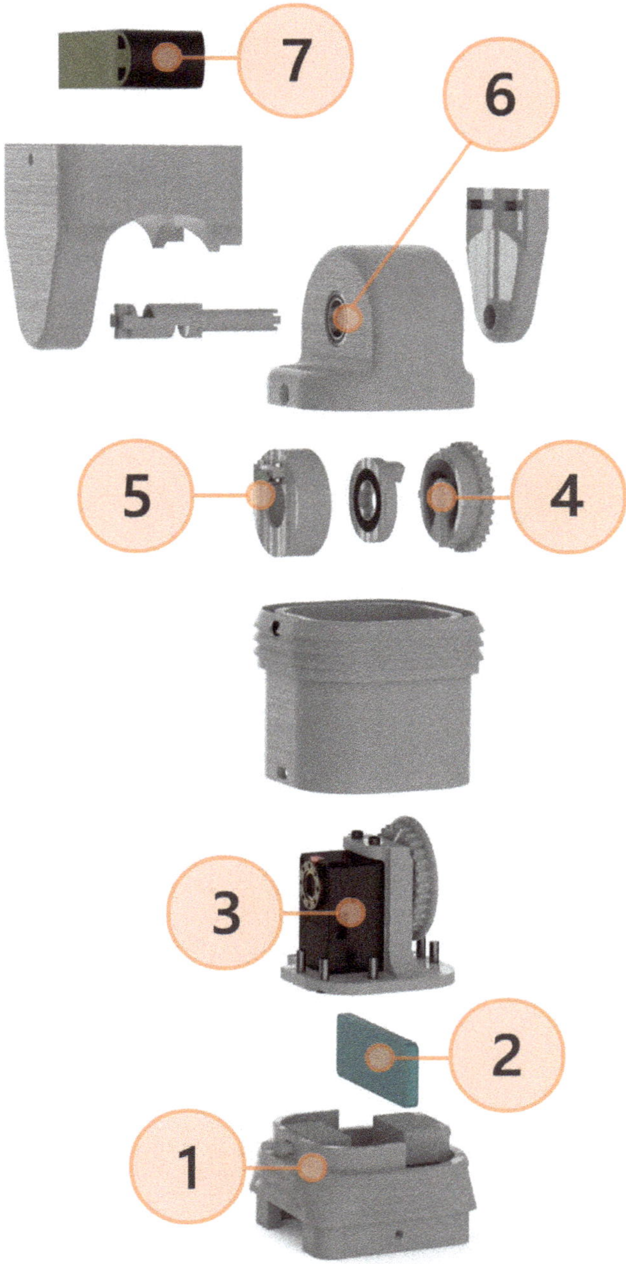

Figure 4. An exploded view of a joint module: (1) screw-less assembly mechanism; (2) micro-controller; (3) actuator; (4) elastic gear; (5) rotary encoder; (6) bearing; and (7) battery-pack.

3.4. Screw-Less Assembly Mechanism

As shown in Figure 5, a screw-less assembly mechanism is proposed for *Serpens* to easily interconnect each joint module through the adoption of specifically designed push-buttons. Each button consists of two springs that locks an oval cylinder in place when triggered. This novel mechanism makes it easier to access the battery-pack and the micro-controller of each module without requiring any tools. A complete dismantling of the modules is easily achievable in a very short time with the removal of just a few screws.

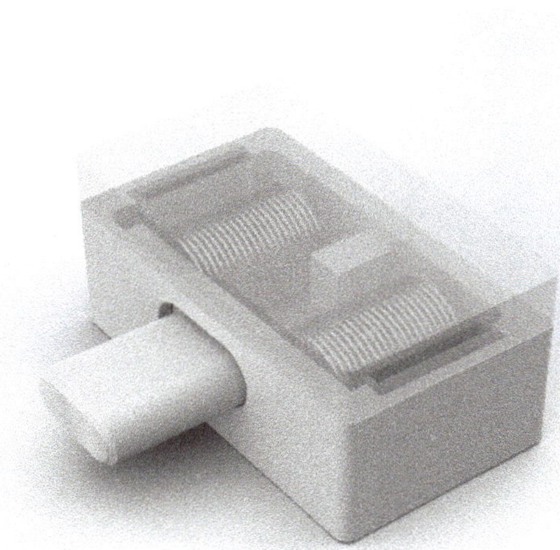

Figure 5. The proposed screw-less assembly mechanism consists of three 3D-printed components (two enclosure parts and a push-button) with two springs orientated in the direction of the pin placed under the enclosure cover.

3.5. Series Elastic Actuator (SEA)

To make *Serpens* highly-compliant with the environment, a newly designed series elastic actuator (SEA) is embedded in each joint module. This makes it possible to achieve passively-compliant motion and precise torque-control. Each SEA deliberately introduces compliance via a spring between the motor-gearbox and the load [27], therefore achieving intrinsic low impedance. As shown in Figure 6, the design of the elastic gear consists of a housing case, a base, a shaft, and a cogwheel. The intermediate element is connected to the shaft and works as a transmission between the cogwheel and the shaft itself. The cogwheel and the actuator are connected through a gear mechanism, where passive-compliance is provided by placing compression springs on each side of the outset of the base, in the chamber of the cogwheel. Even though the spring stiffness can be considered linear within a certain range, an encoder is employed to precisely monitor the misalignment/deviation of the compliant mechanism.

The novel design of *Serpens* is exclusively based on a 3D-printing process with *polylactic acid* (PLA) through FDM with the exception of a limited number of elements such as springs, nuts, bolts, bearings and electrical components. One of the main benefits of using PLA is that a relatively high strength construction can be achieved compared to the production cost. In addition, 3D-printing with PLA allows obtaining a reliable rapid-prototyping process by offering a wide range of customisation for different printing-parameters, such as print-speed, wall-thickness and layer-height. This makes it possible to tune the printing process according to the different parts and the properties the parts should hold.

Figure 6. The proposed design of the elastic gears for *Serpens*. The figure shows: the housing (**left**); the base (**middle**); and the cogwheel (**right**). The shaft (not depicted in this figure) runs through all parts. Compression springs are placed in the chamber of the cogwheel on each side of the outset of the base, providing passive-compliance.

3.6. Heat Dissipation

Within the joint module enclosure, continuous heat generation from the servos might cause overheating and failure. Moreover, the PLA material has poor heat conductivity capabilities (0.13 W/mK) and continuous locomotion could eventually lead to high temperatures within the joint module. To minimise this risk of overheating, the *Dynamixel XM430W-210T*, a COTS actuator produced by *ROBOTIS* [28], is selected to actuate each joint module. Each actuator is enclosed in an aluminium casing providing a good rate of heat transfer. As shown in Figure 4, each actuator is fixed to a specifically designed element which allows the surface area to remain exposed to the climate inside the module. The heat transfer performance of the selected actuator is qualitatively shown in Figure 7 by using a heat map. The motor was run for 50 min at a constant velocity of approximately 57.25 RPM. The servo was run without load continuously in one direction and monitored at 0 min, 25 min and 50 min, respectively. This qualitative experiment shows reasonable heat dissipation even after a medium/long running time. Extensive testing with different loads is still needed to assess the heat performance of *Serpens* in a real operational scenario.

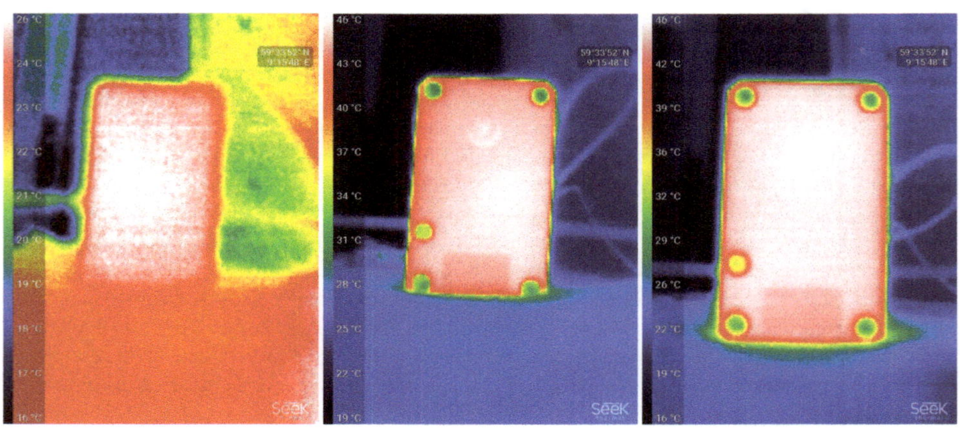

(**a**) (**b**) (**c**)

Figure 7. The heat transfer performance of the selected actuator is qualitatively shown using a heat map. The motor was run for 50 min at a constant velocity of approximately 57.25 RPM. The servo was run without load continuously in one direction and monitored at 0 min, 25 min and 50 min, respectively.

4. Software/Hardware Overview

4.1. Open-Source Software

In line with the overall low-cost approach of *Serpens*, an open-source software framework is designed for the low-level control. To design the software architecture, there are different robotic frameworks and middleware available in recent years [29]. However, the Robot Operating System (ROS) [12] has emerged as a de facto standard for robot software architecture in the research community. ROS is designed as a meta-operating system for robotic applications. The primary goal of ROS is to provide a common platform to make the design of capable robotic applications quicker and easier. Some of the features it provides include hardware abstraction, device drivers, message-passing and package management. In conjunction with ROS, Gazebo 3D simulator [30] can be adopted to accurately and efficiently simulate robots in complex indoor and outdoor environments. Gazebo also provides a robust physics engine, high-quality graphics, and convenient programmatic and graphical interfaces. In this perspective, ROS serves as the interface for the robot model of *Serpens*, while Gazebo is used to simulate both the robot and its operational environment. In addition to ROS and Gazebo, the RViz (ROS visualisation) [31] tool can be adopted to visualise and monitor sensor information retrieved in real-time from both the simulated scenario as well as from the real world. Another benefit for developers are the ROS community-driven support and the stable release-cycle of distributions (a new version is released every year, while a new long-term support (LTS) version is released every second year). Moreover, ROS offers an excellent interface to hardware components such as different micro-controllers and other peripheral hardware, i.e., actuators and sensors. The choice of ROS for the design of the control architecture makes it possible to extend the modular concept to both the hardware as well as the software of *Serpens*.

4.2. Hardware Overview

The control of *Serpens* is dependent on feedback from the actuators regarding position, velocity and torque. This feedback must be provided by the low-level controller of each actuator. Nowadays, there are several COTS actuators available in the market. However, only few commercial options provide precise current-based torque control and profile control for smooth motion planning. For the design of *Serpens*, the *Dynamixel XM430W-210T*, a COTS actuator produced by *ROBOTIS* is selected for each joint module to meet these demanding requirements. This particular actuator provides the aforementioned data as well as additional feedback for temperature and input voltage. In addition to offering the required feedback, the *XM430W-210T* has a sturdy construction with full-metal gears and a metal body, while being able to deliver a stall torque of 3.0 Nm (at 12.0 V, 2.3 A) in a operating temperature of $-5\,^{\circ}$~$80\,^{\circ}$ [28], which is considered sufficient in regards to the applications and the future development of *Serpens*. The chosen actuator communicates through a half duplex asynchronous serial Transistor–Transistor Logic (TTL) communication and also facilitate for daisy-chaining, which provides a simple connection structure for multiple actuators.For implementing the low-level control and the interception of feedback, a micro-controller is required at the joint level. Nowadays, different COTS options are available in the market. To facilitate the integration with the ROS-based architecture of *Serpens*, the *ROBOTIS OpenCM 9.04* micro-controller is embedded in each joint module. The *OpenCM 9.04* is a 32-bit *Cortex-M3* core micro-controller compatible with ROS and with *Arduino* [32] software/hardware. This choice is also motivated by the limited physical space in the presented design of the joint module, therefore the form-factor of the *OpenCM 9.04* (27 mm $\times$ 66.5 mm) is a crucial parameter for the selection of this specific micro-controller for *Serpens*. In addition, since the high-level control can be centralised in a single-board computer (SBC) possibly located either in the head or in an external computer while the low-level control is distributed to the micro-controllers embedded in each joint module, the computing power provided by the *OpenCM 9.04* is adequate for designated applications. The interface between the head module and each generic joint module is shown in Figure 8.

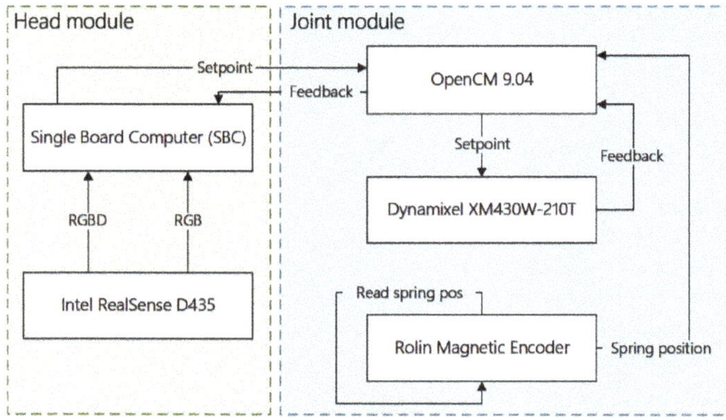

Figure 8. The interface between the head module and each generic joint module.

4.3. Encoders

The proposed SEA are designed for passive-compliance, as described in Section 3. If the spring stiffness k would be constant, the displacement would be linear with respect to the external forces exerted to the load. Thus, the forces acting on a joint could be estimated by applying *Hooke's law*. This can be highlighted in simulation by monitoring the motor- and load-positions and showing that they can be divergent due to external forces acting upon the load, as illustrated in Figure 9. However, in a real-world application, this is not realistic. The spring stiffens can be nonlinear. This is the reason each joint module of *Serpens* is fitted with a rotary incremental encoder that is connected to the *OpenCM 9.04*. The encoder is vital to the control of each SEA, as it provides feedback for the absolute position of the load. In particular, a *RoLin* encoder system is adopted [33] for *Serpens*. The *RoLin* component level encoder system consists of a read-head and a magnetised ring. The actuator is a periodically magnetised ring with a pole length of 2 mm. Axial reading of the ring is adopted for *Serpens*.

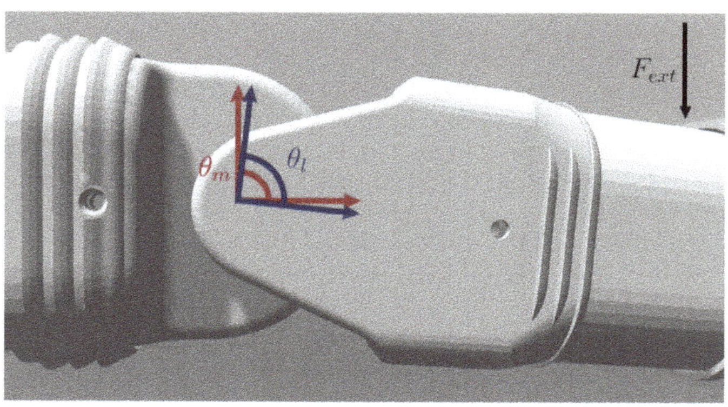

Figure 9. The simulated joint module showing the divergence of the actuator position θ_m, and load position θ_l as a result of the elasticity in the joint when external forces (F_{ext}) are applied to the load.

4.4. Single-Board Computer and Stereoscopic Camera

The head-module is fitted with a *single-board computer* (SBC) and a stereoscopic camera. The SBC is designed to handle all high-level control of *Serpens* in addition to providing the interface to the camera.

To enable visual feedback of the surroundings of *Serpens* while traversing unknown terrains, a camera is fitted on the head module. In particular, a reasonably small stereoscopic vision system is embedded because of the limited space in the design and the need for range detection. The proposed solution utilises a standard COTS *Intel RealSense D435* [25], a low-cost stereo vision camera comprising two depth sensors, a Red-Green-Blue (RGB) sensor, and a infrared projector. A considerable benefit of the *Intel RealSense D435* device is the *realsense2_camera* [34] package available for ROS, which provides a ROS-compatible interface to the *D400-series* from *Intel*, as shown in Figure 10

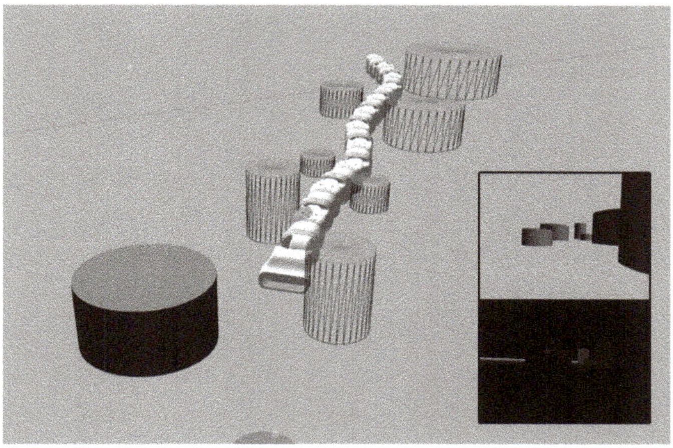

Figure 10. The simulated Gazebo environment showing *Serpens* in a pitch–yaw configuration, and the output of the simulated RGB (**top**) and RGBD (**bottom**) channels from the stereoscopic camera visualised through RVIZ.

4.5. ROS-Based Low-Level Architecture

The proposed ROS-based software architecture is illustrated by using a node-graph in Figure 11. This shows a simplified view of the nodes and topics used to control n joints of *Serpens* in the current implementation. The nodes are represented as ellipses while the topics as rectangles. The arrows represent publishers and subscribers, where arrows directed towards an ellipsis or box indicates a subscriber and an arrow directed outwards indicates a publisher.

The controller node can run either on an external computer or the SBC embedded in the head module. This node provides all high-level control for *Serpens* and acts as a hub for sensory data, such as the depth-sensor data collected by the *Intel RealSense D435*. As described above in this section, each joint module is provided with an embedded micro-controller (*OpenCM 9.04*) that is responsible for low-level control of the designated in the structure. Each of the boards acts as separate nodes in the ROS network architecture. In addition to being responsible for the low-level control, each joint module controller board also collects the feedback from the *XM 430W-210T* actuator and from the *RoLin* rotary incremental encoder. Each micro-controller implement a running ROS-node. This is shown in Figure 11 as */serial_node_n*, where n denotes the micro-controller-index corresponding topics for each joint.

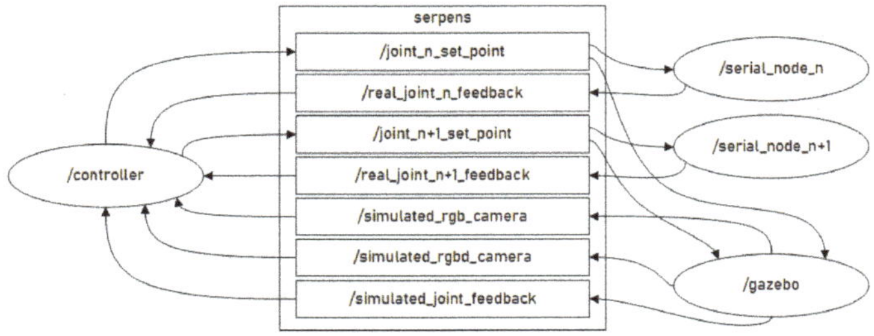

Figure 11. Node-graph showing the structure of nodes and topics contributing to the control to both the simulated and physical robot.

4.6. Guidelines for Designing the Control Framework Architecture to Achieve POAL

In this paper, the fundamental work for implementing the low-level control approach is presented. To practically achieve POAL, a hierarchical control framework is still needed. Possible design guidelines are presented in Figure 12 [2,3]. The following abstraction levels are defined:

- Perception/Mapping: This level is responsible for achieving the functions of sensing, mapping and localisation.
- Motion planning: This level is responsible for decision-making, path-planning and mission planning activities.
- High-level control: This level combines force and torque information with positional data to satisfy simultaneous position and force trajectory constraints.
- Low-level control: This level is responsible for the low-level control of individual joints. This is the level presented in this work.

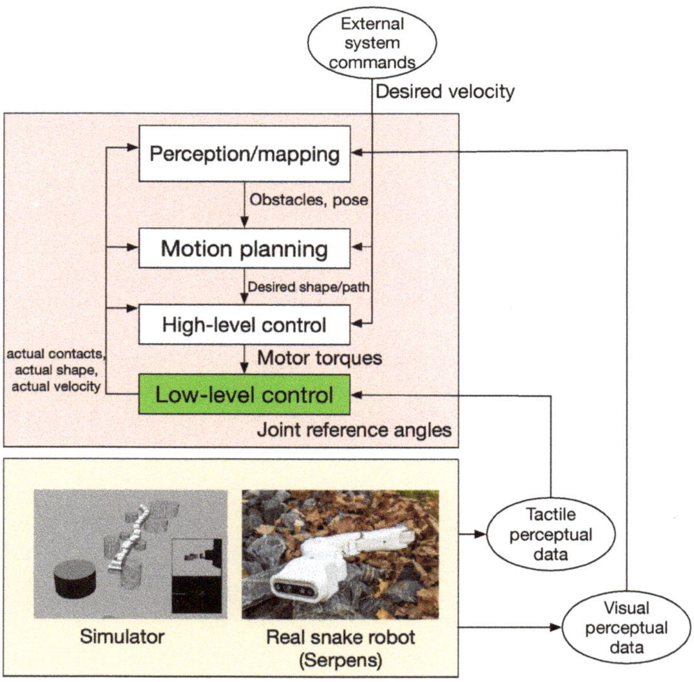

Figure 12. The proposed control framework for achieving POAL.

The abstraction levels above the low-level control layer will be part of our future work.

5. Simulations and Experimental Results

In this section, preliminary experimental simulations are considered with the aim of exploiting and validating the design features of *Serpens*. In particular, the possibility of implementing different characteristics of *Serpens* through the built-in capabilities of the *Universal Robotic Description Format* (URDF) [35] language and the *Gazebo* simulator is considered. In addition, the possibility of replicating the design of the physical model by separating the motor-side and spring-side dynamics in a simulated environment is studied.

To simulate the motor-side and spring-side dynamics, each joint of *Serpens* is modelled as a two-joint structure, as in the physical model. The generic configuration of links and joints for the simulated version of *Serpens* is shown in Figure 13. The motor-side dynamics are simulated with a regular revolute joint that is connected from the link to a shaft. The shaft is connected to the next link through a joint by simulating the spring dynamics. Exploiting the fact that the URDF is converted to *Simulation Description Format* (SDF) when imported to Gazebo, it is possible to use the SDF joint parameter *springStiffness* to simulate the spring dynamics. The capabilities of *macros* in the Extensible Markup Language (XML) Macros (Xacro) language is consistently used to provide an easily configurable setup of links and joints in the simulated environment, thus allowing the implementation of *pitch–yaw*, *pitch–pitch* and *yaw–yaw* configurations.

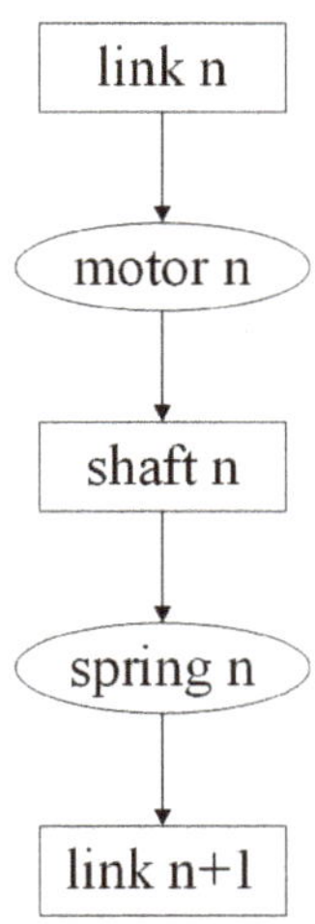

Figure 13. The generic configuration of links and joints in Xacro/URDF. Joints are shown as ellipses, while links are depicted as boxes.

An initial experiment includes the simulated environment and one physical joint module containing the *OpenCM 9.04* and the *XM430W-210T* servo motor. The *OpenCM 9.04* micro-controller is connected to and powered by an external computer running the proposed ROS-based architecture. The communication is achieved by using the *rosserial* package [36]. The *XM430W-210T* actuator is connected through the TTL port of the micro-controller, while the power to the actuator is provided by an external 12V power-supply. The *OpenCM 9.04* micro-controller runs a node with a subscriber for set-points and a publisher for the actuator feedback. The feedback is collected from the control table of the *XM430W-210T* [28] for each execution loop. The controller-node in the external computer intercepts the feedback through a subscriber and continuously publishes the desired position (θ_d) governed by the following equation:

$$\theta_d = Asin(2\pi(t + n\xi)),\tag{1}$$

where t is the time of the ROS-clock, n is the index of the joint to be controlled, and ξ is the spatial frequency. The results of this experiment are depicted in Figure 14, where θ_d denotes the desired position of the actuator, θ_{real} is the actual position of the XM430W-210, and θ_{sim} is the actual position of the simulated joint in Gazebo. The findings of the experiments show a slight delay in terms of the feedback from the *XM430W-210T* actuator. This is to be expected, as the *OpenCM 9.04* subscribes and publishes the set-point, waits for a return feedback signal from the *XM430W-210T* and publishes the feedback intercepted by the controller-node. To characterise the entity of this delay, time measurements were collected during the proposed experiment. In particular, the delay between the instant when data are sent from the controller-node and the time when the corresponding feedback is received was monitored over time. The results are shown in Figure 15. The maximum value is 0.0280 s, the minimum value is 0.0080 s, and the estimated average delay is 0.0152 s.

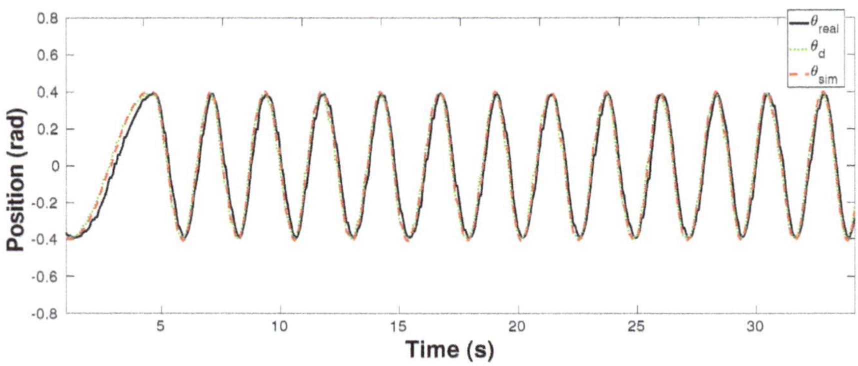

Figure 14. Matlab-plot showing the setpoint θ_d, the joint position feedback from simulation θ_{sim}, and the feedback from a *XM430W-210T* actuator θ_{real}.

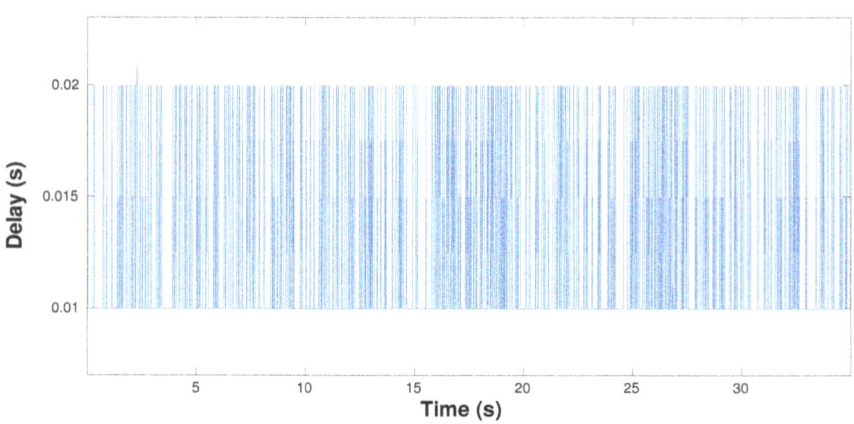

Figure 15. Time plot showing the delay between the instant when data is sent from the controller-node and the time when the corresponding feedback is received. The maximum value is 0.0210 s, the minimum value is 0.0100 s, and the estimated average delay is 0.0129 s.

An additional simulation was performed to highlight the behaviour of the proposed SEA of *Serpens*. As shown in Figure 16, a scenario of a terrain cluttered with cylindrical objects is simulated. The entire body of *Serpens* is constrained by obstacles. The same input signal as outlined in Equation (1) for the desired position (θ_d) is adopted to control the joint module close to the head, as highlighted in Figure 16. The oscillatory motion of the joint module determines collisions of the corresponding link with the adjacent obstacles. These collisions are accommodated through the high level of compliance offered by the SEA of *Serpens*. The motor position θ_m is allowed movement through passive-compliance despite the load position θ_l being blocked by external obstacles. The consequential deviation over time between the motor gear position and the spring reference position is shown in Figure 17.

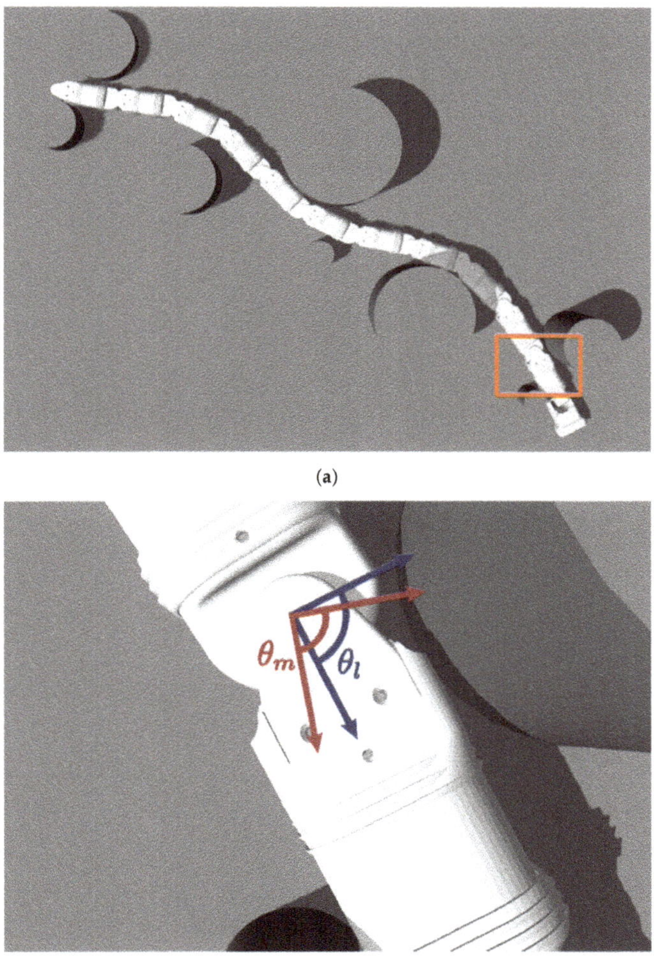

(a)

(b)

Figure 16. The body of *Serpens* is constrained by cylindrical obstacles. (**a**) The same input signal as outlined in Equation (1) for the desired position (θ_d) is adopted to control the highlighted joint module close to the head. (**b**) A zoomed view of one of the joint modules while colliding with obstacles. The motor position θ_m is allowed movement through passive-compliance despite of the load position θ_l being blocked by external obstacles.

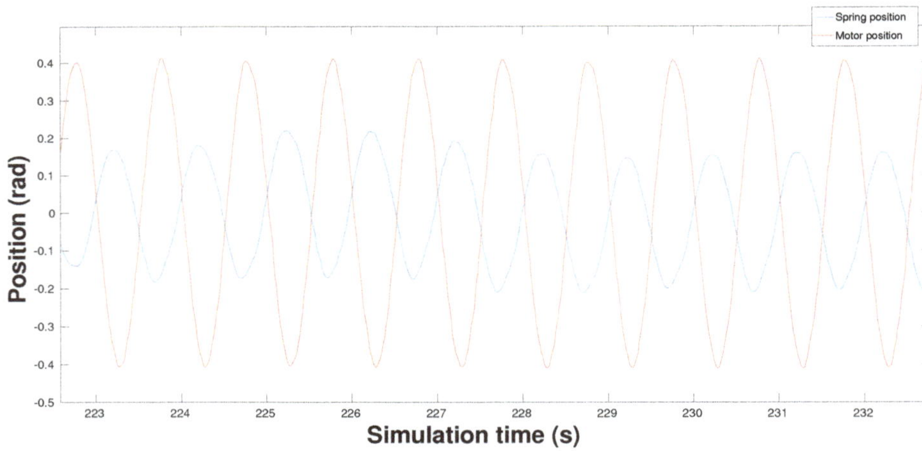

Figure 17. The deviation over time between the motor gear position and the spring reference position.

6. Conclusions and Future Work

Serpens, a low-cost snake robot with elastic joints, torque-controlled actuators and a screw-less assembly mechanism, is presented in this paper based on a modular design and the use of the Robot Operating System (ROS) [12]. The design of the robot relies exclusively on low-cost commercial-off-the-shelf (COTS) components. Fused Deposition Modelling (FDM) manufacturing technology is adopted for 3D-printing the robot modules with *polylactic acid* (PLA), thus making the rapid-prototyping process very fast and economical. A screw-less assembly mechanism makes it possible to assemble the modules and reconfigure the robot in a very reliable, fast and robust manner. A low-cost sensing approach is adopted to allow for torque sensing at the joint level, sensitive collision detection and joint compliant control. These characteristics make *Serpens* very suitable for the interaction with unmapped and dynamic environments or for traversing terrains cluttered with obstacles. The system architecture also follows the concept of modularity on both the software and hardware sides. Each module is independent, being controlled by a self-reliant controller board. The choice of ROS for the implementation of the control framework enables researchers to develop different control algorithms for perception-driven obstacle-aided locomotion (POAL) in a simulated environment with Gazebo. This integration makes the development of control algorithms safe, rapid and efficient. Experimental and simulation results are presented to illustrate the potential of the proposed design.

As future work, the design of reliable low-level control algorithms for the proposed elastic joints will be investigated. Indeed, the design of robust and effective low-level control approaches is essential to enable the achievement of POAL for real-world applications. To achieve this, the current low-level software architecture of *Serpens* must be complemented with a hierarchical organisation by considering the standard functions and capabilities of guidance, navigation, and control (GNC) [37]. Regarding the mechanical design of *Serpens*, the possibility of testing polymers or elastomers and comparing the compliance with our current design based on the use of mechanical springs will also be considered in the future. Moreover, the possibility for *Serpens* to locomote in applications where the gap is narrower than the width of the robot body, i.e., within narrow vertical pipes or walls, will also be explored. More intensive heat dissipation tests are required for practical applications.

Author Contributions: Conceptualization, F.S.; Methodology, F.S.; Software, F.S. and E.H.; Validation, F.S., E.H. and S.L.A.; Formal Analysis, F.S.; Investigation, F.S., E.H., P.A.S., S.L.A.; Resources, F.S.; Data Curation, F.S., E.H., P.A.S., S.L.A.; Writing—Original Draft Preparation, F.S., E.H., P.A.S., S.L.A.; Writing—Review & Editing, F.S.; Visualization, F.S., E.H., P.A.S., S.L.A.; Supervision, F.S.; Project Administration, F.S.; Funding Acquisition, F.S.

Acknowledgments: The authors gratefully acknowledge the contribution of Richard Thue in the 3D-printing prototyping process of this work. The authors also thank Kiran Raja, John Mulholland, Tuan Minh Hua and Thong Ho Sy.

Funding: This work was supported by the Dept. of Science and Industry systems, University of South-Eastern Norway (USN), project title "Secure Multi-sensor Autonomous RoboTs and surveillance operations for Search And Rescue (SMART-SAR) operations in smart buildings".

Conflicts of Interest: The authors declare no conflict of interest.

Abbreviations

The following abbreviations are used in this manuscript:

POAL	perception-driven obstacle-aided locomotion
SEA	series elastic actuator
ROS	Robot Operating System
FDM	Fused Deposition Modelling

References

1. Marvi, H.; Gong, C.; Gravish, N.; Astley, H.; Travers, M.; Hatton, R.L.; Mendelson, J.R.; Choset, H.; Hu, D.L.; Goldman, D.I. Sidewinding with minimal slip: Snake and robot ascent of sandy slopes. *Science* **2014**, *346*, 224–229. [CrossRef] [PubMed]
2. Sanfilippo, F.; Stavdahl, Ø.; Liljebäck, P. SnakeSIM: A ROS-based rapid-prototyping framework for perception-driven obstacle-aided locomotion of snake robots. In Proceedings of the IEEE International Conference on Robotics and Biomimetics (ROBIO), Macau, China, 5–8 December 2017; pp. 1226–1231.
3. Sanfilippo, F.; Stavdahl, Ø.; Liljebäck, P. SnakeSIM: A ROS-based control and simulation framework for perception-driven obstacle-aided locomotion of snake robots. *Artif. Life Robot.* **2018**, *23*, 449–458 [CrossRef]
4. Sanfilippo, F.; Azpiazu, J.; Marafioti, G.; Transeth, A.A.; Stavdahl, Ø.; Liljebäck, P. A review on perception-driven obstacle-aided locomotion for snake robots. In Proceedings of the 14th International Conference on Control, Automation, Robotics and Vision (ICARCV), Phuket, Thailand, 13–15 November 2016; pp. 1–7.
5. Sanfilippo, F.; Azpiazu, J.; Marafioti, G.; Transeth, A.A.; Stavdahl, Ø.; Liljebäck, P. Perception-driven obstacle-aided locomotion for snake robots: The state of the art, challenges and possibilities. *Appl. Sci.* **2017**, *7*, 336. [CrossRef]
6. Sanfilippo, F.; Stavdahl, Ø.; Marafioti, G.; Transeth, A.A.; Liljebäck, P. Virtual functional segmentation of snake robots for perception-driven obstacle-aided locomotion In Proceedings of the IEEE International Conference on Robotics and Biomimetics (ROBIO), Qingdao, China, 3–7 December 2016; pp. 1845–1851.
7. Eitel, E. The Rise of Soft Robots and the Actuators That Drive Them. Available online: https://www.machinedesign.com/robotics/rise-soft-robots-and-actuators-drive-them (accessed on 15 November 2018).
8. Haddadin, S.; Mansfeld, N.; Albu-Schäffer, A. Rigid vs. elastic actuation: Requirements & performance. In Proceedings of the IEEE/RSJ International Conference on Intelligent Robots and Systems (IROS), Algarve, Portugal, 7–12 October 2012; pp. 5097–5104.
9. Rollinson, D.; Bilgen, Y.; Brown, B.; Enner, F.; Ford, S.; Layton, C.; Rembisz, J.; Schwerin, M.; Willig, A.; Velagapudi, P.; et al. Design and architecture of a series elastic snake robot. In Proceedings of the IEEE/RSJ International Conference on Intelligent Robots and Systems (IROS 2014), Chicago, IL, USA, 14–18 September 2014; pp. 4630–4636.
10. Rollinson, D.; Ford, S.; Brown, B.; Choset, H. Design and modeling of a series elastic element for snake robots. In Proceedings of the ASME 2013 Dynamic Systems and Control Conference. American Society of Mechanical Engineers, Palo Alto, CA, USA, 21–23 October 2013; p. V001T08A002.
11. Lipson, H.; Kurman, M. *Fabricated: The New World of 3D Printing*; John Wiley & Sons: New York, NY, USA, 2013.
12. Quigley, M.; Conley, K.; Gerkey, B.; Faust, J.; Foote, T.; Leibs, J.; Wheeler, R.; Ng, A.Y. ROS: An open-source Robot Operating System. In Proceedings of the IEEE International Conference on Robotics and Automation (ICRA), Kobe, Japan, 12–17 May 2009; Volume 3, p. 5.

13. Liljeback, P.; Pettersen, K.Y.; Stavdahl, Ø.; Gravdahl, J.T. Snake robot locomotion in environments with obstacles. *IEEE/ASME Trans. Mech.* **2012**, *17*, 1158–1169. [CrossRef]
14. Hirose, S. *Biologically Inspired Robots: Snake-Like Locomotors and Manipulators*; Oxford University Press: Oxford, UK, 1993.
15. Bayraktaroglu, Z.Y. Snake-like locomotion: Experimentations with a biologically inspired wheel-less snake robot. *Mech. Mach. Theory* **2009**, *44*, 591–602. [CrossRef]
16. Takaoka, S.; Yamada, H.; Hirose, S. Snake-like active wheel robot ACM-R4. 1 with joint torque sensor and limiter. In Proceedings of the IEEE/RSJ International Conference on Intelligent Robots and Systems (IROS), San Francisco, CA, USA, 25–30 September 2011; pp. 1081–1086.
17. Liljebäck, P.; Stavdahl, Ø.; Pettersen, K.Y.; Gravdahl, J.T. A modular and waterproof snake robot joint mechanism with a novel force/torque sensor. In Proceedings of the 2012 IEEE/RSJ International Conference on Intelligent Robots and Systems (IROS), Algarve, Portugal, 7–12 October 2012; pp. 4898–4905.
18. Liljebäck, P.; Stavdahl, Ø.; Pettersen, K.Y.; Gravdahl, J.T. Mamba-A waterproof snake robot with tactile sensing. In Proceedings of the IEEE/RSJ International Conference on Intelligent Robots and Systems (IROS 2014), Chicago, IL, USA, 14–18 September 2014; pp. 294–301.
19. Pratt, G.A.; Williamson, M.M. Series elastic actuators. In Proceedings of the IEEE/RSJ International Conference on Intelligent Robots and Systems, Pittsburgh, PA, USA, 5–9 August 1995; Volume 1, pp. 399–406.
20. Robinson, D.W.; Pratt, J.E.; Paluska, D.J.; Pratt, G.A. Series elastic actuator development for a biomimetic walking robot. In Proceedings of the IEEE/ASME International Conference on Advanced Intelligent Mechatronics, Atlanta, GA, USA, 19–23 September 1999; pp. 561–568.
21. Rouse, E.J.; Mooney, L.M.; Martinez-Villalpando, E.C.; Herr, H.M. Clutchable series-elastic actuator: Design of a robotic knee prosthesis for minimum energy consumption. In Proceedings of the International Conference on Rehabilitation Robotics (ICORR 2013), Bellevue, WA, USA, 20–24 June 2013.
22. Paine, N.; Mehling, J.S.; Holley, J.; Radford, N.A.; Johnson, G.; Fok, C.L.; Sentis, L. Actuator control for the NASA-JSC Valkyrie humanoid robot: A decoupled dynamics approach for torque control of series elastic robots. *J. Field Robot.* **2015**, *32*, 378–396. [CrossRef]
23. Nguyen, M.N.; Tran, D.T.; Ahn, K.K. Robust position and vibration control of an electrohydraulic series elastic manipulator against disturbance generated by a variable stiffness actuator. *Mechatronics* **2018**, *52*, 22–35. [CrossRef]
24. Rollinson, D.; Alwala, K.V.; Zevallos, N.; Choset, H. Torque control strategies for snake robots. In Proceedings of the IEEE/RSJ International Conference on Intelligent Robots and Systems (IROS 2014), Chicago, IL, USA, 14–18 September 2014; pp. 1093–1099.
25. Intel. Intel RealSense D435. Available online: https://click.intel.com/intelr-realsensetm-depth-camera-d435.html (accessed on 15 November 2018).
26. Gonzalez-Gomez, J.; Zhang, H.; Boemo, E. Locomotion principles of 1D topology pitch and pitch-yaw-connecting modular robots. In *Bioinspiration and Robotics Walking and Climbing Robots*; InTech: Vienna, Austria, 2007.
27. Robinson, D.W. Design and Analysis of Series Elasticity in Closed-Loop Actuator Force Control. Ph.D. Thesis, Massachusetts Institute of Technology, Cambridge, MA, USA, 2000.
28. DYNAMIXEL. XM430-W210T. Available online: http://support.robotis.com/en/product/actuator/dynamixel_x/xm_series/xm430-w210.htm (accessed on 15 November 2018).
29. Tsardoulias, E.; Mitkas, P. Robotic frameworks, architectures and middleware comparison. *arXiv* **2017**, arXiv:1711.06842.
30. Koenig, N.; Howard, A. Design and use paradigms for gazebo, an open-source multi-robot simulator. In Proceedings of the IEEE/RSJ International Conference on Intelligent Robots and Systems (IROS), Sendai, Japan, 28 September–2 October 2004; Volume 3, pp. 2149–2154.
31. Kam, H.R.; Lee, S.H.; Park, T.; Kim, C.H. RViz: A toolkit for real domain data visualization. *Telecommun. Syst.* **2015**, *60*, 337–345. [CrossRef]
32. Arduino. Arduino, an Open-Source Electronics Prototyping Platform. Available online: http://arduino.cc/ (accessed on 15 November 2018).
33. RoLin. RoLin Rotary Incremental Encoder. Available online: https://www.rls.si/en/products/rotary-magnetic-encoders/rolin-rotary-incremental-magnetic-encoder-system (accessed on 15 November 2018).

34. Robot Operating System (ROS). realsense2_camera. Available online: http://wiki.ros.org/realsense2_camera (accessed on 15 November 2018).
35. Open Source Robotics Foundation. Tutorial: Using a URDF in Gazebo. Available online: http://gazebosim.org/tutorials/?tut=ros_urdf (accessed on 15 November 2018).
36. Robot Operating System (ROS). Rosserial. Available online: http://wiki.ros.org/rosserial (accessed on 15 November 2018).
37. Kendoul, F. Towards a unified framework for uas autonomy and technology readiness assessment (atra). In *Autonomous Control Systems and Vehicles*; Springer: Berlin, Germany, 2013; pp. 55–71.

applied sciences

MDPI

Article

Head-Raising of Snake Robots Based on a Predefined Spiral Curve Method

Xiaobo Zhang [1,2,3], Jinguo Liu [1,2,*], Zhaojie Ju [1,2,4] and Chenguang Yang [5]

1 State Key Laboratory of Robotics, Shenyang Institute of Automation, Chinese Academy of Sciences, Shenyang 110016, China; zhangxiaobo@sia.cn (X.Z.); zhaojie.ju@port.ac.uk (Z.J.)
2 Institutes for Robotics and Intelligent Manufacturing, Chinese Academy of Sciences, Shenyang 110016, China
3 University of Chinese Academy of Sciences, Beijing 100049, China
4 Intelligent Systems & Biomedical Robotics, School of Creative Technologies, University of Portsmouth, Portsmouth P01 3HE, UK
5 Zienkiewicz Centre for Computational Engineering, Swansea University, Swansea SA1 8EN, UK; cyang@ieee.org
* Correspondence: liujinguo@sia.cn; Tel.: +86-135-0407-9204

Received: 12 September 2018; Accepted: 17 October 2018; Published: 23 October 2018

Abstract: A snake robot has to raise its head to acquire a wide visual space for planning complex tasks such as inspecting unknown environments, tracking a flying object and acting as a manipulator with its raising part. However, only a few researchers currently focus on analyzing the head-raising motion of snake robots. Thus, a predefined spiral curve method is proposed for the head-raising motion of such robots. First, the expression of the predefined spiral curve is designed. Second, with the curve and a line segments model of a snake robot, a shape-fitting algorithm is developed for constraining the robot's macro shape. Third, the coordinate system of the line segments model of the robot is established. Then, phase-shifting and angle-solving algorithms are developed to obtain the angle sequences of roll, pitch, and yaw during the head-raising motion. Finally, the head-raising motion is simulated using the angle sequences to validate the feasibility of this method.

Keywords: snake robots; head-raising; shape-fitting; phase-shifting; spiral curve

1. Introduction

A snake robot is a mobile robot with a long and slender body and plays a crucial role in search and rescue operations, inspection and maintenance [1]. Controlling such robots is difficult, and a review of their modeling, implementation, and control can be found in [2]. Generally, for given tasks such as searching a trapped person in disaster areas, tracking a flying object and repairing a damaged pipe as a manipulator, the snake robot has to move its body with different motion modes, and head-raising motion is one of them. Snake robots' common motion modes include serpentine [3,4], traveling wave [5,6], concertina [7,8], and sidewinding [9,10] locomotion. Additionally, some researchers sought to find other motion modes such as fusion gait [11] and obstacle-aided locomotion [12]. However, these motion modes provide limited visual information. Consequently, snake robots have to raise their heads to observe their surroundings and perform their tasks. Therefore, the motion planning for head-raising of a snake robot is our research object in this paper.

To the best of our knowledge, a few researches on head-raising motion of snake robots have been done, and they can be divided into two types: 2D (two dimensional) head-raising and 3D (three-dimensional) head-raising. This study aims at 3D head-raising. The concepts of 2D and 3D head-raising are illustrated in Figure 1. It should be noted that label 2 and 4 are line segments models of snake robots, which will be utilized in subsequent chapters for simulations. 2D head-raising can be found in [13], the authors proposed an improving serpentine input function to achieve the head-raising motion of a snake robot and analyzed the zero-moment point to assess the stability of the robot. As for 3D head-raising [14,15], to realize it is more complicated. In [14], the works were based on an assumption that the head part of the robot is not contact with the ground and regarded as a manipulator, which is able to achieve several sub-tasks in 3D environment. Therefore, essentially speaking, the researchers mainly considered tracking the trajectory of a snake robot's head rather than the head-raising process. Inspired by the work made by Tanka et al, we currently are working on trajectory tracking of the snake robot head by improving our model. There still exists problems to be solved and a video about the trajectory tracking (Concept_of_trajectory_tracking.avi) can be seen in Supplementary Materials. In [15], the authors regarded a snake robot as a serially linked floating-base robot and divided it into base and active modules. The base modules were used to construct the support polygon for maintaining stability during the head-raising process and can be changed to active modules to allow the head to stretch to a given position and orientation. The trajectory functions of individual joints were set as parameterized cubic spline curves, and the head-raising motion was achieved by optimizing these parameters according to some criterions. In summary, existing works of head-raising attempt to find the input functions of individual joints. However, in this study, the input functions (angle sequences with respect to time) are unknown, and they are obtained by shape-fitting and phase-shifting algorithms, which will be introduced in the subsequent chapters.

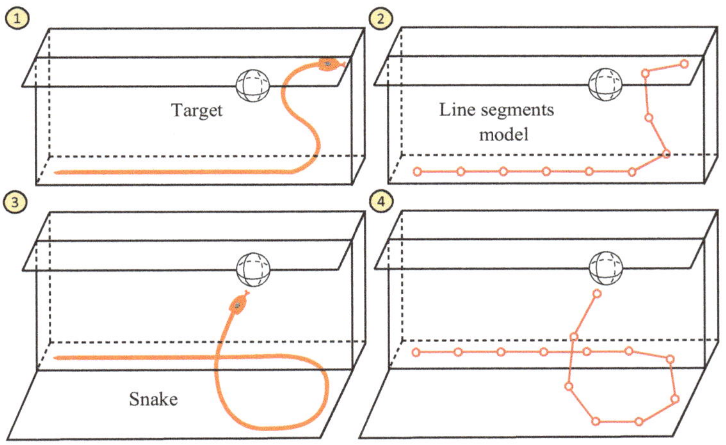

Figure 1. The concepts of 2D and 3D head-raising. Label 1 and 2 represent 2D head-raising, and label 3 and 4 show 3D head-raising.

Achieving motion control for a snake robot is difficult because of the kinematic redundancy, which can be solved by the shape control method. Motion control based on a spiral curve belongs to the category of the shape control. It is necessary to introduce the concept of shape control method. This method can not be described in theory, and it is a kind of idea to control the motion of snake robots or hyper-redundant manipulators. The idea of shape control method is employing a geometric curve to constrain a robot's configuration by fitting the continuous curve and discrete link model as closely as possible. If change the shape of the geometric curve with respect to time, the robot can achieve continuous motion. The key points to use this method include two aspects. First, how to find a proper curve to describe the shape of robots should be considered. Second, if the curve is obtained, there should be an approach to accomplish the fitting process. Examples of using shape control method can be found in [16–25]. In [16], the authors presented a geometric spline method for the shape control of planar manipulators with hyper-degrees of freedom. With the fitting of the robotic macro configuration and the referent spline curve, real-time control could be easily accomplished.

In [17], the authors presented locomotion control using a phase oscillator network, which can produce a smooth transition of the body shape of a snake robot to avoid obstacles. In [18], the researchers introduced a framework to control the shape of robots, and the framework can be divided into four steps, namely, defining the shape control point, using the shape control point to define the Bezier curves, fitting the curves and the shape of the robot, and changing the parameters of the Bezier curves to control the movement of the snake robot according to the desired gait. In [19], the authors used the shape control point method to control the position and orientation of the head of a robot. In [20], the researchers presented a planar manipulator that moves according to a variable-structure polygon function produced by neural networks. In [21], the authors introduced and used a modal approach to generate a backbone curve and thus capture a hyper-redundant manipulator's macroscopic geometric feature. The backbone curve was represented by the linear combination of a set of shape functions with the appropriate coefficients. Then, inverse kinematics reduced to the solution of these coefficients to avoid the heavy computation required by the Jacobian pseudo-inverse method. In [22], the researchers analyzed the curvature of the sidewinding locomotion of a snake robot and developed algorithms to fit the robot's shape to a continuous curve. In [23], the authors proposed the annealed chain fitting algorithm and keyframe wave extraction algorithm; the former maps the shape of the snake robot to a continuous backbone curve and obtains a set of joint angles, whereas the latter obtains a sequence of backbone curves. A similar modeling strategy can also be found in [24,25]. In [25], the researchers considered dynamics during the shape control of hyper-redundant manipulators.

The goal of this paper is to propose a new approach to raise a snake robot's head in 3D environments. This mechanism allows the robot to obtain a 360° field of view and a large working space. In this paper, inspired by the shape control method, a head-raising motion control method based on a predefined spiral curve is proposed. The robot is represented by several line segments, and the corresponding local coordinate system is established. Every line segment corresponds to a module of the snake robot, and the connection point between adjacent line segments corresponds to a joint involving three degrees of freedom, namely, roll, pitch, and yaw angles, which can be obtained by matrix transformation of the local coordinate system. The predefined curve is introduced to constrain the macro shape of the snake robot, and a phase-shifting algorithm is developed to drive the snake robot along the curve. During the phase-shifting process, the angle trajectories for the head-raising motion in a given time can be obtained. With the angle trajectories, the head-raising motion can be achieved. The contributions of this work are summarized by following remarks:

1. A new shape control curve, the predefined spiral curve, is proposed and it is utilized for 3D head-raising of a snake robot;
2. A shape-fitting algorithm is developed for adhering the line segments model of the snake robot to the predefined spiral curve;
3. Establishment rules of coordinate system are given for line segments model of the snake robot;
4. A phase-shifting and an angle-solving algorithms are presented for obtaining angle trajectories used during head-raising motion.

The remainder of this paper is organized as follows. Section 1 reviews previous researches, analyses the necessity for head-raising and introduces our aim, which is raising a snake robot's head based on a predefined spiral curve in 3D environments. Section 2 introduces the modeling of the head-raising motion; the mathematical expression of the predefined spiral curve is discussed in Subsection 2.1, the shape-fitting and phase-shifting methods in Subsection 2.2, and the establishment of the coordinate system and angle-solving algorithm in Subsection 2.3. Section 3 explains the simulation of the head-raising motion. Section 4 presents the conclusions.

2. Modeling of Head-Raising Motion

In this section, a predefined spiral curve and a shape-fitting method are introduced to constrain the macro shape of the snake robot, which is composed of serial links represented by serial line segments. However, the serial line segments do not contain the complete information of the snake robot. Therefore, coordinate system is introduced. The snake robot has to move through the predefined spiral curve from initial posture to final posture, and that process is called phase-shifting. With the proposed phase-shifting and angle-solving algorithms, the angle sequences (or angle trajectories) can be used to accomplish the head-raising motion.

2.1. Predefined Spiral Curve

From the biological perspective, a snake's head-raising movement is irregular, and it is difficult to completely describe its shape using a mathematical model. Based on our observations, the shape of a snake after raising its head is similar to that of a spiral curve, as described in Figure 2. In addition, the spiral curve can ensure the stability of head-raising. Thus, describing the head-raising movement of a snake robot with a spiral curve is reasonable.

$$\begin{aligned} x_s(t) &= at\sin(t+\varphi_0) \\ y_s(t) &= at\cos(t+\varphi_0) \\ z_s(t) &= ct \end{aligned} \quad (1)$$

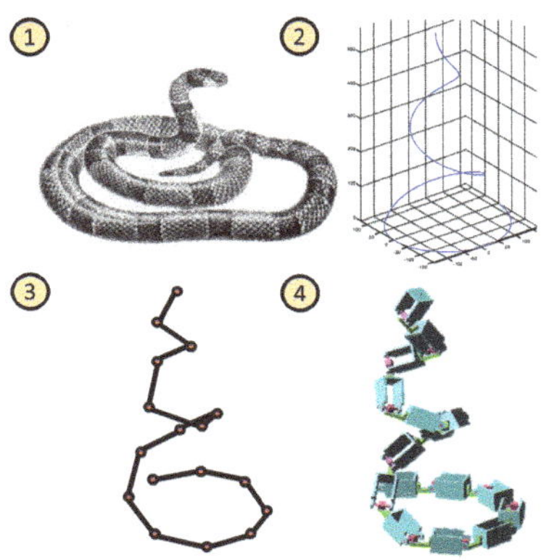

Figure 2. The similar macro shapes of a snake (label 1), a spiral curve (label 2), the line segments model after the shape-fitting algorithm (label 3) and the simplified simulation model (label 4).

A regular spiral curve (Figure 3a) can be expressed as Equation (1). It cannot be utilized to guide the movement of the robot, and has to be modified, because it cannot ensure the stability of head-raising. Then, the predefined spiral curve (Figure 3b) is designed and its mathematical description satisfies Equation (2), where $x_l(t)$, $y_l(t)$, and $z_l(t)$ represent the line part; $x_s(t)$, $y_s(t)$, and $z_s(t)$ describe the spiral curve part; and a, b, and c are the adjustment coefficients of the spiral curve. The amplitudes of $x_s(t)$ and $y_s(t)$ change equally when a is changed and unequally when b is changed. The amplitude of $z_s(t)$ is influenced by c. φ_0 is the initial phase of the sine and cosine functions, which correspond to the slope of the tangent vector for the initial point. n_c is the cycle number of the spiral curve. l_{link} and n are the length and number of the snake robot modules, respectively. t is the independent variable divided into three intervals, which correspond to the line part (defined by LP, interval is $(0, \hat{t}_1)$); the base of the spiral curve part, which is in contact with the ground (defined by BS, interval is $(\hat{t}_1, \hat{t}_2)$); and the rest of the spiral curve (defined by RS, interval is $(\hat{t}_2, \hat{t}_3)$). φ_{base} is the phase value of the second interval.

$$
\begin{array}{l}
x_l(t) = x_s(\hat{t}_1) \quad t \in (0, \hat{t}_1) \\
y_l(t) = y_s(\hat{t}_1) - t + \ \hat{t}_1 \ t \in (0, \hat{t}_1) \\
z_l(t) = 0 \quad t \in (0, \hat{t}_1) \\
x_s(t) = ab(\hat{t}_3 - t)\sin(t + \varphi_0) \quad t \in (\hat{t}_1, \hat{t}_2) \\
y_s(t) = a(\hat{t}_3 - t)\cos\,(t + \varphi_0)_1 \quad t \in (\hat{t}_1, t_2) \\
z_s(t) = 0 \quad t \in (\hat{t}_1, \hat{t}_2) \\
x_s(t) = ab(\hat{t}_3 - t)\sin(t + \varphi_0) \quad t \in (\hat{t}_2, \hat{t}_3) \\
y_s(t) = a(\hat{t}_3 - t)\cos(t + \varphi_0) \quad t \in (\hat{t}_2, \hat{t}_3) \\
z_s(t) = ct - c\hat{t}_2 \quad t \in (\hat{t}_2, \hat{t}_3) \\
\hat{t}_1 = nl_{link} \\
\hat{t}_2 = \hat{t}_1 + \varphi_{base} \\
\hat{t}_3 = \hat{t}_1 + 2n_c\pi
\end{array} \tag{2}
$$

The predefined spiral curve is composed of a line part and a spiral curve part, as shown in Figure 3b. The line and spiral curve parts constrain the initial and final head-raising postures of the snake robot, respectively. As shown in Figure 3, compared with the regular spiral curve expressed in Equation (1), the predefined spiral curve expressed in Equation (2) introduces parameter b to ensure that $x_s(t)$ and $y_s(t)$ change unequally, interval $(0, \hat{t}_1)$ to construct LP, $(\hat{t}_1, \hat{t}_2)$ to construct BS, and $(\hat{t}_2, \hat{t}_3)$ to construct RS. Additionally, by substituting item $(\hat{t}_3 - t)$ for t, it can be ensured that $x_s(t)$ and $y_s(t)$ decrease as $z_s(t)$ increases. Obviously, the predefined spiral curve is appropriate for ensuring stability during head-raising.

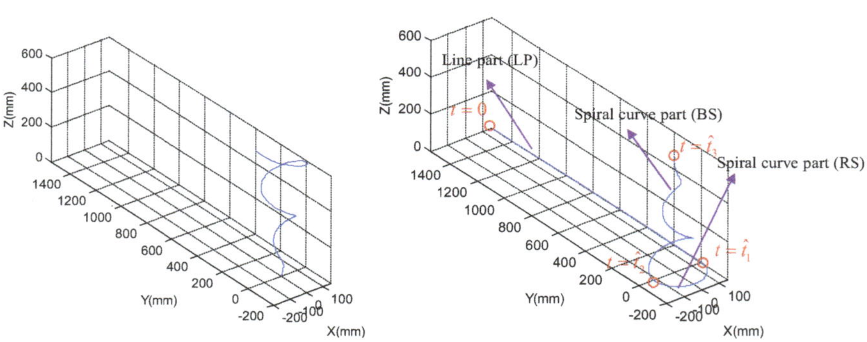

Figure 3. Spiral curves: (**a**) the regular spiral curve, and (**b**) the predefined spiral curve used in this study.

2.2. *Shape-Fitting and Phase-Shifting Methods*

In this section, a shape-fitting and phase-shifting algorithms are developed, and the former ensures that the snake robot adhere to the predefined spiral curve and the latter drive the robot move along with the curve. Figure 4 shows the concepts of these methods. In Figure 4, two links of robot are shown; the red one represents the robot posture at time t_i and the black one corresponds to the robot posture at time t_{i+1}. The robot moves from posture at time t_i to posture at time t_{i+1}, and this process is called phase-shifting. Two links contain three points; that is, the length between adjacent points equals the length of the robot links. Therefore, the snake links are calculated by point sets and represented by line segments. This process is called shape-fitting.

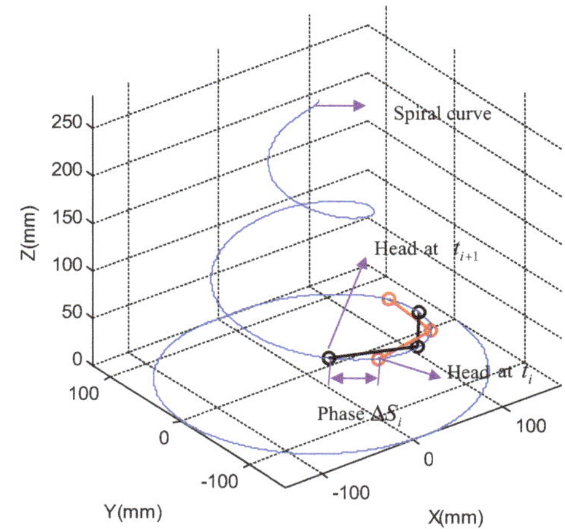

Figure 4. Concepts of shape-fitting and phase-shifting.

The details of how the shape-fitting and phase-shifting algorithms work are described in the following steps.

Step 1: Initialization

The parameters $a, b, c, n_c, n, l_{link}, \hat{t}_1, \hat{t}_2, \hat{t}_3, \varphi_0$, and φ_{base} are initialized. The following constraint equation is used to ensure that line segments model of the robot do not get away from the ground.

$$S_1 + S_2 = \int_{t=\hat{t}_1}^{t=\hat{t}_2} \sqrt{\dot{x}_s(t)^2 + \dot{y}_s(t)^2 + \dot{z}_s(t)^2} dt + \int_{t=\hat{t}_2}^{t=\hat{t}_3} \sqrt{\dot{x}_s(t)^2 + \dot{y}_s(t)^2 + \dot{z}_s(t)^2} dt < n l_{link} \quad (3)$$

where S_1 and S_2 represent the arc lengths of BS and RS, respectively.

Step 2: Determination of basic posture

The snake robot is represented by line segments and the corresponding coordinate system. This section describes how the line segments of the basic posture are determined. The succeeding portion discusses the rules of building the coordinate system.

The point sets used to construct the line segments are defined as follows:

$$\begin{aligned} {}^iP_{snk} &= [{}^iP_1, {}^iP_2, \ldots, {}^iP_j, \ldots, {}^iP_{n+1}] \quad j = 1, 2, \ldots, n+1 \quad i = 1, 2, \ldots, n_{step} \\ {}^iP_j &= ({}^ix_{snk\_j}, {}^iy_{snk\_j}, {}^iz_{snk\_j}) \end{aligned} \quad (4)$$

where ${}^iP_j$ is the position of point j at step i of phase-shifting. n_{step} is the number of phases shifted and satisfies Equation (5).

$$\sum_{i=1}^{i=n_{step}} \Delta S_i = S_1 + S_2 \quad (5)$$

where ΔS_i is the arc length at step i of phase-shifting and defined as "phase" in this study.

Therefore, shape-fitting reduces to find the point set ${}^iP_{snk}$ at step i, and phase-shifting reduces to move all line segments by changing the position of ${}^iP_{snk}$ along with the predefined spiral curve.

The basic posture is considered the initial configuration before the phase-shifting begins, as shown in Figure 5 and calculated as follows.

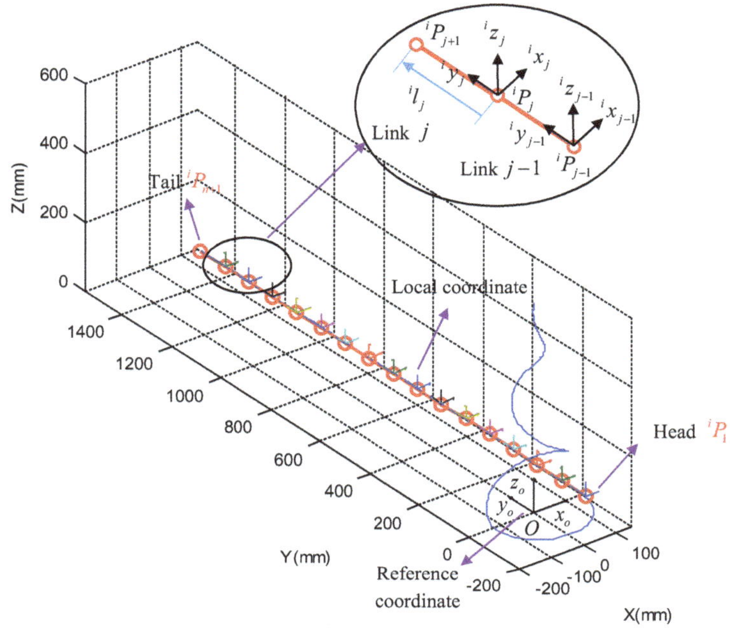

Figure 5. Modeling of the robot's basic posture composed of n line segments and corresponding coordinate system.

We set $i = 1$ to indicate that the line segments are in the basic posture without phase-shifting such that the point sets of the basic posture satisfy Equation (6).

$$
{}^{1}P_{snk} = \begin{bmatrix} (x_s(\hat{t}_1) & y_s(\hat{t}_1) + nl_{link} & 0) \\ (x_s(\hat{t}_1) & y_s(\hat{t}_1) + (n-1)l_{link} & 0) \\ \cdots & \cdots & \cdots \\ (x_s(\hat{t}_1) & y_s(\hat{t}_1) + (n-j+1)l_{link} & 0) \\ \cdots & \cdots & \cdots \\ (x_s(\hat{t}_1) & y_s(\hat{t}_1) & 0) \end{bmatrix}^T \tag{6}
$$

Step 3: One step of phase-shifting

For programming purposes, independent variable t is regarded as a discrete vector t_d and its mathematical description is expressed as follows:

$$
t_d = [0, \Delta t_1, 2\Delta t_1, ..., \hat{t}_1, \hat{t}_1 + \Delta t_2, \hat{t}_1 + 2\Delta t_2, ..., \hat{t}_3] \tag{7}
$$

where Δt_1 and Δt_2 are the step lengths of discretization in intervals $(0, \hat{t}_1)$ and $(\hat{t}_1, \hat{t}_3)$, respectively. After discretization of t, the predefined spiral curve becomes spatial discrete points.

Phase-shifting is achieved by the movement of the snake's head (${}^{i}P_1$) and the refreshing of the positions of the rest of the points (${}^{i}P_2, {}^{i}P_3, \ldots, {}^{i}P_j, \ldots {}^{i}P_{n+1}$) along with the predefined spiral curve. First, given the phase ΔS_i, head ${}^{i}P_1$ changes into a new position ${}^{i+1}P_1$, during which independent variable t is assumed to shift from t_i to t_{i+1}. t_i and ΔS_i are known, and t_{i+1} can be obtained using Equation (8).

$$\Delta S_i = \int_{t=t_i}^{t=t_{i+1}} \sqrt{\dot{x}(t)^2 + \dot{y}(t)^2 + \dot{z}(t)^2}dt \tag{8}$$

Second, we query t_d to identify the element nearest to t_{i+1} and record its index as ${}^{i}Inx_{p_1}$. Finally, the rest of the points (${}^{i}P_2, {}^{i}P_3, \ldots, {}^{i}P_j, \ldots {}^{i}P_{n+1}$) and corresponding indices (${}^{i}Inx_{p_2}, {}^{i}Inx_{p_3}, \ldots, {}^{i}Inx_{p_j}, \ldots, {}^{i}Inx_{p_{n+1}}$) can be obtained by distance judgment going through t_d from index ${}^{i}Inx_{p_1}$ to index 1. Distance judgment is illustrated in Equation (9).

$$l_{link} - eps \leq \left\| {}^{i}P_j - {}^{i}P_{j-1} \right\| \leq l_{link} + eps \tag{9}$$

where eps is utilized to deal with the error between continuous function and discrete approximation.

Step 4: Iteration

Step 3 is one step of phase-shifting from step i to $i+1$. On the basis of Equation (5), all of the values of phase ΔS_i can be computed manually. Then, step 3 is repeated until the head of the snake (${}^{i+1}P_1$) moves to the final point of the discrete predefined spiral curve. Therefore, iteration completes the phase-shifting process from basic posture (${}^{1}P_{snk}$) to final posture (${}^{n_{step}}P_{snk}$) of the snake robot (represented by line segments or point sets without considering the coordinate system in this part).

2.3. Coordinate System and Angle-Solving Algorithm

The line segments cannot contain all the information of the snake robot links or modules. Therefore, we propose an approach on establishing the coordinate system fixed on line segments.

Figure 5 shows the reference coordinate $O_{x_o y_o z_o}$, and its origin is $[0,0,0]$, with x_o-axis $[1,0,0]$, y_o-axis $[0,1,0]$, and z_o-axis $[0,0,1]$. In determining the local coordinates, several rules are needed. Figure 6 shows how these rules are established and illustrates one link at step i during the phase-shifting process. The ${}^{i}y_j$ -axis is calculated as follows:

$${}^{i}y_j = \frac{{}^{i}p_{j+1} - {}^{i}p_j}{\left\| {}^{i}p_{j+1} - {}^{i}p_j \right\|} \tag{10}$$

The ${}^{i}x_j$- and ${}^{i}z_j$-axes are on the plane that is vertical to the link vector shown in the left part of Figure 6. Therefore, the ${}^{i}x_j$- and ${}^{i}z_j$-axes are not unique, and the horizontal plane is introduced to solve this problem. Moving the horizontal plane to point ${}^{i}P_j$, the intersecting line between two planes is used to select the ${}^{i}x_j$-axis, as shown in the right part of Figure 6. The intersecting line has two directions. Thus, the direction is selected by determining whether the angle between ${}^{i}x_j$- and x_o-axes is acute. Afterward, the ${}^{i}z_j$-axis is determined using the right-hand rule. Figure 7 shows the result of the local coordinates during the entire phase-shifting process. The simulation without showing the coordinates is discussed in the subsequent chapter.

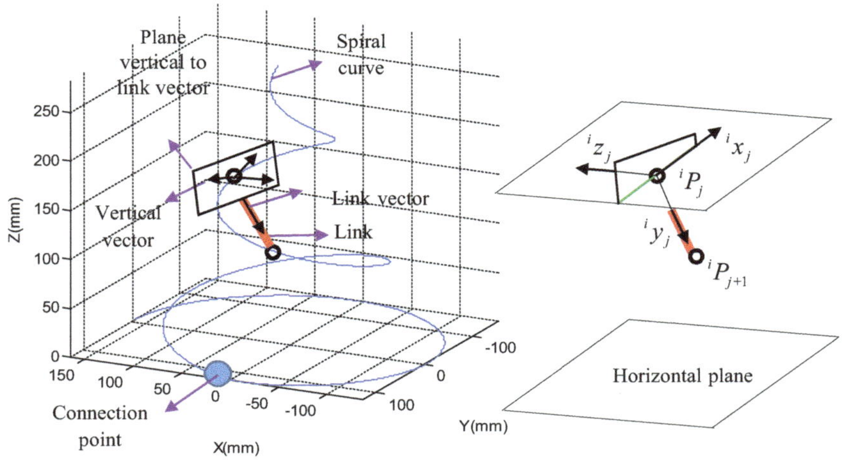

Figure 6. Establishment of local coordinates.

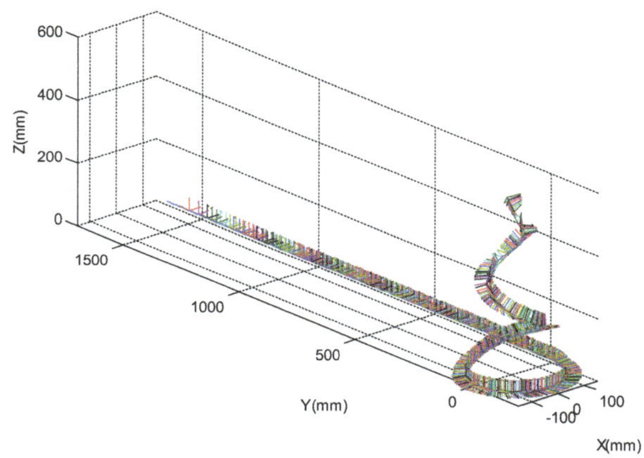

Figure 7. Local coordinates during entire head-raising process.

After the determination of the coordinate system, the relative rotation matrix between adjacent links can be obtained using Equation (11).

$$
\begin{aligned}
&{}^{i}R_{P\_rel_j} = ({}^{i}R_{P_{j+1}})^{-1}({}^{i}R_{P_j}) \quad j = 2,3\ldots,n \quad i = 1,2,\ldots,n_{step} \\
&{}^{i}R_{P_j} = \begin{bmatrix} {}^{i}x_j & {}^{i}y_j & {}^{i}z_j \end{bmatrix} = \begin{bmatrix} r_{11} & r_{12} & r_{13} \\ r_{21} & r_{22} & r_{23} \\ r_{31} & r_{32} & r_{33} \end{bmatrix}
\end{aligned}
\tag{11}
$$

where ${}^{i}R_{P_j} \in R^{3\times3}$ is the orientation matrix of link j relative to the reference coordinate and ${}^{i}R_{P\_rel_j} \in R^{3\times3}$ is the orientation matrix of link j relative to ${}^{i}R_{P_{j+1}}$. Then, the relative roll (${}^{i}\gamma_j$), pitch (${}^{i}\beta_j$), and yaw (${}^{i}\alpha_j$) angle sequences can be calculated using Equation (12).

$$
\begin{aligned}
&{}^{i}\beta_j = A\tan 2(-r_{31}, \sqrt{r_{11}{}^2 + r_{21}{}^2}) \qquad j = 2,3\ldots,n \quad i = 1,2,\ldots,n_{step} \\
&{}^{i}\alpha_j = Atan2(r_{21}/cos({}^{i}\beta_j), r_{11}/cos({}^{i}\beta_j)) \\
&{}^{i}\gamma_j = Atan2(r_{32}/cos({}^{i}\beta_j), r_{33}/cos({}^{i}\beta_j))
\end{aligned}
\tag{12}
$$

The line segments model for the snake robot is abstract, therefore we introduce a simplified Solidworks model for the subsequent simulation in Adams environment, as shown in Figure 8. It is part of Solidworks model and contains three modules. This figure provides us with the line segments model (right part) and Solidworks model (left part), and it is convenient for us to understanding where local coordinates are established and how the roll, pitch, and yaw angles are solved. As mentioned before, a line segment (${}^{i}l_j$) and a local coordinate (${}^{i}P_j - {}^{i}x_j{}^{i}y_j{}^{i}z_j$) represent a snake robot module, and a point (${}^{i}P_j$) represents a joint with three degrees of freedom. The three axes are located at centers of rotating shafts and intersected at the ${}^{i}P_j$. Connecting the ${}^{i}P_j$ and the ${}^{i}P_{j+1}$, the ${}^{i}l_j$ can be determined. Equation (11) and (12) provide us with the solution of roll, pitch and yaw angles. However, it is significant to determine their rotating sequences, which are also illustrated in Figure 8. It should be pointed out that the rotating sequences depend on specific models, otherwise, the rotating sequences and Equation (12) should be modified.

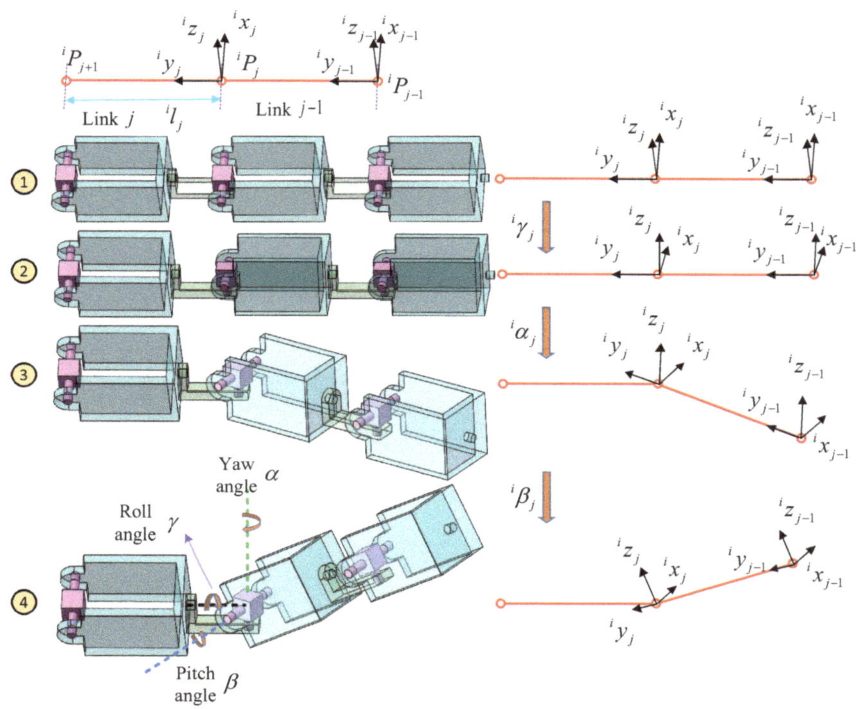

Figure 8. Local coordinates fixed on the simplified Solidworks model and descriptions of roll, pitch and yaw angles. Label 1 corresponds to the initial configuration. Label 2, 3 and 4 describe the configurations after rotating roll angle, yaw angle and pitch angle, respectively.

3. Simulation on Head-Raising Motion

According to the proposed algorithms, the phase-shifting process can be simulated. The parameters of the predefined curve are as follows: $a = 9.7$, $b = 1$, $c = 48.5$, $n_c = 2.5$, $n = 16$, $l_{link} = 97$ (mm), $\hat{t}_1 = 1{,}552$ (rad), $\hat{t}_2 = 1{,}555.1$ (rad), $\hat{t}_3 = 1{,}567.7$ (rad), $\varphi_0 = 1.6022$ (rad), and $\varphi_{base} = pi$ (rad). According to Equation (3), the arc lengths of BS (S_1) and RS (S_2) can be calculated as $S_1 = 431.8891$ (mm) and $S_2 = 1{,}032.6$ (mm). The discrete independent variable is selected as $t_d = [0, 0.1, 0.2, \ldots, \hat{t}_1, \hat{t}_1 + 0.001, \hat{t}_1 + 0.002, \ldots, \hat{t}_3]$ with $\Delta t_1 = 0.1$ and $\Delta t_2 = 0.001$. To facilitate reading, all the symbols utilized in this paper are concluded in Appendix part, as shown in Table A1.

The snake begins to move from basic posture to final posture using the shape-fitting and phase-shifting algorithms. During the phase-shifting process, the phase ΔS_i in every step can be selected manually. In this test, ΔS_i is assumed constant and equal to $(S_1 + S_2)/n_{step}$, where the number of total steps n_{step} equals 500. Table 1 shows the data of ${}^iP_1, {}^iInx_{p_1}, {}^i\alpha_1, {}^i\beta_1$, and ${}^i\gamma_1$ at steps 1, 2, 300, 499, and 500, respectively. Figure 9 shows the phase-shifting process, including the predefined spiral curve (Figure 9a) and snake posture at steps $i = 1$ (Figure 9b), $i = 147$ (Figure 9c), and $i = 500$ (Figure 9d), corresponding to the snake head at $t = \hat{t}_1$, $t = \hat{t}_2$, and $t = \hat{t}_3$, respectively.

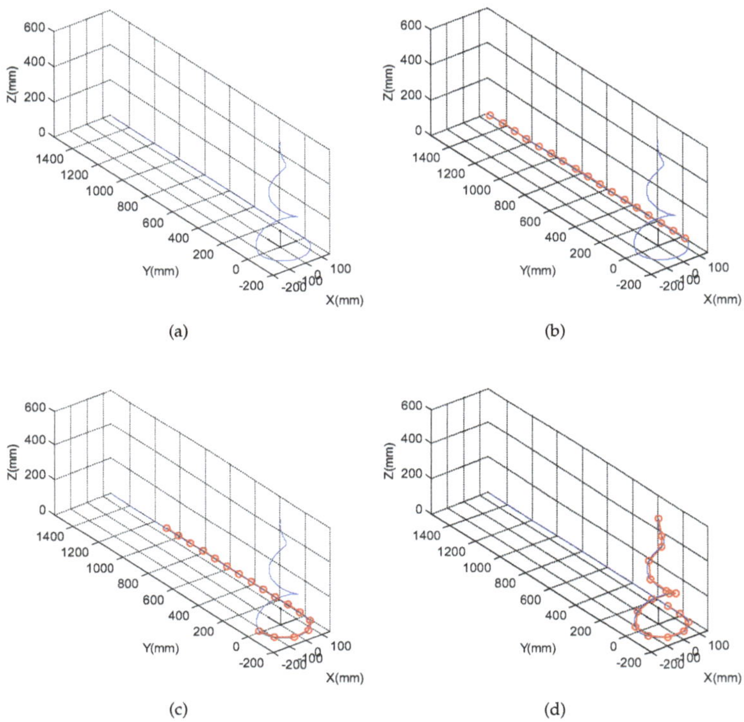

Figure 9. Phase-shifting process: (**a**) predefined spiral curve, (**b**) $i = 1$, $t = \hat{t}_1$, (**c**) $i = 147$, $t = \hat{t}_2$, (**d**) $i = 500$, $t = \hat{t}_3$.

We assume that the head-raising motion time is 50 s, that is, $t_{time} \in (0, 50)$. With Equation (12) and numerical derivation, the roll (${}^i\gamma_j$), pitch (${}^i\beta_j$), and yaw (${}^i\alpha_j$) angle sequences; relative angular velocity (${}^i\dot{\gamma}_j, {}^i\dot{\beta}_j, {}^i\dot{\alpha}_j$); and angular acceleration (${}^i\ddot{\gamma}_j, {}^i\ddot{\beta}_j, {}^i\ddot{\alpha}_j$) can be obtained, as shown in the z-axis of Figure 10. The x- and y-axes represent the joint ID ($ID = 1, 2, \ldots, 15$) and time, respectively. As the snake head moves from $t = \hat{t}_1$ to $t = \hat{t}_2$, the absolute value of ${}^i\gamma_1$ initially increases before the third joint point moves to $t = \hat{t}_1$ and subsequently becomes constant because the first two links are constrained to a semicircle. Meanwhile, the absolute values of ${}^i\alpha_1$ and ${}^i\beta_1$ are zero. Similarly, the motion of the snake head moving from $t = \hat{t}_2$ to $t = \hat{t}_3$ can be divided into two stages. Before the third joint point moves to $t = \hat{t}_2$, the absolute value of ${}^i\gamma_1$ initially decreases in the first stage and then subsequently increases in the second stage. In the meantime, the absolute value of ${}^i\alpha_1$ increases at varying speed. The absolute value of ${}^i\beta_1$ initially increases, subsequently decreases, and finally increases. Unsmooth angles are caused by the unsmooth transition of the piecewise function, which will be improved in future work. Thus, the shapes of the relative velocity and acceleration curve are full of sharp points. Moreover, the curves of the other joint points ($ID = 2, 3, \ldots, 15$) have the same shape with phase lag.

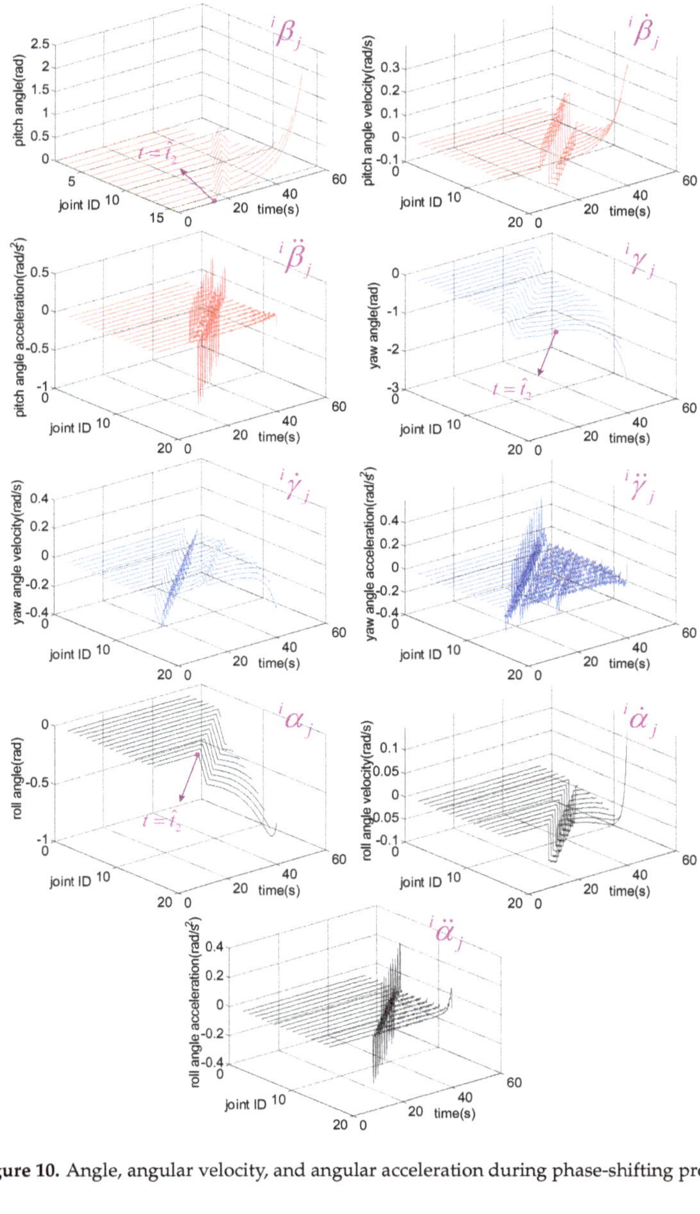

Figure 10. Angle, angular velocity, and angular acceleration during phase-shifting process.

Table 1. Data during phase-shifting process.

Step i	Head ${}^{i}P_1$ (mm)	Index of Head ${}^{i}Inx_{p_1}$	Yaw Angle ${}^{i}\alpha_1$ (rad)	Pitch Angle ${}^{i}\beta_1$ (rad)	Roll Angle ${}^{i}\gamma_1$ (rad)
1	(151.8, −12.8, 0)	15,520	0	0	0
2	(151.3, −15.7, 0)	15,541	−0.0047	0	0
⋮	⋮	⋮	⋮	⋮	⋮
300	(57.3, −61.8, 188.1)	22,544	−0.8964	0.2259	−0.3775
⋮	⋮	⋮	⋮	⋮	⋮
499	(−0.546, 0.015, 606.7)	31,174	−2.5001	2.1308	−0.8348
500	(−0.0036, −0.003, 609.4)	31,230	−2.5309	2.1658	−0.8225

Distinguishing between the phase-shifting and head-raising processes is important. The phase-shifting process is used to obtain the joint trajectory and does not exist in reality if the snake robot has no active wheels with which to move its body along the predefined spiral curve. Meanwhile, the head-raising process is the actual head-raising motion produced using the joint trajectory, and in this case, the tail of the snake robot is fixed. Figure 11 shows the two kinds of simulations of head-raising process. Those simulations are in kinematic level and do not take dynamic factors into consideration. The Simulation in labels 1–5 are implemented in Adams, and the roll, pitch and roll angles are integrated into spline curves to guide the model's movement. The simulation in labels 6–9 are carried out using Matlab, based on the following algorithm:

(1) Determine the basic posture as Equation (6), and set motion step $i = 1$;
(2) Proceed to motion step $i = 2$. The link n is fixed. Calculate the position of link $n-1$ on the basis of angles ${}^{2}\alpha_n$, ${}^{2}\beta_n$, and ${}^{2}\gamma_n$;
(3) Make an iteration to calculate all of the positions of the links, that is, link $j = n-2, n-3, \ldots, 1$, on the basis of angles $({}^{2}\alpha_{n-2}, {}^{2}\beta_{n-2}, {}^{2}\gamma_{n-2}), ({}^{2}\alpha_{n-3}, {}^{2}\beta_{n-3}, {}^{2}\gamma_{n-3}), \ldots, ({}^{2}\alpha_1, {}^{2}\beta_1, {}^{2}\gamma_1)$;
(4) Proceed to motion step $i = i+1$ and repeat steps 2 and 3 until $i = n_{step}$.

In this algorithm, step 2 is important. During the phase-shifting and head-raising processes, their link vectors ${}^{i}l_j$, as shown in Figure 5, are same in orientation and different in position. Therefore, the position of joint point ${}^{i}P_j$ can be obtained as follows:

$$\begin{array}{l}
{}^{i}l_{j-1} = {}^{i}l_j R_z({}^{i}\alpha_j) R_y({}^{i}\alpha_j) R_x({}^{i}\alpha_j) \\
{}^{i}P_j = {}^{i}P_{j-1} + {}^{i}l_{j-1} \\
R_z({}^{i}\alpha_j) = \begin{bmatrix} \cos({}^{i}\alpha_j) & -sin({}^{i}\alpha_j) & 0 \\ \sin({}^{i}\alpha_j) & \cos({}^{i}\alpha_j) & 0 \\ 0 & 0 & 1 \end{bmatrix} R_y({}^{i}\beta_j) = \begin{bmatrix} \cos({}^{i}\beta_j) & 0 & \sin({}^{i}\beta_j) \\ 0 & 1 & 0 \\ -sin({}^{i}\beta_j) & 0 & \cos({}^{i}\beta_j) \end{bmatrix} \\
R_z({}^{i}\gamma_j) = \begin{bmatrix} 1 & 0 & 0 \\ 0 & \cos({}^{i}\gamma_j) & -sin({}^{i}\gamma_j) \\ 0 & \sin({}^{i}\gamma_j) & \cos({}^{i}\gamma_j) \end{bmatrix}
\end{array} \tag{13}$$

Considering that the method of head-raising should be adaptive to deal with different situations, more simulations are implemented as described in Figure 12, where different parameters of the predefined spiral curves are adopted and they are listed in Table 2. In Figure 12, only final postures are shown, and more detailed simulations can be found in Supplementary Materials. Compared with the result in Figures 11 and 12a, the radius of the supporting part BS becomes larger when increasing the value of parameter a (Figure 12b–d); the height of every circle becomes larger if the parameter c increases (Figure 12b–d). When changing the parameter b, the amplitudes of $x_s(t)$ and $y_s(t)$ change unequally. Additionally, the LP part changes accordingly (Figure 12c,d) when changing the parameters l_{link} and n, and it should be noted that φ_0 is set to ensure the initial phases of the sine and cosine functions. All the simulations confirm the feasibility and validity of the proposed method.

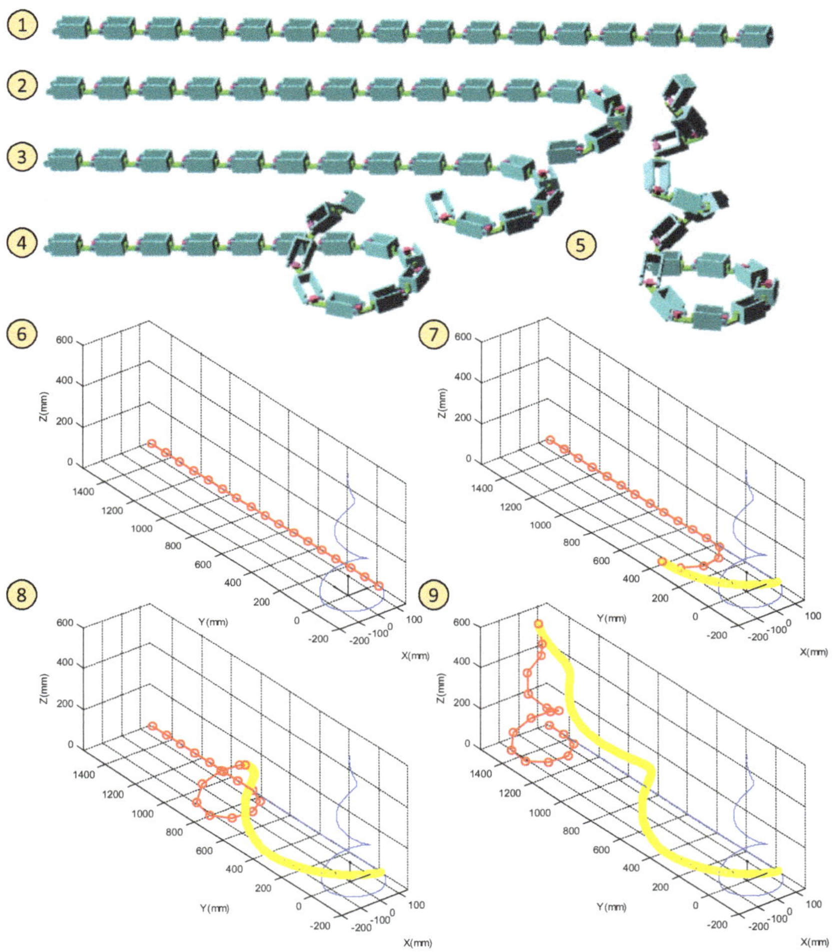

Figure 11. Head-raising motion process. Labels 1–5 show different states from initial posture to final posture of head-raising in Adams environment. Label 1 is the initial posture and label 5 is the final posture. Label 2 corresponds to the case that the robot forms a supporting configuration. Label 3 and 4 are intermediate states. The labels 6–9 are head-raising simulations of line segments model, implemented in Matlab environment. The labels 6–9 correspond to labels 1, 2, 4 and 5, respectively.

Table 2. Parameters of predefined spiral curves used in head-raising simulations shown in Figures 11 and 12.

Figure Number	a	b	c	n_c	φ_0 (rad)	n	l_{link} (mm)
Figure 11	9.7	1	48.5	2.5	0.5 pi	16	97
Figure 12a	9.7	1	20	1.5	0.5 pi	16	97
Figure 12b	60	1	60	0.8	0.5 pi	16	97
Figure 12c	20	1	60	0.9	1.5 pi	12	50
Figure 12d	20	1.5	60	0.9	1.5 pi	12	50

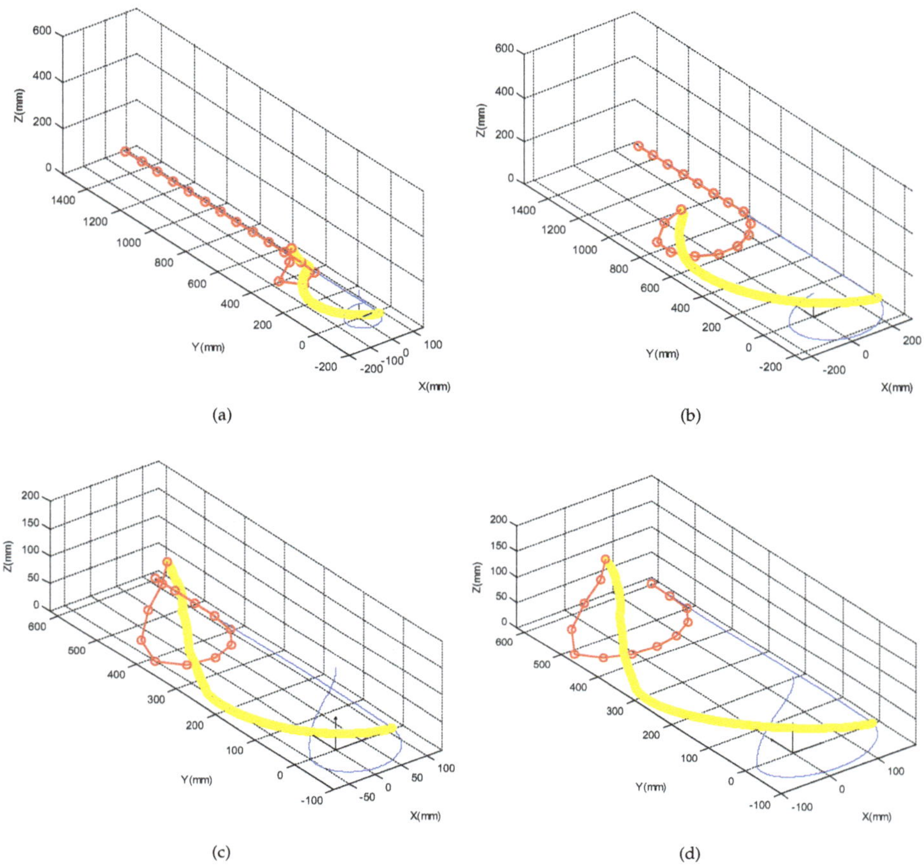

Figure 12. Head-raising motion process with different parameters of the predefined spiral curve, and the parameters are listed in Table 2.

4. Discussion and Conclusions

In this study, head-raising motion of a snake robot based on a predefined spiral curve is proposed and relative simulations are conducted. Obtaining the angle sequences of the head-raising motion of snake robots is difficult under three-dimensional conditions. Thus, the proposed method adopts the predefined spiral curve to guide the robot in moving, thereby improving the performance of the robot and laying the foundation for accomplishing additional tasks. The predefined spiral curve is parameterized. With changes in the parameters, the curves can adapt to different lengths and numbers of the snake robot modules, spiral numbers, and heights of head-raising.

However, there are some problems to be solved in the future work. First, the predefined spiral curve is a piecewise function and has unsmooth points, thus leading to sharp changes in motion trajectories. Second, our model considers only kinematic planning, that is, the dynamic problem is not considered. Dynamic constraints are important factors for successful implementation of our model on a real robot, and will be added into an improved model in the future work. At the same time, experimental verification will be carried out in the next phase of research, after finishing the design and construction of the platform. The greatest challenge to use our model on a real robot is that the robot has to overcome the gravity during the head-raising, and the actuation system must be equipped with motors which have adequate torques to drive the snake modules. The work in [13] is a successful example although they adopted a different method to raise the robot's head. When designing the prototype, lightweight materials can be used to enable the robot with lighter gravity and stronger driving power. More importantly, our model is adaptive. As illustrated in Figure 12, by decreasing the height, the cycle number and increasing the radius of supporting part, it is feasible for the robot to raise its head while satisfying the dynamic constraints. In some specific situations such as exploring the moon, and searching for resources underwater, the gravity will decrease or be compensated. Then, it will be easier for the robot to raise their head. Third, the head-raising motion forms of snake in the natural world can change over time. To accomplish that, the predefined spiral curve should be made time varying without influencing the stability of the robot body, and this principle will also be the focus of the future research.

Supplementary Materials: The following are available online at http://www.mdpi.com/2076-3417/8/11/2011/s1, Video S1: Concept_of_trajectory_tracking.avi; it introduces the concept of trajectory tracking using an improved model in our future work. Video S2: PS_Fig9.avi; it describes the process of phase-shifting (Figure 9). Video S3: HR_Fig11.avi; it shows the process of head-raising using the line segments model of the snake robot (Figure 11). Video S4: SA_Fig11.avi; it shows the simulation of head-raising in Adams environment (Figure 11). Video S5: HR_Fig12a.avi; it displays the head-raising process of Figure 12a. Video S6: HR_Fig12b.avi; it displays the head-raising process of Figure 12b. Video S7: HR_Fig12c.avi; it displays the head-raising process of Figure 12c. Video S8: HR_Fig12d.avi; it displays the head-raising process of Figure 12d.

Author Contributions: Conceptualization, J.L. and Z.J.; Formal analysis, X.Z. and J.L.; Funding acquisition, J.L.; Investigation, J.L., X.Z., Z.J. and C.Y.; Methodology and Simulation, X.Z.; Software and Programming, X.Z.; Supervision, J.L., Z.J. and C.Y.; Validation, J.L.; Writing—original draft, X.Z.; Writing—review and editing, Z.J., C.Y. and J.L.

Funding: This work was partially supported by the National Science Foundation of China (Grant No. 51775541), Research Fund of China Manned Space Engineering (Grant No. 030201), Engineering and Physical Sciences Research Council (EPSRC) (Grant No. EP/S001913/1). And the APC was funded by the National Science Foundation of China (Grant No. 51775541).

Conflicts of Interest: The authors declare no conflict of interest.

Abbreviations

The following abbreviations are used in this manuscript:

2D	Two dimensional
3D	Three dimensional

Appendix A

Table A1 shows all the symbols utilized in this paper.

Table A1. Symbols utilized in this paper.

Symbol	Definition
$x_l(t)$	x value of line part in the predefined spiral curve
$y_l(t)$	y value of line part in the predefined spiral curve
$z_l(t)$	z value of line part in the predefined spiral curve
$x_s(t)$	x value of spiral curve part in the predefined spiral curve
$y_s(t)$	y value of spiral curve part in the predefined spiral curve
$z_s(t)$	z value of spiral curve part in the predefined spiral curve
a	Adjustment coefficient of the spiral curve, which can be used to change the amplitude of $x_s(t)$ and $y_s(t)$ equally
b	Adjustment coefficient of the spiral curve, which can be used to change the amplitude of $x_s(t)$ and $y_s(t)$ unequally
c	Adjustment coefficient of the spiral curve, which can be used to change the amplitude of $z_s(t)$
φ_0	Initial phase of the sine and cosine functions
n_c	Cycle number of the spiral curve
l_{link}	Length of the snake robot modules
n	Number of the snake robot modules
t	Independent variable divided into three intervals: $(0, \hat{t}_1)$, $(\hat{t}_1, \hat{t}_2)$ and $(\hat{t}_2, \hat{t}_3)$
$y_s(t)$	y value of spiral curve part in the predefined spiral curve
LP	Line part of the predefined spiral curve, corresponding to interval $(0, \hat{t}_1)$
BS	The base of the spiral curve part, which is in contact with the ground and corresponds to interval $(\hat{t}_1, \hat{t}_2)$
RS	y The base of the spiral curve part, which is in contact with the ground and corresponds to interval $(\hat{t}_2, \hat{t}_3)$
$\hat{t}_1$	Value of t connecting LP and BS
$\hat{t}_2$	Value of t connecting BS and RS
$\hat{t}_3$	End value of t
φ_{base}	The phase value of interval $(\hat{t}_1, \hat{t}_2)$
t_i	y Discrete value of simulation time
i	Index value of simulation step
j	Index value of point P in line segments model of the snake robot
${}^iP_j$	Position of point j step i of phase-shifting process
n_{step}	Number of phases shifted
${}^ix_{snk\_j}$	x value of ${}^iP_j$
${}^iy_{snk\_j}$	y value of ${}^iP_j$

Table A1. *Cont.*

Symbol	Definition
${}^{i}z_{snk\_j}$	z value of ${}^{i}P_j$
ΔS_i	Arc length at step i and defined as "phase"
${}^{i}P_{snk}$	Point set at step i
t_d	y Discrete vector of t
Δt_1	Step length of discretization in interval $(0, \hat{t}_1)$
Δt_2	Step length of discretization in interval $(\hat{t}_1, \hat{t}_3)$
${}^{i}Inx_{pj}$	Index value of element in t_d, which is nearest to t_{i+1} during phase-shifting process
eps	Threshold utilized to deal with the error between continuous function and discrete approximation
$O_{x_o y_o z_o}$	Reference coordinate
${}^{i}x_j$	x coordinate axis fixed on the snake module j at step i
${}^{i}y_j$	y coordinate axis fixed on the snake module j at step i
${}^{i}z_j$	z coordinate axis fixed on the snake module j at step i
${}^{i}l_j$	Link vector connecting ${}^{i}P_j$ and ${}^{i}P_{j+1}$
${}^{i}P_j - {}^{i}x_j {}^{i}y_j {}^{i}z_j$	y Local coordinate fixed on link $j-1$
${}^{i}\alpha_j$	Yaw angle of joint j at step i
${}^{i}\beta_j$	Pitch angle of joint j at step i
${}^{i}\gamma_j$	Roll angle of joint j at step i
${}^{i}\dot{\alpha}_j$	Yaw angular velocity of joint j at step i
${}^{i}\dot{\beta}_j$	Pitch angular velocity of joint j at step i
${}^{i}\dot{\gamma}_j$	Roll angular velocity of joint j at step i
${}^{i}\ddot{\alpha}_j$	Yaw angular acceleration of joint j at step i
${}^{i}\ddot{\beta}_j$	Pitch angular acceleration of joint j at step i
${}^{i}\ddot{\gamma}_j$	Roll angular acceleration of joint j at step i
t_{time}	Head-raising motion time
${}^{i}R_{P_j}$	Orientation matrix of link j with respect to the reference coordinate at step i
${}^{i}R_{P\_rel_j}$	Orientation matrix of link j with respect to ${}^{i}R_{P_{j+1}}$ at step i
$R_x(.)$	Matrix involving vector rotating around the x axis
$R_y(.)$	Matrix involving vector rotating around the y axis
$R_z(.)$	Matrix involving vector rotating around the z axis

References

1. Transeth, A.A.; Pettersen, K.Y.; Liljebäck, P. A survey on snake robot modeling and locomotion. *Robotica* **2009**, *27*, 999–1015. [CrossRef]
2. Liljebäck, P.; Pettersen, K.Y.; Stavdahl, Ø.; Gravdahl, J.T. A review on modelling, implementation, and control of snake robots. *Robot. Auton. Syst.* **2012**, *60*, 29–40.
3. Liu, J.; Wang, Y.; Li, B.; Ma, S. Path planning of a snake-like robot based on serpenoid curve and genetic algorithms. In Proceedings of the 5th World Congress on Intelligent Control and Automation, Hangzhou, China, 15–19 June 2004; pp. 4860–4864.
4. Saito, M.; Fukaya, M.; Iwasaki, T. Serpentine locomotion with robotic snakes. *IEEE Control Syst. Mag.* **2002**, *22*, 64–81. [CrossRef]
5. Chen, L.; Wang, Y.; Ma, S.; Li, B. Analysis of traveling wave locomotion of snake robot. In Proceedings of the 2003 IEEE International Conference on Robotics, Intelligent Systems and Signal Processing, Changsha, China, 8–13 October 2003; pp. 365–369.
6. Kalani, H.; Akbarzadeh, A.; Safehian, J. Traveling wave locomotion of snake robot along symmetrical and unsymmetrical body shapes. In Proceedings of the 41st International Symposium on Robotics and 6th German Conference on Robotics, Munich, Germany, 7–9 June 2010; pp. 62–68.
7. Liu, J.; Wang, Y.; Li, B.; Chen, L.; Ma, S. Serpentine locomotion with robotic snakes. *Chin. J. Mech. Eng.* **2005**, *41*, 108–113. [CrossRef]
8. Virgala, I.; Dovica, M.; Kelemen, M.; Prada, E.; Bobovsky. Snake robot movement in the pipe using concertina locomotion. *Jixie Gongcheng Xuebao/Chin. J. Mech. Eng.* **2005**, *41*, 108–113. [CrossRef]
9. Chen, L.; Wang, Y.; Ma, S.; Li, B. Study of lateral locomotion of snake robot. *Robot* **2003**, *25*, 246–249.
10. Tanev, I.; Ray, T.; Buller, A. Evolution, robustness, and adaptation of sidewinding locomotion of simulated snake-like robot. In *Genetic and Evolutionary Computation Conference*; Springer: Berlin, Germany, 2004; pp. 627–639.
11. Wang, K.; Gao, W.; Ma, S. Snake-like robot with fusion gait for high environmental adaptability: Design, modeling, and experiment. *Appl. Sci.* **2017**, *7*, 1133. [CrossRef]
12. Sanfilippo, F.; Azpiazu, J.; Marafioti, G.; Transeth A.A.; Stavdahl, Ø.; Liljebäck, P. Study of lateral locomotion of snake robot. *Appl. Sci.* **2017**, *7*, 336. [CrossRef]
13. Ye, C.; Ma, S.; Li, B.; Wang, Y. Head-raising motion of snake-like robots. In Proceedings of the 2004 IEEE International Conference on Robotics and Biomimetics, Shenyang, China, 22–26 August 2004; pp. 595–600.
14. Tanaka, M.; Matsuno, F. Modeling and control of head raising snake robots by using kinematic redundancy. *J. Intell. Robot. Syst.* **2014**, *75*, 53–69. [CrossRef]
15. Cappo, E.A.; Choset, H. Planning end effector trajectories for a serially linked, floating-base robot with changing support polygon. In Proceedings of the 2004 American Control Conference, Portland, OR, USA, 4–6 June 2014; pp. 4038–4043.
16. Wu, W.; Hong, Y.; Wang, G. Geometrical spline approach to shape control of super redundant planar manipulator. *Mach. Tool Hydraul.* **2004**, *4*, 70–72.
17. Nor, N.M.; Ma, S. Body shape control of a snake-like robot based on phase oscillator network. In Proceedings of the 2013 IEEE International Conference on Robotics and Biomimetics, Shenzhen, China, 12–14 December 2013; pp. 274–279.
18. Liljebäck, P.; Pettersen, K.Y.; Stavdahl, O.; Gravdahl, J.T. A control framework for snake robot locomotion based on shape control points interconnected by Bézier curves. In Proceedings of the 2012 IEEE/RSJ International Conference on Intelligent Robots and Systems, Algarve, Portugal, 7–12 October 2012; pp. 3111–3118.
19. Tanaka, M.; Tanaka, K. Shape Control of a snake robot with joint limit and self-collision avoidance. *IEEE Trans. Control Syst. Technol.* **2017**, *25*, 1441–1448. [CrossRef]
20. Liu, J.; Wang, Y.; Ma, S.; Li, B. Shape control of hyper-redundant modularized manipulator using variable structure regular polygon. In Proceedings of the 2004 IEEE/RSJ International Conference on Intelligent Robots and Systems, Sendai, Japan, 28 September–2 October 2004; pp. 3924–3929.
21. Chirikjian, G.S.; Burdick, J.W. A modal approach to hyper-redundant manipulator kinematics. *IEEE Trans. Robot. Autom.* **1994**, *10*, 343–354. [CrossRef]

22. Burdick, J.W.; Radford, J.; Chirikjian, G.S. A "sidewinding" locomotion gait for hyper-redundant robots. *Adv. Robot.* **1995**, *9*, 195–216. [CrossRef]
23. Hatton, R.L.; Choset, H. Generating gaits for snake robots by annealed chain fitting and keyframe wave extraction. *Auton. Robot.* **2010**, *28*, 271–281. [CrossRef]
24. Tavakkoli, S.; Dhande, S.G. Shape synthesis and optimization using intrinsic geometry. *J. Mech. Des.* **1991**, *113*, 379–386. [CrossRef]
25. Mochiyama, H.; Shimemura, E.; Kobayashi, H. Shape control of manipulators with hyper degrees of freedom. *Int. J. Robot. Res.* **1999**, *18*, 584–600. [CrossRef]

applied sciences

MDPI

Article

Generalized Singularity Analysis of Snake-Like Robot

Shunsuke Nansai [1,*], Masami Iwase [2] and Hiroshi Itoh [1]

1 Department of Advanced Machinery Engineering, School of Engineering, Tokyo Denki University, Tokyo 120-8551, Japan; itoh@mail.dendai.ac.jp
2 Department of Robotics and Mechatronics, School of Science and Technology for Future Life, Tokyo Denki University, Tokyo 120-8551, Japan; iwase@fr.dendai.ac.jp
* Correspondence: nansai@mail.dendai.ac.jp; Tel.: +81-3-5284-5120

Received: 13 September 2018; Accepted: 2 October 2018; Published: 10 October 2018

Abstract: The purpose of this paper is to elucidate a generalized singularity analysis of a snake-like robot. The generalized analysis is denoted as analysis of singularity of a model which defines all designable parameters such as the link length and/or the position of the passive wheel as arbitrary variables. The denotation is a key point for a novelty of this study. This paper addresses the above new model denotation, while previous studies have defined the designable parameters as unique one. This difference makes the singularity analysis difficult substantively. To overcome this issue, an analysis method using redundancy of the snake-like robot is proposed. The proposed method contributes to simplify singularity analysis concerned with the designable parameters. The singular configurations of both the model including side-slipping and the one with non side-slipping are analyzed. As the results of the analysis, we show two contributions. The first contribution is that a singular configuration depends on designable parameters such as link length as well as state values such as relative angles. The second contribution is that the singular configuration is characterized by the axials of the passive wheels of all non side-slipping link. This paper proves that the singular configuration is identified as following two conditions even if the designable parameters are chosen as different variables and the model includes side-slipping link. One is that the axials of passive wheels of all non side-slipping links intersect at a common point. Another one is that axials of passive wheels of all non side-slipping links are parallel.

Keywords: snake-like robot; singularity analysis; system design

1. Introduction

A real snake has simple figure like a string, and can locomote by using difference between friction in the propulsive direction and one in the normal direction. It can locomote not only flatland but also irregular terrain such as desert, wildland and grassland, by choosing its motion and posture depending on environments/tasks. Moreover, it can realize skilled locomotions such as swimming, climbing, and squeezing. That is, the real snake possesses highly adaptability corresponding to the environments/tasks. A snake-like robot mimics such highly adaptability of the real snake, and is expected to be an adaptable robot corresponding to the environments/tasks.

Many literatures have reported with respect to locomotion controls of snake-like robots. Hirose, who is a pioneer of the studies for snake-like robots, has found that a curvature of a snake changes sinusoidally along its body axis. He has named it "Serpenoid Curve", and applied it to a trajectory generation for snake-like robots [1]. Endo et al. have implemented a propulsive locomotion control of snake-like robots by using the Serpenoid Curve [2]. As studies extending the Serpenoid Curve, Yamada et al. have proposed the kinematics to stabilize head direction during tracking the robot to the

Serpenoid Curve, and have implemented the kinematics on the robot [3]. Ma et al. have implemented slope climbing control by adjusted amplitude of the Serpenoid Curve via a simulator [4]. In our previous study, we have designed a servo control system for tracking to the Serpenoid Curve [5]. Relative researches with respect to propulsive control use difference of the friction forces between the propulsive direction and the normal direction also have been reported [6,7]. In our previous study, a head position control [8–10], a force control [11] and a slope climbing control [12] have been designed.

These researches [6–12] have been based on a dynamical model of the snake-like robot. The discussions on it has been reported in order to avoid singular configurations as well. In these researches, the model is generally assumed to install non side-slipping passive wheels on its links. Prautsch et al. have derived the dynamical model of the snake-like robot without side-slipping, and have proven some theories and lemmas regarding the dynamics [13,14]. As one of the results of the analyses, they have referred the snake-like robot takes the singular configuration as long as it poses either straight line or arc shape. The result has been shared by a lot of researchers. For example, Date et al. have also reported that the snake-like robot becomes singular when the robot poses straight line or arc shape [15–17]. In addition, Ye et al. have also stated that straight line or arc shape are singular [18]. Matsuno et al. [19] and Tanaka et al. [20–22] also have mentioned that the postures of straight line or arc shape are the singular configuration as well. They [15–22] have led the same results based on the analysis by Prautsch et al. [13,14]. In addition, Dear et al. [23,24] and Guo et al. [25] have reported the snake-like robot is the singular configuration when all relative angles are equivalent. It indicates the posture under this situation is either straight line or arc shape. Liljebäck et al. have addressed a dynamical model supposing the snake-like robot with side-slipping, and have derived the model with viscous friction forces working at the center of each link. They have analyzed nonlinear controllability of the models, and finally have also proven that straight line and arc shape are singular configurations [26–29].

While these studies [13–28] have addressed the either fully non side-slipping or fully side-slipping model, Tanaka et al. have addressed a model including both side-slipping and non side-slipping links [30]. They have referred that either the straight line or the arc shape are still the singular configuration with respect to the fully non side-slipping model based on Prautsch's analysis [14]. Nevertheless, they have concluded that following two conditions are the singular configuration regarding to the model including side-slipping.

Condition

- Axials of passive wheels of all non side-slipping links intersect at a common point.
- Axials of passive wheels of all non side-slipping links are parallel.

These researches [13–30] have a common assumption that the robot has either the passive wheel or the center of gravity located at the center of each link. This common assumption means link length and/or position of the passive wheel are defined as common variable. Hence, the common assumption contributes the analysis simple, because each element of its Jacobian matrix can be factorized by link length or position of passive wheel as common variable. Accordingly, each element is able to be formulated by summation of the trigonometric functions with link length or position of passive wheel as the common variable. For example, Tanaka et al. have defined the length from a front joint to a passive wheel and the one from a passive wheel to hind joint as a same variable for all link. To satisfy this definition, they have supposed that the link length of lifted links is not changed by limiting lifting height infinitesimally, even though their robot allows to lift up it to enough height [30]. Thus, the analyses of previous studies subject to the limitation regarding to the common variable, although these variables could be designable. Conversely, effects of link position of the passive wheel and/or length to its behavior have never discussed yet.—i.e., the analysis without any limitation has never been discussed.

This paper elucidates a generalized singularity analysis of the snake-like robot. The generalized analysis is denoted as analysis of singularity of a model which defines all designable parameters such as the link length and/or the position of the passive wheel as arbitrary variables. From a viewpoint of system design, revealing a relationship between the designable parameters and system behavior affects behavior of the whole system including the locomotion control system as well. In the case of the inverted pendulum for example, it has been reported that changing its designable parameters such as link length results in extending stability margin of the system [31–33]. In another instance, the Jansen linkage mechanism, a representative example of closed link mechanism, is capable of transitioning its gaits significantly by changing length ratio of all link [34,35]. In addition, its characteristic influences robot's morphology as well [36]. The snake-like robot is a kind of non-holonomic system, and its non-holonomic constraints come from the characteristic of the non side-slipping passive wheels. Hence, the position of the passive wheel affects the system behavior. Thus, discussion on effect of the designable parameters is expected to provide some advantages in this area as well.

This paper addresses two main issues. The first main issue is to elucidate dependent relationship between the singular configuration of the snake-like robot and the designable parameters such as link length and position of the passive wheel. As discussed above, the previous studies have identified that the singular configuration of the snake-like robot is either straight line or arc shape. This identification is equivalent to that the singular configuration depends on only its state vector. Tanaka et al. have referred that the postures of straight line or arc shape are the singular configuration regarding to the model with non side-slipping while they have referred that the *Conditions* are the singular configuration regarding to the model including side-slipping [30]. In addition, they have defined all link length as unique parameter. On the other hand, this paper addresses the model which defines the designable parameters as non-unique variables in order to reveal the relation between the singular configuration and the designable parameters. This difference of the assumptions results in a substantial increase of the analysis complexity, because the link length and the position of the passive wheel defined as different parameters are unable to factorized as common variables.

Therefore, in this paper, an analysis method using redundancy of the snake-like robot is proposed. The singular configuration of the snake-like robot is analyzed based on the method. The proposed method contributes to simplify singularity analysis concerned with the designable parameters such as link length. An epitomization of the method is to resolve the snake-like robot into subsystems every three links. Applying the same analysis to all subsystem leads to singularity analysis of whole system. Since the subsystem is absolutely smaller than the whole system, the epitomization reduces complexity coming from different parameters. The subsystem is also able to be composed corresponding to the model including side-slipping as well as the one with non side-slipping. Consequently, the singular configurations of both the model including side-slipping and the one with non side-slipping are analyzed in this paper.

The second main issue is to characterize the singular configuration of the snake-like robot by the axials of the passive wheels of all non side-slipping link. Tanaka et al. [30] have proven that the singular configuration is characterized by the axials of the passive wheels as the *Conditions*. The *Conditions* is novel characteristic to identify the singular configuration. However, they had a limitation for maintaining the uniform link length. By clearing the limitation, this paper proves effectiveness of the novel characteristic even if the designable parameters are defined as different values and the model includes side-slipping links.

This paper is organized as follows: Section 3 derives both kinematics and Jacobian matrices of the snake-like robot. Four kinds of kinematics around a joint are derived corresponding to side-slipping patterns. The Jacobian matrices of both the snake-like robot including side-slipping and the one with non side-slipping are formulated by composing the four kinds of kinematics redundantly. By applying the analysis method using redundancy, Section 4 analyzes the singular configuration of the snake-like robot. Two theories and one lemma are proven with respect to the model with non side-slipping. Also, two theories and two lemmas are proven with respect to the model including side-slipping. Section 5 visualizes the analysis results via two numerical simulations. Section 6 concludes this paper.

2. Notation

Notations used in this paper are defined here for simplicity of description. The first is regarding to representation of side-slipping on schematic figures of the snake-like robot. The second is regarding to product of matrices. The last one is regarding to a relative angle.

Notation 1. *The passive wheel depicted on the schematic figure distinguishes between side-slipping and non side-slipping of a link. Figure 1a represents a non side-slipping link. Figure 1b represents a side-slipping link.*

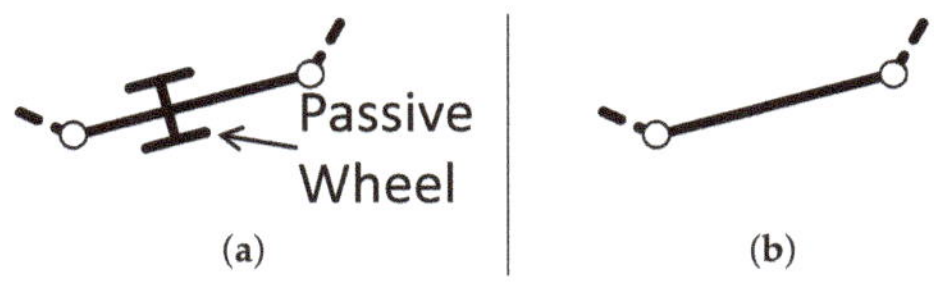

Figure 1. The distinction between a side-slipping link and a non side-slipping link. (**a**): a non side-slipping link. (**b**): a side-slipping link.

Notation 2. *In this paper, production of matrices is represented as follows:*

$$\prod_{i=1}^{n} \boldsymbol{A}_i := \boldsymbol{A}_n \boldsymbol{A}_{n-1} \boldsymbol{A}_{n-2} \cdots \boldsymbol{A}_2 \boldsymbol{A}_1, \qquad \prod_{i=1}^{n} \boldsymbol{A}_i \boldsymbol{B} := \boldsymbol{A}_n \boldsymbol{A}_{n-1} \boldsymbol{A}_{n-2} \cdots \boldsymbol{A}_2 \boldsymbol{A}_1 \boldsymbol{B},$$

where, $\boldsymbol{A}_i \in \mathbb{R}^{n \times n}$, $\boldsymbol{B} \in \mathbb{R}^{n \times m}$.

Notation 3. *The relative angle of adjacent ith link and i − 1th link is represented as follow:*

$$\phi_i := \theta_i - \theta_{i-1}.$$

Whereas, the relative angle of non adjacent ith link and jth link is represented as follow:

$${}^{i}\phi_j := \theta_i - \theta_j.$$

3. Kinematics of the Snake-Like Robot

A kinematics of n-link snake-like robot including some side-slipping links is formulated in this section by deriving velocity relations around each joint. Since the snake-like robot is a kind of redundant robots, its kinematics is also able to be represented by redundant relation. Thus, to derive the kinematics of n-link snake-like robot is equivalent to derive it between adjacent links. In particular, a kinematics around a joint between adjacent non side-slipping links is derived, also, ones including side-slipping link is derived from the derived kinematics. Finally, Jacobian matrices of both the n-link snake-like robot including side-slipping and the one with non side-slipping are formulated, respectively.

3.1. Kinematics around a Joint Adjacent Two Links with Non Side-Slipping

A schematic figure of a joint between adjacent non side-slipping links is shown in Figure 2. Physical parameters and state variables are denoted in Tables 1 and 2, respectively.

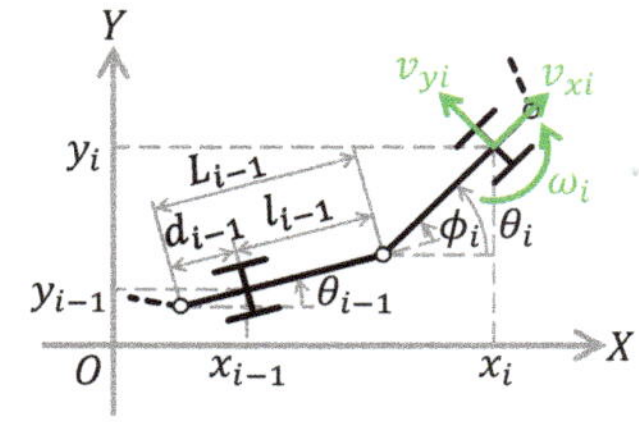

Figure 2. The schematic figure of the joint of two adjacent links both without side-slipping.

Table 1. Physical Parameters$(i = 1, \cdots, n)$.

Parameters	Notation
Link length	L_i [m]
Length from COG to end	l_i [m]
Length from extremity to COG	d_i [m]

Table 2. State Variables$(i = 1, \cdots, n)$.

Variables	Notation
Position in x-coordinate	x_i [m]
Position in y-coordinate	y_i [m]
Absolute angle of each link	θ_i [rad]
Relative angle between ith link and $i-1$th link	ϕ_i [rad]
Velocity in the propulsive direction	v_{xi} [m/s]
Velocity in the normal direction	v_{yi} [m/s]
Angular velocity of each link	ω_i [rad/s]

From Figure 2, the passive wheel, which represents that the link is non side-slipping, is installed on arbitrary position of each link. θ_i $(i = 1, \cdots, n)$ represents absolute angle of each link, ϕ_i $(i = 2, \cdots, n)$ represents relative angle between ith link and $i-1$th link, and $(x_i,\ y_i)$ $(i = 1, \cdots, n)$ represent position of the passive wheel of each link on generalized coordinate. Also, d_i and l_i $(i = 1, \cdots, n)$ represent length from extremity to center of gravity (COG) and length from COG to end, respectively.

The generalized coordinate and the quasi-velocity coordinate are defined. From Figure 2, the generalized coordinate of the system $\boldsymbol{x}_{pi}$ is defined as:

$$\boldsymbol{x}_{pi} := \begin{bmatrix} \theta_i & x_i & y_i \end{bmatrix}^T.$$

Also, the quasi-velocity of the system $\boldsymbol{v}_i$ is defined as:

$$\boldsymbol{v}_i := \begin{bmatrix} \omega_i & v_{xi} & v_{yi} \end{bmatrix}^T.$$

From Figure 2, the velocity transform matrix $\boldsymbol{T}_i$ transforms the generalized coordinate $\boldsymbol{x}_{pi}$ to the quasi-velocity coordinate $\boldsymbol{v}_i$ as below:

$$\dot{\boldsymbol{x}}_{pi} = \boldsymbol{T}_i \boldsymbol{v}_i,$$

$$\boldsymbol{T}_i = \begin{bmatrix} 1 & 0 & 0 \\ 0 & \cos\theta_i & -\sin\theta_i \\ 0 & \sin\theta_i & \cos\theta_i \end{bmatrix}.$$

From Figure 2, ith link is related to $i-1$th link as follows:

$$\begin{cases} \theta_i = \theta_i, \\ x_i = x_{i-1} + l_{i-1}\cos\theta_{i-1} + d_i\cos\theta_i, \\ y_i = y_{i-1} + l_{i-1}\sin\theta_{i-1} + d_i\sin\theta_i. \end{cases}$$

Its derivations are:

$$\begin{cases} \dot{\theta}_i = \dot{\theta}_i, \\ \dot{x}_i = \dot{x}_{i-1} - l_{i-1}\sin\theta_{i-1}\dot{\theta}_{i-1} - d_i\sin\theta_i\dot{\theta}_i, \\ \dot{y}_i = \dot{y}_{i-1} + l_{i-1}\cos\theta_{i-1}\dot{\theta}_{i-1} + d_i\cos\theta_i\dot{\theta}_i, \end{cases} \quad (1)$$

Equation (1) is reformulated as (2) by using the generalized velocity $\dot{\boldsymbol{x}}_{\boldsymbol{pi}} = \mathrm{d}\boldsymbol{x}_{\boldsymbol{pi}}/\mathrm{d}t$.

$$\dot{\boldsymbol{x}}_{\boldsymbol{pi}} = \bar{\boldsymbol{\Lambda}}_{\boldsymbol{i-1}}\dot{\boldsymbol{x}}_{\boldsymbol{pi-1}} + \bar{\boldsymbol{\Delta}}_{\boldsymbol{i}}\dot{\theta}_i, \quad (2)$$

$$\bar{\boldsymbol{\Lambda}}_{\boldsymbol{i-1}} = \begin{bmatrix} 0 & 0 & 0 \\ -l_{i-1}\sin\theta_{i-1} & 1 & 0 \\ l_{i-1}\cos\theta_{i-1} & 0 & 1 \end{bmatrix}, \qquad \bar{\boldsymbol{\Delta}}_{\boldsymbol{i}} = \begin{bmatrix} 1 \\ -d_i\sin\theta_i \\ d_i\cos\theta_i \end{bmatrix}.$$

By transforming (2) to the quasi-velocity coordinate using the velocity transform matrix $\boldsymbol{T_i}$, and we obtain:

$$\boldsymbol{T_i v_i} = \bar{\boldsymbol{\Lambda}}_{\boldsymbol{i-1}}\boldsymbol{T_{i-1}v_{i-1}} + \bar{\boldsymbol{\Delta}}_{\boldsymbol{i}}\omega_i.$$

Please note that the relative angle of each link is defined as $\phi_i = \theta_i - \theta_{i-1}$, $(i = 2, \cdots, n)$ (Notation 3). Then,

$$\boldsymbol{v_i} = \boldsymbol{\Lambda_{i-1}v_{i-1}} + \boldsymbol{\Delta_i}\omega_i,$$

where,

$$\boldsymbol{\Lambda_{i-1}} = \boldsymbol{T}_{\boldsymbol{i}}^{-1}\bar{\boldsymbol{\Lambda}}_{\boldsymbol{i-1}}\boldsymbol{T_{i-1}} = \begin{bmatrix} 0 & 0 & 0 \\ l_{i-1}\sin\phi_i & \cos\phi_i & \sin\phi_i \\ l_{i-1}\cos\phi_i & -\sin\phi_i & \cos\phi_i \end{bmatrix},$$

$$\boldsymbol{\Delta_i} = \boldsymbol{T}_{\boldsymbol{i}}^{-1}\bar{\boldsymbol{\Delta}}_{\boldsymbol{i}} = \begin{bmatrix} 1 \\ 0 \\ d_i \end{bmatrix}.$$

Since adjacent link is supposed non side-slipping, $v_{yi} = v_{yi-1} = 0$. Hence,

$$\omega_i = -\frac{l_{i-1}}{d_i}\cos\phi_i\omega_{i-1} + \frac{1}{d_i}\sin\phi_i v_{xi-1}.$$

Therefore, the kinematics around a joint of two adjacent links both non side-slipping is obtained as:

$$\boldsymbol{v_i} = \boldsymbol{A}'_{\boldsymbol{i}}\boldsymbol{v_{i-1}}, \qquad i = 2, \cdots, n, \quad (3)$$

$$\boldsymbol{A}'_{\boldsymbol{i}} = \begin{bmatrix} -\frac{l_{i-1}}{d_i}\cos\phi_i & \frac{1}{d_i}\sin\phi_i & 0 \\ l_{i-1}\sin\phi_i & \cos\phi_i & 0 \\ 0 & 0 & 0 \end{bmatrix},$$

where, $\boldsymbol{A}'_{\boldsymbol{1}} = \boldsymbol{I}^{3\times 2}$, and $\boldsymbol{I}^{m\times n}$ represents the identity matrix $\in \mathbb{R}^{m\times n}$.

3.2. Kinematics around a Joint Adjacent Two Links with Side-Slipping

From (3) and its derivation process, the kinematics around a joint including side-slipping links are derived as well. In this paper, the passive wheel represents that the link is side-slipping or non side-slipping in order to distinguish clearly (Notation 1). Combinations of side-slipping of two adjacent link regarding around a joint are shown in Figure 3.

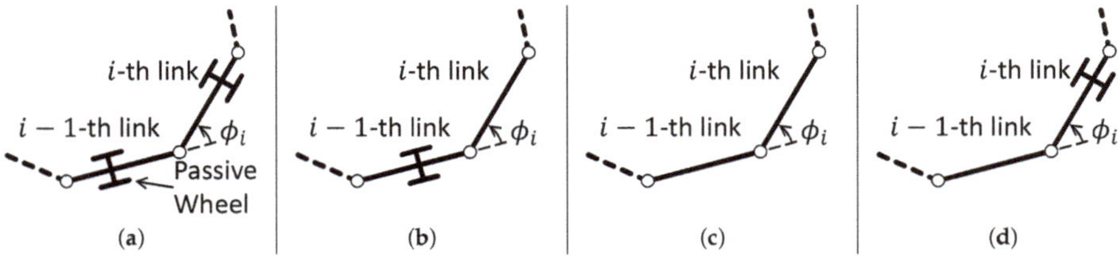

Figure 3. Combinations of side-slipping of ith link and $i-1$th link. (**a**): Both adjacent links are non side-slipping. (**b**): Only hind link is side-slipping. (**c**): Both adjacent links are side-slipping. (**d**): Only front link is side-slipping. A link with the passive wheel represents a non side-lipping link.

The kinematics of Figure 3 are formulated as follows:

(a) Both adjacent links are non side-slipping (Figure 3a).

$$\begin{aligned} \boldsymbol{v}_i &= \boldsymbol{A}'_i \boldsymbol{v}_{i-1}, \\ \boldsymbol{A}'_i &= \begin{bmatrix} -\frac{l_{i-1}}{d_i}\cos\phi_i & \frac{1}{d_i}\sin\phi_i & 0 \\ l_{i-1}\sin\phi_i & \cos\phi_i & 0 \\ 0 & 0 & 0 \end{bmatrix}. \end{aligned} \tag{4}$$

(b) Only hind link is side-slipping (Figure 3b).

$$\begin{aligned} \boldsymbol{v}_i &= \boldsymbol{A}''_{hi} \boldsymbol{v}_{i-1} + \boldsymbol{\Delta}_i \dot{\phi}_i, \\ \boldsymbol{A}''_{hi} &= \boldsymbol{A}'_{hi} \boldsymbol{I}^{2\times 3} \\ \boldsymbol{A}'_{hi} &= \begin{bmatrix} 0 & 0 \\ l_{i-1}\sin\phi_i & \cos\phi_i \\ l_{i-1}\cos\phi_i & -\sin\phi_i \end{bmatrix} + \begin{bmatrix} 1 & 0 \\ 0 & 0 \\ d_i & 0 \end{bmatrix}. \end{aligned} \tag{5}$$

(c) Both adjacent links are side-slipping (Figure 3c).

$$\begin{aligned} \boldsymbol{v}_i &= \boldsymbol{A}'_{si} \boldsymbol{v}_{i-1} + \boldsymbol{\Delta}_i \dot{\phi}_i, \\ \boldsymbol{A}'_{si} &= \begin{bmatrix} 0 & 0 & 0 \\ l_{i-1}\sin\phi_i & \cos\phi_i & \sin\phi_i \\ l_{i-1}\cos\phi_i & -\sin\phi_i & \cos\phi_i \end{bmatrix} + \begin{bmatrix} 1 & 0 & 0 \\ 0 & 0 & 0 \\ d_i & 0 & 0 \end{bmatrix}. \end{aligned} \tag{6}$$

(d) Only front link is side-slipping (Figure 3d).

$$\begin{aligned} \boldsymbol{v}_i &= \boldsymbol{A}''_{fi} \boldsymbol{v}_{i-1}, \\ \boldsymbol{A}''_{fi} &= \begin{bmatrix} -\frac{l_{i-1}}{d_i}\cos\phi_i & \frac{1}{d_i}\sin\phi_i & -\frac{1}{d_1}\cos\phi_i \\ l_{i-1}\sin\phi_i & \cos\phi_i & \sin\phi_i \\ 0 & 0 & 0 \end{bmatrix} = \begin{bmatrix} \boldsymbol{A}'_{fi} \\ \boldsymbol{0}^T \end{bmatrix}. \end{aligned} \tag{7}$$

Submatrices are defined as follows:

$$\boldsymbol{A}_i := \begin{bmatrix} -\frac{l_{i-1}}{d_i}\cos\phi_i & \frac{1}{d_i}\sin\phi_i \\ l_{i-1}\sin\phi_i & \cos\phi_i \end{bmatrix}, \qquad \boldsymbol{\delta}_i := \begin{bmatrix} 0 \\ d_i \end{bmatrix},$$

$$\boldsymbol{A}_{hi} := \begin{bmatrix} l_{i-1}\sin\phi_i & \cos\phi_i \\ l_{i-1}\cos\phi_i & -\sin\phi_i \end{bmatrix}, \qquad \boldsymbol{a}_{hi} := \begin{bmatrix} \sin\phi_i \\ \cos\phi_i \end{bmatrix},$$

$$\boldsymbol{A}_{si} := \begin{bmatrix} \cos\phi_i & \sin\phi_i \\ -\sin\phi_i & \cos\phi_i \end{bmatrix}, \qquad \boldsymbol{a}_{si} := \begin{bmatrix} l_{i-1}\sin\phi_i \\ l_{i-1}\cos\phi_i \end{bmatrix},$$

$$\boldsymbol{A}_{fi} := \begin{bmatrix} \frac{1}{d_i}\sin\phi_i & -\frac{1}{d_i}\cos\phi_i \\ \cos\phi_i & \sin\phi_i \end{bmatrix}, \qquad \boldsymbol{a}_{fi} := \begin{bmatrix} -\frac{l_{i-1}}{d_i}\cos\phi_i \\ l_{i-1}\sin\phi_i \end{bmatrix}.$$

3.3. Jacobian Matrices

The Jacobian matrices of both the n-link snake-like robot including side-slipping and the one with non side-slipping are formulated by using (4)–(7) redundantly. First, the Jacobian matrix $\boldsymbol{J}_g$ of the n-link snake-like robot with non side-slipping is formulated. A tangent speed $\boldsymbol{q}$ is defined as:

$$\boldsymbol{q} := \begin{bmatrix} \omega_1 & v_{x1} \end{bmatrix}^T.$$

The relative angular velocity vector $\dot{\boldsymbol{\phi}}$ is defined as:

$$\dot{\boldsymbol{\phi}} := \begin{bmatrix} \dot{\phi}_2 & \cdots & \dot{\phi}_n \end{bmatrix}^T.$$

Thus, the Jacobian matrix $\boldsymbol{J}_g$ of n-link snake-like robot with non side-slipping is formulated as follow:

$$\begin{aligned} \dot{\boldsymbol{\phi}} &= \boldsymbol{J}_g \boldsymbol{q}, \\ \boldsymbol{J}_g &= \begin{bmatrix} \boldsymbol{e}_1^T(\boldsymbol{A}_2 - \boldsymbol{I})\boldsymbol{A}_1 \\ \boldsymbol{e}_1^T(\boldsymbol{A}_3 - \boldsymbol{I})\boldsymbol{A}_2\boldsymbol{A}_1 \\ \vdots \\ \boldsymbol{e}_1^T(\boldsymbol{A}_n - \boldsymbol{I})\prod_{m=1}^{n-1}\boldsymbol{A}_m \end{bmatrix} \in \mathbb{R}^{n-1\times 2}, \end{aligned} \tag{8}$$

where, $\boldsymbol{e}_1^T = \begin{bmatrix} 1 & 0 \end{bmatrix}$.

Next, the Jacobian matrix $\boldsymbol{J}_s$ of the n-link snake-like robot including side-slipping is formulated. While combinations of side-slipping exist infinite on the n-link snake-like robot, the Jacobian matrix $\boldsymbol{J}_s$ consists of (4)–(7) due to its redundancy, since product of the kinematics formulates it. We denotes a set of non side-slipping links as:

$$\begin{aligned} \mathbb{G} &:= (g_1, \cdots, g_\eta) \in \mathbb{N}, \\ \mathbb{N} &:= (1, \cdots, n). \end{aligned}$$

$\boldsymbol{\phi}$ is decomposed into $\boldsymbol{\phi} = \begin{bmatrix} \boldsymbol{\phi}_s^T & \boldsymbol{\phi}_g^T \end{bmatrix}^T$, where $\boldsymbol{\phi}_g$ and $\boldsymbol{\phi}_s$ are a relative angle vector of non side-slipping links and an one of side-slipping links respectively, and denoted as follows:

$$\begin{aligned} \boldsymbol{\phi}_g &:= \{\phi_{g(i)} | g(i) = \overline{\mathbb{G} \cap 1},\ g(i+1) > g(i)\}, \\ \boldsymbol{\phi}_s &:= \{\phi_{s(i)} | s(i) = \overline{\mathbb{G} \cap \mathbb{N}},\ s(i+1) > s(i)\}, \end{aligned}$$

where $\dot{\phi}_1 := v_{y1}$, since ϕ_1 is unable to be defined as angle. Subscripts with respect to the set of non side-slipping links $g(i)$ are denoted as follow:

$$g(i-1) := \zeta, \qquad g(i) := \iota, \qquad g(i+1) := \kappa.$$

The Jacobian matrix of the n-link snake-like robot including side-slipping is formulated as follows:

$$\begin{bmatrix} \dot{\boldsymbol{\phi}}_s \\ \dot{\boldsymbol{\phi}}_g \end{bmatrix} = \begin{bmatrix} \mathbf{0}^{n-\eta\times 2} & \boldsymbol{I}^{n-\eta\times n-\eta} \\ \boldsymbol{J}_s & *^{\eta-1\times n-\eta} \end{bmatrix} \begin{bmatrix} \boldsymbol{q} \\ \dot{\boldsymbol{\phi}}_s \end{bmatrix}, \tag{9}$$

$$\boldsymbol{J}_s = \begin{bmatrix} \boldsymbol{e}_1^T(\boldsymbol{B}_2 - \boldsymbol{I})\boldsymbol{B}_1 \\ \boldsymbol{e}_1^T(\boldsymbol{B}_3 - \boldsymbol{I})\boldsymbol{B}_2\boldsymbol{B}_1 \\ \vdots \\ \boldsymbol{e}_1^T(\boldsymbol{B}_\eta - \boldsymbol{I})\prod_{m=1}^{\eta-1}\boldsymbol{B}_m \end{bmatrix} \in \mathbb{R}^{\eta-1\times 2}, \tag{10}$$

$$\boldsymbol{B}_i = \begin{cases} \boldsymbol{A}_\iota & \text{if } \iota - \zeta = 1, \\ \boldsymbol{A}'_{f\iota}\boldsymbol{A}'_{h\zeta+1'} & \text{if } \iota - \zeta = 2, \\ \boldsymbol{A}'_{f\iota}\prod_{m=\zeta+2}^{\iota-1}\boldsymbol{A}'_{sm}\boldsymbol{A}'_{h\zeta+1'} & \text{if } \iota - \zeta > 2, \end{cases}$$

where $*$ represents an element of the matrix.

This section has derived the kinematics around a joint corresponding to 4 patterns of side-slipping. By using the derived kinematics redundantly, the Jacobian matrices of both the n-link snake-like robot including side-slipping and the one with non side-slipping have been formulated as (9) and (8) respectively.

4. Singularity Analysis

This section analyzes singularity of n-link snake-like robot. Since elements of matrix to be analyzed is unable to factorize to summation of trigonometric function, it is inappropriate to analyze the singularity by the method of Tanaka et al. [30] in the case of the snake-like robot supposed to non-uniform designable parameters addressed in this paper. The key point of our proposed method is in redundancy of the system which is one of big reason to make the snake-like robot popular. Using redundancy results in avoidance of complexity as well as systematic solution. Our proposed method provides the analysis results as same as Tanaka's analysis [30]. In addition, the singular configuration of the snake-like robot is identified as follows:

- Axials of passive wheel of all non side-slipping links intersect at a common point.
- Axials of passive wheel of all non side-slipping links are parallel.

The singular configuration of the snake-like robot with non side-slipping is analyzed first in order to propose our analysis method using redundancy as well as to prove that the analysis results in the above two conditions. The analysis proves as well that the singular configuration of the snake-like robot depends on the designable parameters. The singular configuration of the snake-like robot including side-slipping is analyzed next. The analysis also proves that the singular configuration of the snake-like robot is identified as the above two even if the side-slipping links are included.

4.1. Snake-Like Robot with non Side-Slipping

The singular configuration of the snake-like robot with non side-slipping is analyzed first. The previous studies have reported that the singular configuration is either straight line or arc shape—i.e., it depends on only state vector. This analysis proves that the singular configuration of the snake-like robot depends on the designable parameters (such as link length and/or the position of the passive wheel) as well as the state vector. With respect to the snake-like robot with non side-slipping, following 2 theories and 1 lemma are obtained.

Theorem 1. *Except for* $\exists\phi_i,\ l_{i-1} = d_i \cap \cos\phi_i = -1$, (8) *is rank deficient if and only if*, $\forall\phi_i$,

$$\left(\frac{d_i}{l_{i-1}}\cos\phi_i + 1\right)\frac{1}{d_{i+1}}\sin\phi_{i+1} - \frac{1}{l_{i-1}}\sin\phi_i\left(\frac{l_i}{d_{i+1}}\cos\phi_{i+1} + 1\right) = 0. \tag{11}$$

Proof. From (8), since $\boldsymbol{J_g} \in \mathbb{R}^{n-1\times 2}$, rank of (8) drops when rank$\boldsymbol{J_g} < 2$, and the snake-like robot poses the singular configuration. Row vectors of (8) are denoted as $\boldsymbol{l}_i^T$ $(i = 1, \cdots, n-1)$. Since (8) possesses one independent row vector when rank$\boldsymbol{J_g}$=1, it can be written by using arbitrary constants α_i $(i = 2, \cdots, n-1)$ as follows:

$$J_g = \begin{bmatrix} \boldsymbol{l}_1^T \\ \alpha_2 \boldsymbol{l}_1^T \\ \vdots \\ \alpha_{n-1}\boldsymbol{l}_1^T \end{bmatrix}, \tag{12}$$

where, $\forall\alpha_i \neq 0$. Rank of (12) equals 1 even if any 2 row vectors are chosen.—i.e., if submatrices $\boldsymbol{J_i}$ $(i = 2, \cdots, n-1)$, consisting of adjacent two row vectors, satisfy $\forall$rank$\boldsymbol{J_i}$=1, rankJ_g=1 and J_g is rank deficient. $\boldsymbol{J_i}$ is

$$\boldsymbol{J_i} = \begin{bmatrix} \boldsymbol{e}_1^T(\boldsymbol{A_i} - \boldsymbol{I})\prod_{m=1}^{i-1}\boldsymbol{A_m} \\ \boldsymbol{e}_1^T(\boldsymbol{A_{i+1}} - \boldsymbol{I})\prod_{m=1}^{i}\boldsymbol{A_m} \end{bmatrix} = \begin{bmatrix} \boldsymbol{e}_1^T(\boldsymbol{I} - \boldsymbol{A}_i^{-1}) \\ \boldsymbol{e}_1^T(\boldsymbol{A_{i+1}} - \boldsymbol{I}) \end{bmatrix}\prod_{m=1}^{i}\boldsymbol{A_m} = \bar{\boldsymbol{J}}_i\prod_{m=1}^{i}\boldsymbol{A_m}.$$

From

$$\left|\boldsymbol{A_i}\right| = -\frac{l_{i-1}}{d_i}\cos^2\phi_i - \frac{l_{i-1}}{d_i}\sin^2\phi_i = -\frac{l_{i-1}}{d_i},$$

since $\boldsymbol{A_i}$ is rank safficient,

$$\text{rank}\boldsymbol{J_i} = \text{rank}\bar{\boldsymbol{J}}_i.$$

By finding rank deficient conditions of $\bar{\boldsymbol{J}}_i$, that is, solving $|\bar{\boldsymbol{J}}_i| = 0$, the conditions of which the snake-like robot poses the singular configuration are identified.

Please note that if $\exists\alpha_i, \alpha_i = 0$, it might be able to rank rank$\boldsymbol{J_g} = 2$ even if $\forall$rank$\boldsymbol{J_i} = 1$ in (12). In particular, when

$$J_g = \begin{bmatrix} \boldsymbol{l}_1^T \\ \alpha_2\boldsymbol{l}_1^T \\ \vdots \\ \alpha_{j-1}\boldsymbol{l}_1^T \\ \boldsymbol{0}^T \\ \boldsymbol{l}_{j+1}^T \\ \alpha_{j+2}\boldsymbol{l}_{j+1}^T \\ \vdots \\ \alpha_{n-1}\boldsymbol{l}_{j+1}^T \end{bmatrix},$$

rank$\boldsymbol{J_g}$ = 2, if $\boldsymbol{l}_1^T$ and $\boldsymbol{l}_{j+1}^T$ are linear independent. However, it is determined as rank deficient in above mentioned condition, because rank$\boldsymbol{J_j}$ = rank$\boldsymbol{J_{j+1}}$ = 1. Since $\forall\boldsymbol{A_i} \neq \boldsymbol{0}$, $\boldsymbol{l_j} = 0$ only when $\boldsymbol{e}_1^T(\boldsymbol{A_j} - \boldsymbol{I}) = \boldsymbol{0}^T$. Because

$$\boldsymbol{e}_1^T(\boldsymbol{A_j} - \boldsymbol{I}) = \begin{bmatrix} -\frac{l_{j-1}}{d_j}\cos\phi_j - 1 & \frac{1}{d_j}\sin\phi_j \end{bmatrix}, \tag{13}$$

$\boldsymbol{e}_1^T(\boldsymbol{A_j} - \boldsymbol{I}) \neq \boldsymbol{0}^T$ as long as $l_{j-1} \neq d_j$. Furthermore, $\boldsymbol{e}_1^T(\boldsymbol{A_j} - \boldsymbol{I}) \neq \boldsymbol{0}^T$ as long as $\cos\phi_j \neq -1$ even if $l_{j-1} = d_j$.

Finally, since

$$\left|\bar{J}_i\right| = \begin{vmatrix} \frac{d_i}{l_{i-1}}\cos\phi_i + 1 & -\frac{1}{l_{i-1}}\sin\phi_i \\ -\frac{l_i}{d_{i+1}}\cos\phi_{i+1} - 1 & \frac{1}{d_{i+1}}\sin\phi_{i+1} \end{vmatrix}$$
$$= \left(\frac{d_i}{l_{i-1}}\cos\phi_i + 1\right)\frac{1}{d_{i+1}}\sin\phi_{i+1} - \frac{1}{l_{i-1}}\sin\phi_i\left(\frac{l_i}{d_{i+1}}\cos\phi_{i+1} + 1\right),$$

except for $l_{i-1} = d_i \cap \cos\phi_i = -1$, (8) is rank deficient if and only if $\forall\phi_i$, (11). □

Equation (11) clearly shows that the singular configuration of the snake-like robot depends on the designable parameters as well as the state vector. Thus, it is limited into neither straight line nor arc shape. Equation (11) indicates to be unable to factorize to summation of trigonometric function due to different coefficients in the premise of different parameter definition. Furthermore, complexity of $|\bar{J}_i A_i|$ is more apparent than (11). This predisposition points difficulty of analyzing the Jacobian matrix at a time, and the method using redundancy has an advantage for avoiding the complexity.

Lemma 1. $l_{i-1} = d_i \cap \cos\phi_i = -1$, *—i.e.,* $e_1^T(A_j - I) = 0^T$ *if and only if the axials of the passive wheel overlap.*

Proof. In (13),

$$-\frac{l_{j-1}}{d_j}\cos\phi_j - 1 = -\frac{1}{d_j}\left(l_{j-1}\cos\phi_j - d_j\right),$$

and equation in parentheses equals x coordinate of jth link corresponding to $j-1$th link. Hence, the wheels of jth link and $j-1$th link overlap when $l_{i-1} = d_i \cap \cos\phi_i = -1$. Thus, the axials of the passive wheel clearly overlap if $e_1^T(A_j - I) = 0^T$. □

The proposed analysis method is capable of analyzing simply even though the designable parameters of each link are defined as different ones. On the other hand, it has been shown the analysis method outputs uncertain result if the special condition ($l_{i-1} = d_i \cap \cos\phi_i = -1$) is satisfied. It also has been proven that the condition is satisfied when the axials of the adjacent passive wheels overlap. Since it is pose of which the links wholly overlaps, the actual snake-like robot generally never take that pose.

Theorem 2. *The snake-like robot is singular if and only if the axials of all passive wheel intersect at a common point or are parallel.*

Proof. ith link is focused on. A coordinate is set the origin at position of the passive wheel of ith link, and let the X axis and the Y axis are on its propulsive direction and normal direction of the passive wheel respectively. Let b_{i-1} and b_{i+1} are an intercept of the axial of the passive wheel of $i-1$th link and the one of $i+1$th link respectively. The axials of the passive wheels of all $i-1$th, ith and $i+1$th link intersect at a common point, if $b_{i-1} = b_{i+1}$. It is proven that equations of the intercepts on that situation equals to (11). Figure 4 depicts the axials of all $i-1$th, ith and $i+1$th link.

The position of the passive wheel of $i-1$th link is

$$\begin{cases} x_{i-1} = -d_i - l_{i-1}\cos\phi_i, \\ y_{i-1} = l_{i-1}\sin\phi_i. \end{cases}$$

Since a slope of the $i-1$th link equals $-\tan\phi_i$, a slope of the axial of the passive wheel is follow:

$$a_{i-1} = \frac{1}{\tan\phi_i} = \frac{\cos\phi_i}{\sin\phi_i}.$$

Hence, the intercept of the passive wheel of the $i-1$th link b_{i-1} is

$$b_{i-1} = \frac{1}{\sin\phi_i}(l_{i-1} + d_i \cos\phi_i).$$

Also, the intercept of the passive wheel of the $i+1$th link b_{i+1} is

$$b_{i+1} = \frac{1}{\sin\phi_{i+1}}(d_{i+1} + l_i \cos\phi_{i+1}).$$

If $b_{i-1} = b_{i+1}$,

$$\frac{l_{i-1}}{\sin\phi_i}\left(\frac{d_i}{l_{i-1}}\cos\phi_i + 1\right) = \frac{d_{i+1}}{\sin\phi_{i+1}}\left(\frac{l_i}{d_{i+1}}\cos\phi_{i+1} + 1\right),$$

By multiplying $\frac{1}{l_{i-1}d_{i+1}}\sin\phi_i \sin\phi_{i+1}$ to the both side,

$$\left(\frac{d_i}{l_{i-1}}\cos\phi_i + 1\right)\frac{1}{d_{i+1}}\sin\phi_{i+1} = \frac{1}{l_{i-1}}\sin\phi_i\left(\frac{l_i}{d_{i+1}}\cos\phi_{i+1} + 1\right). \tag{14}$$

The axials of all passive wheel intersect at a common point when $\forall\phi_i$, (14). Since (14) equals (11), the axials of all passive wheel intersect at a common point when the snake-like robot is singular.

When $\phi_i = \phi_{i+1} = 0$, (11) equals (14) at 0. Since $\phi_i = \phi_{i+1} = 0$ represents the axials of the passive wheel are parallel, the axials of all passive wheel are parallel when the snake-like robot is singular. □

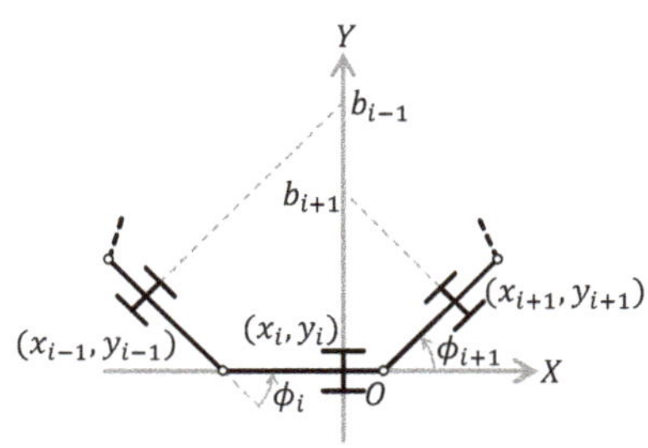

Figure 4. Schematic figure of the axials of the passive wheels in $i-1$th and $i+1$th link, when ith link is focused on.

4.2. Snake-Like Robot with Side-Slipping

The singular configuration of the snake-like robot including side-slipping is analyzed next. It also analyzed as same as the one with non side-slipping as well. An analysis result shows that it satisfies the two conditions mentioned at beginning of this section. With respect to the snake-like robot including side-slipping, following 2 theories and 2 lemmas are obtained.

Theorem 3. *Except for* $\boldsymbol{e}_1^T(\boldsymbol{B}_i - \boldsymbol{I}) = \boldsymbol{0}^T$, *(9) is rank deficient if and only if* $\forall\phi_i$,

$$\left(\frac{d_\iota}{l_\zeta}\cos{}^\iota\phi_\zeta + \sum_{m=\zeta+1}^{\iota-1}\frac{L_m}{l_\zeta}\cos{}^\zeta\phi_m + 1\right)\frac{1}{d_\kappa}\sin{}^\kappa\phi_\zeta - \frac{1}{l_\iota}\sin{}^\iota\phi_\zeta\left(\frac{l_\iota}{d_\kappa}\cos{}^\kappa\phi_\iota + \sum_{m=\iota+1}^{\kappa-1}\frac{L_m}{d_\kappa}\cos{}^\kappa\phi_\iota + 1\right) = 0. \tag{15}$$

Proof. From (9), since columns after 3rd column are clearly rank sufficient and $J_s \in \mathbb{R}^{\eta-1\times 2}$, rank of (9) drops and the snake-like robot is singular, when rank$J_s < 2$. (10) is analyzed as well as (8). Thus, conditions of which rank$\bar{J}_{si}$ drops are formulated. Where,

$$\bar{J}_{si} = \begin{bmatrix} e_1^T(I - B_i^{-1}) \\ e_1^T(B_{i+1}^{-1} + I) \end{bmatrix}.$$

B_i is expanded with supposing $\iota - \zeta > 2$ as the most complex case.

$$B_i = \begin{bmatrix} -\frac{l_\zeta}{d_\iota}\cos{}^\iota\phi_\zeta - \sum_{m=\zeta+1}^{\iota-1}\frac{L_m}{d_\iota}\cos{}^\iota\phi_m & \frac{1}{d_\iota}\sin{}^\iota\phi_\zeta \\ l_\zeta\sin{}^\iota\phi_\zeta + \sum_{m=\zeta+1}^{\iota-1}L_m\sin{}^\iota\phi_m & \cos{}^\iota\phi_\zeta \end{bmatrix}.$$

Thus,

$$B_i^{-1} = \begin{bmatrix} \frac{\cos{}^\iota\phi_\zeta}{|B_i|} & -\frac{\sin{}^\iota\phi_\zeta}{d_\iota|B_i|} \\ * & * \end{bmatrix},$$

$$\because \left|B_i\right| = -\frac{l_\zeta}{d_\iota} - \sum_{m=\zeta+1}^{\iota-1}\frac{L_m}{d_\iota}\cos{}^\zeta\phi_m. \tag{16}$$

$\bar{J}_{si}$ is transformed as:

$$\begin{bmatrix} -\frac{\cos{}^\iota\phi_\zeta}{|B_i|} + 1 & \frac{\sin{}^\iota\phi_\zeta}{d_\iota|B_i|} \\ -\frac{l_\iota}{d_\kappa}\cos{}^\kappa\phi_\iota - \sum_{m=\iota+1}^{\kappa-1}\frac{L_m}{d_\kappa}\cos{}^\kappa\phi_m - 1 & \frac{1}{d_\kappa}\sin{}^\kappa\phi_\iota \end{bmatrix}$$
$$\to \begin{bmatrix} \frac{d_\iota}{l_\zeta}\left(\cos{}^\iota\phi_\zeta - |B_i|\right) & \frac{d_\iota\sin{}^\iota\phi_\zeta}{l_\zeta d_\iota} \\ -\frac{l_\iota}{d_\kappa}\cos{}^\kappa\phi_\iota - \sum_{m=\iota+1}^{\kappa-1}\frac{L_m}{d_\kappa}\cos{}^\kappa\phi_m - 1 & \frac{1}{d_\kappa}\sin{}^\kappa\phi_\iota \end{bmatrix}$$
$$= \begin{bmatrix} \frac{d_\iota}{l_\zeta}\cos{}^\iota\phi_\zeta + \sum_{m=\zeta+1}^{\iota-1}\frac{L_m}{l_\zeta}\cos{}^\zeta\phi_m + 1 & -\frac{1}{l_\iota}\sin{}^\iota\phi_\zeta \\ -\frac{l_\iota}{d_\kappa}\cos{}^\kappa\phi_\iota - \sum_{m=\iota+1}^{\kappa-1}\frac{L_m}{d_\kappa}\cos{}^\kappa\phi_m - 1 & \frac{1}{d_\kappa}\sin{}^\kappa\phi_\iota \end{bmatrix}.$$

Therefore, except for $e_1^T(B_i - I) = 0^T$, (10) is rank deficient if and only if $\forall\phi_i$, (15). □

Equation (15) also indicates as well as (11) that the singular configuration of the snake-like robot depends on the designable parameters as well as the state vector.

Lemma 2. *B_i is singular if and only if the joint of next non side-lipping link is placed on a axial of the passive wheel of current non side-slipping link (see Figure 5).*

Proof. In (16),

$$\left|B_i\right| = -\frac{1}{d_\iota}\left(l_\zeta + \sum_{m=\zeta+1}^{\iota-1}L_m\cos{}^m\phi_\zeta\right),$$

and an equation in the parentheses equals the x coordinate of the joint of ιth link corresponding to ζth link. □

Lemma 3. *$e_1^T(B_i - I) = 0$ if and only if the axials of both current and next non side-slipping link overlap (see Figure 6).*

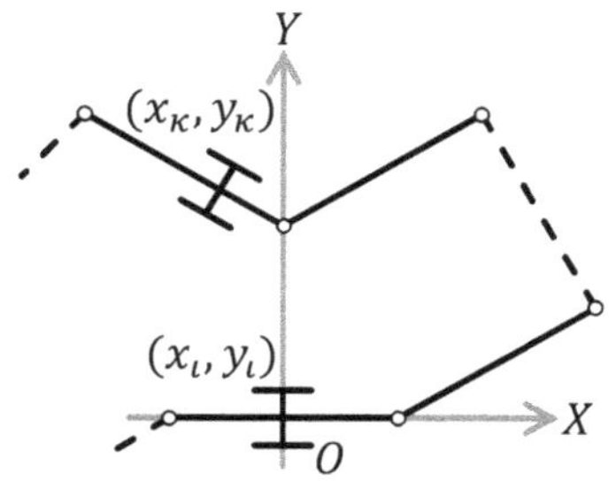

Figure 5. An example of singularity of B_i.

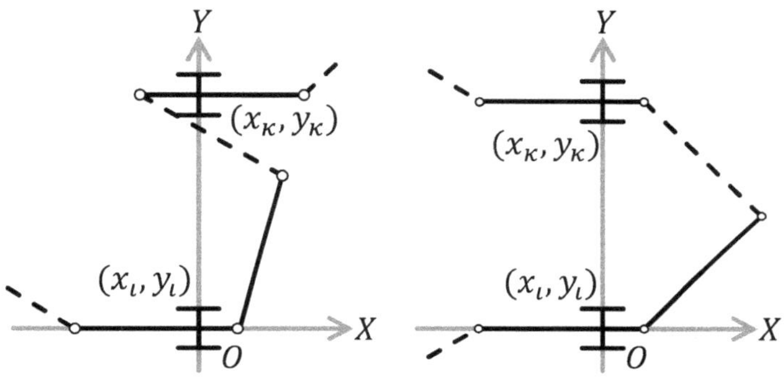

(**a**) ${}^{\iota}\phi_{\zeta} = 0$ (**b**) ${}^{\iota}\phi_{\zeta} = \pi$

Figure 6. Examples of configurations of $\boldsymbol{e}_{\mathbf{1}}^T(\boldsymbol{B_i} - \boldsymbol{I}) = 0$.

Proof.

$$\boldsymbol{e}_{\mathbf{1}}^T(\boldsymbol{B_i} - \boldsymbol{I}) = \begin{bmatrix} -\frac{l_\zeta}{d_\iota}\cos{}^{\iota}\phi_\zeta - \sum\limits_{m=\zeta+1}^{\iota-1} \frac{L_m}{d_\iota}\cos{}^{\iota}\phi_m - 1 & \frac{1}{d_\iota}\sin{}^{\iota}\phi_\zeta \end{bmatrix},$$

$$\theta_\iota = \begin{cases} \theta_\zeta, & \text{if } {}^{\iota}\phi_\zeta = 0, \\ \theta_\zeta + \pi, & \text{if } {}^{\iota}\phi_\zeta = \pi, \end{cases} \qquad \because \frac{1}{d_\iota}\sin{}^{\iota}\phi_\zeta = 0.$$

Thus,

$$-\frac{l_\zeta}{d_\iota}\cos{}^{\iota}\phi_\zeta - \sum_{m=\zeta+1}^{\iota-1} \frac{L_m}{d_\iota}\cos{}^{\iota}\phi_m - 1 = \mp\frac{1}{d_\iota}\left(l_\zeta + \sum_{m=\zeta+1}^{\iota-1} L_m \cos{}^{m}\phi_\zeta \pm d_\iota\right).$$

An equation in the parentheses equals x coordinate of the passive wheel of next non side-slipping link (ζth link) of when ${}^{\iota}\phi_\zeta = 0,\ \pi$. Hence, the axials of the passive wheel of both current and next non side-slipping link overlap. □

Theorem 4. *The snake-like robot is singular if and only if the axials of passive wheel of all non side-slipping links intersect at a common point or are parallel.*

Proof. ιth link is focused on. A coordinate is set the origin at position of the passive wheel of ιth link, and let the X axis and the Y axis are on its propulsive direction and normal direction of the passive wheel respectively. Let b_ζ and b_κ are an intercept of the axial of the passive wheel of ζth link and the one of κth link respectively. The axials of the passive wheels of all ζth, ιth and κth link intersect at a common point, if $b_\zeta = b_\kappa$. It is proven that equations of the intercepts on that situation equals to (15). Figure 7 depicts the axials of all ζth, ιth and κth link.

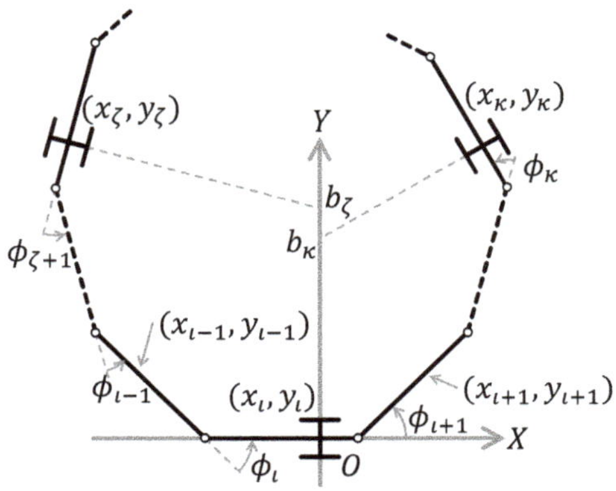

Figure 7. Schematic figure of the axials of the passive wheels in ζth and κth link, when ιth link is focused on.

The position of the passive wheel of κth link is

$$\begin{cases} x_\kappa = l_\iota + \sum\limits_{m=\iota+1}^{\kappa-1} L_m \cos{}^m\phi_\iota + d_\kappa \cos{}^\kappa\phi_\iota, \\ y_\kappa = \sum\limits_{m=\iota+1}^{\kappa-1} L_m \sin{}^m\phi_\iota + d_\kappa \sin{}^\kappa\phi_\iota. \end{cases}$$

Since a slope of the κth link equals $\tan{}^\kappa\phi_\iota$, a slope of the axial of the passive wheel is follow:

$$a_\kappa = -\frac{1}{\tan{}^\kappa\phi_\iota} = -\frac{\cos{}^\kappa\phi_\iota}{\sin{}^\kappa\phi_\iota}.$$

Hence, the intercept of the passive wheel of the κth link b_κ is

$$b_\kappa = \frac{1}{\sin{}^\kappa\phi_\iota}\left(l_\kappa \cos{}^\kappa\phi_\iota + \sum_{m=\iota+1}^{\kappa-1} L_m \cos{}^\kappa\phi_m + d_\kappa\right).$$

Also, the intercept of the passive wheel of the ζth link b_ζ is

$$b_\zeta = \frac{1}{\sin{}^\iota\phi_\zeta}\left(d_\iota \cos{}^\iota\phi_\zeta + \sum_{m=\zeta+1}^{\iota-1} L_m \cos{}^\zeta\phi_m + l_\zeta\right).$$

If $b_\zeta = b_\kappa$,

$$\frac{1}{\sin{}^\iota\phi_\zeta}\left(d_\iota \cos{}^\iota\phi_\zeta + \sum_{m=\zeta+1}^{\iota-1} L_m \cos{}^\zeta\phi_m + l_\zeta\right) = \frac{1}{\sin{}^\kappa\phi_\iota}\left(l_\kappa \cos{}^\kappa\phi_\iota + \sum_{m=\iota+1}^{\kappa-1} L_m \cos{}^\kappa\phi_m + d_\kappa\right).$$

By multiplying $\frac{1}{l_\zeta d_\kappa} \sin{}^\iota\phi_\zeta \sin{}^\kappa\phi_\iota$ to the both side,

$$\left(\frac{d_\iota}{l_\zeta} \cos{}^\iota\phi_\zeta + \sum_{m=\zeta+1}^{\iota-1} \frac{L_m}{l_\zeta} \cos{}^\zeta\phi_m + 1\right) \frac{1}{d_\kappa} \sin{}^\kappa\phi_\zeta = \frac{1}{l_\iota} \sin{}^\iota\phi_\zeta \left(\frac{l_\iota}{d_\kappa} \cos{}^\kappa\phi_\iota + \sum_{m=\iota+1}^{\kappa-1} \frac{L_m}{d_\kappa} \cos{}^\kappa\phi_\iota + 1\right) \quad (17)$$

The axials of all passive wheel intersect at a common point when $\forall \phi_i$, (17). Since (17) equals (15), the axials of passive wheel of all non side-slipping links intersect at a common point when the snake-like robot is singular.

When ${}^{\iota}\phi_{\zeta} = {}^{\kappa}\phi_{\iota} = 0$, (15) equals (17) at 0. Since ${}^{\iota}\phi_{\zeta} = {}^{\kappa}\phi_{\iota} = 0$ represents the axials of the passive wheel of non side-slipping link are parallel, the axials of passive wheel of all non side-slipping links are parallel when the snake-like robot is singular. □

5. Numerical Simulation

It has been so far identified as following two conditions. One is that the axials of all passive wheel intersect at a point. Another is that the axials of all passive wheel are parallel. To show the singular configuration particularly, this section visualizes the singular configuration via two simulations. According to the analysis process in this paper, first simulation verifies that the singaular configuration of the snake-like robot depends on the designable parameters as well as the state vector. Second simulation verifies that the singular configuration of the snake-like robot is able to be characterized by axials of the passive wheels of all non side-slipping links. Both simulations analyze both the snake-like robot including side-slipping and the one with non side-slipping, and performed by MaTX VC version 5.3.45 [37]. The designable parameters used in the both simulations is shown in Table 3.

The first simulation visualizes the state vectors of the singular configuration in the state coordinate space. The state vectors of the singular configuration of the both 3-link and 4-link snake-like robot with non side-slipping are calculated by solving (11). In addition, the state vectors of the singular configuration of the 4-link snake-like robot including side-slipping are calculated by solving (15).

Table 3. Parameters in Numerical Simuations ($i = 1, 2, 3, 4$).

Parameters	Case 1	Case 2	Case 3
d_i [m]	1.0	0.7	0.2
l_i [m]	1.0	0.5	0.8

Simulation results are shown in Figures 8–11.

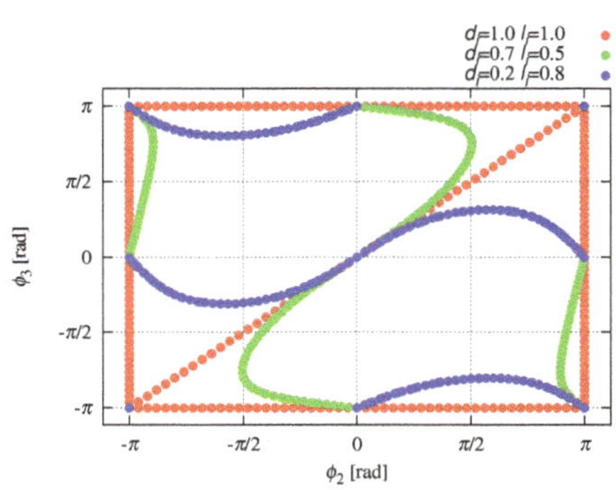

Figure 8. Singular configuration of the 3-link snake-like robot with non-sideslipping. **Red:** Singular configuration with $d_i = 1.0$ m and $l_i = 1.0$ m (Case 1 in Table 3). This configuration has been defined as singular configuration in the previous studies. **Green:** Singular configuration with $d_i = 0.7$ m and $l_i = 0.6$ m (Case 2 in Table 3). **Blue:** Singular configuration with $d_i = 0.2$ m and $l_i = 0.8$ m (Case 3 in Table 3).

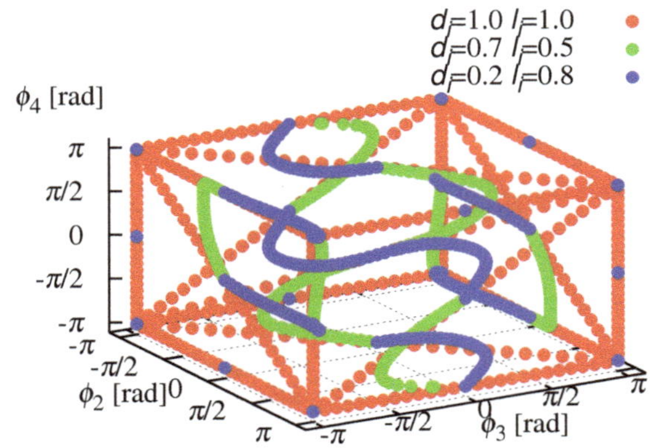

Figure 9. Singular configuration of the 4-link snake-like robot with non-sideslipping. **Red:** Singular configuration with $d_i = 1.0$ m and $l_i = 1.0$ m (Case 1 in Table 3). This configuration has been defined as singular configuration in the previous studies. **Green:** Singular configuration with $d_i = 0.7$ m and $l_i = 0.6$ m (Case 2 in Table 3). **Blue:** Singular configuration with $d_i = 0.2$ m and $l_i = 0.8$ m (Case 3 in Table 3).

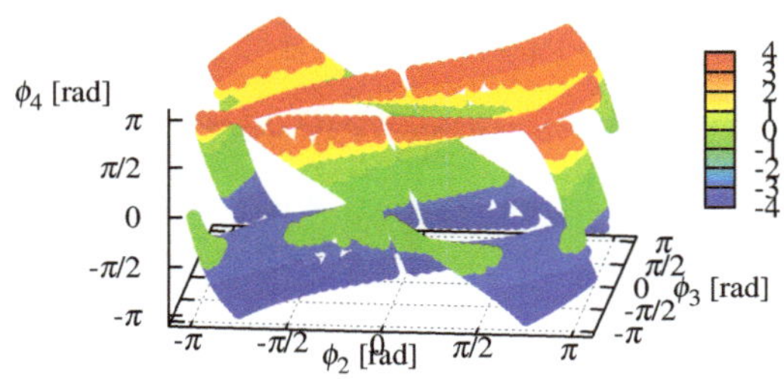

Figure 10. Singular configuration of the 4-link snake-like robot including side-slipping with $d_i = 0.7$ m and $l_i = 0.6$ m (Case 2 in Table 3). The third link is supposed as slipping link.

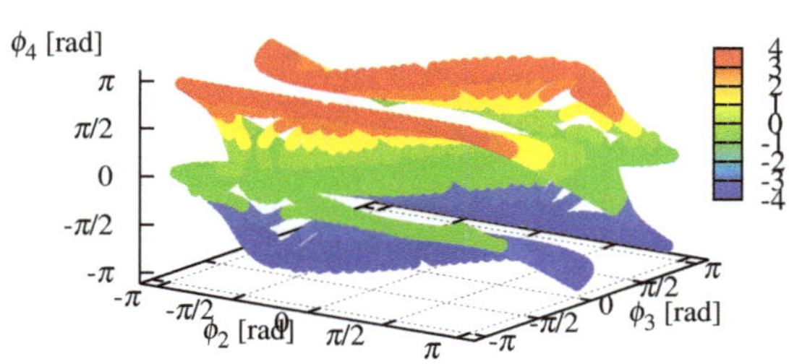

Figure 11. Singular configuration of the 4-link snake-like robot including side-slipping with $d_i = 0.2$ m and $l_i = 0.8$ m (Case 3 in Table 3). The third link is supposed as slipping link.

Figures 8 and 9 show the relative angles on the singular configuration in the state coordinate space. Red shows singular configuration with $d_i = 1.0$ m and $l_i = 1.0$ m (Case 1 in Table 3). This configuration has been defined as singular configuration in the previous studies. In fact, red shows state values of either straight line or arc shape. If the singular configuration is either straight line or arc shape—i.e., it depends on the state values only, the relative angles in the state coordinate space are unchanged even if the designable parameters are changed. However, green and blue in both Figures 8 and 9 are absolutely discord. That is, Figures 8 and 9 clearly show that the singular configuration of the snake-like robot depends on the designable parameters as well as the state values. Figures 10 and 11 show the relative angles on the singular configuration including side-slipping in the state coordinate space, and these are absolutely discord. Hence, Figures 10 and 11 also clearly show that the singular configuration of the snake-like robot depends on the designable parameters as well as the state values. Thus, the designable parameters affect dynamics of the snake-like robot.

Second simulation shows the axials of the passive wheel of all non side-slipping links intersect at a common point when the snake-like robot is singular. The singular configuration of 4-link snake-like robot is analyzed. Case 2 and Case 3 in Table 3 are chosen as the designable parameters. Let $\phi_2 = 1.2$ rad for Case 2 and $\phi_2 = \pi/3$ for Case 3, and the intercepts of the axials of the passive wheel and the singular configurations are obtained by solving (11) and (14). Simulation results are shown in Figures 12–15.

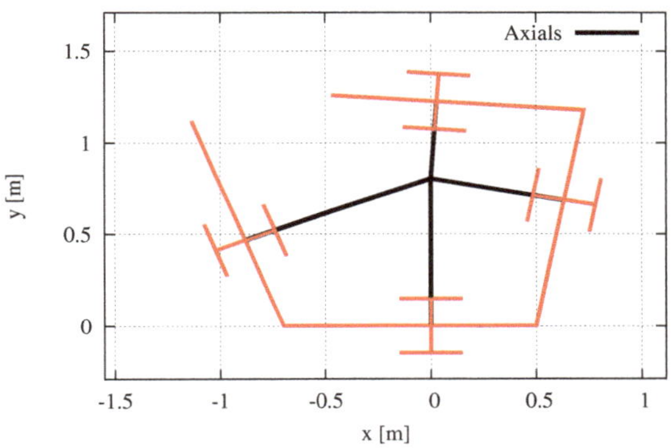

Figure 12. Case 2: The singular configuration of 4-link snake-like robot without side-slipping when setting $\phi_2 = 1.2$ rad, $d_i = 0.7$ m and $l_i = 0.5$ m. $\phi_3 = 1.38$ rad and $\phi_4 = 1.69$ rad on this situation. **Brack:** The axial of the passive wheel. The posture is not arc shape, however, these intersect at a point.

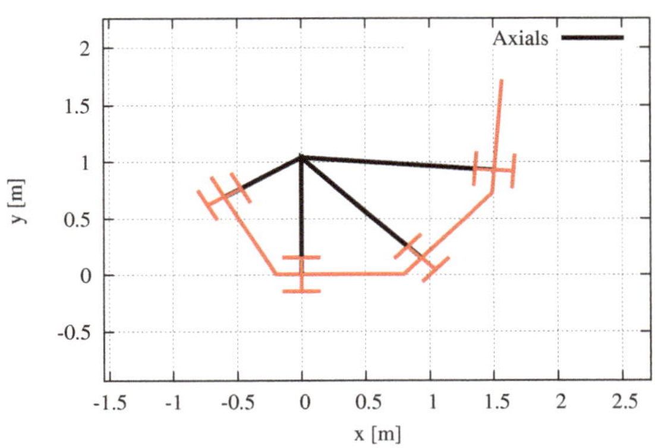

Figure 13. Case 3: The singular configuration of 4-link snake-like robot without side-slipping when setting $\phi_2 = \pi/3$ rad, $d_i = 0.2$ m and $l_i = 0.8$ m. $\phi_3 = 0.81$ rad and $\phi_4 = 0.68$ rad on this situation. **Brack:** The axial of the passive wheel. The posture is not arc shape, however, these intersect at a point.

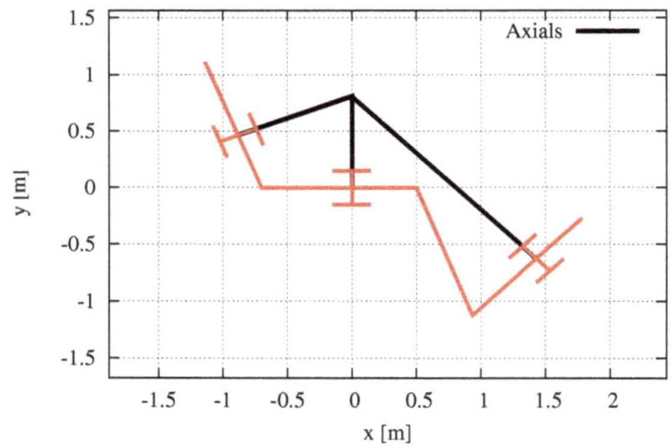

Figure 14. Case 2: The singular configuration of 4-link snake-like robot including side-slipping when setting $\phi_2 = 1.2$ rad, $\phi_3 = -1.2$ rad as slipping link, $d_i = 0.7$ m and $l_i = 0.5$ m. $\phi_4 = 1.98$ rad on this situation. **Brack:** The axial of the passive wheel. The posture is not arc shape, however, these intersect at a point.

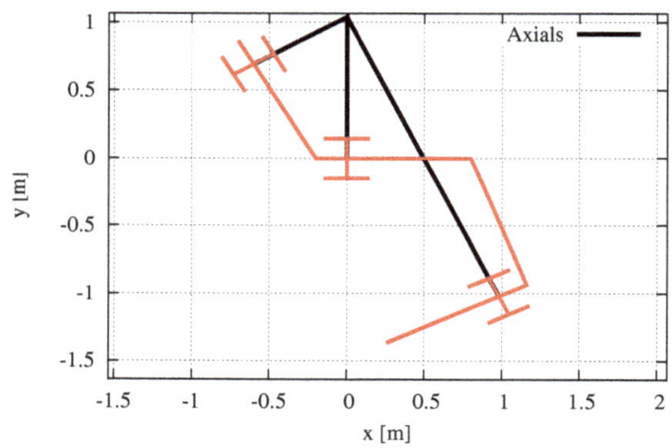

Figure 15. Case 3: The singular configuration of 4-link snake-like robot including side-slipping when setting $\phi_2 = \pi/3$ rad, $\phi_3 = -1.2$ rad as slipping link, $d_i = 0.2$ m and $l_i = 0.8$ m. $\phi_4 = -1.50$ rad on this situation. **Brack:** The axial of the passive wheel. The posture is not arc shape, however, these intersect at a point.

Figures 12–15 shows that the axials of the passive wheel intersect at a common point, when the snake-like robot is singular. Since $\phi_3 = 1.38$ rad and $\phi_4 = 1.69$ rad in Figure 12, $\phi_3 = 0.81$ rad and $\phi_4 = 0.68$ rad in Figure 13, $\phi_4 = 1.98$ rad in Figure 14 and $\phi_4 = -1.50$ rad in Figure 15, the singular configuration of the snake-like robot is not limited to either straight line or arc shape.

This two simulations has proven that the singular configuration of the snake-like robot depends on the designable parameters as well as the state values, and it is characterized by the intersection of the axials rather than the state values.

6. Conclusions

This paper has elucidated the generalized singularity analysis of the snake-like robot affected by the designable parameters such as the link length and/or the position of the passive wheel. This paper has addressed the model installing the passive wheel on arbitrary position, while the previous studies had addressed the model installing the wheel on the center of each link.

By applying the method using redundancy of the snake-like robot, we have reduced high complexity due to different set-up corresponding to the non-unique designable parameters such as link length and/or the position of the passive wheel. First, the kinematics around a joint of two adjacent links installing the passive wheel on arbitrary position have been derived corresponding to four patterns of side-slipping. By composing the joint kinematics redundantly, the kinematics of whole snake-like robot is formulated. Second, the Jacobian matrices of both the one including side-slipping and the one with non side-slipping have been formulated by using the kinematics. Finally, the rank deficient conditions of the Jacobian matrices—i.e., the singular configuration of the snake-like robot have been analyzed.

The analysis creates two contributions. The first contribution is that the singular configuration depends on the designable parameters such as link length and/or the position of the passive wheel as well as the state vector consisted of the relative angles. Theorem 1 and 3 have revealed the relation between the singular configuration of the snake-like robot and the designable parameters, while the previous studies have reported that the singular configuration depends on only the state vector. The second contribution is that the singular configuration of the snake-like robot is able to be characterized by the axials of the passive wheel of all non side-slipping link. Theorem 2 and 4 have proven that the singular configuration of the snake-like robot is identified as the following two conditions regardless of whether the side-slippings actually exist.

- Axials of passive wheels of all non side-slipping links intersect at a common point.
- Axials of passive wheels of all non side-slipping links are parallel.

These analyses lead shows discussions on the system design in the previous studies was insufficient. Conversely, this paper has shown importance, necessity and possibility of the discussion with respect to the system design including the locomotion control system as well as the designable parameters in the area of the snake-like robot. The discussion of the system design has a lot of possibilities to be a new index of mechanical design. For example in the area of the snake-like robot, we will be able to design a robot which is unlikely to result in the singular configuration as well as a novel locomotion control system. For another instance, the discussion will lead to invent a novel snake-like robot which can control the position of passive wheels actively. Also for another robotic system, the discussion helps robot's mechanical design corresponding to its intended use. Especially for unstable system, the discussion contributes to maximize its stability region.

Author Contributions: S.N., M.I. and H.I. conceptualized the paper. S.N. conceived and proofed the theorems and the lemmas. S.N. performed the numerical simulations. S.N., M.I. and H.I. wrote the paper.

Funding: This research received no external funding.

Conflicts of Interest: The authors declare no conflict of interest.

References

1. Hirose, S. *Biologically Inspired Robots: Snake-Like Locomotors and Manipulators*; Oxford University Press: Oxford, UK, 1993.
2. Endo, G.; Togawa, K.; Hirose, S. Study on self-contained and terrain adaptive active cord mechanism. In Proceedings of the IEEE/RSJ International Conference on Intelligent Robots and Systems IROS'99, Kyongju, Korea, 17–21 October 1999; Volume 3, pp. 1399–1405.
3. Yamada, H.; Mori, M.; Hirose, S. Stabilization of the head of an undulating snake-like robot. In Proceedings of the IEEE/RSJ International Conference on Intelligent Robots and Systems, San Diego, CA, USA, 29 October–2 November 2007; pp. 3566–3571.
4. Ma, S.; Tadokoro, N. Analysis of creeping locomotion of a snake-like robot on a slope. *Auton. Robots* **2006**, *20*, 15–23. [CrossRef]
5. Nansai, S.; Iwase, M. Tracking control of snake-like robot with rotational elastic actuators. In Proceedings of the 2012 12th International Conference on Control, Automation and Systems (ICCAS), JeJu Island, Korea, 17–21 October 2012; pp. 678–683.

6. Yamakita, M.; Hashimoto, M.; Yamada, T. Control of locomotion and head configuration of 3D snake robot (SMA). In Proceedings of the IEEE International Conference on Robotics and Automation, ICRA'03, Taipei, Taiwan, 14–19 September 2003; Volume 2, pp. 2055–2060.
7. Matsuno, F.; Sato, H. Trajectory tracking control of snake robots based on dynamic model. In Proceedings of the IEEE International Conference on Robotics and Automation, ICRA 2005, Barcelona, Spain, 18–22 April 2005; pp. 3029–3034.
8. Watanabe, K.; Iwase, M.; Hatakeyama, S.; Maruyama, T. Control strategy for a snake-like robot based on constraint force and its validation. In Proceedings of the 2007 IEEE/ASME International Conference on Advanced Intelligent Mechatronics, Zurich, Switzerland, 4–7 September 2007; pp. 1–6.
9. Watanabe, K.; Iwase, M.; Hatakeyama, S.; Maruyama, T. Control strategy for a snake-like robot based on constraint force and verification by experiment. *Adv. Robot.* **2009**, *23*, 907–937. [CrossRef]
10. Yanagida, T.; Kasahara, M.; Iwase, M. Locomotion Control of Snake-like Robot on Geometrically Smooth Surface. *IFAC-PapersOnLine* **2015**, *48*, 162–167. [CrossRef]
11. Nansai, S.; Elara, M.R.; Iwase, M. Dynamic Hybrid Position Force Control using Virtual Internal Model to realize a cutting task by a snake-like robot. In Proceedings of the 2016 6th IEEE International Conference on Biomedical Robotics and Biomechatronics (BioRob), Singapore, 26–29 June 2016; pp. 151–156.
12. Tashiro, K.; Nansai, S.; Iwase, M.; Hatakeyama, S. Development of snake-like robot climbing up slope in consideration of constraint force. In Proceedings of the IECON 2012-38th Annual Conference on IEEE Industrial Electronics Society, Montreal, QC, Canada, 25–28 October 2012; pp. 5422–5427.
13. Prautsch, P.; Mita, T. Control and analysis of the gait of snake robots. In Proceedings of the 1999 IEEE International Conference on Control Applications, Kohala Coast, HI, USA, 22–27 August 1999; Volume 1, pp. 502–507.
14. Prautsch, P.; Mita, T.; Iwasaki, T. Analysis and control of a gait of snake robot. *IEEJ Trans. Ind. Appl.* **2000**, *120*, 372–381. [CrossRef]
15. Date, H.; Hoshi, Y.; Sampei, M. Locomotion control of a snake-like robot based on dynamic manipulability. In Proceedings of the IEEE/RSJ International Conference on Intelligent Robots and Systems, Takamatsu, Japan, 31 October–5 November 2000; Volume 3, pp. 2236–2241.
16. Date, H.; Hoshi, Y.; Sampei, M.; Nakaura, S. Locomotion control of a snake robot with constraint force attenuation. In Proceedings of the American Control Conference, Arlington, VA, USA, 25–27 June 2001; Volume 1, pp. 113–118.
17. Date, H.; Sampei, M.; Nakaura, S. Control of a snake robot in consideration of constraint force. In Proceedings of the 2001 IEEE International Conference on Control Applications (CCA'01), Mexico City, Mexico, 7 September 2001; pp. 966–971.
18. Ye, C.; Ma, S.; Li, B.; Wang, Y. Locomotion control of a novel snake-like robot. In Proceedings of the 2004 IEEE/RSJ International Conference on Intelligent Robots and Systems (IROS 2004), Sendai, Japan, 28 September–2 October 2004; Volume 1, pp. 925–930.
19. Matsuno, F.; Sato, H. Trajectory tracking control of snake robots based on dynamic model. In Proceedings of the 2005 IEEE International Conference on Robotics and Automation, ICRA 2005, Barcelona, Spain, 18–22 April 2005; pp. 3029–3034.
20. Tanaka, M.; Tanaka, K. Climbing and descending control of a snake robot on step environments based on kinematics. In Proceedings of the 2013 IEEE/RSJ International Conference on Intelligent Robots and Systems (IROS), Tokyo, Japan, 3–7 November 2013; pp. 3285–3290.
21. Tanaka, M.; Matsuno, F. Modeling and control of head raising snake robots by using kinematic redundancy. *J. Intell. Robot. Syst.* **2014**, *75*, 53–69. [CrossRef]
22. Tanaka, M.; Tanaka, K. Control of a snake robot for ascending and descending steps. *IEEE Trans. Robot.* **2015**, *31*, 511–520. [CrossRef]
23. Dear, T.; Kelly, S.D.; Travers, M.; Choset, H. Locomotive analysis of a single-input three-link snake robot. In Proceedings of the 2016 IEEE 55th Conference on Decision and Control (CDC), Las Vegas, NV, USA, 12–14 December 2016; pp. 7542–7547.
24. Dear, T.; Kelly, S.D.; Travers, M.; Choset, H. The three-link nonholonomic snake as a hybrid kinodynamic system. In Proceedings of the American Control Conference (ACC), Boston, MA, USA, 6–8 July 2016; pp. 7269–7274.

25. Guo, X.; Ma, S.; Li, B.; Wang, M.; Wang, Y. Modeling and optimal torque control of a snake-like robot based on the fiber bundle theory. *Sci. China Inf. Sci.* **2015**, *58*, 1–13. [CrossRef]
26. Liljebäck, P.; Pettersen, K.Y.; Stavdahl, Ø.; Gravdahl, J.T. Controllability analysis of planar snake robots influenced by viscous ground friction. In Proceedings of the IEEE/RSJ International Conference on Intelligent Robots and Systems, St. Louis, MO, USA, 10–15 October 2009; pp. 3615–3622.
27. Liljebäck, P.; Pettersen, K.Y.; Stavdahl, Ø.; Gravdahl, J.T. Stability analysis of snake robot locomotion based on averaging theory. In Proceedings of the 2010 49th IEEE Conference on Decision and Control (CDC), Atlanta, GA, USA, 15–17 December 2010; pp. 1977–1984.
28. Liljebäck, P.; Pettersen, K.Y.; Stavdahl, Ø.; Gravdahl, J.T. Controllability and stability analysis of planar snake robot locomotion. *IEEE Trans. Autom. Control* **2011**, *56*, 1365–1380.
29. Pettersen, K.Y.; Liljebäck, P.; Stavdahl, Ø.; Gravdahl, J.T. Snake Robots From Biology to Nonlinear Control. *IFAC Proc. Volumes* **2013**, *46*, 110–115. [CrossRef]
30. Tanaka, M.; Tanaka, K. Singularity analysis of a snake robot and an articulated mobile robot with unconstrained links. *IEEE Trans. Control Syst. Technol.* **2016**, *24*, 2070–2081. [CrossRef]
31. Onishi, Y.; Izutu, M.; Iwase, M. Link lengths search using genetic algorithm for stabilization of quintuple inverted pendulum. In Proceedings of the 2011 SICE Annual Conference (SICE), Tokyo, Japan, 13–18 September 2011; pp. 739–744.
32. Shiraishi, D.; Oonishi, Y.; Izutsu, M.; Hatakeyama, S. Study of system structure with controllability matrix. In Proceedings of the 54th Japan Joint Automatic Control Conference, Tokyohashi, Japan, 19–20 November 2011; Volume 54, p. 29.
33. Nagaki, K.; Izutsu, M.; Hatakeyama, S.; Iwase, M. Structure evaluation applying stabilizable space for physical parameter. In Proceedings of the IECON 2016-42nd Annual Conference of the IEEE Industrial Electronics Society, Florence, Italy, 23–26 October 2016; pp. 642–647.
34. Nansai, S.; Rojas, N.; Elara, M.R.; Sosa, R. Exploration of adaptive gait patterns with a reconfigurable linkage mechanism. In Proceedings of the 2013 IEEE/RSJ International Conference on Intelligent Robots and Systems (IROS), Tokyo, Japan, 3–7 November 2013; pp. 4661–4668.
35. Nansai, S.; Rojas, N.; Elara, M.R.; Sosa, R.; Iwase, M. On a Jansen leg with multiple gait patterns for reconfigurable walking platforms. *Adv. Mech. Eng.* **2015**, *7*. [CrossRef]
36. Nansai, S.; Rojas, N.; Elara, M.R.; Sosa, R.; Iwase, M. A novel approach to gait synchronization and transition for reconfigurable walking platforms. *Digit. Commun. Netw.* **2015**, *1*, 141–151. [CrossRef]
37. Koga, M. MaTX/RtMaTX: A Freeware for Integrated CACSD. In Proceedings of the 1999 IEEE International Symposium on Computer Aided Control System Design, Kohala Coast, HI, USA, 27 August 1991; pp. 452–456.

applied sciences

MDPI

Article

Power Assist Control Based on Human Motion Estimation Using Motion Sensors for Powered Exoskeleton without Binding Legs

Shinnosuke Nomura [1], Yasutake Takahashi [2,*], Katsuya Sahashi [1], Shota Murai [1], Masayuki Kawai [2], Yoshiaki Taniai [2] and Tomohide Naniwa [1]

1 Graduate School of Engineering, University of Fukui, Fukushi 910-8507, Japan; snomura@ir.his.u-fukui.ac.jp (S.N.); ksahashi@ir.his.u-fukui.ac.jp (K.S.); smurai@ir.his.u-fukui.ac.jp (S.M.); naniwa@rbt.his.u-fukui.ac.jp (T.N.)

2 Faculty of Engineering, University of Fukui, Fukushi 910-8507, Japan; m_kawai@u-fukui.ac.jp (M.K.); taniai@u-fukui.ac.jp (Y.T.)

* Correspondence: yasutake@ir.his.u-fukui.ac.jp; Tel.: +81-776-27-8540

Received: 14 November 2018; Accepted: 25 December 2018; Published: 4 January 2019

Abstract: In this study, we propose a novel power assist control method for a powered exoskeleton without binding its legs. The proposed method uses motion sensors on the wearer's torso and legs to estimate his/her motion to enable the powered exoskeleton to assist with the estimated motion. It can detect the start of walking motion quickly because it does not prevent the motion of the wearer's knees at the beginning of the walk. A nine-axis motion sensor on the wearer's body is designed to work robustly in very hot and humid spaces, where an electromyograph is not reliable due to the wearer's sweat. Moreover, the sensor avoids repeated impact during the walk because it is attached to the body of the wearer. Our powered exoskeleton recognizes the motion of the wearer based on a database and accordingly predicts the motion of the powered exoskeleton that supports the wearer. Experiments were conducted to prove the validity of the proposed method.

Keywords: powered exoskeleton; motion sensor; machine learning

1. Introduction

Powered exoskeletons are nowadays used in various fields, such as agriculture, and medical and welfare services [1–4]. They have a wide variety of applications in numerous fields. A powered exoskeleton in the field of rehabilitation [5–8] has low output power for safety assistance. On the other hand, the assisting power used to transport heavy baggage tends to be high [9,10]. Powered exoskeletons have also been developed for workers in a nuclear power plant [11,12]. We have also been developing a powered exoskeleton for workers who transport heavy baggage in a nuclear power plant. In case of a nuclear hazard, workers need to wear radiation-protective equipment that weighs approximately 40 kg. Moreover, the worker is supposed to carry a heavy exploration robot, such as the PackBot [13], around in a nuclear reactor for efficient exploration. The target of our study is to develop a powered exoskeleton that is used to support workers in a nuclear plant who need to wear heavy radiation-protective equipment and carry an exploration robot.

Several approaches have been proposed to control powered exoskeletons. One is based on myoelectric signals measured by electromyography (EMG) sensors [8,14–16]. It estimates human intentions by measuring the action potential of the muscles, and the powered exoskeleton assists human action according to the intention. Since action potential occurs approximately 50 ms before the relevant muscle contracts, the powered exoskeleton enables rapid power assistance. However, it can easily be affected by human sweat in a hot environment. It is thus unsuitable for our intended application because workers often sweat in a radiation-protective equipment [17]. The equipment is composed of highly airtight materials so that temperature and humidity inside the equipment are high.

Another approach is based on a force sensor/switch. Berkeley Lower Extremity Exoskeleton (BLEEX) [18–21] uses pressure sensors that measure the force between the shoe of the powered exoskeleton and the foot of the wearer to control the exoskeleton according to the configuration of the foot relative to the ground. Sano et al. [22,23] also proposed using force sensors attached to the bottom of the wearer's feet to detect the pressure between the shoes of the exoskeleton and the ground. These exoskeletons control joint angle and angular velocity based on the given state of the leg, such as "stance phase" or "swing phase", estimated by the force switches/sensors. We have examined this approach. We place a force sensor on the wearer's back to measure the weight of the load on it. The powered exoskeleton controls itself to generate assist torque based on the floor's reaction force. The powered exoskeleton assists the worker to carry the load according to the measured weight of the load. It has an advantage of not being affected by human sweat. However, we found that this approach cannot distinguish among similar motions, such as "walking forward" and "walking backward". This means that the approach restricts the motion that can be assisted, and the motion needs to be designed in advance. Actually, the potential user needs to do many motions, including walking forward and backward, squat, going up and coming down stairs, run, one-leg standing, and so on. In this research, we focus on only three motions, standing upright position, walking forward, and walking backward. The motion "walking backward" is necessary because we suppose that it is hard to turn around in case that the passageway in the nuclear power plant is narrow. Furthermore, the outputs of a force sensor tend to be noisy, especially at the time of impact. Reactive assist control based on force sensors tends to be jerky due to noise. Low-pass filters can be applied but slow down the assist control. Moreover, they are likely to cause hardware trouble because of repeated impact during the walk because they are likely attached to the bottom of the foot of the powered exoskeleton.

We adopted the PLL-01 [12], designed and developed by Activelink Co., Ltd., Japan, for our study. Its major feature is that it does not bind the legs of the wearer. It binds only the wearer's shoulders and feet so that he/she can move his/her legs freely at the beginning of the motion because there is room to move knees due to the redundancy in the link structure of the human body, even if the joints of the exoskeleton are fixed. Other popular powered exoskeletons often bind the upper and lower legs tightly to links of the exoskeletons. Consequently, the wearer must push the exoskeleton intentionally until it estimates the motion of the wearer and begins assistance according to the estimated motion. Our powered exoskeleton enables the wearer to move his/her legs freely at the beginning of the motion, so that motion sensors on his/her legs and torso can detect motion and quickly start assistance according to the estimated motion. Liu et al. [24] proposed a powered exoskeleton that does not bind the legs of the wearer. However, they bound lightweight, rigid bars to the wearer's legs and measured his/her joint angles using magnetic rotary encoders. Even if the rigid bars are lightweight, they restrict the motion of the wearer. It is well known that the human joint is not a hinge joint. The center of rotation of the human joint changes during bending and extension. Therefore, rigid bars with hinge joints are not suitable for measuring human motion because they restrict human motion. Researchers have also reported motion recognition systems using nine-axis motion sensors for powered exoskeletons [25]. Seven sensors are attached to the wearer's trunk and legs to estimate his/her motion based on hidden semi-Markov models. Such systems can estimate the wearer's motion; however, the only experiments on it were conducted without the user wearing a powered exoskeleton.

We propose a novel approach for power assist control of powered exoskeletons based on human motion estimation using the nine-axis motion sensors. The motion sensor can measure the wearer's motion in a high-temperature and humid environment. Our powered exoskeleton does not bind the wearer's legs, unlike other popular powered exoskeletons, such that he/she can move his/her legs freely at beginning of the motion. Therefore, it can quickly detect the start of walking motion. Our method estimates the wearer's motion using a motion sensor and controls the exoskeleton based on the estimated motion. Our motion estimation and assist control are based on a motion database of the wearer and the powered exoskeleton. The database consists of sequential output data from the motion sensors attached to the wearer and the joint angles at the waist and knees of the powered exoskeleton during specific motions. An advantage of the proposed method is that it can recognize several motions of the wearer that are challenging for other similar methods. Another advantage is its low cost. Motion sensors are cheaper than commercially available load cells and are robust such that they avoid repeated impact during a walk because they are attached to the wearer's limbs. A force sensor or force switch embedded into the bottom of the foot can be easily broken because of the direct impact with the floor during the walk. This paper shows the effectiveness of the proposed method through experiments with a powered exoskeleton.

2. Powered Exoskeleton without Binding Legs

Figures 1 and 2 show the powered exoskeleton, designed and developed by Activelink Co., Ltd., Nara City, Japan [12], used in this research. It consisted of four geared motors and rotary encoders at the knee and hip joints. Their joint angles were controlled by PID controllers. There was no motor at the ankle joints. The degrees of freedom of the joints are shown in Figure 1a. The powered exoskeleton bound a wearer only at his/her shoulders and feet. There was no binding at the limbs of the upper and lower legs, as in conventional powered exoskeletons. The wearer could move his/her knees freely, especially at the beginning of the motion. The powered exoskeleton was designed to have comparatively small torque, at most 50 Nm, at each joint so that the back-drivability ensured safety in case of loss of control. Therefore, the powered exoskeleton was not designed to support all loads on the wearer, but only part of it.

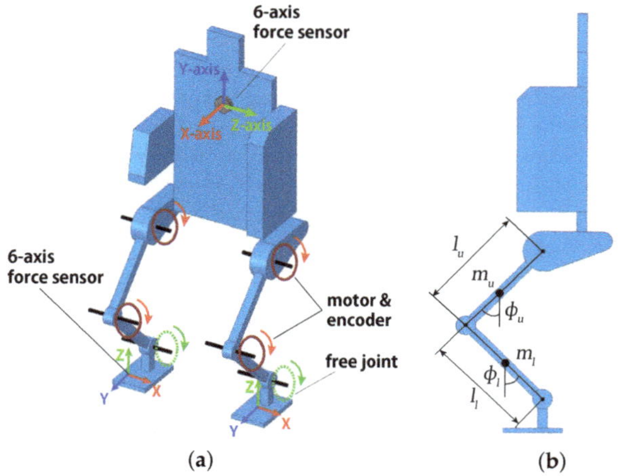

Figure 1. Configuration of powered exoskeleton. (**a**) Axes of force sensors and the degrees of freedom of the joints; (**b**) Side view and zero position of angles.

Figure 2 shows a wearer attaching five nine-axis motion sensors as well as their positions and coordinates. The x-axis was upward, the y-axis was horizontal, and the z-axis was in the forward direction. The wearer attached them to the chest and the upper and lower legs. The powered exoskeleton can distinguish the motions "walking forward" and "walking backward" based on the outputs of the motion sensors. The algorithm proposed by Sebastian Madgwick [26] was adopted to calculate the posture of the motion sensor in this paper. This method used acceleration, angular velocity, and geomagnetism measured by the motion sensors to calculate posture. Three force sensors were attached to the powered exoskeleton. One was on the back, and the others were on the feet.

Figure 3 shows the shoe designed for the wearer, the foot of the powered exoskeleton, and the force sensor attached to both. The force sensor on the back measured the load on the shoulder of the wearer. The force sensors on the feet measured the interactive force between the feet of the wearer and those of the exoskeleton. The axes of force sensors are depicted in Figure 1a. These sensors measured load and moment along the three directions. The force sensors were used only for the evaluation of our proposed method.

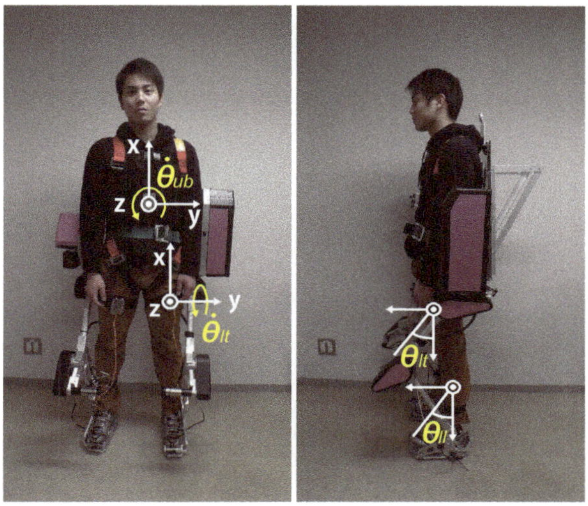

(**a**) front view (**b**) side view

Figure 2. Motion sensors attached to the body of the wearer, and the definition of the axes.

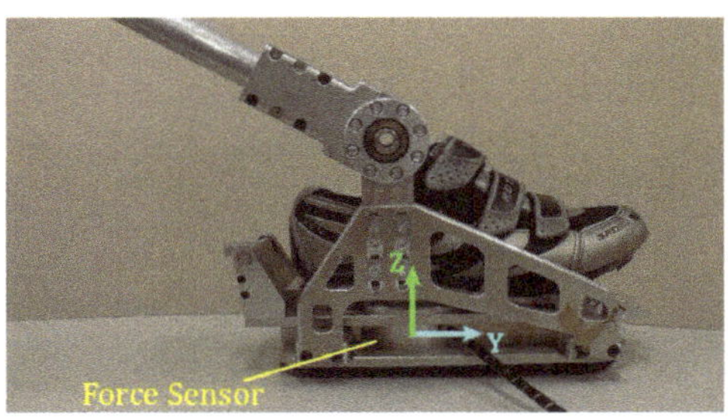

Figure 3. The shoe for the wearer, the foot of the powered exoskeleton, and the force sensor attached to both.

3. Leg Control Based on Human Motion Prediction Using Motion Sensor

Figure 4 shows the overview of the proposed controller using motion sensors for our powered exoskeleton. The powered exoskeleton recognizes the wearer's motion to assist him/her. It estimates in advance by a few hundred milliseconds the future joint angles of the powered exoskeleton according to the recognized motion to assist the wearer in real time. The motion estimation and the calculation of the desired joint angles of the powered exoskeleton are based on a motion database compiled in advance. This database is composed of sequential data of the wearer's motion, the label of the motion, and the corresponding leg motion of the powered exoskeleton. The joint angles of the legs of the powered exoskeleton are controlled to be estimated based on the database by PID controllers.

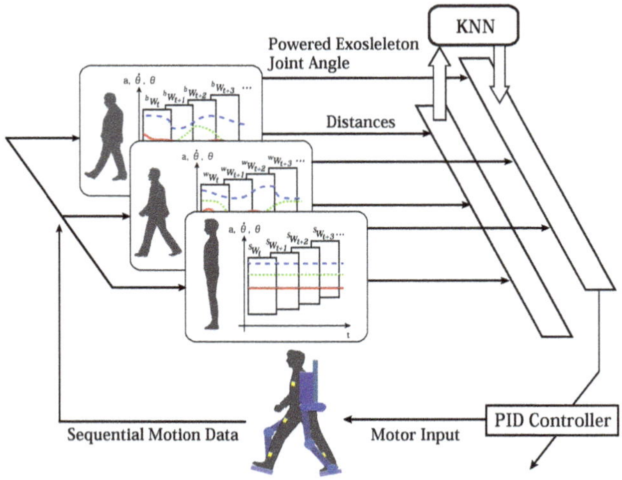

Figure 4. Overview of proposed control system.

The database includes sequential data from motion sensor attached to the wearer as feature vectors, each with motion class label "standing", "walking forward", or "walking backward", and the joint angles of the powered exoskeleton as the wearer exhibited the relevant motion. The data of the motion sensors are angles $\boldsymbol{\theta} = (\theta_{rt}, \theta_{rl}, \theta_{lt}, \theta_{ll})$, angular velocities $\dot{\boldsymbol{\theta}} = (\dot{\theta}_{rt}, \dot{\theta}_{rl}, \dot{\theta}_{lt}, \dot{\theta}_{ll}, \dot{\theta}_{ub})$, and acceleration rates $\boldsymbol{a} = (a_{rt}, a_{rl}, a_{lt}, a_{ll}, a_{ub})$. Indices rt, rl, lt, ll, and ub indicate the upper-right leg, the lower-right leg, the upper-left leg, the lower-left leg, and the torso, respectively. The joint angle of the powered exoskeleton is defined as $\boldsymbol{\phi} = (\phi_{rw}, \phi_{rk}, \phi_{lw}, \phi_{lk})$. Indices rw, rk, lw, and lk indicate the right hip, the right knee, the left hip, and the left knee of the powered exoskeleton, respectively.

The wearer's motion dataset is defined by piecewise sequences of $\boldsymbol{x}_t = (\boldsymbol{a}_t, \dot{\boldsymbol{\theta}}_t, \boldsymbol{\theta}_t)$, where t is the time index. Sequential motion data $(\boldsymbol{x}_1, \boldsymbol{x}_2, \cdots)$ are segmented using a window of size m into sequence data $\boldsymbol{X}_t = (\boldsymbol{x}_t, \boldsymbol{x}_{t+1}, \cdots, \boldsymbol{x}_{t+m})$. The sequence dataset is assigned one of the three motion category indices of "standing" c_s, "walking forward" c_w, and "walking backward" c_b. It is also assigned the joint angles of the powered exoskeleton at time $t + \Delta t$, $\boldsymbol{\phi}_{t+\Delta t}$. A dataset in the motion database is composed of the wearer's motion dataset, the motion category, and the joint angles of the powered exoskeleton, $(\boldsymbol{X}_t, c_i, \boldsymbol{\phi}_{t+\Delta t})$, where c_i is one of c_s, c_w, and c_b. The database D is composed of the set of datasets $D = \{(\boldsymbol{X}_t^0, c_i^0, \boldsymbol{\phi}_{t+\Delta t}^0), (\boldsymbol{X}_t^1, c_i^1, \boldsymbol{\phi}_{t+\Delta t}^1), \cdots\}$.

The powered exoskeleton recognizes the wearer's motion using the k-nearest neighbors method on the database D. We choose the k-nearest neighbors algorithm because it is one of the non-parametric methods that do not make some specific assumption about the motion of the human or the powered exoskeleton and it is the simplest algorithm and works in real time for our application. Motion data of the wearer at time t are defined as $\boldsymbol{x}_t = (\boldsymbol{a}_t, \dot{\boldsymbol{\theta}}_t, \boldsymbol{\theta}_t)$. Query sequential data with window size m are defined as ${}^q\boldsymbol{X} = (\boldsymbol{x}_t, \boldsymbol{x}_{t+1}, \cdots, \boldsymbol{x}_{t+m-1})$. They calculate the normalized Euclidean distance d^i between $\boldsymbol{X}_t^i$ and ${}^q\boldsymbol{X}_t$, where i is the data index in database D. It chooses k datasets from the database D with the smallest distances based on a normalized d_i, and collects the set of motion category IDs $\boldsymbol{c} = (c_1, c_2, \cdots, c_k)$, each of which is one of the motion categories c_s, c_w, and c_b. Then, the k-nearest neighbor algorithm outputs most of the motion category in $\boldsymbol{c}$. The term "majority" indicates the motion category with the maximum number of category indices in the nearest neighbor set of the motion category indices c. For example, if the number of the nearest datasets with the motion category index c_s is higher than that of datasets with motion category indices c_w and c_b, c_s is said to be in the majority. The normalized distance d_i is calculated as below:

$$d_i^2 = ({}^q\boldsymbol{X}_t - {}^c\boldsymbol{X}_t)\Sigma^{-1}({}^q\boldsymbol{X}_t - {}^c\boldsymbol{X}_t)^{\mathrm{T}} \tag{1}$$

where T indicates the transpose and Σ is a variance matrix with variance vector σ on the diagonal. The variance vector σ is the vector whose components are the variances of the corresponding components of $\boldsymbol{X}_t$, which used for database D.

It estimates the appropriate joint angles of the powered exoskeleton at time $t + \Delta t$, $\phi^d_{t+\Delta t}$, according to the estimated motion of the wearer based on the k-nearest neighbor algorithm. For example, if the estimated motion is "standing", it chooses only the datasets whose motion categories c_i is c_s for the estimation of $\phi^d_{t+\Delta t}$. An overview of the estimation of the appropriate joint angles $\phi^d_{t+\Delta t}$ is provided in Figure 5 and the algorithm is shown in Algorithm 1. The input to the joint motor u_t is calculated by Equation (2):

$$u_t = -k_p(\phi^d_{t+\Delta t} - \phi_t) - k_d\left((\phi^d_{t+\Delta t} - \phi_t) - (\phi^d_{t+\Delta t-1} - \phi_{t-1})\right) \tag{2}$$

where ϕ^d_t is the desired joint angle and ϕ_t is the actual joint angle of the powered exoskeleton at time t. k_p, k_i, and k_d are the proportional, integral, and differential gains, respectively.

Algorithm 1 Wearer's motion estimation, and calculation of joint angle of powered exoskeleton

load database $D = \{(\boldsymbol{X}^0_t, c^0_i, \boldsymbol{\phi}^0_{t+\Delta t}), (\boldsymbol{X}^1_t, c^1_i, \boldsymbol{\phi}^1_{t+\Delta t}), \cdots\}$
acquire sequential motion data of the wearer $\boldsymbol{X}_t$
$\boldsymbol{c} = knn(\boldsymbol{X}_t, D)$: components of $\boldsymbol{c}$ is c_s, c_w or c_b
if majority of $\boldsymbol{c}$ is c_s **then**
  recognizes the current motion as "standing" motion
  $\boldsymbol{\Phi} = knn(\boldsymbol{X}_t, D|c_i = c_s) : \boldsymbol{\Phi} = (\boldsymbol{\phi}^1_{t+\Delta t}, \boldsymbol{\phi}^2_{t+\Delta t}, \cdots, \boldsymbol{\phi}^k_{t+\Delta t})$
end if
if majority of $\boldsymbol{c}$ is c_w **then**
  recognizes the "walking forward" motion
  $\boldsymbol{\Phi} = knn(\boldsymbol{X}_t, D|c_i = c_w) : \boldsymbol{\Phi} = (\boldsymbol{\phi}^1_{t+\Delta t}, \boldsymbol{\phi}^2_{t+\Delta t}, \cdots, \boldsymbol{\phi}^k_{t+\Delta t})$
end if
if majority of $\boldsymbol{c}$ is c_b **then**
  recognizes the "walking backward" motion
  $\boldsymbol{\Phi} = knn(\boldsymbol{X}_t, D|c_i = c_b) : \boldsymbol{\Phi} = (\boldsymbol{\phi}^1_{t+\Delta t}, \boldsymbol{\phi}^2_{t+\Delta t}, \cdots, \boldsymbol{\phi}^k_{t+\Delta t})$
end if
$\boldsymbol{\phi}^d_{t+\Delta t} = \frac{1}{k}\sum^k_{i=1}\boldsymbol{\phi}^i_{t+\Delta t}$
return $\boldsymbol{\phi}^d_{t+\Delta t}$

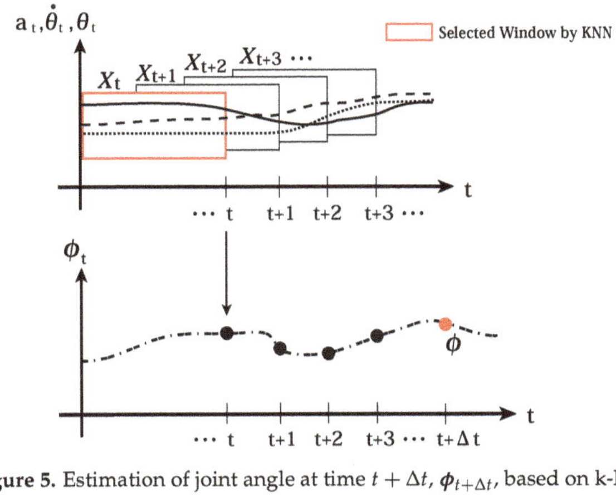

Figure 5. Estimation of joint angle at time $t + \Delta t$, $\boldsymbol{\phi}_{t+\Delta t}$, based on k-NN.

4. Comparative Methods

We evaluated the assistive performance of the proposed method with two comparable methods. We avoided the use of EMG instrument due to its drawbacks for our application, as mentioned in Section 1. We also did not rely on the foot force sensors or foot switches because repeated impact during a walk often breaks them. Therefore, a gravity compensation method was adopted as a comparative method. Unfortunately, we found that it was challenging to reduce the load on the wearer's shoulder in the following experiments. Therefore, we also adopted a foot force switch method as the other comparative method.

4.1. Gravity Compensation Method

The gravity compensation control method was inspired by work by Sano et al. [23]. The powered exoskeleton proposed by Sano et al. [23] had only hip joints and no knee joints. Therefore, we modified the method as follows: The powered exoskeleton generated the torques of the joints of the hips and knees, τ_h and τ_k, so that it counteracted the effect of gravity on the body of the exoskeleton. τ_h and τ_k were controlled by Equations (3) and (4), respectively.

$$\tau_k = k_k \left(\frac{1}{2} l_l m_l g \sin \phi_l\right) \quad (3)$$

$$\tau_h = k_h \left(\frac{1}{2} l_u m_u g \sin \phi_u + m_l g \left(l_u \sin \phi_u + \frac{1}{2} l_l \sin \phi_l\right)\right) \quad (4)$$

where ϕ_u, ϕ_l, l_u, l_l, m_u, and m_l are the posture angles with respect to the force of gravity on, the lengths of, and the masses of the upper and lower legs of the powered exoskeleton, respectively. The definitions are shown in Figure 1. g is gravitation acceleration, and k_k and k_h are the control gains. The method tends to keep the body in upright position to lift the load to the shoulder.

4.2. Force Switch-Based Method

The other method is based on the idea of the force switch algorithm proposed by Sano et al. [22]. They [22] developed a method to estimate the wearer's motion, walking or standing, based on the force switch on the feet of the wearer. The algorithm required two or three steps for estimation and generates torques on the joints based on a pre-defined pattern for the walking motion. Unfortunately, we found that the original method needed time to estimate the walking motion, and the wearer needed to push his/her feet actively while the powered skeleton tried to retain posture before motion estimation was accomplished. Following motion estimation, the powered exoskeleton began assisting the estimated motion of the wearer. However, we found that many parameters needed to be tuned for appropriate power assistance.

Therefore, we designed a simplified algorithm in place of the force switch-based method. Since force sensors are attached to the wearer's feet in our powered exoskeleton, we used them as force switches instead. The powered exoskeleton recognizes the "swing phase" and the "stance phase" according to the difference between the left and right force values. When the difference is small, it recognizes that both feet are in "stance phase". Otherwise, if the force value of the left foot is greater than the right, it recognizes that the left leg is in the "swing phase" and the right leg is in the "stance phase", and vice versa.

Following the recognition of the phase of each foot, the powered exoskeleton controls itself to maintain the posture of the leg as desired according to the recognized phase. The desired angles of the powered skeleton $\boldsymbol{\phi}_t^d = (\phi_{rw}^d, \phi_{lw}^d, \phi_{rk}^d, \phi_{lk}^d)$ are set to the pre-defined desired angles $(\phi_{rw}^{rp}, \phi_{lw}^{lp}, \phi_{rk}^{rp}, \phi_{lk}^{lp})$, where rp and lp indicate the phases of the right and left legs, respectively. ϕ_i^{swing} and ϕ_i^{stand} are the desired angles of joint i for the "swing phase" and the "stance phase", respectively. A PD controller was applied to calculate the input of each joint motor $\boldsymbol{u}_t = (u_{rw}, u_{lw}, u_{rk}, u_{lk})$ by Equation (5):

$$u_t = k_p(\phi_t - \phi_t^d) + k_d\left((\phi_t - \phi_t^d) - (\phi_{t-1} - \phi_{t-1}^d)\right) \tag{5}$$

where $\phi_t = (\phi_{rw}, \phi_{lw}\ \phi_{rk}, \phi_{lk})$ are the given joint angles of the powered exoskeleton. The algorithm for this process is shown in Algorithm 2. Thresholds τ_A and τ_B are determined by trial and error.

The two competing methods introduced above struggle to distinguish forward and backward walks. Therefore, we apply and tune the parameters of these methods only for the forward walking motion.

Algorithm 2 Foot Force Switch-based Walk Assistance System

Obtain the floor reaction force f_l, f_r from the force sensors on the wearer's feet
$f_{diff} = f_l - f_r$
if $\tau_A \leq f_{diff} \leq \tau_B$ **then**
  recognizes that both legs are in "stance phase"
  $\phi_{rw}^d, \phi_{lw}^d, \phi_{rk}^d$ and ϕ_{lk}^d is set to $\phi_{rw}^{stand}, \phi_{lw}^{stand}, \phi_{rk}^{stand}$ and ϕ_{lk}^{stand}
end if
if $f_{diff} < \tau_A$ **then**
  recognizes that the left leg is in "swing phase" and the right leg is in "stance phase"
  $\phi_{rw}^d, \phi_{lw}^d, \phi_{rk}^d$ and ϕ_{lk}^d is set to $\phi_{rw}^{stand}, \phi_{lw}^{swing}, \phi_{rk}^{stand}$ and ϕ_{lk}^{swing}
end if
if $\tau_B < f_{diff}$ **then**
  recognizes that the right leg is in "swing phase" and the left leg is in "stance phase"
  $\phi_{rw}^d, \phi_{lw}^d, \phi_{rk}^d$ and ϕ_{lk}^d is set to $\phi_{rw}^{swing}, \phi_{lw}^{stand}, \phi_{rk}^{swing}$ and ϕ_{lk}^{stand}
end if

---

5. Experiments

Experiments were conducted to test the proposed method by comparing it with two comparative methods (The experiments were approved as No. H2016001 by the Research Ethics Committee, Department of Human and Artificial Intelligent Systems, Graduate School of Engineering, University of Fukui.). One wearer was a male student in his early 20s. In this experiment, he walked forward and backward for approximately 5 m wearing the powered exoskeleton. The data for the database were obtained while the powered exoskeleton was lifted by a gantry, and the wearer walked with the gantry so that he did not have any payload from the powered exoskeleton while his motion was restricted by the kinematics of the exoskeleton. Figure 6 shows how the data for the database were obtained for (a) "walking forward" and (b) "walking backward". The datasets for "standing" were also obtained when the wearer stood in upright position. The sampling time was approximately 80 milliseconds. The window size of the dataset was 10 steps. The number of datasets for each motion category was approximately 50.

Once the database had been constructed, the proposed method was applied. The k of the k-NN was set to 5 for motion category recognition and 10 for appropriate joint angle estimation in the experiments.

Figure 7 shows the results of the estimation of the wearer's motion based on the proposed method. The wearer first stood in the upright position, started walking forward, stopped, and stayed still there for a while; he then walked backward, and stopped. The figure shows that the proposed method successfully recognized the wearer's motion. The sampling time of the control system was approximately 80 milliseconds. The calculation of the motion recognition takes only about 20 milliseconds on the controller. The calculation of the whole control system including sensor value acquisition and motor control takes less than 80 milliseconds so that the powered exoskeleton assists the wearer's motion in real time.

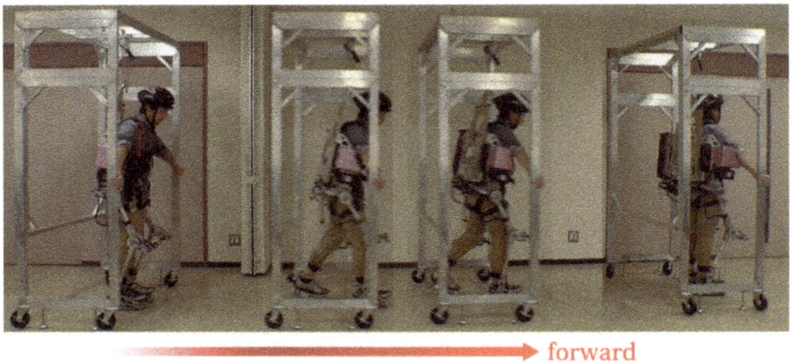

(a) Walk forward

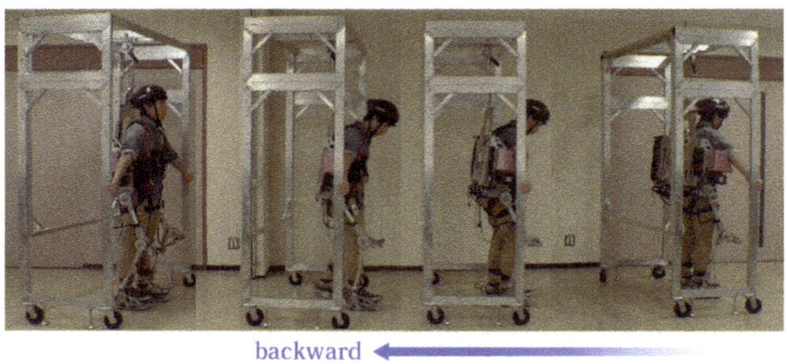

(b) Walk backward

Figure 6. Data acquisition for database construction: the wearer walks forward and backward wearing the powered exoskeleton while it is lifted by a gantry and pushes the gantry himself.

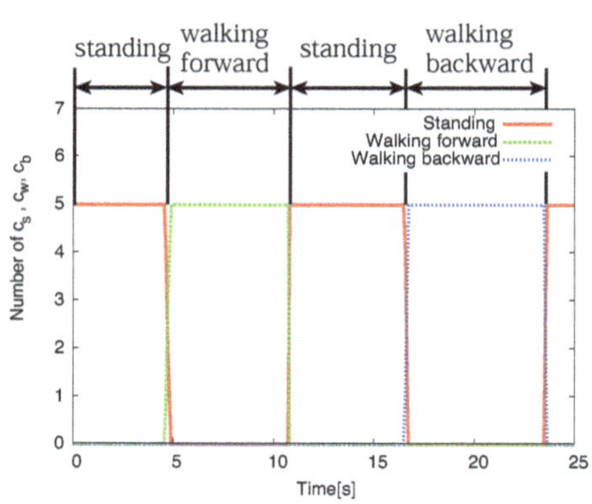

Figure 7. Results of motion estimation based on k-nearest neighbor algorithm.

An additional 15 kg weight was placed on the powered exoskeleton in the experiments. The proposed method and the competitive methods described in Sections 4.1 and 4.2 were applied to the powered exoskeleton one by one across enough breaks for the wearer. Figures 8 and 9 show the image sequences of the motion "walking forward" and "walking backward" based on each method. The images were captured from a side. The motion category recognition worked perfectly based on the proposed method and the foot force switch-based method. Table 1 shows the walking speeds for "walking forward" and "walking backward" based on each method.

Figure 8 shows that the proposed method and the foot force switch-based method enabled the wearer to walk smoothly while the gravity compensation-based method did not. The sampling time of image capture was approximately 1.8 s. The gravity compensation-based method caused the wearer to walk more slowly than the other methods. Table 1 also shows that the proposed and the force switch-based methods supported "walking forward". The wearer supported by the gravity compensation method slowly walked forward because this method does not actively support walking.

Table 1. Walking speed for each method.

Method	Walking Forward [km/h]	Walking Backward [km/h]
Proposed method	2.70	2.16
Gravity compensation method	2.37	1.65
Force Switch-based method	2.70	0.75

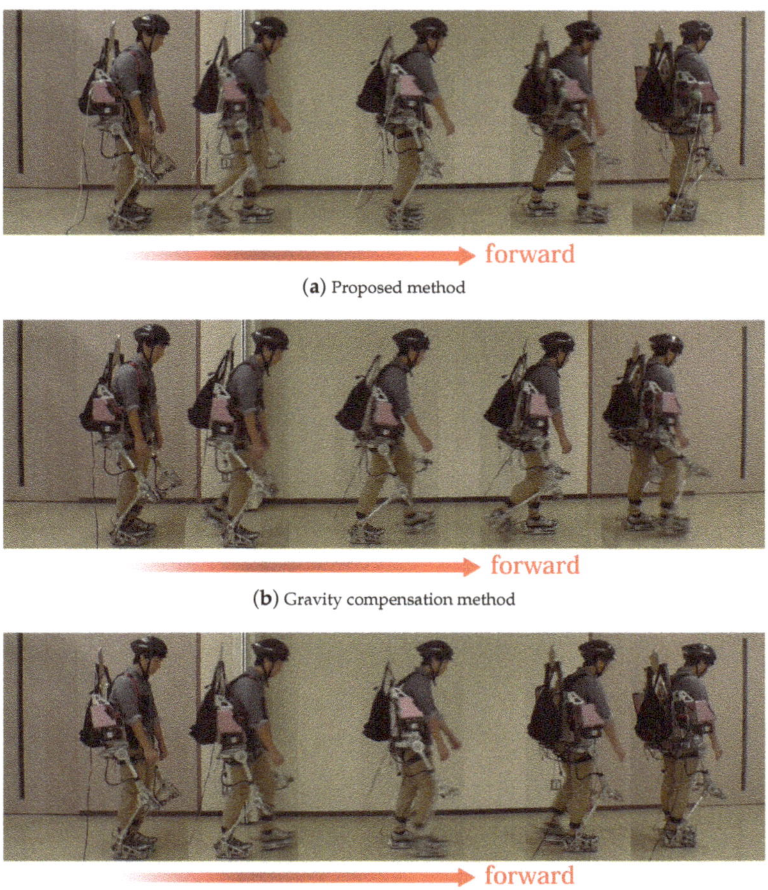

(a) Proposed method

(b) Gravity compensation method

(c) Foot force switch-based method

Figure 8. Image sequences of walking forward based on each method: the sampling time of image capture was approximately 1.8 s.

Figure 9 shows that the proposed method allowed the wearer to walk backward faster than other methods. The proposed method recognized the wearer's motion of walking backward correctly and supported it appropriately. On the other hand, the foot force switch-based method caused the wearer to walk backward slowly because it tried to support him in walking forward even though he was walking backward. Eventually, the wearer needed to exert a strong force to push his leg backward and walk slowly. It was difficult for the foot force switch-based method to recognize walking forward and backward based only on the outputs of the force sensors of the feet. This was one of the drawbacks of the method. The gravity compensation method showed good result, but the walk tended to be slow because it did not assist the horizontal motion of the leg, even though it assisted vertical leg motion, such that the wearer had to firmly push his leg back. Table 1 supports the analysis in terms of the walking speed for the motion "walking backward".

(a) Proposed method

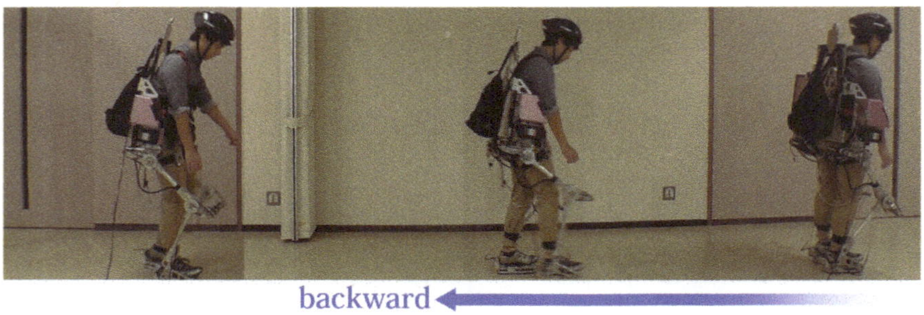

(b) Gravity compensation method

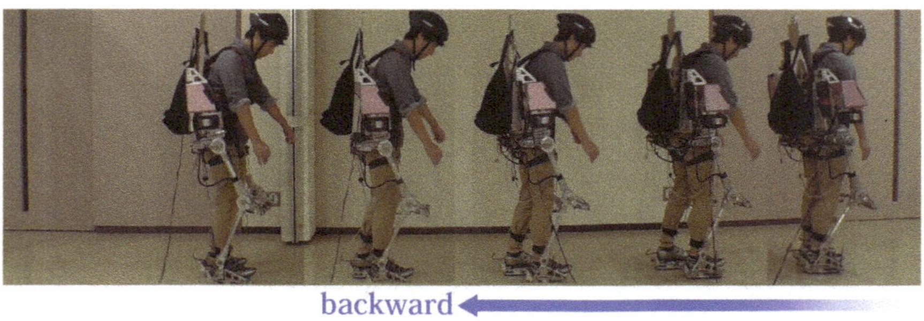

(c) Foot force switch-based method

Figure 9. Image sequences of walking backward based on each method: the sampling time of the image capture was approximately 5.4 s.

Figure 10 shows the average load on the wearer's shoulder while walking forward and backward. The proposed method and the foot force switch method maintained a load of approximately 100 N whereas the gravity compensation method maintained one of 350 N. If there was no assist control, the wearer had approximately 350 N on his shoulders. The proposed method successfully reduced the load. It depends on the motion database D. When the datasets for the database were sampled, the powered exoskeleton was hung up on the gantry so that the wearer had no load due to the exoskeleton. Therefore, the proposed method lifted the exoskeleton. To maintain the back-drivability of the powered exoskeleton, we kept the control gain as small as possible. An approximately 100 N load on the shoulder was imposed because of the small control gain for back-drivability. The gravity compensation method did not reduce the load on the wearer's shoulder. If the control gains k_k and k_h in Equations (3) and (4) became large, the load on the wearer's shoulder in the upright position became small, but it became challenging for the wearer to swing the leg because the powered exoskeleton tried to keep the leg as vertical as possible. It eventually lost back-drivability. To retain back-drivability, the control gains k_k and k_h needed to be small, in which case the system failed to reduce the load on the wearer's shoulder. The foot force switch-based method was as good as the proposed method according to Figure 10.

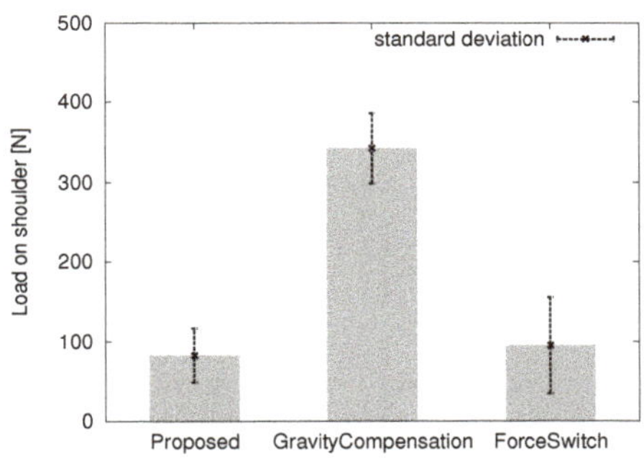

(a) Walking forward

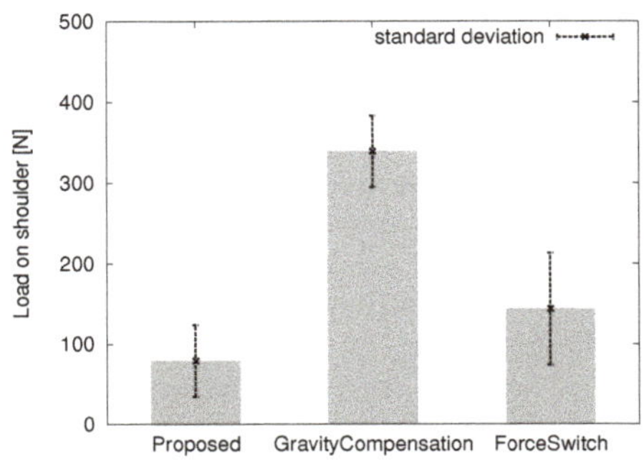

(b) Walking backward

Figure 10. Average load on the shoulders while walking.

Figure 11 shows the horizontal front-back reaction force measured by foot force sensors while the wearer walked based on each control method. The reaction forces on the left leg under the proposed method and the force switch-based method were smaller than that for the gravity compensation method. This was because the gravity compensation method did not consider the motion of the feet in the horizontal direction. The proposed method used motion sensors to predict the posture of the powered exoskeleton and successfully reduced the reaction forces on the feet in the horizontal direction. The force switch-based method also reduced the reaction forces because the pre-defined motion for the method fed the swinging leg forward and the standing leg backward. The reaction force on the right leg was comparatively small when the gravity compensation was applied because of the wearer's gait.

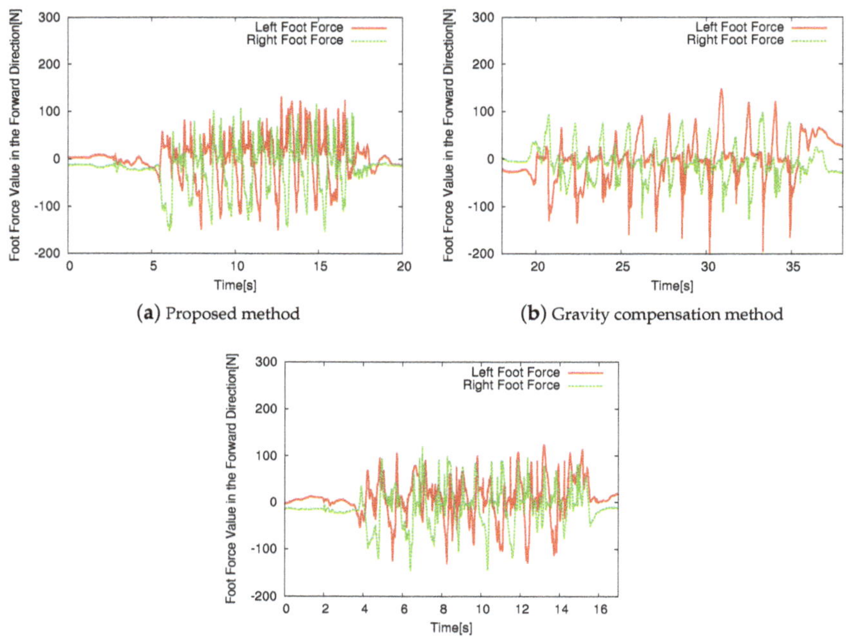

(**a**) Proposed method (**b**) Gravity compensation method (**c**) Force switch-based method

Figure 11. Reaction force on feet in horizontal front-back direction while walking forward based on each control method.

Figure 12 shows the horizontal front-back force measured by the foot force sensors while the wearer walked backward based on each control method. The proposed method showed the smallest magnitudes of forces during this. This was because it appropriately recognized the wearer's motion and controlled the legs of the powered exoskeleton based on the estimated motion. The gravity compensation method yielded the highest resistance force to the wearer's legs because it did not consider the motion of the feet in the horizontal direction, as mentioned above. The force switch-based method failed to support backward walking because it could not distinguish between walking forward and backward, and the pre-defined motion for the method was designed for forward walking. Eventually, the wearer had to push the swinging leg more strongly. The reaction force on the right foot was small because of the manner of the wearer's walk.

Figures 11 and 12 show that there are differences of the gate frequencies of the walks. The wearer with the gravity compensation method (b) walks slower than the other methods (a) and (c). The wearer with the gravity compensation method (b) needs to push the powered exoskeleton forward and backward by his legs intentionally because the gravity compensation method (b) just compensates for the gravity effect of the powered exoskeleton and does not support the current human motion. On the other hand, the proposed method (a) and the force switch-based method (c) support the walk motion actively so that the wearer can walk faster. Figures 11 and 12 show that there were differences in the gate frequencies of the walks. The wearer using the gravity compensation method (b) walked slower than with methods (a) and (c). The wearer using the gravity compensation method (b) needed to push the powered exoskeleton forward and backward using his legs because this method (b) only compensates for the effect of gravity due to the powered exoskeleton and does not support the human motion. On the other hand, the proposed method (a) and the force switch-based method (c) supported the walking motion actively such that the wearer could walk faster.

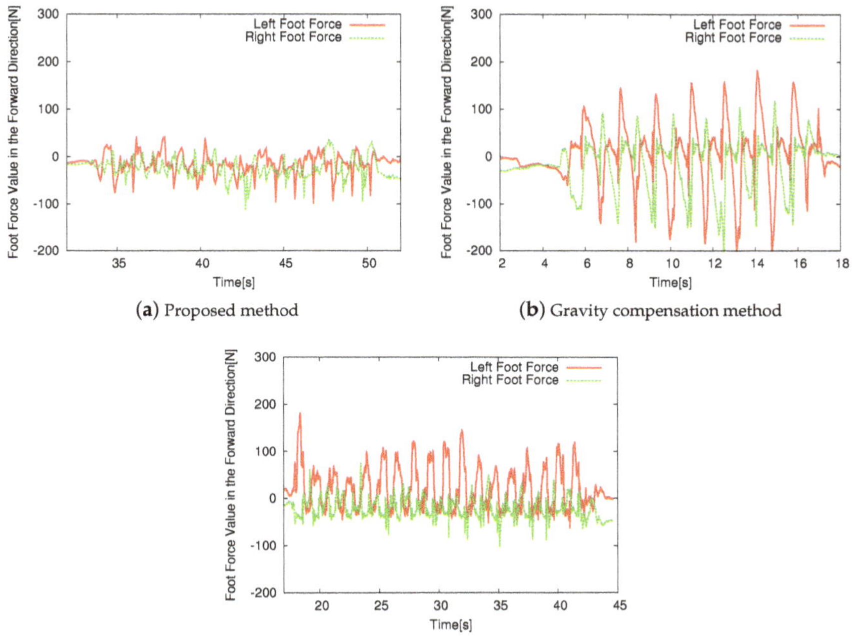

(**a**) Proposed method (**b**) Gravity compensation method (**c**) Force switch method

Figure 12. Reaction force on feet in horizontal front-back direction while walking backward based on each control method.

To evaluate the usability of the proposed powered exoskeleton, we had a questionnaire on the powered exoskeleton controlled by each method. Three users wore the powered exoskeleton controlled by each method, the proposed method, the gravity compensation method, and the force switch method. After they walked forward, stopped, and walked backward, and repeated them a few times, they answered the questions on the lightness of the shoulder load, lightness of the reaction force to feet, and how freely they could move. The users answered these questions with numbers from 1 to 5; 1 is for the lowest and the 5 is the highest.

Figure 13 shows the results of the questionnaire. According to Figure 13a, they were aware of the lightness of the shoulder load if the proposed and the force switch methods applied. The gravity compensation method failed to reduce the shoulder load. The answers are consistent with the discussions on Figure 10.

Figure 13b suggests that the proposed and force switch methods successfully support the feet of the users when they walk forward but the force switch method failed to support them when they walk backward. The evaluation of the gravity compensation method depends on the user's preference. This result is also consistent with the discussion on Figures 11 and 12.

The gravity compensation method received high scores on how freely they can move according to the Figure 13c. The reason is that the gravity compensation method does not assist the power actively and just follows the motion of the user while the other methods try to assist the motion actively, but the assistant becomes against the user's intention occasionally. The force switch method has a low evaluation from the users especially when they walk backward. The reason is that the method was designed for walking forward.

Figure 13 indicates that the overall evaluation of the proposed method is better than the other while it has room to improve the power assistant abilities. It is one of the future works to improve them.

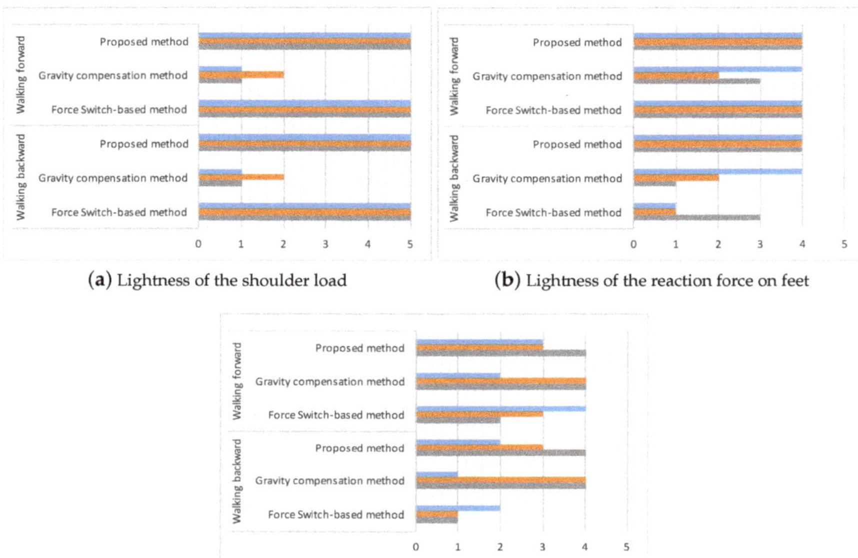

(**a**) Lightness of the shoulder load (**b**) Lightness of the reaction force on feet

(**c**) How freely they can move

Figure 13. Answers to questionnaire on the powered exoskeleton controlled by the methods.

The experimental results show that the proposed method outperformed the other competitive methods comprehensively, as it enabled the wearer to walk faster with a smaller reaction force than the other methods.

6. Conclusions and Discussions

This study proposed a power assist control system based on the wearer's estimated motion using motion sensors for a powered exoskeleton without leg binding. It recognizes the wearer's motion using motion sensors, estimates appropriate joint angles for the powered exoskeleton based on a motion database compiled in advance, and assists the wearer's motion in real time. The experimental results exhibited the effectiveness of the proposed assistive system.

The major feature of our powered exoskeleton is that it does not bind the legs of the wearer. It allows the wearer to move his/her legs freely at the beginning of the motion even if the joints of the exoskeleton are fixed because of the room to move knees. The feature enables us to use motion sensors to recognize the wearer's motion and give feedback on the power assist. It supports only hip and knee joints rotating on the lateral direction. The other joints, for example, hip joints rotating in different directions and ankle joints are passive. The wearer needs power assists on those joints, too, if the load on the wearer increases more. It is one of the future works from the viewpoint of the mechanical design of the powered exoskeleton to strengthen the existing active joints and replace the passive joints to the active one.

In principle, the proposed method simply replays the pre-recorded joint angles from the database. However, even if the walking speed changes, it tries to find the best matching motion from the database to support it. If the wearer walks more slowly than the pre-recorded walk, the system tries to find the best matching phase of the walk and assists faster walking. If the wearer walks more quickly than the pre-recorded walk, it assists in walking slower but does not prevent the human walk because it always follows the walk to find the best matching phase based on the database. Therefore, it can adapt to a certain degree. If the walk is too far from the pre-recorded datasets, the proposed method fails to support it and needs a new dataset for walks at different speeds. This will form part of our future work.

In this paper, we employed the k-nearest neighbors algorithm to deal with the motion database. The reason is that it does not make some specific assumptions on the motions of the human or powered exoskeleton and it is the simplest method among the various machine learning technique. However, there is a possibility to employ the other sophisticated algorithm. We are investigating more effective algorithms for motion learning [27,28]. Another part of our future work in this area will involve increasing the variety of motions that can be assisted, that is, not only standing and walking motions and forward-backward motions, but also sideways walking, squatting, swinging the body, climbing stairs, and so on. We also intend to investigate online updates of the motion database.

Author Contributions: conceptualization, S.N. and Y.T. (Yasutake Takahashi); methodology, S.N. and Y.T. (Yasutake Takahashi); software, S.N.; validation, S.N., K.S. and S.M.; formal analysis, S.N. and S.M.; investigation, S.N., K.S., S.M., and Y.T. (Yasutake Takahashi); resources, Y.T. (Yasutake Takahashi), M.K., Y.T. (Yoshiaki Taniai), T.N.; data curation, S.N., K.S. and S.M.; writing—original draft preparation, Y.T. (Yasutake Takahashi); writing—review and editing, Y.T. (Yasutake Takahashi); visualization, S.N., K.S. and S.M.; supervision, Y.T. (Yasutake Takahashi), Y.T. (Yoshiaki Taniai), and T.N.; project administration, M.K.; funding acquisition, M.K.

Funding: This work was supported by research grants from the Fukui Prefectural Government, Japan.

Conflicts of Interest: The authors declare no conflict of interest.

References

1. Dollar, A.M.; Herr, H. Lower Extremity Exoskeletons and Active Orthoses: Challenges and State-of-the-Art. *IEEE Trans. Robot.* **2008**, *24*, 144–158. [CrossRef]
2. Satoh, H.; Kawabata, T.; Sankai, Y. Bathing Care Assistance with Robot Suit HAL. In Proceedings of the International Conference on Robotics and Biomimetics, Guilin, China, 19–23 December 2009; pp. 498–503.
3. Kawamoto, H.; Hayashi, T.; Sakurai, T.; Eguchi, K.; Sankai, Y. Development of Single Leg Version of HAL for Hemiplegia. In Proceedings of the 31st Annual International Conference of the IEEE EMBS, Minneapolis, MN, USA, 3–6 Septtmber 2009; pp. 5038–5043.
4. Kawabata, T.; Satoh, H.; Sankai, Y. Working Posture Control of Robot Suit HAL for Reducing Structual Stress. In Proceedings of the International Conference on Robotics and Biomimetics, Guilin, China, 19–23 December 2009; pp. 2013–2018.
5. Lu, R.; Li, Z.; Su, C.Y.; Xue, A. Development and Learning Control of a Human Limb With a Rehabilitation Exoskeleton. *IEEE Trans. Ind. Electron.* **2014**, *61*, 3776–3785. [CrossRef]
6. Stegall, P.; Winfree, K.; Zanotto, D.; Agrawal, S.K. Rehabilitaion Exoskeleton Design: Exploring the Effect of the Anterior Lunge Degree of Freedom. *IEEE Trans. Robot.* **2013**, *29*, 838–846. [CrossRef]
7. Zhang, X.; Xiang, Z.; Lin, Q.; Zhou, Q. The Design and Development of a Lower Limbs Rehabilitation Exoskeleton Suit. In Proceedings of the International Conference on Complex Medical Engineering, Beijing, China, 25–28 May 2013; pp. 307–312.
8. Ma, W.; Zhang, X.; Yin, G. Design on Intelligent Perception System for Lower Limb Rehabilitation Exoskeleton Robot. In Proceedings of the International Conference on Ubiquitous Robots and Ambient Intelligence, Xi'an, China, 19–22 August 2016; pp. 587–592.
9. Seungnam, Y.; Changsoo, H.; Ilje, C. Design considerations of a lower limb exoskeleton system to assist walking and load-carrying of infantry soldiers. *Appl. Bionics Biomechan.* **2014**, *11*, 119–134.
10. Gui, L.; Yang, Z.; Yang, X.; Gu, W.; Zhang, Y. Design Control Technique Research of Exoskeleton Suit. In Proceedings of the International Conference on Automation and Logistics, Jinan, China, 18–21 August 2007; pp. 541–546.

11. Cyberdyne. Hal for Disaster Recovery <on the Research and Development Stage>. 2016. Available online: http://www.cyberdyne.jp/english/products/supporting.html (accessed on 21 December 2016).
12. Activelink Co. Ltd. Powerloader Linght "PLL". 2014. Available online: http://activelink.co.jp/doc/668.html (accessed on 21 December 2016).
13. Yamauchi, B. PackBot: A versatile platform for military robotics. In Proceedings of the Unmanned Ground Vehicle Technology VI, Orlando, FL, USA, 12–16 April 2004; Volume 5422, pp. 228–237.
14. Fleischer, C.; Hommel, G. A Human-Exoskeleton Interface Utilizing Electromyography. *IEEE Trans. Robot.* **2008**, *24*, 872–882. [CrossRef]
15. Takahashi, J.; Meraz, N.S.; Suezawa, S.; Hasegawa, Y.; Sankai, Y. Alternative Interface System by Using Surface EMG of Residual Muscles for Physically Challenged Person. In Proceedings of the International Conference on Robotics and Biomimetics, Phuket, Thailand, 7–11 December 2011; pp. 1349–1354.
16. Chandrapal, M.; Chen, X.; Wang, W. Preminary evaluation of a lower-limb exoskeleton—Stair climbing. In Proceedings of the International Conference on Advanced Intelligent Mechatronics, Wollongong, NSW, Australia, 9–12 July 2013; pp. 1458–1463.
17. Tsuji, M.; Kakamu, T.; Hayakawa, T.; Kumagai, T.; Hidaka, T.; Kanda, H.; Fukushima, T. Risk and preventive factors for heat illness in radiation decontamination workers after the Fukushima Daiichi Nuclear Power Plant accident. *J. Occup. Health* **2013**, *55*, 53–58.
18. Kazerooni, H.; Racine, J.L.; Huang, L.; Steger, R. On the Control of the Berkeley Lower Extremity Exoskeleton (BLEEX). In Proceedings of the International Conference on Robotics and Automation, Barcelona, Spain, 18–22 April 2005; pp. 4353–4360.
19. Huang, L.; Steger, J.R.; Kazerooni, H. Hybrid Control of the Berkeley Lower Extremity Exoskeleton (BLEEX). In Proceedings of the 2005 ASME International Mechanical Engineering Congress and Exposition, Orlando, FL, USA, 5–11 November 2005; pp. 1–8.
20. Zoss, A.; Kazerooni, H.; Chu, A. Hybrid Control of the Berkeley Lower Extremity Exoskeleton (BLEEX). In Proceedings of the 2005 IEEE/RSJ International Conference on Intelligent Robots and Systems, Edmonton, AB, Canada, 2–6 August 2005; pp. 3132–3139.
21. Steger, R.; Kim, S.H.; Kazerooni, H. Control Scheme and Networked Control Architecture for the Berkeley Lower Extremity Exoskeleton (BLEEX). In Proceedings of the 2006 IEEE International Conference on Robotics and Automation, Orlando, FL, USA, 15–19 May 2006; pp. 3469–3476.
22. Sano, K.; Yagi, E.; Sato, M. Estimation of Walking Intention of Non-Handicapped Persons Using Foot Switches and Hip Joint Angles. *J. Jpn. Soc. Mechan.* **2013**, *79*, 3487–3500.
23. Sano, K.; Yagi, E.; Sato, M. Development of a Wearable Assist Suit for Walking and Lifting-Up Motion Using Electric Motors. *J. Robot. Mechatron.* **2013**, *25*, 923–930. [CrossRef]
24. Liu, D.X.; Wu, X.; Wang, M.; Chen, C.; Zhang, T.; Fu, R. Non-Binding Lower Extremity Exoskeleton (NextExo) for Load-Bearing. In Proceedings of the IEEE Conference on Robotics and Biomimetics, Zhuhai, China, 6–9 December 2015; pp. 2312–2317.
25. Wang, M.; Wu, X.; Liu, D.X.; Wang, C. A Human Motion Prediction Algorithm Based on HSMM for SIAT's Exoskeleton. In Proceedings of the 35th chinese Control Conference, Chengdu, China, 27–29 July 2016; pp. 3891–3896.
26. Madgwick, S.O. *An Effcient Orientation Filter for Inertial And Inertial/Magnetic Sensor Arrays*; Report x-io and University of Bristol; University of Bristol: Bristol, UK, 2010; Volume 25, pp. 113–118.
27. Ishizuka, Y.; Murai, S.; Takahashi, Y.; Kawai, M.; Taniai, Y.; Naniwa, T. Modeling Walking Behavior of Powered Exoskeleton based on Complex-Valued Neural Network. In Proceedings of the 2018 IEEE International Conference on Systems, Man, and Cybernetics, Miyazaki, Japan, 7–10 October 2018; pp. 1923–1928. [CrossRef]
28. Murata, F.; Takahashi, Y. Walking Motion Model based on Quaternion-valued Recurrent Neural Network for Powered Exoskeleton. In Proceedings of the 2018 Joint 10th International Conference on Soft Computing and Intelligent Systems and 19th International Symposium on Advanced Intelligent Systems, Toyama, Japan, 5–8 December 2018; pp. 1354–1359.

applied sciences

MDPI

Article

Expanded Douglas–Peucker Polygonal Approximation and Opposite Angle-Based Exact Cell Decomposition for Path Planning with Curvilinear Obstacles

Jin-Woo Jung *, Byung-Chul So, Jin-Gu Kang, Dong-Woo Lim and Yunsik Son

Department of Computer Science and Engineering, Dongguk University, Seoul 04620, Korea; sbc10620@naver.com (B.-C.S.); kanggu12@dongguk.edu (J.-G.K.); aehddn@gmail.com (D.-W.L.); sonbug@dongguk.edu (Y.S.)

* Correspondence: jwjung@dongguk.edu; Tel.: +82-2-2260-3812

Received: 16 January 2019; Accepted: 11 February 2019; Published: 14 February 2019

Abstract: The Expanded Douglas–Peucker (EDP) polygonal approximation algorithm and its application method for the Opposite Angle-Based Exact Cell Decomposition (OAECD) are proposed for the mobile robot path-planning problem with curvilinear obstacles. The performance of the proposed algorithm is compared with the existing Douglas–Peucker (DP) polygonal approximation and vertical cell decomposition algorithm. The experimental results show that the path generated by the OAECD algorithm with EDP approximation appears much more natural and efficient than the path generated by the vertical cell decomposition algorithm with DP approximation.

Keywords: curvilinear obstacle; douglas–peuker polygonal approximation; opposite angle-based exact cell decomposition; path planning; mobile robot

1. Introduction

The path-planning process of a mobile robot aims at finding a collision-free path to move the robot from the current posture to the goal posture [1–3]. If there are multiple available paths, the optimal path in the sense of an objective function, such as the minimum distance, can be chosen. The algorithms for mobile-robot path planning can be grouped into four categories: roadmap approaches, such as the visibility graph or generalized Voronoi graph; cell decomposition approaches, such as the vertical cell decomposition (VCD) or approximate cell decomposition; sampling-based planning methods, such as the rapidly exploring random tree (RRT) or probabilistic roadmap (PRM), and potential field methods [2–5]. Among these methods, roadmap approaches are generally fast and easy to implement, but an intrinsic way to describe environmental information is not provided [1–3]. Sampling-based planning methods are more practical, but they do not provide completeness so we cannot recognize the non-existence of a path [2,3]. Potential field methods are useful to control the robot by generating a differentiable smooth path, but they cannot give explicit information on the roadmap and easily fall into a local minimum [1–3]. There are also some hybrid approaches, such as a potential field with RRT [6] or potential field with cell decomposition [7].

Cell decomposition, which is a classical and representative method for mobile-robot path planning, decomposes the given environment into several cells and finds a collision-free path based on the connectivity graph of these cells [2–10]. Here, each node of a connectivity graph is made by the representative point of each cell or its border line. Each link of the connectivity graph between the nodes indicates that the corresponding cell is adjacent to each other [11]. In various cell decomposition algorithms [12–17], one of the most widely known algorithms is vertical cell decomposition (VCD), but it does not generate an efficient path because it uses too many cells [8]. Figure 1 shows an example of the previous exact cell decomposition using VCD. VCD makes vertical lines from the convex vertices to decompose the environment into cells. The adjacency relationship among the cells is represented by the connectivity graph and used to find a path using graph search algorithms. VCD does not guarantee the optimality of the number of decomposed cells since there is no consideration of the shape of the obstacle. Reducing the number of decomposed cells directly increases the efficiency in path planning, but finding the optimal decomposition case is known as an NP-hard problem [1–3].

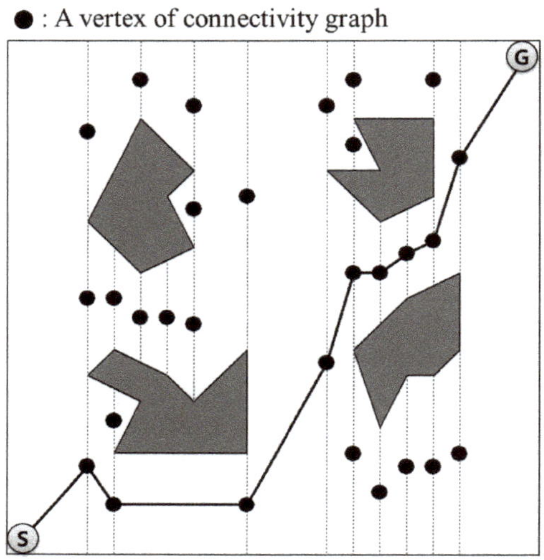

Figure 1. An example of Vertical Cell Decomposition (VCD) and its path.

To supplement this drawback, the Opposite Angle-based Exact Cell Decomposition (OAECD) [18] was proposed by using a type of greedy approach to minimize the number of cells and increase efficiency. However, the OAECD can only be applied for polygonal obstacles, since it is operated based on the relationship among the vertices of the obstacles. In other words, path planning in an environment with curvilinear obstacles is impossible by itself. Therefore, in this paper, a novel expanded polygonal approximation method based on Douglas–Peucker (DP) algorithm is proposed to apply OAECD path planning to the cases with curvilinear obstacles.

In Chapter 2, a novel polygonal approximation algorithm for curvilinear obstacles is addressed. Then, the algorithm is applied to a modified OAECD algorithm in Chapter 3. The experimental results are shown in Chapter 4, and the paper is concluded in Chapter 5.

2. Polygonal Approximation Algorithm of Curvilinear Obstacle

In this chapter, the Douglas–Peucker (DP) algorithm [19], which is one of the most popular algorithms for polygonal approximation, is reviewed. An Expanded Douglas–Peucker (EDP) algorithm for the application with curvilinear obstacles is proposed with mathematical validation on the circumscription of the EDP Algorithm.

2.1. Douglas-Peucker Algorithm

The DP algorithm is a representative method of polygonal approximation. The purpose of a DP algorithm is to find a similar piecewise linear curve with fewer points given a closed curve. The algorithm uses the concept of dissimilarity based on the maximum distance between the original curve points and their simplified piecewise linear curve [17,18].

The algorithm (Algorithm 1) measures the distance between each point of a curve and the base line, which is the line segment with the same first and last points with the curve, to find the farthest point from the line segment with the maximum perpendicular distance.

Here, C is the set of obstacle contours, which is the set of point lists of the obstacle; ε is the threshold value for the maximum dissimilarity tolerance; R is the final result of the polygonal approximation of all obstacles; c is a point list of the obstacle contour, which is arranged in counter-clockwise order; pl is a point list; r_1 and r_2 indicate each result of the recursiveDP procedure; r indicates the final result of the polygonal approximation of an obstacle by DP algorithm.

Algorithm 1. Pseudo Code of the Initial DP.

Input:
$C \leftarrow$ Set of point lists of the obstacle
$\varepsilon \leftarrow$ Threshold value for maximum dissimilarity tolerance
Output:
$R \leftarrow$ Final result of polygonal approximation of all obstacles represented as a set of point lists

Begin ***DP Procedure***
for each c **in** C **do**
  find two points in c, p_1 and p_2, which have the maximum distance from each other and p_1 is in front of p_2
  $r_1 \leftarrow$ ***recursiveDP***$(pl[p_1 \ldots p_2], \varepsilon)$ // pl: point list of a segment of obstacle contour
  $r_2 \leftarrow$ ***recursiveDP***$(pl[p_2 \ldots$ starting point of $c \ldots p_1], \varepsilon)$
  $r \leftarrow$ point list$[r_1[0] \ldots r_1[\text{end}_1 - 1]\ r_2[0] \ldots r_2[\text{end}_2 - 1]]$ // end_i: number of points in r_i
  insert r ***to*** R
end for
End ***DP Procedure***

In the recursive procedure of DP algorithm (Algorithm 2), the line segment is further divided into two sub-line segments using the farthest point as the via-point whenever the maximum perpendicular distance is greater than or equal to the threshold value for maximum dissimilarity tolerance, ε. This process is recursively repeated until the maximum perpendicular distance is less than ε.

Here, r_1 and r_2 are each the result of the recursiveDP procedure, and r is the result of the polygonal approximation of the given curve.

Figure 2 shows the process of the DP algorithm. l_b is the base line, which is identical to $line_{base}$ in Algorithm 2. l_{b1} is determined with the starting point P_1 and ending point P_2, which makes the widest width of the given obstacle. Then, we check whether $dist_{max}$ is less than ε. If $dist_{max}$ is less than ε, two vertices of l_{b1} are inserted into the point list of the polygonal approximation. If $dist_{max}$ is greater than or equal to ε, a new base line l_{b2} with P_1 and P_3 is made, new $dist_{max}$ is found again based on l_{b2}, and the above procedures are repeated.

The obstacle approximation result by DP algorithm does not guarantee the circumscription of the original obstacle. In other words, some interior points of the obstacle region may not be included inside the polygon by the DP algorithm. Similarly, some exterior points of the obstacle region may be included inside the polygon by the DP algorithm.

Thus, if the result of the DP algorithm on a curvilinear obstacle is directly used for path planning, the generated path may penetrate inside the real obstacle region, and the robot may easily collide with the obstacle. To overcome this problem, a modified DP algorithm is proposed in this paper.

Algorithm 2. Pseudo Code of the Recursive DP.

Input:
$pl \leftarrow$ Point list of the given curve
$\varepsilon \leftarrow$ Threshold value for maximum dissimilarity tolerance
Output:
$r \leftarrow$ Result of the polygonal approximation of the given curve represented as a point list
Initialization:
$line_{base} \leftarrow$ line segment connected from $pl[0]$ to $pl[\text{end}]$ // end: number of points in pl
$dist_{max} \leftarrow -1$
$i_{max} \leftarrow -1$

```
Begin recursiveDP Procedure
    for each point p in pl[1 ... end−1] do
        dist ← perpendicular distance between line_base and p
        if dist > dist_max
            dist_max ← dist
            i_max ← index of p
        end if
    end for
    if dist_max >= ε then
        r1 ← recursiveDP(pl[0 ... i_max], ε) // pl: point list of a segment of the given curve
        r2 ← recursiveDP(pl[i_max ... end], ε)
        r ← point list[r1[0] ... r1[end1 − 1] r2[0] ... r2[end2]] // end_i: # of points in r_i
    else
        insert pl[0] to r
        insert pl[end] to r
    end if
End recursiveDP Procedure
```

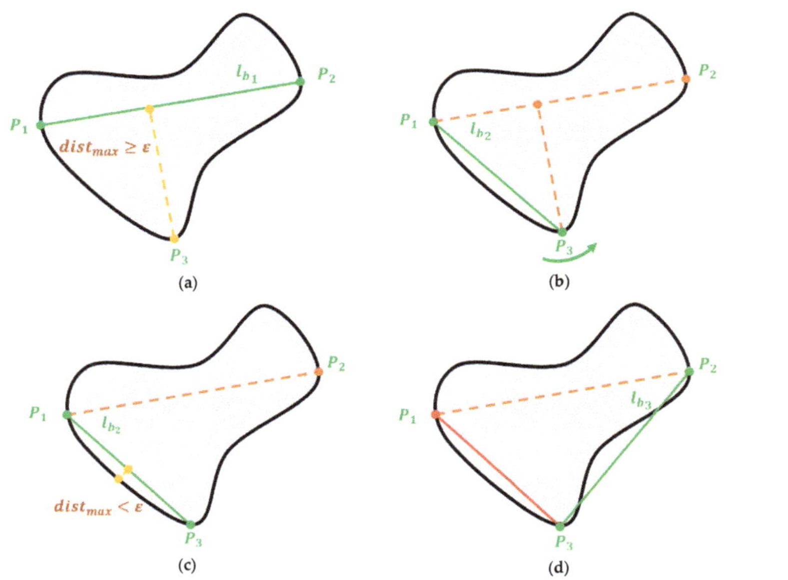

Figure 2. Douglas–Peucker (DP) process between P_1 and P_2: (**a**) If $dist_{max} \geq \varepsilon$, then (**b**) l_{b2} between P_1 and $dist_{max}$ point P_3 is used for the next base line; (**c**) If $dist_{max} < \varepsilon$, then (**d**) l_{b3} between P_3 and P_2 is used for the next base line.

2.2. Proposed Expanded Douglas-Peucker (EDP) Algorithm

The path planning for curvilinear obstacles using DP algorithm may not be collision-free. The basic philosophy of the EDP algorithm is to guarantee the circumscription of obstacles by expanding the polygon of the DP algorithm with the maximum dissimilarity tolerance. In addition, by appending additional points near the convex corner, the convex corner clearance is considered, where the robot control may become more difficult during path following.

Figure 3 shows the operation of EDP based on the counter-clockwise traversal. The first step is to perform the DP algorithm. If the half angle between line segments l_0 and l_1 is 0–90°, i.e., the corner of DP polygon is convex, the perpendicular half-lined to l_0 and l_1 are v_0 and v_1 at the corner point, respectively. Then, the points on the arc, with the center point as the convex corner and the radius of ε, are appended starting from v_0 and rotating with angle θ_{rot} until they meet v_1 in the counter-clockwise direction. These points are added for the convex corner clearance. If the half angle between line segments l_0 and l_1 is 90° or more, i.e., the corner of DP polygon is concave, the point far from the cross point of l_0 and l_1 with distance ε to the outer direction of the obstacle is appended. The above procedures are repeated for the remaining corner points of the DP polygon.

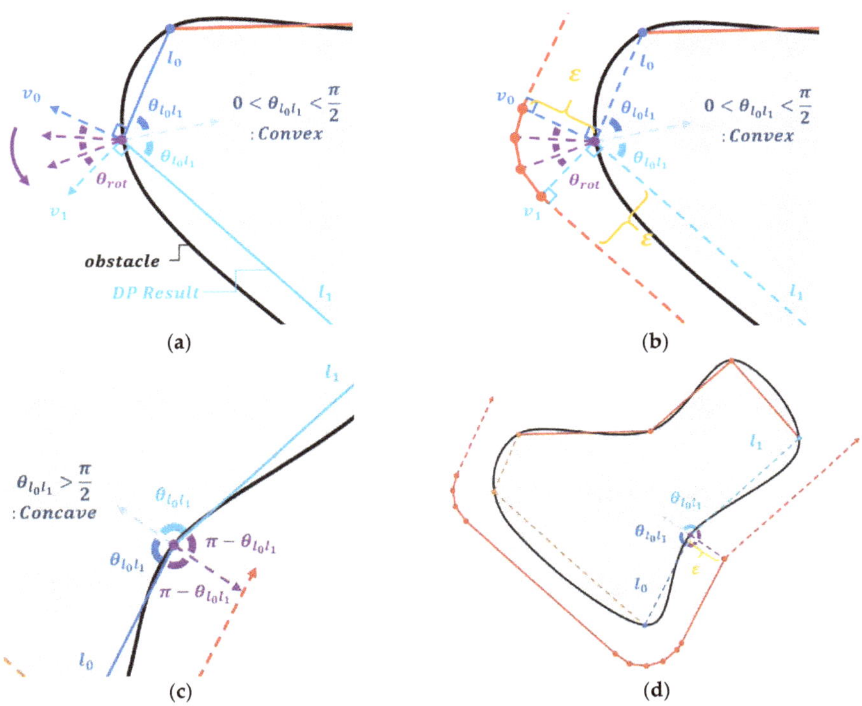

Figure 3. Abstract process of Expanded Douglas–Peucker (EDP) algorithm: (**a**) Before applying EDP for a convex corner; (**b**) After applying EDP for a convex corner; (**c**) Before applying EDP for a concave corner; (**d**) Overall appearance after applying EDP for a concave corner.

Algorithm 3 shows the abstract version of the pseudo code of the EDP algorithm.

Algorithm 3. Pseudo Code of EDP (Abstract version).

Input:
$R_{DP} \leftarrow$ Result of DP polygonal approximation of all obstacles represented as a set of point lists
$\varepsilon \leftarrow$ Threshold value for maximum dissimilarity tolerance
$\theta_{rot} \leftarrow$ Constant angle for obstacle corner clearance
Output:
$R \leftarrow$ Final result of the EDP polygonal approximation represented as a set of point lists

```
Begin Expanded-DP Procedure
    for each r_DP in R_DP do
        R ← null
        for each point p in r_DP do
            if p is convex vertex then
                θ ← half inner angle of p
                dist ← ε
                l_0 ← line segment connected from the previous point of p to p
                l_1 ← line segment connected from p to the next point of p
                v_0 ← perpendicular half-line to l_1 started from p
                v_1 ← perpendicular half-line to l_2 started from p
                loop
                    insert points p_cv on the arc with center point p and radius ε, which is consisted of the points according to θ_rot angle from v_0 to v_1, to R
                end loop
            else
                θ ← half outer angle of p
                dist ← ε/sinθ
                insert point p_cc far from p with distance dist to the outer direction of the obstacle to R
            end if
        end for
    end for
End Expanded-DP Procedure
```

Here, R_{DP} is the set of point lists of the polygonal approximation of the obstacles by the DP algorithm; ε is the threshold value for maximum dissimilarity tolerance; θ_{rot} is a constant angle for obstacle corner clearance; R is the final result of the polygonal approximation of all obstacles by the EDP algorithm, and r_{DP} is a point list of the obstacle polygonal approximation by the DP algorithm, which is arranged in counter-clockwise order.

Figure 4 illustrates the detailed version of the EDP algorithm (Algorithm 4).

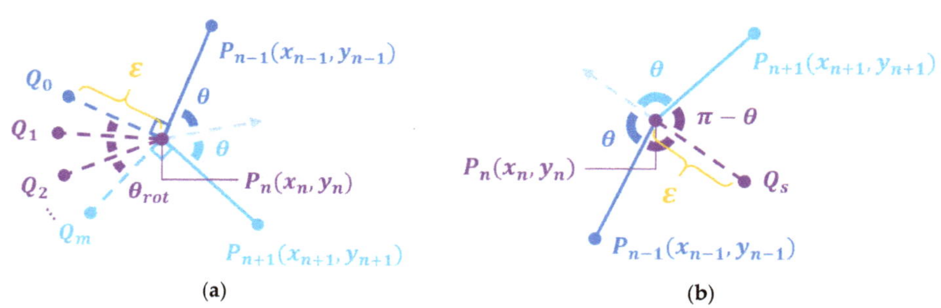

Figure 4. Detailed process of the EDP: (**a**) Convex; (**b**) Concave.

Since $\overrightarrow{P_{n-1}P_n} = (x_n - x_{n-1}, y_n - y_{n-1})$ and $\overrightarrow{P_{n-1}P_n} \perp \overrightarrow{P_n Q_0}$,
unit vector of

$$\overrightarrow{P_n Q_0} = \frac{1}{\sqrt{(y_n - y_{n-1})^2 + (x_n - x_{n-1})^2}} (y_n - y_{n-1}, -(x_n - x_{n-1}))^T \tag{1}$$

Therefore,

$$Q_0 = \frac{\varepsilon}{\sqrt{(y_n - y_{n-1})^2 + (x_n - x_{n-1})^2}} (y_n - y_{n-1}, -(x_n - x_{n-1}))^T \tag{2}$$

In addition, since Q_i ($i = 1, 2, \ldots, m-1$) are on the arc with center point P_n and radius ε, rotating Q_i with angle θ_{rot} in the counter-clockwise direction yields

$$Q_i = \begin{pmatrix} \cos\theta_{rot} & -\sin\theta_{rot} \\ \sin\theta_{rot} & \cos\theta_{rot} \end{pmatrix} (Q_{i-1} - P_n) + P_n \tag{3}$$

Similarly, in the case of a concave corner,

$$\overrightarrow{OP_n} + \overrightarrow{P_n Q_s} = (x_n, y_n)^T + \begin{pmatrix} \cos(\pi - \theta) & -\sin(\pi - \theta) \\ \sin(\pi - \theta) & \cos(\pi - \theta) \end{pmatrix} (x_{n-1} - x_n, y_{n-1} - y_n)^T \tag{4}$$

Around the convex vertex of DP polygon, the concept of θ_{rot} is used for the convex corner clearance. When $\theta_{rot} \geq \angle Q_0 P_n Q_m$, there is no additionally appending vertex, and $Q_m = Q_1$; when $\theta_{rot} < \angle Q_0 P_n Q_m$, there are additionally appending vertices $Q_1, Q_2, \ldots$, which makes the effect of securing additional free space near the obstacle corner relatively difficult to path following.

Algorithm 4 shows the detailed version of the pseudo code of the EDP algorithm.

Algorithm 4. Pseudo Code of EDP (Detailed version).

Input:
$R_{DP} \leftarrow$ Result of DP polygonal approximation of all obstacles represented as a set of point lists
$\varepsilon \leftarrow$ Threshold value for maximum dissimilarity tolerance
$\theta_{rot} \leftarrow$ Constant angle for obstacle corner clearance
Output:
$R \leftarrow$ Final result of the EDP polygonal approximation represented as a set of point lists

Begin ***Expanded-DP Procedure***
  for each r_{DP} in R_{DP} **do**
    for each point $P_n(x_n, y_n)$ in point list r_{DP} ***do***
      if $\angle P_{n+1}P_nP_{n+1} < \pi$ ***then***
        $\theta \leftarrow \frac{1}{2}\angle P_{n+1}P_nP_{n+1}$
        $dist \leftarrow \varepsilon$
        $Q_0 \leftarrow (x_n, y_n)^T + \frac{\varepsilon}{\sqrt{(y_n - y_{n-1})^2 + (x_n - x_{n-1})^2}}(y_n - y_{n-1}, -(x_n - x_{n-1}))^T$
        $Q_m \leftarrow (x_{n+1}, y_{n+1})^T + \frac{\varepsilon}{\sqrt{(y_{n+1} - y_n)^2 + (x_{n+1} - x_n)^2}}(y_{n+1} - y_n, -(x_{n+1} - x_n))^T$
        loop until $i \leq \frac{\pi - 2\theta}{\theta_{rot}}$
          $Q_i \leftarrow \begin{pmatrix} \cos\theta_{rot} & -\sin\theta_{rot} \\ \sin\theta_{rot} & \cos\theta_{rot} \end{pmatrix}(Q_{i-1} - P_n) + P_n$
          insertQ_i to R
          $i \leftarrow i + 1$
        end loop
      else
        $\theta \leftarrow \frac{1}{2}\angle P_{n+1}P_nP_{n+1}$
        $dist \leftarrow \varepsilon/\sin\theta$
        $Q_s \leftarrow \overrightarrow{OP_n} + \overrightarrow{P_nQ_s} = (x_n, y_n)^T + \begin{pmatrix} -\cos(\theta) & -\sin(\theta) \\ \sin(\theta) & -\cos(\theta) \end{pmatrix}(x_{n-1} - x_n, y_{n-1} - y_n)^T$
        insert Q_s ***to*** R
      end if
    end for
  end for
End ***Expanded-DP Procedure***

2.3. Mathematical Validation on the Circumscription of the EDP Algorithm

[*Theorem*] The polygon that consists of the resulting points from the expanded DP (EDP) algorithm on an obstacle includes all points of the original obstacle, i.e., no point of the obstacle region is outside of the polygon that consists of the resulting points from EDP on that obstacle.

(Proof) Let P_n be the n-th point from the DP polygonal approximation of an obstacle. Then, the following propositions are always true by the DP algorithm.

[P1] P_{n-1}, P_n, P_{n+1} are points included in the boundary line of the obstacle.

[P2] P_n is the farthest point from the line segment $\overline{P_{n-1}P_{n+1}}$ among the boundary points of the obstacle in $[P_{n-1}, P_{n+1}]$.

[P3] Every boundary point of the obstacle in $[P_{n-1}, P_n]$ has a shorter distance than ε from the line segment $\overline{P_{n-1}P_n}$.

[P4] Every boundary point of the obstacle in $[P_n, P_{n+1}]$ has a shorter distance than ε from the line segment $\overline{P_nP_{n+1}}$

If P_n is a convex point.

Let us assume that there is a boundary point $c \in [P_{n-1}, P_{n+1}]$ of the obstacle such that c is located outside the polygon Q_m, Q'_0, Q'_1, ..., Q'_m, Q''_0 such as Figure 5a, which consists of the resulting points from the EDP algorithm on the obstacle. Since P_{n-1}, P_n, and P_{n+1} are points on the boundary line of the obstacle by proposition [P1], boundary point c should be identical to P_n or included in the range of $[P_{n-1}, P_n]$ or (P_n, P_{n+1}).

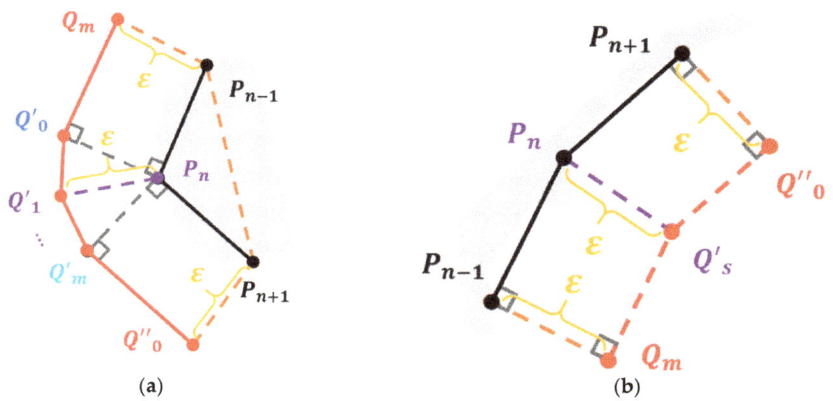

Figure 5. Circumscription of EDP: (**a**) Convex; (**b**) Concave.

If point c is identical to P_n, it is trivial that c cannot be located outside of the polygon with vertices Q_m, Q'_0, Q'_1, ... , Q'_m, and Q''_0, since P_n is inside the polygon.

If point c is included in the range of $[P_{n-1}, P_n)$, the distance from point c to the line segment $\overline{P_{n-1}P_n}$ should be less than ε by proposition [P3]. Therefore, c cannot be located outside of the polygon with vertices $Q_m \sim Q'_0$.

If point c is included in the range of $(P_n, P_{n+1}]$, the distance from point c to the line segment $\overline{P_nP_{n+1}}$ should be less than ε by proposition [P4]. Therefore, c cannot be located outside of the polygon with vertices $Q'_m \sim Q''_0$.

Therefore, no boundary point of the obstacle region is outside of the polygon with vertices Q_m, Q'_0, Q'_1, ... , Q'_m, and Q''_0 in the case of convex P_n.

If P_n is a concave point,

Let us assume that there is a boundary point $c \in [P_{n-1}, P_{n+1}]$ of the obstacle such that c is located outside of the polygon Q_m, Q'_s, Q''_0 such as Figure 5b, which consists of the resulting points from the EDP algorithm on the obstacle. Since P_{n-1}, P_n, and P_{n+1} are points on the boundary line of the obstacle by proposition [P1], boundary point c should be identical to P_n or included in the range of $[P_{n-1}, P_n]$ or (P_n, P_{n+1}).

If point c is identical to P_n, it is trivial that c cannot be located outside of the polygon with vertices Q_m, Q'_s, and Q''_0, since P_n is inside the polygon.

If point c is included in the range of $[P_{n-1}, P_n)$, the distance from point c to the line segment $\overline{P_{n-1}P_n}$ should be less than ε by proposition [P3]. Therefore, c cannot be located outside of the polygon with vertices Q_m and Q'_s.

If point c is included in the range of $(P_n, P_{n+1}]$, the distance from point c to the line segment $\overline{P_nP_{n+1}}$ should be less than ε by proposition [P4]. Therefore, c cannot be located outside of the polygon with vertices Q'_s and Q''_0.

Therefore, no boundary point of the obstacle region is outside the polygon with vertices Q_m, Q'_s, and Q''_0 in the case of concave P_n (*Q.E.D.*).

Once the polygonal approximation with the EDP algorithm is completed, it is possible to solve the obstacle collision problem that may occur when DP algorithm is used for the polygonal approximation of a curvilinear obstacle. Since the polygon created by the EDP algorithm can guarantee the circumscription of obstacles, this result of EDP can be used to apply OAECD in path planning with curvilinear obstacles.

3. Modified Opposite Angle-Based Exact Cell Decomposition (OAECD) Algorithm for Path Planning with Curvilinear Obstacles

The basic concept of the modified OAECD algorithm is to reduce as possible as many cells after the execution of the EDP algorithm and increase the calculation efficiency. Using the concept of inclusion of an opposite angle, OAECD first tries to connect the closest neighboring cell in the opposite angle region. OAECD does not randomly or sequentially process the points but processes in the order from a close pair to a far pair through three consecutive steps, which are similar to the human method. In addition, OAECD does not try to make an additional decomposition if the shape of the generating cell is convex.

The detailed algorithm for the modified OAECD consists of three steps [18]. Figure 6 shows how the decomposing lines are formed between the obstacles for each step of the modified OAECD algorithm. In the first step, the modified OAECD finds the closest neighboring vertex in the set of all vertices of other obstacles for every convex vertex (CV) of each obstacle. For every CV of each obstacle and its closest neighboring vertex (NV) in other obstacles, a new decomposing line is drawn if there is no existing decomposing line with the current CV, its closest NV is inside the region of the opposite angle of the current CV, and no intersection is made with other obstacles or other decomposing lines.

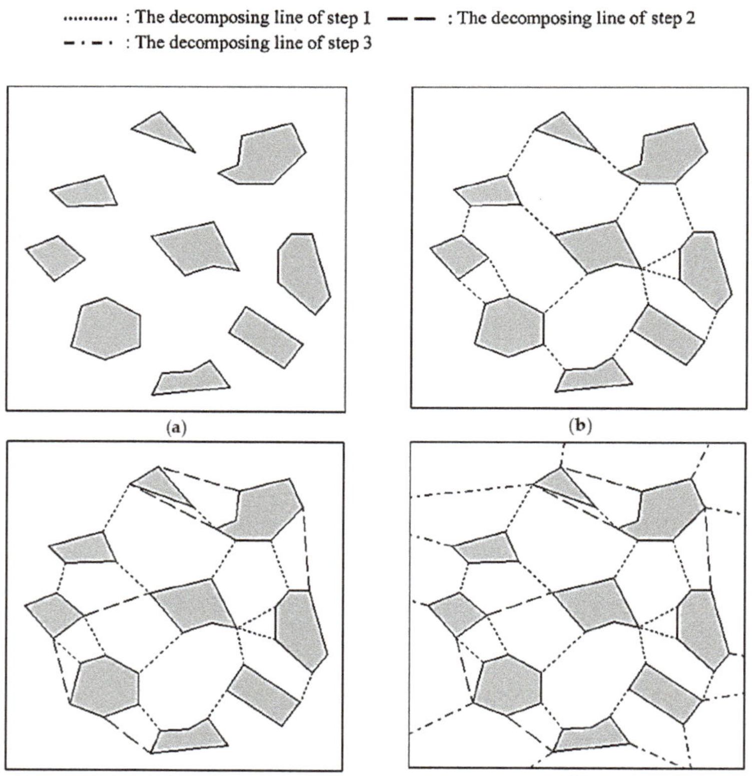

Figure 6. Decomposition results of each step in the modified Opposite Angle-Based Exact Cell Decomposition (OAECD) algorithm after the EDP algorithm: (**a**) Environment; (**b**) Result of step 1; (**c**) Result of step 2; (**d**) Result of step 3.

In the second step, for every CV among the remaining vertices and a vertex (AV) in another obstacle inside the region of the opposite angle of the current CV, a new decomposing line is drawn if there is no decomposing line with the current CV, the AV in another obstacle is the shortest one from the current CV among all available AVs, and there is no intersection with other obstacles or other decomposing lines.

Finally, in the third step, for every CV of the remaining vertices, a new decomposing line is drawn in the direction of the equiangular line of the opposite angle of the current CV until it intersects with the boundary of the environment, obstacle, or another decomposing line, if no decomposing line is connected with the current CV or more than one decomposing line is already connected from another CV, and the created angle near the current CV is larger than 180°.

The OAECD can be properly applied in a static environment. Although a single-sensed map may generate noise, it can sense multiple times for the same environment in a short period and remove the noise on the map by the moving average method [20] or the median method [21]. Based on these methods, one can plan the route. However, this process is only accurate in the static environment. In the dynamic environment, it is difficult to resolve the sensor noise completely by the moving average method or the median method for the same position because the obstacle moves.

Algorithms 5–7 show the pseudo code of the modified OAECD algorithm for each step in the process. Here, v_{c1} is a convex vertex of an obstacle; V_c is the set of point lists, which is the result of the EDP polygonal approximation of all obstacles; v_{a1} is the closest neighboring vertex from v_{c1}; V is the set of all vertices of the obstacles; V_{check} is the point set to check for processing.

Algorithm 5. Pseudo Code of the modified OAECD (Step 1).

Input:
$V_c \leftarrow R$ // R: Result of the EDP polygonal approximation of all obstacles represented as a set of point lists
$M \leftarrow$ Environment map
Output:
$V_{chec} \leftarrow$ Point set to check for processing
$M \leftarrow$ Environment map with decomposing lines of step 1

Begin ***OAECD-Algorithm Procedure***
for every convex vertex of each obstacle in V_c ***do***
    find the closest vertex in the set of all vertices of other obstacles by comparing the distances between the current vertex and other vertices
end for
for each v_{c1} in V_c ***do***
    if there is no decomposing line that is already connected with v_{c1} ***then***
        if v_{a1} is included in the region of the opposite angle of v_{c1} and ***line***(v_{c1}, v_{a1} in *V*) is not intersected with other obstacles or other lines ***then***
            draw decomposing line from v_{c1} ***to*** v_{a1} and ***insert*** v_{c1} ***to*** V_{check}
        else if
    else if all angles near v_{c1} are less than or equal to 180° ***then***
        insert v_{c1} ***to*** V_{check}
    end if
end for
End ***OAECD-Algorithm Procedure***

Generally, the cells created by the cell decomposition methods are shaped as a convex polygon to avoid being penetrated by obstacles. Therefore, only the convex vertices of the obstacles are considered in this step because the concave vertices of the obstacles obviously result in convex vertices of the free cell. In other words, there is no need to make an additional decomposing line from some vertices if they preserve the concaveness. In addition, the closest neighboring vertex is chosen to reduce the number of possible cells and the possibility of crossing with other decomposing lines. Here, v_{c2} is a convex vertex of an obstacle in V_c-V_{check}, v_{a2} is a vertex in the region of the opposite angle of v_{c2}, and V_i is the set of all vertices inside the region of the opposite angle of v_{c2}.

Algorithm 6. Pseudo Code of the modified OAECD (Step 2).

Input:
$V_c \leftarrow R//R$: Result of the EDP polygonal approximation of all obstacles represented as a set of point lists
$M \leftarrow$ Environment map
$V_{check} \leftarrow$ Point set to check for processing (Result of previous step 1)
Output:
$V_{check} \leftarrow$ Point set to check for processing (Result of this step)
$M \leftarrow$ Environment map with the decomposing lines of step 1 and 2

Begin ***OAECD-Algorithm Procedure***
for each v_{c2} in V_c - V_{check} ***do***
  if there is no decomposing line that is already connected with v_{c2} ***then***
    if v_{a2} has the shortest distance with v_{c2} compared with those in V_i - $\{v_{a2}\}$ and ***line***(v_{c2}, v_{a2} in V_i) is not intersected with the obstacle and other lines ***then***
      draw decomposing line from v_{c2} ***to*** v_{a2} and ***insert*** v_{c2} ***to*** V_{check}
    end if
  else if all angles near v_{c2} are less than or equal to 180° ***then***
    insert v_{c2} ***to*** V_{check}
  end if
end for
End ***OAECD-Algorithm Procedure***

From Step 2, the decomposing line from v_{c2} to v_{a2} is drawn, and v_{c2} is inserted into V_{check} although v_{a2} is not the closest vertex in V. Here, v_{c3} is an unchecked convex vertex of each obstacle.

Algorithm 7. Pseudo Code of the modified OAECD (Step 3).

Input:
$V_c \leftarrow R//R$: Result of the EDP polygonal approximation of all obstacles represented as a set of point lists
$M \leftarrow$ Environment map
$V_{check} \leftarrow$ Point set to check for processing (Result of previous step 2)
Output:
$M \leftarrow$ The final environment map with decomposing lines of steps 1, 2, and 3

Begin ***OAECD-Algorithm Procedure***
for each v_{c3} in V_c - V_{check} ***do***
  if there is no decomposing line that is already connected with v_{c3} ***then***
  draw decomposing line from v_{c3} in the direction of half angle of the opposite angle of v_{c3} until it intersects with the boundary of the environment, other obstacles, or other lines and ***insert*** v_{c3} ***to*** V_{check}
  else if all angles near v_{c3} are less than or equal to 180° ***then***
    insert v_{c3} ***to*** V_{check}
  end if
end for
End ***OAECD-Algorithm Procedure***

The time complexity of the modified OAECD is $O(n^2)$ because the sub-time complexity is basically all $O(n^2)$ for Step 1, Step 2, and Step 3. In addition, the time complexity for Step 3 can be reduced to $O(nlogn)$ approximately when the sweep-line method [22,23] is applied.

4. Experimental Results

Two types of experiments were conducted to find the performance of the proposed EDP and modified OAECD algorithm. An additional simulation has been performed to verify the feasibility of the modified OAECD algorithm for a map with curvilinear obstacles. The first experiment is conducted to compare the approximation error and the number of vertices of the approximated polygon made by EDP and DP. In each experiment, twenty maps were chosen in the pre-created one hundred fifty random maps, each of which is assumed to be 20 × 20 m^2 in size and randomly have the position of the obstacle, position of the vertices of each obstacle, and area of the obstacle by using the random function in the math library of the MS Visual C++ compiler. The number of obstacles and vertices of each obstacle were fixed at fifteen. Bezier curve was used for the curvilinear obstacle representation by interconnecting the vertices of each obstacle and creating naturally curved obstacles. Each map was created by an image in bitmap format.

The second experiment was conducted to compare the performance of the OAECD algorithm and the VCD algorithm. Similar to the first experiment, the size of the map was assumed as 20 × 20 m^2, and the position of the obstacle, positions of the vertices of each obstacle, number of vertices of each obstacle, and number of obstacles in the map were random variables.

Table 1 shows the experimental results by averaging the results of each experiment. Here, IA is the percentage area of the inner space of the approximated polygons outside the obstacle regions in comparison with the area of the original obstacle region. OA is the percentage area of the outer space of the approximated polygons inside the obstacle regions in comparison with the area of the original obstacle region. SA is the sum of IA and OA. AVG is the average values for various ε.

Table 1. Performance comparison between Expanded Douglas–Peucker (EDP) algorithm and Douglas–Peucker (DP) alogrithm for various ε values (The number of obstacles and vertices of each obstacle were fixed as 15, and θ_{rot} for EDP was fixed as 30°).

ε (m)	EDP (%)			DP (%)		
	IA	OA	SA	IA	OA	SA
0.05	19.05	0.00	19.05	2.45	0.70	3.15
0.08	28.46	0.00	28.46	4.89	1.17	6.06
0.11	37.95	0.00	37.95	7.19	1.54	8.73
0.14	47.63	0.00	47.63	9.26	1.83	11.08
0.17	57.34	0.00	57.34	11.50	2.08	13.58
0.20	67.17	0.00	67.17	13.38	2.48	15.86
0.23	77.11	0.00	77.11	15.52	2.76	18.28
0.26	87.01	0.00	87.01	17.23	3.20	20.43
0.29	96.91	0.00	96.91	19.55	3.56	23.10
0.32	106.7	0.00	106.7	22.00	3.86	25.85
0.35	116.7	0.00	116.7	23.90	4.40	28.30
0.38	127.3	0.00	127.3	26.91	4.83	31.74
0.41	137.9	0.00	137.9	27.99	5.27	33.26
0.44	148.0	0.00	148.0	30.57	5.62	36.19
0.47	158.8	0.00	158.8	32.13	5.86	37.99
0.50	169.5	0.00	169.5	33.11	6.60	39.72
AVG	92.72	0.00	92.72	15.53	2.99	18.52

From Table 1, the average approximation error of the result of the EDP algorithm is approximately 5 times larger than that of DP algorithm to cover the entire area of the original obstacle regions.

Table 2 shows the experimental results by averaging the results of ten experiments with 15 random obstacles and 15 random vertices for each obstacle.

From Table 2, on average, the result of the EDP algorithm has 2.88 times more vertices than the result of the DP algorithm when θ_{rot} for EDP is 30°.

Figure 7 shows a graphical representation of Tables 1 and 2. The approximation performance of EDP is worse than DP in the aspects of IA, SA, and average number of vertices. Nonetheless, OA of

EDP is completely zero and better than DP since the result of the EDP can cover the entire area of the original obstacle regions.

Table 2. Average number of vertices created by EDP and DP (The numbers of obstacles and vertices of each obstacle were fixed as 15, and θ_{rot} for EDP was fixed as 30°).

ε (m)	Average Number of Vertices	
	EDP	**DP**
0.05	31.08	15.92
0.08	27.87	11.92
0.11	25.89	10.19
0.14	24.68	9.18
0.17	23.67	8.33
0.20	22.85	7.66
0.23	22.13	7.32
0.26	21.55	6.83
0.29	20.97	6.56
0.32	20.15	6.20
0.35	19.32	5.91
0.38	18.97	5.58
0.41	18.63	5.40
0.44	18.06	5.06
0.47	17.65	5.00
0.50	17.16	4.78
AVG	21.91	7.65

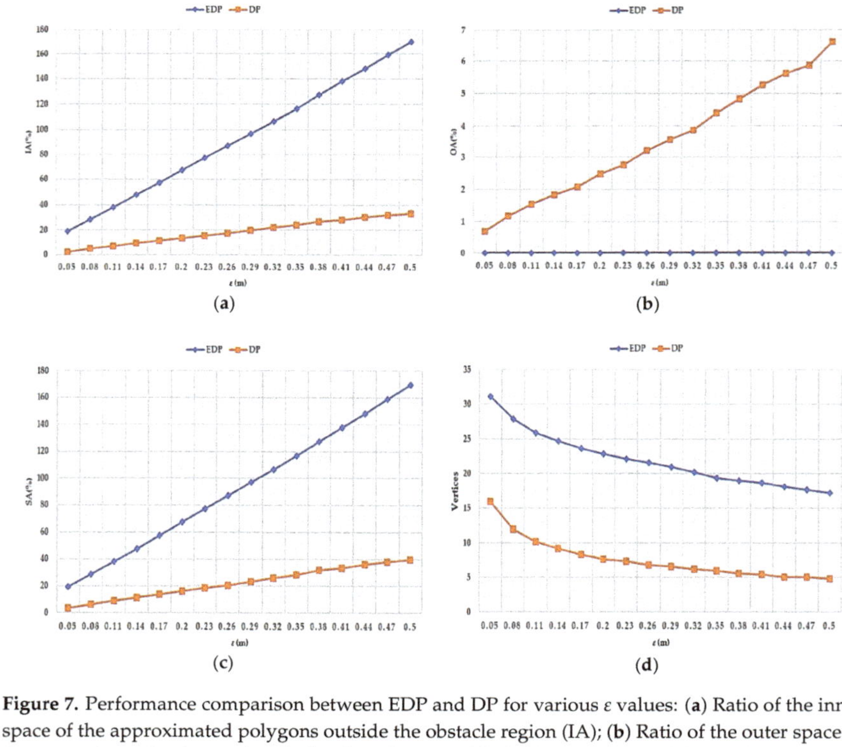

Figure 7. Performance comparison between EDP and DP for various ε values: (**a**) Ratio of the inner space of the approximated polygons outside the obstacle region (IA); (**b**) Ratio of the outer space of the approximated polygons inside the obstacle region (OA); (**c**) Sum of IA and OA (SA); (**d**) Average number of vertices created by EDP and DP (The numbers of obstacles and vertices of each obstacle were fixed as 15, and θ_{rot} for EDP was fixed as 30°).

Table 3 compares the performance of the modified OAECD algorithm and VCD algorithm. Here, NV is the total number of vertices in the connectivity graph, VV is the number of vertices that composes the generated path by the A* algorithm, and PL is the path length.

Table 3. Performance comparison between modified Opposite Angle-Based Exact Cell Decomposition (OAECD) and Vertical Cell Decomposition (VCD) (The numbers of obstacles and vertices of each obstacle were fixed as 15).

(NO, NV)	OAECD			VCD		
	NV (ea)	VV (ea)	PL (m)	NV (ea)	VV (ea)	PL (m)
(1,10)	7.60	4.30	28.43	12.00	7.00	29.27
(3,10)	19.80	6.00	29.16	36.60	16.50	33.43
(5,10)	31.70	10.20	29.85	61.80	24.70	37.84
(7,10)	44.10	10.60	30.37	88.20	29.90	45.89
(9,10)	56.80	11.90	29.68	111.80	34.60	47.90
(11,10)	68.50	13.40	30.88	139.00	36.80	53.64
(13,10)	79.20	14.90	30.62	164.70	43.80	56.38
(15,10)	92.50	17.60	30.92	192.00	45.70	48.31
(15,3)	40.70	9.80	28.88	76.60	28.90	57.03
(15,5)	54.90	11.10	29.72	108.00	29.80	51.94
(15,7)	69.20	13.80	29.83	141.20	37.70	59.94
(15,9)	84.70	15.70	30.48	174.60	43.50	54.79
(15,11)	100.60	17.60	30.82	207.80	43.00	53.86
(15,13)	116.10	21.40	31.26	249.50	46.90	58.88
(15,15)	130.60	20.30	31.32	279.20	43.30	55.28
AVG	66.47	13.24	30.15	136.20	34.14	49.63

From Table 3, the path length by the modified OAECD algorithm is 60.7% shorter than the path length by VCD algorithm on average.

Figure 8 shows the comparison of VCD with DP and OAECD with EDP to show the feasibility of the modified OAECD algorithm for maps with static curvilinear obstacles.

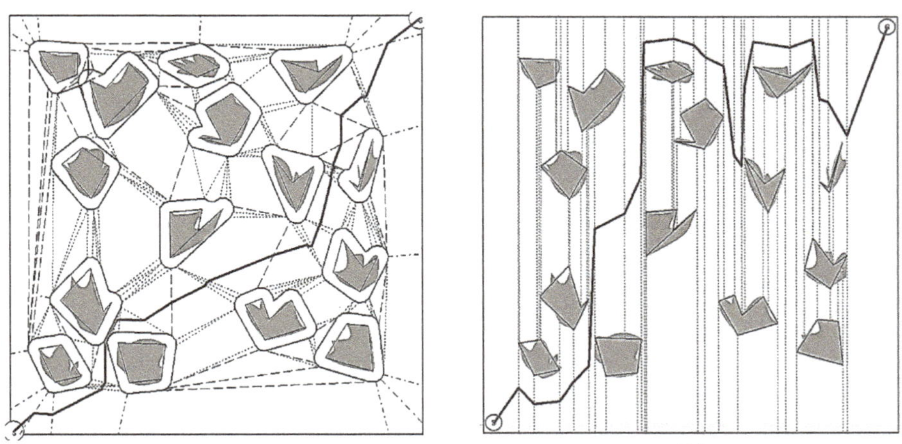

(a) **(b)**

Figure 8. Comparison between OAECD with EDP and VCD with DP (ε = 0.05 m): (**a**) Path planning using EDP and OAECD; (**b**) Path planning using DP and VCD (The numbers of obstacles and vertices of each obstacle were fixed as 15, and θ_{rot} for EDP was fixed as 30°).

The EDP shows that the polygonal approximation completely covers each obstacle. In addition, the OAECD shows a more natural-looking and efficient path than VCD in Figure 8a. The path from VCD with the DP algorithm may also easily collide with the obstacle, such as Figure 8b.

Even though the path in Figure 8a is more natural looking than that in Figure 8b, there are still some sharp corners which may occur generating an additional problem of motion control for real-world robots due to kinematic constraints. These corners can be smoothed by some additional path smoothing techniques [5,24].

5. Conclusions

In this paper, the Expanded Douglas–Peucker (EDP) polygonal approximation algorithm and its application method for the Opposite Angle-based Exact Cell Decomposition (OAECD) are proposed for the mobile-robot path-planning problem with curvilinear obstacles. In addition, mathematical analysis has been conducted to guarantee the circumscription of obstacle regions by the EDP approximation. Since OAECD is basically focused on the static environment, OAECD is not robust enough to sensor noises in the dynamic environment. Nonetheless, the proposed method is useful with both polygonal and curvilinear obstacles, since the EDP approximation can guarantee the circumscription of obstacle regions, and the modified OAECD algorithm can effectively reduce the number of decomposing cells.

The experimental results show that the path generated by the OAECD algorithm with the EDP approximation looks much more natural and is collision-free. That path is also more efficient than the path generated by the VCD algorithm with DP approximation, although on average, the EDP approximation may induce a larger approximation error and more approximation vertices than DP approximation.

Author Contributions: Idea and conceptualization: J.-W.J. and S.-B.C.; methodology: J.-W.J. and S.-B.C.; software: S.-B.C. and Y.S.; experiment: S.-B.C., J.-.G.K. and D.-W.L.; validation: J.-W.J.; investigation: S.-B.C. and J.-G.K.; resources: J.-W.J.; writing: J.-W.J., S.-B.C., J.-G.K., D.-W.L. and Y.S.; visualization: S.-B.C. and J.-G.K.; project administration: J.-W.J.

Funding: This research was supported by the Ministry of Science and ICT, Korea, under the National Program for Excellence in Software supervised by the Institute for Information & Communications Technology Promotion (2016-0-00017), the KIAT (Korea *Institute* for Advancement of Technology) grant funded by the Korea Government (MOTIE: Ministry of Trade Industry and Energy) (No. N0001884, HRD program for Embedded Software R&D) and the National Research Foundation of Korea (NRF) grant funded by the Korea government (MSIT) (No. 2018R1A5A7023490).

Conflicts of Interest: The authors declare no conflicts of interest.

References

1. Latombe, J.-C. *Robot Motion Planning;* Kluwer Academic Publishers: Dordrecht, The Netherlands, 1991.
2. Choset, H.; Lynch, K.M.; Hutchinson, S.; Kantor, G.; Burgard, W.; Kavraki, L.E.; Thrun, S. *Principles of Robot Motion: Theory, Algorithms, and Implementations;* MIT Press: Boston, MA, USA, 2005.
3. La Valle, S.M. *Planning Algorithms;* Cambridge University Press: Cambridge, UK, 2006.
4. Yang, L.; Qi, J.; Song, D.; Xiao, J.; Han, J.; Xia, Y. Survey of robot 3D path planning algorithms. *J. Control Sci. Eng.* **2016**, *5*, 22–44. [CrossRef]
5. Gonzalez, D.; Perez, J.; Milanes, V.; Nashashibi, F. A Review of Motion Planning Techniques for Automated Vehicles. *IEEE Trans. Intell. Transp. Syst.* **2016**, *17*, 1135–1145. [CrossRef]
6. Yang, H.; Jia, Q.; Zhang, W. An Environmental Potential Field Based RRT Algorithm for UAV Path Planning. In Proceedings of the 2018 37th Chinese Control Conference (CCC), Wuhan, China, 25–27 July 2018; pp. 9922–9927.
7. Yanyi, Y.; Yingming, Z.; Xingchen, L. An improved artificial potential field algorithm based on nonuniform cell decomposition. In Proceedings of the 2017 6th International Conference on Measurement, Instrumentation and Automation (ICMIA 2017), Zhuhai, China, 29–30 June 2017.

8. Avnaim, F.; Boissonnat, J.D.; Faverjon, B. A practical exact motion planning algorithm for polygonal objects amidst polygonal obstacles. In Proceedings of the IEEE International Conference on Robotics and Automation, Philadelphia, PA, USA, 24–29 April 1988.
9. Brooks, R.A.; Lozano-Perez, T. A subdivision algorithm in configuration space for find path with rotation. *IEEE Trans. Syst.* **1985**, *15*, 224–233.
10. Iswanto, I.; Oyas, W.; Imam, C.A. Quadrotor Path Planning Based on Modified Fuzzy Cell Decomposition Algorithm. *Telecommun. Comput. Electron. Control* **2016**, *14*, 655–664. [CrossRef]
11. Nora, S.; Tschichold-Gurmann, N. *Exact Cell Decomposition of Arrangements Used for Path Planning in Robotics;* Technical Report; ETH Zürich, Department of Computer Science: Zürich, Switzerland, 1999.
12. Zhang, L.; Kim, Y.J.; Manocha, D. A hybrid approach for complete motion planning. In Proceedings of the International Conference on Intelligent Robots and Systems, San Diego, CA, USA, 29 October–2 November 2007.
13. Kim, J.-T.; Kim, D.-J. New Path Planning Algorithm based on the Visibility Checking using a Quad-tree on a Quantized Space, and its improvements. *J. Inst. Control Robot. Syst.* **2010**, *16*, 48. [CrossRef]
14. Arney, T. An efficient solution to autonomous path planning by Approximate Cell Decomposition. In Proceedings of the 3rd International Conference on Information and Automation for Sustainability, Melbourne, Australia, 4–6 December 2007.
15. Rosell, J.; Iniguez, P. Path planning using Harmonic Functions and Probabilistic Cell Decomposition. In Proceedings of the IEEE International Conference on Robotics and Automation, Barcelona, Spain, 18–22 April 2005.
16. Cowlagi, R.V.; Tsiotras, P. Beyond quadtrees: Cell decompositions for path planning using wavelet transforms. In Proceedings of the 46th IEEE Conference on Decision and Control, New Orleans, LA, USA, 12–14 December 2007.
17. Ghita, N.; Kloetzer, M. Cell Decomposition-Based Strategy for Planning and Controlling a Car-like Robot. In Proceedings of the 14th International Conference on System Theory and Control, Sinaia, Romania, 17–19 October 2010.
18. So, B.-C.; Jung, J.-W. Mobile Robot Path Planning with Opposite Angle-Based Exact Cell Decomposition. *Adv. Sci. Lett.* **2012**, *15*, 144–148. [CrossRef]
19. Ramer, U. An iterative procedure for the polygonal approximation of plane curves. *Comput. Graph. Image Process.* **1972**, *1*, 244–256. [CrossRef]
20. Hwang, Y.-S.; Lee, J. Robust 3D map building for a mobile robot moving on the floor. In Proceedings of the 2015 IEEE International Conference on Advanced Intelligent Mechatronics (AIM), Busan, Korea, 7–11 July 2015; pp. 1388–1393.
21. Gao, H.; Hu, M.; Gao, T. Robust detection of median filtering based on combined features of difference image. *Signal Process. Image Commun.* **2017**, *72*, 126–133. [CrossRef]
22. Douglas, D.; Peucker, T. Algorithms for the reduction of the number of points required to represent a digitized line or its caricature. *Can. Cartogr.* **1973**, *10*, 112–122. [CrossRef]
23. Chazelle, B.; Edelsbrunner, H. An optimal algorithm for intersecting line segments in the plane. In Proceedings of the 29th Annual Symposium on Foundations of Computer Science, White Plains, NY, USA, 24–26 October 1988.
24. Ravankar, A.; Ravankar, A.; Kobauashi, Y.; Hishino, Y.; Peng, C.-C. Path smoothing techniques in robot navigation: State-of-the-art, current and future challenges. *Sensors* **2018**, *18*, 3170. [CrossRef] [PubMed]

MDPI
St. Alban-Anlage 66
4052 Basel
Switzerland
Tel. +41 61 683 77 34
Fax +41 61 302 89 18
www.mdpi.com

Applied Sciences Editorial Office
E-mail: applsci@mdpi.com
www.mdpi.com/journal/applsci

www.ingramcontent.com/pod-product-compliance
Lightning Source LLC
Chambersburg PA
CBHW041750220926
51847CB00014B/156

* 9 7 8 3 0 3 9 2 1 9 4 4 5 *